C0-AYS-017

Springer
Tokyo
Berlin
Heidelberg
New York
Barcelona
Hong Kong
London
Milan
Paris
Singapore

N. Ae, J. Arihara, K. Okada, A. Srinivasan (Eds.)

Plant Nutrient Acquisition

New Perspectives

With 230 Figures, Including 6 in Color

Noriharu Ae, Ph.D.
Laboratory of Soil Biochemistry
National Institute for Agro-Environmental Sciences
3-1-3 Kannondai, Tsukuba, Ibaraki 305-8604, Japan

Joji Arihara, Ph.D.
Legume Physiology Laboratory
National Institute of Crop Science
2-1-18 Kannondai, Tsukuba, Ibaraki 305-8518, Japan

Kensuke Okada, Ph.D.
Crop Production and Environment Division
Japan International Research Center for Agricultural Sciences
1-1 Owashi, Tsukuba, Ibaraki 305-8686, Japan

Ancha Srinivasan, Ph.D.
Regional Science Institute
4-13, Kita-24 Nishi-2, Kita-ku, Sapporo 001-0024, Japan

AGR
QK
867
.P53
2001

ISBN 4-431-70281-4 Springer-Verlag Tokyo Berlin Heidelberg New York

Library of Congress Cataloging-in-Publication Data
Plant nutrient acquisition : new perspectives / N. Ae, ...[et al.], (eds.)
p. cm.
Paper presented at the International Workshop held in Tsukuba, Japan, March 24-27, 1998.
Includes bibliographical references (p.).
ISBN 4431702814 (alk. paper)
1. Plants--Nutrition--Congresses. 2. Plant-soil relationships--Congresses. 3. Roots (Botany)--Congresses. 4. Plant nutrients--Congresses. I. Ae, N. (Noriharu), 1946-

QK867 .P53 2001
575.7'6--dc21

2001020401

Printed on acid-free paper

© Springer-Verlag Tokyo 2001
Printed in Japan
This work is subject to copyright. All rights are reserved, whether the whole or part of the material is concerned, specifically the rights of translation, reprinting, reuse of illustrations, recitation, broadcasting, reproduction on microfilms or in other ways, and storage in data banks.
The use of registered names, trademarks, etc. in this publication does not imply, even in the absence of a specific statement, that such names are exempt from the relevant protective laws and regulations and therefore free for general use.

Printing and binding: Hicom, Japan
SPIN: 10691853

Foreword

Katsuyuki Minami
Director General, National Institute for Agro-Environmental Sciences (NIAES)

Plant Nutrient Acquisition: New Perspectives is a compilation of research papers presented at the International Workshop held in Tsukuba, Japan, March 24–27, 1998. The Workshop was organized by the National Institute of Agro-Environmental Sciences (NIAES) and was sponsored by the Secretariat of the Research Council of the Ministry of Agriculture, Forestry and Fisheries, Japan.

The objectives of the Workshop were to discuss newly emerging concepts of the mechanisms that plants use to acquire soil nutrients, and to relate those concepts in the context of applied agriculture. Some 40 papers and posters highlighting recent advances in this exciting field were presented. I greatly appreciate the efforts of the authors in critically evaluating the past work and presenting new experimental findings and concepts of both theoretical and applied significance.

The scope of the Workshop was defined by the Organizing Committee—the members of which are the editors of this book—by holding regular discussions for one year prior to the Workshop. The Committee also received excellent professional support from several staff members and researchers of NIAES in organizing the Workshop. I express my sincere thanks to all of them.

I am confident that this book will serve as a valuable reference for researchers, students, and extension staff in their efforts to develop more effective approaches to sustainable nutrient management.

May 2001

Welcoming Address

Michinori Nishio
Professor, University of Tsukuba
Former Director General, National Institute for Agro-Environmental Sciences (NIAES)

On behalf of the Organizing Committee and the National Institute of Agro-Environmental Sciences, I extend a warm welcome to all of you. The concept of sustainable agriculture has become the most important concern in the global farming sector since the Earth Summit in Rio de Janeiro in 1992. The concept itself is beautiful, and no one denies that. However, there are many interpretations of this concept. It is quite natural to have such diverse interpretations, as types of agriculture vary widely across countries, regions, and areas, depending on different natural and social conditions.

Broadly speaking, there are two types of agriculture: extensive and intensive. Different factors are threatening sustainability of both types. For example, the deficiency of nutrients is threatening extensive agriculture while the surplus of nutrients is damaging intensive agriculture. One of the most important concerns for sustainable agriculture, therefore, is the management of plant nutrients. In intensive agriculture, overuse of chemical fertilizers is a matter of concern, while in extensive agriculture, acquisition of nutrients from alternatives to chemical fertilizers is of main interest. There are many ways to approach sustainable nutrient management, including the utilization of slow-release fertilizers, improvement of fertilization methods, utilization of renewable resources, recycling of farm wastes, and so on. But I am convinced that the essential approach to sustainable nutrient management is the utilization of biological functions of plants and soil organisms. While studies on the role of microorganisms in relation to nutrient supply are relatively many, there are few studies on plant nutrient acquisition. I believe that effective utilization of nutrient acquisition mechanisms by plants themselves and by plant–soil organism combinations is a fundamental and reliable approach to sustainable nutrient management in both intensive and extensive agriculture.

With these considerations, the Organizing Committee planned this Workshop. The Committee made special efforts to identify and send invitations to distinguished international experts in various fields. I am happy to learn that almost all invited researchers kindly agreed to participate and share their views with us. I am especially glad to see Dr. Swindale and Dr. Johansen, who extended valuable help to the Committee members in their research at International Crops Research Institute for the Semi-Arid Tropics (ICRISAT).

Let me once again welcome all of you to Tsukuba and wish you success in your research endeavors. I am confident that discussions held here will lead to new collaborations that will guide plant nutrition research in the years ahead.

March 1998

Preface

Leslie D. Swindale
Director General (Emeritus), International Crops Research Institute for the Semi-Arid Tropics (ICRISAT)

Plants, like people, need food.

The scientific study of plant nutrients commenced in the mid-nineteenth century with the work of Justus von Liebig and his application of scientific principles to "the means of nutrition of vegetables and with the influence of soils and action of manures upon them." Liebig proposed the Law of the Minimum, that is, that plant growth was dependent upon the least available nutrient. Soon after the turn of the century it was accepted that the primary means of nutrient uptake was by ion exchange at root surfaces with H^+ ions being released from the roots in cation uptake and OH^- and HCO_3^- being released in the uptake of anions. By the early 1920s it was recognized that nutrient supply to any plant was dependent upon both the amount of nutrient available in the surrounding soil and the rate at which the nutrient was supplied to the plant, the so-called capacity and intensity factors. After it was realized that nutrients would first need to pass into the soil solution that existed between the soil and the plant roots—although some adhered to the idea of nutrient supply through direct contact between the solid phases of soil and plant—a physico-chemical model of nutrient uptake was developed and held sway for several decades. In this model there were basically three rate-determining steps: the rate at which a nutrient was removed from the soil particle into the soil solution, the rate at which the nutrient diffused through the solution, and the rate at which it was subsequently taken up by the plant.

The model, to this stage, was essentially a physical one. Jacobsen and Overstreet in the 1960s were among the first to introduce a biological dimension by proposing that organic chelating compounds occurred in the soil solution and affected both the removal of ions from the soil and the activities of ions in solution. The metabolic products of the microbial breakdown of carbohydrates in soils include citrate, malate, and malonate, all of which have chelating abilities. In more recent years the possible role of enzymes in nutrient release and particularly in nutrient uptake has received attention.

Chelating compounds would have particular importance on the availability of phosphate (P), which is one of the three most important nutrients that plants derive from soils. It is often one of the least available because it can be fixed in unavailable forms and because it is not very mobile in soil solutions. Plants forage for P. Several authors have shown how P uptake is related to the total mass and volume of roots. Nitrogen can influence P uptake not only by increasing the root mass but also by altering the nature of root systems.

The Workshop on New Concepts of Plant Nutrient Acquisition, held at Tsukuba, Japan, March 24–27, 1998, the proceedings of which are reported in this publication, brought together people from all over the world to exchange views on old and new perspectives of plant nutrient uptake and to show how the newer perspectives might be applied to practical agriculture. It was to a large extent an outgrowth of a 15-year-long project between International Crops Research Institute for the Semi-Arid Tropics (ICRISAT) and the Japan International Research Center for Agricultural Sciences (JIRCAS), with core support from the Consultative Group on International Agricultural Research (CGIAR) and special project support from the Government of Japan.

The project, entitled "Development of Sustainable Cultivation of Upland Crops in the Semi-Arid Tropics," was conducted in three phases (I: 1984–1989, II: 1989–1994, III: 1995–1999). During that time Japanese scientists worked at ICRISAT with ICRISAT scientists and fellows. The project addressed the issues of food security and poverty alleviation by looking for ways to increase the availability of nutrients without additions of fertilizers in food crops commonly used by poor people in the semi-arid tropics. The project first tried to elucidate the mechanisms by which some of these crops were more efficient than others in extracting nutrients from soils, then studied the role of roots and root development, and, in the final phase, moved on to learning how plant breeding could be combined with management options to improve nutrient availability in soils with low nutrient-supplying powers.

We have known for a long time that pigeonpea seemed to be able to extract P from soils in ways that other crops could not. Project scientists obtained evidence that suggested that root exudates of pigeonpea might be the cause, or at least a cause, of the plant's ability to extract P from soil. Piscidic acid was identified as one of the constituents of the exudates. Mechanisms involving root exudates may make limited contributions to the P nutrition of most plants in most soils, but they seemed to be important in the case of pigeonpea growing in soils with low nutrient-supplying power. This finding was the first report indicating that crop plants are also able to modify rhizosphere conditions, thereby solubilizing P from Fe compounds, which are regarded as the most insoluble form of P.

Several papers at the Workshop in Tsukuba further explored the role of plant exudates in P uptake. Otani and Ae found that pigeonpea, groundnut, and rice had greater abilities to release P from Fe- and Al-fixed forms in Andosols than did sorghum if root volume was constrained. Pigeonpea exudates were shown to be able to dissolve $FePO_4$ and $AlPO_4$ in artificial conditions. In addition to piscidic acid, malonates and citrates were also found as the components of the exudates. P uptake by maize and soybean from soils with low available P was enhanced when artificial pigeonpea exudates were added to the rhizosphere soil. Their effect, in fact, was to create conditions through which P can be solubilized from the soil so that absorption by the plant becomes possible.

Papers presented at this workshop highlight various new aspects of plant nutrient acquisition. One set of papers discusses recent models regarding nutrient uptake that place greater emphasis on the active role of plants in that process. This is followed by several papers describing management options designed to reduce the reliance on chemical fertilizer inputs. The important role of soil microorganisms for nutrient acquisition is stressed and evidence presented that shows that organic fertilizers can be used directly by plants as a result of their ability to take up organic nitrogen forms. Finally, the potential for improving the nutrient acquisition ability of crops by plant breeding methods is explored. Results presented indicate that breeding effective crops is possible even for tropical soils, which have long been considered to be problem soils due to low P-fertility. It was also shown that another important problem of tropical soils—soil acidity—can be overcome by exploring the genetic diversity present in crop plants. Genetic improvements in combination with adjustments in cropping systems will have a significant impact not only in developing countries characterized by low fertilizer use but also in developed countries, where they will contribute to making agricultural systems more environmentally sound.

The organizers of the workshop in an epilogue made a plea for reducing dependence upon mineral fertilizers to increase food crop production in and for the developing world through greater use of natural soil components and processes. I believe that the publication of these proceedings will stimulate research into these more sustainable forms of plant nutrient acquisition.

Contents

Contributors

Ae, Noriharu
Laboratory of Soil Biochemistry, National Institute for Agro-Environmental Sciences (NIAES), 3-1-3 Kannondai, Tsukuba, Ibaraki 305-8604, Japan

Arias, Maria P.
Maize Program, International Center for the Improvement of Wheat and Maize (CIMMYT/CIAT), Apdo Aéreo 6713, Cali, Colombia

Arihara, Joji
Legume Physiology Laboratory, National Institute of Crop Science, 2-1-18 Kannondai, Tsukuba, Ibaraki 305-8518, Japan

Blamey, Frederick Pax C.
Department of Agriculture, School of Land and Food, The University of Queensland, Brisbane, Queensland 4072, Australia

De León, Carlos
Maize Program, International Center for the Improvement of Wheat and Maize (CIMMYT/CIAT), Apdo Aéreo 6713, Cali, Colombia

Delhaize, Emmanuel
Commonwealth Scientific and Industrial Research Organization (CSIRO) Plant Industry, GPO Box 1600, Clunies Ross Street, Canberra, ACT 2601, Australia

Fischer, Albert J.
Weed Science Program—Vegetable Crops Department, University of California, Davis, CA 95616, USA

Gleddie, Sanford
Philom Bios Inc., 318-111 Research Drive, Innovation Place, Saskatoon, Saskatchewan, S7N 3R2, Canada

Hayes, Julie E.
Commonwealth Scientific and Industrial Research Organization (CSIRO) Plant Industry, GPO Box 1600, Clunies Ross Street, Canberra, ACT 2601, Australia

Hocking, Peter J.
Commonwealth Scientific and Industrial Research Organization (CSIRO) Plant Industry, GPO Box 1600, Clunies Ross Street, Canberra, ACT 2601, Australia

Holloway, Greg
Philom Bios Inc., 318-111 Research Drive, Innovation Place, Saskatoon, Saskatchewan, S7N 3R2, Canada

Ishikawa, Satoru
Faculty of Agriculture, Yamagata University, 1-23 Wakaba-cho, Tsuruoka, Yamagata 997-8555, Japan

Karasawa, Toshihiko
Laboratory of Plant Nutrition, Department of Agro-Environment Sciences, National Agricultural Research Center for Hokkaido Region, 1 Hitsujigaoka, Toyohira-ku, Sapporo, Hokkaido 062-8555, Japan

Kato, Yoji
Laboratory of Food Science, Faculty of Education, Hirosaki University, 1 Bunkyo-cho, Hirosaki, Aomori 036-8224, Japan

Kielland, Knut
Institute of Arctic Biology, University of Alaska, Fairbanks, Alaska 99775-0180, USA

Leggett, Mary
Philom Bios Inc., 318-111 Research Drive, Innovation Place, Saskatoon, Saskatchewan, S7N 3R2, Canada

Matoh, Toru
Laboratory of Plant Nutrition, Division of Applied Life Science, Graduate School of Agriculture, Faculty of Agriculture, Kyoto University, Oiwake-cho, Kita-Shirakawa, Sakyo-ku, Kyoto 606-8502, Japan

Matsumoto, Shingo
Department of Agricultural Environment, Shimane Agricultural Experiment Station, 2440 Towatari-cho, Izumo, Shimane 693-0035, Japan

Matsuo, Kazuyuki
Crop Production and Environment Division, Japan International Research Center for Agricultural Sciences (JIRCAS), 1-1 Owashi, Tsukuba, Ibaraki 305-8686, Japan

Minami, Katsuyuki
National Institute for Agro-Environmental Sciences (NIAES), 3-1-3 Kannondai, Tsukuba, Ibaraki 305-8604, Japan

Mori, Satoshi
Department of Applied Biological Chemistry, Division of Agriculture and Agricultural Life Science, Faculty of Agriculture, University of Tokyo, 1-1-1 Yayoi, Bunkyo-ku, Tokyo 113-0032, Japan

Narro, Luis
Maize Program, International Center for the Improvement of Wheat and Maize (CIMMYT/CIAT), Apdo Aéreo 6713, Cali, Colombia

Nishio, Michinori
Institute of Agriculture and Forestry, University of Tsukuba, 1-1-1 Tennodai, Tsukuba, Ibaraki 305-8572, Japan

Nishizawa, Naoko K.
Department of Global Agricultural Sciences, Faculty of Agriculture, University of Tokyo, 1-1-1 Yayoi, Bunkyo-ku, Tokyo 113-0032, Japan

Ofei-Manu, Paul
Faculty of Agriculture, Yamagata University, 1-23 Wakaba-cho, Tsuruoka, Yamagata 997-8555, Japan

Okada, Kensuke
Crop Production and Environment Division, Japan International Research Center for Agricultural Sciences (JIRCAS), 1-1 Owashi, Tsukuba, Ibaraki 305-8686, Japan

Okajima, Hideo
Emeritus Professor of Hokkaido University, 4-9-30 4-jo, Nishino, Nishi-ku, Sapporo, Hokkaido 063-0034, Japan

Otani, Takashi
Regional Research Promotion Division, Agriculture, Forestry and Fisheries Research Council Secretariat, Ministry of Agriculture, Forestry and Fisheries, 1-2-1 Kasumigaseki, Chiyoda-ku, Tokyo 100-8950, Japan

Pandey, Shivaji
International Center for the Improvement of Wheat and Maize (CIMMYT), Lisboa, 06600 Mexico D.F., Mexico

Poulton, Paul R.
Soil Science Department, IACR-Rothamsted, Harpenden, Hertfordshire, AL5 2JQ, UK

Randall, Peter J.
Commonwealth Scientific and Industrial Research Organization (CSIRO) Plant Industry, GPO Box 1600, Clunies Ross Street, Canberra, ACT 2601, Australia

Richardson, Alan E.
Commonwealth Scientific and Industrial Research Organization (CSIRO) Plant Industry, GPO Box 1600, Clunies Ross Street, Canberra, ACT 2601, Australia

Salazar, Fredy
Maize Program, International Center for the Improvement of Wheat and Maize (CIMMYT/CIAT), Apdo Aéreo 6713, Cali, Colombia

Sanchez, Pedro A.
International Centre for Research in Agroforestry (ICRAF), P.O.Box 30677, Nairobi, Kenya

Shen, Renfang
Laboratory of Soil Biochemistry, National Institute for Agro-Environmental Sciences (NIAES), 3-1-3 Kannondai, Tsukuba, Ibaraki 305-8604, Japan

Smithson, Paul C.
International Centre for Research in Agroforestry (ICRAF), P.O.Box 30677, Nairobi, Kenya

Srinivasan, Ancha
Regional Science Institute, 4-13, Kita-24 Nishi-2, Kita-ku, Sapporo 001-0024, Japan

Swindale, Leslie D.
Director General (Emeritus) of International Crops Research Institute for the Semi-Arid Tropics (ICRISAT), 2910 Nanihale Pl, Honolulu, HI 96822, USA

Wagatsuma, Tadao
Faculty of Agriculture, Yamagata University, 1-23 Wakaba-cho, Tsuruoka, Yamagata 997-8555, Japan

Wissuwa, Matthias
Laboratory of Soil Biochemistry, National Institute for Agro-Environmental Sciences (NIAES), 3-1-3 Kannondai, Tsukuba, Ibaraki 305-8604, Japan

Yamagata, Makoto
Department of Integrated Research for Agriculture, National Agricultural Research Center for Hokkaido Region, Shinsei, Memuro-cho, Kasai-gun, Hokkaido 082-0071, Japan

Part I.
Historical Review and Global Concerns

Historical Significance of Nutrient Acquisition in Plant Nutrition Research

Hideo Okajima

Summary. To highlight the significance of plant nutrient acquisition, which lately has been the focus of much attention, this chapter reviews the history of studies on plant nutrient absorption. It is found that the field has greatly progressed based on concepts developed in the first half of the 20th century. By synthesizing this progress, scientists in the 1970s developed mathematical models to describe the relationship between nutrient availability and plant nutrient absorption. These models have had a major impact on the understanding of nutrient absorption and have profoundly contributed both conceptually and practically. The models also have identified mechanisms which are not well comprehended or areas where further experiments are required. Nutrient acquisition is one of these subjects requiring further study. Plants have an autocontrolled feedback system in nutrient absorption, whereby the nutrient absorption power is induced and increased by its nutrient stress; thus, roots actively "mine" their nutrients. This nutrient acquisition may have been seen in a new light after the discovery of mugineic acid (a phytosiderophore) which had been discovered in a careful inquiry into the significance of flooding for rice growth. Such an adaptive feature of nutrient absorption to the environment has received attention toward efficient use of plant potential under the paradigm of sustainable agriculture. Thus, in the 1990s we have come to a new age in these studies whereby the term "nutrient acquisition" has become common usage.

Key words. Discovery of mugineic acid, Efficient use of plant potential for sustainable agriculture, Excretion of organic acid, Mechanisms of nutrient acquisition, Mobility concept, Models for nutrient absorption, Nutrient availability, Studies in Japan

1. Introduction

Plants cannot select their surroundings, nor can they pursue food. Thus, they have adapted evolutionarily to their environments such as to climatic or edaphic conditions. To meet the demand for nutrients, plants actively "mine" their nutrients from their given surroundings. Under the paradigm of sustainable agriculture, plant nutrient acquisition, which is an aspect of active nutrient absorption to counter adverse conditions, is now receiving special attention in the study of plant nutrient absorption.

The main mechanisms of active absorption to counter adverse conditions seem to be: (i) root extension for increasing the root surface to gather nutrients, (ii) selective nutrient absorption by roots, (iii) root action to make rhizosphere conditions favorable for root growth, and, (iv) root action to solubilize more of the limiting nutrients. All of these mechanisms substantially influence plant nutrient absorption. Yet their causal relations in the acquisition are so complex that the understanding of them is still limited. This chapter reviews the history of study on plant nutrient absorption related to nutrient acquisition, to provide a basis for future study.

2. Beginnings of Study

Scientists have been interested in the study of available mineral nutrients in soils and plant growth since Liebig proposed the Mineral Theory (1840) and further added what came to be known as the Law of the Minimum. The Law of the Minimum states that the nutrient present in the lowest amount relative to the others is the limiting nutrient. It means that all the other nutrients are present in excess until the deficient or limiting nutrient is made adequate. It was the Law of the Minimum, in particular, that prompted the search, by chemical analysis, for the deficient nutrient or limiting nutrient in soils that governed plant growth. A typical example is Dyer's one per cent citric acid extraction method (1894) for estimating available soil phosphorus and potassium (cited by Russell 1915; Russell 1973). This reagent was intended to imitate the acidity of plant root excretions. Dyer had assumed that the reason different species of plants differed in ability to extract nutrients was because of differences in the total acidity of root sap. The strength of 1% citric acid solution was supposed to represent the average acidity of root sap.

Dyer's method contributed to the development of chemical methods for estimating available nutrients more simply and quickly, under the expectation that extraction can be more closely related to the fraction used by crops. Along this line, many kinds of chemical analyses of soils (so-called soil tests) have been proposed based on behavior of each nutrient in soils and traits of nutrient acquisition of crops. In fact, they have provided guiding information for fertilization in farming.

Soil solution is a natural medium for nursing plants, as proved by the Sachs and Knop solution culture experiments of the 1800s. Against the trend of predicting plant nutrient availability by the chemical analysis of a handful of soil, Cameron (1911) stressed that much more attention should be given to the study of soil solution. He stated that there should be a general recognition that soil phenomena are essentially dynamic in character, and that the investigation of the properties of the soil solution and its relation to crop production is a procedure certain to yield results of positive value. By comparing results of solution culture experiments with the soil solutions obtained from various soils, it was assumed that the soil solution is normally of a concentration amply sufficient to support ordinary crop plants, and that it is maintained at sufficient concentration of mineral plant nutrients. It seemed also that different crop plants require different amount of these minerals. However, it was unclear whether different plants require different concentrations of the constituents in the nutrient solution for their respective best growths. Furthermore, Cameron demonstrated that the organic substances in the soil solution affect plant growth. Some organic compounds may disturb plant growth while the others promote growth. The former was found in infertile soils and the latter in fertile soils.

Hoagland (1922) conducted pot experiments with oats for seven years to investigate the relation between acid soluble nutrients and nutrient uptake. He could not find, however, a consistent relationship, and finally concluded that nutrient supplying ability of soils or availability of soil nutrients is not simply determined by the nutrient content of soil or solution but may be determined by the replenishment ability of nutrients in the soil solution which are reduced by plant absorption. The idea later became the base for the term "buffer power" of soil nutrient availability.

Johnston and Hoagland (1929) also pointed out a consideration for estimating nutrient availability, that is, that concentration and volume are two distinct characteristics of a solution. These two factors are in part analogous to the "intensity" and "capacity" factors of energy. The energy of a falling stream of water depends on the height of fall (the intensity factor) and on the quantity of water falling (the capacity factor). In the soil system, the intensity factor is regarded as the concentration of the soil solution, whereas the capacity factor is determined by the nature of solid phases and by the volume of soil physiologically available for root dispersion.

It was suggested, therefore, that experiments on the effect of nutrient concentration on nutrient uptake or plant growth should be conducted under conditions without the limitation of capacity. Otherwise, one would not be able to properly evaluate the results obtained from the solution culture experiments. Using a flowing culture technique without the limitation of capacity, they demonstrated that an optimum potassium concentration, that is, the minimum concentration of potassium giving maximum yield for tomato, is only 5 ppm. Their work clarified that nutrient availability depends on intensity and capacity, including buffer power.

3. Concept of Active Absorption

Meanwhile, Hoagland and his collaborators, who had studied giant kelp as a potential potassium resource during World War I, presented a novel concept of mineral nutrient absorption by plants. They found the remarkable selective accumulation by these plants of potassium, iodide and bromide. It did not appear from their results that the uptake of the nutrient ions was merely a diffusion process proceeding to achievement of equal concentration of a solute in internal and external phases, as many texts taught.

To simplify the study, they selected the fresh water alga *Nitella*, which produces internodal cells, often several inches in length, from which most of the sap can be recovered. The analyses of *Nitella* sap and pond water indicated clearly that the cells must have absorbed all the principal ions against concentration and activity gradients. This distribution of ions could not be explained according to the Donnan equilibrium, which comes into play when an ion of one polarity cannot pass through a membrane. The phenomenon was confirmed in experiments of excised barley roots (Fig. 1). It is notable that remarkable accumulation of potassium, halide and nitrate ions was observed only in aerobic conditions.

They also observed marked increase in the capacity for accumulation of potassium with the addition of sugar to the culture medium. Based on these observations, they (Hoagland and Broyer 1936; Hoagland 1944) considered that selection or accumulation of nutrients depends on aerobic metabolic process of roots, that is, to absorb nutrients, plants require the expenditure of energy against concentration and activity gradients. In other words, aerobic respiration supplies the energy required for the selective absorption. Such selective accumulation has been called "active or metabolic absorption," distinguished from non-metabolic passive absorption such as diffusion of the nutrient through the surface of plant roots.

However, there was a major controversy on respiration for nutrient absorption between Hoagland's and Lundegårdh's concepts . Lundegårdh (1937) stressed that most of the energy is required for anion absorption because the anion needs to be absorbed against a cell charge gradient. He postulated that the anion was absorbed in conjunction with an electron transfer system in an oxidation process of aerobic respiration. To differentiate, he called such respiration "anion respiration." It was considered later that anion respiration can be a process which ameliorates the thermodynamically unfavorable conditions for anion uptake (Lundegårdh 1951). In any case, the debate contributed to the recognition of the significance of respiration in nutrient acquisition. Thus, it can be said that the basic concepts of nutrient availability and plant nutrient absorption had been established by the mid-20th century.

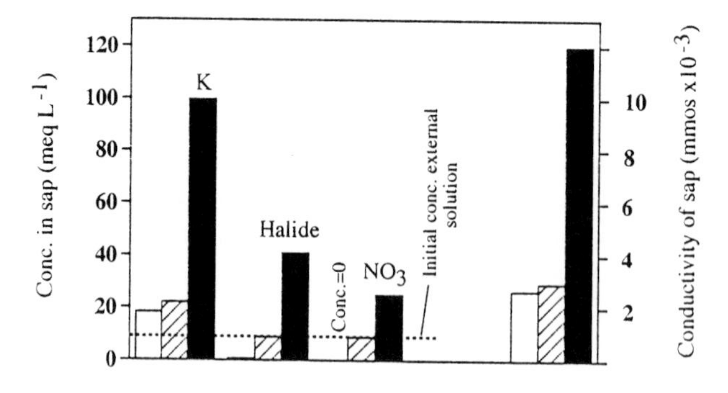

Fig. 1. Effect of aeration on accumulation of potassium, halide, and nitrate ions by excised roots of barley plants. During the absorption period (10 h), a rapid stream of air was passed through one solution and purified nitrogen gas through the other. (Hoagland and Broyer 1936). *White bars,* initial concentration in sap; *oblique line bars,* concentration after absorption with N_2; *black bars,* concentration after absorption with air

4. Studies in Japan

The Japanese Society of Soil Science and Plant Nutrition was founded in 1927. Attempts have been made for the timely introduction of concepts or techniques developed in Europe and the United States into our Society. Despite frequent and unexpected difficulties in the introduction, efforts to overcome the difficulties have yielded undeniable knowledge of plant nutrient absorption. For instance, at one time, ammonium sulfate, a well-known chemical fertilizer produced in Europe, was found to be ineffective in waterlogged paddy fields. Investigation indicated only about 20% to 30% of the nitrogen in the fertilizer applied was used by plants. Improving the efficiency of ammonium sulfate in paddy fields, thus, was recognized as an urgent issue.

In order to clarify the cause of the low efficiency of nitrogen fertilizer, the following factors were tested: leaching loss of nitrogen, change of inorganic nitrogen to organic nitrogen, and evaporative loss of ammonia nitrogen from soils. The results of the investigations, however, showed that none of the factors described above contributed to the low efficiency of ammonium sulfate. After intensive studies by a number of researchers, Shioiri (1942) concluded that the low efficiency of ammonium sulfate was due to a denitrification characteristic of paddy fields with development of an oxidative and a reductive soil layer.

Waterlogged paddy soils differentiate into an oxidized layer and a reduced layer. Ammonium nitrogen changes to nitrate nitrogen which is applied to the oxidized layer of the soil surface, a thin surface film of soil. The nitrate nitrogen formed in the oxidative layer moves down into the reduced layer, a thicker layer of reduced soil just below the oxidized one. There, the nitrate nitrogen is reduced to nitrogen gas which is lost from the soil. Thus it became evident that the low efficiency of ammonium sulfate was caused by the application of fertilizer to the surface soil, oxidized layer. To avoid nitrogen loss, it was suggested that the nitrogen fertilizer should be applied to the reductive soil layer. Thus, application of ammonium fertilizer in the overall soil layers has been recommended in rice farming.

Shioiri's main evidence for denitrification was an observation in incubation experiments that the decrease of ammonium nitrogen (5.66 mg/30g soil) and the increase of N_2 gas in air (4.5 mg) were nearly equal in amount under submerged conditions. The evidence may be not have been conclusive, but the improvement of efficiency of ammonium nitrogen fertilizers by the recommended fertilization was so remarkable that his insight has been extensively accepted.

Concerning the development of a reduced layer, naturally the study was pushed forward to concern changes in the forms of iron, manganese and sulfur in submerged paddy soils. Sometimes the reducing layer will have a redoxy potential sufficiently low for hydrogen sulfide to be formed which is an inhibitor of aerobic respiration for active nutrient absorption. Hydrogen sulfide was found to be produced, particularly in so-called degraded paddy soils which are characterized by low levels of iron. *Akiochi* (autumn decline in rice) had been the most serious physiological disease of rice in degraded paddy field. *Akiochi* had been considered to be induced primarily by root rot caused by hydrogen sulfide and its related unfavorable rhizosphere condition.

Following on this, researchers focused on practical methods for improving low-productivity degraded paddy fields. It was recognized that a heavy dressing of soil high in iron, adequate application of major mineral nutrients, application of silica-containing materials, application of non-sulfur-containing fertilizers, and water management to keep suitable redoxy potentials are quite effective. These findings not only revealed the dynamic aspect of plant nutrient absorption but also contributed to the amelioration of low-productivity paddy fields and the increase of rice production desperately needed after World War II.

5. Performance of Solution Culture in Japan

Ishizuka and Kimura in about1930 pioneered the solution culture technique in Japan for studies of plant nutrition under a simplified controlled condition. Their studies on nutrient uptake and rice plant growth stages under solution culture led to the fundamental concept of rational fertilization in crop production, leading to subsequent study in Japan (Ishizuka 1932; Kimura and Chiba 1943). In the beginning, Kimura could not properly grow rice plants raised with the Sachs or Knop nutrient culture solutions for upland crops. He succeeded, however, in rice culture with lower solution concentration than that for upland crops. He demonstrated that the dilution effect of the solution was caused by the lowering of phosphorus concentration (Fig. 2). Rice growth was favorable at low P concentration of the solution while retarded at high P concentration due to iron chlorosis. In contrast, the growth of barley was normal over the whole range of the concentrations and increased with the increase of P concentration (Kimura 1931, 1932). It was also observed that rice plants hardly absorbed iron from sparingly soluble ferric phosphate, while barley readily absorbed it under the same condition. Furthermore, rice took up P more readily than did barley.

From the results Kimura assumed the iron absorption ability of rice to be lower than that of barley, acclimatized to the rich availability of iron in the paddy field. Gile and Carrero (1914) also noted the low ability of rice in iron absorption. Therefore, in the case of high P application, P absorbed excessively in rice may act on iron in the tissue of leaves such that the iron existing in the cell is turned into an inactive form and causes chlorosis. In contrast, barley, which readily absorbs iron to adapt to the low availability of iron in upland conditions, keeps an active form of iron in the cell even at high P application. This seemed to be why rice required a low P concentration of the culture solution. Kimura's hypothesis presents a concept whereby traits of nutrient acquisition differ with measures adapted to their environments. However, Kimura (1952) stated also that such specific traits of rice in Fe and P nutrition could be observed only in the nutrient solution culture. The phenomenon could not be observed in the field because of low solubility of phosphorus compounds in soil systems. To generalize the results obtained from the solution culture experiments, therefore, we must pay special attention to their conditions.

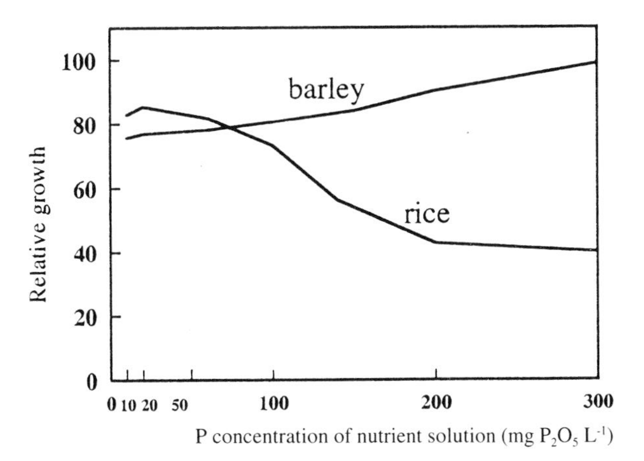

Fig. 2. Effect of P concentrations of culture solution on plant growth in rice and barley (modified from Kimura 1931)

6. Chemical Potential of Nutrients and Required Energy for Uptake

The concept of intensity and capacity factors in nutrient availability proposed by Hoagland and his colleagues has been developed in several ways since the 1950s. Among them, at first, it is natural to introduce the achievements of Schofield and Woodruff, who have tried to apply thermodynamics in the study of nutrient availability. The idea is that the intensity of the availability can be measured by the work needed to withdraw nutrients from an equipotential pool of soils. The work can be expressed as chemical potential, such as the chemical potential of $Ca(H_2PO_4)_2$, the phosphate potential (Schofield 1955) or changes in free energy (ΔF) for potassium (Woodruff 1955).

Phosphate potential is calculated from the pH and concentration of calcium and phosphorus in a solution of 0.01M-$CaCl_2$ which is in equilibrium with the soil. Calculation is done using the formula $1/2\ pCa + pH_2PO_4$ (p: negative logarithm, as pH). Schofield explained the significance of phosphate potential as the measurement of the intensity of its supply, by the analogy of pumping water from a well. The potential is a measure of the depth of the level of soil phosphate just as the depth of water in the well determines the work needed to get the water to the top. As water is taken from a well it is replaced by lateral movement of ground water. In the same way, when phosphorus is taken from the soil solution it is replaced by more dissolution. The withdrawals generally cause some of the level lowering, depending on the capacity of the system.

The energy of exchange for any pair of cations in the exchange complex and in the solution should be the same under the equilibrium condition. If most of the cationic nutrients are held on the exchange complex, the energy of exchange may indicate the work needed by plants to withdraw the nutrients. Woodruff (1955) was the first to calculate the exchange energy for potassium from the cationic composition of the soil solution , based on the equation, changes in free energy $\Delta F = 1364 \log aK/\sqrt{aCa}$ (in calories/g equivalent, a: activity) as the intensity of potassium availability. Arnold (1962), cited by Cooke (1967), also calculated the energy as $\Delta G_{Ca,K}$=2.303RT[pK -1/2p(Ca+Mg)]. Arnold found a close relation between $\Delta G_{Ca,K}$ and potassium uptake by perennial ryegrass for most of the soils tested. Along the same lines, Beckett (1964) adduced a concept of potential buffering capacity of soil nutrients, called the "Q/I relation." This was the first attempt to describe numerically the buffer power of nutrient supply which represents the dynamic relation between intensity and capacity factors.

These proposals demonstrated the genuine usefulness of thermodynamic aspect for simplifying and clarifying the complicated phenomena of soil systems, thus contributing to advances in soil science. However, they were not enough to describe nutrient uptake. Wild (1964) pointed out from a pot experiment that in solutions which are as dilute as soil solutions it is phosphate concentration and not phosphate potential that determines uptake. $\Delta G_{Ca,K}$ is a function of [pK-1/2p(Ca+Mg)] which are the ratios of activities of ions in the soil solution. Hence, the activity ratio for potassium, AR^k also has been used for studying potassium availability. Figure 3 shows the relationship between potassium uptake of alfalfa and AR^k (Okajima and Matsunaka 1972). Potassium uptake increases with the decrease of AR^k, indicating that AR^k governs potassium uptake. The relations, however, differ in different soils used: that is, the regression lines differ with soils. For the criterion of potassium availability, the AR^k should be universal. On the AR^k as a measure for potassium availability, it has been suggested that in any attempt to apply equilibrium solution methods to potassium testing, a careful choice of experimental conditions will be necessary (Taylor 1958). Nye (1968) suggested that energy provided from root respiration greatly exceeds the energy for potassium exchange.

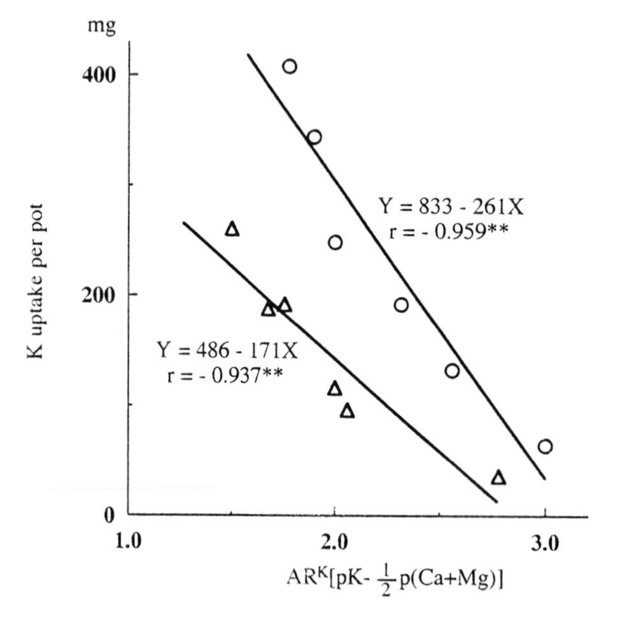

Fig. 3. Relationship between potassium uptake and AR^k in alfalfa (modified from Okajima and Matsunaka 1972). *Circles*, Hokudai-soil, a Fluvisol; *triangles*, Hokunoushi-soil, an Andosol

7. Mobility Concept

Bray (1954) pointed out that studies of availability had neglected the mobility of nutrients. He proposed a concept on this matter in a paper entitled "A nutrient mobility concept of soil-plant relationships." He stated: "The mobility of nutrients in soils is one of the most important single factors in soil fertility relationships. The term mobility, as used here, means the overall process whereby nutrient ions reach sorbing root surfaces, thereby making possible their sorption into the plant."

On the nature of nutrient availability, besides the Law of the Minimum the Mitscherlich-Baule Percentage Sufficiency Theory was proposed in which plant growth is said to follow a diminishing increment type of curve well known as the "yield curve." Originally the mobility concept was proposed for understanding both the theories without contradiction. The concept is illustrated in Fig. 4. There are two kinds of root sorption zones for plants. One is the large volume of soil occupied by the major part of plant root system. It is called the "root system sorption zone." From this zone plant roots can remove almost quantitatively the highly mobile nutrients such as nitrate nitrogen. Such nutrients may act as limiting nutrients, under the Law of Minimum. The other is a relatively thin layer of soil adjacent to each root surface. It is called the "root surface sorption zone." From this zone the roots effectively absorb the relatively immobile nutrients such as exchangeable potassium or hardly soluble phosphate. The sum of these root surface sorption zones represents a small part of soil, hence the roots absorb only a small part of the nutrients present. Thus, the amounts that must be present for maximum yields are many times larger than plant requirement. Such nutrients may act not according to the Law of the Minimum but according to the Percentage Sufficiency Theory.

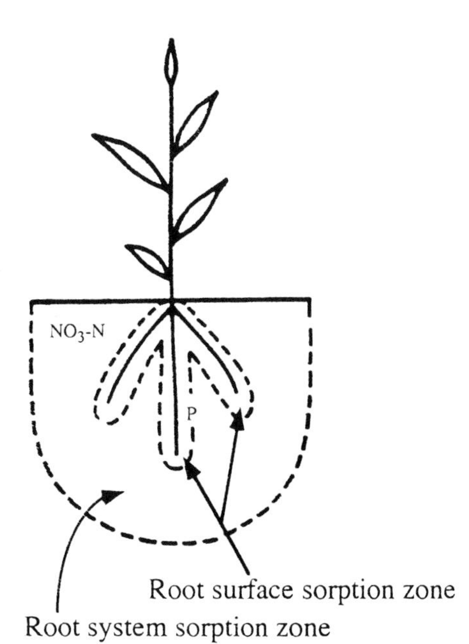

Fig. 4. Root system sorption zone for mobile nutrients and root surface zone for immobile nutrients (modified from Bray 1954)

The concept had a major impact on the field and opened the way for quantitative estimation of nutrient availability. The mobility of ions is governed by two processes: mass flow of the soil solution to the root induced by transpiration, and diffusion of ions to the root induced by their concentration gradient near the root surface (Barber 1962). The relative importance of mass flow and diffusion on the nutrient transfer from sources of supply to sites of absorption primarily depends on nutrient concentration of the soil solution and demand by plants. Because the concentrations of potassium and phosphorus are relatively low, much of these nutrients reach the root surface by diffusion. In contrast, the concentration of calcium, magnesium, and nitrate nitrogen is usually high enough for mass flow to be a dominant factor in the transport of these nutrients. However, the contribution of mass flow and diffusion on the nutrient supply is relative. Even for nitrate nitrogen, which is a highly mobile nutrient, the contribution rate of mass flow on its supply varies within a day, depending on diurnal changes of water uptake and the concentration (Okajima and Hatano 1988).

A number of studies addressed nutrient movement in relation to plant nutrient uptake (see Nye and Tinker 1977). Among them Olsen et al. have applied the diffusion equation to a mathematical model for phosphorus movement to the root surface and phosphorus uptake of plants. For example, Olsen and Watanabe (1963) introduced an equation for the steady-state diffusion of ions in porous media proposed by Porter et al. (1960) into their phosphorus uptake model, in order to demonstrate the significance of soil texture on P supply and P uptake, as shown in the following equation.

$$Q = ab(C_o - C_r)[2\pi^{-1/2}T^{1/2}+1/2T-1/6T3/2\pi^{-1/2}+1/16T^2\ldots]$$

where Q is the amount of uptake of P per unit of root surface area in time, t, a is the root radius, b is a capacity factor, C_o is the initial concentration of P, C_r is the concentration of P at the root surface, and π is 3.14. T is $D_p t/ba^2$, where D_p is the diffusion coefficient of P in the soil.

The results in Fig. 5 and Table 1 show that actual P uptake of corn seedlings in a 24-h period was approximately a linear function of P concentration between 0.05 and 0.3 ppm, and P uptake was one-third as much from sandy soil as from the clay soil at equal concentrations of P in the soil solution. This result agreed closely with the predicted values. As the clay content of three soils increased from 17 to 51 %, the diffusion coefficient for P increased from 1.1×10^{-7} to 6.7×10^{-7} cm^2 per second. Transport of P by diffusion in soil to a plant root depends on the concentration gradient between the soil and the root surface and the capacity of the soil to replenish P taken up by the plants. Textural variations influence all of these factors. Hence, they suggested that an evaluation of fertilizer P availability on soils differing in clay content must account for the effects of these variables.

Table 1. Observed and calculated relative rates of P uptake by corn roots (from Olsen and Watanabe 1963)

Soil	P concentration (ppm)	Relative uptake rate	
		Observed	Calculated
Pierre clay	0.15	1	1
	0.20	1	1
Apishapa silty clay loam	0.15	0.73	0.70
	0.20	0.69	0.70
Tucumcari fine sandy loam	0.15	0.50	0.30
	0.20	0.49	0.30

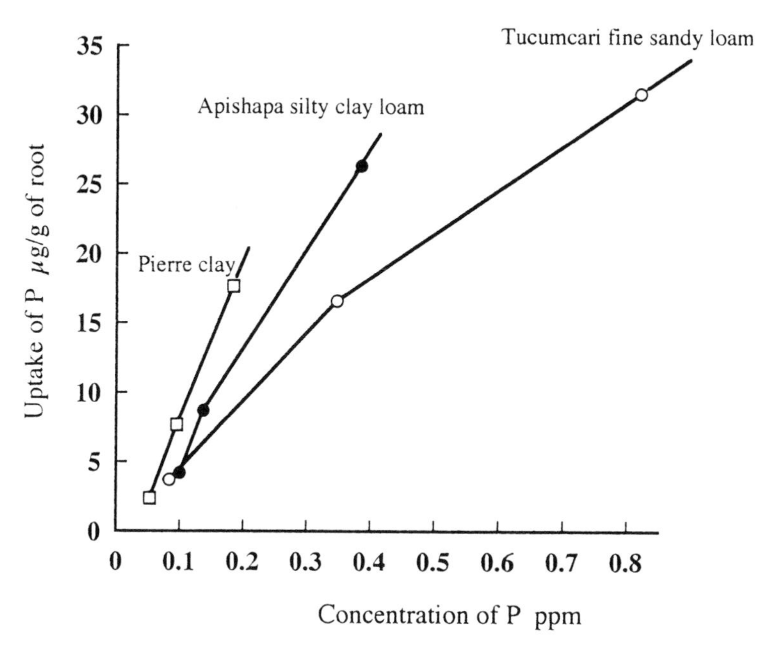

Fig. 5. Absorption of P by corn seedlings as a function of concentration of P in the solution on three soils varying in clay content. Uptake period of 24 h (modified from Olsen and Watanabe 1963)

8. Prediction Model for Nutrient Uptake

Since the nutrient uptake model was proposed by Nye and Marriott(1969), flux by mass flow and diffusion to the root (which can be a main limiting factor for the replenishment ability of nutrients in the soil solution) have been the focus of other models. Various models have been developed using computer technology, such as the Claassen-Barber model for K uptake (1976), the Nye-Brewster-Baht model for P (1975), the Barber-Cushman model for N (1981), etc. A model is very effective in integrating the knowledge obtained, as well as indicating whether proposed mechanisms are proper and defining areas where further experiments are required. The models which have been developed since the1970s, tell this story. The Nye-Brewster-Bhat model (Brewster et al. 1975) for prediction of P uptake is an example. It is characterized by introducing a plant growth model to the P uptake model. The parameters include the net assimilation rate, the ratio of leaf surface area to shoot weight, and the ratio of shoot weight to root weight which determine the plant growth rate.

The results obtained are given in Fig. 6. At intermediate levels of soil P there was close agreement between experimental and predicted uptake. This confirmed previous developments in this field. However, the experimental P uptake slightly exceeded that predicted for the soil lowest in P content and was considerably lower than that predicted for heavily fertilized soil. Regarding the unexpectedly high P,

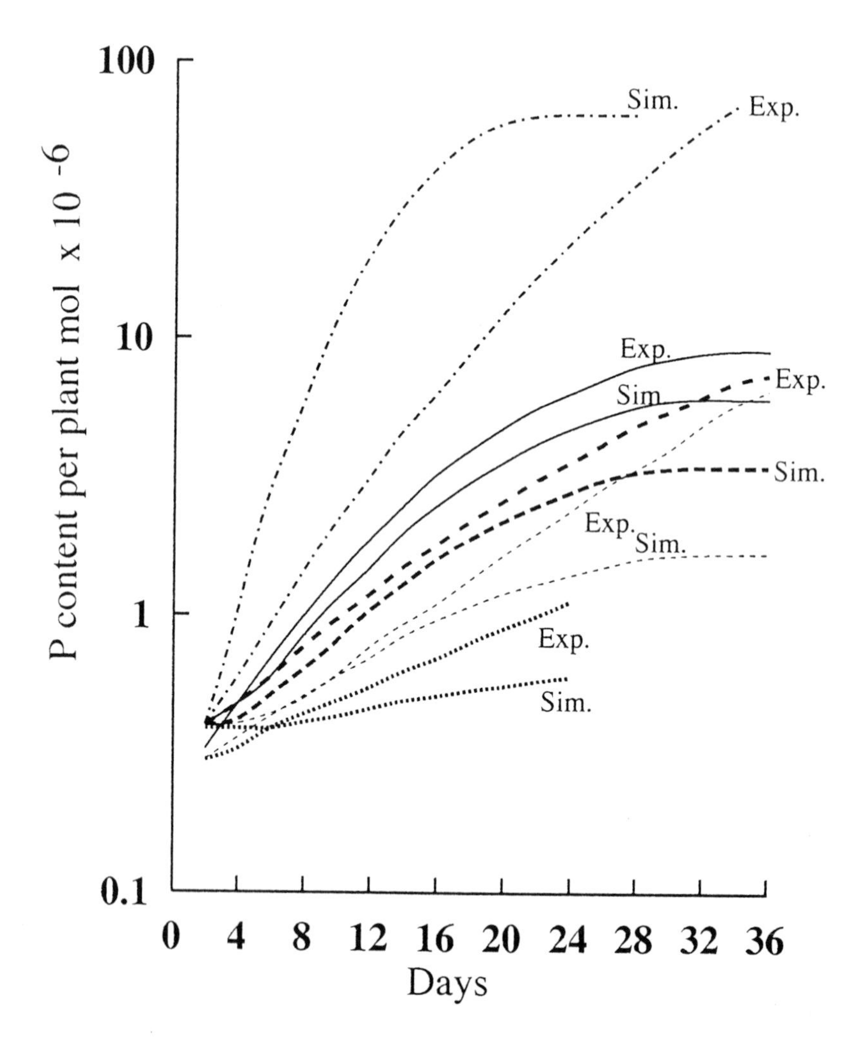

Fig. 6. Measured and simulated phosphorus absorption of onion plants using Nye-Brewster-Bhat whole plant phosphate nutrition models (from Brewster et al. 1975)

they suggested that structured soil appeared to have a lower phosphate buffering capacity than was measured using suspension for prediction parameter; it was thought this lower buffering capacity may influence onion seedlings growing in soils of high P concentration. Also they assumed that in low P soil the roots may be acting as rather more than simple sinks for P. These were the issues, suggested by the models, which would be clarified by further study.

9. Study of Rice Plant Nutrient Absorption

A flooded condition is highly favorable for rice plant growth compared with an upland water condition. The primary factors favorable for growth are abundance of water and of nutrients. Irrigation water supplies nutrients and anaerobic reductive soil conditions increase the nutrient availability and control soil pH. Also, movement of nutrients to the root surface such as by diffusion may be favorable due to water saturation. The advantages, however, are only enjoyed by wetland plants, such as rice, which have a specific mechanism for growth in anaerobic submerged conditions. The main mechanism for this is a transport system of oxygen from shoot to root. Also, rice roots have an enzymatic oxidation system. These functions make their rhizosphere condition favorable through oxidizing harmful hydrogen sulfide or excess ferrous iron produced in the reductive submerged conditions. Thus, they support nutrient acquisition.

Clarification of these relations were attempted during the 1950s to find a way to control rice root environments and nutrient acquisition. The harmful effect of hydrogen sulfide on nutrient uptake of rice had been investigated mostly in solution culture experiments (Mitsui et al. 1951; Okajima and Takagi 1953). The inhibition rate of hydrogen sulfide to uptake was calculated from the change of nutrient concentration in the culture solution. The inhibition rate appears to follow the order: $P>K>Si>NH_4\text{-}N>Mn>H_2O>Mg>Ca$. Yamane et al. (1956) observed in a pot experiment using degraded soils that hydrogen sulfide inhibited nutrient uptake of rice, an inhibition which was prevented by the addition of iron oxide. Besides the inhibition of nutrient uptake, hydrogen sulfide enters the shoot through the roots, and disturbs rice growth in acidic conditions of the medium (Takagi and Okajima 1956a). The concept of active nutrient absorption was thus introduced in Japan.

Depression of aerobic respiration of rice roots treated with hydrogen sulfide was observed in an experiment using Warburg's manometer (Okajima and Takagi 1955; Takagi and Okajima 1956b). Hydrogen sulfide consumes oxygen in the process of its oxidation. The root also consumes oxygen. Hence it seems likely that competition for oxygen consumption occurs between rice root and hydrogen sulfide in the rice rhizosphere. Figure 7 shows the effect of Fe and pH on the oxygen consumption of sulfide. The rate of oxidation of sulfide was accelerated by iron, and the maximum consumption of oxygen was observed in the solution of about pH 7. The oxidation of hydrogen sulfide by roots in pot experiments showed the same tendency. From these results it was assumed that the oxidation of hydrogen sulfide occurs through the formation of $Fe\text{-}H_2S$ intermediate complex on the root surface, and pH controls the formation. The phenomenon is complex but this may be one reason why H_2S enters the shoot in the acidic condition. H_2S-oxidation-Fe-pH may be an issue for study in nutrient acquisition of rice.

Although knowledge has been accumulated on the transport system of molecular oxygen from the shoot to the root and their relation to root oxidizing power, details of their mechanism are limited, because of technical difficulties investigating the complex rhizosphere phenomena in chemical, physical, and biological reactions.

Recently the simulation model study has demonstrated its power in this field. For example, based on the kinetics of Fe^{2+} in submerged soil condition, Kirk et al. (1990) developed a simulation model of the simultaneous diffusion and reaction of oxygen and ferrous iron in soil, with which to investigate conditions near rice releasing oxygen. They predicted that large amounts of iron would be transferred towards the oxidizing surface, leading to a well defined zone of ferric hydroxide accumulation, and that the pH in this zone would be more than two units lower than in the bulk soil. The results give numerous implications for rhizosphere processes. For instance, the phosphate buffer of fresh precipitate at low pH would be very high, resulting in reduction of the diffusive supply of P to roots. At the same time the low pH would favor release of soil phosphorus. Also, nitrification in oxi-

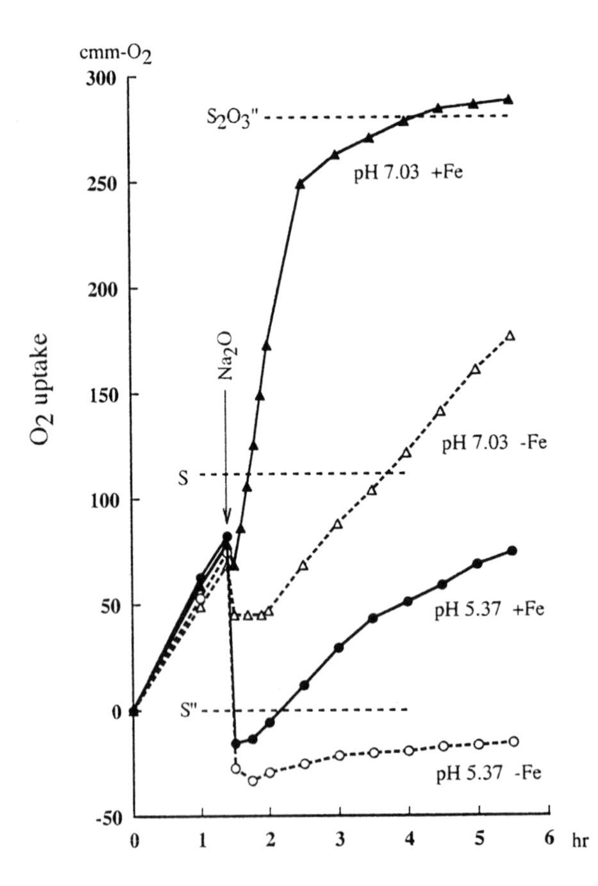

Fig. 7. Effect of Fe and pH on oxygen uptake by Na_2S in phosphate buffer solution by Warburg's manometric method. Phosphate buffer in main chamber, 1 mL; 10^{-2} M Na_2S in sidearm, 25°C (from Takagi and Okajima 1956b). *+Fe*, with Fe-citrate; *–Fe*, without Fe

dized rhizosphere would be very low at low pH, and a low pH would hinder biological nitrogen fixation.

The models described above are called micro-models. In contrast, there is a macro-model which describes such mechanisms as nutrient flow or nutrient uptake in fields, covering special growth stages or entire growth period. Integrating the knowledge obtained, Itoh and Awasaki (1996) developed a macro-model for predicting nitrogen dynamics and rice growth. The main processes taken up were: diffusion and mass flow of ammonium N, mineralization, immobilization and remineralization of immobilized N, denitrification, eluviation and N uptake by plants. N dynamics are so complex that they used a quite large number of parameters, that is, 34 for their model, while the Claassen-Barber micro-model for potassium uptake described above uses 11 parameters.

They presented fruitful results. One is given in Table 2. The table shows reasonably close agreement between the predicted and measured values for N uptake and dry matter of rice with successive growth after top dressing of N. Such a process-oriented simulation model is very useful not only to get a better understanding of N economy but also to improve rice farming practice such as by improving the decision-making process on fertilizer management.

10. Discovery of Mugineic Acid

Study of traits of rice plant growth under flooding condition also has resulted in significant advances in knowledge of plant nutrition for plants other than rice. This point demands mention here. As an example, the story of mugineic acid's discovery is introduced. Takagi (1960) studied the physiological significance of flooded conditions in rice plant growth. He raised rice seedlings in an artificial soil consisting of sand-clay minerals without organic materials. No chemical changes of nutrients were expected to occur in the medium without organic matter even in the flooded condition. Hence, only the effect of soil water of the flooded condition would be observed in the artificial soil.

Table 2. Stimulated and experimental values for N uptake and dry matter of rice at 21 days after topdressing of N at panicle formation (from Itoh and Awazaki 1996)

Variable	Stimulated value	Experimental value
Dry matter (t ha^{-1})	0.641	0.837
N uptake (kg ha^{-1})	86.1	119.0
^{15}N uptake (kg ha^{-1})	11.0	12.3
NH_4-N in soil (kg ha^{-1})	11.5	3.8
NH_4-^{15}N in soil (kg ha^{-1})	0.11	0.08
Immobilized ^{15}N (kg ha^{-1})	4.0	3.2
Lost ^{15}N (kg ha^{-1})	4.9	4.4

Rice seedling growth was promoted proportionally to the increase of soil water over the range from 40% of maximum water holding capacity to the flooded condition. The flooded moisture condition is the best for rice. However, as shown in Fig. 8, even for rice, the growth was depressed in flooded conditions when pH > 6.0. This was clearly due to iron deficient chlorosis, later called "flooding-induced chlorosis." In order to elucidate the cause of the phenomenon, a simulated upland water condition "root submergence intermission" was treated in the water culture experiments. The treatment method was brief removal of nutrient solution until the first signs of wilting. The treatment was repeated periodically once or twice per day. The results distinctly showed no iron deficient chlorosis occurring with the treatment even in pH > 6.0 medium, whereas serious chlorosis occurred without the treatment (Takagi 1966).

From these results Takagi intuited that rice has a characteristic iron acquisition system. That is, rice root may excrete a chelator-like iron solubilizing substance. If so, the substance would be leached from the rhizosphere under the flooding condition of artificial soils, whereas the substance would accumulate in the rhizosphere in the upland water condition. He thought this might be why the flooding-induced chlorosis occurred. His hypothesis was proved correct by the discovery of a chelating substance with iron produced from rice roots under iron deficiency, which was later identified as mugineic acid (MA) (Takagi 1976; Takemoto et al. 1978; Takagi et al. 1984)

Excitement greeted the finding that mugineic acid governed iron acquisition not only in rice but also in other crops. They demonstrated that Fe uptake of grasses is a direct function of the amount of Fe(III)-MA complexes formed at the root-soil interface, and that any failure of Fe(III)-MA formation in the rhizosphere

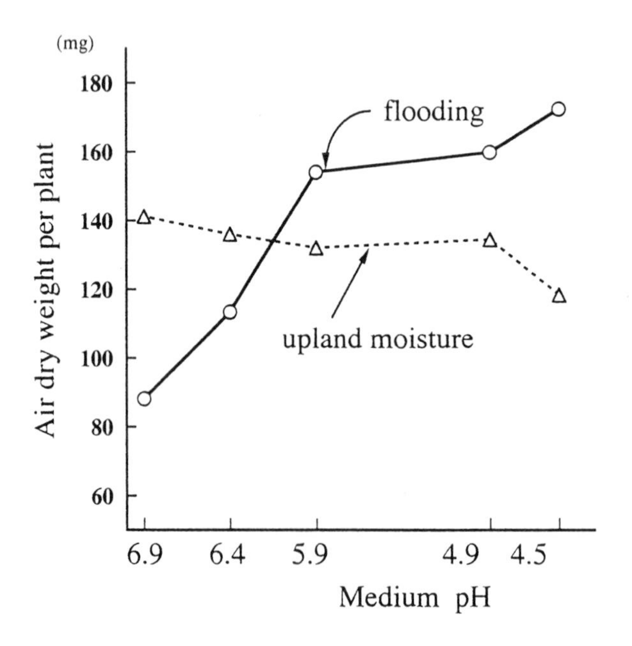

Fig. 8. Effect of flooding and upland moisture conditions on rice seedling growth in artificial soils varying in pH (from Takagi 1960)

causes the plants to develop Fe-deficiency chlorosis. In fact, there is a good correlation between the degree of Fe efficiency of major cereal crops and the ability of their roots to liberate mugineic acid under Fe deficiency; both of these traits being in the order: barley > wheat, rye > oats > maize > sorghum > rice (Marschner et al. 1986; Kawai et al. 1988). Consequently, the low ability of rice in iron acquisition was found to be due primarily to a low ability to secrete mugineic acid from the roots.

Today, the MAs (mugineic acid and its family of phytosiderophores) production system has come to be called Strategy II for graminaceous crops, as distinguished from Strategy I for iron efficient dicotyledonous crops (Marschner et al. 1986). Strategy I has been clarified in intensive studies which were triggered by the work of Brown et al. (1961), on enzymatic reduction of ferric iron to ferrous by roots. The studies prompted a reconsideration of the study of nutrient acquisition in plant species and provided a stimulus to addressing the serious worldwide problem of iron deficient chlorosis.

11. Significance of Root Extension and Root Hairs

Extension of the root system into new soil volumes results in a continued absorption of nutrients to meet the demand of plants. It can be said that nutrient availability and nutrient uptake are primarily functions of root extension as well as root activity. A number of studies have pointed out that nutrient uptake is closely related to morphological properties such a root length, surface area, fineness and root hair. Barber et al. (1963) have estimated the significance of root extension quantitatively. Barley (1970), who reviewed the subjects in detail, stressed the significance of these root properties in nutrient uptake.

Extension of the root system becomes particularly significant in less mobile and hardly soluble nutrients such as phosphorus, as implied in Bray's mobile concept. If the source of phosphorus is limited in the root surface sorption zone, root extension would be a main mechanism of gathering P from soil. The significance of root on P and other nutrient absorption has been proved by the prediction models introduced in previous sections in which root length or root surface is included as a main parameter.

Recently Otani and Ae (1996) clarified the significance of root length in P absorption in six plant species which were grown in fields with different P statuses. One site (P0) was poor in P fertility (Truog P: 0.5 mg P kg^{-1}) and the other (P1) was fertile (Truog P: 7.3 mg P kg^{-1}). The P uptake was in the order of: sorghum > soybean > groundnut > castor bean > pigeon pea > buckwheat at the P1 site. The P uptake by sorghum which had the longest roots, was the highest in both the sites. A strong relation ($r = 0.846$) between root length and P uptake at the P1 site was observed but not at the P0 site. In the pot experiments, it was also observed that the root length of sorghum was the longest, but P uptake was much less than in peanut. It was suggested from the results that root length mostly governs P uptake but another mechanism operates when P supply is limited. Crops such as sorghum in which P uptake mainly depends on root length, are able to take up a large amount of P by developing root systems where the capacity in P supply is large enough to meet the demand, as in a field. However, when the capacity is limited, as in a pot, the root length is not able to work properly in the uptake. In contrast, crops such as groundnut in which P uptake depends on not only root length but also other mechanisms, are able to absorb P reasonably well even in P limited conditions, as in a pot. This may be one reason why P uptake was higher in sorghum than in groundnut in the field, while the reverse was true in the pot.

Regarding the root hairs, they were long held by some to be significant in nutrient absorption, although others held their significance to be low. These differences among researchers probably depend on soil nutritional conditions and plant species. Using the modified Barber-Cushman model, Itoh and Barber (1983) calculated that the contribution of root hairs to P uptake was high in Russian thistle and tomato. On the other hand, the role of root hairs was very limited in carrot and onion, and was absent in wheat. They pointed out that the differing effect of root hair on P uptake was mainly due to morphological differences of root hairs between crop species. These results were confirmed by experiments utilizing a prediction model for potassium uptake (Meyer and Jungk 1993).

Mengel and Steffents (1985) pointed out that plant species differ in their ability to utilize nonexchangeable K, and this has been attributed to differences in root length. Also, Mitsios and Rowell (1987) observed that the contribution of nonexchangeable K increased with a corresponding increase in root density. Nair (1996) mentioned that root hairs contribute to an increase in root absorption surface. Therefore, reducing the distance of diffusion from the site of K release to the site of K uptake, and increasing the K concentration gradient by root hairs can be expected to exert a pronounced effect on K availability from the less mobile K fraction. Gabelman and Gerloff (1982) cited by Graham (1984) indicated a root extension response to phosphorus stress which had quite high heritability.

12. New Approach to Root Growth and Nutrient Absorption

Root arrangement in soils, which governs the efficiency of plant nutrient uptake, is not spatially uniform; it is usually uneven. Nevertheless, to simplify the relations, almost all models of nutrient absorption are based on regular root arrangement. Several studies, however, show that this simplification can be questioned, because root spatial arrangement may be significantly uneven.

Based on systematic study of spatial variability of root density, Tardieu (1988) reported that the assumption of regularity can be unrealistic. From the viewpoints of heterogeneous porous soil system, it was suggested that the distribution patterns of the micropores reflect the mechanical strength, bulk density, and other factors in the soil which affect root elongation (Hatano et al. 1988). Noordwijk et al. (1993) discussed the estimation of nutrient and water uptake efficiency of nonregular root distribution and incomplete root-soil contact, toward reducing negative environmental impacts of agriculture. Using fractal dimensions, Hatano and Sakuma (1990) proposed a way to describe the role of unequal macropore distribution in soils on the solute transport and root growth. These recent works may suggest a paradigm shift in the study of nutrient absorption: from one of homogeneity to heterogeneity.

13. Feedback System in Nutrient Absorption

Active nutrient absorption against a nutrient gradient has been described quantitatively by the Carrier Theory. The theory, supported by Epstein who proposed the enzyme-kinetic hypothesis on ion absorption, is pertinent to understanding active absorption (Epstein and Hagen 1952). Since the carrier resembles an enzyme in action, it is possible to describe the absorption mathematically by using Michaelis-Menten kinetics. Most of the simulation models of the absorption described above in substance consist of the nutrient concentration on root surface and coefficient of root absorption power obtained from the Michaelis-Menten equation.

The models, however, are not complete. For example, as already mentioned, in the prediction model for phosphorus absorption (Fig. 6) the simulation was satisfactory for intermediate levels of phosphorus, but the simulation was lower than the observed results at low levels of phosphorus. Other researchers frequently have confirmed this tendency. For example, underestimation of the prediction for phosphorus was reported under phosphorus limiting conditions in rape and sugar beet (Hoffland et al. 1990; Claassen 1990). These observations can be understood by assuming that plants are capable of increasing their nutrient absorption power automatically in response to nutrient stress. To improve the prediction models, therefore, it will be required to devise a way to introduce such an autocontrolled system of nutrient uptake into the models.

MAs biosynthesis has presented a perspective on autocontrolled feedback regulation in nutrient acquisition. Takagi (1991) described the mechanism as follows: "With the onset of Fe deficiency, synthesis of MAs in barley roots is accelerated remarkably, to cause a rapid and large accumulation in the roots. Most of the MAs thus accumulated are then secreted collectively through a special transport system, which operates periodically every morning (Fig. 9). It was disclosed also that this MAs secretory transport system is highly dependent on metabolic energy, that the MAs liberation by this system is attended by symport of equimolar K^+, and that its operation time is under the control of an endogenous circadian clock." The periodical excretion every morning means an economical energy-saving strategy of plants for minimizing loss of MAs by microbial degradation in the rhizosphere and by outflow into the bulk soil.

These dramatic findings open the way for a new approach to disclosing the mechanism of feedback system in plant nutrient acquisition. Along this line, investigations on the regulation mechanism of MAs production have been conducted intensively from the macro level to the micro level in molecular biology (Mori and Nishizawa 1987; Kawai et al. 1988; Shojima et al. 1990; Marschner et al. 1990).

The feedback system observed in nutrient acquisition can be diagrammed to present the plant adaptation mechanism and plant principal survival strategy. High-input-intensive agriculture has a tendency to mask such a latent plant faculty through the application of fertilizers or additives. Hence, scientists have paid much more attention lately to searching for such kinds of plant adaptive faculties to adverse conditions which may still be hidden from sight at present. Such investigation promises to best make efficient use of great plant potential through low-input technology for sustainable agriculture. To this day we find a great number of studies titled along the lines of "tolerance to nutrient deficiency", "tolerance to toxicity of heavy metals", "tolerance to acid soil, saline soil, or calcareous soil" based on various leading studies concerning efficient genotypes, such as for Mn (Graham 1984), Cu (Nambiar 1976), and Zn (Randhawa and Takkar 1976). To show the trend and to gain a perspective on these studies, several recent papers on nitrogen acquisition, phosphorus acquisition, tolerance of aluminum toxicity and related problems are introduced here.

13.1 Nitrogen Acquisition

Roots supply energy to microbes, thus generating a typical rhizosphere society in which microbes influence root activity in soils. The important role of microbe activity in the rhizosphere for nutrient acquisition is well-known, such as the role of the root growth promoting factor produced by microbes or mycorrhizal association.

Higher green plants are able to live and grow by using inorganic nutrients. They are called "autotrophs" indicating that they can grow without depending on or-

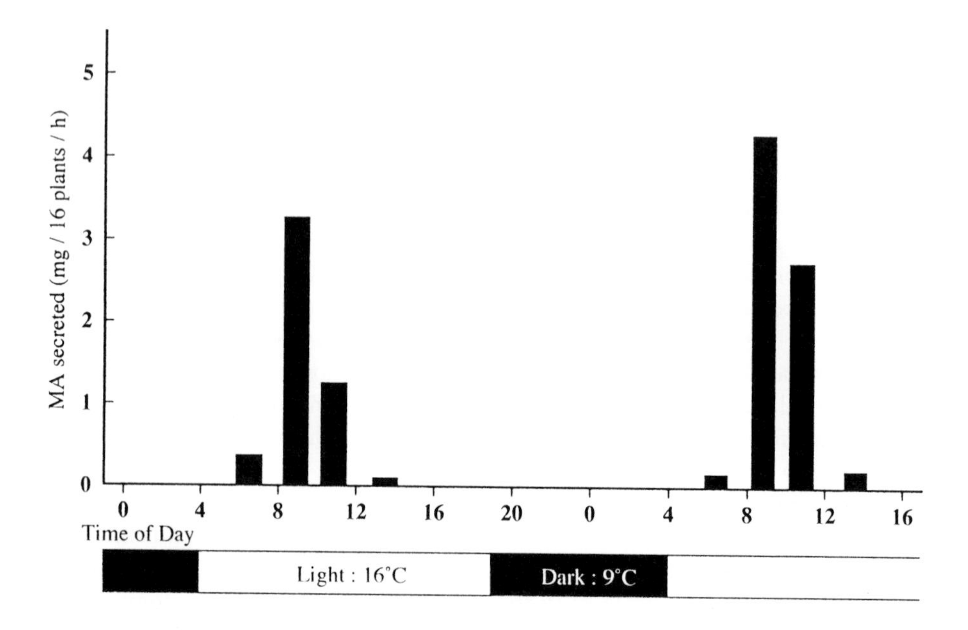

Fig. 9. Diurnal variation in mugineic acid secretion by Fe-stressed barley roots under 16-h photoperiod (from Takagi et al. 1984)

ganic foods synthesized by other organisms. However, the source of mineral nitrogen, nitrate or ammonium ions is primarily derived from the biological fixation of atmospheric nitrogen. Nitrogen nutrition of plants depends on the nitrogen fixation of symbiotic and free-living organisms. Furthermore, in modern agriculture, artificially fixed nitrogen also is a source of nitrogen for plant requirements. In order to introduce the study of this unique nitrogen cycle and plant nitrogen uptake which had progressed to that time, Epstein (1972) put to use the term "the acquisition of nitrogen" in his book. It can be said that nitrogen absorption in plants with root nodules which fix actively inert gas nitrogen is a typical example of nutrient acquisition.

Recently, several papers demonstrated that arctic tundra plants absorb organic nitrogen directly. Kielland (1994) observed that those plants take up free amino acids preferentially to mineral nitrogen. Arctic soils have small pools and slow fluxes of inorganic nitrogen, despite having large pools of organic nitrogen including free amino acids. Thus, amino acid acquisition may be a strategy of adaptation to the adverse soil condition of limited mineral nitrogen. Most of the plants studied are graminoids, deciduous and evergreen shrubs with mycorrhizal association, lichens, and mosses. Some are saprophytic.

On the direct uptake of organic nitrogen, Yamagata and Ae (1996) observed that several crops absorbed much more nitrogen in conditions adverse to the release of mineral nitrogen from organic nitrogen under the application of high C/N organic materials than in high application of mineral nitrogen. That is, some plants could take up more nitrogen in a small amount of inorganic nitrogen with large amount of organic nitrogen than in a high amount of mineral nitrogen in soils, suggesting the existence of organic nitrogen acquisition. Upland rice showed a particularly clear and high response to organic nitrogen. The phenomenon could not be understood by a priming effect or N mineralization of roots. Thus, they concluded that upland rice has quite a high ability to absorb organic nitrogen. It has already been reported that upland rice and barley absorb organic nitrogen, even proteins such as albumin and hemoglobin (Mori and Nishizawa 1979).

The study of organic nitrogen acquisition developed in the last decade provides new perspectives of mineral nutrition and nitrogen cycle and on ways to minimize the negative environmental impacts of cropping system under good yields, which is an acute need of society.

13.2 Phosphorus Acquisition

Phosphorus minerals, the source of phosphorus, are distributed in an extremely limited area in the world. Hence, the cycle of phosphorus differs primarily from the cycle of nitrogen. Furthermore, phosphorus availability is very low, due to chemical fixation, though the demand of plants is large, as described in the famous story of Lawes who produced the commercial fertilizer superphosphate for a more soluble form in 1843 (cited by Johnston 1994).

Japan is a typical volcanic country. Most of the fields for upland crops are soils derived from volcanic ash, Andosols. The main growth limiting factor is the low availability of phosphorus, due to the soils' allitic character. Without phosphorus application, crop growth is very poor. Thus, phosphorus-bearing fertilizers are looked upon as a soil ameliorant, like limestone. To produce good crops, heavy applications of phosphorus are needed initially and also as cropping proceeds. Accordingly, considerable amounts of phosphorus imported from other countries have accumulated in soils in sparingly soluble form. Currently there is awareness of using such kinds of sparingly soluble phosphorus in soils. Consequently, in Japan, phosphorus acquisition has been the focus of study (Suzuki and Hirata 1997).

Summarizing the knowledge at present, it can be said that the strategy of phosphorus acquisition depends on the following mechanisms: In nutrient solution culture, (1) root surface area and (2) P uptake power per unit root length. In a soil system, in addition to (1) and (2), (3) release of protons, (4) release of organic acids, (5) release of enzyme, and (6) mycorrhizal association. In a nutrient solution culture, because the rate of phosphorus movement to root surface is controlled artificially, phosphorus absorption is mainly enhanced by root surface and root activity. In contrast, in a soil system a continuous renewal of phosphorus at the location of root absorption is more likely an important limiting factor in phosphorus absorption. That is, the rate of diffusion which governs phosphorus transport to the roots enhances phosphorus absorption, as already discussed in Olsen's works.

The phosphorus concentration of the soil solution is so low in P limited condition that the rate of diffusion can be increased by raising the concentration of the soil solution. This is one reason why a root action of increasing phosphorus concentration by solubilizing more of the sparingly soluble phosphorus has been attractive as a strategy of phosphorus acquisition in efficient plant species. That is, root exudation, whose contents are capable of solubilizing sparingly soluble phosphorus such as chelating, reducing, and hydrolytic enzymes have been a particular focus of the study of phosphorus acquisition since the 1980's. Fungal mycelia can also solubilize the phosphorus.

Gardner et al. (1983) reported that the roots of lupins, a species efficient in P utilization, excreted a large amount of citrate ion. They postulated the action sequence as follows: the citrate ions react in the soil to form ferric hydroxy phosphate polymers which diffuse to the root surface where they are degraded by the action of reducing agents in the presence of an Fe^{2+} uptake mechanism balanced by hydrogen ion secretion, resulting in Fe-P availability. Thus, they consider phosphorus acquisition to be related to the concept of nutrient mobility and strategy of iron acquisition I. Hoffland (1992) concluded that organic acid exudation was a highly effective strategy to increase phosphate uptake in rape from rock phosphate. It has also been observed that phosphatase is released under P deficiency. Tadano and Sakai (1991) found the liberation of acid phosphatase significantly increased by P deficiency in all nine plant species tested. They are working to clarify the induction mechanism of acid phosphatase secretion at the molecular or cellular level (Wasaki et al.1997).

Ae et al. (1990) also found citric acid production to be effective for use of sparingly soluble Ca-P in some crops. However, this mechanism could not explain the excellent ability of the pigeon pea to extract P from Alfisols which contain Fe-P. They discovered the pigeon pea root exudate characteristically contains piscidic acid and its derivatives, which have very high capacity to solubilize ferric phosphate. These acids release P from $FePO_4$ by chelating with Fe (III). Thus, they solved the mystery of a traditional practice in Deccan, India in which the pigeon pea has been cultivated for thousands of years as an intercrop with cereals in low P Alfisol with low input of P fertilizer.

Groundnut has a superior ability to take up P from a soil with low P infertility as mentioned previously. Ae et al. (1996) and Ae and Otani (1997) have presented the evidence that this is not due to better root development or production of root exudate but due to the P-solubilizing ability of the root cell wall. They have reported that cell wall preparations from groundnut roots show a superior P solubilizing ability than those of soybean and sorghum, and that the active site of the cell is located on the root epidermal cell surface. These works emphasize the significance of the Contact Exchange Theory of nutrient uptake, proposed by Jenny et al. in the late 1930s (Jenny and Overstreet 1939), which was dismissed in those days. The theory suggests that root exudate can exchange directly with nutrients held on the soil surface. Groundnut behavior in phosphorus absorption indicated that there is a mechanism of the contact exchange without regulation by nutrient mobility. Thus, they provided a new perspective for studies on nutrient acquisition and cell wall function.

13.3 Tolerance Strategy for Al Toxicity

Acid soils comprise up to 40% of the world's arable land (Kochian 1995). Their low productivity is a serious problem for global food production. Al phytotoxicity is the primary growth limiting factor in acid soils. Yet there are many other such factors, including high acidity, low P, and low base saturation. Therefore, the subject is complex, and there is also a mutual relation in tolerance for these limiting factors. For example, the root exudate of organic acid is a promising mechanism in Al toxicity tolerance. Ojima et al. (1984, 1989) observed in cell culture experiments that carrot and tobacco selected for Al tolerance had an enhanced ability to excrete citric acid induced by Al treatment . However, they consider there to be a possibility that formation of Al-phosphate precipitates causes P deficiency that in turn may trigger citric acid excretion because liberation of organic acid is induced by P deficiency.

Tanaka and his group (Tanaka 1984) have conducted comprehensive studies on plant growth limiting factors of acid soils in the tropics and tolerance of plant species. Among their works, they demonstrated the relation between tolerance of Al and tolerance of low P. For example, in the case of acid tolerant crops which are tolerant to Al, such as soybean, maize and wheat, phosphorus was found to be a growth limiting factor in acid soils, due to these crops' intolerance to low P. Therefore, efficient use of these crops may be expected, when P is properly applied.

The literature has long documented Al toxicity and the mechanism of A1 tolerance. Miyake (1916) discussed Al toxicity on rice seedling growth regarding Al forms in the culture solutions. There has been much research and speculation from various aspects, through which much knowledge has been accumulated, resulting in a remarkable recent advance (see Kochian 1995).

Wagatsuma (1983) has reported that different Al tolerance among plant species might be associated with the barrier resistance of the cell membranes which controls the degree of passive Al movement into the root cells. Several scientists have hypothesized that Al resistance should be favored by low cell-wall cation-exchange capacity which would result in the binding of lower amounts of Al within the cell wall. Blamey et al. (1990, 1993) suggested that the binding of A1 in the apoplasm played an important role in the expression of Al toxicity to plant roots. They also demonstrated by using synthetic Ca pectate that the presence of A1 in the root environment may reduce the porosity of the cell wall, and thereby limit solution movement via the apoplastic pathway. This, in turn, may influence plant water relation and nutrient uptake.

Excretion of organic acids that chelate and detoxify Al in the rhizosphere has been implicated in the Al tolerance mechanism as mentioned above. It has been suggested that organic-Al complexes are not toxic to plant roots (Bartlett and Regio 1972). Release of citric acid from the roots of snapbeans was stimulated by Al, and an Al-tolerant genotype was shown to excrete 10 times more citric acid than an Al-sensitive genotype (Miyasaka et al. 1991). On the other hand, Delhaize et al. (1993) reported there was a consistent correlation of Al tolerance with high rates of malic acid excretion stimulated by Al in the tolerant genotype of wheat seedlings and that malic acid excretion was specifically stimulated by Al, and neither La, Fe, nor the absence of P was able to elicit this response.

Even though the problem is complex, an intriguing paper, "Aluminum tolerance in transgenic plants by alteration of citrate synthesis," (de la Fuente et al. 1997) suggests a way to proceed. To examine whether increased citrate production and release in transgenic plants could confer Al tolerance, they produced transgenic tobacco and papaya plants that overexpress a citrate synthase gene from *Pseudomonas aeruginosa* in their cytoplasm. They showed that production of citrate resulted in Al tolerance in the transgenic plants. Thus, they provide direct evidence that organic acid excretion is one of the important mechanisms of Al tolerance, which opens the possibility of applying this technology to improve crop production.

14. Conclusions

Based on the concepts developed in the first half of the 20th century, studies on plant nutrient absorption have shown great progress: Deeply and broadly tied to other disciplines, plant nutrient acquisition has come to receive special attention. History tells us also that studies have developed along with the needs of society. Such studies include soil tests for soil fertility, effective use of fertilizer for maximum yield, tolerance ability of crops to adverse soil conditions, and optimized nutrient cycle to minimize external inputs. Agricultural scientists can take pride in having responded to the demands of society.

Notwithstanding, it is very impressive to notice that throughout history, great progress has come primarily from the observation of a scientist in his particular field of interest, progress which has come through intuition of the hidden problems behind the phenomenon on which he focuses independently. Such observations can be seen regarding potassium and iodine accumulation in giant kelp, denitrification in paddy fields, nutrient mobility as a function of availability, flooding-induced chlorosis, and the substantial difference in solution culture, pot, and field. Indeed, these observations have provided new knowledge and added to the fundamental information available on plant nutrient absorption, thus opening new fields of study.

Under the paradigm of sustainable agriculture, one step starts from the observations of dynamic and complex phenomena in nature: once again through the naked eye, we see beautifully organized regularity. We are at the very beginning of a quest for treasures.

References

Ae N, Otani T (1997) Groundnut roots in solubilizing sparingly soluble phosphorus in low fertility soils. In: Ando T, Fujita K, Mae T, Matsumoto H, Mori S, Sekiya J (Eds) Plant nutrition for sustainable food production and environment. Kluwer Academic, Dordrecht, pp 309-314

Ae N, Arihara J, Okada K, Yoshihara T, Johansen C (1990) Phosphorus uptake by pigeon pea and its role in cropping systems of the Indian subcontinent. Science 248:477-480

Ae N, Otani T, Makino T, Tazawa J (1996) Role of cell wall of groundnut roots in solubilizing sparingly soluble phosphorus in soil. Plant and Soil 186:197-204

Arnold PW (1962) The potassium status of some English soils considered as a problem of energy relationships. In: Cook GW (1967)(Eds) The control of soil fertility. Crosby Lockwood, London, p 46

Barber SA (1962) A diffusion and mass-flow concept of soil nutrient availability. Soil Sci 93:39-49

Barber SA, Cushman JH (1981) Nitrogen uptake model for agronomic crops. In: Iskandar IK (Ed) Modeling waste water renovation-land treatment. Wiley-Interscience, New York, pp 382-409

Barber SA, Walker JM, Vasey EH (1963) Mechanisms for movement of plant nutrients from the soil and fertilizer to the plant root. J Agr Food Chem 11:204-207

Barley KP (1970) The configuration of the root system in relation to nutrient uptake. Adv Agron 22:159-301

Bartlett RJ, Regio DC (1972) Effects of chelation on toxicity of aluminum. Plant and Soil 37:419 -423

Beckett PHT (1964) Studies on soil potassium. II. The immediate Q/I relations of labile potassium in the soil. J Soil Sci 15:9-23

Blamey FPC, Edmeades DC, Wheeler DM (1990) Role of root cation- exchange capacity in different aluminum tolerance of Lotus Species. J Plant Nutr 13:729-744

Blamey FPC, Asher CJ, Edward DG, Kerven GL (1993) In vitro evidence of aluminum effects on solution movement through root cell walls. J Plant Nutr 16:555-562

Bray RH (1954) A nutrient mobility concept of soil-plant relationships. Soil Sci 78:9-22

Brewster JL, Bhat KKS, Nye PH (1975) The possibility of predicting solute uptake and plant characteristics. III. The growth and uptake of onions in a soil fertilized to different initial levels of phosphate and a comparison of the results with model predictions. Plant Soil 42:197-226

Brown JC, Holmes RS, Tiffin LO (1961) Iron chlorosis in soybeans as related to the genotype of rootstalk. 3. Chlorosis susceptibility and reductive capacity at the root. Soil Sci 91:127-132

Cameron FK (1911) The soil solution. Williams Norgate, London, 136pp

Claassen N (1990) Fundamentals of soil-plant interactions as derived from nutrient diffusion in soil, uptake kinetics and morphology of roots. Trans 14th Int Congress Soil Sci 2:118-123

Claassen N, Barber SA (1976) Simulation model for nutrient uptake from soil by a growing plant root system. Agro J 63:961-964

de la Fuente JM, Ramírez-Rodriquez V, Cabrera-Ponce JL, Herrera-Estrella L (1997) Aluminum tolerance in transgenic plants by alteration of citrate synthesis. Science 276:1566-1568

Delhaize E, Ryan PR, Randall PJ (1993) Aluminum tolerance in wheat (*Triticum aestium* L.). II. Aluminum-stimulated excretion of malic acid from root apices. Plant Physiol 103:695-702

Epstein E (1972) Mineral nutrition of plants: Principle and perspectives. John Wiley, New York, pp 257-283

Epstein E, Hagen CE (1952) A kinetic study of the absorption of alkali cations by barley roots. Plant Physiol 27:457-474

Gabelman WH, Gerloff GC (1982) The search for, and interpretation of, genetic controls that enhance plant growth under deficiency levels of a macronutrient. Cited by Graham RD Breeding for nutritional characteristics in cereals. In: Tinker PB, Läuchli A (1984) (Eds) Advances in plant nutrition 1. Praeger, New York, pp 57-102

Gardner WK, Barber DA, Paebery DG (1983) The acquisition of phosphorus by *Lupinus albus* L. III. The portable mechanism by which phosphorus movement in the soil/root interface is enhanced. Plant and Soil 70:107-124

Gile PL, Carrero JO (1914) Assimilation of colloidal iron by rice. J Agr Res 3:205-210

Graham RD (1984) Breeding for nutritional characteristics in cereals. In: Tinker PB, Läuchli A (Eds) Advances in plant nutrition. I. Praeger, New York, pp 57-102

Hatano R, Sakuma T (1990) The role of macropore on rooting pattern and movement of water and solutes in various field soils. Trans 14th Int Congress Soil Sci 2:130-135

Hatano R, Iwanaga K, Okajima H, Sakuma T (1988) Relationship between the distribution of soil macropores and root elongation. Soil Sci Plant Nutr 34:535-546

Hoagland DR (1922) Soil analysis and soil and plant interrelations. Agr Exp Sta Univ California, Circular: No 235:1-7

Hoagland DR (1944) Lectures on the inorganic nutrition of plants. Chronica Botanica, Mass, pp 48-71

Hoagland DR, Broyer TC (1936) General nature of the process of salt accumulation by roots with description of experiment methods. Plant Physiol 11:471-507

Hoffland E (1992) Quantitative evaluation of the role of organic acid exudation in the mobilization of rock phosphate by rape. Plant Soil 140:279-289

Hoffland E, Findenegg GR, Leffelaar PA, Nelemans JA (1990) Use of a simulation model to quantify the amount of phosphate released from rock phosphate by rape. Trans 14th Int Congress Soil Sci 2:170-175

Ishizuka Y (1932) Absorption and utilization of nutrients at different stages of rice plants by means of water culture (in Japanese). Bull Agr Chem, Jpn 8:849-867

Itoh S, Awazaki H (1996) The development of a process-oriented simulation model for nitrogen dynamics and crop growth in a paddy soil. Proceedings of the International Symposium on maximizing sustainable rice and environmental management. pp 637-647

Itoh S, Barber SA (1983) Phosphate uptake by six plant species as related to root hairs. Agr J 75:457-461

Jenny H, Overstreet R (1939) Cation interchange between plant roots and soil colloids. Soil Sci 47:257-272

Johnston AE (1994) The Rothamsted classical experiments. In Leigh RA, Johnston AE (Eds) Long-term experiments in agricultural and ecological sciences, CAB International, pp 9-37

Johnston ES, Hoagland DR (1929) Minimum potassium level required by tomato plants grown in water cultures. Soil Sci 27:89-108

Kawai S, Ito K, Takagi S, Iwashita T, Nomoto K (1988) Studies of phytosiderphores: Biosynthesis of mugineic acid and 2'-deoxymugineic acid in *Hordeum vulgare* L. var. Minorimugi. Tetrahedron Lett 29:1053-1056

Kielland K (1994) Amino acid absorption by arctic plants: implications for plant nutrition and nitrogen cycling. Ecology 75:2373-2383

Kimura J (1931) On the behavior of rice to mineral nutrients in solution culture, especially compared with those of barley and wheat (in Japanese). J Agr Exp Sta 1:375-402

Kimura J (1932) Further studies on the specific habits of rice in regard to its nutritive behavior in solution culture (in Japanese). J Agr Expt Sta 2:1-32

Kimura J (1952) On the characteristic behavior of crops in mineral nutrition. In: New aspects on soil and plant nutrition (in Japanese). Yokendo, Tokyo, pp 172-182

Kimura J, Chiba H (1943) Studies on the efficiency of nitrogen absorbed by the rice plants for the yields of grain and straw (in Japanese). J Sci Soil Manure, Jpn 17:479-497

Kirk GJD, Nye PH, Ahmad AR (1990) Diffusion and oxidation of iron in the rice rhizosphere. Trans 14th Int Congress Soil Sci 2:153-157
Kochian LV (1995) Cellular mechanisms of aluminum toxicity and resistance in plants. Ann Rev Plant Physiol Plant Mol Biol 46:237-260
Lundegårdh H (1937) Untersuchungen über die Anionenatmung. Biochem Zeitschr 290:104-124
Lundegårdh H (1951) Leaf analysis, Hilger and Watts, London, 176 pp
Marschner H, Römheld V, Kissel M (1986) Different strategies in higher plants in mobilization and uptake of iron. J Plant Nutr 9:695-713
Marschner H, Römheld V, Zhang FS (1990) Mobilization of mineral nutrients in the rhizosphere by root exudates. Trans 14th Int Congress Soil Sci 2:158-163
Mengel K, Stetffens D (1985) Potassium uptake of ryegrass (*Lolium perenne*) and red clover (*Trifolium pratense*) as related to root parameters. Biol Fert Soils I:53-58
Meyer D, Jungk A (1993) A new approach to quantify the utilization of non exchangeable soil potassium by plants. Plant Soil 149:235-243
Mitsios IK, Rowell DL (1987) Plant uptake from exchangeable and non-exchangeable potassium. 1. Measuring and modeling for onion roots in a chalky boulder clay soil. J Soil Sci 38:53-63
Mitsui S, Aso S, Kumazawa K (1951) Dynamic studies on the nutrient uptake by crop plants. 1. The nutrient uptake by rice roots as influenced by hydrogen sulfide (in Japanese). J Sci Soil Manure, Jpn 22:46-52
Miyake K (1916) The toxic action of soluble aluminum salts upon the growth of the rice plant. J Biol Chem 25:23-28
Miyasaka SC, Buta JG, Howell PK, Foy CD (1991) Mechanism of aluminum tolerance in snapbeans: root exudation of citric acid. Plant Physiol 96:737-743
Mori S, Nishizawa N (1979) Nitrogen absorption by plant roots from the culture medium where organic and inorganic nitrogen coexist. II. Which nitrogen is preferentially absorbed among [U-^{14}C]GluNH$_2$, [2,3-^{3}H]Arg and Na^{15}NO$_3$? Soil Sci Plant Nutr 25:51-58
Mori S, Nishizawa N (1987) Methionine as a dominant precursor of phytosiderophores in *Graminaceae* plants. Plant Cell Physiol 28:1081-1092
Nair KPP (1996) The buffering power of plant nutrients and effects on availability. Adv Agron 57:238-287
Nambiar EKS (1976) Genetic differences in the copper nutrition of cereals. Aust J Agric Res 27:465-477
Noordwijk MV, Brouwer G, Harmanny K (1993) Concepts and methods for studying interactions of roots and soil structure. Geoderma 56:351-375
Nye PH (1968) Processes in the root environment. J Soil Sci 19:205-215
Nye PH, Marriott FHC (1969) A theoretical study of the distribution of substances around roots resulting from simultaneous diffusion and mass flow. Plant Soil 30:459-472
Nye PH, Tinker PB (1977) Solute movement in the soil-root system. Blackwell Scientific, Oxford, 342 pp
Ojima K, Abe H, Ohira K (1984) Release of citric acid into the medium by aluminum-tolerant carrot cells. Plant Cell Physiol 25:855-858
Ojima K, Koyama H, Suzuki R, Yamaya T (1989) Characterization of tobacco cell lined selected to grow in the presence of either ionic Al or insoluble Al-phosphate. Soil Sci Plant Nutr 35:545-551
Okajima H, Hatano R (1988) An estimation of NO_3-N transport to the plant roots and the minimum NO_3-N concentration of the soil solution by means of the single root model (in Japanese). Jpn J Soil Sci Plant Nutr 59:172-177
Okajima H, Matsunaka T (1972) Evaluation of AR^K and PBC^K as the measurement of potassium supplying power of soils. Part 1 On AR^K (in Japanese) J Sci Soil Manure, Jpn 43:409-416
Okajima H, Takagi S (1953) Physiological behavior of hydrogen sulfide in the rice plant. Part 1 Effect of hydrogen sulfide on the absorption of nutrients. The Science Reports of the Research Institute, Tohoku Univ, Series D 5:21-31
Okajima H, Takagi S (1955) Physiological behavior of hydrogen sulfide in the rice plant. Part 5 Effect of hydrogen sulfide on respiration of rice roots (in Japanese). J Sci Soil Manure, Jpn 26:323-326
Olsen SR, Watanabe FS (1963) Diffusion of phosphorus as related to soil texture and plant uptake. Soil Sci Amer Proc 27:648-653
Otani T, Ae N (1996) Sensitivity of phosphorus uptake to changes in root length and soil volume. Agron J 88:371-375

Porter LK, Kemper WD, Jackson RD, Stewart BA (1960) Chloride diffusion in soils as influenced by moisture content. Soil Sci Soc Am Proc 24:460-463

Randhawa NS, Takkar PN (1976) Screening of crop varieties with respect to micronutrient stresses in India. In: Wright MJ (Ed) Plant adaptation to mineral stress in problem soils. Cornell Univ Agric Exp Sta, pp 393-400

Russell EJ (1915) Soil conditions and plant growth, 2nd ed. Longmans, London, 190pp

Russell EW (1973) Soil conditions and plant Growth, 10th ed. Longman, London, 549pp

Schofield RK (1955) Can a precise meaning be given to available soil phosphorus? Soils Fert18:373-375

Shioiri M (1942) Denitrification in the paddy field (in Japanese). J Sci Soil Manure, Jpn 16:104-116

Shojima S, Nishizawa N, Fushiya S, Nozoe S, Irifune T, Mori S (1990) Biosynthesis of phytosiderophores. In vitro biosynthesis of 2'-deoxymugineic acid from L-methionine and nicotiamine. Plant Physiol 93:1497-1503

Suzuki N, Hirata H (1997) Changes in phosphorus compounds around the rhizosphere of groundnut and pigeonpea grown in an Andosol with arbuscular mycorrhizal fungi. In: Ando T, Fujita K, Nae T, Matsumoto H, Mori S, Sekiya J (Eds) Plant nutrition for sustainable food production and environment. Kluwer Academic, Dordrecht, pp 513-514

Tadano T, Sakai H (1991) Secretion of acid phosphatase by roots of several crop species under phosphate deficiency conditions. Soil Sci Plant Nutr 37:129-140

Takagi S (1960) Physiological study on the adaptability of rice plants to the water logged condition. Part 1. On the flooding induced chlorosis of rice seedlings. The Science Reports of the Research Institutes, Tohoku University, Series D 11:77-96

Takagi S (1966) Studies on the physiological significance of flooded soil condition in rice plant growth - with special reference to flooding induced chlorosis of rice seedlings (in Japanese). Bull Inst Agri Res Tohoku Univ 18:1-158

Takagi S (1976) Naturally occurring iron-chelating compounds in oat-and rice-root washing. 1. Activity, measurement and preliminary characterization. Soil Sci Plant Nutr 22:423-433

Takagi S (1991) Mugineic acids as examples of root exudates which play an important role in nutrient uptake by plant roots. In: Johansen C, Kee KK, Sahrawat KL (Eds) Phosphorus nutrition of grain legumes in the semi-arid tropics. ICRISAT, India, pp 77-90

Takagi S, Okajima H (1956a) Physiological behavior of hydrogen sulfide in the rice plant. Part 3, Detection of sulfide in the rice plant. The Science Reports of the Research Institute, Tohoku Univ, Series D 7:17-26

Takagi S, Okajima H (1956b) Physiological behavior of hydrogen sulfide in the rice plant. Part 6 Influence of iron, pH and rice plant root on oxidation of H_2S in water culture solution (in Japanese). J Sci Soil Manure, Jpn 26:455-458

Takagi S, Nomoto T, Takemoto T (1984) Physiological aspect of mugineic acid, a possible phytosiderophore of graminaceous plants. J Plant Nutr 7: 469-477

Takemoto T, Nomoto K, Fushiya S, Ouchi R, Kusano G, Hikino H, Takagi S, Matsuura Y, Kondo M (1978) Structure of mugineic acid, a new amino acid possessing an iron-chelating activity from root washings of water-cultured *Hordeum vulgare* L. Proc Jpn Acad 54 B:469-473

Tanaka A (1984) In: Tanaka A (Ed) Acid soils and their use for agriculture: especially present situation and perspectives in the tropics (in Japanese). The Japanese Society of Soil Science and Plant Nutrition, Hakuyu-Sha, Tokyo, 310pp

Tardieu F (1988) Analysis of the spatial variability of maize root density. I. Effect of wheel compaction on the spatial arrangement of roots. Plant Soil 107:259-266

Taylor AW (1958) Some equilibrium solutions on Rothamsted soils. Soil Sci Soc Am Proc 22:511-513

Wagatsuma T (1983) Effect of non-metabolic conditions of the uptake of aluminum by plant roots. Soil Sci Plant Nutr 29:323-333

Wasaki J, Ando M, Ozawa K, Omura M, Osaki M, Ito H, Matsui H, Tadano T (1997) Properties of secretory acid phosphatase from lupine roots under phosphorus-deficient conditions. In: Ando T, Fujita K, Mae T, Matsumoto H, Mori S, Sekiya J (Eds) Plant nutrition for sustainable food production and environment, Kluwer Academic, Dordrecht, pp 295-300

Wild A (1964) Soluble phosphate in soil and uptake by plants. Nature 203:326-327

Woodruff CM (1955) The energies of replacement of calcium by potassium in soils. Soil Sci Soc Proc 19:167-171

Yamagata M, Ae N (1996) Nitrogen uptake response of crops to organic nitrogen. Soil Sci Plant Nutr 42:389-394

Yamane I, Usami T, Ikeda K, Omukai S (1956) Effect of root injury caused by hydrogen sulfide upon the rice plant (in Japanese). Rep Tohoku Agr Exp Sta 10:134-155

Plant Nutritional Problems in Marginal Soils of Developing Countries

Paul C. Smithson and Pedro A. Sanchez

Summary. Tropical regions have both the highest population growth rates in the world and a preponderance of soils which are less suitable for agricultural production. The relatively infertile Oxisols and Ultisols occupy about 43% of the tropics but only 7% of the temperate zone, while fertile Mollisols account for 2% of the tropics and 11% of temperate regions. Within the tropics, soils with chemical deficiencies or toxicities total about 47% of South America, 59% of Southeast Asia and 18% of Africa. In Africa, nutrient depletion of originally fertile soils has been caused by decades of continuous cropping with few external inputs, which in turn has resulted in stagnant or declining yields in many areas. Aluminum toxicity is among the most widespread problems in Latin America and Asia. Phosphorus deficiency (often accompanied by high phosphorus fixation) usually accompanies Al toxicity, and also occurs in many otherwise relatively fertile soils. Nitrogen deficiency can be considered as a given in almost all agroecosystems. Appropriate management practices vary according to the major deficiency or toxicity. Nitrogen deficiency may be alleviated in part by improved organic matter management. Leguminous intercrops or fallows, or biomass transfer from off-site, are some options for N management, though it must be recognized that in many situations the loss of time or area to fallows is untenable due to land pressure. For P management, inorganic additions are necessary. Soluble P sources, phosphate rocks and organic amendments should be used in an integrated fashion to recapitalize soil P in nutrient depleted soils. For Al-toxic soils, required lime rates of up to several t ha^{-1} and high transport costs often preclude liming. Selection and breeding of Al-tolerant and agronomically acceptable varieties are probably the best options for agriculture in acid soils. Along with integrated agricultural systems, newer methods of soil characterization need to be developed and standardized. Technological advances in remote sensing will allow better characterization over large areas in the near future. In the laboratory, newer methods as well as reassessments of older techniques, will make possible better indices of nutrient availability and potential yields.

Key words. Active soil C, Agroforestry, Al toxicity, N deficiency, Organic matter, P deficiency, P solubilization, Rock phosphate, Soil constraints, Tropical soils

1. Introduction

This chapter presents the global distribution of the major problem soils of the world, with emphasis on soil chemical constraints to agricultural production. The focus is on developing countries of the tropics, particularly in sub-Saharan Africa, where soil constraints endanger food security now. Particular emphasis will be placed on two of the most widespread constraints, aluminum toxicity and phosphorus deficiency. Aluminum toxicity occurs in the acid, infertile soils of the humid tropics, and is among the most widespread problems in Latin America and Asia. Phosphorus deficiency is a common problem accompanying acid infertility, and also occurs in many otherwise relatively fertile soils. Phosphorus deficiency may be accompanied by high phosphorus fixation, creating other problems and opportunities in P management.

The prevalence of problem soils in the humid tropics is contrasted with the situation in much of sub-Saharan Africa, which has a relatively smaller area of inherently infertile soils. In sub-Saharan Africa, where per capita food production has continued to decline over the past several decades, we argue that the major cause is not inherent infertility, but rather soil fertility depletion resulting from years of continuous cropping with few external inputs.

Various options for agricultural development of problem soils are presented, contrasting "Green Revolution"-type development with newer approaches that integrate organic and inorganic inputs, in combination with selection for adapted species and varieties. The use of organic inputs is highlighted as a low cost, if partial solution to several problems. In addition to the main goal of mitigating the nitrogen deficiency which is a given in most low input cropping systems, organic inputs can also help alleviate aluminum toxicity and in some cases enhance phosphorus availability.

Finally, approaches currently under study at ICRAF are detailed, in which newer techniques suitable for assessment of soil fertility status under organic and low-input management regimes are being tested. The emerging issue of soil quality is a yet-nebulous but appropriate concept which incorporates the complexities of organic-based management systems. Potential methods include soil P indices suitable for soils receiving organic or low-solubility P additions, including methods to estimate the active portion of soil organic P fractions. For N management, measurement of inorganic N accumulations, mineralizable fractions, and various measures of active soil carbon show promise. Many of the studies are still in the preliminary stages, but already show promise for improved prediction of soil quality and fertilizer needs under integrated management schemes.

2. Worldwide and Tropical Distribution of Marginal Soils

Increasing world population has placed ever-increasing demands on agricultural production to keep up with even basic food needs, especially in tropical regions, where population growth rates are the highest in the world. At the same time, soils which are less suitable for agricultural production are relatively more abundant in tropical regions compared to the temperate zone, while the chemical inputs needed for high productivity are often scarce and expensive. Table 1 gives a generalized picture of the relative distribution of soil orders (in U.S. Taxonomy) worldwide, based on the FAO 1:5 million Soils Map of the World (FAO/UNESCO, 1977). The relatively infertile Oxisols (FAO Ferralsols) and Ultisols (FAO Acrisols) occupy about 43% of the tropics and only 7% of the temperate zone, while fertile Mollisols account for only 2% of the tropics and 11% of temperate regions. Other soil orders with wide distribution, such as the younger Entisols and Inceptisols, are not as easy to generalize as to their suitability for agriculture, varying according to parent material mineralogy.

Table 2 gives a breakdown of major limitations to agriculture, including chemical, physical and climatic factors. Again, a relatively larger percentage of tropical soils suffer from one or more serious constraints. The physical constraints such as stoniness, shallow depth or steep slopes, or climatic constraints such as droughtiness, do not readily lend themselves to management interventions at the farm scale, while chemical constraints can in most cases be ameliorated with appropriate management. This chapter will focus on soil chemical constraints in the developing world, with emphasis on examples from sub-Saharan Africa and Latin America, where the authors have the most experience.

Soils with chemical constraints generally occupy a greater percentage of land area in the tropics than in temperate regions, accounting for 47% of South American and 59% of Southeast Asian soils, compared to 22% of North America and 33% of Europe. Even so, the inherent chemical constraints of African soils are often less severe than those of other tropical regions. Table 3 gives a breakdown of the major soil chemical constraints to crop production by major regions of the developing

Table 1. Approximate distribution of soil orders in the temperate and tropical regions of the world. (From Sanchez and Logan 1992, based on FAO 1971-79)

	Tropics[a]		Temperate regions	
	10^6 ha	%	10^6 ha	%
Soil associations dominated by:				
Oxisols	833	23	7	—
Ultisols	749	20	598	7
Alfisols	559	15	1231	13
Mollisols	74	2	1026	11
Entisols	574	16	2156	24
Inceptisols	532	14	1015	11
Vertisols	163	5	148	2
Aridisols	87	2	2189	24
Andisols	43	1	101	1
Histosols	36	1	204	2
Spodosols	20	1	458	5

[a]Latitude below 23°

Table 2. Major limitations for agriculture worldwide (From Dent 1980, based on FAO/UNESCO 1977)

Region	Percent of total land area with limitation					
	Drought	Mineral stress[a]	Shallow depth	Water excess	Permafrost	No serious limitation
North America	20	22	10	10	16	22
Central America	32	16	17	10	—	25
South America	17	47	11	10	—	15
Europe	8	33	12	8	3	36
Africa	44	18	13	9	—	16
South Asia	43	5	23	11	—	18
North and Central Asia	17	9	38	13	13	10
Southeast Asia	2	59	6	19	—	14
Australia	55	6	8	16	—	15
World	28	23	22	10	6	11

[a] Nutrient deficiencies or toxicities

world (including both tropical and subtropical regions), and again it is notable that Africa has relatively less area which suffers from severe soil chemical constraints. Here it is useful to distinguish between inherent infertility, characteristic of deep sands and leached soils on old geologic surfaces, and fertility depletion, resulting from human activities, mostly agricultural. For example, vast areas of the Amazon basin in South America have never been exploited for agriculture, since due to inherent infertility of the soils human settlement has tended to concentrate elsewhere. In contrast, much of eastern and southern Africa has adequate rainfall and moderate inherent soil fertility, so that during several waves of migration from the north and west, substantial human populations have developed in these areas.

Sanchez and Buol (1975) detailed the distribution of animal and human populations in relation to soil fertility. They note that the vast herds of ungulates found in the relatively fertile African and North American savannahs are missing from the extensive but infertile savannahs of South America. Human settlement has followed the same pattern, and in recent decades the increased population in sub-Saharan Africa has exceeded the limits of the soils to sustain indigenous low-input agriculture.

The problem in sub-Saharan Africa is primarily one of nutrient depletion due to continuous cropping, and recent studies give at least a rough estimate of the magnitude of the problem. Smaling et al. (1997) have made estimates of nutrient depletion at farm, national and continental scales in sub-Saharan Africa (Fig. 1), where vast areas are now suffering from serious soil degradation. Estimated annual per-hectare losses in Africa are 22 kg N, 2.5 kg P and 15 kg K. Loss estimates vary widely, but are uniformly negative throughout sub-Saharan Africa, excluding South Africa. In the Kisii District in western Kenya, a potentially highly productive area, the loss estimates are 112 kg N, 2.5 kg P and 70 kg K ha^{-1} yr^{-1}.

Table 3. Main chemical soil constraints in the developing world by geographic region (includes tropical and subtropical regions). (From Sanchez and Logan 1992, based on FAO 1971-79)

Soil constraint	Latin America		Africa		South and Southeast Asia		Developing world	
	10^6 ha (% of total)							
Low nutrient reserves	941	(43)	615	(20)	261	(16)	1817	(27)
Aluminum toxicity	821	(38)	479	(16)	236	(15)	1527	(23)
High P fixation (iron)	615	(28)	205	(7)	192	(12)	1012	(15)
Acidity (not Al toxic)	313	(14)	471	(16)	320	(20)	1104	(16)
Calcareous reaction	96	(4)	331	(11)	360	(23)	788	(12)
Low CEC	118	(5)	397	(13)	67	(4)	582	(9)
Salinity	62	(3)	75	(3)	97	(6)	234	(3)
Alkalinity	35	(2)	18	(—)	7	(—)	60	(1)
High P fixation (allophane)	44	(2)	5	(—)	7	(—)	56	(1)
High soil OM	9	(—)	12	(—)	23	(1)	44	(—)
Total area	2172		3011		1575		6758	

2.1 Specific Chemical Constraints

Table 3 is based on a conversion of the FAO soils map to the fertility capability classification system (Sanchez et al. 1982), which separates soils based on constraints to agricultural production. As such, it is largely based on inferences, since the FAO and other classification systems do not explicitly consider topsoil characteristics for most soil properties. The system is useful at least in a broad sense, however, and allows some notion of the extent of particular problems. As noted above, emphasis will be placed on two particularly widespread constraints, with brief mention of other important chemical factors.

2.1.1 Soil Acidity and Aluminum Toxicity

The problem of acid soil infertility has been recognized since early times, and the value of liming for centuries. The involvement of aluminum in the poor growth of plants on acid soils was identified by Hartwell and Pember (1918) and by McLean and Gilbert (1927) (see also references in the chapter by Okada and Fischer, this volume). Emphasis later shifted to H^+ as the active agent in soil acidity, and research on Al as the toxic agent waned until the 1960's (see the review on changing fashions in soil acidity research by Jenny 1961).

In general, Ultisols, and most Oxisols can be assumed to be Al toxic as well as some Alfisols, Entisols, and Inceptisols. The problem is characteristic of high leaching environments and soils derived from acidic parent materials, with low native reserves of basic cations. Large areas of southeastern USA and southeastern China are comprised of acidic Ultisols. Aluminum toxicity is particularly severe in humid tropical regions. Some 38% of Latin American soils are Al toxic, with lesser areas in other regions, but overall nearly one-fourth of the developing world's

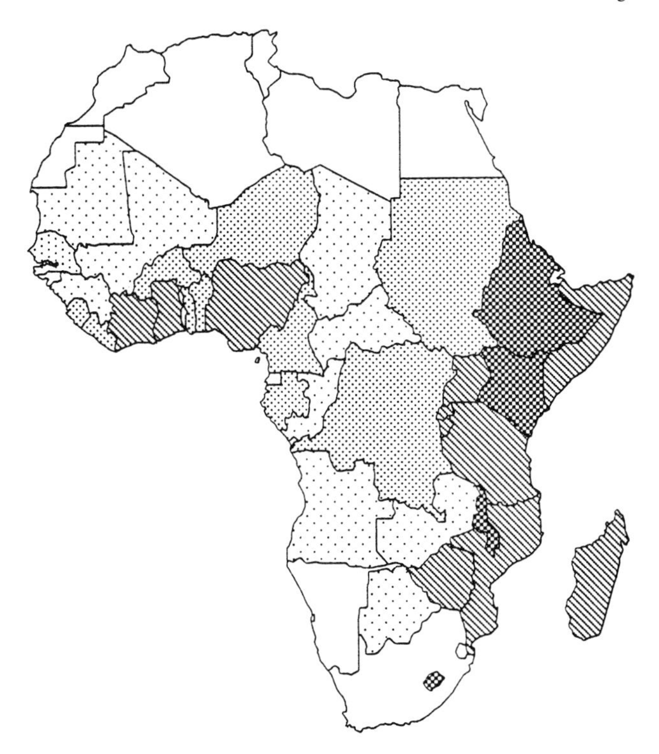

Nutrient depletion (kg ha^{-1} yr^{-1})

	N	P	K
Low	<10	<1.7	<8.3
Moderate	10-20	1.7-3.5	8.3-16.6
High	20-40	3.5-6.6	16.6-33.2
Very high	≥40	≥6.6	≥33.2

Fig. 1. Estimated nutrient depletion rates in sub-Saharan Africa (From Smaling et al. 1997, based on Stoorvogel and Smaling 1990)

soils suffer from acid infertility (Table 3).

While the existence and extent of acid infertility, and the identification of Al as the primary toxic agent are well accepted, the exact mechanisms and sites of the toxic action of Al are still debated after decades of research. One of the problems in detecting Al toxicity in the field is the relative lack of specific above-ground symptoms, with plants showing generally poor growth and low yields. Symptoms often are similar to those of P or Ca deficiency (Foy 1984). The physiology of Al toxicity was reviewed by Foy (1984) among others, and more recent research and references are given in other chapters of this volume.

There are many studies showing the root cell wall as a primary site of attack, for example Clarkson (1967) and Mugwira and Elgawhary (1979). Recent work and interpretation is reported by Blamey (this volume) and others in this volume on Al interactions with root cell wall components. Other evidence points to the plasma membrane as the primary site of Al injury. Wagatsuma and Ishikawa (this volume) present data on the cell plasma membrane as the primary site of damage, showing higher rigidity and depolarization of the plasma membrane. A similar result was reported by Vierstra and Haug (1978), and the subject of Al interaction with the plasma membrane was reviewed by Haug (1984). Plant species and varieties within species show differential tolerance to Al injury. Delhaize (this volume) examines organic acid exudation by roots as an inducible Al tolerance mechanism, and shows varietal differences in the quantity of root exudates, which is correlated with Al tolerance.

Among the consequences of Al injury is the reduced uptake of nutrients, and Ca has been identified as key among these, as it is involved in enzyme activation and in the maintenance of membrane and cell wall integrity (Haug 1984). Added Ca has long been known to partially alleviate the symptoms of Al toxicity (Munns 1965), and the Al:Ca or Al:CEC ratio is an important determinant of plant response, both in soil solution and on the soil exchange complex (Foy 1984; Kamprath 1984; Shortle and Smith 1988). In fact, in an Al-toxic and Ca-deficient Colombian soil, Okada and Fischer (this volume) found rice (*Oryza sativa*) performance to be better related to efficiency of Ca uptake than to Al tolerance per se.

Elevated soil solution Al (or perhaps the Al:Ca ratio) is the determinant of toxic effects, but its routine estimation is difficult. Aluminum saturation of the exchange complex has been used to predict solution Al, but the increase in solution levels is not linear, and the Al saturation at which solution concentration increases rapidly is variable (Kamprath 1984). Numerous titration and buffer-pH schemes have been devised for estimation of lime requirements, none of which have proved universally applicable. However, for soils with mostly kaolinitic clays and therefore mostly pH-dependent charge, which includes many acidic tropical soils, KCl-exchangeable acidity has been successful in predicting lime needs (Kamprath 1970).

2.1.2 Phosphorus Deficiency and Phosphorus Fixation

Phosphorus is required for all life, being a structural constituent of nucleic acids, as well as being involved in metabolic energy transfers through adenosine triphosphate (Ozanne 1980). After nitrogen, phosphorus is usually the most limiting nutrient for crop production, though crop uptake and harvest removal of P is only a fraction that of N or K. Much of the reason for continued need for P additions is due to the slow conversion of P to plant-unavailable forms, or P fixation.

Phosphorus fixation is not explicitly considered in soil classification schemes, but can be related fairly reliably to specific soil properties. Phosphorus deficiency and/or severe P fixation are likely in acidic soils dominated by 1:1 clay minerals, particularly those with clayey or loamy surfaces, and with substantial Fe and Al oxide contents. These soil types correspond to Oxisols, as well as clayey or loamy Ultisols and clayey, rhodic subgroups of Alfisols (portions of Ferralsols, Acrisols and Nitosols in the FAO legend). High P fixation due to allophane is primarily limited to soils of volcanic origin, and occurs in only about 1% of the developing world. These soils and their management problems are extremely important locally, but will not be discussed further due to their limited areal extent.

While global or continental estimates of P fixation are useful in a general sense, they cannot provide any useful information at the level of the management unit. The FAO soils map on which these estimates are based is itself based largely on reconnaissance surveys or generalizations from the underlying geology, especially in Africa. At the 1:5 000 000 scale at which the map was produced, the smallest unit which can be reasonably represented on a map is 100 000 ha (Eswaran et al. 1997). In addition, estimates of fertility problems made from the FAO data are themselves based on inferences rather than directly measured properties. This is particularly true for P, and estimates vary several-fold according to the assumptions used.

Using Africa as an example and considering only clayey Oxisols (Ferralsols) as potentially P-fixing (S.W. Buol, personal communication) gives an estimate of about 7% of the area as potentially P-fixing (Table 3). On the other hand, including other acidic soil types with medium or fine textured surfaces nearly triples the estimate, to about 18% (Fig. 2). In any event, P deficiency rather than P fixation per se is the important factor in soil management, but again soil P parameters are not usually considered in soil classification schemes, so prediction of P deficiency is subject to error. As a first approximation, however, one can consider essentially all Oxisols and Ultisols as likely to be P deficient, representing 43% of the area of the tropics.

As with resistance to Al toxicity, there are species and varietal differences in tolerance to low soil P, and in the ability to utilize poorly soluble P sources. In at least some cases the Al tolerance and P efficiency are related, which is not surprising since both soil problems often occur together. A classic example is tea (*Camellia sinensis*), which is an Al accumulator. Jayman and Sivasubramaniam (1975) showed that roots of tea plants produced malic acid, which solubilized both Al and the P associated with it. Organic acid root exudates have already been mentioned as an Al tolerance mechanism, and Otani and Ae (this volume) show that the root exudation of organic acids also explains many differences in P uptake from poorly soluble P compounds.

As with Al, however, the mechanism is not always clear, and again the cell wall seems to be involved in some cases. Ae and Otani (this volume) studied P acquisition by groundnut, and localized the cell wall as the site of a yet unexplained P solubilizing and uptake mechanism. Soil and plant factors important in using poorly soluble P sources will be discussed further in a later section.

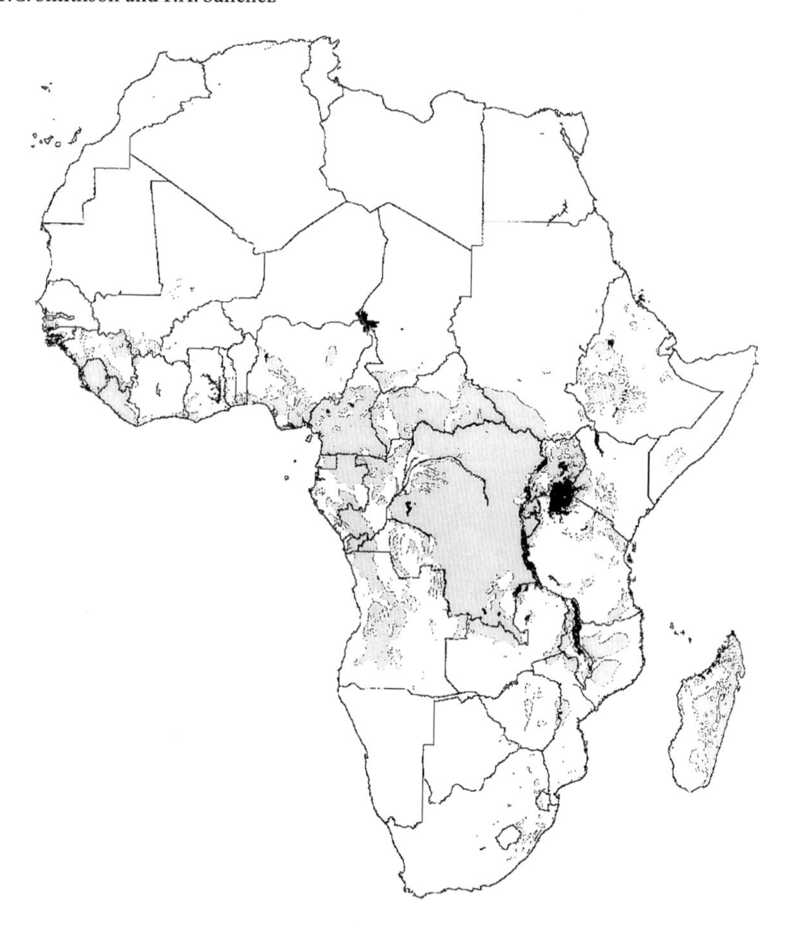

Fig. 2. Estimated area of potentially high P-fixing soils in Africa, totaling 530 million ha (From Sanchez et al. 1997)

2.1.3 Other Chemical Constraints

Again referring to Table 3, 27% of the developing world suffers from low soil nutrient reserves, defined as < 10% weatherable minerals in the sand-plus-silt fraction. This category represents highly weathered soils or those formed from acidic parent materials. These soils will require regular additions of nutrients in order to be productive. The problem of low CEC (9% of the developing world) will often accompany the low nutrient category. These soils will require careful fertility management in order to avoid leaching problems, due to their limited ability to retain nutrients.

Mild acidity without Al toxicity covers 16% of the developing world, but except for a few particularly sensitive crops will require no special management. Further weathering in the long term, or heavy fertilization and crop offtake in the short term, will cause these soils to acidify further, requiring liming or other ameliorative action.

Calcareous soils occupy a relatively small percentage of the developing world, though they are locally important and often intensively managed. The primary agricultural constraint is the probability of micronutrient deficiencies.

Saline soils, and the alkalinity which often accompanies salinity, occupy only a few percent of the area of the developing world. Poor physical structure, salt damage to crop plants, and element toxicities or deficiencies are common. These soils require either salt-tolerant crops or costly reclamation procedures to make them productive.

Nitrogen deficiency is not considered in any soil classification scheme, and so does not appear in most maps or tables. However, essentially all soils will suffer from N deficiency under all but the most extensive production systems, and N deficiency is a primary soil constraint in most of the developing world.

The following sections will emphasize what are perhaps the most widespread soil chemical constraints: N deficiency, P deficiency/fixation and Al toxicity. These problems are both widespread and amenable to cost-effective solutions. Many of the examples will come from recent studies in Africa, and will highlight organic matter management as a part of improving overall soil quality and resilience.

3. Options for Sustainable Development of Tropical Soils

Since the development of improved crop varieties over the past few decades, food production has increased substantially in Asia and to a lesser extent in Latin America, where impending food shortages seemed certain in the 1960's. Both regions are now largely self-sufficient in basic foodstuffs. In contrast, per capita food production continues to decline in sub-Saharan Africa (Fig. 3), where yields of many staple crops remain low or are even declining. Nutrient depletion is the primary biophysical cause of agricultural stagnation in sub-Saharan Africa (Sanchez et al. 1997).

Sanchez (1994, 1997) has referred to the use of industrial technology to solve soil fertility problems as the "first paradigm" of soil management. In this approach, soil fertility constraints are removed by application of lime and fertilizer inputs. Highly successful in the industrialized nations, this paradigm was also the basis of agricultural development in the Cerrado region of central Brazil, which is dominated by Oxisols. Large initial applications of soluble P fertilizers, phosphate rock and lime, followed by lower maintenance applications, along with adapted crop and pasture germplasm, sound agronomic practices and an enabling policy environment have transformed much of the Cerrado from a "fertility desert" into a world-class commercial agricultural success (Goedert 1987; Abelson and Rowe 1987; Sanchez 1997).

First-paradigm development has been less successful in most of the tropics, due to the dominance of smallholder rather than commercial farms, along with other social and economic constraints. Therefore, a "second paradigm" has developed, in which biological processes are emphasized (Sanchez 1994). This includes use of adapted local species, development of Al-tolerant or P-efficient varieties, use of integrated nutrient management including organic manures, and optimization of nutrient cycling.

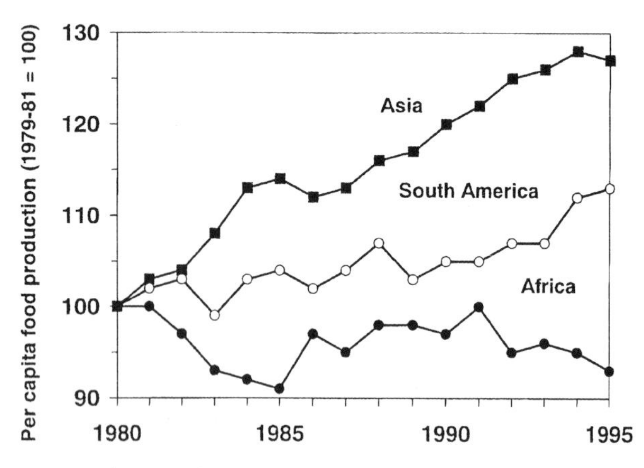

Fig. 3. Per capita food production in Africa, Asia, and South America, 1980-1995 (From World Bank 1996a, b)

Sanchez et al. (1996) consider that fertility replenishment should be viewed as a capital investment, much as irrigation projects or other large development schemes. In some areas of the tropics, Sanchez (1997) further observes that tropical development is now moving towards a "third paradigm", in which a moderate use of fertilizers, adapted species and varieties, and traditional local practices, are all combined to develop problem soils of the tropics in a sustainable manner. In this paradigm, yields will not reach their full genetic potential for some time. Rather, yield increases will be incremental, far below yields typical of modern industrial agriculture, but double and even triple those of current smallholder farms in the tropics. With increased grain yield and biomass production, the downward spiral of yield declines and soil degradation can be reversed, and the soil's natural resiliency restored.

3.1 Nitrogen and Organic Matter Management

Historically, most soil fertility management was based on addition of organic materials such as animal or green manures, or cleared fallow vegetation. The same is true today in much of the developing world, but land scarcity caused by population increase has caused these formerly viable systems to break down. The following sections highlight some possible options for improved organic matter management, both for the primary purpose of alleviating N deficiency and for secondary benefits such as reduction in Al toxicity and P fixation.

Known reserves of raw materials for N fertilizer production are extremely limited in sub-Saharan Africa, and production facilities are also rare in the region (Mudahar 1986). Except for Nigeria and South Africa, essentially all inorganic N fertilizer is imported, which due to poor infrastructure and high transport costs, costs 2 to 3 times the world market price at the farm gate (Donovan 1996; Sanchez et al. 1997). For these reasons, production and management of organic N sources is a more attractive option for short-term yield increases. While a portion of crop N needs may be met biologically, sufficient land for grazing or for fallow biomass production is generally not available in most areas of Africa (Giller et al. 1997). Quiñones et al. (1997) maintain that to achieve the necessary growth in agricultural production, inorganic sources of N will also be required. The emphasis here on organic sources of N is not meant to be exclusive or dogmatic, but represents an immediately possible set of technologies which can provide critically needed yield increases in the near term. These technologies may involve in situ systems such as intercrops or fallows, or biomass transfers, where materials are grown elsewhere and transferred to the field.

3.1.1 Organic Inputs and Quality

In organic-based systems, both the quantity and quality of inputs will be important in determining their benefits to agricultural production. Organic matter quality is important but ill-defined. While total C and N contents of organic materials can be measured unequivocally, and in a broad sense can be related to a given material's N-supplying capacity, these measures often fail to predict actual N release patterns. The C:N ratio, often used as a predictor of potential N mineralization of organics, fails with many tropical legumes, many of which contain substantial amounts of polyphenolics, which can retard N release (Palm and Sanchez 1991). Lignin:N ratio (Melillo et al. 1982), polyphenolic:N ratio (Palm and Sanchez 1991), or (lignin + polyphenolic):N ratio (Tian et al. 1992; Mafongoya et al. 1997) often are better predictors of potential N release from tropical legumes. Standardization of methods is needed, however, before valid comparisons among studies can be made; Palm and Rowland (1997) present a list of suggested parameters and methods in an attempt to move toward standardization.

Good quality organic residues common in tropical areas may supply up to about 150 kg N ha^{-1} at a rate of addition of 4 t dry matter ha^{-1} (Palm 1995). However, crop recovery of N from organic sources is generally low, around 10% to 20% (Palm 1995), compared to an average of about 50% recovery from inorganic sources (Baligar and Bennett 1986). Part of the low recovery of N from organic sources is due to the fact that N release may occur at a time when crops are unable to utilize it. The timing of N release to coincide with crop needs is now generally referred to as "synchrony" (Swift 1987). In practical terms, organic additions on the farm must be added according to availability of materials, usually at planting time when crop uptake is minimal. Nitrogen release therefore often occurs before high crop demand, thus losing much of the potential yield advantage due to leaching and other losses. In southern Malawi, rapid release of N from *Gliricidia sepium* prunings after the onset of rains caused soil inorganic N levels to fall to low levels within one month after planting, well before crop uptake was substantial (Fig. 4; ICRAF 1998). Predicted N mineralization and crop yields in climate based models, presented by Duxbury et al. (1989), suggest that nonsynchrony of N release may be a greater problem in tropical than in temperate areas. Likewise, synchrony of N release may be more difficult to attain with a basal application of organics than with properly split applications of mineral fertilizers.

Research has therefore been directed toward the regulation of N release to coincide with plant demand. Mixing of different quality materials can delay N release so that it better coincides with crop needs (Handayanto et al. 1997). Becker and Ladha (1997) adjusted N mineralization rates by mixing *Sesbania rostrata* with rice straw in unplanted fields, compared with sesbania alone, which released N too rapidly to coincide with predicted crop uptake.

The relevance of synchrony to actual farming systems is still debated. While it is easy to manipulate input quality experimentally, it may not easily translate into an adoptable technology. Availability of organic materials on-farm is often limited, and most materials have competing uses, which may take precedence over their use as crop supplements. As Myers et al. (1997) point out, synchrony must take into account many factors influencing farmers' decisions if a technology is to be adopted.

3.1.2 Other Effects of Organic Inputs

In addition to the main purpose of N supply, organic inputs can have other beneficial effects on soil health and crop performance. Some of the potential benefits in terms of increased P availability or decreased P sorption will be covered in following sections on P management.

Other effects include the amelioration of Al toxicity by organic anions, which has been documented for decades (Eisenmenger 1935; Bartlett and Riego 1972). The effects seem to occur whether the substances are soil humic materials, decomposition products of added organic matter, or are excreted directly by roots of living plants.

Organics also have beneficial physical effects such as improved soil tilth, better aeration and water infiltration and improved aggregate stability. These and other benefits are important but often difficult to quantify, and serve to highlight the relatively recent and still ill-defined concepts of soil resilience and soil quality, considered to be important for sustainable management of soils (Bezdicek et al. 1996). In low input systems of the developing world, lack of organic inputs over several decades is postulated to be a major factor in the decreasing resilience of these soils to environmental perturbations. Organic-based management systems highlighted in the following sections are an attempt to restore these soils to their former health, and some laboratory methods which attempt to quantify the soil quality concept are covered in the final section.

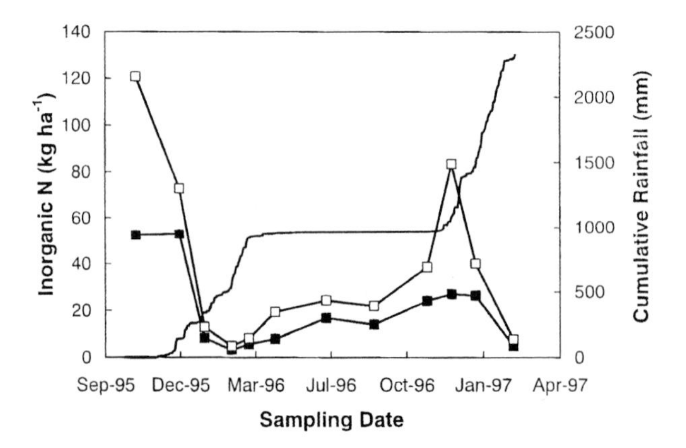

Fig. 4. Soil inorganic N (ammonium + nitrate) in top 20 cm soil layer over two growing seasons on an Alfisol in southern Malawi, comparing *Gliricidia sepium*—maize intercrop with sole maize. *Filled symbols,* sole maize; *open symbols,* gliricidia intercrop; *no symbols,* cumulative rainfall. (From ICRAF 1998)

3.1.3 Organic-Based Management Systems

Improved Fallows in Eastern Zambia

The Eastern Province of Zambia is representative of the southern African miombo woodlands, characterized by a unimodal rainy season averaging 800 to 1200 mm annual rainfall. Nitrogen deficiency is widespread and severe, and large yield responses to N fertilizers are common. Recent price increases have forced most farmers to abandon use of fertilizers except on high-value cash crops, and the traditional long fallows of up to several decades have shortened to 1 to 5 years in eastern Zambia due to population increase. Unfertilized maize (*Zea mays*) yields averaged 1.5 t ha^{-1} over 5 years in one study, while N at 112 kg ha^{-1} increased average yields to 4.9 t ha^{-1} (Kwesiga and Coe 1994; Kwesiga et al. 1999; ICRAF 1996).

Studies aimed at improving the existing bush fallow systems have been conducted, using *Sesbania sesban*, *Tephrosia vogelii* and pigeonpea (*Cajanus cajan*), among others. The following summary deals specifically with the sesbania trials on-station. One- or 2-year fallows were installed in a phased entry design (Table 4), to minimize effects of variable rainfall in different years. Two-year fallows were found to be the most successful, maintaining better yields for at least 2 seasons after the fallow period. Economic analysis showed that the 2-year fallow became more profitable by the 4th year than continuous unfertilized maize, despite the labor requirements for fallow establishment and loss of 2 years' yield (Kwesiga et al. 1999).

Another potential effect of fallows is the capture of N leached beyond the rooting zone of maize, and brought to the surface by the deeper-rooted trees during the fallow. In Oxisols and oxic Alfisols of western Kenya, subsoil N accumulation may be as high as 70 to 315 kg N ha^{-1}, and 18-month sesbania fallows significantly reduced the subsoil nitrate (Hartemink et al. 1996). This effect is only likely where substantial subsoil anion exchange capacity exists, allowing accumulation of leached nitrate. In cases where N accumulation occurs, however, fallows provide a means of recycling N which would otherwise be lost, and therefore represent an increase in efficiency of the system.

Nonleguminous Fallows and Biomass Transfer

The weedy shrub *Tithonia diversifolia* (Mexican sunflower) of the family Asteraceae is widespread in the tropics, including much of Kenya and elsewhere in sub-Saharan Africa with humid to sub-humid climates (Jama et al. 1999). Tithonia grows rapidly and accumulates substantial biomass with high concentrations of major nutrients, about 3% to 4% N, 0.25% to 0.4% P and 3% to 5% K. In western Kenya, tithonia biomass additions of 2 to 4 t dry weight ha^{-1} gave around 60% maize yield increases relative to unfertilized controls (ICRAF 1997). Since this technology involves substantial biomass transfer, availability of both labor and sufficient nearby biomass are issues which affect the area of applicability. Neither of these are serious constraints in western Kenya where thousands of farmers are now using tithonia green manures as nutrient inputs.

There may also be an opportunity for nutrient recycling by using short-term

Table 4. Maize yield data following 1- and 2-year *Sesbania sesban* fallows in eastern Zambia. (From ICRAF 1996)

Treatment	Maize yield (t ha^{-1})					
	1990/91	1991/92	1992/93	1993/94	1994/95	Cumulative total yield
Continuous maize (- fert.)	2.72	1.07	1.86	1.03	0.94	7.62
2-year fallow	F[a]	F	4.67	2.84	2.28	9.79
1-year fallow	2.72	F	2.89	1.44	1.15	8.20
2-year fallow	2.72	F	F	3.36	2.30	8.38
1-year fallow	2.72	1.07	F	1.87	1.16	6.82
2-year fallow	2.72	1.07	F	F	3.06	6.85
1-year fallow	2.72	1.07	1.86	F	1.75	7.40
Continuous maize (+ fert.)	7.15	1.95	7.68	3.83	3.75	24.36
SED	0.615	0.239	0.293	0.446	0.289	

[a]F = period under fallow in phased-entry design
SED = standard error of the difference in means

fallows of tithonia or various other nonlegumes. In Southeast Asia, for example, farmers practice short-rotation fallows using another composite, *Austroeupatorium inulifolium*. Dry matter, N and P accumulations were all about 3 to 4 times those produced by fallows of native ferns or the invasive grass *Imperata cylindrica* (ICRAF 1996). Similarly, tithonia is used by Philippine farmers to smother *Imperata* and restore soil fertility (ICRAF 1996).

All of the above systems should be considered as transitional between traditional shifting cultivation, which is no longer sustainable, and high-yield "first-paradigm" management. The intent is not to stagnate at less than potential yields, but to provide incremental gains which can be built upon as farmers realize greater profits and then can afford more expensive purchased inputs.

3.2 Managing Phosphorus Deficiency

Management of P deficiency and high P fixation are generally similar wherever they occur, in general requiring additions of inorganic P, whether from processed fertilizers or regionally available phosphate rocks. This section will emphasize the East African highlands, where much of ICRAF's soil fertility research is located.

The East African highlands comprises portions of Ethiopia, Kenya, Uganda, Tanzania, Rwanda and Burundi, and an estimated 26% of the arable land in the region suffers from P deficiency (KARI 1994). Within the densely populated Lake Victoria region of western Kenya, data from several hundred farmers' fields show that more than 80% are seriously P-deficient, indicated by < 5 mg kg^{-1} bicarbonate-extractable P (Jama et al., unpublished data; Niang et al., unpublished data). Yields of the staple crop maize average < 1 t ha^{-1} under usual smallholder conditions in this area (Sanchez et al. 1997). Responses to P are significant even at rates as low as 10 kg P ha^{-1} (Jama et al. 1997). Various agricultural options for increasing P supply to crop plants are covered in the following sections.

3.2.1 Native Soil Phosphorus

Plant-available P has generally been considered to be inorganic orthophosphate (Pi), but in recent years more emphasis has been placed on various fractions of soil organic P (Po) as a potential source of P for crops by mineralization of Po over longer time periods, particularly in tropical soils, which often have a greater percentage of total P as Po. Sharpley et al. (1987), for example, found an increasing percentage of bicarbonate-extractable P as Po in a weathering sequence, from 30% in calcareous soils to 74% in highly weathered soils.

In unfertilized systems, bicarbonate- or NaOH-extractable Po has been correlated with yields. Linquist et al. (1997) found that both of these Po fractions were correlated with soybean (*Glycine max*) yields in an unfertilized Hawaiian Ultisol, while Beck and Sanchez (1994) found NaOH Po the most important P fraction in explaining crop yields under long-term unfertilized cultivation of a Peruvian Ultisol. The potential of native soil Po to increase crop yields in western Kenya is so far unclear. Some fraction of the Po in western Kenyan soils might be mobilized by microbial activity following the addition of high quality organic materials, known as a "priming" effect (Palm et al. 1997), but this can be considered only as a temporary measure. Recycling of the limited resource cannot solve the fertility depletion problem, and some external sources of P are essential for recapitalization of soil P stocks.

3.2.2 Organic Sources of Phosphorus

Unlike plant N, plant P content is generally too low to provide enough P for annual crops. Organic P sources may be important as supplements, however. Palm et al. (1997) list over 20 crop wastes and manures available for use in eastern and southern Africa, which range in P content from < 0.1% for some crop wastes to 1.8% for high quality poultry manure (average 0.35%). In China, Zhu and Xi (1990) estimate that in 1986 about 42% of China's agricultural P needs were provided by organic inputs, still leaving more than half of P requirements to be met by inorganic inputs.

In western Kenya, Jama et al. (1997) obtained maize yield increases from as little as 10 kg P ha^{-1} added as cattle manure on a P-deficient Oxisol. Manure is the most common soil amendment in the area, but a majority of farmers own no livestock and therefore have no source of manure (Shepherd and Soule 1998). Thus low P content, low availability and competing uses of organic sources and labor will generally preclude exclusive use of organics for P fertilization requirements, though they are certainly important as supplements. Organic materials added in integrated systems may also increase P availability by reducing P adsorption by the soil or enhancing solubility of poorly soluble P sources, as detailed in a later section.

3.2.3 Inorganic Sources of Phosphorus

Phosphate Rock

Due to limited fertilizer production facilities in sub-Saharan Africa, most soluble P fertilizer currently used in the region is imported, high priced and often unavailable when needed (Donovan 1996; Sanchez et al. 1997). While high-analysis P fertilizers are probably essential for the needed growth in African agricultural production (Quiñones et al. 1997), better use of indigenous sources is also necessary.

There are a number of phosphate rock (PR) deposits in sub-Saharan Africa, and in eastern and southern Africa most are of low reactivity (Table 5). There is at least one PR deposit, at Minjingu Hill near Lake Manyara in Tanzania, which is suitable for direct application to row crops in acidic soils (Buresh et al. 1997; Sanchez et al. 1997). The Minjingu deposit contains 13% total P and 3% neutral ammonium citrate (NAC)-soluble P (Smithson, unpublished data). The Minjingu deposit is small, containing perhaps only 2.2×10^6 tonnes or less of easily processed ore. The deposit also appears to be unique, of biogenic origin, consisting of Pliocene era deposits of bird and fish bones (P. van Straaten, personal communication), although similar ones may occur in other Rift Valley areas.

There is thus a need to exploit the more abundant but poorly reactive PR deposits which dominate the region. Most PR deposits of eastern and southern Africa are weathered carbonatite deposits of igneous origin and of generally low solubility, and often with substantial Fe + Al oxide content. Typical of the deposits of the region are the Busumbu and Sukulu deposits in eastern Uganda. The Busumbu deposit contains from 4.8% to 14% total P and 0.5 to 0.9 % NAC-soluble P (Smithson, unpublished data), while the Sukulu deposit contains from 5.7% to 18% total P and 0.7% to 1.2% NAC-soluble P (Van Kauwenbergh 1991). These and most other sub-Saharan PR deposits will require some sort of processing to increase their solubility for use with agricultural crops. Possibilities include production of TSP, simple superphosphate (SSP) or partially acidulated PRs (PAPRs). Due to the sensitivity of the TSP process to Fe and Al oxide impurities (more than 2% to 3% Fe + Al oxide content is unacceptable), most sub-Saharan PRs may be better suited for SSP or PAPR production with current industrial technologies (Roy and McClellan 1986).

Biosolubilization of Phosphate Rock

Just as some plants are able to access poorly soluble forms of P, so certain groups of bacteria and fungi can solubilize soil P or PRs (see historical review by Goldstein (1986), and the chapter by Leggett et al., this volume). Gram-negative bacteria of the genus *Pseudomonas* and others have been studied, and the gene responsible for the solubilizing ability has been identified and cloned (Liu et al. 1992). Production of gluconic acid in particular seems to be responsible for the solubilization of apatites, and selection of mutant strains offers possibilities for enhancement of solubilization potential (Goldstein et al. 1993).

Table 5. Estimated resources of some phosphate rocks (PR) in sub-Saharan Africa (From Buresh et al. 1997)

Country	Name of deposit	Type of PR	Reactivity[a]	Estimated reserves of PR (10^6 tonnes)	Total P content (g P kg^{-1})	1994 production[b] (10^6 tonnes PR)
Angola	Cabinda and Lucunga	Sedimentary	Medium	28	109	
Burkina Faso	Kodjari	Sedimentary	Low to medium	63[c]	109 to 140	
Burundi	Matongo	Igneous	Low	25	48	
Congo D.R.	Lueshe	Igneous	NA	30	17 to 44	
Guinea Bissau	Saliquinhe	Sedimentary	NA[d]	112	131	
Madagascar	Iles Barren	Biogenic	Medium to high	0.6[C]	122 to 166	
Mali	Tilemsi	Sedimentary	Medium	20 to 25	66 to 140	
Mauritania	Bofal-Loubboira	Sedimentary	NA	94	83 to 87	
Niger	Tahoua and Parc W	Sedimentary	Low to medium	100	79 to 153	
Senegal	Taiba and Thies	Sedimentary	Medium to high	155	122 to 166	1.59
South Africa	Phaloborwa	Igneous	Low	1800	26	2.55
Tanzania	Minjingu	Biogenic	Medium to high	10	87 to 109	0.10[e]
Tanzania	Panda Hill	Igneous	Low	125	26	
Togo	Hahotoe	Sedimentary	Low	100	122 to 140	2.15
Uganda	Sukulu Hill	Igneous	Low	230	48 to 57	
Zambia	Nkombwa-Rufunsa	Igneous	Low	700	17	
Zimbabwe	Dorowa	Igneous	Low	47[C]	35	0.15

Data sources: McClellan and Notholt (1986), Sheldon (1987), and Van Kauwenbergh (1995) unless indicated otherwise.

[a]Reactivity of a PR is determined by its inherent mineralogy and is correlated to the amount of citrate soluble (CS) P in the PR: High reactivity > 25 g CS P kg^{-1}, medium reactivity = 15 to 25 g CS P kg^{-1}, and low reactivity < 15 g CS P kg^{-1}

[b]From FAO (1996), except for Minjingu PR

[c]From IFDC/CIRAD/ICRAF/NORAGRIC (1996)

[d]NA = Not available

[e]Extraction in 1995. Production halted in 1996/97 because of excess stock and closure of Tanga fertilizer factory (E.P. Mwaluko, Mipco Ltd, Nairobi, 1997, personal communication)

Historically, there has been interest in inoculating soil or seed with P solubilizing organisms, but early results from many studies using so-called "phosphobacterin" fertilizer were unsuccessful (Tinker 1980). Recent work has therefore shifted towards industrial bioprocessing of ores or incorporation of organisms into fertilizers or seed treatments. Rogers and Wolfram (1993) have developed a continuous-process bioreactor which has been successful at laboratory scale, and may have promise for industrial processing of ores. Another option under development is the incorporation of suitable organisms along with PR ore and a soluble C source into a granular fertilizer product (a "bio-bead"; R.D. Rogers, personal communication). Both processes are still experimental and their industrial utility speculative, but may offer a possibility of processing PR ores at lower energy cost, with cheaper raw materials, lower percentage of waste materials, and with the ability to economically process ores of lower quality than is economically feasible with current industrial processes (Goldstein et al. 1993).

Leggett et al. (this volume) discuss an alternative approach, using a fungal seed inoculum based on *Penicillium bilaii*, which gives modest yield increases of several crops and has been commercialized in Canada. The relative merits of bacterial vs. fungal solubilizers are also discussed by Leggett et al.

Some or all of the above industrial processes may have future potential in the developing world, but all these forms of processing require currently nonexistent industrial capacity. Possible exceptions are the biosolubilization processes, which may be amenable to down-scaling and incorporation into a more appropriate technology for smallholder farmers (R.D. Rogers, personal communication).

Direct Use of Phosphate Rocks

Even relatively soluble PRs require acidic soils to be effective, generally with pH less than 5.5 (Sanchez 1976; Rajan et al. 1996). In the lake region of western Kenya, about 47% of nearly 1800 soils sampled from the area meet this criterion (Smithson, unpublished data). Though acidic, 60% of those soils with pH < 5.5 have $< 20\%$ Al saturation of the effective CEC, and only 4% have $> 40\%$ Al saturation (Smithson, unpublished data). Therefore, P deficiency rather than Al toxicity is the major problem over most of the area, and the relatively low pH may be considered an asset for the purpose of PR application.

Soil P sorption capacity also affects PR solubility (Smyth and Sanchez 1982; Chien et al. 1990; Rajan et al. 1996), with higher P sorption capacity causing greater dissolution by removing dissolved P from solution. High P sorption capacity has generally been considered as a liability hindering crop production (Sanchez and Uehara 1980), but moderate sorption capacity may also be considered as an asset, providing a reservoir of residual P available to later crops (Sanchez et al. 1997). As Rajan et al. (1996) discuss in their review, the situation is a complex interaction of two competing reactions, the dissolution of PR and its adsorption and conversion to less available forms. Differences in soil pH and P buffering capacity make each situation hard to predict. The same factors are involved in deciding the relative merits of large capital additions of P or smaller annual additions.

ICRAF began work with Minjingu PR in Western Kenya in 1996, comparing PR with TSP in terms of first-year responses and residual effects over time, using both annual maintenance applications of 50 kg P ha^{-1} and a large one-time application of 250 kg P ha^{-1}. Results at the initial site (a Eutrudox, pH 5.1, clay content 30% to 35%) were encouraging, with Minjingu PR averaging around 75% to 85% as effective as TSP, depending upon P rate (Sanchez et al. 1996; Mutuo et al. 1999).

Similar experiments have been installed at a number of other sites in western Kenya, in order to determine the likely range of applicability of direct PR use. Soils in the region are mostly Eutrudox and Kandiudalfs, with pH < 5.5 at 53% of 34 sampled sites, and surface clay content ranging from 31% to 58%. From the above 34 sites, 26 (76%) were determined to be P-responsive, indicated by at least 0.5 t ha^{-1} yield increment with 100 kg P (as TSP) ha^{-1}, with inorganic N and K supplied at recommended rates (Lijzenga 1998). At these 26 sites, Minjingu PR at 100 kg P ha^{-1} gave first-season response relative to TSP of greater than 60% at over half of the sites (Fig. 5).

Agronomic Effectiveness and Residual Effects of Phosphorus

Relative agronomic effectiveness (RAE) of a PR or other poorly soluble P source can be defined as 100*(yield increase with PR/yield increase with TSP). Rajan et al. (1996) summarize the results of many studies of PR application, and in general in the first season after PR application, the RAE of PR is less than 100%, but due to slow dissolution may exceed 100% in later years, as the rapidly dissolved TSP is subsequently adsorbed by soil sesquioxides and becomes less available. Barrow (1980) and Sanchez and Uehara (1980) summarize several studies in which early PR performance was inferior to that of TSP, but in later years was superior to TSP, though in many cases never attaining equivalent cumulative yields. This is an important consideration in terms of sustainability, because the overall effects of several years are more important than the initial effects. The temporal scale deserves increasing attention nowadays.

Regardless of P source, with increasing time since P application, labile P is converted to less available forms, therefore P fertilizer recommendations are normally much greater than harvest removal and other P exports. For example, Baligar and Bennett (1986) estimated that only 10% to 30% of applied P is taken up by the first crop after P fertilization. Some of the remaining P will be available for subsequent crops, but generally at an ever-decreasing rate, so that continued annual P applications are usually necessary. There is an approximate exponential decline

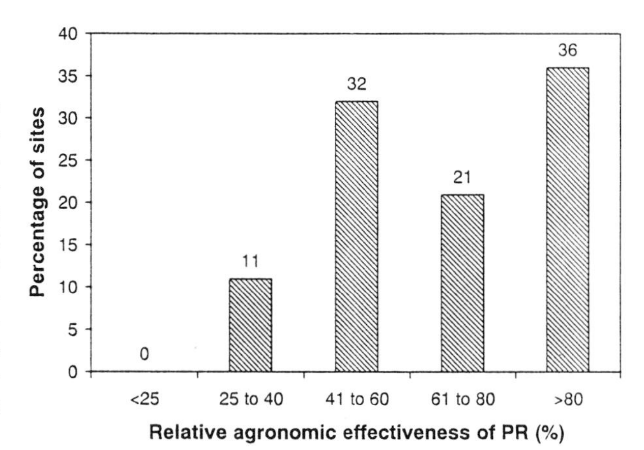

Fig. 5. Relative agronomic effectiveness (RAE) of Minjingu rock phosphate compared to triple superphosphate on 26 P-responsive sites in western Kenya. RAE = 100*(yield increase with PR/yield increase with TSP). Percentage of sites in given category are above bars. All sites received N and K at recommended rates (From Lijzenga 1998)

in soil P, and the early loss rates are generally higher for higher initial soil P levels (McCollum 1991). Cox et al. (1981) modeled soil P buildup and decline in several USA soils, and also found an approximate exponential decrease in available P. Figure 6 compares modeled soil P decline with one-year decline in bicarbonate-extractable P in a western Kenyan Oxisol, suggesting that the basic form of the model is reasonable.

Even though available P appears to decline rapidly after P additions, reports of residual effects of high P applications are abundant. Barrow (1980) reports studies which measured crop recoveries of 80% or greater after 9 years, varying with initial rate of P application. Kamprath (1967) reported sustained maximum maize yields on a North Carolina Ultisol of high P sorption capacity for 7 to 9 years after a single addition of 545 kg P ha^{-1}.

The residual effect is not universal, however. Linquist et al. (1996) found decreasing soybean (*Glycine max*) yields during a residual cropping phase following cumulative P additions of up to 930 kg P ha^{-1} over 2 years on a high P-fixing Oxisol in Hawaii. They recommended more frequent small applications of P as a more economical choice. Modeling efforts by Barrow (1980), and modeled experimental data of Cox et al. (1981) show that frequent small additions eventually result in higher soil P than large one-time additions (Fig. 6).

The problem with small additions in severely depleted or high P-fixing soils is that acceptable yield levels may not be achieved in the first few years. The Cerrado of Brazil was a case of this sort, in which large initial P additions were required to obtain economic yields (Yost et al. 1979), and the above study of Kamprath obtained low yields with only small annual additions. Maximal yields were obtained with a moderately large initial application followed by small annual additions. In summary, while large capital replenishment additions will usually have substantial residual effects, and in fact may be necessary for economic returns, the need for "new" vs. "old" soil P will require continual maintenance additions of P (see discussion by Fox et al. 1990).

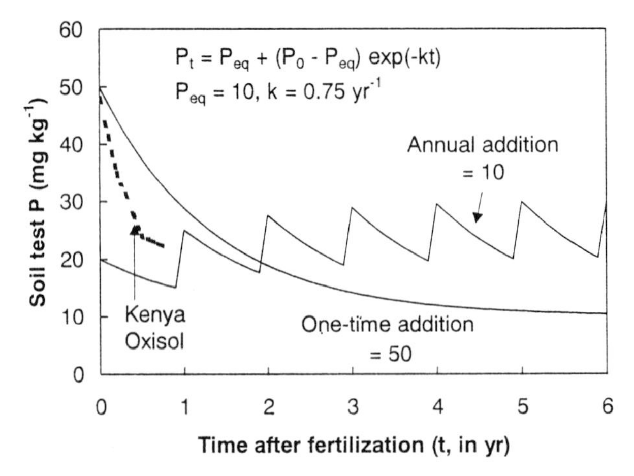

Fig. 6. Modeled soil P build-up and decline under single vs. annual additions of P compared to actual one-year decline in bicarbonate-extractable P over one year in a western Kenya Oxisol (Model from Cox et al. 1981; data from ICRAF unpublished)

Enhancing PR Availability with Organic Additions

Besides direct additions of P, organic inputs may also increase P availability by reducing P sorption, presumably by competition for P adsorption sites by organic anions. Organic additions may also directly enhance PR solubility, due both to lowered pH and to decomposition-related production of metal-complexing organic acids. Reduction in P sorption has been shown for pure organic acids such as oxalic or malic acids (Lopez-Hernandez et al. 1986), and with incorporated green manures, crop residues or animal manures (Syers and Lu 1990; Nziguheba et al. 1998).

Direct enhancement of Minjingu PR solubility was demonstrated by Ikerra et al. (1994) with compost- and manure-amended soil in Tanzania. Cellulose addition enhanced PR dissolution in an acid soil in a study by He et al. (1996), presumably by enhancing microbial activity. On the other hand, there have been cases of decreased solubility as well. Zaharah and Bah (1997) found that green manures tended to enhance the plant P uptake from low solubility PRs, but decreased that from more soluble ones.

Effects of Crop Species and Varieties

Besides additions of P fertilizers, selection of crops or varieties for efficiency of P uptake or utilization is also an option for integrated management. Certain groups of plants can solubilize P from Al and Fe phosphates or other poorly soluble sources by exudation of various complexing organic acids. Some examples are tea (*Camellia sinensis*) whose root exudates contain malic acid (Jayman and Sivasubramaniam 1975); lupins (*Lupinus* spp.) with citric acid (Gardner et al. 1983); and pigeon pea (*Cajanus cajan*), with piscidic acid (Ae et al. 1990). Within most crop species there are also varietal differences in P efficiency (Hedley et al. 1990). There has been substantial work on breeding for Al and/or low P tolerance in major crops (see the volume edited by Johansen et al. 1995; Narro et al. this volume; and Bahia et al. 1997 for some recent examples). While crop and varietal selection may have a future role in P management in Africa, current work at ICRAF is concentrating on soil replenishment of P stocks using existing locally accepted crops and varieties.

3.3 Management of Acid Soils

Large areas of the humid tropics suffer from soil acidity, and permanent agricultural development in these areas is hindered by the problem. Much indigenous agriculture in humid areas depends on slash-and-burn or swidden agriculture, where the ash from burned vegetation provides both nutrients and the alkalinity necessary to neutralize toxic aluminum. Increasing acidity and weed encroachment force the rapid abandonment of fields, yet population increases make the development of a more permanent and sustainable system a priority.

Organic matter management is one low cost but high labor means of reducing soil Al saturation and increasing base cation availability. Increasing organic content of soils decreases both exchangeable and solution Al, with accompanying decrease in Al toxicity (Evans and Kamprath 1970). Forms of Al in solution are also altered, forming less toxic Al-organic complexes instead of the more toxic Al^{3+}. Green manure additions or organic inputs from fallow clearance have similar effects as native soil organic matter, but the effects are more temporary as the fresh organic matter oxidizes (Wade and Sanchez 1983). In acid soil areas, a side benefit of some of the fallow and biomass transfer systems discussed above may be to reduce both the absolute amount and the relative toxicity of Al in solution.

Liming as a practice to ameliorate soil acidity has been known since ancient times (see Barber 1984 for some historical references). The agronomic benefits of liming are well accepted, but in much of the humid tropics, liming is not economic for various reasons. In extremely acid soils, lime requirements may be several t ha^{-1}, making transport costs prohibitive in many areas. Suitable liming materials may not be available locally, or may be of poor quality, further increasing costs. Various alternative options for the management of acid soils are covered briefly here, with no attempt at a comprehensive review. Sanchez and Salinas (1981) and Helyar (1991) provide good reviews of the topic.

Where suitable local liming materials occur, liming may be economical. In many cases liming to a predetermined pH, or to a pH which maximizes crop yield, as is recommended in high-input agriculture, is not feasible. More moderate additions of lime sufficient to neutralize most of the soluble Al are often more cost-effective. Liming at a Ca rate of about two times the KCl-exchangeable acidity has been shown to be effective on many variable-charge soils of the tropics (Kamprath 1984). Other low-cost liming options which may be appropriate include pelleting lime and Mo fertilizer with legume seed, solving the problem of poor Mo availability in acid soils (Helyar 1991).

Even if surface soils are limed, subsoil acidity can still reduce production due to shallow rooting, which in turn reduces deep nutrient uptake and increases drought susceptibility. Helyar (1991) gives some possible methods for reducing subsoil acidity. These include use of a more soluble Ca source such as gypsum ($CaSO_4$). Besides causing more rapid movement of Ca into subsoil, gypsum can also alter the chemical speciation of Al, reducing Al^{3+} and increasing the less toxic $AlSO_4^+$. Ritchey et al. (1980) demonstrated accelerated downward movement of Ca in an Oxisol when more soluble sources of Ca such as $CaSO_4$ or $CaCl_2$ were applied.

Plant architecture may also contribute to alleviating subsoil acidity. Deep-rooted species can recycle subsoil nitrate (Hartemink et al. 1996), which reduces the acidification caused by anion leaching (Helyar 1991) as well as contributing alkalinity to the subsoil by the bicarbonate excretion which accompanies nitrate uptake. This is another potential benefit of the improved fallows detailed above.

In extremely leached soils, Ca deficiency may be a greater problem than Al toxicity per se. Small additions of Ca sufficient to increase plant Ca uptake may be adequate, even though Al saturation may remain fairly high. Kamprath (1984) reviews data of Andrew and Norris (1961) on tropical and temperate legume growth in a low-Ca soil; most species showed maximum growth at around 26% Ca saturation of the CEC. The same study also highlighted the differential Al tolerance (or Ca efficiency; see Okada and Fischer, this volume) of different species.

Both temperate and tropical species showed maximum growth at similar Ca saturation, but the tropical legumes showed moderate growth at a Ca saturation of only 3%, while the temperate species grew hardly at all.

The species differences in the above study highlight the importance of plant factors in the management of acid soils, in that breeding for Al tolerance or Ca efficiency may be the best option for successful agriculture in acid soil areas. Both species and varietal differences in acid tolerance are well known, and can be exploited in breeding programs. Sanchez and Salinas (1981) considered selection of Al tolerant crop varieties as the main component of acid soil management. That statement is still true today, though simple selection of tolerant varieties has progressed to active breeding programs to increase both Al tolerance and yield potential (see, for example, Narro et al., this volume). Howeler (1991) reviewed the progress of the 1980's in breeding for Al tolerance in eleven important tropical crops. Large germplasm collections maintained by many of the International Agricultural Research Centers provide a reservoir of breeding lines for continued improvement in acid tolerance.

4. Laboratory Methods for Integrated Agricultural Systems

New technologies combining organic and inorganic inputs often require new accompanying methodologies. Some examples of field and laboratory methods are given in the section below; most of them are at the development stage.

4.1 Soil Mapping with Remote Sensing

For providing broad estimates of the areal extent of problem soils and their suitability for various management interventions, the FAO soils map of the world was and is a valuable tool. The small scale (1:5 million), however, is useful only at the regional or continental scale. At the other end of the spectrum, detailed soil surveys are unlikely to be available for much of the developing world in the near term, if ever. Some means of filling in detail at acceptable cost is needed, and recent developments in remote sensing technology offer such a possibility.

The most exciting recent possibility for improving our knowledge about soils over broad regions is that of near infrared (NIR) reflectance spectroscopy. Near IR reflectance spectroscopy is a well-developed technique for plant and litter analysis (McLellan et al. 1991), and current laboratory NIR techniques can provide rapid, accurate and cost-effective prediction of organic matter and clay content of soils (Stenberg 1997; Stenberg et al. 1995) Near infrared spectroscopy can also identify many important soil minerals, including 1:1 versus 2:1 clay minerals, several Fe-oxides and other minerals diagnostic for various soil properties (Kruse et al. 1993), from which inferences about soil genesis and management properties may be drawn with some reliability.

Near infrared spectroscopy is well developed in the laboratory and in ground-based field applications, and remote sensing technology holds promise to provide similar information at larger scales. Laboratory spectral signatures can be used to calibrate field (ground truth)-based spectra, which in turn have been used to identify mineral signatures obtained from airborne imaging systems, in particular the airborne visible and infrared imaging system (AVIRIS) (Kruse et al. 1993). The technique has so far been used mostly in regions of sparse to no vegetation cover, but new techniques of spectral unmixing show promise in areas of at least moderate vegetation cover.

At present, NIR technology is restricted to ground-based or airborne systems. Spaceborne imaging has until recently been restricted to fairly broadband spectra, such as those provided by the USA's Landsat series of satellites, because existing detectors were of insufficient sensitivity to obtain useful signal strength in narrower bandwidths. Advances in detector technology, however, have allowed narrow bandwidth spectral imaging systems to be mounted on space-based platforms, and the first of these was launched in November 2000 (M. Walsh, personal communication). The next few years promise information about soils over large areas and at a level of detail and reliability heretofore unattainable.

4.2 Laboratory Methods

The basic chemical principles behind the experimental agricultural systems described above are no different than for any other agricultural practice. However, in these lower input integrated systems, traditional laboratory methods are less successful in predicting yield responses or fertilizer needs than in the higher input systems for which most such methods were developed. In particular, organic sources of N present a problem in analysis. Availability of organic materials dictates the timing of their addition, and crop N availability is dependent upon both inorganic N and potentially mineralizable N fractions. For phosphorus, use of inorganic P sources of varying solubility presents challenges in estimating plant availability. Integration of organics and inorganics complicates the situation further, since organic P fractions are likely to play an increasing role in crop nutrition. The following sections detail work in the ICRAF laboratories and elsewhere which attempt to unravel these issues and provide appropriate and simple analyses for soils under integrated management systems.

4.2.1 Organic Matter and Soil Quality

In a broad sense, total C and N contents of soils are related to potential nutrient supplying ability, but these measures are rather insensitive to short-term organic additions to soils, and to subsequent crop productivity. Therefore, there has been increasing interest in finding simple measures of the active portion of soil organic matter, that which turns over on a time scale considered relevant to crop nutrient availability, or larger particulate fractions which are temporarily affected by recent organic additions.

Size and density fractionation of SOM may be useful to separate management-sensitive fractions from the large recalcitrant pool of humified SOM. Various methods involving simply sieving to separate particulate organic matter (POM) and/or flotation on dense liquids or salt solutions to separate various "light fractions" have been used. Christensen (1992) lists 22 different density separation methods, dating from as early as 1962, using various heavy organic liquids and salts of I and Br, among others.

Magid et al. (1977) found that the fraction of SOM which floated on sodium polytungstate solution (1.4 Mg m^{-3}) were most affected by field additions of crop residues. Similarly, Barrios et al. (1977) found that the fractions of SOM which floated on a silica suspension (< 1.13 Mg m^{-3}; Meijboom et al. 1995) or on a sodium iodide solution (< 1.7 Mg m^{-3}; Strickland and Sollins 1987) were significantly higher under maize-legume rotation or *Gliricidia sepium* additions compared to continuous maize in semiarid central Kenya. Total soil C and N were not significantly affected by the different cropping systems in this study.

Using the method of Meijboom et al. (1995), Barrios et al. (1997) also studied experimental improved fallow systems in eastern Zambia. Again, total C and N under various 2- to 3-year fallow systems or continuous maize were not significantly different, but light fraction weight, and total soil content of light fraction C, N and P were all significantly higher under various fallow systems than under continuous maize. Nutrient contents in light fractions may also provide good predictions of potential yields, as discussed in following sections.

The various fractionation methods for soil organic matter are useful but time-consuming, and a more rapid technique which can be adapted to soil testing programs is also needed. For example, Islam (1997) has attempted to find a rapid method which is well related to his "soil quality index", but which does not require extensive and tedious quantification of the many biological parameters necessary for determining the soil quality index. Islam tested total C, K_2SO_4-extractable C for intact and partially sterilized soil, H_2SO_4- and $NaHCO_3$-extractable C, and anthrone-reactive C in K_2SO_4 extracts of partially sterilized soil. Of these, the anthrone-reactive C (Brink et al. 1960) was by far the best single estimator related to the soil quality index. Anthrone reacts with reducing sugars, generally referenced as "glucose equivalents", and thus presumably measures only easily oxidized and therefore "active" carbon.

Another recent method suggested by Blair et al. (1995), modified from Loginow et al. (1987) measures only a portion of total organic C ("labile" C) which is oxidized by 0.333 M $KMnO_4$ in 1 hour. From the labile C and total C measurements, these authors have developed a "Carbon Management Index" (CMI) to characterize the relative lability (and thus presumably nutrient supplying capacity) of C in soils under different management regimes. Murage (1998) sampled fields from 12 farms in the agriculturally important central Kenyan Highlands, which were identified by farmers as productive and nonproductive. Total C was about 25% higher in productive fields, significant at $p < 0.001$, and a majority of the difference was due to higher labile ($KMnO_4$-reactive) C in productive soils. Labile C was 41% higher in productive fields, while nonlabile C (total C – $KMnO_4$-reactive) was only 15% higher. At 2 of the sites, where permanent pasture or remnants of virgin forest existed, relatively undisturbed soils could be compared with the cultivated soils. At each site, there was a clear decline in CMI from undisturbed to productive to nonproductive soils (Fig. 7). This suggests that at least part of the documented decline in yields in the Kenyan Central Highlands is due to loss of

SOM stocks, particularly "young" or labile carbon.

Another rapid method recommended by Franzluebbers et al. (1996) is the CO_2 evolution in the first 24 hours after rewetting air-dry soil. This method is relatively rapid, and does not require field moist samples, which many methods require but which is often difficult to accomplish. Their findings were that the CO_2 evolution correlated well with potential N mineralization and a number of other measurements, including microbial biomass, arginine ammonification, substrate-induced respiration and others. The authors caution that soils which have had recent large additions of fresh organic material of high C:N ratio may overestimate potential N mineralization, due to the rapid C flush as well as possible N immobilization due to the excess C.

4.2.2 Nitrogen Availability Indices

Just as total C of a soil is not a very sensitive indicator of soil quality, so total N is rather insensitive to short term management. Total N in POM or light fraction SOM, microbial biomass N, extractable inorganic N or various measures of potential N mineralization, have all been more successful in detecting treatment differences and predicting yields. In two studies, Barrios et al. (1996, 1997) found that total N was not significantly affected by fallow or green manure treatments, but as noted above, light fraction SOM N content was significantly higher under leguminous fallows than under continuous maize. In a study in eastern Zambia, Barrios et al. (1997) also found good correlations of fallow treatment with soil inorganic N and both aerobic and anaerobic N mineralization potential. Maroko et al. (1998) at two locations in Kenya, found that nitrate N, total inorganic N and anaerobic N mineralization were correlated with maize grain yield at both sites.

Extractable soil nitrate as an N availability index is not a new idea, and dates at least to the beginning of this century (King and Whitson 1901, 1902, cited in Dahnke and Johnson 1990). The predictive power of the measurement is usually better

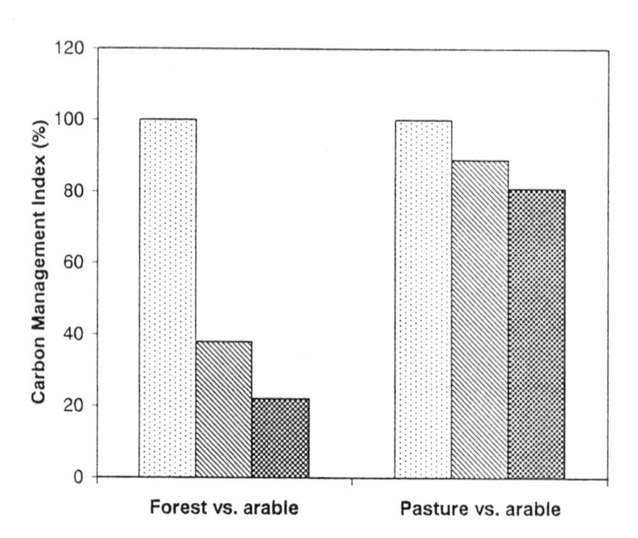

Fig 7. Carbon Management Index (CMI) as estimated by $KMnO_4$-reactive (labile) C vs. total C at undisturbed and adjacent arable sites in the Central Highlands of Kenya. CMI of undisturbed site is set to 100. *Dotted bars,* undisturbed; *hatched bars,* productive arable; *cross-hatched bars,* unproductive arable. (CMI after Blair et al. 1995; data from Murage 1998)

when soils are sampled to a depth of about 1 meter, currently not commonly done in routine soil testing (Dahnke and Johnson 1990). The situation may be complicated, however, by soils with significant anion exchange capacity in the subsoil, where substantial nitrate may accumulate (Hartemink et al. 1996). Maroko et al. (1998) found that correlations of inorganic N with maize yield were improved on an Alfisol by including depths to 1 m, but on an Oxisol with significant subsoil anion exchange capacity, correlations were poorer when deeper layers were included.

In a study of a *Gliricidia sepium* – maize intercrop in the unimodal rainfall regime of southern Malawi, Ikerra (1997) found that inorganic N and nitrate N in the top 20 cm of soil, measured just prior to the onset of rains, was highly correlated with maize yields over 2 seasons ($r^2 = 0.71$, $p < 0.001$). The correlation was slightly improved when anaerobic N mineralization potential was included in the model ($R^2 = 0.82$). Potential N mineralization is likely to be a valuable component of N availability in agroforestry and other organic-based low-input farming systems, but may be less useful when inorganic sources of N are used in higher input systems. For example, Kapkiyai (1996) found that anaerobic N mineralization was correlated with unfertilized common bean (*Phaseolus vulgaris*) yields, but not with the following fertilized maize crop. The correlation became significant when inorganic N-fertilized treatments were excluded from the analysis. A combination of readily and potentially available N will probably be the best predictors of N availability in integrated systems.

4.2.3 Phosphorus Availability Indices

Organic Phosphorus

Chemical extractions have been used as indices of plant available P for more than a century, and historical reviews of the subject appeared as early as 1894 (Dyer, cited by Kamprath and Watson 1980). Most research until recently has tended to emphasize the inorganic P (Pi) fractions and processes such as sorption/desorption reactions while neglecting to a large extent organic forms of P and biochemical processes (Gressel and McColl 1997). The same authors maintain that current views of the controls on P mineralization from soil organic P (Po), which is based on phosphatase production in response to plant P demand, are inadequate to explain observed plant uptake from added Po sources. While there is recent increased interest in Po fractions in soil, understanding has lagged behind.

As noted above, the highly weathered soils which dominate the tropics have a generally higher percentage of Po than less weathered soils of the temperate zone (Sharpley et al. 1987). This reservoir of Po may serve as a slowly available P source for plant growth, but the Po fraction in depleted western Kenya soils seems to be relatively unavailable in the short term, and the question then becomes one of which estimate of soil Po provides the best index to short- and longer-term P availability.

Many studies have been based on various sequential chemical extraction schemes, e.g., Hedley et al. (1982) or Tiessen and Moir (1993). Most studies have identified NaOH Pi as an important slowly available P fraction (e.g., Schmidt et al.

1996), but as noted above, Po fractions have been shown to be important especially in unfertilized systems. Recent work in western Kenya indicates that 18-month fallows with *Sesbania sesban* increased NaOH Po (Buresh et al. 1997), but the differences were small, and have not yet been consistently related to crop yields. In an ongoing field study of mineral phosphate sources the NaOH Pi rather than Po fraction shows the greatest changes during the first 18 months after P additions (Mutuo et al. 1999). To date no strong relationships between yields and extractable Po fractions have been found.

In contrast to extractable Po, which under our field conditions apparently contains a large proportion of recalcitrant forms of P, several other measures of organic P fractions have proved useful. Maroko et al. (1998) found that P in particulate OM and light fraction OM, and chloroform-extractable (microbial biomass) P were significantly increased in *Sesbania sesban* or natural weedy fallows compared to continuous maize or bare uncultivated fallow, on two contrasting soils in Kenya. Unfertilized maize yields after the fallows were best related to chloroform-extractable P and P in particulate OM on these sites, which were not N limited in the absence of P fertilization. Other organic or inorganic P fractions were unaffected or inconsistently affected by the land use systems.

Inorganic Phosphorus

Since most real increases in crop yields will require inorganic P additions, Pi fractions assume greater importance in the prediction of yields and fertilization needs. Direct use of PRs in particular will require a rapid and generally applicable soil test to predict plant P availability for P sources of varying solubility.

In the initial field study comparing Minjingu PR with TSP on maize in Western Kenya, Boening (1997) compared two chemical and two P-sink type extractions. Chemical extractants were $NaHCO_3$ + EDTA (modified Olsen) and HCl + NH_4F (Bray-1); P-sink extractions were anion resin (Sibbesen 1978) and Fe-oxide coated paper (Menon et al. 1990). Overall, P-sink methods were far superior to chemical extractants in estimating plant-available P in PR-fertilized soils, with the resin method slightly superior to Fe-oxide paper. At equal P additions as TSP or PR, Fe-oxide paper estimated availability of PR as about 55% the availability of TSP and resin extraction about 68% that of TSP-P; these ratios are comparable to the 87% RAE of the Minjingu PR for the first season. On the other hand, both chemical extractants underestimated available P from Minjingu PR, extracting only 16 to 19% as much as from TSP-treated soils.

These results illustrate a point that in soils fertilized with PR materials, the assumptions about relative lability of P in sequential fractionation schemes may not be relevant. Resin extraction is often the first step in the fractionations and is therefore generally considered as the most labile fraction (e.g., Linquist et al. 1997). In further work in western Kenya, Mutuo et al. (1999) found that in soils fertilized with Minjingu PR, resin-extractable P was often higher than NaOH-Pi. Recovery tests indicated the resin was able to solubilize 37% to 40% of the solid PR, though an estimate of dissolution (by ΔNaOH-P) indicated no more than 50% dissolution after 10 months. As noted above, however, the resin-extractable P gave the best estimate of plant availability, indicating a need for caution in the interpretation of P fractionation data. Khasawneh and Doll (1978) found that substantial amounts of P are extracted from PRs by anion exchange resins, even in the absence of soil. Thus lability, solubility and plant availability are not necessarily the same, though they are most certainly interrelated.

Saggar et al. (1992, 1995) found that a mixed resin system which included a cation as well as anion exchange resin was the best of several acidic, alkaline and resin extractions in predicting yields when several PRs of varying solubility were used. Mutuo et al. (1999) in western Kenya, found a slight increase in P availability estimates compared to anion resin only, even slightly overestimating P availability, based on relative yields. Other more poorly soluble PR materials such as the Busumbu PR are also being tested, to determine whether a simple and generally applicable soil test can be developed for a variety of soil and fertilizer combinations. Preliminary results indicate that extractability with mixed resin roughly approximates plant performance for a given P source (Savini 1999).

5. Conclusions

Soils presenting chemical problems are more common in the tropics than in temperate regions, which in turn presents difficulties for sustainable agricultural development. In Africa, inherent soil fertility problems are less common than in Asia or Latin America, because many of these soils were originally fertile, and supported rapidly growing human populations in arable areas. The problems in much of sub-Saharan Africa have more to do with depletion of inherent soil fertility through decades of continuous cropping with few external inputs.

In order to reverse the decades-long decline in per capita food production in Africa, and to maintain gains of the last few decades elsewhere, interventions must be targeted to agronomic, social and economic realities on smallholder farms. Phosphorus recapitalization by use of inorganic P sources, whether soluble sources or phosphate rock, should be viewed as a capital investment, with national and global benefits as well as increased income for smallholder farmers. Nitrogen replenishment should strive to better utilize and manage existing organic resources, with appropriate improvements in existing indigenous systems, for instance by incorporation of leguminous trees in fallow systems. This is not to ignore the use of organic sources for P replenishment or inorganic sources for N additions, for these will both be useful and even necessary for continued agricultural growth in sub-Saharan Africa. The technologies detailed above are merely the first step.

Along with integrated agricultural systems, newer methods of soil characterization need to be developed and standardized. Technological advances in remote sensing will allow better characterization over large areas in the near future. In the laboratory, newer methods as well as reassessments of older techniques, will make possible better indices of nutrient availability and potential yields.

References

Abelson PH, Rowe JW (1987) A new agricultural frontier. Science 235:1450–1451

Ae N, Arihara J, Okada K, Yoshinara T, Johansen C (1990) Phosphorus uptake by pigeon pea and its role in cropping systems in the Indian subcontinent. Science 248:477-480

Andrew CS, Norris DO (1961) Comparative responses to calcium of five tropical and four temperate pasture legume species. Austr J Agric Res 12:40-55

Bahia Filho AFC, Magnavaca R, Shaffert RE, Alves VCM (1997) Identification, utilization and economic impact of maize germplasm tolerant to low levels of phosphorus and toxic levels of exchangeable aluminum in Brazilian soils. In: Moniz AC, Fulani AMC, Shaffert RE, Fageria NK, Rosolem CA, Cantarella H (Eds) Plant-soil interactions at low pH: Sustainable agriculture and forestry production. Brazilian Soil Sci Soc, Vićosa, Minas Gerais, Brazil, pp 59-70

Baligar VC, Bennett OL (1986) NPK-fertilizer efficiency—a situation analysis for the tropics. Fert Res 10:147-164

Barber SA (1984) Liming materials and practices. In: Adams F (Ed) Soil acidity and liming. Agronomy Series No 12, Soil Sci Soc America, Madison, Wisconsin, USA, pp 171-209

Barrios E, Buresh RJ, Sprent JI (1996) Organic matter in soil particle size and density fractions from maize and legume cropping systems Soil Biol Biochem 28:185-193

Barrios E, Kwesiga F, Buresh RJ, Sprent JI (1997) Light fraction soil organic matter and available nitrogen following trees and maize. Soil Sci Soc Am J 61:826-831

Barrow, NJ (1980) Evaluation and utilization of residual phosphorus in soils In: Khasawneh FE, Sample EC, Kamprath EJ (Eds) The role of phosphorus in agriculture. Soil Sci Soc America, Madison, Wisconsin, USA, pp 333-359

Bartlett RJ, Riego DC (1972) Effect of chelation on the toxicity of aluminum. Plant Soil 37:419-423

Beck MA, Sanchez PA (1994) Soil phosphorus fraction dynamics during 18 years of cultivation on a Typic Paleudult. Soil Sci Soc Am J 58:1424-1431

Becker M, Ladha JK (1997) Synchronizing residue N mineralization with rice N demand in flooded conditions. In: Cadisch G, Giller KE (Eds) Driven by nature: Plant litter quality and decomposition. CAB International, Wallingford, UK, pp 231-238

Bezdicek DF, Papendick RI, Lal R (1996) Introduction: Importance of soil quality to health and sustainable land management. In: Doran JW, Jones AJ (Eds) Methods for assessing soil quality. SSSA Special Publication No 49, Soil Sci Soc America, Madison, Wisconsin, USA, pp 1-8

Blair GJ, Lefroy RDB, Lisle L (1995) Soil carbon fractions based on their degree of oxidation, and the development of a carbon management index for agricultural systems. Austr J Agric Res 46:1459-1466

Boening MI (1997) Evaluation of P extraction methods in western Kenya soils fertilized with phosphate rock or soluble P source. MSc thesis, University of Goettingen, Goettingen, Germany

Brink RH Jr, Dubach P, Lynch DL (1960) Measurement of carbohydrates in soil hydrolyzates with anthrone. Soil Sci 89:157-166

Buresh RJ, Smithson PC, Hellums DT (1997) Building soil phosphorus capital in Africa. In: Buresh RJ, Sanchez PA, Calhoun F (Eds) Replenishing soil fertility in Africa. SSSA Special Publication No 51, Soil Sci Soc America, Madison, Wisconsin, USA, pp 111-149

Chien SH, Sale PWG, Hammond LL (1990) Comparison of the effectiveness of phosphorus fertilizer products. In: Phosphorus requirements for sustainable agriculture in Asia and Oceania. International Rice Research Institute, Los Baños, the Philippines, pp 143-156

Christensen BT (1992) Physical fractionation of soil and organic matter in primary particle size and density separates. Adv Soil Sci 20:1-90

Clarkson DT (1967) Interactions between aluminum and phosphorus on root surfaces and cell wall material. Plant Soil 27:347-356

Cox FR, Kamprath EJ, McCollum RE (1981) A descriptive model of soil test nutrient levels following fertilization. Soil Sci Soc Am J 45:529-532

Curtin D, Syers JK, Smillie GW (1987) The importance of exchangeable cations and resin-sink characteristics in the release of soil phosphorus. J Soil Sci 38:711-716

Dahnke WC, Johnson GV (1990) Testing soils for available nitrogen. In: Westerman RL (Ed) Soil testing and plant analysis. SSSA Book Series No 3, Soil Sci Soc America, Madison, Wisconsin, USA, pp 127-139

Dent, FJ (1980) Major production systems and soil-related constraints in Southeast Asia. In: Soil-related constraints to food production in the tropics. International Rice Research Institute, Los Baños, Philippines, pp 79-106

Donovan W G (1996) The role of inputs and marketing systems in the modernization of agriculture. In: Breth SA (Ed) Achieving greater impact from research investments in Africa. Sasakawa Africa Association, Mexico City, pp 178–194

Doran JW, Parkin TB (1996) Quantitative indicators of soil quality: A minimum data set. In: Doran JW, Jones AJ (Eds) Methods for assessing soil quality. SSSA Special Publication No 49, Soil Sci Soc America, Madison, Wisconsin, USA, pp 25-37

Duxbury JM, Smith MS, Doran JW (1989) Soil organic matter as a source and a sink of plant nutrients. In: Coleman DC, Oades JM, Uehara G (Eds) Dynamics of soil organic matter in tropical ecosystems. NifTAL Project, University of Hawaii Press, Honolulu, Hawaii, USA, pp 33-67

Eisenmenger WS (1935) Toxicity of aluminum on seedlings and action of certain ions in the elimination of toxic effects. Plant Physiol 10:1-25

Eswaran H, Reich P, Beinroth F (1997) Global distribution of soils with acidity. In: Moniz AC, Fulani AMC, Shaffert RE, Fageria NK, Rosolem CA, Cantarella H (Eds) Plant-soil interactions at low pH: Sustainable agriculture and forestry production. Brazilian Soil Sci Soc, Vićosa, Minas Gerais, Brazil, pp 159-164

Evans CE, Kamprath EJ (1970) Lime response as related to percent Al saturation, solution Al, and organic matter content. Soil Sci Soc Am Proc 34:893-896

FAO (1996) FAO fertilizer yearbook volume 45—1995. Food and Agriculture Organization of the United Nations, Rome, Italy

FAO/UNESCO (1977) FAO-UNESCO soil map of the world, 1:5 000 000. United Nations Educational, Scientific and Cultural Organization, Paris

Fox RL, Bosshart RP, Sompongse D, Lin Mu-lien (1990) Phosphorus requirements and management of sugarcane, pineapple, and banana. In: Phosphorus requirements for sustainable agriculture in Asia and Oceania. International Rice Research Institute, Los Baños, Philippines, pp 409-425

Foy CD (1984) Physiological effects of hydrogen, aluminum, and manganese toxicities in acid soil. In: Adams F (Ed) Soil acidity and liming. Agronomy Series No 12, Soil Sci Soc America, Madison, Wisconsin, USA, pp 57-97

Franzluebbers AJ, Haney RL, Hons FM, Zuberer DA (1996) Determination of microbial biomass and nitrogen mineralization following rewetting of dried soil. Soil Sci Soc Am J 60:1133-1139

Gardner WK, Barber DD, Parbery, DG (1983) The acquisition of phosphorus by *Lupinus albus* L. III. The probable mechanisms by which phosphorus movement in the soil/root interface are enhanced. Plant Soil 70, 107-124

Giller KE, Cadisch G, Ehaliotis C, Adams E, Sakala WD, Mafongoya PL (1997) Building soil nitrogen capital in Africa. In: Buresh RJ, Sanchez PA, Calhoun F (Eds) Replenishing soil fertility in Africa. SSSA Special Publication No 51, Soil Sci Soc America, Madison, Wisconsin, USA, pp 151-192

Goedert WG (1987) Management of acid tropical soils in the savannas of South America. In: Management of acid tropical soils for sustainable agriculture: Proceedings of an inaugural workshop. IBSRAM, Bangkok, Thailand, pp 109–127

Goldstein AH (1986) Bacterial mineral phosphate solubilization: historical perspectives and future prospects. Am J Alt Agric 1:57-65

Goldstein AH, Rogers RD, Mead G (1993) Mining by microbe. Bio/Technology 11:1250-1254

Gressel N, McColl JG (1997) Phosphorus mineralization and organic matter decomposition: A critical review. In: Cadisch G, Giller KE (Eds) Driven by nature: Plant litter quality and decomposition. CAB International, Wallingford, UK, pp 297-309

Handayanto E, Cadisch G, Giller KE (1997) Regulating N mineralization from plant residues by manipulation of quality. In: Cadisch G, Giller KE (Eds) Driven by nature: Plant litter quality and decomposition. CAB International, Wallingford, UK, pp 175-185

Hartemink AE, Buresh RJ, Jama B, Janssen BH (1996) Soil nitrate and water dynamics in sesbania fallows, weed fallows, and maize. Soil Sci Soc Am J 60:568-574

Hartwell BL, Pember FR (1918) The presence of aluminum as a reason for the difference in the effect of so-called acid soil on barley and rye. Soil Sci 6:259-279

Haug A (1984) Molecular aspects of aluminum toxicity. CRC Crit Rev Plant Sci 1:345-373.

He ZL, Baligar VC, Martens DC, Ritchey KD, Kemper WD (1996) Kinetics of phosphate rock dissolution in acidic soil amended with liming materials and cellulose. Soil Sci Soc Am J 60:1589-1595

Hedley MJ, Stewart JWB, Chauhan BS (1982) Changes in inorganic and organic soil phosphorus fractions induced by cultivation practices and laboratory incubations. Soil Sci Soc Am J 46:970-976

Hedley MJ, Hussin A, Bolan NS (1990) New approaches to phosphorus fertilization. In: Phosphorus requirements for sustainable agriculture in Asia and Oceania. International Rice Research Institute, Los Baños, Philippines, pp 125-142

Helyar KR (1991) The management of acid soils. In: Wright RJ, Baligar VC, Murrmann RP (Eds) Plant-soil interactions at low pH. Kluwer Academic Publishers, Dordrecht, The Netherlands, pp 365-382

Howeler RH (1991) Identifying plants adaptable to low pH conditions. In: Wright RJ, Baligar VC, Murrmann RP (Eds) Plant-soil interactions at low pH. Kluwer Academic Publishers, Dordrecht, The Netherlands, pp 885-904

ICRAF (1996) ICRAF Annual Report 1995. International Centre for Research in Agroforestry, Nairobi, Kenya

ICRAF (1997) ICRAF Annual Report 1996. International Centre for Research in Agroforestry, Nairobi, Kenya

ICRAF (1998) ICRAF Annual Report 1997. International Centre for Research in Agroforestry, Nairobi, Kenya

IFDC/CIRAD/ICRAF/NORAGRIC (1996) PR feasibility case studies: synthesis report. An assessment of phosphate rock as a capital investment: Evidence from Burkina Faso, Madagascar, and Zimbabwe. Unpublished report prepared for the World Bank and Africa Environment. The World Bank, Washington, DC

Ikerra ST (1997) Soil nitrogen dynamics in agroforestry systems at Makoka in southern Malawi. Final report for ICRAF/Ford Foundation postgraduate research fellowship in agroforestry project for women, International Centre for Research in Agroforestry, Nairobi, Kenya

Ikerra TWD, Mnkeni PNS, Singh BR (1994) Effects of added compost and farmyard manure on P release from Minjingu phosphate rock and its uptake by maize. Norwegian J Agric Sci 8:13-23

Islam KR (1997) Test for active soil carbon as a measure of soil quality. PhD dissertation, University of Maryland at College Park, College Park, Maryland, USA

Jama B, Swinkels RA, Buresh RJ (1997) Agronomic and economic evaluation of organic and inorganic phosphorus in western Kenya. Agron J 89:597-604

Jama B, Palm CA, Buresh RJ, Niang A, Gachengo C, Nziguheba G, Amadalo B (2000) *Tithonia diversifolia* as a green manure for soil fertility improvement in western Kenya: A review. Agroforestry Systems 49:201-221

Jayman TCZ, Sivasubramaniam S (1975) Release of bound iron and aluminium from soils by the root exudates of tea (*Camellia sinensis*) plants. J Sci Food Agric 26:1895-1898

Jenny H (1961) Reflections on the soil acidity merry-go-round. Soil Sci Soc Am Proc 25:428-432

Johansen C, Lee KK, Sharma KK, Subbarao GV, Kueneman EA (Eds) (1995) Genetic manipulation of crop plants to enhance integrated nutrient management in cropping systems—1 Phosphorus: proceedings of an FAO-ICRISAT expert consultancy workshop. International Crops Research Institute for the Semi-Arid Tropics, Patancheru, Andhra Pradesh, India

Kamprath EJ (1967) Residual effect of large applications of phosphorus on high phosphorus fixing soils. Agron J 59:25-27

Kamprath EJ (1970) Exchangeable aluminum as a criterion for liming leached mineral soils. Soil Sci Soc Am Proc 34:252-254

Kamprath EJ (1984) Crop response to lime on soils in the tropics. In: Adams F (Ed) Soil acidity and liming. Agronomy Series No 12, Soil Sci Soc America, Madison, Wisconsin, USA, pp 349-368

Kamprath EJ, Watson ME (1980) Conventional soil and tissue tests for assessing the phosphorus status of soils. In: Khasawneh FE, Sample EC, Kamprath EJ (Eds) The role of phosphorus in agriculture. Soil Sci Soc America, Madison, Wisconsin, USA, pp 433-469

Kapkiyai, JJ (1996) Dynamics of soil organic carbon, nitrogen and microbial biomass in a long-term experiment as affected by inorganic and organic fertilizers. MSc thesis, University of Nairobi, Nairobi, Kenya

KARI (1994) Fertilizer use recommendations Vol 1-22. Fertilizer Use Recommendation Project, Kenya Agricultural Research Institute, Nairobi, Kenya

Khasawneh FE, Doll EC (1978) The use of phosphate rock for direct application to soils. Adv Agron 30:159-206

Kruse FA, Lefkoff AB, Dietz JB (1993) Expert system-based mineral mapping in northern Death Valley, California/Nevada, using the airborne visible/infrared imaging spectrometer (AVIRIS). Remote Sensing of the Environment 44:309-336

Kwesiga FR, Coe R (1994) The effect of short rotation *Sesbania sesban* planted fallows on maize yields. Forest Ecology and Management 64:199-208

Kwesiga FR, Franzel S, Place F, Phiri D, Simwanza CP (1999) *Sesbania sesban* improved fallows in eastern Zambia: Their inception, development and farmer enthusiasm. Agroforestry Systems 47:49-66

Lijzenga M (1998) Maize response to NPK in relation to fertility indices in western Kenya. MSc thesis, Wageningen Agricultural University, The Netherlands

Linquist BA, Singleton PW, Cassman KG, Keane K (1996) Residual phosphorus and long-term management strategies for an Ultisol. Plant Soil 184:47-55

Linquist BA, Singleton PW, Cassman KG (1997) Inorganic and organic phosphorus dynamics during a build-up and decline of available phosphorus in an Ultisol. Soil Sci 162:254-264

Lins IDG, Cox FR (1989) Effect of extractant and selected soil properties on predicting the correct phosphorus fertilization of soybean. Soil Sci Soc Am J 53:813-816

Liu S-T, Lee L-Y, Tai C-Y, Hung C-H, Chang Y-S, Wolfram JH, Rogers RD, Goldstein AH (1992) Cloning of an *Erwinia herbicola* gene necessary for gluconic acid production and enhanced mineral phosphate solubilization in *Escherichia coli* HB101: Nucleotide sequence and probable involvement in biosynthesis of the coenzyme pyrroloquinoline quinone. J Bacteriol 174:5814-5819

Loginow W, Wisniewski W, Gonet SS, Ciescinska B (1987) Fractionation of organic carbon based on susceptibility to oxidation. Polish J Soil Sci 20:47-52

Lopez-Hernandez D, Siegert G, Rodriguez JV (1986) Competitive adsorption of phosphate with malate and oxalate by tropical soils. Soil Sci Soc Am J 50:1460-1462

Mafongoya PL, Dzowela BH, Nair PK (1997) Effect of multipurpose trees, age of cutting and drying method on pruning quality. In: Cadisch G, Giller KE (Eds) Driven by nature: Plant litter quality and decomposition. CAB International, Wallingford, UK, pp 167-174

Magid J, Mueller T, Jensen LS, Nielsen NE (1977) Modelling the measurable: Interpretation of field-scale CO_2, N-mineralization, soil microbial biomass and light fractions as indicators of oilseed rape, maize and barley straw decomposition. In: Cadisch G, Giller KE (Eds) Driven by nature: Plant litter quality and decomposition. CAB International, Wallingford, UK, pp 349-362

Maroko JB, Buresh RJ, Smithson PC (1998) Soil nitrogen availability as affected by fallow-maize systems on two soils in Kenya. Biol Fertil Soils 26:229-234

Maroko JB, Buresh RJ, Smithson PC (1999) Soil phosphorus fractions in unfertilized fallow-maize systems on two soils in Kenya. Soil Sci Soc Am J 63:320-326

McClellan GH, Notholt AJG (1986) Phosphate deposits of tropical sub-Saharan Africa. In: Mokwunye AU, Vlek PLG (Eds) Management of nitrogen and phosphorus fertilizers in sub-Saharan Africa. Martinus Nijhoff, Dordrecht, The Netherlands, pp 173–224

McCollum RE (1991) Buildup and decline in soil phosphorus: 30-year trends on a Typic Umprabuult. Agron J 83:77-85

McLean FT, Gilbert BE (1927) The relative aluminum tolerance of crop plants. Soil Sci 24:163-175

Meijboom FW, Hassink J, van Noordwijk M (1995) Density fractionation of soil organic matter using silica suspensions. Soil Biol Biochem 27:1109-1111

Melillo JM, Aber JD, Muratore JF (1982) Nitrogen and lignin control of hardwood litter decomposition rates. Ecology 63:621-626

Menon RG, Chien SH, Hammond LL (1990) Development and evaluation of the Pi soil test for plant-available phosphorus. Commun Soil Sci Plant Anal 21:13-16

Mudahar MS (1986) Fertilizer problems and policies in sub-Saharan Africa. In: Mokwunye AU, Vlek PLG (Eds) Management of nitrogen and phosphorus fertilizers in sub-Saharan Africa. Martinus Nijhoff Publishers, Dordrecht, The Netherlands, pp 1-32

Mugwira LM, Elgawhary SM (1979) Aluminum accumulation and tolerance of triticale and wheat in relation to root cation exchange capacity. Soil Sci Soc Am J 43:736-740

Munns DN (1965) Soil acidity and growth of a legume II. Reactions of aluminum and phosphate in solution and effects of aluminum, phosphate, calcium and pH on *Medicago sativa* L. in solution culture. Austr J Agric Res 16:743-755

Murage EW (1998) Soil carbon fractions based on physical, chemical and biological fractionation and their relationship to fertility of Humic Nitisols in the central Kenya highlands. MSc thesis, University of Nairobi, Nairobi, Kenya

Mutuo PK (1997) Soil phosphorus pools following phosphorus fertilization, their relationships to maize yield in western Kenya. MSc thesis, Moi University, Eldoret, Kenya

Mutuo PK, Smithson PC, Buresh RJ, Okalebo RJ (1999) Comparison of phosphate rock and triple superphosphate on a phosphorus-deficient Kenyan soil. Commun Soil Sci Plant Anal, 30:1091-1103

Myers RJK, van Noordwijk M, Vityakon P (1997) Synchrony of nutrient release and plant demand: Plant litter quality, soil environment and farmer management options. In: Cadisch G, Giller KE (Eds) Driven by nature: Plant litter quality and decomposition. CAB International, Wallingford, UK, pp 215-229

Novais R, Kamprath EJ (1978) Phosphorus supplying capacities of previously heavily fertilized soils. Soil Sci Soc Am J 42:931-935

Nziguheba G, Palm CA, Buresh RJ, Smithson PC (1998) Soil phosphorus fractions and adsorption as affected by organic and inorganic sources. Plant Soil 198:159-168

Ozanne PG (1980) Phosphate nutrition of plants—a general treatise. In: Khasawneh FE, Sample EC, Kamprath EJ (Eds) The role of phosphorus in agriculture. Soil Sci Soc America, Madison, Wisconsin, USA, pp 559-589

Palm CA (1995) Contribution of agroforestry trees to nutrient requirements of intercropped plants. Agroforestry Systems 30:105-124

Palm CA, Rowland AP (1997) A minimum dataset for characterization of plant quality for decomposition In: Cadisch G, Giller KE (Eds) Driven by nature: Plant litter quality and decomposition. CAB International, Wallingford, UK, pp 379-392

Palm CA, Sanchez PA (1991) Nitrogen release from the leaves of some tropical legumes as affected by their lignin and polyphenolic contents. Soil Biol Biochem 23:83-88

Palm CA, Myers RJK, Nandwa SM (1997) Combined use of organic and inorganic nutrient sources for soil fertility maintenance and replenishment. In: Buresh RJ, Sanchez PA, Calhoun F (Eds) Replenishing soil fertility in Africa. SSSA Special Publication No 51, Soil Sci Soc America, Madison, Wisconsin, USA, pp 193-217

Quiñones MA, Borlaug NE, Dowswell CR (1997) A fertilizer-based green revolution for Africa. In: Buresh RJ, Sanchez PA, Calhoun F (Eds) Replenishing soil fertility in Africa. SSSA Special Publication No 51, Soil Sci Soc America, Madison, Wisconsin, USA, pp 81-95

Rajan SSS, Watkinson JH, Sinclair AG (1996) Phosphate rocks for direct application to soils. Adv Agron 57:77-159

Ritchey KD, Souza DMG, Lobato E, Correa O (1980) Calcium leaching to increase rooting depth in a Brazilian savannah Oxisol Agron J 72:40-44

Rogers RD, Wolfram JH (1993) Biological separation of phosphate from ore. Phosphorus, Sulfur, Silicon 77:137-140

Roy AH, McClellan GH (1986) Processing phosphate ores into fertilizers In: Mokwunye AU, Vlek PLG (Eds) Management of nitrogen and phosphorus fertilizers in sub-Saharan Africa. Martinus Nijhoff Publishers, Dordrecht, The Netherlands, pp 225-252

Saggar S, Hedley MJ, White RE, Gregg PEH, Perrott KW, Cornforth IS (1992) Development and evaluation of an improved soil test for phosphorus. 2. Comparison of the Olsen and mixed cation-anion exchange resin tests for predicting the yield of ryegrass grown in pots. Fert Res 33:135-144

Saggar S, Hedley MJ, White RE, Gregg PEH, Perrott KW, Cornforth IS (1995) Evaluation of soil tests for predicting pasture response on acid soils to phosphate supplied in water soluble and sparingly soluble forms. In: Date RA, Grundon NJ, Rayment GE, Probert ME (Eds) Plant-soil interactions at low pH: Principles and management. Kluwer Academic Publishers, Dordrecht, The Netherlands, pp 657-660

Sanchez PA (1976) Properties and management of soils in the tropics. John Wiley and Sons, New York, NY, USA

Sanchez PA (1994) Tropical soil fertility research, towards the second paradigm. Trans 15th World Congress of Soil Sci, Acapulco, Mexico 1:65–88

Sanchez PA (1997) Changing tropical soil fertility paradigms: from Brazil to Africa and back. In: Moniz AC, Fulani AMC, Shaffert RE, Fageria NK, Rosolem CA, Cantarella H (Eds) Plant-soil interactions at low pH: Sustainable agriculture and forestry production. Brazilian Soil Sci Soc, Vićosa, Minas Gerais, Brazil, pp 19-28

Sanchez PA, Buol SW (1975) Soils of the tropics and the world food crisis. Science 188:598-603

Sanchez PA, Logan TJ (1992) Myths and science about the chemistry and fertility of soils in the tropics. In: Lal R, Sanchez PA (Eds) Myths and science of soils of the tropics. SSSA Special Publication No 29. Soil Sci Soc of America, Madison, Wisconsin, USA, pp 35-46

Sanchez PA, Salinas JG (1981) Low input technology for managing Oxisols and Ultisols in tropical America. Adv Agron 34:279-406

Sanchez PA, Uehara G (1980) Management considerations for acid soils with high phosphorus fixation capacity. In: Khasawneh FE, Sample EC, Kamprath EJ (Eds) The role of phosphorus in agriculture. Soil Sci Soc America, Madison, Wisconsin, USA, pp 471-514

Sanchez PA, Couto W, Buol SW (1982) The fertility capability soil classification system: Interpretation, applicability and modification. Geoderma 27:283-309

Sanchez PA, Izac A-M N, Valencia I, Pieri C (1996) Soil fertility replenishment in Africa: a concept note. ICRAF, Nairobi, Kenya

Sanchez PA, Shepherd KD, Soule MJ, Place FM, Buresh RJ, Izac A-M N, Mokwunye AU, Kwesiga FR, Ndiritu CG, Woomer PL (1997) Soil fertility replenishment in Africa: An investment in natural resource capital. In: Buresh RJ, Sanchez PA, Calhoun F (Eds) Replenishing soil fertility in Africa. SSSA Special Publication No 51, Soil Sci Soc America, Madison, Wisconsin, USA, pp 1-46

Savini I (1999) The effect of organic and inorganic amendments on phosphorus release and availability from two phosphate rocks and triple superphosphate in phosphorus fixing soils. MSc thesis, Dept of Soil Science, University of Nairobi, Nairobi, Kenya

Schmidt JP, Buol SW, Kamprath EJ (1996) Soil phosphorus dynamics during seventeen years of continuous cultivation: Fractionation analyses. Soil Sci Soc Am J 60:1168–1172

Sharpley AN, Tiessen H, Cole CV (1987) Soil phosphorus forms extracted by soil tests as a function of pedogenesis. Soil Sci Soc Am J 51:362-365

Sharpley AN, Troeger WW, Smith SJ (1991) The measurement of bioavailable phosphorus in agricultural runoff. J Environ Qual 20:235-238

Sheldon RP (1987) Industrial minerals—emphasis on phosphate rock. In: McLaren DJ, Skinner BJ (Eds) Resources and world development. John Wiley & Sons, New York, pp 347–361

Shepherd KD, Soule MJ (1998) Assessment of the economic and ecological impacts of agroforestry and other soil management options on western Kenyan farms using a dynamic simulation model. Agric Ecosys Environ 71:131-145

Shortle WC, Smith KT (1988) Aluminum-induced calcium deficiency syndrome in declining red spruce. Science 240:1017-1018

Sibbesen E (1978) A simple ion-exchange resin procedure for extracting plant-available elements from soil. Plant Soil 46:665-669

Singh BB, Jones JP (1976) Phosphorus sorption and desorption characteristics of soil as affected by organic residues. Soil Sci Soc Am J 40:389-394

Smaling EMA, Nandwa SM, Janssen BH (1997) Soil fertility in Africa is at stake. In: Buresh RJ, Sanchez PA, Calhoun F (Eds) Replenishing soil fertility in Africa. SSSA Special Publication No 51, Soil Sci Soc America, Madison, Wisconsin, USA, pp 47-61

Smyth TJ, Sanchez PA (1982) Phosphate rock dissolution and availability in Cerrado soils as affected by phosphorus sorption capacity. Soil Sci Soc Am J 46:339-245

Stenberg B (1997) Integrated evaluation of physical, chemical and biological properties of agricultural soil. PhD thesis, Swedish University of Agricultural Sciences, Uppsala

Stenberg B, Nordkvist E, Salomonsson L (1995) Use of near infrared reflectance spectra of soils for objective selection of samples. Soil Sci 159:109-114

Stewart JWB, Thiessen H (1987) Dynamics of soil organic phosphorus. Biogeochemistry 4:41-60

Stoorvogel JJ, Smaling EMA (1990) Assessment of soil nutrient depletion in sub-Saharan Africa, 1983-2000. Report 28, DLO Winand Staring Center for Integrated Land, Soil, Water Research, Wageningen, The Netherlands

Strickland TC, Sollins P (1987) Improved method for separating light- and heavy-fraction organic material from soil. Soil Sci Soc Am J 51:1390-1393

Swift MJ (1987) Tropical soil biology and fertility: Interregional research planning workshop. Biology International Special Issue 13 IUBS, Paris, France

Syers JK, Lu Ru-kun (1990) Inorganic reactions influencing phosphorus cycling in soils. In: Phosphorus requirements for sustainable agriculture in Asia and Oceania. International Rice Research Institute, Los Baños, the Philippines, pp 191-197

Tian G, Kang BT, Brussaard L (1992) Effects of chemical composition on N, Ca, and Mg release during incubation of leaves from selected agroforestry and fallow plant species. Biogeochemistry 16:103–119

Tiessen H, Moir JO (1993) Characterization of available P by sequential extraction. In: Carter MR (Ed) Soil sampling and methods of analysis. Can Soc Soil Sci, Lewis Publishers, pp 75-86

Tinker PB (1980) Role of rhizosphere microorganisms in phosphorus uptake by plants. In: Khasawneh FE, Sample EC, Kamprath EJ (Eds) The role of phosphorus in agriculture. Soil Sci Soc America, Madison, Wisconsin, USA, pp 617-654

van Kauwenbergh SJ (1991) Overview of phosphate deposits in East and Southeast Africa. Fert Res 30:127-150

van Kauwenbergh SJ (1995) Overview of the global phosphate rock production situation. In: Dahanayake K, Van Kauwenbergh SJ, Hellums DT (Eds) Direct application of phosphate rock and appropriate technology fertilizers in Asia—what hinders growth and acceptance. Institute of Fundamental Studies, International Fertilizer Development Center, Kandy, Sri Lanka, pp 15–28

Vierstra R, Haug A (1978) The effect of Al^{3+} on the physical properties of membrane lipids in *Thermoplasma acidophilum*. Biochem Biophys Res Comm 84:138-143

Wade MK, Sanchez PA (1983) Mulching and green manure applications for continuous crop production in the Amazon basin. Agron J 75:39-45

World Bank (1996a) African development indicators. 1996 World Bank, Washington, DC

World Bank (1996b) Natural resources degradation in sub-Saharan Africa: Restoration of soil fertility. Africa Region World Bank, Washington, DC

Yost RS, Kamprath EJ, Lobato E, Naderman GC (1979) Phosphorous response on an Oxisol as influenced by rates and placement. Soil Sci Soc Am J 43:338–343

Yost RS, Kamprath EJ, Naderman GC, Lobato E (1981) Residual effects of phosphorus applications on a high phosphorus adsorbing Oxisol of central Brazil. Soil Sci Soc Am J 45:540-543

Zaharah AR, Bah AR (1997) Effect of green manures on P solubilization and uptake from phosphate rocks. Nutr Cycl Agroecosys 48:247-255

Zhu Zhao-ling, Xi Zhen-bang (1990) Recycling phosphorus from crop and animal wastes in China. In: Phosphorus requirements for sustainable agriculture in Asia and Oceania. International Rice Research Institute, Los Baños, the Philippines, pp 116-123

Part II.
Root Exudates in Nutrient Acquisition and Metal Tolerance

Root Exudates in Phosphorus Acquisition by Plants

Peter J. Randall, Julie E. Hayes, Peter J. Hocking, and Alan E. Richardson

Summary. This chapter discusses the processes in the rhizosphere that are determined by exudates from roots and that in turn affect the availability of phosphorus to plants. These include control of rhizosphere pH, exudation of organic acids and root phosphatases. Possibilities for manipulating these processes in order to improve phosphorus acquisition in agricultural plants are discussed in the context of the phosphorus cycle in agriculture.

Key words. Cation excess, Citrate, Malate, Organic acids, Phosphatase, Phosphorus acquisition, Phosphorus nutrition, Phytase, Phytate, Proteoid roots, Rhizosphere, Rhizosphere pH, Root exudates, Soil phosphorus

1. Introduction and Background

1.1 Introduction

A wide range of organic compounds are exuded from healthy, undamaged roots and are found in the rhizosphere of plants (Rovira 1962; Rovira and Davey 1974; Curl and Trueglove 1986). Early studies of the root-soil interface focused on the microbial populations which inhabit this zone (Bowen 1980), and interest in root exudates focused more on their effects on microbial populations rather than possible direct effects on nutrient availability (Rovira 1969). Although the influence of exudates on plant nutrition was recognized, it was considered to be largely indirect through effects on the microflora (Rovira and Davey 1974).

By the early 1980s, definitive evidence had emerged for a direct role of root exudates in increasing the availability in soil of three plant nutrients. These were manganese (Mn), iron (Fe) and phosphorus (P), each of which is relatively immobile in soil (Marschner 1986).

The first evidence was for Mn when Bromfield (1958) and Uren (1982) showed that root exudates could solubilize MnO_2 and release Mn^{2+}, the form available to plants. Subsequently, Reisenauer and colleagues (Godo and Reisenauer 1980; Jauregui and Reisenauer 1982) demonstrated that malic acid in root exudates could chelate Mn^{2+}, preventing its reoxidation and maintaining Mn in a form available to plants.

Work on Fe acquisition, showed that dicotyledonous plants could improve the availability of Fe^{2+} around their roots by releasing H^+ and phenolic compounds

(Romheld and Marschner 1981). By contrast, graminaceous species use an alternative strategy to obtain Fe. This involves the release of non-protein amino acids, such as mugeneic acid (Tagaki 1976; Takagi et al. 1984), which mobilize Fe by chelation and act as phytosiderophores.

For P, Gardner et al. (1982a,b, 1983) provided the first strong evidence linking organic acid exudation from roots with solubilization of poorly-available soil P and enhanced P uptake. The roots of white lupin (*Lupinus albus*) were shown to exude citric acid (citrate), and it was proposed that citrate improved the P nutrition of the plant by forming a ferric-hydroxy-phosphate polymer in the rhizosphere which diffused to the root and there released P following reduction (Gardner et al. 1983).

Since these early studies, there has been considerable interest worldwide in the direct role of root exudates in enhancing nutrient acquisition by plants. General aspects of this work have been reviewed by Marschner et al. (1986), Curl and Trueglove (1986), Uren and Reisenauer (1988), Bar-Yosef (1991), Jungk (1991), Darrah (1993), and Jones (1998). In this chapter, we discuss how organic acids and phosphatases exuded from roots, and pH changes in the rhizosphere affect the P nutrition of plants. We concentrate on P because it plays a major role in agricultural production and P fertilizer is a significant input cost for farmers worldwide. We begin by mentioning aspects of the P nutrition of plants in natural communities and agricultural systems in Australia, as background to a discussion of root exudates and their roles in solubilizing inorganic and organic forms of P in soils. We conclude by assessing some possibilities for developing plants with an improved ability to acquire P from poorly-available sources in soils and fertilizers.

1.2 Plant Nutrition in the Australian Context

1.2.1 The Importance of P in Agriculture

Most Australian soils are old, highly weathered, leached and therefore deficient in the nutrients required by agricultural plants. In reviewing fertilizer responses in Australian soils, Stephens and Donald (1958) wrote: "Artificial fertilizers...have produced some of the most spectacular crop and pasture responses in the world. These responses have been given by phosphatic fertilizers, and by the trace elements...." Studies of soil fertility and plant nutrition have therefore been crucial to the development of the fertilizer technology and management practices that underpin modern-day agriculture. The use of superphosphate and one or more trace elements, together with the establishment of pasture legumes, particularly subterranean clover in southern Australia and a range of tropical legume species in northern Australia, raised the levels of available nitrogen (N) in the soil and increased pasture growth and the productivity of grazing animals. The extensive development of improved pastures on what were formerly nutrient-impoverished soils has been of great economic significance to Australia (Williams and Andrew 1970). Phosphorus fertilizer is a key input, both for its direct effects on pasture and crop productivity and indirectly through stimulating N fixation by pasture and grain legumes, thus making an invaluable contribution to the N economy of Australian agriculture.

1.2.2 Plants with Improved Ability to Acquire P as a Strategy for Managing P in Australian Agriculture

While fertilizer application to correct P deficiency has received most attention, there has also been considerable interest in improving the P acquisition of agricultural plants by selection and plant breeding (Graham 1984). The case for this approach is based on several arguments. Firstly, it is known that agricultural species differ in their ability to obtain P from the same soil (e.g., McLachlan 1976). Secondly, in most soils total P greatly exceeds crop and pasture requirements even though available P is inadequate (Graham 1984). Thirdly, considerable amounts of P have accumulated in soils from past fertilizer applications, because P added in fertilizer generally reacts rapidly with soil constituents. The fertilizer P is "fixed" by becoming bound to charged surfaces, precipitated in insoluble forms or immobilized by soil microorganisms and incorporated into organic forms (Barrow 1982). Because of this "fixation", the initial recovery of fertilizer P by plants is usually poor, often in the order of only 20% during the first season (Sharpley 1986; McLaughlin et al. 1988). While more of the applied P is recovered in subsequent seasons and is thus said to have a residual value, appreciable quantities of P from fertilizer continue to accumulate in Australian agricultural soils. Estimates of P budgets for major agricultural enterprises in Australia indicate that inputs of P in fertilizer considerably exceed removals in produce (McLaughlin et al. 1992, Table 1). Cropping, typified by wheat production, has an input to output ratio of about 2 to 1. By comparison, the animal production industries have a higher input to output ratio (circa 5 to 1) as the P is applied to grow pasture, whilst relatively small amounts are removed in animal products. Horticultural industries such as vegetable and fruit production also have high ratios, reflecting their more intensive nature and the need to ensure high levels of productivity (Table 1). The P accumulating in "fixed" forms in agricultural soils represents a substantial investment by farmers. Considerable research has been conducted to understand the processes involved in the fixation of P, in an effort to improve the return on this investment. Phosphorus fertilizer is a significant cost of production to farmers, particularly in grazing enterprises where it may comprise 20% of variable costs, so improvement in its efficiency of use is highly desirable.

Table 1. Fertilizer input(I) and product output(O) of phosphorus for the major agricultural enterprises in Australia (extracted from McLaughlin et al. 1992)

Enterprise type	P applied in fertilizer (I) (kt)	P in harvested product (O) (kt)	I/O ratio
Wheat	85	42.2	2.0
Wool, meat, milk, and live animals	179	35.8	5.0
Vegetables	8	1.3	6.2
Fresh fruit	6	0.4	15.0

1.2.3 Adaptations by Australian Native Plants to Low Nutrient Soils

Despite the infertility of many Australian soils, an extensive and diverse native flora is able to flourish where agricultural species fail because they are unable to acquire sufficient P, unless given fertilizer (Specht and Groves 1966). The adaptive mechanisms that allow these native plants to thrive on soils low in available P include those that economize on the internal use of P and those that maximize acquisition of P from the soil. Amongst the latter are adaptations that increase the effective surface area of roots for nutrient uptake. These include association with mycorrhizae and with P-solubilizing bacteria, the production of roots taking priority over above-ground growth, preferential root development in nutrient-rich, organic layers near the soil surface, and development of specialized roots including proteoid root clusters (Handreck 1996). Other adaptations involve the ability of roots to modify the chemistry of the rhizosphere. The best-described example is the exudation of citrate by proteoid roots of *Banksia integrifolia* (Grierson 1992), which enhances uptake of P (Jescke and Pate 1995), a mechanism similar to that described for the agricultural plant, white lupin (*Lupinus albus*) (Gardner et al. 1983). It is likely that further studies of rhizosphere processes in Australian native plants that have evolved on infertile soils will reveal other strategies. An understanding of the range of adaptive mechanisms of the Australian flora to infertile soils may provide conceptual models and possibly genes for improvement of nutrient efficiency in agricultural plants.

2. Phosphorus in Soils and P Uptake by Plants

2.1 Soil P and the P Cycle in Agricultural Systems

As a background to discussing the role of root exudates in P acquisition, it is useful to consider briefly the forms of P in soils and the factors affecting their availability to plants. Phosphorus in soils occurs in both organic forms, including plant and animal residues and microbial P, and inorganic forms. Most of the P is soil-bound or precipitated and only a small fraction is in solution (Fig. 1). This is important for plant nutrition because only the inorganic phosphate (P_i) in the soil solution is directly available for uptake by roots. In most soils, and despite the high content of total soil P, it is necessary to apply fertilizer P to maintain a source of soluble P_i close to the roots in order to maximize agricultural plant production. Some of the fertilizer P not taken up by plants is used by soil microbes and immobilized in organic forms. However, a major portion is adsorbed to charged surfaces or reacts with soil minerals and precipitates in sparingly-soluble inorganic forms. Some of the organic P (P_o) can return to the soluble pool as a result of microbial turnover, and this may be in organic or inorganic forms. Although P is removed from farms in harvested products, over the longer term, P accumulates in both the inorganic and organic soil P fractions with the rate and form accumulated depending on soil type, land use and fertilizer history (Harrison 1987).

The inorganic fractions of soil P include phosphate ions adsorbed to hydrous oxides of aluminium (Al) and Fe, and P precipitated as amorphous and crystal-

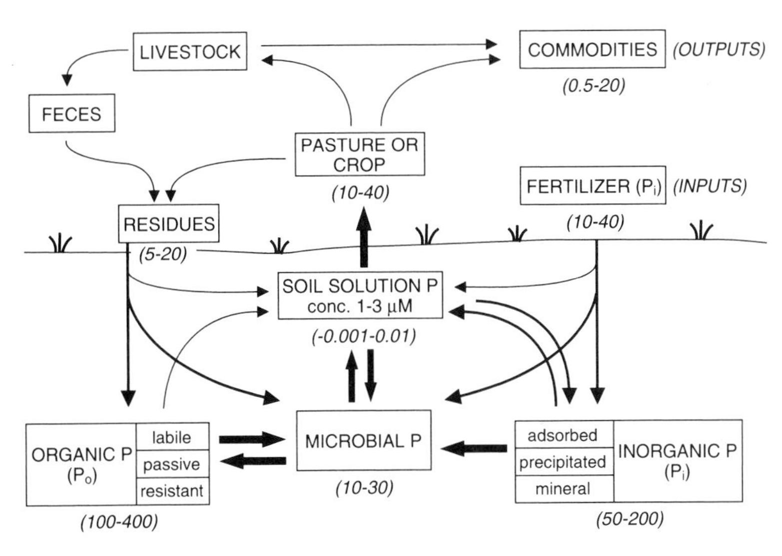

Fig. 1. Schematic representation of the major components of the soil–plant phosphorus cycle in agriculture. Major P fractions and the amounts of P (kg ha^{-1} ; soil 0-10 cm) typically associated with each fraction are shown in *parentheses*. The passage of P between different fractions is shown by *arrows*. Fertilizer P (*INPUTS*) dissolves in soil water and is usually rapidly immobilized in the inorganic (P_i) pool or via microbial activity, in the organic (P_o) pool. The immobilized P_i is in equilibrium with P_i in the soil solution, the source of P for plants. P in crops or pasture is either removed in harvested commodities (*OUTPUTS*) or returned to the soil in residues. Note the low concentration and amount of P in the soil solution. (after Richardson 1994)

line Al and Fe minerals, and as calcium (Ca) phosphates (Sample et al. 1980). The actual compounds formed depend on the soil chemistry, and solubilization-precipitation reactions involving these compounds are strongly pH-dependent. The inorganic forms of P are in equilibrium with the P_i in soil solution.

Organic P may account for up to 80% of the total soil P in some soils (McLaughlin et al. 1990). Constituents of soil P_o include inositol phosphates (phytate), representing up to 50% of the total organic P (Anderson 1980), and lesser amounts (1% to 5%) of other P esters (phospholipids, sugar phosphates, nucleic acids), polyphosphates, pyrophosphates and phosphonates (Newman and Tate 1980). However, some 50% or more of the soil P_o is in unidentified high molecular weight compounds (Dalal 1977).

Organic P is found largely in the bulk soil but also exists in soil solution, often at concentrations greater than soil solution P_i (Ron Vaz et al. 1993; Seeling and Zasoski 1993; Seeling and Jungk 1996). For example, 80% of the P in dilute $CaCl_2$ extracts of a luvisol, and a similar proportion in the soil solution, was in the organic form (Seeling and Jungk 1996). Despite its abundance, the identity of the P_o in solution remains largely unknown, although inositol monophosphate has been identified (Wild and Oke 1966).

Accumulation of P_o in soil occurs by a combination of immobilization of fertilizer P in soil organisms and the build-up of plant residues that are resistant to mineralization (Dalal 1977). Soil P_o compounds differ somewhat in their susceptibility to mineralization. Phytate and the high molecular weight compounds are relatively resistant, whilst other P esters, polyphosphates and phosphonates appear more susceptible to hydrolysis.

2.2 Phosphate Uptake by Plants in Relation to Rhizosphere Processes

Plants obtain their P from the soil solution in contact with their roots. Diffusion is the major transport process supplying P_i to plant roots, with mass flow making a comparatively small contribution (Barber 1984). The concentration of P_i in the soil solution is very low (typically 1 to 3 mmol m^{-3} ([μM], Fig. 2) and, as plants deplete the P_i in solution, it is replenished from labile P sources. This occurs by desorption of phosphate from charged surfaces, solubilization of P-containing minerals and hydrolysis of organic P compounds (Fig. 2). Despite this, rates of diffusion of P_i in the soil are low (~0.13 mm day^{-1}) (Jungk 1991) and are generally insufficient to match the rates of uptake occurring at the root surface. Consequently, P_i in the soil solution immediately adjacent to the root surface is rapidly depleted and a marked concentration gradient is established in the rhizosphere (Hendriks et al. 1981). This concentration gradient creates the conditions for the net diffusion of P towards the root, which in turn increases the rate of desorption of P_i from soil. Both the rates of replenishment and the diffusion of P_i towards the root can be enhanced by root-mediated processes, as will be discussed later.

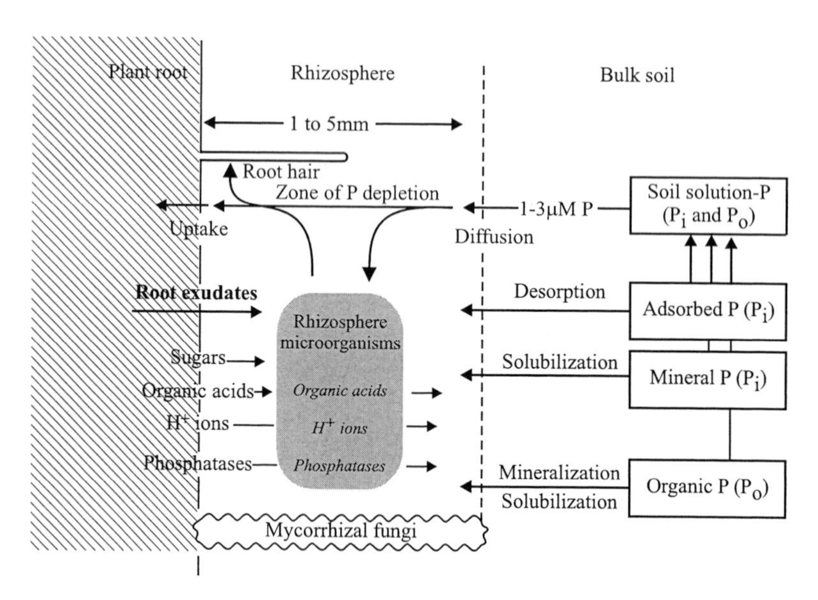

Fig. 2. Diagram of the major processes affecting P availability in the rhizosphere and P supply to plant roots. Plant roots take up P_i from the soil solution. This is replenished from P_i and P_o in the solid phase. Root exudates increase the rates of solubilization, desorption, and mineralization either directly or via effects on microorganisms. Root hairs and mycorrhiza increase the effective absorbing area of the root and extend the volume of soil explored. (after Richardson 1994)

2.3 Role of Root Exudates in P Solubilization in the Rhizosphere

Features of the rhizosphere that are important for this discussion are summarized in Fig. 2. Root hairs and mycorrhizae increase the volume of soil explored, improve contact between root and soil and increase the effective absorbing area of the root. The other processes depicted are concerned with modifying the chemistry in the rhizosphere and thus altering the availability of P compounds. This can be mediated directly by root exudates, or indirectly through promoting the growth of rhizosphere microorganisms, which may assist in both solubilizing and mineralizing P, as reviewed by Richardson (1994) and Kucey et al. (1989). A wide range of soil bacteria and fungi have been shown to solubilize soil P through the production of organic acids (Moghimi et al. 1978; Whitelaw et al. 1999). Studies of microbial influences on P availability in soils have resulted in commercial products containing microbial inoculants for application to seeds in order to improve the P nutrition of crops (see chapter by Leggett, this volume).

Root exudates also affect P availability in the rhizosphere more directly. The mechanisms include: (1) changing conditions in the soil solution (e.g., pH), thus promoting the dissolution of sparingly soluble P-containing minerals; (2) altering surface characteristics of soil particles; (3) competing with phosphate ions for adsorption sites; (4) complexing cations which are bound to phosphate; and (5) enzymic hydrolysis of organic forms of P, and there is a detailed discussion of these factors in Bar-Yosef (1991). The importance of each of these mechanisms will differ depending on the soil, the sources of P it contains, and other factors. For example, an increase in acidity would increase solution P by dissolution of calcium phosphates in neutral to alkaline soils, and by desorption reactions in acid soils where P solubility is controlled by ion-exchange equilibria involving charged clay minerals and organic matter (White 1980).

3. Direct Effects of Root Exudates on Rhizosphere pH

Rhizosphere pH is determined largely by the balance between cation and anion uptake (Hedley et al. 1983; Haynes 1990). Plants usually absorb unequal quantities of nutrient cations ($Ca^{2+} + Mg^{2+} + K^{+} + Na^{+} + NH_4^{+}$) and anions ($SO_4^{2-} + H_2PO_4^{-} + Cl^{-} + NO_3^{-}$), and maintain charge balance through excretion of H^+ or $RCOO^-$ in net amounts that are stoichiometrically equal to the excess cation or anion uptake. Because of the greater uptake of N relative to other nutrients, the form of N (NH_4^+ or NO_3^-) available to the plant has a major effect on conditions in the rhizosphere, with NH_4^+ nutrition resulting in a lower pH and NO_3^- nutrition leading to an increase in pH (Jarvis and Robson 1983c). Narrow-leaved lupin (*Lupinus angustifolius*) provides an example of a species which excretes H^+ to balance excess cation uptake (Loss et al. 1993), and excretes citrate to balance excess anion uptake without increasing pH. Citrate excretion rates of 1.2 μmol h^{-1} g^{-1} (dry wt of root) were measured for plants supplied with nitrate (Loss et al. 1994).

It has been shown that the form of the N supply can affect P uptake and that this effect depends on soil properties. For example, ammonium sulfate decreased the root surface pH of corn, and this increased P uptake from monocalcium phosphate fertilizer, in soils above pH 5.5 (Blair et al. 1971). Gahoonia et al. (1992) showed that ammonium nutrition lowered the pH in the rhizosphere of ryegrass and this resulted in increased P mobilization from a luvisol in which calcium phosphates were dissolved by acidification, but had no effect in an oxisol. Conversely, nitrate nutrition raised the pH of the rhizosphere, and this increased P mobilization in the oxisol, probably by ligand exchange, but had no effect on P in the luvisol. McLachlan (1976) concluded that buckwheat was superior to rye and clover in obtaining P from superphosphate in soil, due to its ability to acidify the rhizosphere. This was attributed to excess cation uptake, particularly Ca^{2+}, by buckwheat. Bekele et al. (1983), using rock phosphate as a P source, confirmed this ability of buckwheat. They suggested a second mechanism as a consequence of excess cation uptake: the lowering of Ca^{2+} in the soil solution shifted the mass-action equilibrium and increased the rate of dissolution of the calcium phosphate, thus releasing P.

The rhizosphere of P-deficient oilseed rape plants becomes acidified in restricted zones. While initially this was attributed to excess cation uptake (Hedley et al. 1983; Moorby et al. 1988), later work showed that exudation of malate and citrate was important (Hoffland et al. 1989). Calculations show that this mechanism can account for the ability of oilseed rape to obtain P from calcium phosphates contained in rock phosphate in an unbuffered system (Hoffland 1992). Experiments with rock phosphate added to alumina sand, an acid, P-fixing medium, showed that rape plants substantially raised the pH in the rhizosphere and increased solubilization of rock phosphate and P uptake. It was suggested that P was released due to the acid nature of the medium and because roots depleted Ca in the rhizosphere. The released P immediately reacted with the strongly P-sorbing alumina, but as the rhizosphere pH increased during plant growth, P was desorbed and became available for uptake (Hinsinger and Gilkes 1997).

4. Organic Acids in the Rhizosphere

4.1 Organic Acids Excreted from Roots and Uptake of P

Many studies have shown that plants excrete organic acids from their roots. Citric, malic and malonic acids are commonly found, and oxalic, succinic, tartaric, piscidic, aconitic and fumaric acids have also been reported (Table 2). The effectiveness of an organic acid in mobilizing soil P will vary depending on its ability to complex metals and displace P from charged surfaces. Organic acids decreased the adsorption of P in soils in the order tricarboxylic > dicarboxylic > monocarboxylic acids, and the amount of P released was proportional to the ability to complex Al (Bar-Yosef 1991; Bolan et al. 1994). Table 3 summarizes the formation constants of the major organic acids reported in root exudates with Al, Fe and Ca. The tricarboxylic acid, citric acid is often singled out for attention because it binds more strongly than the di- or monocarboxylic acids to metals that are important in P chemistry in soils (Hue et al. 1986; Bar-Yosef 1991). Studies by Jones and

Table 2. Organic acids measured in exudates from whole root systems of selected species

Plant species	Growth conditions	Organic acid released from roots (whole root systems; nmol h^{-1} g^{-1} dry wt[a])				Reference
		Citric	Malic	Malonic	Others	
Pigeon pea	Complete nutrient solution	2	tr	580	Oxalic (186), piscidic (90), tartaric (1)	(Otani et al. 1996)
Chick pea	Fe deficient solution	1	tr	6	Tartaric (2), fumaric (1)	(Ohwaki and Sugahara 1997)
Rice	Soil, low P	2300			Oxalic, malic, lactic, fumaric.	(Kirk et al. 1999)
Maize	Nutrient-deficient	1300	6000			(Jones and Darrah 1995)
Narrow-leaved lupin	Solution with nitrate	1200				(Loss et al. 1994)
White lupin	P-deficient solution	11000	8000			(Johnson et al. 1996)

tr, trace detected
[a]Assumed dry weight is 7% of fresh weight where conversion required

Table 3. Log formation constants with Al, Fe, and Ca for various organic acids found in root exudates

Organic acid	Log formation constants		
	For Al-ligand	For Fe^{3+}-ligand	For Ca-ligand
$Citrate^{3-}$	9.6[a]	12.5[a]	3.1[a]
$Oxalate^{2-}$	7.3[a]	8.9[a]	2.9[a]
$Tartrate^{2-}$	5.3[a]	6.5[a]	2.8[a]
$Malate^{2-}$	5.4[b]	—	2.7[c]
$Malonate^{2-}$	—	7.5[a]	2.3[a]
$Acetate^{1-}$	1.5[a]	3.4[a]	1.2[a]

[a](Bar-Yosef 1991)
[b](Bolan et al. 1994)
[c](Hue et al. 1986)

Darrah (1994) indicate that citric acid is likely to be most effective at releasing P in soils containing acid-soluble calcium-phosphates. Citric acid is excreted in considerable quantities by some plants such as white lupin (Table 2) and has been detected in rhizosphere soils (Dinkelaker et al. 1989; Grierson 1992; Gerke et al. 1994; Gerke et al. 1995).

There are several lines of evidence that indicate the importance of organic acid exudation from roots in the acquisition of soil and fertilizer P by plants. For example, additions of organic acids, particularly citrate, to soils may solubilize significant quantities of P (Trainia et al. 1986; Bar-Yosef 1991). Responses to P deficiency by a number of plant species include increased rates of organic acid excretion from roots (Table 4) and it appears that this mechanism benefits these species under P-limiting conditions. A contrary result was reported for pigeon pea (*Cajanus cajan*) grown in sand culture, where exudation rates of malonic, citric and piscidic acids were higher from P-adequate than P-stressed plants (Otani and Ae 1997). This is surprising because the Fe-P solubilizing activity of exudates from pigeon pea roots increases with P stress (Subbarao et al. 1997b).

Root exudates containing citrate, malate and other organic acids have been shown to solubilize P in in vitro experiments. For example, Gardner et al. (1983) showed that exudate collected from white lupin roots, which contained citrate, could solubilize P bound to Fe. Similarly, Otani et al. (1996) found that exudates from pigeon pea, containing malonic and piscidic acids, solubilized P bound to Fe or Al. Further, measurements in rhizosphere soil of proteoid roots of white lupin showed elevated levels of both citrate and of soluble P, Al and Fe (Gerke et al. 1994). Dinkelaker et al. (1989) reported that P was depleted from soil in the rhizosphere of proteoid roots of white lupin grown in calcareous soil, and deposits of calcium citrate were clearly visible near the roots. Proteoid roots of *Banksia integrifolia* excrete citrate (Grierson 1992), and xylem sap collected from proteoid roots contained 1 mmol m^{-3} P compared to < 0.1 mmol m^{-3} P in sap from non-proteoid roots (Jescke and Pate 1995). At the agronomic level, correlative evidence

Table 4. Selected data for a range of plant species illustrating increases in organic acid excretion due to low P nutrition

Plant species	Organic acid	P status and amount exuded	Comments	Reference
Sorghum	Total dicarboxylic acids (predominantly aconitic and succinic)	9-fold higher in exudates from low P compared to high P plants	Whole roots washed out of soil after 6 weeks growth, treated with antibiotic, then exudates collected for 12 h	(Schwab et al. 1983)
		μmol d^{-1} g^{-1} root dry wt		
Lucerne	Citrate	+P 1.4, -P 2.6[a]	Whole root, aseptic sand culture, 24 days	(Lipton et al. 1987)
Tomato	Citrate	+P 0.2, -P 0.8	Whole roots, solution culture, 6-h collection	(Imas et al. 1997)
White lupin	Citrate malate	+P 12, -P 260[b] +P 12, -P 200[b]	Whole roots, sand culture, 22-h collection	(Johnson et al. 1996)

[a]Assumes citrate accumulated over final 12 days of growth
[b]Assumes dry weight is 7% of fresh weight

shows that species which exude appreciable quantities of organic acids can obtain more P from a deficient soil than plants without this attribute (Fig. 3; Ae et al. 1991; Hocking et al. 1997). Finally, calculations and mathematical models indicate the potential importance of organic acid exudates for increasing soluble P in the rhizosphere and consequently rates of P uptake (Bar-Yosef 1991; Hoffland 1992; Kirk et al. 1999).

4.2 Does Organic Acid Excretion Allow Plants to Draw on Different Pools of P in the Soil?

While it is clear that species differ in the amounts of P they can obtain from the same soil (e.g., McLachlan 1976), it has been more difficult to determine if they draw on different P pools in the soil or the same pools but at different rates. Smith (1981) grew buffel grass and four tropical pasture legumes (*Macroptilium* sp., *Desmodium* sp. and two *Stylosanthes* spp.) in two alkaline clay soils and compared P uptake and the ability to obtain P from different soil P pools using the ^{32}P isotopic dilution technique. Although the species differed considerably in their rates of removal of P and in total P uptake, L-values did not differ significantly. He concluded that these species drew on the same pools of P and that differences in P acquisition were due to differences in the ability of the plants to locate soluble P. As an alternative to labeling the soluble P pool in the soil, Armstrong et al. (1993) added ^{32}P-labelled amorphous Fe and Al phosphates, the crystalline forms strengite and variscite, and potassium phosphate to soils, on which were grown maize, sorghum, mung bean, cowpea and soybean. Their conclusions were similar to those of Smith (1981). Both studies indicated that the P-efficient and P-inefficient plants drew on the same pools of soil P.

It should be pointed out that none of the species studied by Smith (1981) or Armstrong et al. (1993) are known to excrete significant amounts of organic acids, although detailed studies are lacking. In contrast, studies including species known to exude organic acids suggest that plants vary considerably in their abil-

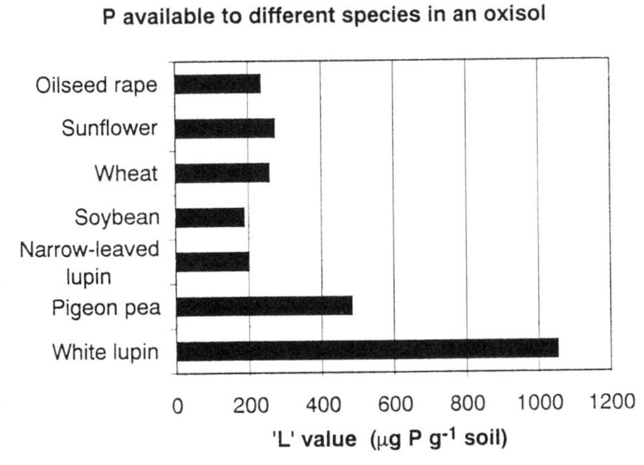

Fig. 3. Estimates of soil P available to seven crop species grown in an oxisol with 1.8 mg kg^{-1} Bray-1 P and 3.1 g kg^{-1} total P. Plants were sown after the soil was thoroughly mixed with $^{32}P_i$ to label the isotopically exchangeable P pool. 'L' values were calculated as a measure of the amount of P available to the plants. 'L'value = ((shoot ^{31}P-seed ^{31}P)* ^{32}P added to soil)/ shoot^{32}P (drawn from data of Hocking et al. 1997)

ity to take up P from different minerals in the soil. For example, Ae and his colleagues (Ae et al. 1991, 1993) found that maize, sorghum, millet, soybean and chickpea obtained more P from Ca-phosphate than from Fe- or Al-phosphates, whereas pigeon pea took up P equally well from Ca- and Fe-bound P. More recently, two groups (Braum and Helmke 1995; Hocking et al. 1997) used the ^{32}P isotope dilution technique to show that white lupin in particular can draw on soil P which is not available to other species. Hocking et al. (1997) compared the growth of seven species, including white lupin and pigeon pea, in an oxisol labeled with ^{32}P. After 5 weeks of growth the plants were all deficient in P and varied 8-fold in the specific activity of ^{32}P in the shoots, with oilseed rape having the highest specific activity and white lupin the lowest, indicating that the latter species had taken up proportionately more P from unlabelled P pools in the soil. Calculations of L-values were consistent with this, indicating that the pool of soil P available to white lupin was substantially larger than that available to the other species (Fig. 3). This may be explained by the high rates of citrate exudation by white lupin. The L-value for pigeon pea, while lower than that of white lupin, was higher than values for narrow-leaved lupin, wheat, sunflower, soybean or oilseed rape. This finding is consistent with the conclusions of Ae and colleagues (Ae et al. 1991) and may be due to root exudation by pigeon pea of malonic and possibly piscidic acids (Otani et al. 1996). It is interesting that oilseed rape appeared to be inefficient at obtaining P from insoluble sources in this soil (Fe- and Al-bound P) (Fig. 3) while being able to solubilize Ca-bound P in rock phosphate (Hoffland 1992; Hinsinger and Gilkes 1996). This may be due to the comparatively low rates of exudation, particularly of citrate, from roots of P-deficient oilseed rape (~3% of the rates from proteoid roots of white lupin; Table 5).

4.3 Quantities of Organic Acid Exudates from Roots

4.3.1 Problems of Measurement

Root exudates are usually collected from roots growing in nutrient solution because it facilitates the collection of exudates and also experimental manipulations such as control of the nutrient supply or applications of inhibitors (Ryan et al. 1995). Also, aseptic conditions can be imposed to avoid the complication of consumption of exudates by microorganisms. Use of solution-grown plants suffers from the criticism that the chemical and physical conditions are not comparable with those in soil, with the result that the roots may be morphologically and physiologically different, and consequently the rhizosphere conditions may be different. The presence of bacteria is claimed to enhance the development of proteoid roots of white lupin (Dinkelaker et al. 1995), which is a consideration if aseptically-grown roots are studied. In other studies, soil- or sand-grown roots have been washed out and placed in aqueous solutions for collection of exudates (e.g., Schwab et al. 1983; Ae et al. 1993). There is potential to damage roots in this procedure, with possible leakage of organic compounds from ruptured cells.

Attempts have been made to estimate exudation from roots in soil by measuring accumulation of organic acids in the rhizosphere. In a study of white lupin plants growing in a calcareous soil for 13 weeks, Dinkelaker et al. (1989) extracted

Table 5. Summary of rates of exudation of malate or citrate from different parts of root systems of white lupin and oilseed rape as affected by P supply

Plant species	Location of exudation	Nutrient status	Organic acid		Reference
			Malate	Citrate	
			(nmol h^{-1} cm^{-1})		
White lupin	Young proteoid roots	P-deficient		5.4	(Keerthisinghe et al. 1998)
		P-adequate		1.5	
	Lateral roots (non-proteoid)	P-deficient		0.7	
		P-adequate		0.2	
Oilseed rape	Root tips	P-deficient	0.44	0.14	(Hoffland et al.1989)
		P-adequate	0.08	0.03	
	Older root	P-deficient	0.10	0.07	
		P-adequate	0.01	0.01	

citrate from the rhizosphere soil with water for analysis. The soil was autoclaved but not maintained aseptic during plant growth. On the basis of the measured quantities of citrate, it was calculated that citrate exudation represented over 20% of the total plant dry weight at harvest. Li et al. (1997c) used irradiated soil in a box divided into 0.5 mm sections with polyester cloth so that rhizosphere soil at different distances from the roots of lupin could be extracted and analyzed. Under their conditions organic acid levels were elevated within 2 mm of the root and levels were similar in sterilized and unsterilized treatments. However, citrate is not readily extracted from all soils, and little was extractable in water from an oxisol studied by Gerke et al. (1994), and in situ analysis methods had to be used. Other workers have used experimentally derived correction factors to account for losses due to incomplete extraction and degradation by microorganisms when calculating rates of organic acid exudation (Kirk et al. 1999).

Exudation rates have often been expressed on a whole root fresh or dry weight or length basis, but this may underestimate localized rates if the exudation is not uniform along the root. Hoffland (1992) attempted to correct for this by using pH indicators to estimate the proportion of the root system exuding citric acid in studies with P-deficient oilseed rape plants. Alternatively, root compartments have been used to collect exudates from defined parts of roots (Hoffland et al. 1989; Delhaize et al. 1993b; Keerthisinghe et al. 1998).

4.3.2 Localization of Organic Acid Exudation from Roots

Rates of exudation of organic acids vary greatly between different parts of root systems (Table 5). In oilseed rape, young regions of the root exude more organic acids than older parts (Hoffland et al. 1989). In white lupin, proteoid roots are the major sites of citrate exudation, with low amounts exuded from other roots. The highest rates of citrate exudation occur in the recently-developed portions of the proteoid root (Keerthisinghe et al. 1998; Watt and Evans 1999). Exudation ceases after a few days, although the mature proteoid roots are still metabolically active and retain the ability to take up P. Concentrating exudates along a limited portion of the root allows intensive chemical extraction of a small volume of soil and is the most effective strategy for P mobilization per unit of organic acid (Dinkelaker et al. 1995).

The rate of citrate exudation in proteoid roots measured in the zone of active excretion of 5.4 nmol citric acid h^{-1} cm^{-1} root (Table 5) compares with rates of 6.7 (Neumann et al. 1999) and 14 nmol citric acid h^{-1} cm^{-1} root (Watt and Evans 1999) in similar regions of the root. These rates are an order of magnitude greater than those from root tips of P-deficient oilseed rape, 0.14 and 0.44 nmol h^{-1} cm^{-1}, for citric and malic acids, respectively (Table 5). For comparison, rates of organic acid exudation from Al tolerant cereal root tips in the presence of Al were 0.3 nmol citric acid h^{-1} cm^{-1} root for maize (Pellet et al. 1995) and 13 nmol malic acid h^{-1} cm^{-1} root for wheat (Delhaize et al. 1993b).

4.3.3 Effects of P Supply on Organic Acid Exudation from Roots

Early evidence using standard solution cultures suggested that proteoid root formation in white lupin was a specific response to P deficiency (Marschner et al. 1986). However, field observations indicate that it is normal to find abundant proteoid roots on white lupin growing in a range of fertilized agricultural soils (P Hocking unpublished). Recent work has shown strong development of proteoid roots in P-adequate white lupin plants growing in solutions with P maintained at 10 mmol m^{-3}, and proteoid roots were only suppressed at levels unlikely to be reached in soil solutions (Keerthisinghe et al. 1998). Rates of citrate excretion were, however, nearly 3-fold higher in the plants deprived of P than those given 10 mmol m^{-3}. Although exudation rates were much lower from non-proteoid roots, there was a similar proportionate increase due to P deficiency (Table 5). In oilseed rape, P stress increased exudation from root tips by approximately 5 fold, from 0.03 to 0.14 nmol h^{-1} cm^{-1} (Table 5).

4.3.4 What Controls Organic Acid Excretion?

Two sets of processes, the synthesis of organic acids within the root cells and their release into the apoplast, have been implicated in determining the rate of exudation from roots. Direct evidence for the role of synthesis comes from experiments with tobacco transformed with a bacterial citrate synthase gene. In the transformed plants, internal concentrations of citrate were up to 10-fold higher than in the wild type, while exudation rates were increased 4-fold (de la Fuente et al. 1997). This supports earlier data showing a correlation between organic acid exudation by roots of oilseed rape and white lupin and the activities of key enzymes involved in organic acid biosynthesis. In rape, phosphoenolpyruvate carboxylase (PEPC) was increased by P stress in root tips but not in older roots (Hoffland et al. 1992), in keeping with the differences in organic acid exudation between young and old root tissue (Table 5). In white lupin, both PEPC and malate dehydrogenase (MDH) were higher in citrate-excreting proteoid roots compared to non-proteoid roots (Johnson et al. 1994). Detailed studies of internal citrate levels, citrate excretion and enzyme activity in young and old sections of proteoid roots showed elevated PEPC and MDH activities in the recently-developed section, coinciding with high citrate excretion in that zone. However, as the proteoid roots aged, citrate excretion declined sharply followed by a decline in PEPC activity (Keerthisinghe et al. 1998), indicating that elevated citrate efflux is not necessarily a direct consequence of increased organic acid metabolism. In addition, internal citrate concentrations and rates of excretion along the proteoid roots were not correlated.

The transport mechanisms that control release of organic acids from roots have been most extensively studied in wheat, where malate release from root tips is controlled by anion channels (see chapter by E. Delhaize, this volume). These open in response to Al^{3+} and allow malate efflux down the electrochemical gradient (Delhaize and Ryan 1995; Ryan et al. 1995). Calculations show that diffusion through the plasmamembrane would be adequate to account only for the observed background rates (Jones 1998). It is likely that organic acid-permeable channels are involved in the substantial citrate excretion shown in proteoid roots and in P-deficient plants. When these channels open, organic acids diffuse down the concentration and electrical potential gradients between the cytoplasm and the apoplast. Figure 4 is a schematic diagram showing synthesis of organic acids and their efflux from root cells. Recent work supports the notion that channels are involved; in the case of white lupin proteoid roots, a decrease of approximately 50% in citrate efflux occurred following treatment with known inhibitors of anion channels, anthracene-9-carboxylic acid and ethacrynic acid (Neumann et al. 1999). There is little information available on ion fluxes required to balance the exudation of organic acids, except in the case of malate efflux in wheat root tips where K^+ was shown to be the counter ion (Ryan et al. 1995). Other possibilities include H^+ efflux (Dinkelaker et al. 1989) or anion influx. Further work is needed to understand the transport mechanisms and their regulation involved in the release of organic acids under low P conditions in localized regions of the root, and at particular stages of root development.

5. Acid Phosphatases of Plant Roots

5.1 Organic P as a Substrate for Plant Growth

The organic P fractions in both the soil solution and in solid phases are an important potential source of P for plants (Richardson 1994). Organic P is generally considered to be accessible to roots only after hydrolysis (mineralization) to yield soluble inorganic P. The availability of P_o to plants has been investigated in a number of studies. For example, seedlings of barley could use a portion of the P_o from $CaCl_2$ extracts of soil for growth (Seeling and Jungk 1996). Tarafdar and Jungk (1987) showed that a number of plant species substantially depleted P within the rhizosphere, with up to 86% of the total extractable P_o removed from the rhizo-

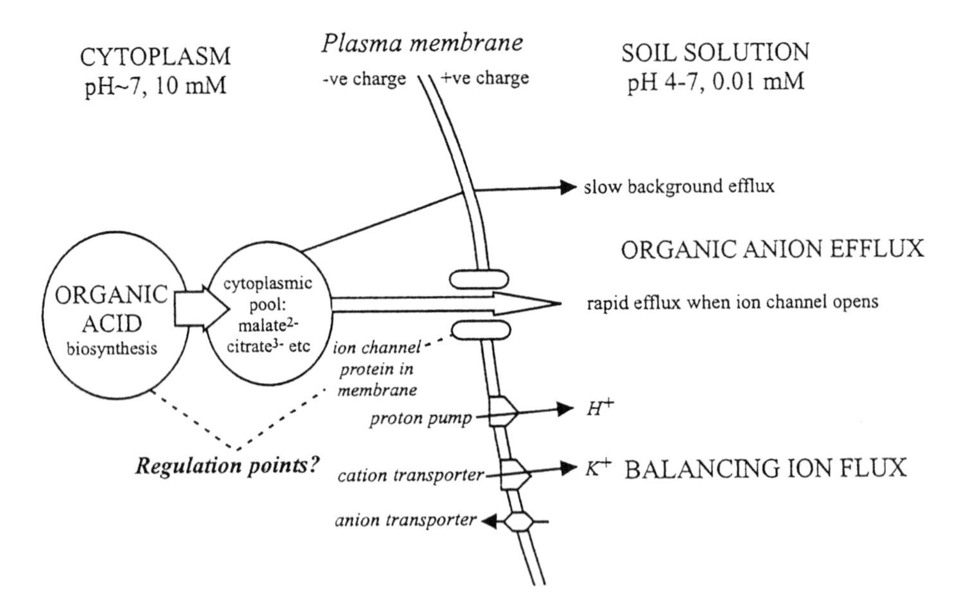

Fig. 4. Representation of efflux of organic acids from living cells in the root. Organic acid-permeable channels rather than slow diffusion through the plasma membrane is likely to account for the observed rates of efflux in P-deficient plants or proteoid roots. A proton pump maintains membrane polarization. Proton or cation efflux or anion uptake balances the efflux of organic acids. Control points for regulating organic acid efflux could be the organic acid-permeable channels and/or the net synthesis of organic acids

sphere soil. This depletion was associated with a markedly higher phosphatase activity in the rhizosphere relative to the bulk soil.

The capacity for plants to use specific P_o substrates for growth has also been assessed. Wheat seedlings grown under aseptic conditions were able to use a range of monoester P substrates (including glucose 1-phosphate, glycerophosphate and ATP) for plant growth, equally well as P_i. By contrast, plants were less able to obtain their P requirements from diester substrates (RNA and cyclic AMP) or from inositol hexaphosphate (IHP or phytate), despite IHP being a monoester P compound (Table 6; Richardson et al., unpublished). In other work, β-glycerophosphate and AMP have been shown to be effective P sources, both in soil and solution culture, while RNA appears to be of variable availability for plant growth (Table 6; Adams and Pate 1992; Jayachandran et al. 1992; Barrett-Lennard et al. 1993).

Phytate is an important P_o substrate because of its abundance in many soils. The effectiveness of phytate as a source of P for plant growth has been investigated by many groups, but with conflicting results. Phytate has been shown to be completely available (Rogers et al. 1940; Kroehler and Linkins 1991), of variable availability (Martin and Cartwright 1971; Helal 1990; Helal and Sauerbeck 1991; Adams and Pate 1992; Findenegg and Nelemans 1993) or unavailable to plants (Table 6; Barrett-Lennard et al. 1993; Hübel and Beck 1993). Using radio-labeled substrate, Martin (1970, 1973) concluded that the availability of IHP to plants depended on the amount added to soil and also on the P-retention capacity of the growing media. While

Table 6. Accumulation of shoot dry weight and phosphorus uptake by wheat seedlings supplied with P either as inorganic phosphate, a range of organic P substrates (all supplied at 1240 μg P plant^{-1}), or grown in the absence of added P. The seedlings were grown for 19 days under sterile conditions in agar culture (Richardson et al. unpublished)

Substrate	Shoot dry weight (mg plant^{-1})	Shoot P content (μg P plant^{-1})
Inorganic P (H_2PO_4)	68	371
Adenosine 3':5'-cyclic monophosphate	59	225
Ribonucleic acid	70	238
D(-)3-phosphoglycerate	67	338
D-glucose 1-phosphate	68	381
Adenosine-5'-triphosphate	73	407
myo-Inositol hexaphosphate	41	74
No P	35	27

perennial ryegrass grew well in sand with phytate as the P source, the species could not obtain P from phytate when added to soil (Martin and Cartwright 1971). This observation is supported by similar studies using non-labeled IHP (Adams and Pate 1992; Findenegg and Nelemans 1993). Hübel and Beck (1993) also used radio-labeled IHP and found that it was not depleted across the rhizosphere of soil-grown maize. However, excised roots from *Eriophorum vaginatum* were able to take up ^{32}P from radio-labeled IHP in solution (Kroehler and Linkins 1991).

Collectively, these studies indicate that IHP is a poorer source of P for plant growth than most other organic P substrates found in soils. This observation is consistent with the accumulation of phytate P in soils, particularly after application of P fertilizer, and its apparent resistance to mineralization (Williams and Anderson 1968).

Several studies have indicated that soil microorganisms (including bacteria, fungi and mycorrhizal fungi) improve the ability of plants to access P from IHP (Heinrich et al. 1988; Jayachandran et al. 1992; Tarafdar and Marschner 1994, 1995; Richardson et al. unpublished). Microorganisms may solubilize IHP either indirectly, through modification of the rhizosphere, or by direct hydrolysis of substrate. Many organisms capable of hydrolyzing phytate have been isolated from soil (Cosgrove et al. 1970; Hilger and Krause 1989; Richardson and Hadobas 1997). However, the relative contribution of phosphatase enzymes from either soil microorganisms, or directly from plant roots, for increasing the availability of P_o in soils, remains unclear. For example, in the work by Tarafder and Jungk (1987), where increases in phosphatase activity and the depletion of soil P_o were observed in the rhizosphere, the origin of the phosphatase activity could not be determined because large increases in both bacterial and fungal populations also occurred within the rhizosphere.

5.2 Phosphatases in Plant Roots

Phosphatases are a diverse group of hydrolase enzymes that catalyze the release of P_i from P_o compounds containing monoester, diester, triester, pyrophosphate or phosphonate linkages (Bieleski 1973). In broad terms, phosphatases may be classified as either acid (E.C. 3.1.3.2) or alkaline phosphatases (E.C. 3.1.3.1), depending on whether their optimal pH for catalysis is below or above pH 7.0. Acid phosphatases are generally considered to be more important for plant P nutrition, although both acid and alkaline phosphatase activities were correlated with decreases in P_o across the rhizosphere (Tarafdar and Jungk 1987).

Phosphatases participate in a wide range of cellular functions and have been characterized from many organisms and tissues, including plant roots (McLachlan 1980; Panara et al. 1990; Ozawa et al. 1995). These enzymes may function either to hydrolyze external soil P_o compounds, or in the recycling and recovery of P_i from internal plant metabolic products and storage compounds (Hübel and Beck 1996; Hunter et al. 1999). A further role proposed for root phosphatases is to conserve P by converting ester P compounds leaked from the root to P_i which can then be reabsorbed (Barrett-Lennard et al. 1993).

Phosphatases show considerable variability in their affinity for P_o substrates; some possess broad substrate specificity while others have activity only for specific P_o substrates (Duff et al. 1994). Phytases are a class of phosphatase that have a high affinity for phytate (Gibson and Ullah 1988; Laboure et al. 1993; Konietzny et al. 1995; Li et al. 1997a), and are of particular interest for plant P nutrition given the abundance of phytate in most soils. Phytases catalyze the removal of up to five of the six phosphate moieties from IHP. These enzymes have been isolated and characterized from a range of plant tissues including pollen, germinated seeds and grains and cotyledon tissue (see review by Wodzinski and Ullah 1996). Recently, phytases with high substrate affinity (23 to 117 μM) have been isolated and characterized from the roots of maize (Hübel and Beck 1996) and tomato (Li et al. 1997a).

5.3 Cellular Localization of Phosphatases and Response to P Nutrition

Phosphatases located within root cells are unlikely to hydrolyze P_o in the soil. Rather, the root phosphatases of potential importance in affecting P supply to plant roots are the extracellular forms. These are either associated with the cell wall (Dracup et al. 1984; Hunter et al. 1999) or secreted by the root into the rhizosphere (Tadano et al. 1993).

The activity of root extracellular acid phosphatases and their response to plant P status have been extensively studied. In general, P-deficient plants possess higher levels of extracellular root acid phosphatase activity than plants grown where P is not limiting (Ridge and Rovira 1971; McLachlan 1980; Helal and Sauerbeck 1991; Adams and Pate 1992; Tadano et al. 1993; Ozawa et al. 1995; Ascencio 1996, 1997; Barrett et al. 1998; Hayes et al. 1999). Enhanced phosphatase activity due to P deficiency was similarly observed for suspension-cultured cells (Goldstein et al. 1988; Lefebvre et al. 1990), isolated cell wall fractions (Dracup et al. 1984; Barrett-Lennard et al. 1993; Hunter et al. 1999), soluble cell components (Hunter et al. 1999) and extracts from whole roots (Dracup et al. 1984; Hayes et al. 1999). More recently, an extracellular phosphatase that is responsive to plant P nutrition has been isolated and extensively characterized (Ozawa et al. 1995; Wasaki et al. 1999). With a few exceptions where phosphatase activity did not change with P status (Ascencio 1997; Hayes et al. 1999), the general rule appears to be that P deficiency increases phosphatase often by 30% or more, and even up to 20-fold in a range of plant species. This is presumptive evidence for their implicit role in plant P nutrition.

Changes in phytase activity in roots in response to plant P status have also been observed for a wide range of species (Asmar 1997; Li et al. 1997b; Hayes et al. 1999). Li et al. (1997b) demonstrated that the levels of phytase secreted by roots of P-deficient plants increased up to ten-fold, and that the extracellular phytase activities of the roots were equivalent to, and in some cases, greater than acid phosphatase activity. This is in contrast to Barrett-Lennard et al. (1993), who showed that whilst acid phosphatase activity in subterranean clover was enhanced by P deficiency, levels of phytase activity were insignificant and did not respond to plant P status. More recently, Hayes et al. (1999) have showed that whilst phytase activity was increased by P deficiency, phytase activity constitutes a small component (less than 5%) of the total acid phosphatase activity in extracts from plant roots. Moreover, the extracellular phytase activity, or the fraction that was eluted from the roots of subterranean clover with dilute NaCl, represented only a small proportion of the total activity of the root (Table 7). Extracellular acid phosphatases that can hydrolyze a wide range of organic P substrates, other than IHP, have been reported in both cell wall-associated and secreted forms (Dracup et al. 1984; Tadano et al. 1993; Hunter et al. 1999).

Histochemical localization of acid phosphatase in mature maize roots has indicated that enzyme activity is predominantly located within epidermal layers and throughout the outer cortex (Hübel and Beck 1996). Phosphatases in root meristems are also associated primarily with the epidermal cell layers for a wide range of species (Shaykh and Roberts 1974). The cellular localization of phytase enzymes in roots is less clear. Hübel and Beck (1996), detected phytase activity only in the endodermal and pericycle layers of adult roots of maize and suggested that phytase was not involved in the hydrolysis of external phytate substrates.

Table 7. Acid phosphatase and phytase activities (nkat g^{-1} root fresh wt, measured against *p*-nitrophenolphosphate and inositol hexaphosphate, respectively) of roots from 21-day-old seedlings of subterranean clover. Plants were grown under conditions of either P deficiency or were supplied with P. Enzyme activities were determined for soluble extracts from ground roots, as extracellular components that were eluted for 6 h into 0.1 M NaCl, or as total extracellular acid phosphatase activities for intact roots. Each value is the mean (±SE) of between three and six independent observations (adapted from Hayes et al. 1999)

Activity component		Enzyme activity	
		P-adequate	P-deficient
Acid phosphatase (*p*NPP)	Soluble extracts	20.1 ± 1.3	34.0 ± 1.8
	Eluted into 0.1 M NaCl	1.2 ± 0.16	1.8 ± 0.02
	Extracellular activity	22.9 ± 1.0	34.0 ± 1.1
Phytase (IHP)	Soluble extracts	0.79 ± 0.02	0.97 ± 0.02
	Eluted into 0.1 M NaCl	0.025 ± 0.01	0.042 ± 0.001

However, Li et al. (1997b) clearly demonstrated the presence of an extracellular phytase activity in exudates collected from a wide range of plant species. We have similarly measured an extracellular component of phytase activity from the roots of subterranean clover (Table 7; Hayes et al. 1999) and wheat (Richardson et al. unpublished) but, in contrast to Li et al. (1997b), the phytase activity was only a small proportion of the extracellular acid phosphatase activity, and represented a small proportion only of the total root phytase activity (Table 7). Clearly, further studies are required to establish the role of phytases in the P nutrition of plants, in view of this conflicting evidence concerning the activity of extracellular phytases and observations that plants have limited ability to utilize P from phytate (Table 6).

5.4 Genetic Studies and Regulation of Expression of Root Phosphatases

While species differences in root phosphatase activities are known (McLachlan 1976), as yet there are few reports relating intra-specific differences in root phosphatase activity to P uptake. Helal (1990) showed a strong positive relationship between the uptake of P from phytate and external phosphatase activity of roots among four common bean varieties. Cultivars of wheat varied 2 to 3-fold in external phosphatase activity (McLachlan and De Marco 1982), but the significance of this for P uptake from organic sources was not tested and differences may have reflected internal plant P status rather than genotype. However, genotypic differences in the ability to increase root phosphatase at low P may be an important mechanism by which different plants acquire P from organic sources.

Factors involved in the regulation of expression of phosphatase in plant roots are receiving increased attention. Enzyme responses to P deficiency generally occur prior to observed effects of P deficiency on plant growth (Bosse and Kock 1998), indicating that they are not generalized plant stress responses, but specific to P stress. The trigger for exudation of phosphatases from roots may either be a low supply of external P, or a decrease in internal P concentration. While cytoplasmic concentrations of P_i in plant cells remain relatively constant, vacuolar P_i can fluctuate dramatically, depending on the P status of the plant, and may therefore be an important sensor for the P deficiency-induced responses (Mimura et al. 1996). Duff et al. (1991) observed a considerable delay between the imposition of P deficiency and the synthesis of phosphatase protein in black mustard cells in suspension culture, which probably coincided with a depletion of the vacuolar P_i pool. Moreover, the de novo synthesis of phosphatase protein in the cells was correlated with enhanced phosphatase activity (Duff et al. 1991), indicating that the activity was regulated by enzyme synthesis. Levels of extracellular acid phosphatase activity of suspension-cultured cells of black mustard (Lefebvre et al. 1990) and tomato (Goldstein et al. 1988) also increased following removal of P from the growth medium.

Molecular studies of plant phosphatase genes and their contribution to plant P nutrition have only recently been undertaken. An acid phosphatase gene (*Aps-1*) was first cloned from tomato, as a marker linked to a gene for nematode resistance (Aarts et al. 1991; Williamson and Colwell 1991). However, the function of this phosphatase in plants has not been investigated in detail, although it does not appear to be induced by P deficiency. More recently, a cDNA encoding an inducible acid phosphatase (*LASAP1*), has been isolated from P-deficient roots of white lupin (Wasaki et al. 1999). The polypeptide encoded by *LASAP1* shared strong homology with deduced amino acid sequences of purified purple acid phosphatases from bean (Klabunde et al. 1994). Northern blot analysis of the phosphatase showed that expression of the gene was induced in P-deficient roots but not in other plant tissues. A cDNA coding for phytase (*phy S11*) has similarly been cloned from germinated seedlings of maize (Maugenest et al. 1997). Although this gene was predominantly expressed during seed germination, where it is most likely to be involved in the mobilization of P from seed phytate, mRNA was also detected in adult roots (Maugenest et al. 1999). In situ hybridization and immunolocalization indicated the presence of the phytase in the root epidermis, the endodermis and the pericycle, although the involvement of the phytase in plant P nutrition remains to be investigated.

6. Potential for Improving the Ability of Plants to Acquire Phosphorus More Efficiently

6.1 Breeding Objectives

The efficiency with which plants take up P during their growth cycle depends on many factors including root growth, morphology and function, as well as the efficiency with which they use P for growth and distribute assimilate between roots and shoots. Progress towards improving P acquisition may be slow using selection criteria based on plant performance, because many factors are involved, they interact, and their relative importance alters with changes in P supply (see Blair 1993). More rapid progress is likely if particular plant attributes can be identified that are important in the acquisition of P, that are measurable and under genetic control. Selection for root morphological characters such as root hair number and length, although difficult, has given promising results (Caradus 1995). There has been less interest in selection on the basis of root function, largely due to lack of information that can be used to identify suitable selection criteria. Current knowledge of root function, particularly aspects that relate to the capacity of roots to increase the rate of P mobilization in soil, suggests several possibilities where significant progress might now be made and where molecular approaches might be feasible.

6.2 Manipulating Rhizosphere pH Through Changing Cation/Anion Balance

Plants with an enhanced capacity for net H^+ excretion may have greater ability to solubilize and acquire P, particularly in soils where the P chemistry is controlled by calcium phosphates (with the proviso that soil pH buffer capacity will determine the pH change in the rhizosphere for a given H^+ output). Net H^+ excretion from roots is determined largely by the balance between uptake of cations and anions (Haynes 1990), and it is worth considering if manipulating this character in plants is feasible and if a practical screening technique can be devised. Also, it should be noted that increasing the excess cation content of plants would result in greater rates of soil acidification if this were reflected in the ash alkalinity of the products removed from the land.

Diversity among plant genotypes in tissue concentration of the major cations and anions has been reported (Clark 1983), and the "adequate" range of concentrations of Ca, Mg, K, Na, P, S and Cl in healthy plants is relatively wide (Reuter and Robinson 1997), suggesting alteration may be possible without detrimental effects on growth, although effects on product quality may have to be monitored. Heritability studies suggest scope for alterations in cation content of herbage (Smith et al. 1999).

Two approaches might be considered. Firstly, measuring changes in pH around roots is readily done using pH dyes or electrodes (Marschner et al. 1987), but it would be difficult to scale such techniques to the demands of a breeding program. Plant analysis may offer an alternative approach. This would rely on the relationship between excess cations in the plant and net proton excretion from the roots (Pierre and Banwart 1973; Jarvis and Robson 1983a). Comparisons would need to be made under standard conditions because the form of nitrogen taken up by the plant root - cationic NH_4^+, anionic NO_3^- or symbiotically fixed uncharged N_2 -will affect the net H^+ excretion (Jarvis and Robson 1983b). Jarvis and Robson's work (1983a) is encouraging, with 10 cultivars of subterranean clover differing significantly in content of excess cations. Calculations showed that the greater the excess of cations, the lower the soil pH at harvest and the higher the total P uptake. Technically, such an approach is feasible because excess cations in plant tissue can be easily calculated from elemental analyses by x-ray fluorescence spectrometry (XRFS) (Noble et al. 1996), and XRFS has been adapted for screening large numbers of plants for leaf nutrient content (Delhaize et al. 1993a).

6.3 Organic Acid Excretion

There is limited information on intra-specific variation in organic acid excretion from plant roots. Experimentally, it may be difficult to demonstrate consistent differences because organic acid excretion rates are altered by plant P status, age of root and environment (Table 5). Recent work with cultivars of pigeon pea did not attempt to directly measure organic acid excretion, but instead used a functional test to measure the ability of concentrated exudates to solubilize P bound to either Fe or Al (Subbarao et al. 1997b). Phosphorus uptake by pigeon pea from Fe-P in vermiculite culture was also measured (Subbarao et al. 1997a). Differences between genotypes were found by both methods and the two approaches were not inconsistent. There is need to understand the effects of internal P status and plant age in the cultivar comparisons. Nevertheless, these approaches are promising and offer hope for the development of a screening procedure to select efficient plants

Although a number of laboratories have studied factors that control the synthesis and release of organic acids from roots, the only reported attempt to increase rates by plant transformation has involved the enzyme citrate synthase. In tobacco, expression of a bacterial citrate synthase gene significantly increased both the internal concentration and exudation of citrate from the roots (de la Fuente et al. 1997), and enhanced Al tolerance. Although this finding needs to be confirmed and the significance for P nutrition of the changes engineered carefully evaluated, it is encouraging for future work as it demonstrates the potential of a molecular approach. A further step would be to characterize the genes controlling release of organic acids from roots so that the interaction between biosynthetic capacity and rates of release of organic acids can be examined and optimum strategies for plant improvement established.

6.4 Root Phosphatases

Elucidation of the putative roles of plant phosphatases in making soil P_o available to plants has not been straightforward. This is partly because the relative importance of root and microbial phosphatase activities in the rhizosphere have not always been quantified, and partly because the identity and plant-availability of P_o in soil solution is still not clearly understood. The question of location and activity of phospatases and phytases in and around plant roots is presently unclear and a more detailed understanding at the physiological and molecular level is required. However, good progress is now being made in characterizing root phosphatases. The recent success of Wasaki et al. (1999) in cloning the root-specific inducible *LASAP1* gene now provides an excellent opportunity for further work to define the roles of root phosphatases in plant P nutrition.

7. Concluding Remarks

Efforts to improve the efficiency of P acquisition by plants have passed through several stages. Initially, researchers sought to explain the observed differences in P uptake and use efficiency and looked for variation within agricultural species that might be exploited by breeding and selection. However, it proved difficult to define practical selection criteria for P-efficiency in plants and breeders were reluctant to include P efficiency as an aim in their breeding programs in addition to more readily achievable objectives such as disease resistance, improved yield and product quality. Work to explain mechanisms proceeded at the physiological level, particularly following the landmark discovery of high rates of citric acid excretion by roots of white lupin (Gardner et al. 1982a). This understanding at the physiological level laid the foundation for the current progress at the molecular level. We are now in the position of being able to move ahead on several fronts.

There is sufficient understanding of some P efficiency mechanisms to be able to design useable selection criteria to begin screening populations of plants. However, more work is needed to assess the efficacy of the mechanisms of rhizosphere acidification, organic acid exudation and root phosphatases in different soils and for different forms of P fertilizer. At the agronomic level, the roles of P-efficient plants in the P economy of crop and pasture sequences need to be evaluated.

The first report of genetic engineering to increase citrate exudation from roots (de la Fuente et al. 1997) is encouraging for future attempts to improve P acquisition. The rapid advances in understanding of P efficiency at the molecular level will enhance both conventional and genetic engineering approaches aimed at improving the P acquisition efficiency of agricultural plants.

References

Aarts JMMJG, Hontelez JGJ, Fischer P, Verkerk R, van Kammen A, Zabel P (1991) Acid phosphatase-1, a tightly linked molecular maker for root-knot nematode resistance in tomato: from protein to gene, using PCR and degenerate primers containing deoxyinosine. Plant Mol Biol 16:647-661

Adams MA, Pate JS (1992) Availability of organic and inorganic forms of phosphorus to lupins (*Lupinus* spp.). Plant Soil 145:107-113

Ae N, Arihara J, Okada K (1991) Phosphorus uptake mechanisms of Pigeon pea grown in Alfisols and Vertisols. In: Johansen C, Lee KK, Sahrawat KL (Eds) Phosphorus nutrition of grain legumes in the semi-arid Tropics. ICRISAT, Patancheru, AP India, pp 91-98

Ae N, Arihara J, Okada K, Yoshihara T, Otani T, Johansen C (1993) The role of piscidic acid secreted by pigeon pea roots grown on an Alfisol with low-P fertility. In: Randall PJ, Delhaize E, Richards RA, Munns R (Eds) Genetic aspects of plant mineral nutrition. Kluwer Academic, Dordrecht, pp 279-288

Anderson G (1980) Assessing organic phosphorus in soils. In: Khasawneh FE, Sample EC, Kamprath EJ (Eds) The role of phosphorus in agriculture. American Society of Agronomy, Madison, Wisconsin, pp 411-431

Armstrong RD, Helyar KR, Prangnell R (1993) Direct assessment of mineral phosphorus availability to tropical crops using ^{32}P labeled compounds. Plant Soil 150:279-287

Ascencio J (1996) Growth strategies and utilization of phosphorus in *Cajanus cajan* L. Millsp. and *Desmodium tortuosum* (Sw.) DC under phosphorus deficiency. Commun Soil Sci Plant Anal 27:1971-1993

Ascencio J (1997) Root secreted acid phosphatase kinetics as a physiological marker for phosphorus deficiency. J Plant Nut 20:9-26

Asmar F (1997) Variation in activity of root extracellular phytase between genotypes of barley. Plant Soil 195:61-64

Barber SA (1984) Soil nutrient bioavailability. Wiley, New York.

Barrett DJ, Richardson AE, Gifford RM (1998) Elevated atmospheric CO_2 concentrations increase wheat root phosphatase activity when growth is limited by phosphorus. Aust J Plant Physiol 25:87-93

Barrett-Lennard EG, Dracup M, Greenway H (1993) Role of extracellular phosphatases in the phosphorus nutrition of clover. J Exp Bot 44:1595-1600

Barrow NJ (1982) Soil fertility changes. In: Costin AB, Williams CH (Eds) Phosphorus in Australia. Centre for Resource and Environment Studies, Australian National University, Canberra, pp 197-220

Bar-Yosef B (1991) Root excretions and their environmental effects. Influence on availability of phosphorus. In: Waisel Y, Eshel A, Kafkafi U (Eds) Plant roots. The hidden half. Marcel Dekker, New York, pp 529-557

Bekele T, Cino BJ, Ehlert PAI, van der Maas AA, van Diest A (1983) An evaluation of plant-borne factors promoting the solubilization of alkaline rock phosphates. Plant Soil 75:361-378

Bieleski RL (1973) Phosphate pools, phosphate transport and phosphate availability. Ann Rev Plant Physiol 24:225-252

Blair G (1993) Nutrient efficiency - what do we really mean? In: Randall PJ, Delhaize E, Richards RA, Munns R (Eds) Genetic aspects of plant mineral nutrition. Kluwer Academic, Dordrecht, pp 205-214

Blair GJ, Mamaril CP, Miller MH (1971) Influence of nitrogen source on phosphorus uptake by corn from soils differing in pH. Agron J 63:235-238

Bolan NS, Naidu R, Mahimairaja S, Baskaran S (1994) Influence of low molecular-weight organic acids on solubilization of phosphates. Biol Fertil Soils 18:311-319

Bosse D, Kock M (1998) Influence of phosphate starvation on phosphohydrolases during development of tomato seedlings. Plant Cell Environ 21:325-332

Bowen GD (1980) Misconceptions, concepts and approaches in rhizosphere biology. In: Elwood DG, (Ed) Contemporary microbial ecology. Academic, London, pp 283-304

Braum SM, Helmke PA (1995) White lupin utilizes soil phosphorus that is unavailable to soybean. Plant Soil 176:95-100

Bromfield SM (1958) The solution of γ-MnO_2 by substances released from soil and roots of oats and vetch in relation to manganese availability. Plant Soil 10:147-160

Caradus JR (1995) Genetic control of phosphorus uptake and phosphorus status in plants. In: Johansen C, Lee KK, Sharma KK, Subbarao GV, Kueneman AE (Eds) Genetic manipulation of crop plants to enhance integrated nutrient management in cropping systems. 1. Phosphorus. ICRISAT, Patancheru, AP India, pp 55-74

Clark RB (1983) Plant genotype differences to uptake, translocation, accumulation, and use of mineral elements. In: Saric MR (Ed) Genetic specificity of mineral nutrition of plants. Serbian Academy of Sciences and Arts, Belgrade, pp 41-55

Cosgrove DJ, Irving GCJ, Bromfield SM (1970) Inositol phosphate phosphatases of microbiological origin. The isolation of soil bacteria having inositol phosphate phosphatase activity. Aust J Biol Sci 23:339-343

Curl EA, Trueglove B (1986) The rhizosphere. Springer, Berlin

Dalal RC (1977) Soil organic phosphorus. Adv Agron 29:85-117

Darrah PR (1993) The rhizosphere and plant nutrition: a quantitative approach. In: Barrow NJ (Ed) Plant nutrition - from genetic engineering to field practice. Kluwer Academic, Dordrecht, pp 3-22

de la Fuente JM, Ramirez-Rodriguez V, Cabrera-Ponce JL, Herrera-Estrella L (1997) Aluminum tolerance in transgenic plants by alteration of citrate synthesis. Science 276: 1566-1588

Delhaize E, and Ryan PR (1995) Aluminum toxicity and tolerance in plants. Plant Physiol 107: 315-321

Delhaize E, Randall PJ, Wallace PA, Pinkerton A (1993a) Screening *Arabidopsis* for mutants in mineral nutrition. Plant Soil 155/156:131-134

Delhaize E, Ryan PR, Randall PJ (1993b) Aluminum tolerance in wheat *(Triticum aestivum* L.). II. Aluminum-stimulated excretion of malic acid from root apices. Plant Physiol 103:695-702

Dinkelaker B, Romheld V, Marschner H (1989) Citric acid excretion and precipitation of calcium citrate in the rhizosphere of white lupin (*Lupinus albus* L.). Plant Cell Environ 12:285-292

Dinkelaker B, Hengeler C, Marschner H (1995) Distribution and function of proteoid roots and other root clusters. Bot Acta 108:183-200

Dracup MNH, Barrett-Lennard EG, Greenway H, Robson AD (1984) Effect of phosphorus deficiency on phosphatase activity of cell walls from roots of subterranean clover. J Exp Bot 35:466-480

Duff SMG, Plaxton WC, Lefebvre DD (1991) Phosphate-starvation response in plant cells: *De novo* synthesis and degradation of acid phosphatases. Proc Natl Acad Sci USA 88:9538-9542

Duff SMG, Sarath G, Plaxton WC (1994) The role of acid phosphatases in plant phosphorus metabolism. Physiol Plant 90:791-800

Findenegg GR, Nelemans JA (1993) The effect of phytase on the availability of P from *myo*-inositol hexaphosphate (phytate) for maize roots. Plant Soil 154:189-196

Gahoonia TS, Claassen N, Junk A (1992) Mobilization of phosphate in different soils by ryegrass supplied with ammonium or nitrate. Plant Soil 140:241-248

Gardner WK, Barber DA, Parbery DG (1982a) The acquisition of phosphorus by *Lupinus albus* L. I. Some characteristics of the soil/root interface. Plant Soil 68:19-32

Gardner WK, Parbery DG, Barber DA (1982b) The acquisition of phosphorus by *Lupinus albus* L. II. The effect of varying phosphorus supply and soil type on some characteristics of the soil-root interface. Plant Soil 68:33-41

Gardner WK, Barber DD, Parbery DG (1983) The acquisition of phosphorus by *Lupinus albus* L. III. The probable mechanisms by which phosphorus movement in the soil/root interface is enhanced. Plant Soil 70:107-124

Gerke J, Romer W, Jungk A (1994) The excretion of citric and malic acid by proteoid roots of *Lupinus albus* L.: effects on soil solution concentrations of phosphate, iron, and aluminum in the proteoid rhizosphere in samples of an oxisol and a luvisol. Z Pflanzenernahr Bodenkd 157:289-294

Gerke J, Meyer U, Romer W (1995) Phosphate, Fe and Mn uptake of N_2 fixing red clover and ryegrass from an Oxisol as affected by P and model humic substances application. 1. Plant parameters and soil solution composition. Z Pflanzenernahr Bodenkd 158:261-268

Gibson DM, Ullah AHJ (1988) Purification and characterization of phytase from cotyledons of germinating soybean seeds. Arch Biochem Biophys 260:503-513

Godo GH, Reisenauer HM (1980) Plant effects on soil manganese availability. Soil Sci Soc Am J 34:993-995

Goldstein AH, Baertlein DA, McDaniel RG (1988) Phosphate starvation inducible metabolism in *Lycopersicon esculentum*. I. Excretion of acid phosphatase by tomato plants and suspension-cultured cells. Plant Physiol 87:711-715

Graham RD (1984) Breeding for nutritional characteristics. In: Tinker PB, Lauchli A (Eds) Advances in plant nutrition. Praeger, New York, Vol 1, pp 57-102

Grierson PF (1992) Organic acids in the rhizosphere of *Banksia integrifolia* L.f. Plant Soil 144:259-265

Handreck KA (1996) Phosphorus requirements of Australian native plants. Aust J Soil Res 35:241-289

Harrison AF (1987) Soil organic phosphorus: A world review. Commonwealth Agricultural Bureau, International, Wallingford, UK

Hayes JE, Richardson AE, Simpson RJ (1999) Phytase and acid phosphatase activities in extracts from roots of temperate pasture grass and legume seedlings. Aust J Plant Physiol 26:801-809

Haynes RJ (1990) Active ion uptake and maintenance of cation-anion balance: A critical examination of their role in regulating rhizosphere pH. Plant Soil 126:247-264

Hedley MJ, Nye PH, White RE (1983) Plant induced changes in the rhizosphere of rape (*Brassica napus* var. Emerald) seedlings. IV. The effect of rhizosphere phosphorus status on the pH, phosphatase activity and depletion of soil phosphorus fractions in the rhizosphere and on cation-anion balance in the plants. New Phytol 95:69-82

Heinrich PA, Mulligan DR, Patrick JW (1988) The effect of ectomycorrhizas on the phosphorus and dry weight acquisition of Eucalyptus seedlings. Plant Soil 109:147-149

Helal HM (1990) Varietal differences in root phosphatase activity as related to the utilization of organic phosphates. In: El Bassam N, Dambroth M, Loughman BC (Eds) Genetic aspects of plant mineral nutrition. Kluwer Academic, Dordrecht, pp 103-105

Helal HM, Sauerbeck D (1991) Soil and root phosphatase activity and the utilization of inositol phosphates as dependent on phosphorus supply. In: McMichael BL, Persson H (Eds) Plant roots and their environment. Elsevier Science, Amsterdam, pp 93-97

Hendriks L, Claassen N, Jungk A (1981) Phosphatverarmung des wurzelnahen Bodens und Phosphataufnahme von Mais und Raps. Z Pflanzenernahr Bodenkd 144:486-499

Hilger AB, Krause HH (1989) Growth characteristics of *Laccaria laccata* and *Paxillus involutus* in liquid culture media with inorganic and organic phosphorus sources. Can J Bot 67:1782-1789

Hinsinger P, Gilkes RJ (1996) Mobilization of phosphate from phosphate rock and alumina-sorbed phosphate by the roots of ryegrass and clover as related to rhizosphere pH. Eur J Soil Sci 47:533-544

Hinsinger P, Gilkes RJ (1997) Dissolution of phosphate rock in the rhizosphere of five plant species grown in an acid, P-fixing mineral substrate. Geoderma 75:231-249

Hocking PJ, Keerthisinghe G, Smith FW, Randall PJ (1997) Comparison of the ability of different crop species to access poorly-available soil phosphorus. In: Ando T, Fujita K, Mae T, Matsumoto H, Mori S, Sekiya J (Eds) Proceedings XIII International plant nutrition colloquium. Kluwer Academic, Dordrecht, pp 305-308

Hoffland E (1992) Quantitative evaluation of the role of organic acid exudation in the mobilization of rock phosphate by rape. Plant Soil 140:279-289

Hoffland E, Findenegg GR, Nelemans JA (1989) Solubilization of rock phosphate by rape. II. Local root exudation of organic acids as a response to P-starvation. Plant Soil 113:161-165

Hoffland E, Van den Boogaard R, Nelemans J, Findenegg G (1992) Biosynthesis and root exudation of citric and malic acids in phosphate-starved rape plants. New Phytol 122:675-680

Hübel F, Beck E (1993) In-situ determination of the P-relations around the primary root of maize with respect to inorganic and phytate-P. Plant Soil 157:1-9

Hübel F, Beck E (1996) Maize root phytase. Purification, characterization, and localization of enzyme activity and its putative substrate. Plant Physiol 112:1429-1436

Hue NV, Craddock GR, Adams F (1986) Effect of organic acids on aluminum toxicity in subsoils. Soil Sci Soc Am J 50:28-34

Hunter DA, Watson LM, McManus MT (1999) Cell wall proteins in white clover: influence of plant phosphate status. Plant Physiol Biochem 37:25-32

Imas P, Bar-Yosef B, Kafkafi U, Ganmore-Neumann R (1997) Phosphate induced carboxylate and proton release by tomato roots. Plant Soil 191:35-39

Jarvis SC, Robson AD (1983a) A comparison of cation/anion balance of ten cultivars of *Trifolium subterraneum* L. and their effects on soil acidity. Plant Soil 75:235-243

Jarvis SC, Robson AD (1983b) The effects of nitrogen nutrition of plants on the development of acidity in Western Australian soils. I. Effects with subterranean clover grown under leaching conditions. Aust J Agric Res 34:341-353

Jarvis SC, Robson AD (1983c) The effects of nitrogen nutrition of plants on the development of acidity in Western Australian soils. II. Effects of differences in cation-anion balance between plant species grown under non-leaching conditions. Aust J Agric Res 34:354-365

Jauregui MA, Reisenauer HM (1982) Dissolution of oxides of manganese by root exudate components. Soil Sci Soc Am J 46:314-317

Jayachandran K, Schwab AP, Hetrick BAD (1992) Mineralization of organic phosphorus by vesicular-arbuscular mycorrhizal fungi. Soil Biol Biochem 24:897-903

Jescke WD, Pate JS (1995) Mineral nutrition and transport in xylem and phloem of *Banksia prionotes* (Proteaceae), a tree with dimorphic root morphology. J Exp Bot 46:895-905

Johnson JF, Allan DL, Vance CP (1994) Phosphorus stress-induced proteoid roots show altered metabolism in *Lupinus albus*. Plant Physiol 104:657-665

Johnson JF, Allan DL, Vance CP, Weiblen G (1996) Root carbon dioxide fixation by phosphorus-deficient *Lupinus albus*. Plant Physiol 112:19-30

Jones DL (1998) Organic acids in the rhizosphere - a critical review. Plant Soil 205:25-44

Jones DL, Darrah PR (1995) Influx and efflux of organic acids across the soil-root interface of *Zea mays* L. and its implications in rhizosphere C flow. Plant Soil 173:103-109

Jones LJ, Darrah PR (1994) Role of root derived organic acids in the mobilization of nutrients from the rhizosphere. Plant Soil 166:247-2570

Jungk AO (1991) Dynamics of nutrient movement at the soil-root interface. In: Waisel Y, Eshel A, Kafkafi U (Eds) Plant roots: the hidden half. Marcel Dekker, New York, pp 455-481

Keerthisinghe G, Hocking PJ, Ryan PR, Delhaize E (1998) Effect of phosphorus supply on the formation and function of proteoid roots of white lupin (*Lupinus albus* L.). Plant Cell Environ 21:467-478

Kirk GJD, Santos EE, Santos MB (1999) Phosphate solubilization by organic anion excretion from rice growing in aerobic soil: rates of excretion and decomposition, effects on rhizosphere pH and effects on phosphate solubility and uptake. New Phytol 142:185-200

Klabunde T, Stahl B, Suerbaum H, Hahner S, Karas M, Hillenkamp F, Krebs B, Witzel H (1994) The amino acid sequence of the red kidney bean Fe(III)-Zn(II) purple acid phosphatase. Determination of the amino acid sequence by a combination of matrix-assisted laser desorption/ionization mass spectrometry and automated Edman sequencing. Eur J Biochem 226:369-375

Konietzny U, Greiner R, Jany K-D (1995) Purification and characterization of a phytase from spelt. J Food Biochem 18:165-183

Kroehler CJ, Linkins AE (1991) The absorption of inorganic phosphate from ^{32}P-labeled inositol hexaphosphate by *Eriophorum vaginatum*. Oecologia 85:424-428

Kucey RMN, Janzen HH, Leggett ME (1989) Microbially mediated increase in plant available phosphorus. Adv Agron 42:199-228

Laboure A-M, Gagnon J, Lescure A-M (1993) Purification and characterization of a phytase (*myo*-inositol hexakisphosphate phosphohydrolase) accumulated in maize (*Zea mays*) seedlings during germination. Biochem J 295:413-419

Lefebvre DD, Duff SMG, Fife CA, Julien-Inalsingh C, Plaxton WC (1990) Response to phosphate deprivation in *Brassica nigra* suspension cells. Enhancement of intracellular, cell surface, and secreted phosphatase activities compared to increases in Pi-absorption rate. Plant Physiol 93:504-511

Li M, Osaki M, Honma M, Tadano T (1997a) Purification and characterization of phytase induced in tomato roots under phosphorus-deficient conditions. Soil Sci Plant Nutr (Tokyo) 43:179-190

Li M, Osaki M, Rao IM, Tadano T (1997b) Secretion of phytase from the roots of several plant species under phosphorus-deficient conditions. Plant Soil 195:161-169

Li M, Shinano T, Tadano T (1997c) Distribution of exudates of lupin roots in the rhizosphere under phosphorus deficient conditions. Soil Sci Plant Nutr (Tokyo) 43:237-245

Lipton DS, Blanchar RW, Blevins DG (1987) Citrate, malate, and succinate concentration in exudates from P-sufficient and P-stressed *Medicago sativa* L. seedlings. Plant Physiol 85:315-317

Loss SP, Robson AD, Ritchie GSP (1993) H^+/OH^- excretion and nutrient uptake in upper and lower parts of lupin (*Lupinus angustifolius* L.) root systems. Ann Bot 72:315-320

Loss SP, Robson AD, Ritchie GSP (1994) Nutrient uptake and organic acid anion metabolism in lupins and peas supplied with nitrate. Ann Bot 74:69-74

Marschner H (1986) Mineral nutrition of higher plants. Academic, London

Marschner H, Romheld V, Horst WJ, Martin P (1986) Root induced changes in the rhizosphere: Importance for the mineral nutrition of plants. Z Pflanzenernahr Bodenkd 149:441-456

Marschner H, Romheld V, Cakmak I (1987) Root-induced changes of nutrient availability in the rhizosphere. J Plant Nut 10:1175-1184

Martin JK (1970) Preparation of ^{32}P-labeled inositol hexaphosphate. Anal Biochem 36:233-237

Martin JK (1973) The influence of rhizosphere microflora on the availability of ^{32}P-*myo*-inositol hexaphosphate phosphorus to wheat. Soil Biol Biochem 5:473-483

Martin JK, Cartwright B (1971) The comparative plant availability of ^{32}P *myo*-inositol hexaphosphate and $KH_2{}^{32}PO_4$ added to soils. Soil Sci Plant Anal 2:375-381

Maugenest S, Martinez I, Lescure A-M (1997) Cloning and characterization of a cDNA encoding a maize seedling phytase. Biochem. J. 322:511-517

Maugenest S, Martinez I, Godin B, Perez P, Lescure A-M (1999) Structure of two maize phytase genes and their spatio-temporal expression during seedling development. Plant Mol Biol 39:503-514

McLachlan KD (1976) Comparative phosphorus response in plants to a range of available phosphorus situations. Aust J Agric Res 27:323-341

McLachlan KD (1980) Acid phosphatase activity of intact roots and phosphorus nutrition in plants. I. Assay conditions and phosphatase activity. Aust J Agric Res 31:429-440

McLachlan KD, De Marco DG (1982) Acid phosphatase activity of intact roots and phosphorus nutrition of plants. 3. Its relation to phosphorus garnering by wheat and a comparison with leaf activity as a measure of phosphorus status. Aust J Agric Res 33:1-12

McLaughlin MJ, Alston AM, Martin JK (1988) Phosphorus cycling in wheat-pasture rotations. 1. The source of phosphorus taken up by wheat. Aust J Soil Res 26:323-331

McLaughlin MJ, Baker TG, James TR, Rundle JA (1990) Distribution and forms of phosphorus and aluminium in acidic topsoils under pastures in south-eastern Australia. Aust J Soil Res 28:371-385

McLaughlin MJ, Fillery IR, Till AR (1992) Operations of the phosphorus, sulphur and nitrogen cycles. In: Gifford RM, Barson MM (Eds) Australia's renewable resources: sustainability and global change. Bureau of Rural Resources Proceedings No. 14, Australian Government Publishing Service, Canberra

Mimura T, Sakano K, Shimmen T (1996) Studies on the distribution, re-translocation and homeostasis of inorganic phosphate in barley leaves. Plant Cell Environ 19:311-320

Moghimi A, Lewis DG, Oades JM (1978) Release of phosphate from calcium phosphates by rhizosphere products. Soil Biol Biochem 10:277-281

Moorby H, White RE, Nye PH (1988) The influence of phosphate nutrition on H ion efflux from the roots of young rape plants. Plant Soil 105:247-256

Neumann G, Massonneau A, Martinoia E, Romheld V (1999) Physiological adaptations to phosphorus deficiency during proteoid root development in white lupin. Planta 208:373-382

Newman RH, Tate KR (1980) Soil phosphorus characterisation by ^{31}P nuclear magnetic resonance. Commun Soil Sci Plant Anal 11:835-842

Noble AD, Zenneck I, Randall PJ (1996) Leaf litter ash alkalinity and neutralisation of soil acidity. Plant Soil 179:293-302

Ohwaki Y, Sugahara K (1997) Active extrusion of protons and exudation of carboxylic acids in response to iron deficiency by roots of chickpea (*Cicer arietinium* L). Plant Soil 189:49-55

Otani T, Ae N (1997) The exudation of organic acids by pigeonpea roots for solubilizing iron- and aluminum-bound phosphorus. In: Ando T, Fujita K, Mae T, Matsumoto H, Mori S, Sekiya J (Eds) Proceedings XIII International plant nutrition colloquium. Kluwer Academic, Dordrecht, pp 325-326

Otani T, Ae N, Tanaka H (1996) Phosphorus (P) uptake mechanisms of crops grown in soils with low P status II Significance of organic acids in root exudates of pigeon pea. Soil Sci Plant Nutr (Tokyo) 42:553-560

Ozawa K, Osaki M, Matsui H, Honma M, Tadano T (1995) Purification and properties of acid phosphatase secreted from lupin roots under phosphorus-deficiency conditions. Soil Sci Plant Nutr (Tokyo) 41:461-469

Panara F, Pasqualini S, Antonielli M (1990) Multiple forms of barley root acid phosphatase: purification and some characteristics of the major cytoplasmic isoenzyme. Biochim Biophys Acta 1037:73-80

Pellet DM, Grunes DL, Kochian LV (1995) Organic acid exudation as an aluminum-tolerance mechanism in maize (*Zea mays* L.). Planta 164:788-795

Pierre WH, Banwart WL (1973) Excess-base and excess-base/ nitrogen ratio of various crop species and parts of plants. Agron J 65:91-96

Reuter DG, Robinson JB (1997) Plant analysis: an interpretation manual. CSIRO, Melbourne

Richardson AE (1994) Soil microorganisms and phosphorus availability. In: Pankhurst CE, Doube BM, Gupta VVSR Grace PR (Eds) Management of the soil biota in sustainable farming systems. CSIRO Publishing, Melbourne, pp 50-62

Richardson AE, Hadobas PA (1997) Soil isolates of *Pseudomonas* spp. that utilize inositol phosphates. Can J Microbiol 43:509-516

Ridge EH, Rovira AD (1971) Phosphatase activity of intact young wheat roots under sterile and non-sterile conditions. New Phytol 70:1017-1026

Rogers HT, Pearson RW, Pierre WH (1940) Absorption of organic phosphorus by corn and tomato plants and the mineralizing action of exo-enzyme systems of growing roots. Soil Sci Soc Am Proc 285-291

Romheld V, Marschner H (1981) Rhythmic iron stress reactions in sunflower at suboptimal iron supply. Physiol Plant 53:347-353

Ron Vaz MD, Edwards AC, Shand CA, Cresser MS (1993) Phosphorus fractions in soil solution: Influence of soil acidity and fertilizer additions. Plant Soil 148:175-183

Rovira AD (1962) Plant-root exudates in relation to the rhizosphere microflora. Soils Fert 25:167-172

Rovira AD (1969) Plant root exudates. Bot Rev 35:35-57

Rovira AD, Davey CB (1974) Biology of the rhizosphere. In: Carson EW (Ed) The plant root and its environment. University of Virginia Press, Charlottesville, pp 153-204

Ryan PR, Delhaize E, Randall PJ (1995) Characterisation of Al-stimulated efflux of malate from the apices of Al-tolerant wheat roots. Planta 196:103-110

Sample EC, Soper RJ, Racz GJ (1980) Reactions of phosphate fertilizers in soils. In: Khasawneh EC, Sample CE, Kamprath EJ (Eds) The role of phosphorus in agriculture. American Society of Agronomy, Madison, Wis, pp 263-310

Schwab SM, Menge JA, Leonard RT (1983) Quantitative and qualitative effects of phosphorus on extracts and exudates of sudan grass roots in relation to vesicular-arbuscular mycorrhiza formation. Plant Physiol 73:761-765

Seeling B, Jungk A (1996) Utilization of organic phosphorus in calcium chloride extracts of soil by barley plants and hydrolysis by acid and alkaline phosphatases. Plant Soil 178:179-184

Seeling B, Zasoski RJ (1993) Microbial effects in maintaining organic and inorganic solution phosphorus concentrations in a grassland topsoil. Plant Soil 148:277-284

Sharpley AN (1986) Disposition of fertilizer phosphorus applied to winter wheat. Soil Sci Soc Am J 50:953-958

Shaykh MM, Roberts LW (1974) A histochemical study of phosphatases in root apical meristems. Ann Bot 38:165-174

Smith FW (1981) Availability of soil phosphate to tropical pasture species. In: Proceedings of XIV International Grassland Congress, Lexington, USA, pp 282-285

Smith KF, Rebetzke GJ, Eagles HA, Anderson MW, Easton HS (1999) Genetic control of mineral concentration and yield in perennial ryegrass (*Lolium perenne* L.), with special emphasis on minerals related to grass tetany. Aust J Agric Res 50:79-86

Specht RL, Groves RH (1966) A comparison of the phosphorus nutrition of Australian heath plants and introduced economic plants. Aust J Bot 14:201-221

Stephens CG, Donald CM (1958) Australian soils and their responses to fertilizer. Adv Agron 10:167-256

Subbarao GV, Ae N, Otani T (1997a) Genetic variation in acquisition, and utilisation of phosphorus from iron-bound phosphorus in pigeonpea. Soil Sci Plant Nutr (Tokyo) 43:511-519

Subbarao GV, Ae N, Otani T (1997b) Genotypic variation in iron- and aluminum-phosphate solubilizing activity of pigeonpea root exudates under P deficient conditions. Soil Sci Plant Nutr (Tokyo) 43:295-305

Tadano T, Ozawa K, Sakai H, Osaki M, Matsui H (1993) Secretion of acid phosphatase by the roots of crop plants under phosphorus-deficient conditions and some properties of the enzyme secreted by lupin roots. Plant Soil 155/156:95-98

Tagaki S (1976) Naturally occurring iron-chelating compound in oat- and rice-root washings. 1. Activity measurements and preliminary characterisation. Soil Sci Plant Nutr (Tokyo) 22:423-433

Takagi S, Nomoto K, Takemoto T (1984) Physiological aspects of mugeneic acid, a possible phytosiderophore of graminaceous plants. J Plant Nut 7:469-477

Tarafdar JC, Jungk A (1987) Phosphatase activity in the rhizosphere and its relation to the depletion of soil organic phosphorus. Biol Fertil Soils 3:199-204

Tarafdar JC, Marschner H (1994) Efficiency of VAM hyphae in utilisation of organic phosphorus by wheat plants. Soil Sci Plant Nutr (Tokyo) 40:593-600

Tarafdar JC, Marschner H (1995) Dual inoculation with *Aspergillus fumigatus* and *Glomus mossae* enhances biomass production and nutrient uptake in wheat (*Triticum aestivum* L.) supplied with organic phosphorus as Na-phytate. Plant Soil 173:97-102

Trainia SJ, Sposito G, Hesterberg D, Kafkafi U (1986) Effects of organic acids on orthosphosphate solubility in an acidic, montmorillonitic soil. Soil Sci Soc Am J 50:45-52

Uren NC (1982) Chemical reduction at the root surface. J Plant Nut 5:515-520

Uren NC, Reisenauer HM (1988) The role of root exudates in nutrient acquisition. In: Tinker B, Lauchli A (Eds) Advances in plant nutrition. Praeger, New York, Vol 3, pp 79-114

Wasaki J, Omura M, Osaki M, Ito H, Matsui H, Shinano T, Tadano T (1999) Structure of a cDNA for an acid phosphatase from phosphate-deficient lupin (*Lupinus albus* L.) roots. Soil Sci Plant Nutr (Tokyo) 45:439-449

Watt M, Evans JR (1999) Linking development and determinacy with organic acid efflux from proteoid roots of white lupin grown with low phosphorus and ambient or elevated atmospheric CO_2 concentration. Plant Physiol 120:705-716

White RE (1980) Retention and release of phosphate by soil and soil constituents. In: Tinker PB (Ed) Soils and agriculture. Blackwell Scientific, Oxford, Vol 2, pp 71-114

Whitelaw MA, Harden TJ, Helyar KR (1999) Phosphate solubilization in solution culture by the soil fungus *Penicillium radicum*. Soil Biol Biochem 31:655-665

Wild A, Oke OL (1966) Organic phosphate compounds in calcium chloride extracts of soils: identification and availability to plants. J Soil Sci 17:356-371

Williams CH, Anderson G (1968) Inositol phosphates in some Australian soils. Aust J Soil Res 6:121-130

Williams CH, Andrew CS (1970) Mineral nutrition of pastures. In: Moore MR(Ed) Australian grasslands. ANU, Canberra, pp 321-338

Williamson VM, Colwell G (1991) Acid phosphatase-1 from nematode resistant tomato. Isolation and characterization of its gene. Plant Physiol 97:139-146

Wodzinski RJ, Ullah AHJ (1996) Phytase. Adv Appl Microbiol 42:263-302

Interspecific Differences in the Role of Root Exudates in Phosphorus Acquisition

Takashi Otani and Noriharu Ae

Summary. Phosphorus (P) is highly fixed with aluminum (Al) and iron (Fe), and present as sparingly soluble forms in Andosols with low fertility. We tried to select crops which had superior P uptake ability in P-limiting soils, and examined their P uptake mechanisms. Field-grown sorghum took up large amounts of P in the low P soil. This ability depended on the development of a larger root system, and sorghum failed to survive when soil volume was limited. On the other hand, pigeonpea, groundnut, and rice were selected as crops that had higher P uptake ability in low P soils in a pot experiment compared with soybean, sorghum or maize. These selected crops seemed to have some mechanisms to take up P from Al- and Fe-bound forms. The ability was not related to their root development, microorganisms association, rhizosphere pH, P absorption characteristics (I_{max}, K_m, C_{min}), or Fe^{III}-reducing activity. In the examination of root exudate, only pigeonpea exudate could dissolve $FePO_4$ and $AlPO_4$. The solubilizing ability was attributed to the organic acids (malonic, citric, malic, piscidic and oxalic acids) which can make chelate rings with Al and Fe. The exudation of these active acids was not enhanced by plant P-deficiency, but may be controlled by plant growth stage. The application of artificial pigeonpea-root exudate, consisting of malonic and citric acid, increased the P uptake amounts of crops with low P uptake ability, such as soybean and maize in a low P Andosol. Therefore, it was demonstrated that organic acids exuded from roots play an important role for P uptake of pigeonpea from Al- and Fe-bound sparingly soluble forms in Andosols.

Key words. Aluminum, Andosol, Iron, Organic acid, Phosphorus, Pigeonpea, Root exudate

1. Introduction

Phosphorus (P) is the second most important nutrient for plants next to nitrogen, and is taken up from soils by roots. Phosphorus is contained in significant amounts in the earth's crust (0.24% as P_2O_5 on average; Taylor 1964), and the constituent of various rock and soil minerals. However, most amounts of P in soils are usually in unavailable forms for plants through bonding with aluminum (Al), iron (Fe), and calcium (Ca). It was pointed out that P deficiency in soils is the main limiting factor for crop production in developing countries (Johansen et al. 1994). Price increases of P fertilizer are considered to be imminent due to the rapid depletion of high grade, low cost phosphate rock reserves throughout the world (Cathcart 1980). Therefore, searching for crops that are able to utilize less-available P, studying the reactions between roots and soils for superior P uptake, and stimulating this ability are important strategies to improve the sustainability of agriculture, not only on marginal lands, but also in the developed countries of the world.

Many scientists have tried to select crops or genotypes with greater ability to absorb P under P-limiting conditions, and have proposed that factors such as root morphology and physiology (Schenk and Barber 1979; Itoh and Barber 1983), rhizosphere pH (van Ray and van Diest 1979), root exudates (Gardner et al. 1983; Hoffland et al. 1989), and mycorrhizal association (Yost and Fox 1979), are important in superior P uptake. However, the effectiveness of each mechanism may depend on the chemical status of P in soils.

An appealing mechanism for P uptake is that plant roots exude compounds dissolving Al- and Fe-bound P. In Andosols, major upland soils in Japan, it is known that soil P is strongly fixed with Al and Fe. Ae et al. (1990) showed that pigeonpea (*Cajanus cajan* L.) has a superior ability to utilize P bound with Fe in Indian Alfisols, and that this ability is attributed to organic acid in root exudate. It was necessary to confirm the effectiveness of this root exudate mechanism in Andosols for utilizing P, which is strongly fixed and present as organic forms in large amounts (Amano 1981). Moreover, if we can find some crops which have a superior P uptake ability in Andosols, it is of interest to investigate whether this ability also owes to organic acids exuded from the roots or not.

We studied superior P uptake mechanisms of upland crops from P-deficient soils as follows: researching P status in low fertile soils, mainly in Andosols which have high P fixing capacity; screening crops, including pigeonpea, for their ability to take up P in these soils; and examining several hypothetical factors by which plants can take up P advantageously from unavailable forms. As the result of these studies, we were able to show an important role for the root exudates of pigeonpea for superior P uptake from sparingly soluble (Al- and Fe-bound) P forms.

2. The Status of "Unavailable" P in Soils

In studies on the P uptake mechanisms of crops, it is important to identify the P forms present in soils. Phosphate anions are fixed to soil minerals more selectively and irreversibly, compared to other anions. It is said that Al and Fe in acidic conditions and Ca in alkaline conditions mainly participate in this reaction, and make sparingly soluble forms (Tan 1993). Andosols fix large amounts of P, and characteristically also contain much of this P in organic forms (Amano 1981). Therefore, we determined soil P as Ca-, Al-, and Fe-bound forms of both inorganic (Pi) and organic (Po) phosphorus by a combination of P fractionation and wet digestion techniques (Otani and Ae 1997a). In the tested soils, i.e., three Andosols and a yellow soil (another major upland soil next in abundance to Andosols in Japan), most P, Pi as well as Po, were determined as Al- and Fe-bound forms in low fertile soils (Table 1).

High fixation of Pi with Al and Fe in Andosols is commonly known because there are large amounts of active (acid oxalate extractable) Al and Fe present in the soils (Wada and Gunjigake 1979). It is also noteworthy that most Po was also bonded with Al and Fe. It was pointed out that Po in the rhizosphere may be mineralized by phosphatases released from plant roots (Caradus and Snaydon 1987; Tadano and Sakai 1991) or microorganisms (Bishop et al. 1994). However, the Po probably must be present as the released form in the soil solution for hydrolyzation by the enzymes. Pant et al. (1994) pointed out that Al and Fe produced a stable complex with Po, and it could provide protection from the action of the enzymes.

To examine the significance of Po in soils for plant nutrition, we determined the amounts of Po extracted from two Andosols using several methods for the evaluation of available P (Truog, Bray II, Olsen, and citric methods), and then allowed them to react with the three phosphatases (acid phosphatase, alkaline phosphatase, and phytase). Among the extraction methods, the Olsen and citrate methods could extract large amounts of Po (Table 2). Citrate-Po was hydrolyzed to a certain degree (30%–40% by phytase, or 12%–13% by acid phosphatase), while Olsen-Po was not hydrolyzed. The citrate method could extract superior amounts of both Al and Fe compared to the other methods (Otani and Ae 1998). These results suggest that releasing Po from Al- and Fe-bound forms is necessary for hydrolyzation by phosphatases. Tarafdar and Claassen (1988) pointed out that phosphatase activity is not the limiting factor in the use of Po but rather the availability of phosphatase hydrolysable P compounds. Therefore, even if the phosphatases produced by plant roots or microorganisms contribute to hydrolyzation of organic phosphates in the rhizosphere, the most important strategy for enhancement of plant P uptake from what are generally considered as non-available forms may be to release P from Al- and Fe-bound forms of Po and Pi in Andosols.

Table 1. Chemical properties and the amounts of soil P extracted in the different fractions of the tested soils

Soil	pH (H_2O)	Total C (gc kg^{-1})	Total P (mg P kg^{-1})	P abs. coeff[e] (mg P_2O_5 kg^{-1})	Oxalate soluble (g kg^{-1})		Inorganic P (mg P kg^{-1})			Organic P		
					Al	Fe	Al-P_i	Fe-P_i	Ca-P_i	Al-P_0	Fe-P_0	Ca-P_0
Tsukuba[a]	6.1	39.3	1097	2574	63.6	57.1	1	263	181	2	144	236
Utsunomiya[b]	4.9	127.1	1622	2790	57.1	18.0	5	287	167	7	326	562
Kawatabi[c]	4.9	108.1	809	1943	22.0	10.1	1	41	51	1	91	539
Ishigaki[d]	4.9	2.4	246	426	1.5	2.3	0	2	27	0	4	7

[a] Haplic Andosol
[b] Cumulic Andosol
[c] Cumulic Non-Allophanic Andosol
[d] Terrace Yellow Soil
[e] Phosphorus absorption coefficient

Table 2. The amounts of orthophosphate released by the phosphatases in the extractants from four different extraction methods in the Andosols

Ext. Method	Soil[a]	Extractable P		Released Pi by Enzyme					
		P_i	P_o	Acid Pase		Phytase		Alkaline Pase	
		(μg P mL^{-1})		μg Pi mL^{-1}(8h)$^{-1}$	%	μg Pi mL^{-1}(8h)$^{-1}$	%	μg Pi mL^{-1}(8h)$^{-1}$	%
Truog	U	0.11	0.36	0.00 ± 0.02[c]	0[d]	0.02 ± 0.02	6	-0.03 ± 0.02	-8
(×10)[b]	K	0.02	0.10	0.00 ± 0.00	0	0.00 ± 0.00	0	0.00 ± 0.03	0
Bray II	U	0.71	0.45	0.00 ± 0.00	0	0.05 ± 0.00	11	-0.05 ± 0.03	-11
	K	0.20	0.18	0.00 ± 0.00	0	0.02 ± 0.02	11	-0.02 ± 0.02	-11
Olsen	U	0.27	1.70	0.05 ± 0.03	3	0.04 ± 0.00	2	0.09 ± 0.03	5
	K	0.06	2.23	0.00 ± 0.00	0	0.04 ± 0.03	2	0.03 ± 0.02	1
Citrate	U	0.66	2.13	0.27 ± 0.06	13	0.63 ± 0.03	30	0.18 ± 0.04	8
	K	0.12	1.60	0.19 ± 0.00	12	0.61 ± 0.02	38	0.02 ± 0.02	1

[a]Soil samples from uncultivated plots in Utsunomiya (U-Uc), and Kawatabi (K-Uc)
[b]10 times concentration of the conventional extractant
[c]Mean ± standard error (n=2)
[d]Proportion of P_I released in each enzyme application to the amount of P_o

3. Screening of Crops for Efficient P Uptake

3.1 Comparison of P Uptake Ability Among Crops in Low P Fields

We compared P uptake ability among various upland crops in some P-deficient soils. The selected crops that took up much P can be thought to have superior P uptake mechanism(s) from Al- and Fe-bound P.

First, we cultivated 17 crops (7 families) in two sites in Andosol fields of different P status (Table 3). One site (P0) had received no P fertilizer for 12 years and the other (P1) received 90 kg P ha^{-1} yr^{-1} every year. All crops received enough N, K, and lime, but P was not applied to either site for this test. Therefore, available P was less not only in P0 but also P1 site compared with the fields fertilized conventionally. Phosphorus uptake from the fields of each crop was compared (Fig. 1). At the P0 site, P uptake by sorghum was highest, with groundnut, pigeonpea, cotton, and sunflower following. At the P1 site, sorghum also took up the greatest amounts of P. However, the ratio of P uptake in P0 to P1 was lower in sorghum than in pigeonpea, groundnut, and rice. Therefore it is doubtful that sorghum can show superior P uptake ability under further P-limiting conditions.

3.2 Relationship Between P Uptake and Root Length

Morphological characters such as root length, surface area, fineness (radius), and intensity of root hairs are considered to strongly influence P uptake because soil P is supplied to plants mainly by diffusion and the P diffusion coefficient is very low (Barber 1984). Several mathematical models have been developed for estimating P uptake (e.g., Nye and Marriott 1969; Claassen and Barber 1976; Itoh and Barber 1983). Factors such as growth-rate and fineness of roots, and root hairs were considered to be the most important variables influencing P uptake

Table 3. Phosphorus status of soils used in the field and pot experiments

Site	PH (H_2O)	Soil P status (mg P kg^{-1})							
		Total P[a]	Available P as measured by				Inorganic P[b]		
			Truog	Bray II	Olsen	Citrate	Ca-P	Al-P	Fe-P
P0	6.1	988	0.5	4.5	7.4	3.2	0.0	255	134
P1	6.1	1472	7.3	25.6	28.2	14.6	2.5	563	261

[a]Measured by digestion
[b]Measured by the Sekiya method

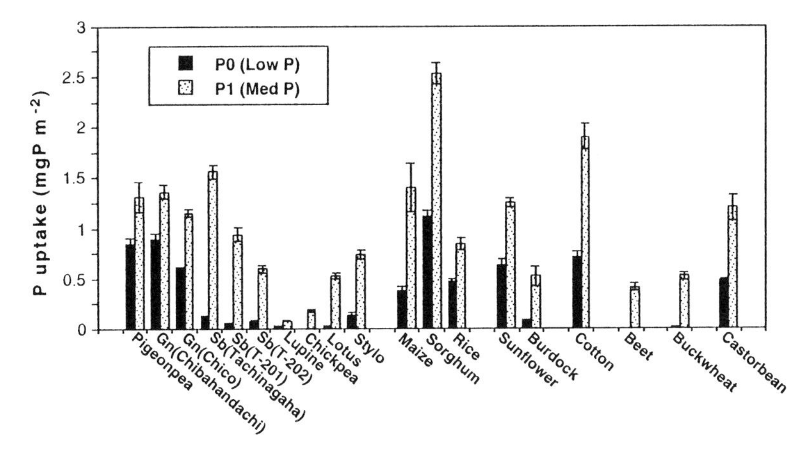

Fig. 1. Phosphorus (*P*) uptake by crops grown in low (*P0*) and medium (*P1*) fertility under field conditions. *Gn*, groundnut; *Sb*, soybean. *Bars*, standard errors of means (n=3)

(Silberbush and Barber 1983). The models successfully predict P uptake in soils with adequate P status, but tended to underestimate P uptake by species such as rape (Hoffland et al. 1990) and sugar beet (Claassen 1990) grown under P-limiting conditions. Consequently, we investigated the relationship between P uptake and root length, as affected by soil volume, and soil P status.

Under field conditions, sorghum, in which P uptake was the highest, had the longest roots at both P0 and P1 sites (Table 4). A strong correlation (r=0.846**) between root length and P uptake among crops was observed in the relatively high-P (P1) site, but not (r=0.235) in the low-P (P0) site. On the other hand, using the same soils in a pot experiment, though root length per pot was greatest in sorghum (a similar response to that in the field), P uptake was much less than in groundnut in the P1 soil. Moreover, in the P0 soil, P uptake by sorghum was 6 times less than that of groundnut while root length was very close. In a further experiment where pot size was changed, differences in P uptake were correlated to changes in soil volume rather than root length per pot (root density) (Otani and Ae 1996a). Greater root elongation under field conditions also implies an increased soil volume from which roots can absorb easily soluble P. At the P1 site,

Table 4. Phosphorus uptake and root length in different crops grown in field and pot conditions with low (P0) and medium (P1) P status

Crops	Field				Pot			
	P0		P 1		P 0		P 1	
	P uptake (g P m^{-2})	Root length (mm^{-2})	P uptake (g P m^{-2})	Root length (mm^{-2})	P uptake (mg P pot^{-1})	Root length (m pot^{-1})	P uptake (mg P pot^{-1})	Root length (m pot^{-1})
Buckwheat	0.02 d[a]	484 bc	0.57 d	588 bc	0.77 b	41 b	9.47 c	151 b
Castorbean	0.49 b	262 c	1.32 ab	360 cd				
Groundnut	1.01 a	300 c	1.43 a	424 cd	15.61 a	194 a	61.29 a	208 a
Pigeonpea	0.97 a	350 bc	1.20 a	248 d				
Sorghum	1.27 a	1258 a	2.86 b	1747 a	2.47 b	148 a	39.12 b	976 a
Soybean	0.12 c	814 b	1.81 c	1077 ab				
	r[b]=0.235		r=0.846*					

*Statistically significant (P<0.01)

[a] Means within colums, followed by the same letter are not significantly different at P<0.05 according to an ANOVA-protected Tukey's Multiple Range Test

[b] r:coefficient of correlation between P uptake and root length (n=18)

sorghum, which had a long root system, seemed to absorb large amounts of available P from a larger volume of soil. In such a case, the strong correlation between P uptake and root length is not surprising. In other words, the ranking of P uptake ability under field conditions is probably similar to the ranking of root elongation ability. The P uptake strategy of sorghum owing to its greater root elongation ability is effective in low P fertile fields, but the P uptake ability may be weak in cases in which the supply of available P from soils is severely limited. Indeed, sorghum plants failed to survive up to the heading stage in small volume (1.6 l) pots with P0 soil. In this situation, crops like groundnut, in which P uptake per unit of soil was high, can be expected to absorb much P from sparingly soluble P forms. The conceptional figure described above is shown in Fig. 2. Based on this concept, we chose a pot experiment to identify crops that can take up P efficiently in P-deficient soils.

3.3 Screening of Crops for Efficient P Uptake from Sparingly Soluble P Forms

Phosphorus uptake by 10 pot-grown crops was compared by using two kinds of Andosols (Minori, Kawatabi), and a yellow soil (Taketoyo) without P application. In all the soils, the content of available P was very low, and most P was in both Fe- and Al-bound forms (Table 5). Among 10 crops, P uptake of pigeonpea, groundnut, and rice was much higher than that of other crops (Fig. 3). Upland rice took

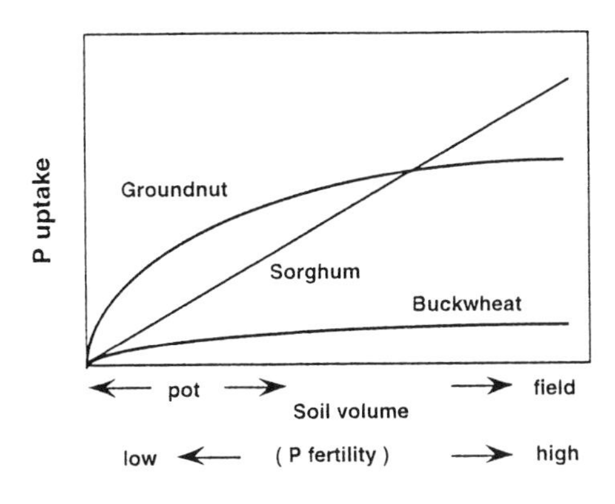

Fig. 2. A conceptional diagram of the relationship between P uptake of some crops and soil volume or soil P fertility

Table 5. Phosphorus status of three soils used for screening of superior P uptake crops in pot experiment

Soil	PH (H_2O)	Total C ($\frac{gc}{kg^{-1}}$)	Exch. Al ($\frac{cmol(+)}{kg^{-1}}$)	Total P ($\frac{mg\ P}{kg^{-1}}$)	Inorg P ($\frac{mg\ P}{kg^{-1}}$) Ca-P	Al-P	Fe-P	Org P ($\frac{mg\ P}{kg^{-1}}$)	Available P ($\frac{mg\ P}{kg^{-1}}$) Truog	BrayII	Olsen	Citrate
Minori	5.3	40	0.18	612	trace	77	60	104	0.5	1.3	1.5	0.6
Kawatabi	4.9	92	3.33	752	trace	36	23	295	1.1	12.2	1.2	2.6
Taketoyo	4.7	3	2.46	132	trace	3	11	16	1.8	2.9	1.6	trace

up more P than other crops especially in the Kawatabi soil (non-allophanic Andosol). Such a high P uptake ability in rice was presumably associated with its tolerance to Al. To avoid the influence of Al toxicity, P uptake was again compared among 6 crops with liming of the Kawatabi soil. Liming enhanced the growth of each plant, but higher P uptake of pigeonpea, groundnut, and rice was still observed compared with soybean, maize, and sorghum (Otani and Ae 1996b). These results suggest that pigeonpea, groundnut and rice have superior ability to take up P from Fe- or Al-bound forms in the soil. The superior P uptake ability of pigeonpea and rice was confirmed in an experiment in which $FePO_4$ or $AlPO_4$ was applied as a P source in a vermiculite potting medium (Otani and Ae 1996b). This experiment was not in perfectly sterile conditions, but the possibility of participation by soil microorganisms, like mycorrhizae, seemed to be low. Ae et al. (1990) pointed out that the higher P uptake efficiency in mycorrhizal associated plants was not due to a higher solubilizing ability of Fe-P but that it was largely due to a greater efficiency in taking up easily soluble P.

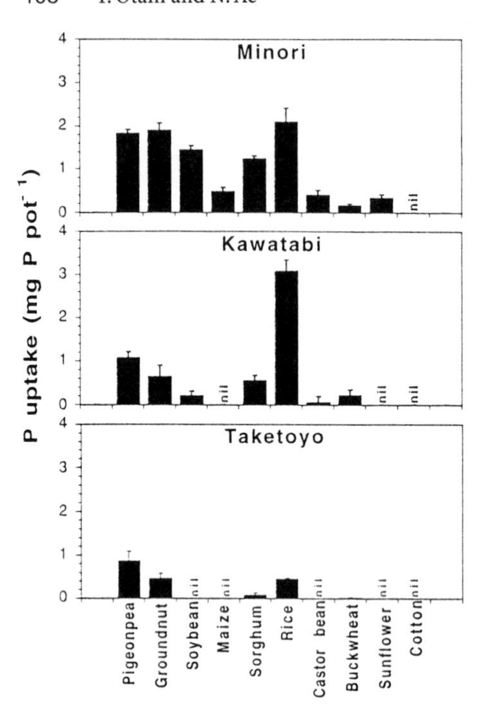

Fig. 3. Phosphorus uptake by pot-grown crops in three soils with low P availability. *Bars*, standard errors of means (n=3)

4. Examination of P Uptake Mechanisms from Sparingly Soluble P Forms

4.1 Hypothetical Factors Advantageous for P Uptake from Sparingly Soluble P Forms

By the examination of P status in low P soils, and the screening of superior P uptake crops in these soils, pigeonpea, groundnut, and rice were considered to possess mechanisms to take up P efficiently from sparingly soluble Al- and Fe-bound P, probably by an ability to dissolve them. Many researchers have suggested that factors such as root morphology and physiology, rhizosphere pH, root exudates, and mycorrhizal association play an important role in P uptake under P-limiting conditions. Among these factors, root morphology (root length, or density) and mycorrhizal association seemed to play a role in efficient take up of easily soluble P, but did not account for the superior ability of pigeonpea, groundnut, and rice to take up P from sparingly soluble P forms in the soil. Rhizosphere pH (ability of H^+ releasing from roots) is a significant factor in the case where the P source is in Ca-bound forms, like rock phosphate or apatite in calcareous soils, but also seems to be ineffective in dissolving P from Al- and Fe-bound P.

We proposed three hypothetical factors that could be considered to be advantageous for P uptake from Al- and Fe-bound P, that is, P absorption characteristics (I_{max}, K_m, C_{min}), Fe^{III} reducing ability of roots, and root exudates. We compared the effect of each factor on P uptake among the crops selected from screening (pigeonpea, groundnut, and rice as superior P uptake ability plants; soybean, sorghum and maize as inferior plants).

4.1.1 P Absorption Characteristics

In general, the I_{max}, K_m and C_{min} values in cereals were higher than in legumes (Table 6). Rice showed higher I_{max}, K_m and C_{min} values than sorghum at 6 weeks after sowing but lower values at 8 weeks. Among the 3 legumes, soybean showed a higher I_{max} value than pigeonpea and groundnut. In contrast, the K_m and C_{min} values were higher in pigeonpea and groundnut than in soybean. It was suggested that a plant with high I_{max}, low K_m, and low C_{min} values would take up P efficiently (Nielsen and Barber 1978). However, the data showed that the P absorption characteristics could not account for the higher ability of pigeonpea, groundnut and rice to take up P from sparingly soluble P forms. Itoh (1987) also pointed out that these kinetic characteristics could not adequately account for the P uptake ability of chickpea on Vertisols with low P fertility.

4.1.2 Fe^{III} Reduction by Roots

It was reported that iron deficiency resulted in an increase in Fe^{III} reducing activity by roots in dicots and non-graminaceous monocots (Römheld and Marschner 1983). If iron in $Fe^{III}PO_4$ were reduced to Fe^{II}, HPO_4^{2-} would be released. In order to test this possibility, Fe^{III} reduction activity of roots was measured under P and/or Fe-deficient conditions. Among all crops, Fe^{III} reduction activity was lowest in rice despite its high P uptake ability (Fig. 4). Among legumes, Fe^{III} reduction activity was highest in groundnut, but similar in pigeonpea and soybean. Therefore, the superior P uptake ability of pigeonpea and rice could not be explained by differences in Fe^{III} reduction activity, but which may partly contribute to P uptake efficiency in groundnut. Additionally, Fe^{III} reduction activity was high under Fe deficiency, but low under P deficiency. If plants can acquire P from $Fe^{III}PO_4$ by Fe^{III} reduction activity, this mechanism could be effective in both P and Fe deficiency conditions. Even in this case, plants may recover from Fe deficiency first, and this activity may decrease even though P deficiency continues because plants usually require 10 times more P than Fe (mole base). Therefore, this mechanism may play a limited role even in groundnut. Furthermore, this hypothesis may not be valid

Table 6. Phosphorus absorption characteristics in the P depletion experiment

Crop	I_{max} (μmol h^{-1} g^{-1} root dry wt.)		K_m (μM)		C_{min} (μM)	
	Time of measuring					
	6w[a]	8w	6w	8w	6w	8w
Rice	7.7	5.9	14.4	19.5	1.95	ND[b]
Sorghum	6.4	13.1	13.7	21.3	1.70	1.85
Pigeonpea	4.2	3.4	4.4	6.9	1.03	0.82
Groundnut	3.2	3.9	2.5	8.2	1.06	0.64
Soybean	6.3	13.0	2.1	6.3	0.76	0.75

[a]Weeks after sowing
[b]Not determined

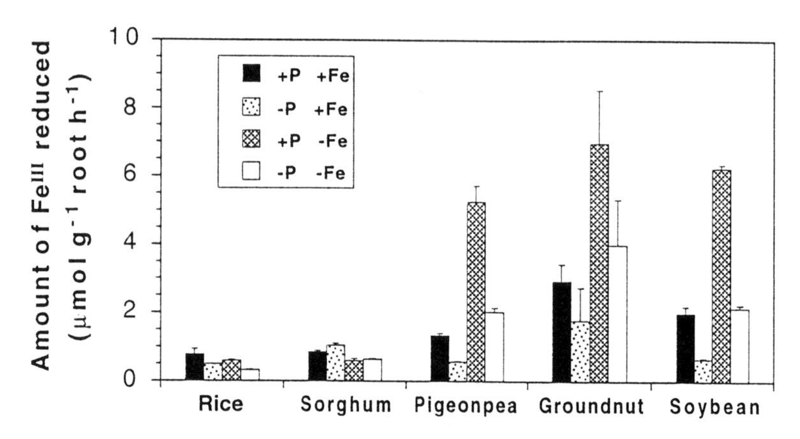

Fig. 4. Effect of P and Fe status on the ability of Fe^{III} reduction by roots. *Bars*, standard errors of means (n=4)

in the case of Al-bound P, which, together with Fe-bound P, was a major part of insoluble P in soil in acid soils in Japan.

4.1.3 Root Exudates

Root exudates were collected from 6 crops grown in water culture, and the solubilizing ability of $FePO_4$ and $AlPO_4$ were compared. Root exudate from pigeonpea had a higher ability to dissolve Fe- and Al-P than exudates from other crops, especially at 4 and 5 weeks after transfer to the nutrient solution (Table 7). The ability to dissolve $FePO_4$ and $AlPO_4$ is not due to the changing of pH, but probably due to the Fe and Al chelating chemicals included in the root exudate of pigeonpea. This is because the solubility measurements were carried out under constant pH (5.6) using acetate buffer. These results suggest that root exudate may play an important role in the superior P uptake ability of pigeonpea from sparingly soluble P forms in the soils.

4.2 Significance of Organic Acids in Root Exudates of Pigeonpea

4.2.1 Components in Root Exudates of Pigeonpea

We tried both qualitative and quantitative analysis of the root exudates of pigeonpea. After the separation of crude root exudates of pigeonpea into cationic, anionic, and neutral fractions, only the anionic fraction showed a notable ability to dissolve $FePO_4$ and $AlPO_4$ (Otani et al. 1996). This result suggested that the active components in pigeonpea root exudates seemed to be organic acids. Further analysis of the anionic fractions using a GC-MS technique after trimethylsilylation, revealed that larger amounts of malonic, oxalic, and piscidic acid were included in pigeonpea root exudates compared with other crops (Table 8).

The ability to release P from $FePO_4$ and $AlPO_4$ of several organic acids, including the components exuded from pigeonpea roots, were compared (Fig. 5). Malonic, oxalic, and piscidic acid, the components exuded from pigeonpea roots, were

Table 7. Effect of root exudates of different crops at various stages on the amount of P released from $FePO_4$ or $AlPO_4$

Crop	P release from $FePO_4$ (μg P mL^{-1} [a])				P release from $AlPO_4$ (μg P mL^{-1})			
	Time of sampling							
	2w[b]	3w	4w	5w	2w	3w	4w	5w
Rice	0.1	0.1	ND[c]	ND	0.3	ND	0.1	ND
Sorghum	0.1	ND	0.5	ND	ND	ND	ND	ND
Maize	—[d]	—	ND	ND	—	—	ND	ND
Pigeonpea	ND	ND	1.2	1.1	ND	ND	7.5	9.3
Groundnut	ND	0.4	ND	0.2	0.4	0.1	ND	ND
Soybean	ND	ND	ND	ND	0.1	ND	ND	ND

[a]The amount of P released from 5 mg $FePO_4$ or $AlPO_4$ by 500μL of root exudates.
[b]Weeks in the solution culture
[c]Not detected
[d]Not determined

Table 8. Organic acids exuded from roots of six crops

Crop	Organic acids [μmol (g^{-1} root) (22 h)$^{-1}$]					
	Oxalic	Malonic	Malic	Tartaric	Citric	Piscidic
Rice	Tr	Tr	Tr	0.02	Tr	ND[a]
Sorghum	Tr	0.02	Tr	Tr	0.02	ND
Maize	Tr	0.01	Tr	Tr	0.01	ND
Pigeonpea	3.87	8.85	0.01	0.01	0.04	2.10
Groundnut	Tr	Tr	Tr	Tr	Tr	ND
Soybean	Tr	0.02	Tr	0.02	Tr	ND

[a]Not detected

confirmed to have the ability to dissolve Fe-P and Al-P. Malic acid and citric acid could also dissolve both P forms, but succinic acid could not. It was noteworthy that differences in the affinity of organic acids to P forms were observed. That is, malonic acid dissolved Al-P more than Fe-P, while solubility by piscidic acid was opposite. Oxalic, malic, and citric acid dissolved both P forms similarly. The higher solubilizing ability of Al-P by the root exudate of pigeonpea (Table 7) could be explained by the characteristic of malonic acid, the major component in the exudate, to dissolve Al-P preferentially. These different solubilizing abilities of organic acids can be explained from the molecular structure of each acid. The organic acids which can dissolve Fe-P and Al-P have a favorable structure to make stable 5- or 6-membered chelate rings with Fe or Al (Hue et al. 1986), by the carboxyl- and/or hydroxyl-groups in their structure (Fig. 6). Therefore, it was suggested that the organic acids (malonic, oxalic, and piscidic) exuded from pigeonpea roots can release orthophosphate from Fe- and Al-bound P by their ability to make stable chelate rings with Fe and Al.

Many researchers have suggested that organic acids in root exudate play an

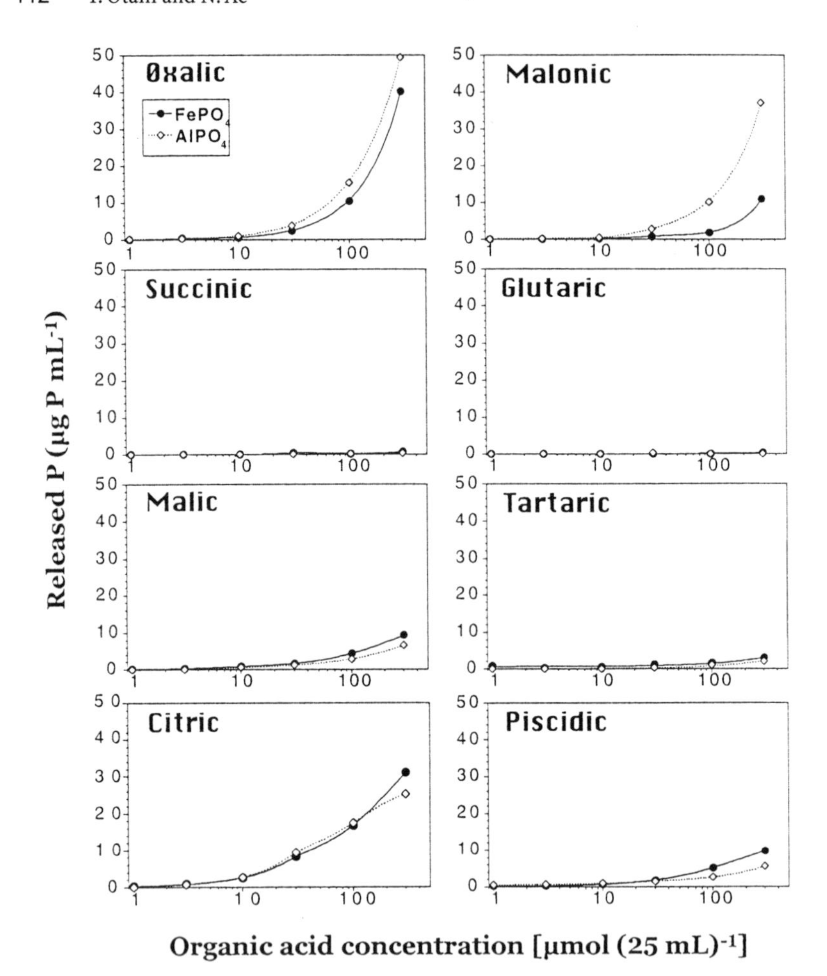

Fig. 5. Effects of organic acids on P release from $FePO_4$ and $AlPO_4$

important role in P uptake under P-limiting conditions. Gardner et al. (1983) indicated that citric acid from lupin can dissolve Ca-P. Several crops were reported to exude organic acids in root exudates. For instance, citric and/or malic acids exuded by maize (Kraffczyk et al. 1984), alfalfa (Lipton et al. 1987), and rape (Hoffland et al. 1989), and oxalic acid exuded by *Eucalyptus gummifera* (Mullete et al. 1974), were reported as the major components in root exudates. Ae et al. (1990) already postulated that piscidic acid exuded from the roots of pigeonpea promoted the release of P from $FePO_4$. However, there were few reports that malonic acid was a major component in root exudates. Because of its high affinity to Al-P, malonic acid exuded from roots seems to play an important role in dissolving sparingly soluble P forms, especially in low fertile Andosols, in which P was mainly fixed with active Al.

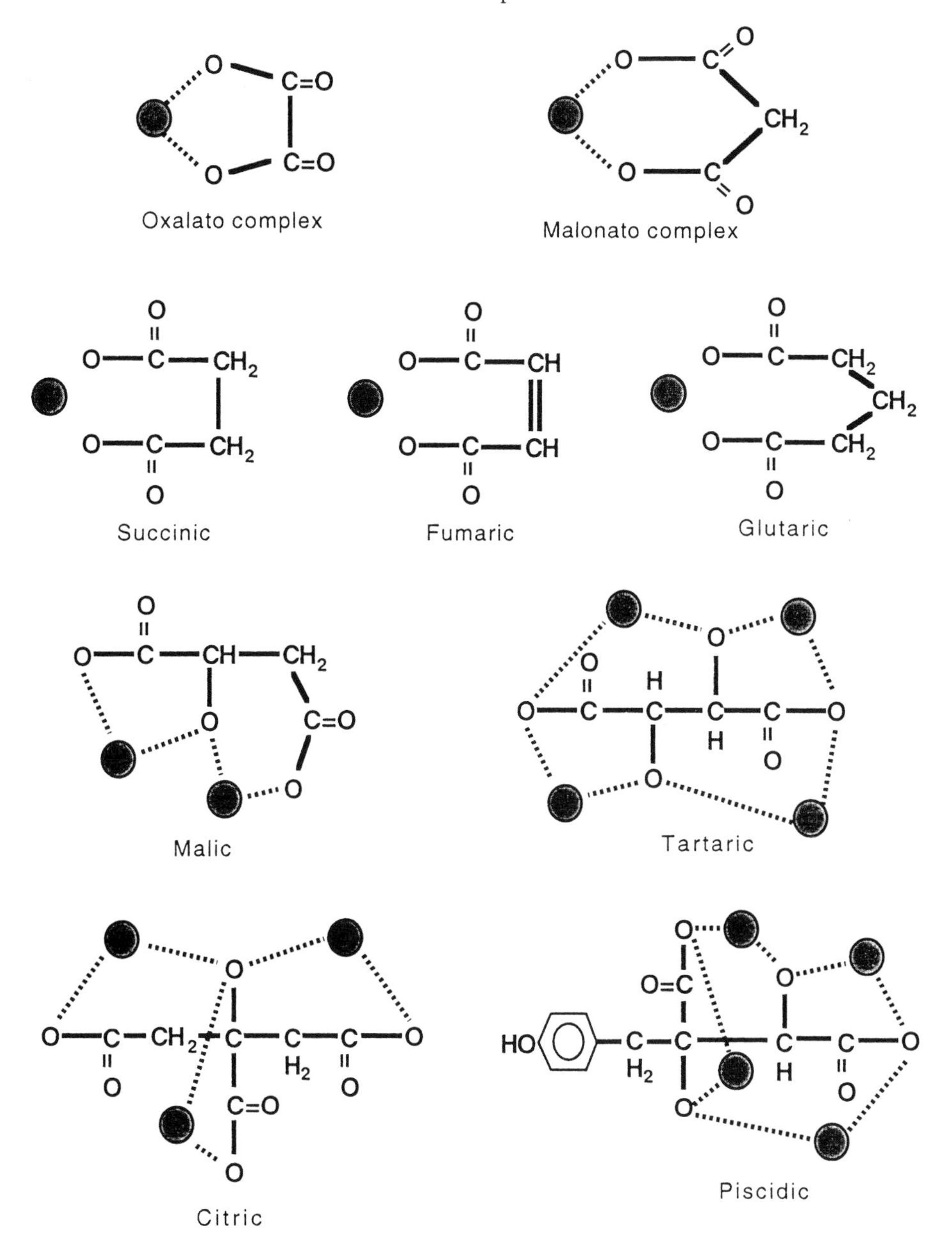

Fig. 6. A scheme of chelate rings formed by organic acids with Al or Fe. *Filled circles*, metals (Al or Fe)

4.2.2 Factors Controlling the Exudation of Organic Acids

We have shown in this chapter that the active organic acids that can dissolve Al-P and Fe-P exist in the root exudate of pigeonpea. It is important to understand the controlling factors of this exudation with regard to the components and their amounts for the enhancement of P uptake in crop breeding. The chemical components and their amounts in root exudates may be affected by factors such as plant species, plant age, temperature, and nutrient condition, etc. (Rovira 1969). It was pointed out that the amounts of citric and malic acids in root exudates were increased by P-starvation in alfalfa (Lipton et al. 1987), and rape (Hoffland et al. 1989). Therefore, we determined the active organic acids in root exudates of pigeonpea under different P nutritional conditions, and examined the possible factors controlling their exudation.

Pigeonpea plants were grown in pots filled with sand. Phosphorus was applied at 3 levels, low (Pl), medium (Pm), and high (Ph), and other nutrients were applied in surplus. The dry weight and P concentration of the tested plants increased with increasing P application. In practice, plants in the Ph treatment showed P-excess, while those in Pl showed P deficiency (Otani and Ae 1997b). The ability to dissolve $FePO_4$ and $AlPO_4$ appeared suddenly after 9 (Ph) to 11 (Pm, Pl) weeks after sowing (WAS), and was greater at higher P levels (Table 9). The amounts of active organic acids, that is, malonic, citric, malic, and piscidic acids in the root exudates could explain the changes in P-solubilizing ability. In this experiment, the plants developed faster at higher P applications. Malonic and citric acids began to exude at 9 WAS in Ph, 10 WAS in Pm, and 11 WAS in Pl, respectively, and the starting time seemed to correspond to the particular plant growth stage, i.e., at development of the 12-14th leaves and the making flower buds in each P treatment. These results suggest that the exudation of active organic acids from pigeonpea roots seems to be controlled not by P-deficiency, but is related to plant growth stage.

In this experiment, malonic and citric acids were the main components of the organic acids in the root exudates. Despite using the same variety (ICPL 87), the components of the organic acids exuded from pigeonpea roots differed compared to the result described above (Table 8). The reason for such differences was not entirely clear. Differences in growth culture (water vs. sand) may be possible. Ae et al. (1993) reported that piscidic acid was the major component, and that malic or citric acids were present only in small amounts, in the root exudates of a different variety of pigeonpea (C-11). It was suggested that the P-solubilizing ability of root exudates by pigeonpea differed among varieties (Subbarao et al. 1997). Further studies on the controlling factors of the active organic acids exudation from plant roots are necessary.

4.2.3 Effectiveness of Exuded Organic Acids from Roots for the P Uptake of Plants

The organic acids exuded from pigeonpea roots can release P from $FePO_4$ and $AlPO_4$ under artificial conditions. However, it is uncertain how much organic acids contribute to P uptake from sparingly soluble P forms in the actual rhizosphere, the plant-soil interaction system, because of the following possibilities; i) decomposition of exuded acids by microorganisms, and, ii) resorption or re-fixation of released P with Al and Fe on the soil surface. Therefore, it is necessary to demonstrate the effectiveness of the organic acids for plant P uptake from soils.

Pigeonpea, soybean, and maize were grown in plastic pots filled with P-deficient Andosols (P0 and P1 soil). Of these plants, soybean and maize have a lower ability to exude organic acids from their roots (Table 8). A notional solution of pigeonpea root exudates was applied to each pot. The components and amounts of applied organic acids, and the periods of their application were determined from the results of the experiment using pigeonpea grown in sand culture as described above. The solution consisted of malonic and citric acids adjusted to a pH of 5.6 by adding KOH, and the amounts of the applied organic acids were decided based on the amounts in the root exudates in Pl, Pm, Ph treatments at 11 WAS. Using a syringe, the solution was applied to each rhizosphere every day for 3 weeks at the flower bud stage (Table 9).

Table 9. Amounts of released P from $FePO_4$ and $AlPO_4$ with the root exudates of pigeonpea, the exuded organic acids,and the plant growth stage at different P application levels

Weeks After Sowing	Applied P level[a]	Released P from (μg P mL^{-1})		Exuded organic acids [μmol $(plant)^{-1}$ $(24h)^{-1}$]				Leaf stage
		$FePO_4$	$AlPO_4$	Malonic	Malic	Citric	Piscidic	
8w	Pl	0.5	0.3	0.3	0.2	0.1	0.0	6 - 7
	Pm	0.5	0.1	0.0	0.1	0.1	0.0	6 - 8
	Ph	0.4	0.1	0.0	0.0	0.1	0.0	9 - 10
9w	Pl	0.6	0.3	0.1	0.1	0.1	0.0	7 - 8
	Pm	1.6	1.1	0.1	0.3	0.6	0.0	9 - 10
	Ph	2.4	2.3	2.1	0.3	1.0	0.1	13
10w	Pl	0.8	0.4	0.0	0.1	0.1	0.0	11 - 12
	Pm	2.2	2.1	1.5	0.1	1.1	0.1	12 - 13
	Ph	8.1	13.2	10.1	4.2	3.5	0.5	15 - 17
11w	Pl	8.7	9.9	1.8	1.1	2.7	0.1	12 - 14
	Pm	13.9	25.8	7.0	4.3	8.0	0.8	13 - 17
	Ph	9.8	30.0	23.8	2.6	12.7	1.6	17 - 20
12w	Pl	8.8	9.9	6.7	0.3	2.3	0.2	15 - 17
	Pm	9.0	9.2	8.4	1.7	3.4	0.6	18 - 20
	Ph	9.4	11.4	5.7	1.4	3.9	0.3	20 - 22

[a]Phosphorus was applied at three levels, low(Pl), medium(Pm), and high(Ph) as NaH_2PO_4 solution

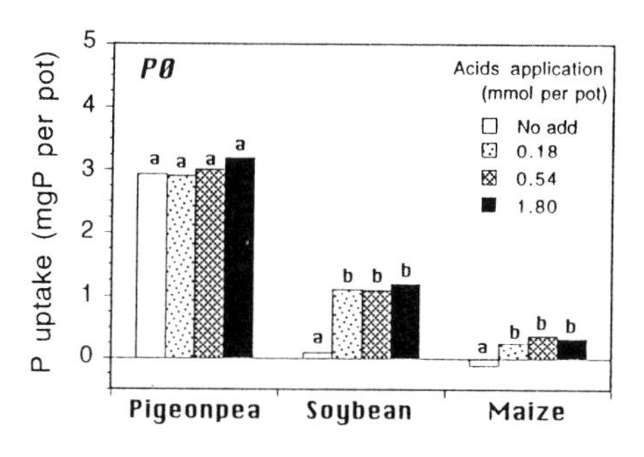

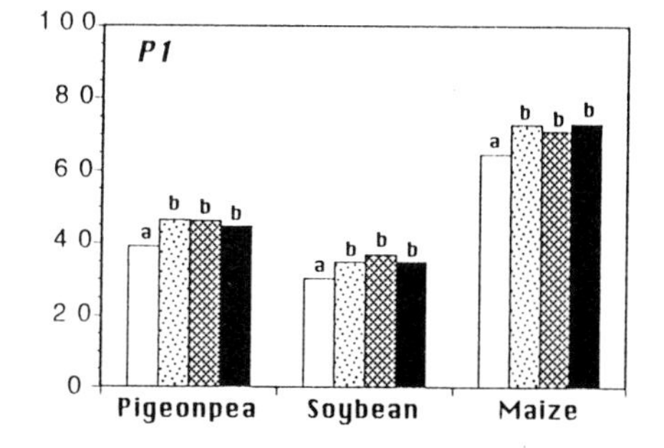

Fig. 7. Effects of organic acid application to the rhizosphere on P uptake of crops. Letters (*a, b*) within the crops denote significant differences at $P<0.05$ according to an ANOVA-protected Tukey's Multiple Range Test. *No add,* no addition

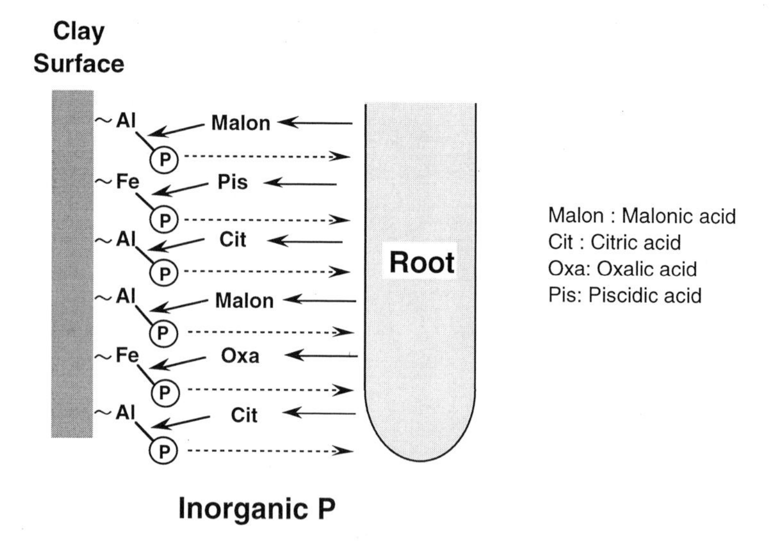

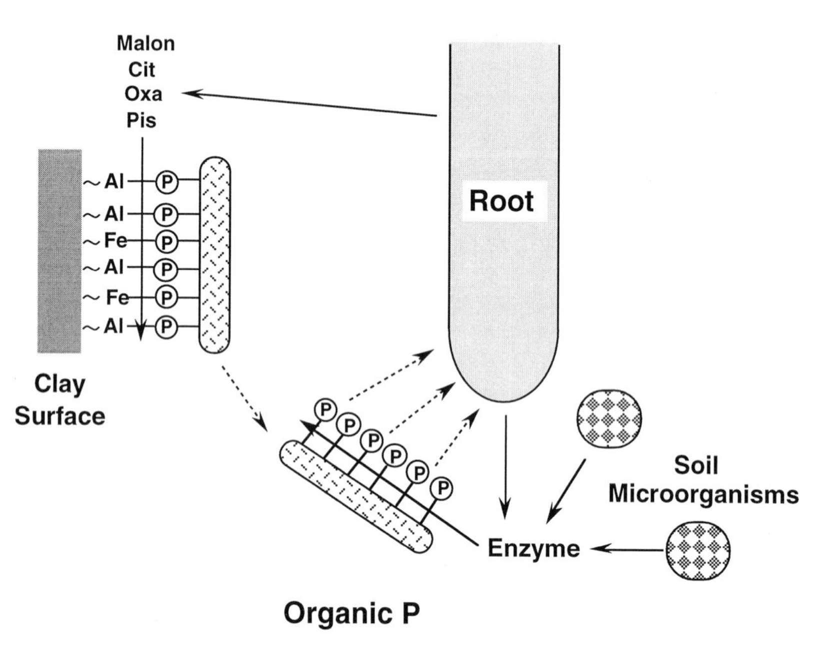

Fig. 8. A scheme of P fixed in Andosols and the release and utilization mechanisms of pigeonpea by organic acid exudation from roots

The P uptake amounts of soybean and maize increased with the application of the artificial root exudates. In particular the effect of the application was clear in the P0 soil, which had little P (Fig. 7). Therefore, it was demonstrated that the organic acids exuded from plant roots contributed to P uptake and survival of crops under P-deficient soil conditions. On the other hand, an effect of organic acid application was not observed in pigeonpea plants. In addition, a further increase in the amount of the acids applied did not increase P uptake of all tested crops. That is, the amount of P which can be released by the exuded organic acids from sparingly soluble forms per unit of soil may be limited under pot culture conditions with restricted total soil volume. However, under field conditions, plants can develop a greater root area and so can be expected to gain new soil sites to be contacted by roots, thus dissolving more P with the organic acids exuded from the roots.

5. Conclusions

We showed a superior P uptake mechanism of pigeonpea from sparingly soluble forms in Andosols by the plant's use of organic acids in root exudates, which can release P from Fe- and Al-bound P (Fig. 8). In the case of organic phosphates, it is necessary for one more hydrolyzation step by enzymes to take place in the rhizosphere for P utilization by plants. But if Po is released by organic acids exuded from roots, this seems to be an easier step than Po release with enzymes produced by microorganisms or plant roots, as in the usual upland soil condition.

We would like to emphasize here that this result was achieved by a combination studies in soil science, crop science, and plant physiology. In this manner, it seems to be important to cooperate with various scientific and agronomic fields in studying plant nutrition. For producing crop species superior in P uptake under P-limiting conditions, breeding work will advance efficiently by using the exudation ability of active organic acids from the roots as an indicator in selection. Moreover, after clarifying the controlling mechanisms of active organic acids exudation from roots qualitatively and quantitatively and by enhancing the function of root exudates through genetic manipulation, it can be expected that new species can be made which have a higher tolerance of P-limiting conditions. Therefore, further studies on the controlling factors and the genetic mechanisms for this action are necessary. Likewise, studies on the mechanisms responsible for superior P uptake ability in rice and groundnut are also required. As for groundnut, a new P uptake mechanism is proposed in the chapter by Ae N et al.

References

Ae N, Arihara J, Okada K, Yoshihara T, Johansen C (1990) Phosphorus uptake by pigeonpea and its role in cropping systems of Indian subcontinent. Science 248:477-480

Ae N, Arihara J, Okada K, Yoshihara T, Otani T, Johansen C (1993) The role of piscidic acid secreted by pigeonpea roots grown in an alfisol with low P fertility. In: Randall PJ et al. (Eds) Genetic aspects of plant mineral nutrition. Kluwer Academic, Dordrecht, pp 279-288

Amano Y (1981) Phosphorus status of some Andosols in Japan. JARQ 15:14-21

Barber SA (1984) Phosphorus. In: Soil nutrient bioavailability – a mechanistic approach. John Wiley & Sons, New York , pp 201-228

Bishop ML, Chang AC, Lee EWK (1994) Enzymatic mineralization of organic phosphorus in a volcanic soil in Chile. Soil Sci 157:238-243

Caradus JR, Snaydon RW (1987) Aspects of the phosphorus nutrition of white clover populations. II Root exocellular acid phosphatase activity. J Plant Physiol 10:287-301

Cathcart JB (1980) World phosphorus reserves and resources. In: Khasawneh FE et al. (Eds) The role of phosphorus in agriculture, Am Soc Agron, Crop Sci Soc Am, Soil Sci Soc Am, Madison, Wisconsin , pp 1-18

Claassen N (1990) Fundamentals of soil-plant interactions as derived from nutrient diffusion in soil, uptake kinetics and morphology of roots. In: Koshino M et al. (Eds) Trans 14th Int Congr Soil Sci, Kyoto, Japan, Vol II, pp 118-123

Claassen N, Barber SA (1976) Simulation model for nutrient uptake from soil by growing plant root system. Agron J 68:961-964

Gardner WK, Barber DA, Parbery DG (1983) The acquisition of phosphorus by *Lupinus albus* L. III The probable mechanisms by which phosphorus movement in the soil/root interface is enhanced. Plant Soil 70:107-109

Hoffland E, Findenegg GR, Nelemans JA (1989) Solubilization of rock phosphate by rape. II Local root exudation by organic acids as a response to P-starvation. Plant Soil 113:161-165

Hoffland E, Findenegg PA, Leffelaar PA, Nelemans JA (1990) Use of a simulation model to quantify the amount of phosphorus released from rock phosphate by rape. In: Koshino M et al. (Eds) Trans 14th Int Congr Soil Sci, Kyoto, Japan. Vol II, pp 170-175

Hue NV, Craddock GR, Adams F (1986) Effect of organic acids on aluminum toxicity in subsoils. Soil Sci Soc Am J 50:28-34

Itoh S (1987) Characteristics of phosphorus uptake of chickpea in comparison with pigeonpea, soybean, and maize. Soil Sci Plant Nutr 33:417-422

Itoh S, Barber SA (1983) Phosphorus uptake by six plant species as related to root hairs. Agron J 75:457-461

Johansen C, Subbarao GV, Lee KK, Sharma KK (1994) Genetic manipulation of crop plants to enhance integrated nutrient management in cropping system – the case of phosphorus. In: Johansen C et al. (Eds) Proc FAO-ICRISAT Expert Consultancy Workshop, Patancheru, India, pp 9-29

Kraffczyk I, Trolldenier G, Beringer H (1984) Soluble root exudates of maize: influence of potassium supply, and rhizosphere microorganisms. Soil Biol Biochem 16:315-322

Lipton DS, Blanchar RW, Blevins DG (1987) Citrate, malate, and succinate concentration in exudates from P-sufficient and P-stressed *Medicago sativa* L. seedlings. Plant Physiol 85:315-317

Mullette KJ, Hannon NJ, Elliott AGL (1974) Insoluble phosphorus usage by *Eucalyptus*. Plant Soil 41:199-205

Nielsen NE, Barber SA (1978) Differences among genotypes of corn in the kinetics of P uptake. Agron J 70:695-698

Nye PH, Marriott FHC (1969) A theoretical study of the distribution of substances around roots resulting from simultaneous diffusion and mass flow. Plant Soil 30:459-472

Otani T, Ae N (1996a) Sensitivity of phosphorus uptake to changes in root length and soil volume. Agron J 88:371-375

Otani T, Ae N (1996b) Phosphorus (P) uptake mechanisms of crops grown in soils with low P status. I. Screening of crops for efficient P uptake. Soil Sci Plant Nutr 42:155-163

Otani T, Ae N (1997a) The status of inorganic and organic phosphors in some soils in relation to plant availability. Soil Sci Plant Nutr 43:419-429

Otani T, Ae N (1997b) The exudation of organic acids by pigeonpea roots for solubilizing iron- and aluminum-bound phosphorus. In: Ando T et al. (Eds) Plant nutrition – for sustainable food production and environment. Kluwer Academic, Dordrecht, pp325-326

Otani T, Ae N (1998) Extraction of organic phosphorus in Andosols by various methods. Soil Sci Plant Nutr, 45:151-161

Otani T, Ae N, Tanaka H (1996) Phosphorus (P) uptake mechanisms of crops grown in soils with low P status. II. Significance of organic acids in root exudates of pigeonpea. Soil Sci Plant Nutr 42:553-56

Pant HK, Edwards AC, Vaughan D (1994) Extraction, molecular fractionation and enzyme degradation of organically associated phosphorus in soil solutions. Biol Fertil Soils 17:196-200

Rovira AD (1969) Plant root exudates. Bot Rev 35:35-58

Römheld V, Marschner H (1983) Mechanisms of iron uptake by peanut plants. I. Fe^{III} reduction, chelate splitting, and release of phenolics. Plant Physiol 71:949-954

Schenk MK, Barber SA (1979) Root characteristics of corn genotypes as related to P uptake. Agron J 71:921-924

Silberbush M, Barber SA (1983) Sensitivity of simulated phosphorus uptake to parameters used by mechanistic-mathematical model. Plant Soil 74:93-100

Subbarao GV, Ae N, Otani T (1997) Genotypic variation in iron-, and aluminum-phosphate solubilizing activity of pigeonpea root exudates under P deficient conditions. Soil Sci Plant Nutr 43:295-305

Tadano T, Sakai H (1991) Secretion of acid phosphatase by the roots of several crop species under phosphorus-deficient conditions. Soil Sci Plant Nutr 37:129-140

Tan KH (1993) Anion exchange. In: Principles of soil chemistry, 2nd Ed, Marcel Dekker Inc, New York, pp 245-254

Tarafdar JC, Claassen N (1988) Phosphatase activity in the rhizosphere and its relation to the depletion of soil organic phosphorus. Biol Fertil Soils 3:199-204

Taylor SR (1964) Abundance of chemical elements in the continental crust: A new table. Geochem Cosmochim Acta 28:1273-1285

van Ray B, van Diest A (1979) Utilization of phosphate from different sources by six plant species. Plant Soil 51:577-589

Wada K, Gunjigake N (1979) Active aluminum and iron and phosphate adsorption in Ando soils. Soil Sci 128:331-336

Yost RS, Fox RL (1979) Contribution of mycorrhizae to P nutrition crops growing on an Oxisol. Agron J 71:903-908

The Role of Mugineic Acid in Iron Acquisition: Progress in Cloning the Genes for Transgenic Rice

Satoshi Mori

Summary. In order to make transgenic rice tolerant to Fe-deficiency on calcareous soils, we cloned several genes from barley roots that contribute to the biosynthesis of the mugineic acid family phytosiderophores (MAs) from the precursor methionine. Of these, adenine phosphoribosyltransferase (*aprt*), nicotianamine synthase (*nas*), nicotianamine aminotransferase (*naat*), and a putative mugineic acid synthase (*Ids3*) were induced specifically in the roots of Fe-deficient plants. The hypothetical 'Fe^{III}-MA transporter' gene has not yet been cloned successfully. We are now introducing *nas* and *naat* genes into rice plants.

Key words. *Aprt*, Chlorosis, *Ids3*, Iron, Mugineic acid, *Naat*, *Nas*, Strategy-I, Strategy-II, Transgenic rice

1. An Advanced Concept of the Two Iron Acquisition Mechanisms

Lime-induced chlorosis in calcareous soil is one of the major agricultural problems leading to reduced crop yields, considering that about 30% of the cultivated soils in the world are calcareous. Under high pH conditions, most of the iron in the soil forms sparingly soluble Fe^{III}-precipitates, mainly as derivatives of $Fe(OH)_3$ (Inoue et al. 1993) that are unavailable to plants, according to the following equations:

$$K_s = [Fe^{3+}][OH]^3 = 3 \times 10^{-38} \quad (1)$$
$$[Fe^{3+}] = 3 \times 10^{-38}/[OH]^3 \quad (2)$$

For instance, only 3×10^{-17} M Fe^{3+} ions are theoretically soluble at pH 7, which is far less than plants demand (10^{-4} – 10^{-8} M) for normal growth.

Graminaceous monocots release Fe-chelating substances, the mugineic acid family of phytosiderophores (MAs), in response to Fe-deficiency stress (Takagi 1976; Takagi et al. 1984). These phytosiderophores solubilize inorganic Fe^{III}-compounds by chelating them, and the Fe^{III}-MAs complexes are taken up in the plasma membrane of root cells by a specific transport system (Römheld and Marschner 1986; Mihashi and Mori 1989). This mechanism is called the Strategy-II mechanism (Marschner et al.1986) and it resembles the microbial siderophore system

(Neilands 1981). In contrast, Strategy-I is an iron acquisition mechanism that is found in all higher plants except graminaceous monocots. Under Fe-deficient conditions, nongraminaceous plants a) increase their ferric reduction capacity at the root surface, b) enhance proton excretion in the rhizosphere, and c) release reductants or chelators in the rhizosphere (Römheld and Marschner 1983). Consequently, plants elevate the iron availability in the rhizosphere and enhance their iron acquisition (Chaney et al.1972). In calcareous soils, however, this reduction

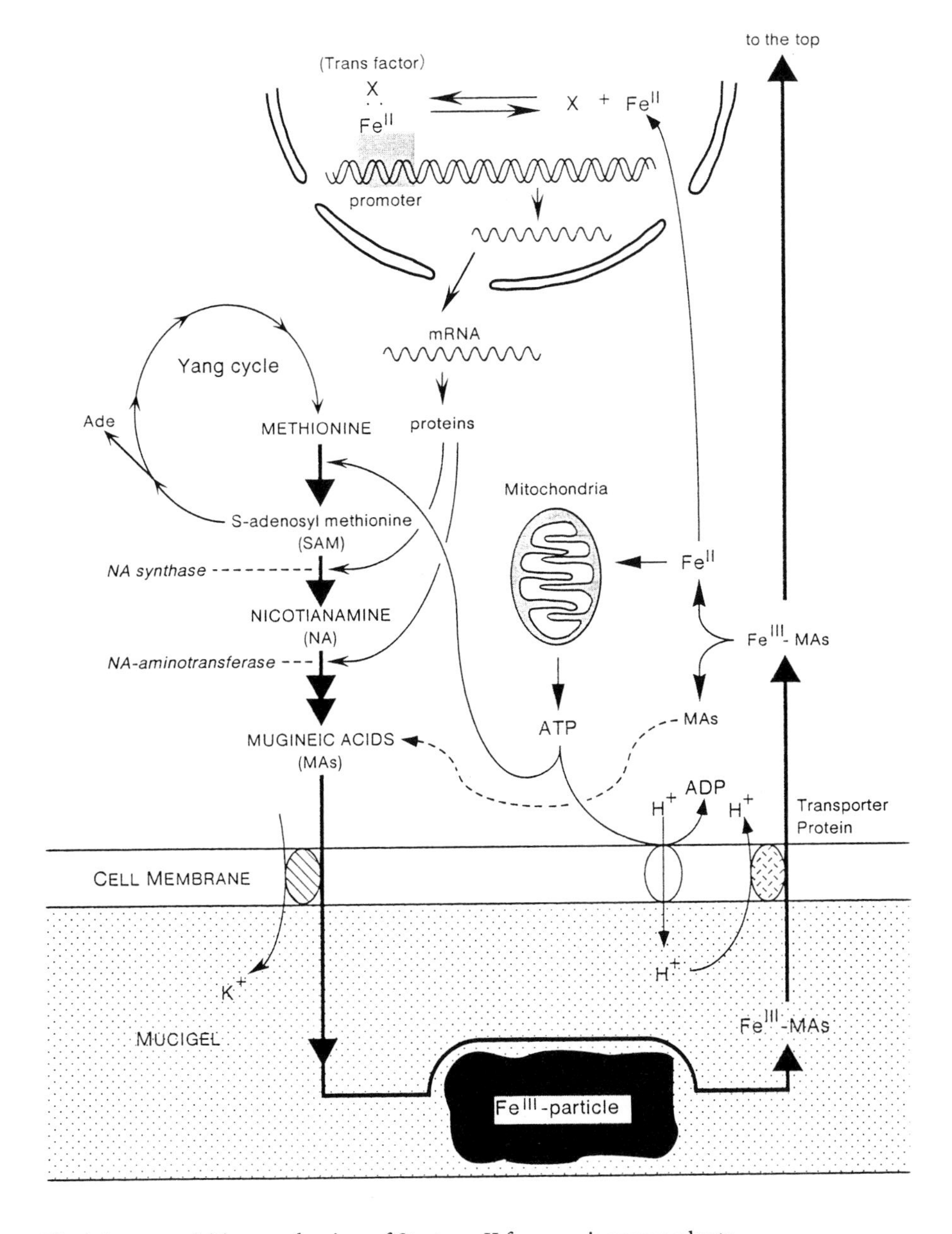

Fig. 1. Iron acquisition mechanism of Strategy-II for graminaceous plants

step is inhibited by the high pH of the soil, and subsequently iron uptake is limited. Since the Fe^{II}-transporter is the only iron transport system in the plasma membrane of the root cells of nongraminaceous plants, Fe^{3+} must be reduced to Fe^{2+} before absorption.

Following the discovery of mugineic acid by Takagi (1976), the Strategy-II iron acquisition mechanism of graminaceous monocots (Fig. 1) and the Strategy-I mechanism of dicots and nongraminaceous monocots (Fig. 2) were proposed by Römheld and Marschner (1986). Their model is still plant iron scientists' leading working hypothesis.

This paper mainly describes our recent progress in cloning the genes relating to the Strategy-II plants in order to make transgenic cultivars tolerant to Fe-deficiency on calcareous soils.

1.1 Strategy-I

As will be discussed later, progress in purifying the proteins involved in the Strategy-I mechanism has been much slower than with those of the Strategy-II mechanism. These proteins include ferric reductase, ATPase, and Fe^{II}-transporter. Until now, only a putative Fe^{II}-transporter from *Arabidopsis thaliana* has been cloned (Eide et al. 1996). Figure 2 shows my modification of Römheld's (1987) original figure illustrating Strategy-I.

1.1.1 Ferric Reductase

Ferric reductase is the most important enzyme to be highly correlated with Fe-deficiency tolerance in many non-graminaceous plants (Römheld 1987). Ferric reductase has been partly purified by many scientists, but has never been purified sufficiently to determine a partial amino acid sequence to allow cloning the

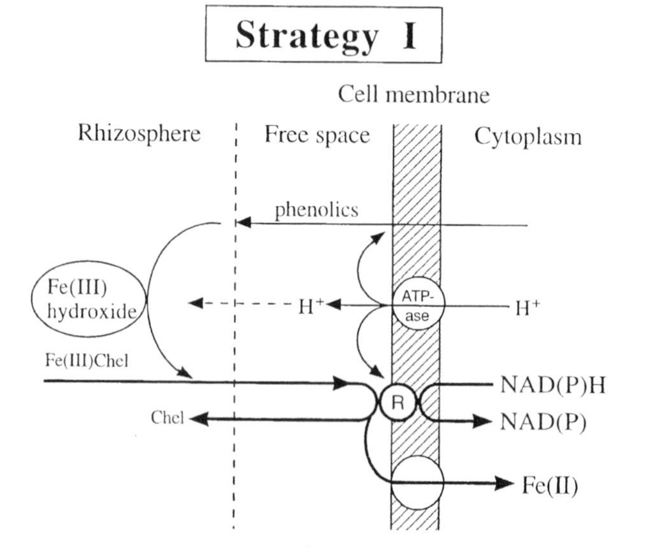

Fig. 2. Iron acquisition mechanism of Strategy-I for nongraminaceous plants

ferric reductase genes. This is probably because ferric reductase is a plasma membrane protein and the amounts recovered in purification are too low for protein analysis (Chaney et al.1972; Holden et al.1991).

Dancis et al. (1990, 1992) cloned two genes (*FRE1* and *FRE2*) from *Saccharomyces cerevisiae* by the yeast complementation method. They introduced yeast cDNA into a yeast mutant with a defect in ferric reductase and cloned the functional genes from the rescued yeast. Both gene products, FRE1 and FRE2, can also reduce Cu^{2+} to Cu^{1+} in yeast (Fig. 3).

1.1.2 ATPase

In some bean plants, Fe-deficiency inducible ATPase activity in the cell membrane of the plant roots correlates with Fe-deficiency tolerance (Loeppert et al. 1995; Ohwaki et al. 1997). If the pH of the rhizosphere falls with increased H^+-extrusion by ATPase from root cells, the number of soluble Fe^{3+} ions released by dissolution of sparingly soluble ferric hydroxide in the soil increases. The solubility follows equations (1) and (2) shown above and 10^3 times more Fe^{3+} ions are released into the rhizosphere with a decrease in the pH of 1. The pH decrease on the root surface also optimizes the enzymatic activity of ferric reductase in the root epidermal cells, because the optimum pH of the plant membrane reductase is generally around 6 (Chaney et al.1972). The increased ferric reductase activity increases the available Fe^{2+} ions on the cell surface, and these ions are absorbed into the root cells by the Fe^{II}-transporter. Therefore, many scientists have been trying to isolate an ATPase that is either specifically induced by iron deficiency or is found in the plasma membrane of Fe-deficient plant roots. Until now, many ATPase genes have been cloned in plants, but nobody has successfully cloned the specific ATPase genes regulated by Fe-deficiency.

1.1.3 Fe^{II}-Transporter

In Strategy-I plants, the plasma membrane Fe^{II}-transporter is thought to transport Fe^{2+} ions into root cells. For a long time, nobody succeeded in cloning plant Fe^{II}-transporter genes. While plant nutrition scientists were bewildered, yeast

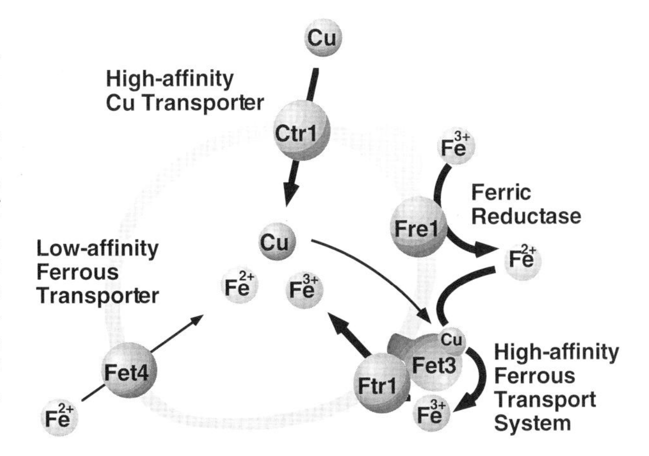

Fig.3. Iron acquisition mechanism of yeast. Reductase (FRE1 and FRE2), low-affinity Fe^{II}-transporter (FET4), and high-affinity Fe-transporter system (FET3 plus FTR1). When the amount of external Fe^{II} ions is very low, the high-affinity system works, requiring endogenous Cu^{2+}. In the high-affinity Fe-transporter system Fe^{II} ion is oxidized by FTR1, then Fe^{III} ions are transported by Fe^{III}-transporter (FET3)

scientists discovered the cDNA of two types of Fe^{II}-transporter in yeast. Fet4 protein is a low-affinity Fe^{II}-transporter (Dix et al. 1994) and Fet3 protein is a multicopper oxidase (Stearman et al. 1996), which is regulated by copper and forms a complex with Ftr1 protein, a high-affinity Fe-transporter. Ftr1 protein is a postulated Fe^{III}-transporter. Moreover, two types of Cu transporter, low (Ctr2) and high-affinity (Ctr1) transporters, were found in yeast. Therefore, the endogenous Cu content, which is dependent on the high-affinity Cu transporter (Ctr1), regulates the high-affinity Fe-transporter (Ftr1+Fet4) (Fig. 3).

Recently, Eide et al. (1996) introduced an *Arabidopsis thaliana* cDNA library into a yeast double mutant (*Δ fet3 Δ fet4*), and successfully cloned a plant Fe^{II}-transporter cDNA by screening the clones that complimented the defects. They designated the putative Fe^{II}-transporter IRT1 (Iron-Regulated Transporter). The *IRT1* gene was only induced in the roots of Fe-deficient *Arabidopsis*. Introduction of high copy number *IRT1* genes into *Saccharomyces cerevisiae* increased the iron uptake. A different type of Fe-transporter gene has been cloned from mammals (Gunshin et al.1997).

1.2 Strategy-II

1.2.1 Discovery of MAs

Knowledge of the Strategy-II iron acquisition mechanism has increased considerably since the discovery of mugineic acid by Takagi (1976, 1993). The discovery of Fe-solubilizing substances in washings from the roots of Fe-deficient rice and oats has been briefly described elsewhere by the author (Mori 1995). The Takemoto group (Nomoto et al. 1987) determined the chemical structure of MAs, and between 1986 and 1993 almost all of the biosynthetic pathways were clarified, as mentioned later (Fig. 4).

1.2.2 Evidence that Methionine is the Predominant Precursor of MAs

I developed an amino acid-radioanalyzer (Mori 1981) and used it to screen amino acids and organic acids to determine the precursors of MAs in Fe-deficient barley root tips. Methionine was identified as the sole precursor (Mori and Nishizawa 1987). Subsequently, we tried to identify the chemical intermediates in the steps from methionine to MAs. S-adenosylmethionine (SAM) is synthesized from methionine by SAM synthetase (Shojima et al. 1989). Then nicotianamine synthase (NAS) combines three molecules of SAM to form one molecule of nicotianamine (NA) (Shojima et al. 1989; Higuchi et al. 1996). NA is then converted to [3"-keto acid] by nicotianamine-aminotransferase (NAAT) (Shojima et al. 1990; Ohata et al. 1993; Kanazawa et al. 1994, 1995b). Finally, deoxymugineic acid (DMA) is synthesized by an unknown reductase (Shojima et al. 1990). The hydroxylation of DMA to form the other MAs was proposed from indirect evidence using wheat (cv. Chinese Spring)-barley (cv. Betzes) and wheat (cv. Chinese Spring)-rye (cv. Imperial) addition lines (Fig. 4; Mori and Nishizawa 1989; Mori et al. 1990). Ma and Nomoto (1994) obtained direct proof from NMR studies using ^{13}C-methionine and ^{2}H-methionine as precursors.

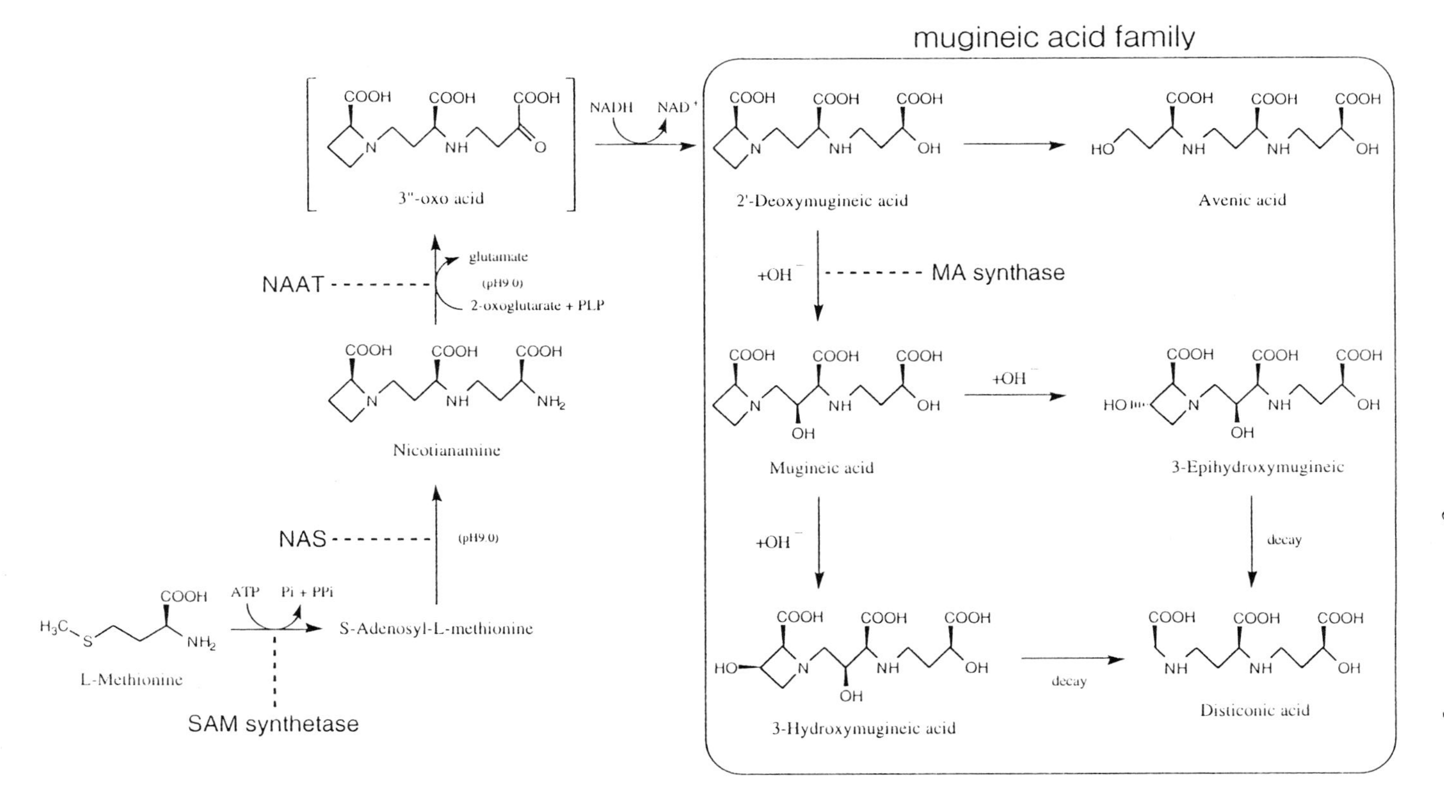

Fig. 4. Biosysnthetic pathway of mugineic acid family phytosiderophores for strategy-II plants

1.2.3 What is the Source of Methionine?

Ma et al. (1995) proposed the methionine cycle (Yang Cycle) as one of the methionine sources, based on NMR studies using ^{2}H-methyl thioribose and ^{13}C-ribose as precursors (see Fig. 9). The Yang Cycle generates methionine as a precursor for ethylene synthesis when environmental stress occurs or in the climacteric stage of fruit (Yang et al.1984). However, methionine has been detected in the phloem of rice (0.08-0.3 mol% of total amino acid) and wheat (1.8 mM/261.7 mM of total amino acid) (Hayashi and Chino 1986, 1990). The question still remained as to whether methionine transported from the shoot to the roots was used for the direct synthesis of MAs in the roots. To answer this, the rate of translocation of methionine from the shoot to the roots needed to be determined. We used real time imaging of ^{11}C-methionine transport in barley to address this question. The results showed that methionine from a cut leaf of iron deficient barley did not move to the roots, but moved to other chlorotic leaves preferentially. This suggests that methionine from the upper part of the plant is not used to synthesize MAs. In conclusion, ATP molecules that are either produced from photoassimilates or transported to iron deficient roots directly from the top of the plant may be the precursor of the methionine skeleton in the Yang Cycle (Fig. 4) (Nakanishi et al. 1999).

1.2.4 Gene Cloning for MAs-Synthesis

S-adenosylmethionine Synthetase Genes (sam1, sam2, sam3)

Since SAM is a precursor of MAs, the SAM synthetase activity in Fe-deficient and Fe-sufficient barley roots was compared. There was, however, no difference (Takizawa et al. 1996). We then cloned the cDNA of three *sam* genes (they are deposited in the DDBJ database as D63835, D85237, and D85238) from the commercial barley-cDNA library (λZapII), using the consensus region from all the *sam* genes listed in the database that have ever been cloned from living organisms. However, we could not clone the *sam* genes that were specifically expressed during Fe-deficiency, although those three genes were specifically expressed in barley roots. Therefore, we have tentatively concluded that there is no specific Fe-deficiency *sam* gene and that constitutive SAM synthetase activity is sufficient for MAs-synthesis.

Nicotianamine Synthase Genes (nas)

Nicotianamine synthase (NAS) catalyzes trimerization of S-adenosyl-methionine to form one molecule of nicotianamine as an intermediate substrate in the biosynthetic pathway of MAs (Fig. 5). The method used for the enzyme assay was first proposed by Shojima et al. (1989, 1990) and subsequently developed by Higuchi et al. (1994). For a long time, we have been purifying NAS by trial and error (Higuchi et al.1994, 1996) and we have finally succeeded. The NAS protein was degraded with CNBr and several peptides were obtained. Using synthetic degenerate primers for the partial amino acid sequences of those peptides, we isolated seven NAS cDNAs, designated as *nas*, from a cDNA library constructed

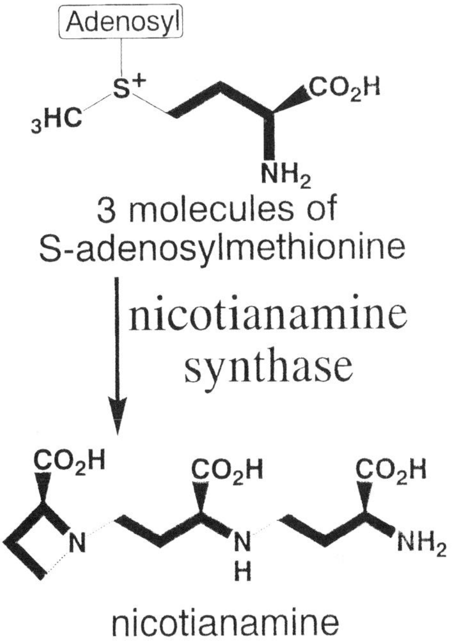

Fig. 5. Two functions of nicotianamine synthase (NAS): trimerization of S-adenosylmethionine and azetidine ring formation

from mRNA from Fe-deficient barley roots. Nucleotide sequence analysis of *nas1* revealed an open reading frame of 985 bp encoding a polypeptide with a predicted molecular weight of 35,144, which corresponds to the size of NAS estimated using SDS-PAGE. *Nas1* hybridized to the root and Fe-deficiency-specific 1.3 kb transcripts. Southern hybridization revealed that *nas* genes form a multigene family in both the barley and rice genomes. The barley *nas* gene family is deposited in the DDBJ database as *nas1* (AB010086), *nas2* (AB011265), *nas3* (AB011264), *nas4* (AB011266), *nas5-1* (AB011267), *nas5-2* (AB011268) and *nas6* (AB011269). Each gene was introduced into *E. coli* and enzyme activity was obtained with all except *nas5-1* and *nas5-2.* 3'-DNA sequences that correspond to 53 amino acids are deleted in these two genes, which may cause their lack of NAS activity (Higuchi et al. 1998).

Nicotianamine Aminotransferase (NAAT-A, NAAT-B)

Nicotianamine aminotransferase (NAAT) converts nicotianamine to 3”-keto acid (Fig. 6). A specific method for measuring enzyme activity was proposed by Ohata et al. (1993) and developed by Kanazawa et al. (1994). Like NAS activity, NAAT enzyme activity is strongly induced by Fe-deficiency in barley roots. Using the enzyme assay system, a crude enzyme fraction from Fe-deficient barley roots has been purified after several purification steps using column chromatography. Kanazawa et al. (1995a, 1995b) found at least two NAATs in Fe-deficient barley roots; one is strongly induced by Fe-deficiency and the other is constitutively expressed in Fe-sufficient barley roots, but is also inducible by Fe-deficiency. The elutes were separated using 2D-SDS PAGE after affinity chromatography using

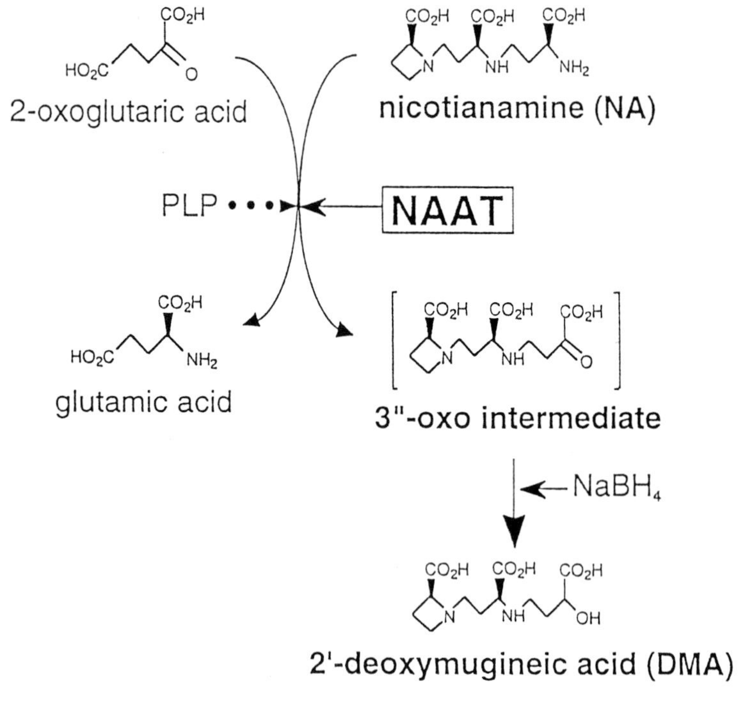

2-oxoglutaric acid
nicotianamine (NA)
PLP
NAAT
glutamic acid
3"-oxo intermediate
NaBH$_4$
2'-deoxymugineic acid (DMA)

Fig. 6. Conditions for the enzymatic analysis of nicotianamine aminotransferase (NAAT)

nicotianamine as a ligand, and eluting with nicotianamine. All the peptide spots stained with Coomassie brilliant blue that appeared specifically in the sample from Fe-deficient barley were analyzed. Each peptide was digested by CNBr and developed on SDS-PAGE. Several distinct bands were blotted on PVDF membrane, and then the amino acid sequence of each band was determined. One of these proteins had the partial amino acid sequence of an amino-group transferase that binds with pyridoxal phosphate. We thought that this protein might be nicotianamine aminotransferase. After degrading this protein, we conducted PCR using several of the peptide fragments. The PCR product was used as a probe to clone two genes (designated as *naat-a* and *naat-b*) from a cDNA library made from Fe-deficient barley roots. The amino acid sequence of the pyridoxal phosphate binding site was identified in both clones and in other peptide fragments resulting from CNBr digestion. We introduced both cDNAs into two genotypes of *Saccharomyces cerevisiae* and both yeasts showed NAAT activity. Northern analysis showed that *naat-a* was specifically induced by Fe-deficiency, while *naat-b* showed a different expression pattern. Therefore, we concluded that we have cloned *naat* genes (Takahashi et al. 1999). *Naat-a* and *naat-b* are deposited in the DDBJ database, as D88273 and AB005788, respectively.

Putative Mugineic Acid Synthase (Ids3)

We cloned *Ids1* (Okumura et al. 1991, GSDB/DDBJ/EMBL/NCBI, X58540, D50640), *Ids2* (Okumura et al.1994) and *Ids3* (Nakanishi et al.1993, GSDB/DDBJ/EMBL/NCBI, D37796) using the differential hybridization method with mRNA from Fe-deficient and Fe-sufficient barley roots. These genes are expressed in both an Fe-deficiency-specific and a root-specific manner. The genomic DNA was also cloned and sequenced. Northern analysis using full length *Ids3* as a probe showed that it was expressed in the Fe-deficient roots of *Hordeum vulgare* cv. Ehimehadaka (this variety produces DMA, MA, and epiHMA), *Hordeum vulgare* cv. Minorimugi (produces DMA and MA), *Add-4* in wheat (cv. Chinese Spring)-barley (cv. Betzes) addition lines (*Add-4* produces DMA and MA) and rye (produces DMA, MA, and HMA). Western analysis using anti-IDS3-antibody showed the same trend as the Northern analysis. *Ids3* was never expressed in wheat, rice, or corn (those species only produce DMA) (see Fig. 4). Homology research showed that *Ids3* has a dioxygenase nature. Dioxygenases oxidize the substrate with one O from O_2 and the other O is trapped with 2-oxoglutarate, which is then changed to succinate. Since the synthesis of MAs has been identified to proceed DMA → MA → epiHMA/HMA (Mori and Nishizawa 1989; Mori et al.1990; Ma et al.1994) it is speculated that *Ids3* is a putative mugineic acid synthase (deoxymugineic acid hydroxylase) gene (Fig. 7). To prove this hypothesis, we tried producing enzyme activity using a crude homogenate of *E. coli* into which the *Ids3* gene was introduced, but we were not successful. To function as a dioxygenase, this enzyme needs Fe^{2+} ions as an enzyme activator. Fe^{2+} ions are easily converted into Fe^{3+} ions autocatalytically during the enzyme reaction in vitro. The Fe^{3+} ions formed are easily chelated with DMA, a specific chelator of Fe^{3+} ions, which is also the substrate of this enzyme. This means that DMA is consumed by Fe^{3+} ions and Fe^{III}-DMA cannot be hydroxylated by this enzyme, possibly because the targeting site (2'-C) by this enzyme is masked stereochemically in the structure of Fe^{III}-DMA. These sequential events may occur during the reaction, and this may be the reason that we cannot detect IDS3 enzyme activity. We introduced *Ids3* into rice as an alternative way of proving our hypothesis that it is an MA-synthase. Our hypothesis was proved by obtaining transgenic rice that produces MA under Fe-deficiency, because wild rice can produce only DMA (see Fig. 4)(Kobayashi et al. 2001).

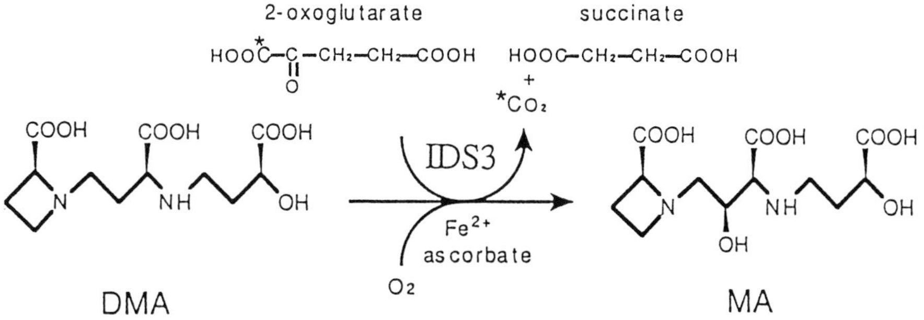

Fig. 7. A plausible function of IDS3 as mugineic acid synthase

Adenine Phosphoribosyltransferase (APRT)

APRT protein was identified by sequence homology research as a specific protein from Fe-deficient barley roots that is not seen in Fe-sufficient roots with 2D SDS-PAGE. PCR was conducted with degenerate primers constructed from the amino acid sequences of the peptide fragments produced by CNBr degradation of the protein. The PCR product was used as a probe for screening a cDNA library made from Fe-deficient barley roots and one positive gene was cloned. This cDNA is deposited in DDBJ (AB012046). Gene expression and APRT enzyme activity were induced in a root-specific and Fe-deficiency dependent manner. When we cloned this gene, we could not imagine the role of this gene in MAs-synthesis. APRT plays a role in the production of AMP from adenine (Fig. 8). Adenine is released during the Yang Cycle, and in turn should be recovered as a precursor of ADP. ATP, which is the important precursor supplying the ribose moiety for methionine synthesis in the Yang Cycle (Fig. 9), is synthesized from ADP. Both APRT activity and transcription of the *aprt* gene are clearly enhanced by Fe-deficiency in barley roots (Itai et al. 2000). Therefore, we now speculate that this gene is involved in salvaging adenine from the Yang Cycle as shown in Fig. 9.

Fe^{III}-Mugineic Acid Transporter

A hypothetical "Fe^{III}-MA transporter" is based on the findings: 1. Translocation of ^{59}Fe to the top of a rice plant supplied with $^{59}Fe^{III}$-MA is much faster than with $^{59}Fe^{III}$-EDTA (Fig. 10); 2. The rate of absorption of $^{59}Fe^{III}$-MA by Fe-deficient barley roots is markedly higher than by Fe-sufficient barley roots (Fig. 11), and 3. BPDS does not inhibit the absorption of $^{59}Fe^{III}$-MA (Römheld and Marschner 1986). There was, however, no clear difference in the silver staining pattern of membrane proteins seen with SDS-PAGE between Fe-deficient and Fe-sufficient barley roots (Mihashi and Mori 1989). von Wirén et al. (1997) identified four proteins labeled with ^{35}S-methionine within the membrane fraction of maize, wild

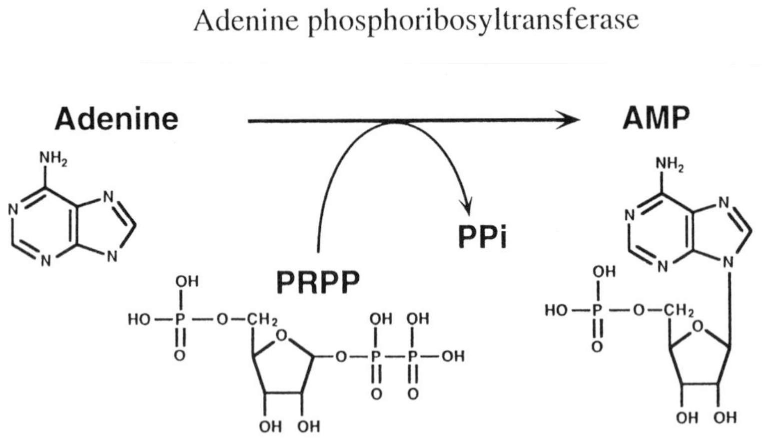

Fig. 8. The function of adenine phosphoribosyltransferase (APRT) for purine salvage

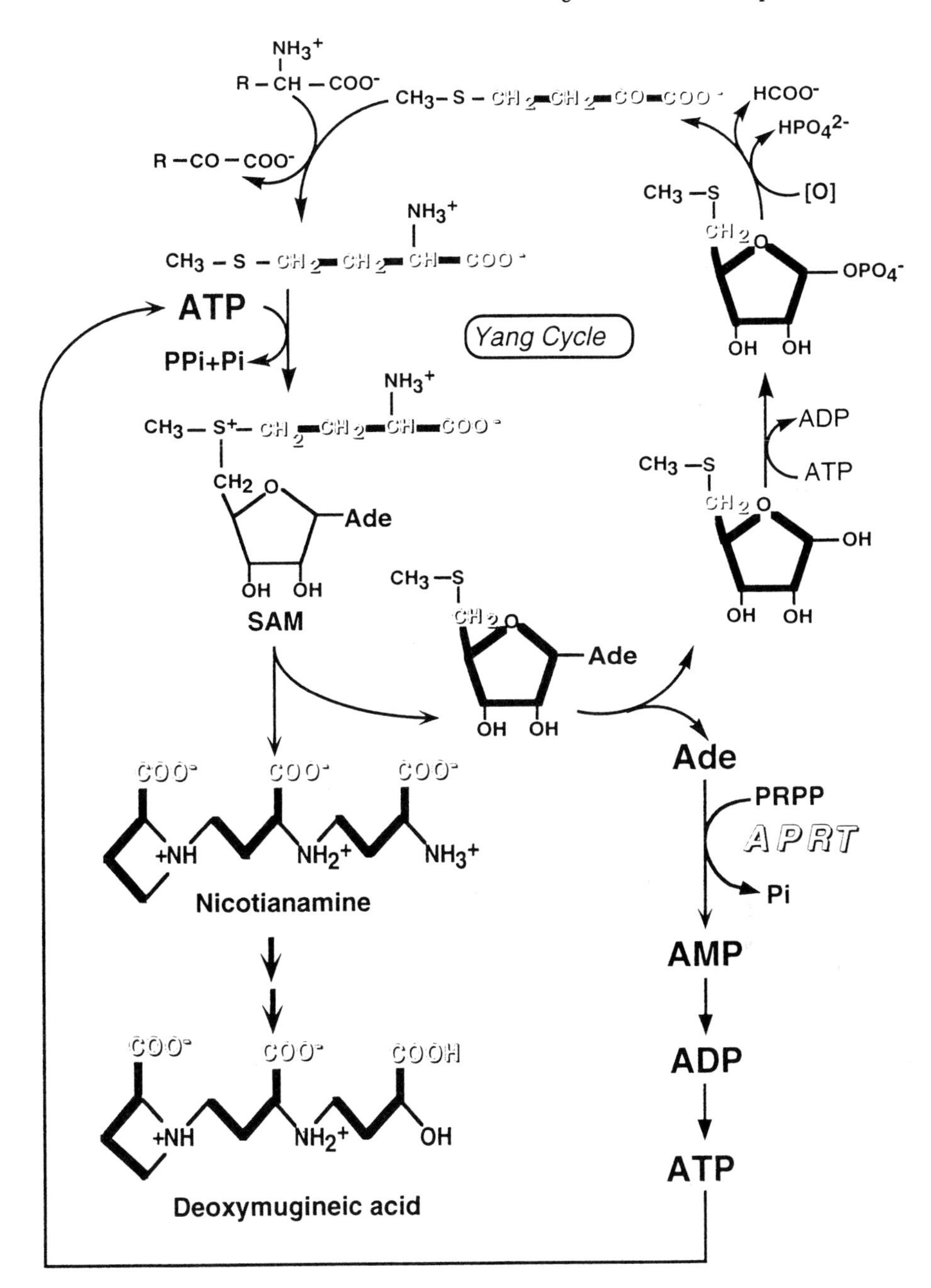

Fig. 9. The Yang cycle and APRT for methionine supply, and ATP recycling for the production of MAs in Fe-deficient roots

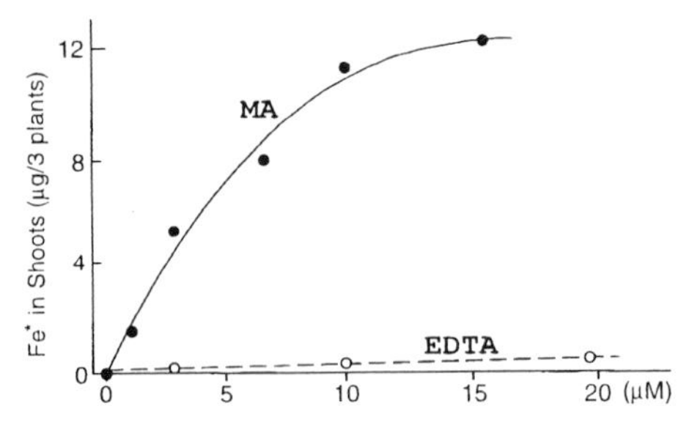

Fig. 10. The effect of mugineic acid (MA, *solid line*) or EDTA (*broken line*) on the absorption of iron by young rice roots (*Oryza sativa* cv. Norin 16, pH 7.0). The data were obtained 24 h after $^{59}FeCl_3$ supply (Takagi et al. 1984)

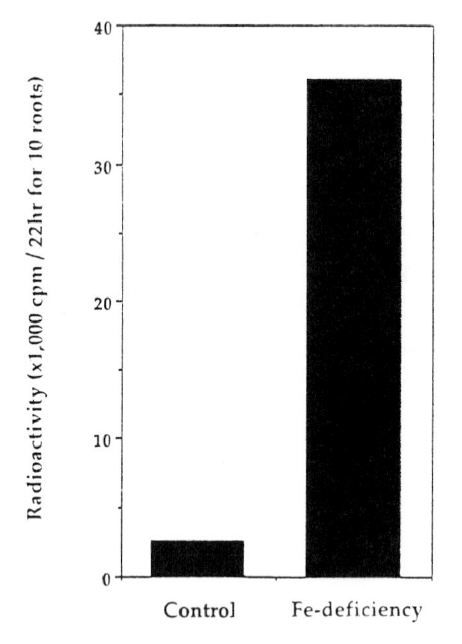

Fig. 11. Iron uptake activity measured by the multicompartment root box method (Mihashi and Mori 1989)

type Alice, when compared with mutant *ys1*, which has a defect in Fe^{III}-MA transporter activity. Whether one of these is a transporter protein remains to be determined.

We then changed the strategy used to clone the Fe^{III}-MA transporter gene to the "yeast complementation method". A cDNA library made from Fe-deficient barley roots was introduced into the mutant yeast *Δfre1* (a gift of Dr. H. Dancis, NIH), in which ferric reductase is defective. They were screened under severe Fe^{II}-deficiency caused by an excess of BPDS (a synthetic chelator of Fe^{2+} ions) in the culture medium, with Fe^{III}-MA as the sole source of Fe. However, no gene was successfully cloned. Next, we tried to compliment the mutant yeast, *Δctr1* (a defect in the high-affinity Cu-transporter, also a gift of Dr. H. Dancis, NIH), by introducing the cDNA of Fe-deficient barley roots. In yeast cells, Cu serves as an essential catalyzer of the high-affinity Fe^{III}-transporter (FET3), as mentioned earlier (Fig. 3). We hypothesized that if the Cu content in the yeast cell is deficient because of the defect in the *ctr1* gene, FTR1 does not oxidize FeII to FeIII. So the function of FET3 (Fe^{III}-transporter) will not work, resulting in Fe-deficiency in yeast. If an Fe^{III}-MA transporter gene from the cDNA library made from Fe-deficient barley roots is introduced into this yeast *ctr1* mutant, expressed, transported to the yeast cell membranes, and Fe^{III}-MAs outside the yeast cells are absorbed through this Fe^{III}-MA transporter, then this yeast will be rescued. As a result, however, the transporter was not cloned successfully using this screening method. Instead, we cloned a novel gene designated the gene *SFD1* (Suppressor of Ferric uptake Defect). This novel gene may be involved in Cu homeostasis in plant cells. This gene is deposited in DDBJ (AB006591). Next, we will try to clone the Fe^{III}-MA transporter from graminaceous plants after constructing yeast double mutants such as (Δ*fre1* Δ*fre2*), (Δ*fet3* Δ*fet4*), and (Δ*fre1* Δ*ftr1*) or a triple mutant (Δ*fre1* Δ*ftr1* Δ*fet4*).

2. Transgenic Plants

The foliar application of iron chelators such as Fe-EDTA or Fe-EDDHA has been recommended to cure "Fe-chlorosis" (Chen 1997). These chemicals are, however, very expensive for use by farmers. Until now, however, no effective iron fertilizers have been invented for basal application, and the best alternative technology has been to develop tolerant cultivars by conventional methods involving crossbreeding and grafting. Now plant biotechnology may be able to solve this problem as an alternative technology, since the Strategy-I and Strategy-II iron acquisition mechanisms have been clarified and genes relating to these mechanisms are now being cloned.

2.1 Introduction of the Refre1 Gene into Tobacco

The ferric reductase gene from plants has not yet been cloned, so we introduced the yeast *FRE1* gene into tobacco plants (*Nicotiana tabacum* L. cv. SR1). However, the transgenic tobacco showed no additional reductase activity, because the *FRE1* transcripts from these transgenic tobaccos were shorter than expected. Further investigation revealed that polyadenylation had occurred in the coding region of the introduced *FRE1* gene product. We then reconstructed the entire sequence of the *FRE1* gene by the PCR method and named it *refre1* (Reconstructed *FRE1*). We introduced the *refre1* gene into tobacco plants. The transgenic tobaccos carrying the *refre1* gene produced full-length mRNA and showed constitutive ferric reductase activity under Fe-sufficient conditions (Fig. 12) (Oki et al. 1999). Under Fe-deficient conditions, however, wild type tobacco plants had ferric reductase activity that was ten times higher than that of the transgenic plants. Therefore, in order to produce an enhanced variety of tobacco that is tolerant to Fe-deficiency a more potent promoter than the CaMV35S promoter should be ligated in front of the *refre1* gene. Alternatively, other unknown factors, such as the amount of NADPH, should be increased to act as a cofactor for the full function of the *refre1* gene.

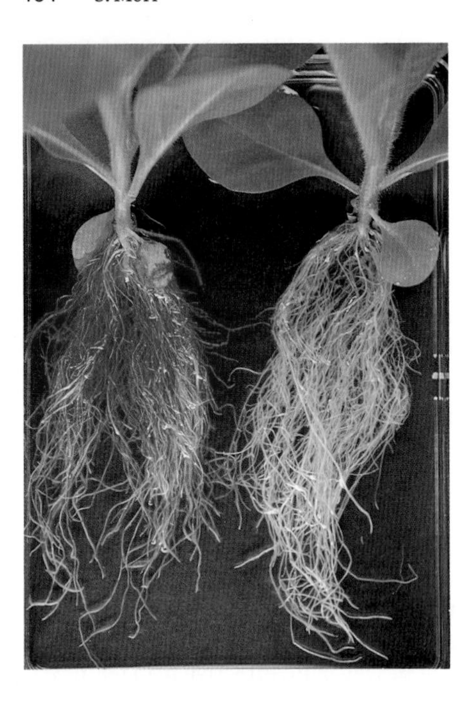

Fig. 12. Fe^{III}-reductase activity was obtained in the transgenic tobacco introduced with *refre 1* gene (*left*) under Fe-sufficient conditions. The *red* color is Fe^{II}-bathophenathroline disulfonate (BPDS)

Although the Fe^{II}-transporter has been cloned (Eide et al. 1996), no one has reported transgenic plants into which this gene has been introduced and overexpressed. The ATPase gene that is induced by Fe-deficiency has not yet been cloned (Fig. 2). The development of transgenic plants containing these genes remains for future study.

2.2 Introduction of Strategy-II Genes into Rice

In many graminaceous species, the ability to secrete total MAs under Fe-deficiency is reported to be correlated with the tolerance of Fe-deficiency. The order of Fe-deficiency tolerance is barley > rye > oats > wheat > sorghum = corn = rice (Römheld 1987). Transgenic plants that produce high amounts of MAs are expected to tolerate Fe-deficiency on calcareous soils.

Since we were unable to clone Fe-deficiency inducible *sam* genes from barley, the activity of SAM synthetase is thought to be sufficient to fulfill the demand for SAM in the production of MAs under Fe-deficiency. Moreover, the phenotype of a transgenic *Arabidopsis* that overexpressed the *sam* gene was sometimes "chlorotic" (Boerjan et al. 1994). This implies that suppression or gene silencing of the *sam* gene had occurred and therefore the amount of nicotianamine was reduced, resulting in the inhibition of iron transport in the plants, causing chlorosis. In dicots, nicotianamine is supposed to mediate iron transport in the phloem (Stephan and Schulz 1993). Therefore, we did not try to make transgenic plants that overexpress the *sam* gene(s).

On the other hand, the introduction of one or both of the *nas* (*nas1, 2, 3, 4, 6*) or *naat* genes (*naat-a* or *naat-b*) into an Fe-deficiency susceptible cultivar of graminaceous plants is a very important idea, because those genes are strongly induced by Fe-deficiency stress (Fig. 13B, C), and their enzyme activity is highly correlated with the amount of MAs secreted (Fig. 13A). The expression of these genes might be one of the limiting factors for the biosynthesis of MAs.

Making transgenic graminaceous plants with a strong Fe^{III}-MA transporter gene (if the transporter gene can be cloned in future) is also an interesting idea. If Fe^{III}-MAs are formed on the root surface just after secretion of MAs and then immediately absorbed through the introduced potent Fe^{III}-MA transporter, the plants might be tolerant to Fe-deficiency. Usually, more than 100 times as much as MAs are secreted from the roots each day than the number of Fe^{3+} ions that must be absorbed each day. Therefore, if all the secreted MAs are efficiently uptaken by the powerful transporter without loss, it is not necessary for the plants to increase the secreted amount of MAs. In plants like sorghum, maize, and rice in which the potency of the secreted MAs is low, transgenic plants overexpressing the Fe^{III}-MAs transporter gene could be a new cultivar tolerant to Fe-deficiency even if the secreted amount of MAs is not increased. In this case, overexpressing only Fe^{III}-MAs transporter gene is sufficient to compensate for the low amount of secreted MAs.

Among graminaceous species, rice is the most important food grain in developing countries, where populations are increasing rapidly . Recently, successful ways to develop "*Agrobacterium*-mediated" transgenic rice have been advanced considerably by Hiei et al. (1994). Therefore, we are focusing on a way to introduce useful genes that are specifically expressed in the roots of Fe-deficient plants into rice. Since barley is the cultivar most tolerant of Fe-deficiency, barley *nas*, *naat*, or Fe^{III}-MA transporter genes may have strong promoters in the roots that are triggered by Fe-deficiency. We are now going to introduce 10-20 kb barley genomic DNA including those genes directly into the rice genome, using a large capacity binary vector developed by Thunoda et al. (1998). This vector is recommended for the easy stabilization of the transgenic genome into the host genome, because of the genomic synteny between rice and barley. To get transgenic rice that produces a higher amount of MAs than a single gene introduction, we are also attempting to introduce the genes that are linearly combined with "*nas* and *naat*" or "*nas*, *naat* and *Ids3*".

3. Conclusion

Based on the Strategy-I and Strategy-II mechanism hypotheses for Fe acquisition in plants, many advances have been made during the past two decades, from the practical field level to the molecular level. These advances, however, are still insufficient for the development of new transgenic cultivars that will not develop Fe-chlorosis when grown on calcareous soils.

In particular, more intensive attempts to clone the Strategy-I genes should be made. The Strategy-II genes should be studied in finer detail at the genomic DNA level. The promoter region for Fe-deficiency-specific and root-specific regula-

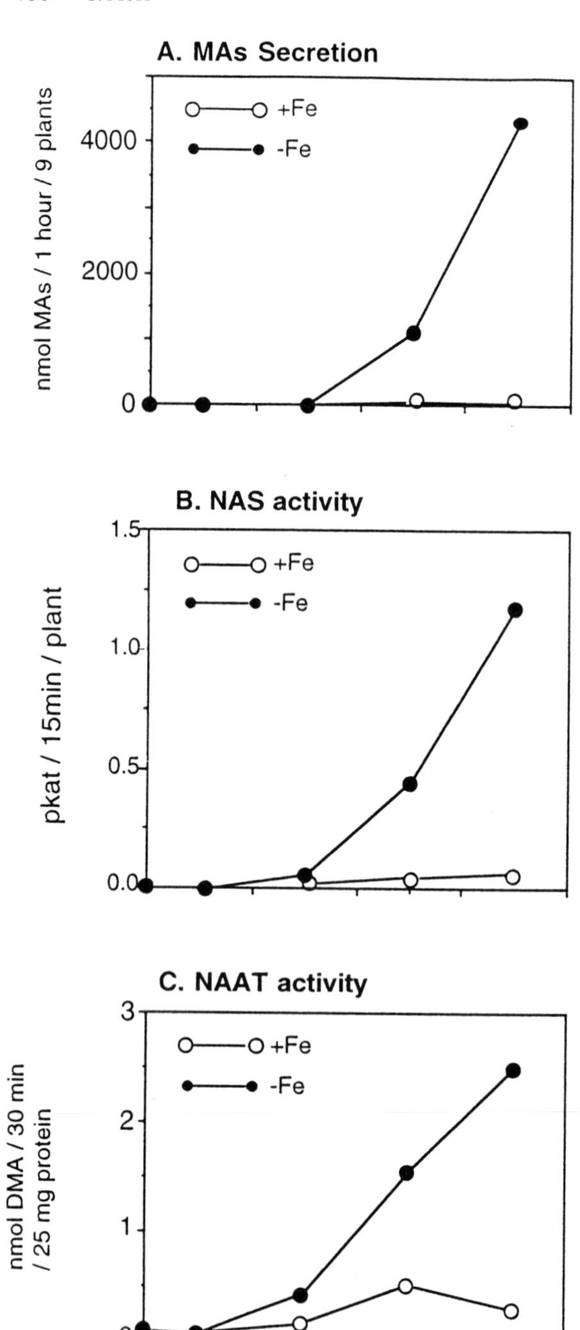

Fig. 13. Time-course study of MAs secretion (**A**), NAS activity (**B**), and NAAT activity (**C**) under Fe-deficiency treatment

tion of the *naat*, *nas*, and *Ids3* genes needs to be understood in detail to allow the effective expression of these genes in transgenic plants. Moreover, the trans factors that regulate Fe-deficiency-specific cis-element(s) have not yet been cloned. Manipulation of these trans factors may be another way to make Fe-deficiency-tolerant transgenic plants, as demonstrated by the example of the overexpression of a transfactor of cold-stress-induced genes, *CBF1*, inducing freezing tolerance in *Arabidopsis* (Jagro-Ottsen KR et al. 1998). Cloning the Fe^{III}-MA transporter gene may pave the way to identifying new types of transporter genes that have been acquired specifically in the evolution of graminaceous monocots.

Addendum. During the editing of this book, our work advanced to generate transgenic rice tolerant to iron deficiency in calcareous soil, as reported in Takahashi M, Nakanishi H, Kawasaki S, Nishizawa NK, Mori S (2001). Enhanced tolerance of rice to low iron availability in alkaline soils using barley nicotianamine aminotransferase genes. Nature Biotechnol 19:466–469.

Acknowledgments. This work was supported by the Ministry of Education, Science and Culture as a Specific Research Project on Priority Areas, entitled "Perspectives of an Advanced Technologied Society". The work also was supported by CREST (Core Research for Evolutional Science and Technology) of the Japan Science and Technology Corporation(JST).

References

Boerjan W, Bauw G, Montagu MV, Inze D (1994) Distinct phenotypes generated by overexpression and suppression of S-adenosyl-methionine synthetase reveal developmental patterns of gene silencing in tobacco. Plant Cell 6:1401-1414

Chaney RL, Brown JC, Tiffin LO (1972) Obligatory reduction of ferric chelates in iron uptake by soybeans. Plant Physiol 50:208-213

Chen Y (1997) Remedy of iron deficiency - present and future. 9th International Symposium on Iron Nutrition and Interactions in Plants. 20-25, July 1997, University Hohenheim, Stüttgart, p 51

Dancis A, Klausner RG, Anderson GJ, Hinnebush AG, Barriocanal JC (1990) Genetic evidence that ferric reductase is required for iron uptake in *Saccharomyces cerevisiae*. Mol Cell Biol 10:2294-2301

Dancis A, Roman DG, Anderson GJ, Hinnebusch AG, Klausner RD (1992) Ferric reductase of *Saccharomyces cerevisiae*: molecular characterization, role in iron uptake, and transcriptional control by iron. Pro Natl Acad Sci USA 89:3869-3873

Dix DR, Bridgham JT, Broderius MA, Byerdorfer CA, Eide DJ (1994) The *Fet4* gene encodes the low affinity Fe(II) transport protein of *Saccharomyces cerevisiae*. J Biol Chem 269:26092-26099

Eide D, Broderius M, Fett J, Guerinot ML (1996) A novel iron-regulated metal transporter from plants identified by functional expression in yeast. Proc Natl Acad Sci USA 93:5624-5628

Gunshin H, Mackenzie B, Berger UV, Gunshin Y, Romero MF, Boron WF, Nussgerger S, Gollan JL, Hediger MA (1997) Cloning and characterization of a mammalian proton-coupled metal-ion transporter. Nature 388:482-488

Hayashi H, Chino M (1986) Collection of pure phloem sap from wheat and its chemical composition. Plant Cell Physiol 27:1387-1393

Hayashi H, Chino M (1990) Chemical composition of phloem sap from the uppermost internode of the rice plant. Plant Cell Physiol 31:247-251

Hiei Y, Ohta S, Komari T, Kumashiro T (1994) Efficient transformation of rice (*Oryza sativa* L.) mediated by *Agrobacterium* and sequence analysis of the boundaries of the T-DNA. Plant J 6:271-282

Higuchi K, Kanazawa K, Nishizawa NK, Chino M, Mori S (1994) Purification and characterization of nicotianamine synthase from Fe deficient barley roots. Plant and Soil 176:173-179

Higuchi K, Kanazawa K, Nishizawa NK, Mori S (1996) The role of nicotianamine synthase in response to Fe nutrition status in Gramineae. Plant and Soil 178:171-177

Higuchi K, Suzuki K, Nakanishi H, Yamaguchi H, Nishizawa NK, Mori S (1999) Cloning of nicotianamine synthase genes, novel genes involved in the biosynthesis of phytosiderophores. Plant Physiol 119:471-480

Holden MJ, Luster DG, Chaney RL, Buckout TJ, Robinson C (1991) Fe^{3+}-chelate reductase activity of plasma membranes isolated from tomato (*Lycopersicon esculentum* Mill) roots. Plant Physiol 97:537-544

Inoue K, Hiradate S, Takagi S (1993) Interaction of mugineic acid with synthetically produced iron oxides. Soil Sci Soc Am J 57:1254-1260

Itai R, Suzuki K, Yamaguchi H, Nakanishi H, Nishizawa NK, Yoshimura E, Mori S (2000) Induced activity of adenine phosphoribosyltransferase (APRT) in iron deficient barley roots: a possible role for phytosiderophore production. J Exp Bot 51:1179-1188

Jaglo-Ottosen KR, Gilmour SJ, Zarka DG, Schabenberger O, Thomashow MF (1998) Arabidopsis CBF1 overexpression induces COR genes and enhances freezing tolerance. Science 280:104-106

Kanazawa K, Higuchi K, Nishizawa NK, Fushiya S, Mitsuo C, Mori S (1994) Nicotianamine aminotransferase activities are correlated to the phytosiderophore secretions under Fe-deficient conditions in Gramineae. J Exp Botany 45:1903-1906

Kanazawa K, Higuchi K, Fushiya S, Nishizawa NK, Chino M, Mori S (1995a) Induction of two enzyme activities involved in the biosynthesis of mugineic acid in Fe-deficient barley roots. In: Abadia J (Ed) Iron nutrition in soils and plants. Kluwer, Dordrecht, pp 37-41

Kanazawa K, Higuchi K, Nishizawa NK, Fushiya S, Mori S (1995b) Detection of two distinct isozymes of nicotianamine aminotransferases in Fe-deficient barley roots. J Exp Botany 46:1241-1244

Kobayashi T, Nakanishi H, Takahashi M, Kawasaki S, Nishizawa NK, Mori S (2001) In vivo evidence that *Ids3* from *Hordeum vulgare* encodes a dioxygenase that converts 2'-deoxymugineic acid to mugineic acid in transgenic rice. Planta 212:864-871

Loeppert R, Wei LC, Ocumpaugh W (1995) Variations in the strategy of iron uptake in plants - importance of the Fe-deficiency induced H^+-pump. International Conference on BioIron (ICBI), April 16-21, 1995, Asheville, North Carolina, p 22

Ma JF, Nomoto K (1994) Incorporation of label from ^{13}C, ^{2}H-, and ^{15}N-labeled methionine molecules during the biosynthesis of 2'-deoxymugineic acid in roots of wheat. Plant Physiol 105:607-610

Ma JF, Shinada T, Matsuda C, Kyosuke N (1995) Biosynthesis of phytosiderophores, mugineic acid, associated with methionine cycle. J Biol Chem 270:16549-16554

Marschner H, Römheld V, Kissel M (1986) Different strategies in higher plants in mobilization and uptake of iron. J Plant Nutr 9:695-713

Mihashi S, Mori S (1989) Characterization of mugineic acid-Fe transporter in Fe-deficient barley roots using multi-compartment transport box method. BioMetals 2:146-154

Mori S (1981) A new continuous flow monitoring system for radioactive amino acids. Agrc Biol Chem 45:1880-1881

Mori S (1995) A strategy for cloning the genes in the synthetic pathway of mugineic acid-family phytosiderophores. In: Johansen C, Lee KK, Sharma KK, Subbarao GV, Kueneman EA (Eds) Genetic manipulation of crop plants to enhance integrated nutrient management in cropping systems. - Phosphorus: Proceedings of an FAO/ICRISAT Expert Consultancy Workshop, 15-18 Mar 1994, ICRISAT Asia Center, India, pp 129-144

Mori S, Nishizawa N (1987) Methionine as a dominant precursor of phytosiderophores in Gramineae plants. Plant Cell Physiol 28:1081-1092

Mori S, Nishizawa N (1989) Identification of barley chromosome no.4, possible encoder of genes of mugineic acid synthesis from 2'-deoxymugineic acid using wheat-barley addition lines. Plant Cell Physiol 30:1057-1060

Mori S, Nishizawa N, Fujigaki K (1990) Identification of rye chromosome 5R as a carrier of the genes for mugineic acid and related compounds. Jpn J Genet 65:343-352

Nakanishi H, Okumura N, Umehara Y, Nishizawa NK, Chino M, Mori S (1993) Expression of a gene specific for iron deficiency (*Ids3*) in the roots of *Hordeum vulgare*. Plant Cell Physiol 34:401-410

Nakanishi H, Bughio N, Matsuhashi S, Ishioka NS, Uchida H, Tsuji A, Osa A, Sekine T, Kume T, Mori S (1999) Visualizing real time [^{11}C] methtonine translocation in Fe-sufficient and Fe-deficient barley using a positron emitting tracer imaging system (PETIS). J Exp Bot 50:637-643

Neilands JB (1981) Iron absorption and transport in microorganisms. Annu Rev Nutr 1:27-46

Nomoto K, Sugiura Y, Takagi S (1987) Mugineic acids, studies on phytosiderophores. In: Winkelman G, van der Helm D and Neilands JB (Eds) Iron transport in microbes, plants and animals. VCH, Weinheim New York, pp 401-425

Ohata T, Mihashi S, Nishizawa NK, Fushiya S, Nozoe S, Chino M, Mori S (1993) Biosynthetic pathway of phytosiderophores in iron-deficient gramineous plants. Soil Sci Plant Nutr 39:745-749

Ohwaki Y, Sugahara K, Suzuki K, Moriyama H, Fukuhara T, Nitta T (1997) Iron stress response in chick pea (*Cicer arietinum* L): morphological changes and levels of H^+-ATPase mRNA in root tips. In: Ando T et al. (Eds) Plant nutrition for sustainable food production and environment. Kluwer, Dordrecht, pp 261-262

Oki H, Yamaguchi H, Nakanishi H, Mori S (1999) Introduction of the reconstructed yeast ferric reductase gene, *refre1*, into tobacco. Plant Soil 215:211-220

Okumura N, Nishizawa NK, Umehara Y, Mori S (1991) An iron deficiency specific cDNA (*Ids1*) with two homologous systein-rich domains. Plant Mol Biol 17:531-533

Okumura N, Nishizawa NK, Umehara Y, Ohata T, Nakanishi H, Yamaguchi T, Chino M, Mori S (1994) A dioxygenase gene (*Ids2*) expressed under iron deficiency conditions in the roots of *Hordeum vulgare*. Plant Mol Biol 25:705-719

Römheld V (1987) Existence of two different strategies for the acquisition of iron in higher plants. In: Winkelman G van Herm G Neilands JB (Eds) Iron transport in microbe, plants and animals. VCH ,Weinhelm, pp 353-374

Römheld V, Marschner H (1983) Mechanisms of iron uptake by peanut plant. I. Fe^{III} reduction, chelate splitting, and release of phenolics. Plant Physiol 71:949-954

Römheld V, Marschner H (1986) Evidence for a specific uptake system for iron phytosiderophore in roots of grasses. Plant Physiol 80:175-180

Shojima S, Nishizawa N, Mori S (1989) Establishment of a cell free system for the biosynthesis of nicotianamine. Plant Cell Physiol 30:673-677

Shojima S, Nishizawa NK, Fushiya S, Nozoe S, Irifune T, Mori S (1990) Biosynthetic pathway of phytosiderophores in iron-deficient gramineous plants. Plant Physiol 93:1497-1503

Stearman R, Yuan DS, Yamaguchi-Iwai Y, Klausner RD, Dancis A. (1996) A permiase-oxidase complex involved in high-affinity iron uptake in yeast. Science 271:1552-1557

Stephan U, Schulz G (1993) Nicotianamine: mediator of transport of iron and heavy metals in the phloem? Physiologia Plantarum 88:522-529

Takagi S (1976) Naturally occurring iron-chelating compounds in oat-and rice-root washing. I. Soil Sci Plant Nutr 22:423-433

Takagi S (1993) Production of Phytosiderophores. In: Barton L, Hemming B (Eds) Iron chelation in plants and soil microorganisms. Academic, pp 111-131

Takagi S, Nomoto K, Takemoto S (1984) Physiological aspect of mugineic acid, a possible phytosiderophore of graminaceous plants. J Plant Nutr 7:469-477

Takahashi M, Yamaguchi H, Nakanishi H, Shioiri T, Nishizawa K, Mori S (1999) Cloning two genes for nicotianamine aminotransferase, a critical enzyme in iron acquisition (strategy II) in graminaceous plants. Plant Physiol 121:947-956

Takizawa R, Nishizawa NK, Nakanishi H, Mori S (1996) Effect of iron deficiency on S-adenosylmethionine in barley roots. J Plant Nutr 19:1189-1200

Thunoda Y, NamSoo JWA, Akiyama K, Sasaki N, Kawasaki S (1998) Construction of a functional high capacity genome library with complementation ability of plant genome in rice. Plant Cell Physiol 39: supplement s116.

von Wirén N, Peltier JB, Rouquie D, Rossignol M, Briat JF (1997) Four root plasmalemma polypeptides under-represented in the maize mutant *ys1* accumulate in a Fe-efficient genotype in response to iron deficiency. Plant Physiol Biochm 35:945-950

Yang SF, Hoffman NE (1984) Ethylene biosynthesis and its regulation in higher plants. Annu Rev Plant Physiol 35:155-189

The Role of Root Exudates in Aluminum Tolerance

Emmanuel Delhaize

Summary. Aluminum is prevalent in soil and under acidic conditions is solubilized into the toxic Al^{3+} form. There are now numerous reports to show that Al stimulates the efflux of organic acids from roots of a range of Al-tolerant plant species. These organic acids are able to chelate Al to render it non-toxic and are therefore implicated in Al-tolerance mechanisms. The efflux of phosphate and the secretion of proteins from root tips have also been implicated in Al-tolerance mechanisms of some species. In this chapter I discuss the literature regarding the various compounds that are secreted from roots and their role in Al tolerance. Recent research is moving on from the initial observation that Al activates or induces organic acid efflux towards a better understanding of the mechanisms occurring at the cellular level.

Key words. Acid-soils, Aluminum, Organic acids, Tolerance, Toxicity, Transport

1. Introduction

Soils that are naturally acid or have become acid through the activities of humans account for large areas of the earth's arable land (Von Uexküll and Mutert 1995). Generally it is not the acidity itself that poses a problem for growth of plants but the metals that are released by the acidity. Aluminum solubilized into the trivalent form (Al^{3+}) under the acidic conditions represents a major reason for poor growth on these soils. The primary effect of Al^{3+} is to inhibit root growth with subsequent deleterious effects on uptake of water and nutrients (Delhaize and Ryan 1995; Kochian 1995). The loss of production due to acid soils in Australia is considerable with both pastures and crops being affected. The management of acid soils in agriculture involves a combination of liming to neutralize the acidity, the use of Al-tolerant plants and the implementation of agricultural practices that reduce inputs of acid. In the Australian context the application of lime in itself is not a feasible strategy to reduce the effects of soil acidity. Because of the large areas involved and the low input nature of some aspects of Australian agriculture, the application of sufficient lime to correct soil acidity is not economical in many regions. Furthermore, when acidity occurs at depth, lime takes many years to move down the soil profile to neutralize the acidity. For these reasons Al-tolerant plants are an important component for managing acid soils.

There is great variability between and within plant species in Al tolerance and this has been exploited in some cases to produce cultivars for improved growth on acid soils. For example, in Australia there are successful breeding programs that have incorporated Al-tolerance genes into wheat (*Triticum aestivum* L.; Fisher and Scott 1993) and phalaris (*Phalaris aquatica* L.; Culvernor et al.1986). For these species there is sufficient Al tolerance in the existing range of genotypes to allow relatively easy incorporation of Al tolerance into elite cultivars. Selection for Al tolerance has been based on root growth in the presence of Al or on the hematoxylin stain (see below). However, in some other species such as lucerne (*Medicago sativum* L.) and rape (*Brassica napus* L.), breeding programs aimed at enhancing Al tolerance have been less successful, reflecting the limited range of Al tolerance in the germplasm of these species. For these species enhanced Al tolerance might be accomplished by genetic engineering and much of the research on Al tolerance in plants has its ultimate aim of identifying and isolating genes that would be useful for this purpose. The identification and isolation of Al-tolerance genes is not restricted to using germplasm of related species. To illustrate this point, recent exciting work has identified a gene from a bacterium with the potential of increasing Al tolerance of plant species used in agriculture (de la Fuente et al. 1997). Recent progress in plant transformation now makes it possible to introduce and express foreign genes into many crop and pasture species illustrating the generic nature of some potential Al-tolerance genes.

There has been considerable research undertaken to understand the mechanisms of Al tolerance in plants. Aluminum tolerance mechanisms can be broadly divided into those that keep Al out of cells and those that detoxify Al internally. Root exudates containing ligands that bind Al could clearly contribute to the former category. To date the most convincing evidence points to mechanisms based on the secretion of compounds from roots that bind Al to render it non-toxic. For the most part these compounds are organic acids but there is also evidence implicating the efflux of phosphate and the secretion of proteins contributing to Al tolerance. It has been known for some time that organic acids added to the external medium can prevent Al toxicity (Bartlett and Riego 1972; Hue et al. 1986) or reverse its effects (Ownby and Popham 1989). The ability of different organic acids to chelate Al varies considerably. A study by Hue et al. (1986) found that, of the common organic acids, citric, oxalic and tartaric were the most effective in protecting cotton roots from Al toxicity. Malic, malonic and salicylic acids were of moderate effectiveness while succinic, lactic, formic, acetic, and phthalic acids provided little protection against Al toxicity. The effectiveness of organic acids in protecting roots was related to the relative positions of OH/COOH groups on the main carbon chain. Organic acids able to form stable 5- or 6-bond ring structures (Fig. 1) with Al provide the best protection. In this chapter I arrange the various Al-tolerance mechanisms based on the identity of the secreted substance.

2. Malate

Wheat shows a large range in Al tolerance between genotypes and for this reason has been the focus of studies on Al tolerance in many laboratories. The genetics of Al tolerance in wheat has been well studied and in several genotypes much of the tolerance can be accounted for by one or a few major genes (Aniol and Gustafson 1984; Berzonsky and Kimber 1989; Kerridge and Kronstad 1968). Because of the existence of major dominant genes in wheat germplasm, it represents a potential source for the isolation of Al-tolerance genes and their subsequent introduction into other species that lack Al tolerance. As a precursor to isolating Al-tolerance genes, much work has been aimed at understanding the mechanisms of Al tolerance in wheat. The development of near-isogenic and isogenic lines that differ in Al tolerance at a single dominant locus, consistent with the presence of a single gene, provided a simplified system for physiological, biochemical and molecular studies (Fig. 2; Delhaize et al.1993a). These lines were derived from genotypes developed in a breeding program and their original purpose was for estimating the economic benefit that a major dominant gene conferring Al tolerance had on wheat yields (Fisher and Scott 1993; Fisher and Scott 1987). The selection of Al-tolerant lines was based on the hematoxylin assay previously developed by Polle et al. (1978). Hematoxylin is a dye that binds Al to develop a reddish-purple color. Sensitive genotypes accumulate Al in the root tips and stain heavily in this region with hematoxylin whereas root tips of toler-

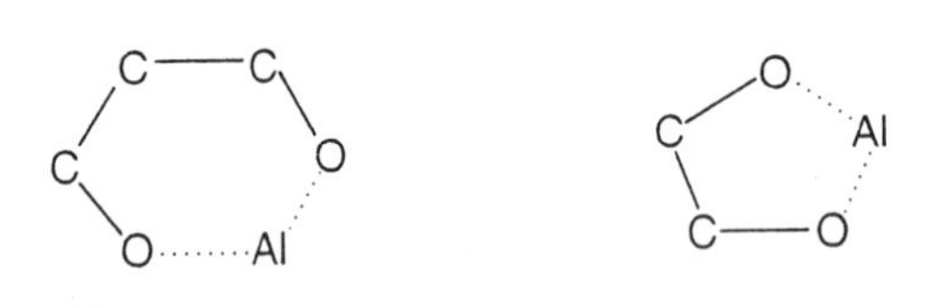

Fig. 1. Organic acids with relative positions of OH/COOH groups that allow the formation of stable 5- or 6-bond ring structures with Al are effective at protecting plants from Al toxicity (Hue et al.1986)

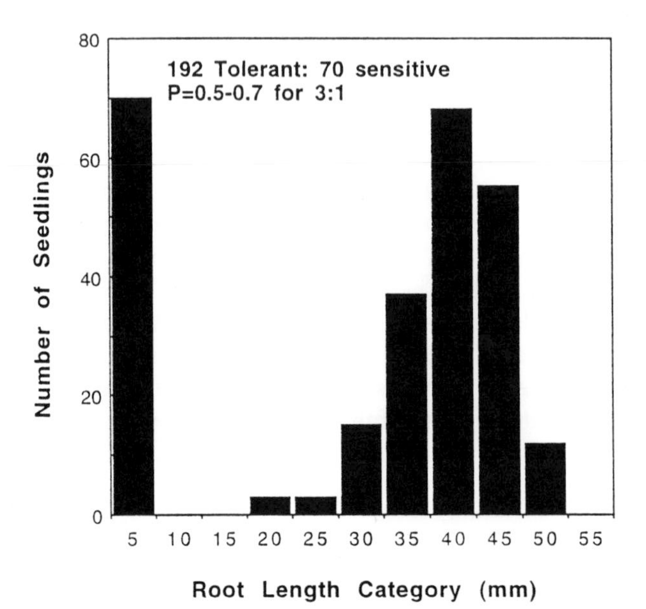

Fig. 2. Al tolerance segregating as a single gene in F2 progeny resulting from a cross between near isogenic wheat lines that differ in Al tolerance. The root length of seedlings of the F2 population grown in 20 μ M Al are shown as categories 5 (0-5mm), 10 (6-10mm), etc. Al sensitive seedlings are denoted by the 0-5mm category while other seedlings able to grow in the solution are designated as been Al-tolerant. From Delhaize et al. (1993a), with permission

ant genotypes remain unstained. Subsequently the effectiveness of the selection strategy was confirmed by field trials that showed enhanced yields of Al-tolerant selections relative to sensitive genotypes when grown on acid soils (Fisher and Scott 1987). Furthermore, it was shown that on non-acid soils, there was no yield penalty from the incorporation of Al-tolerance genes into wheat cultivars.

Earlier work on wheat and other species was hampered by comparisons between cultivars of different genetic backgrounds. Quite apart from the difficulties of comparing cultivars with different attributes such as vigor and root structure, the conclusions drawn from these studies were confounded by the possibility of several different mechanisms contributing to the tolerance. A number of studies provided evidence to show that Al tolerance in wheat is associated with the ability to exclude Al from root apices (Delhaize et al.1993a; Rincón and Gonzales 1992; Tice et al.1992) and it was hypothesized that this may have been related to the secretion of organic acids. Coupled with the observation that the root apex is the part of the plant particularly sensitive to Al toxicity (Ryan et al.1993), this was an attractive hypothesis.

The first evidence to suggest that malate secreted by roots had a role in Al tolerance came from the work of Christiansen-Weniger et al. (1992). In this study, it was found that an Al-tolerant wheat cultivar secreted more malate than an unrelated sensitive cultivar. The malate was secreted without Al being added to the external solution and therefore appeared to be constitutive. This work suffered from the above-mentioned problems of comparing genetically unrelated genotypes but it did provide the impetus for others to investigate the potential of malate as a mechanism for Al tolerance in wheat. Delhaize et al. (1993b) using near-isogenic wheat lines provided strong evidence to indicate that malate efflux was indeed involved in Al tolerance. The key findings from that study were: (i) Al stimulated the efflux of malate from roots of an Al-tolerant genotype and this was 5-10 times greater than the corresponding efflux from the near-isogenic sensitive genotype (Fig. 3), (ii) most of the malate was derived from the terminal 3mm of the root which is the part of the root most susceptible to Al toxicity (Ryan et al.1993) (iii) only Al, and not other trivalent metals or P-deficiency, stimulated the efflux (iv) high malate efflux in seedlings co-segregated with the Al-tolerance phenotype in a population of plants segregating for Al tolerance and (v) malate added to the growth solution protected sensitive seedlings from Al toxicity. In the absence of Al there were no differences in malate efflux which suggests that the experiments of Christiansen-Weniger et al. (1992), where Al was not deliberately added, had small amounts of contaminating Al that stimulated the efflux of malate. Subsequently others found similar differences in malate efflux between wheat cultivars of differing Al tolerance (Basu et al.1994b; de Andrade et al.1997; Pellet et al.1996), and a strong correlation between malate efflux and Al tolerance was found amongst over 30 wheat cultivars that covered a range of Al tolerances (Fig.4; Ryan et al.1995b). These wheat cultivars were derived from diverse sources and this result suggests that genes encoding the malate efflux mechanism accounts for a large proportion of the Al tolerance found in the germplasm of hexaploid wheat. This finding does not preclude the existence of other Al-tolerance mechanisms occurring in wheat and it is possible that other mechanisms, such as the

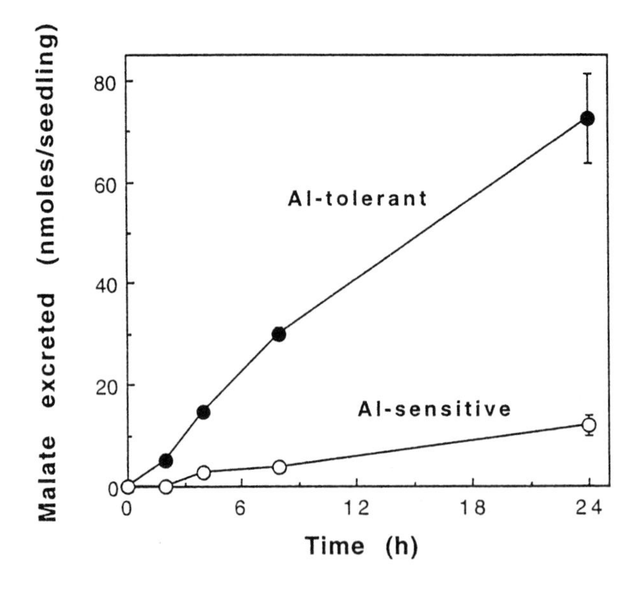

Fig. 3. Aluminum stimulates a 10-fold greater secretion of malate from roots of Al-tolerant than Al-sensitive wheat seedlings. In the absence of Al, both genotypes secreted low amounts of malate. Seedlings were incubated in nutrient solution that contained 50 μM Al and samples of nutrient solution were removed at the indicated times for malate assay (*Vertical bars* denote ± range, *n*=2). From Delhaize et al. (1993b), with permission

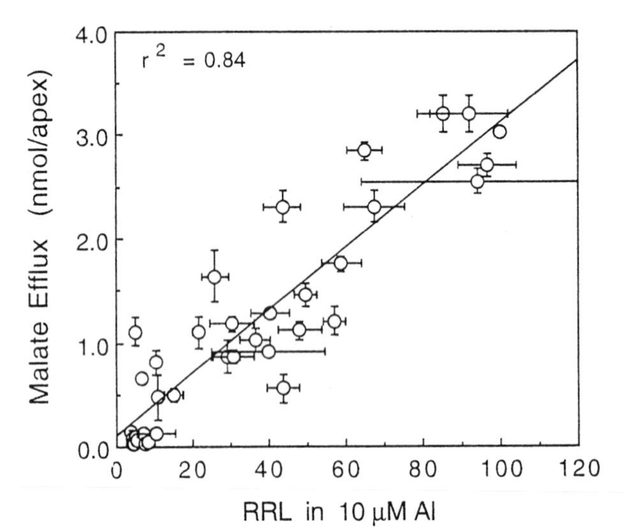

Fig. 4. Aluminum-activated malate efflux and Al tolerance are highly correlated in 36 wheat genotypes. The relationship between relative root length (RRL) of intact seedlings grown in 10 μM Al and malate efflux from excised root apices is shown. Values for RRL show the means and standard errors from 12 to 16 roots. Malate efflux was measured over 80 min in 200 μM $AlCl_3$ with 200 μM $CaCl_2$. The means and range for two to three replicates are shown. From Ryan et al. (1995b), with permission

one discussed below dealing with phosphate efflux, can provide an additional level of tolerance when combined with high malate efflux. The implication of this for breeding programs is that cultivars could be developed that include several mechanisms with additive effects on Al tolerance.

Ryan et al. (1995) used excised root segments to further study the mechanism of malate efflux. This work confirmed that the root apices secreted the bulk of the malate and showed that the rate of secretion was constant with no discernible lag phase after the addition of Al to the external medium (Fig. 5). This result suggested that a pre-existing mechanism for the transport of malate was being activated by Al. Malate exists as a divalent anion in the cytoplasm and its concentration (about 1 mM in the cytoplasm) coupled to a negative membrane potential (typically -150mV) results in a strong electrochemical gradient for malate that is directed out of the cell. This means that malate could flow out of the cell through a malate-permeable anion-channel. Ion channels are proteins found in membranes that allow the transport of specific ions down their electrochemical gradient. In plants, anion channels have been found to be a critical component of many processes that include control of stomatal closure, signal transduction and mineral nutrient uptake (Schroeder 1995). Indirect evidence pointing towards the existence of an anion channel involved in malate efflux came from the use of anion-channel inhibitors that blocked the Al-stimulated efflux of malate (Ryan et al.1995a). If malate flows out of the cell then the movement of negative charge needs to be balanced by an equivalent flow of cations out of the cell or by an equivalent flow of anions into the cell. It was found that K^+ was associated with the flow of malate out of the cell suggesting a role for K^+ channels in balancing charge. Recent electrophysiological work by Papernik and Kochian (1997) showed that the plasmamembrane of cells in root apices of an Al-tolerant wheat genotype was depolarized on exposure to Al. This result is consistent with Al activating the efflux of malate since there is a loss of divalent anion from the cytoplasm to the external medium. Furthermore, the absence of a comparable depolarization across the membrane of an Al-sensitive genotype would appear to provide additional evidence for this interpretation. However, because the Al-induced depolarization was observed in the presence of an inhibitor of malate efflux and the response was also found in mature parts of the root that did not secrete malate, Papernik and Kochian (1997) concluded that the depolarization was probably not due to the transport of malate out of the cell and that some other explanation must be sought. One suggestion put forward by the authors is that the depolarization is the initial step in a signal transduction pathway that is involved in activating malate efflux (see Fig. 6 and below).

The difference in malate efflux between wheat genotypes could be due to differences in biosynthesis of malate or transport of malate out of the cell or a combina-

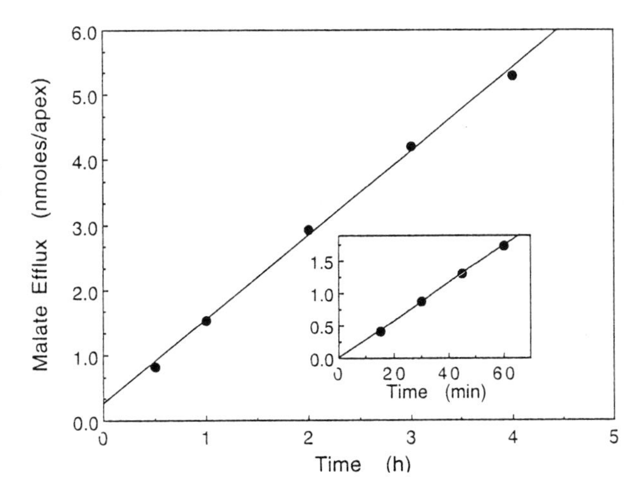

Fig. 5. Aluminum-stimulated efflux of malate from excised root apices of an Al-tolerant wheat genotype (*Vertical bars*, ±SE, n=3). The excised root apices were washed in 200 µM $CaCl_2$ before being exposed to 200 µM Al in 200 µM $CaCl_2$. The *inset* shows malate release over the initial 60 min. From Ryan et al. (1995a), with permission

tion of both factors. Work by Basu et al. (1994b) and de Andrade et al. (1997) using ^{14}C labeled acetate showed that *de novo* synthesis of malate occurred in roots of Al-tolerant genotypes during Al exposure. Basu et al. (1994b) concluded that differences in malate efflux were likely to be due to differences in capacity to synthesize malate. By contrast, Ryan et al. (1995a) showed that there were no differences in the activity of malate dehydrogenase and phosphenolpyruvate carboxylase, two key enzymes involved in malate biosynthesis, between root apices of Al-tolerant and Al-sensitive genotypes. This indicated that although the Al-tolerant genotype synthesized more malate during exposure to Al, both genotypes had equal capacity for malate synthesis and the difference was more likely to be found in transport of malate across the plasmamembrane. It is known that internal malate concentrations are tightly regulated and the loss of cytoplasmic malate by Al-activated transport is likely to have triggered increased synthesis of malate.

Direct evidence for the existence of an Al-activated anion channel was obtained using the patch-clamp technique (Ryan et al.1997). Several properties of an Al-activated anion-channel found in root apices of Al-tolerant wheat are similar to the properties of malate efflux observed in intact root apices. These properties include: (i) the channel is specifically activated by Al and not by La, (ii) the channel is found in protoplasts derived from root apical cells and not in protoplasts derived from mature parts of the root and, (iii) compounds that inhibit malate efflux in intact root tips also inhibit the channel. However, one apparent inconsistency is the finding that protoplasts derived from both Al-tolerant and Al-sensitive wheat genotypes show an Al-activated anion channel (Ryan et al. personal communication) whereas the amount of malate secreted from intact root apices differs markedly between the genotypes. The patch-clamping data were obtained using chloride as the anion transported and it is possible that the Al-activated

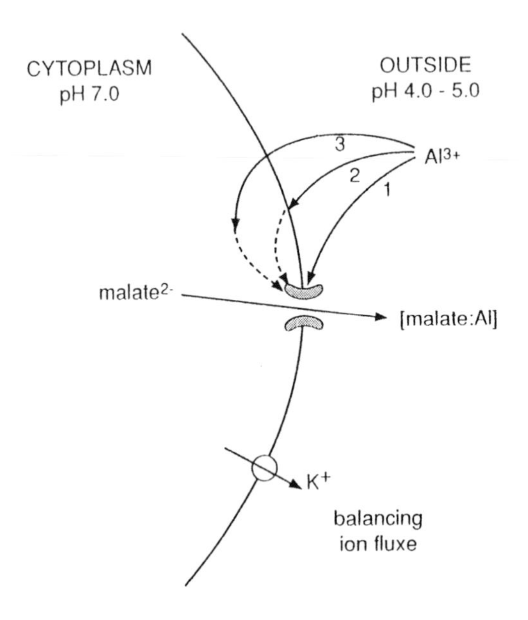

Fig. 6. A hypothetical model showing how Al^{3+} might interact with a malate-permeable channel (*hatched structure*) in the plasma membrane of Al-tolerant root apices of wheat to activate malate efflux. Al could interact directly with the channel (*1*) or use signal transduction pathways to activate the channel indirectly (*2* and *3*). Papernik and Kochian (1997) recently identified Al-induced depolarizations of the plasma membrane from Al-tolerant wheat which may be involved as an initial step in a signal transduction pathway (for example, pathway *2*) which activates malate efflux. From Delhaize and Ryan (1995), with permission

anion-channels of the different genotypes have different malate permeabilities. Further work will be needed determine the relative permeabilities to malate of these channels found in the different wheat genotypes. Figure 6 shows a model summarizing properties of the Al-activated efflux of malate at the cellular level. Aspects of the model remain hypothetical such as the mechanism by which Al activates the channel. Data from patch clamp studies indicate that Al may be acting through secondary messengers to trigger opening of the channel (Ryan et al.1997). If secondary messengers are involved in a signal transduction pathway from Al to channel gating, then this implies the existence of an Al-specific receptor as the primary activator of the pathway. As noted above, a depolarization of the membrane by Al in Al-tolerant wheat may represent an initial component of a signal transduction pathway.

Despite the mounting evidence in support of malate having a role in the Al tolerance of wheat, some researchers remain skeptical. Calculations that estimate the concentration of malate at the membrane surface of apical root cells indicate that there may be only a tenth of the malate needed to detoxify a given amount of Al (Ryan et al.1995b). However, other calculations yield concentrations for malate that are of the same order expected for malate to confer Al tolerance (Pellet et al.1997). Recent work by Parker and Pedler (1998) using La and Cu as "proxy" ions in combination with Al also cast some doubt on the malate hypothesis for Al tolerance in wheat. The idea behind these experiments is that if malate is capable of protecting roots from metals, then an Al-tolerant genotype should show an enhanced tolerance to the proxy ion when added in combination with Al. By contrast an Al-sensitive genotype should not show a comparable tolerance to the proxy ion because it is not secreting malate. Since there was no cross-tolerance conferred to the ions by the Al-stimulated efflux in the tolerant genotypes, the authors concluded that there is insufficient malate to be effective in conferring tolerance. However, other interpretations of these data are possible that do not conflict with the hypothesis that malate efflux confers Al tolerance and additional evidence will be needed to contradict the mounting evidence for organic acids in Al-tolerance mechanisms.

3. Citrate

Of the common organic acids found in plants, citrate is one of the most effective chelators of Al. Citrate secretion as an Al-tolerance mechanism was implicated in several early studies using cell cultures (Ojima et al.1984, 1989) and intact plants (Miyasaka et al.1991). Ojima et al. (1984) showed that cell cultures of carrot (*Daucus carota* L.) selected for enhanced Al tolerance secreted more citrate than a corresponding sensitive line and similar results were found for tobacco (*Nicotiana tabacum* L.) cell lines differing in Al tolerance (Ojima et al.1989). Miyasaka et al.(1991) showed that an Al-tolerant cultivar of snapbean (*Phaseolus vulgaris* L.) secreted 10-fold more citrate than an Al-sensitive cultivar. In retrospect, these studies were probably not given the credence that they deserved in view of the current data providing support for citrate efflux as an Al-tolerance mechanism.

The experiments were conducted over several days where it was difficult to unravel cause from effect and there were uncertainties regarding the P-nutrition of these cultures. It is well known that P-deficiency induces the efflux of organic acids from roots of many species (discussed in other chapters) and because Al can form a poorly-soluble precipitate with phosphate, the high Al concentrations used in the experiments may have inadvertently resulted in P-deficiency.

Convincing data to show a role for citrate in Al tolerance came from the work of Pellet et al. (1995) on maize (*Zea mays* L.). In this work, an Al-tolerant maize cultivar secreted 10-fold more citrate than a sensitive cultivar over relatively short-term experiments (48 h). In addition, the efflux of smaller amounts of malate and phosphate occurred only from the Al-tolerant genotype. Like the wheat system, the efflux of citrate in maize occurs primarily from the root apex. Unlike the immediate Al-activated response found in wheat for malate efflux (Fig. 5), in maize there is a considerable lag phase before maximal citrate efflux is obtained (Fig.7). This suggests that Al is inducing the secretion of citrate as opposed to activating a pre-existing system. In this chapter, I use the words "induced" or "induction" to denote a mechanism that relies on the synthesis of protein after exposure to Al. This induction may result from enhanced synthesis of proteins involved in citrate biosynthesis or transport of citrate. Jorge and Arruda (1997) subsequently showed, using a different pair of maize cultivars differing in Al tolerance, similar differences in citrate efflux. Furthermore they were able to show that citrate efflux was not a consequence of an Al-induced P-deficiency. In this study they followed the accumulation of Al into root apices with the dye hematoxylin. Hematoxylin binds Al to form a purple compound and has been used to identify Al-tolerant genotypes from Al-sensitive genotypes by the intensity of staining at the root apex (Polle et al. 1978). Root apices of both the Al-tolerant and Al-sensitive genotypes stained with hematoxylin after 24 h exposure to Al indicating accumulation of Al. After a further 48 h while the sensitive genotype still showed heavy staining, the staining on root apices of the tolerant genotype was considerably reduced. The

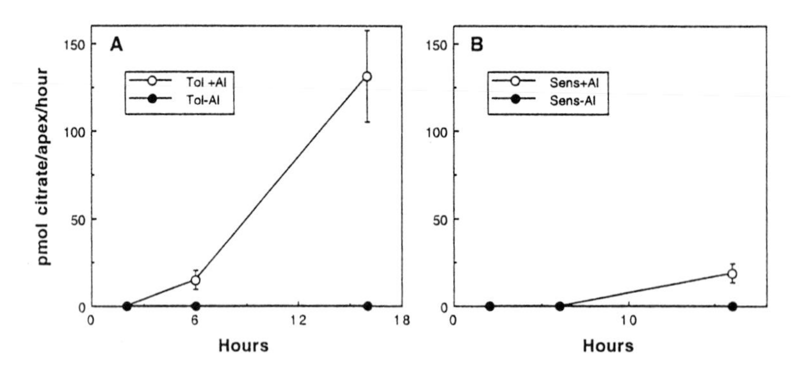

Fig. 7. Time course of Al (9 μM) on citrate efflux from root apices of (**A**) Al-tolerant and (**B**) Al-sensitive maize cultivars (*Vertical bars*, ± SE, *n*=4). Note that the *y-axis* shows a rate of citrate secretion and that this rate increases dramatically after 6 h. By contrast the rates of malate efflux shown in Figs. 3 and 4 are relatively constant over the entire time course. From Pellet et al. (1995), with permission

authors concluded that the reduced staining in the tolerant genotype was a consequence of citrate chelating the Al and reducing its accumulation into the root apex. Curiously, in their study they showed that the Al-tolerant cultivar was also more tolerant to La than the Al-sensitive cultivar. However, they did not present data regarding the ability of La to induce citrate efflux, a question worth investigating in the future.

Ma et al. (1997c) recently showed that Al caused a similar secretion of citrate from roots of Al-resistant *Cassia tora* L. There was a marked lag phase of over 4 h after addition of Al before citrate efflux could be detected (Fig. 8). The authors concluded that this lag phase may represent an induction period as noted above for the maize. Ma et al. (1997c) showed that exposing *C. tora* roots to a 3 h pulse of Al caused citrate efflux over the following 6 h where Al was absent (Fig. 9). This result is once again markedly different to the Al-activated efflux of malate from wheat where efflux responds relatively quickly to either addition or removal of Al (Ryan et al.1995a). The implication of this result is two-fold: (i) it provides further evidence of an induction of citrate secretion by Al and, (ii) it suggests that the mechanism transporting citrate, possibly a citrate-specific anion channel, does not require the continual presence of Al to be active. The secretion of citrate from *C. tora* roots shows a degree of specificity for Al with other trivalent cations or P-deficiency not inducing a comparable secretion. Comparison of *C. tora* with the related, more Al-sensitive species *C. occidentalis*, showed that for a given concentration of Al, *C. tora* secreted 2.5 - 3 times more citrate (Ma et al.1997b).

In some remarkable work, de la Fuente et al. (1997) showed that for some species the synthesis of citrate appears to limit its secretion. In this work they expressed a *Pseudomonas aureginosa* citrate synthase gene in tobacco under the control of the 35S-cauliflower mosaic virus promoter. This is a constitutive promoter that is active throughout the plant. As a result of expressing the citrate synthase gene, the transgenic plants showed enhanced Al tolerance which was associated with an increase in citrate efflux. Greater total citrate synthase activity in the transgenic plants was found with up to 10-fold greater internal citrate concentrations than wild type plants. The cellular location of the bacterial citrate synthase in the transgenic plants was not certain but is likely to be found in the

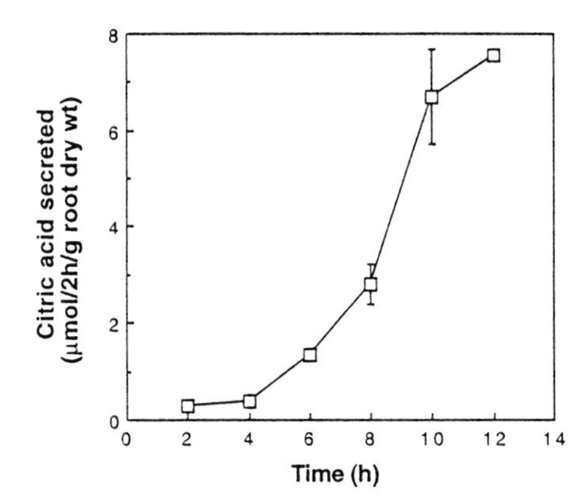

Fig. 8. Time course of citric acid secretion by *Cassia tora* L. roots exposed to 50 μM Al (*Vertical bars*, ± SE, n=3). From Ma et al. (1997c), with permission

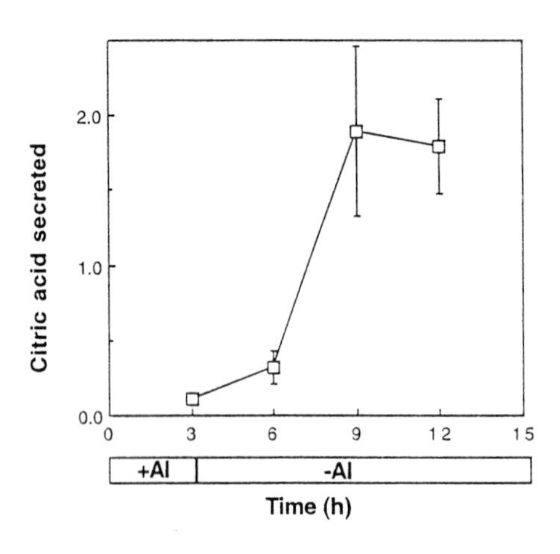

Fig. 9. Effect of a 3-h pulse of 50 μM Al on the secretion of citric acid by *Cassia tora* L. Seedlings were exposed to Al for 3 h and then subsequently to a solution without Al (*Vertical bars*, ± SE, n=3). From Ma et al. (1997c), with permission

cytosol. Citrate synthases endogenous to the tobacco are found in either mitochondria or in glyoxymes. In the mitochondria, citrate synthase plays a key role in the tricarboxylic acid cycle while the glyoxysomal form is involved in the glyoxylate cycle that metabolizes fatty acids. The introduction of a bacterial citrate synthase into the cytosol may have circumvented the usual controls imposed on the endogenous enzymes to limit internal citrate concentrations. An important aspect of this work was the finding that there were no apparent deleterious effects of increasing internal citrate concentrations in the transgenic plants. Similar results were found when the citrate synthase gene was expressed in papaya (*Carica papaya* L.) suggesting that this gene could be used as a general method to enhance the Al tolerance of plants by genetic engineering (de la Fuente et al.1997). The observation that citrate efflux in the absence of Al exposure was increased about 4-fold despite internal concentrations being increased over 10-fold for the highest expressing transgenic, suggests that the transport of citrate out of roots is limiting in these plants. Further work is needed to provide an understanding of how the increased internal citrate concentrations leads to enhanced efflux. As noted above, citrate is likely to be transported out of the cell through an anion channel and in the highest expressing transgenic plants, transport through the channel may have become limiting. In addition, it is possible that an increased internal citrate concentration in itself may confer a degree of Al tolerance. For instance, Ma et al. (1997a) have shown that hydrangea (*Hydrangea macrophylla* L.) accumulates Al as an Al:citrate complex. Presumably Al complexed by citrate provides protection from the potential deleterious effects of ionic Al inside as well as outside the cell.

There is evidence that Al-tolerance mechanisms based on the secretion of organic acids are also present in tree species. Aluminum increased the secretion of citrate from roots of three leguminous trees and this enhanced secretion was not related to P-deficiency stress (Osawa et al. 1997). For one species (*Paraserianthes falcataria* L. Neilson), citrate secretion in the absence of Al was undetectable over 30 days but showed the greatest secretion of citrate in the presence of Al amongst the three species studied. The ability to secrete citrate was correlated with the degree of Al tolerance amongst the three species examined.

4. Oxalate

A recent report has implicated a role for oxalate in Al-tolerance of buckwheat (*Fagopyrum esculentum* Moench). Exposure of roots to Al elicited secretion of oxalate (Ma et al. 1997d) with similar kinetics to the Al-stimulated efflux of malate from wheat roots (Fig. 10). Unlike wheat, buckwheat accumulated high concentrations of Al in leaves and much of this Al was complexed with oxalate. The mechanism for the Al-stimulated efflux of oxalate is not known but the absence of a lag phase suggests that, as hypothesized for wheat, an anion-channel is responsible. The efflux of oxalate and the internal chelation of Al by oxalate are likely to represent two related mechanisms that confer Al tolerance to buckwheat.

5. Exudates Other Than Organic Acids

Pellet et al. (1996) found that Atlas, an Al-tolerant wheat cultivar, in addition to secreting large amounts of malate, also secreted phosphate (Fig. 11). The efflux of phosphate was not activated by Al but occurred continuously from root apices. By contrast, a sensitive wheat genotype did not show comparable phosphate efflux from its root apices. When Pellet et al. (1996) compared the Al-tolerant genotype ET3 with Atlas they found (i) that although ET3 showed Al-stimulated efflux of malate, it did not show phosphate efflux and (ii) that Atlas was more Al-tolerant than ET3. From these observations they concluded that the additional Al tolerance of Atlas may have been due to the efflux of phosphate. Phosphate will complex Al and Pellet et al (1996) showed that increasing the phosphate concentration in the growth medium alleviated the toxicity of Al. The efflux of phosphate as a contributor to Al-tolerance might not be restricted to wheat as Pellet et al. (1995) also found greater efflux of phosphate from root tips of Al-tolerant maize but unlike the wheat, this efflux was Al-stimulated. A detailed study of the likely contributions of malate and phosphate efflux was undertaken by Pellet et al. (1997). In this study they showed that the Al-tolerant cultivar was able to maintain a higher pH in the rhizosphere of its root apices compared to the Al-sensitive genotype in the presence of Al. The combined effects of malate, phosphate and higher pH

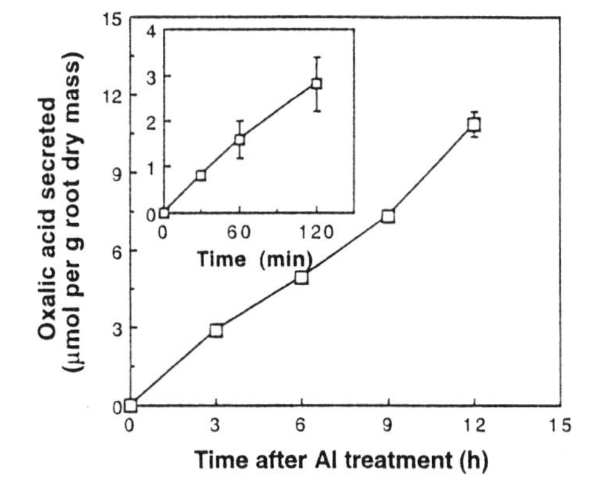

Fig. 10. Aluminum-stimulated secretion of oxalic acid from roots of buckwheat (*Vertical bars*, ±SE, n=3). The *inset* shows the response over the initial 2 h of Al treatment. Note that the *y-axis* shows the cumulative amount of oxalic acid at each time and the rate is therefore approximately constant over 12 h. This is similar to the malate efflux for wheat as shown in Figs. 3 and 4 but differs to citrate efflux from maize (Fig. 6) and *Cassia tora* L. (Fig. 7). From Ma et al. (1997d), with permission

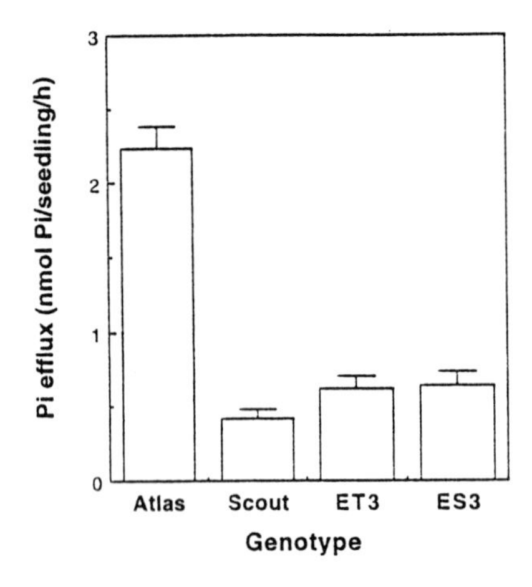

Fig. 11. The constitutive efflux of phosphate from two Al-tolerant (Atlas and ET3) and two Al-sensitive (Scout and ES3) wheat genotypes (*Vertical bars*, ± SE, n=5 to 7). Seedlings were exposed to Al for 7 h but similar results were obtained in the absence of Al. From Pellet et al. (1995), with permission

Table 1. Ability of poly-L-glutamic acid (poly-glu) to protect pollen tube growth from Al toxicity. The poly-glu concentrations are given on the basis of glutamic acid residues. From Konishi et al. (1988)

Treatment	Pollen tube growth (%)
+ 0 μM Al	100
+ 20 μM Al	19.3
+ 20 μM Al + 20 μM poly-glu	23.2
+ 20 μM Al + 40 μM poly-glu	43.3
+ 20 μM Al + 200 μM poly-glu	99.7

could all contribute to the observed Al tolerance of Atlas and this was supported by calculations that estimated the concentrations of these components at the surface of the root apex. These observations are consistent with the genetics of Al tolerance in Atlas. Unlike the near-isogenic lines ET3 and ES3 which were deliberately bred for a single gene difference in Al tolerance, Atlas contains a number of major genes for Al tolerance (Berzonsky 1992). Although there is evidence for phosphate efflux contributing to Al tolerance in laboratory experiments, the role of phosphate efflux in acid soils is less certain. Acid soils are generally low in soluble phosphate so it is more likely that plants will be low in their P-nutrition and would take up rather than secrete phosphate.

Proteins or polypeptides that contain a large proportion of the acidic amino acids glutamate and aspartate can chelate Al (Putterill and Gardner 1988). Poly-L-glutamate and poly-L-aspartate are able to protect germinating pollen from Al toxicity (Table 1; Konishi et al. 1988) demonstrating that, in principle, the exudation of proteins from roots could confer a degree of Al tolerance. Basu et al. (1994a) showed that an Al-tolerant wheat cultivar secreted a range of polypeptides in response to Al. These polypeptides had greater Al-binding capacity than those secreted from a sensitive genotype. From a metabolic point of view it is likely to be more costly to synthesize protein to bind Al rather than to use organic acids or phosphate. In addition the secretion of protein in response to Al is likely to be considerably slower than the secretion of organic acids. However, despite these reservations other factors may be involved which results in exuded protein providing a degree of Al tolerance.

6. Conclusions

There are numerous reports that provide evidence for the hypothesis that plants secrete substances to protect their roots from Al toxicity. The strongest evidence supports a role for organic acids such as malate, citrate and oxalate as well as for the inorganic anion, phosphate. Further work is needed to identify the mechanisms involved in the transport of these substances out of the cell, the mechanisms involved in activating or inducing the efflux of organic acid by Al and the cloning of genes involved in these processes. The expression of a bacterial citrate synthase gene in plants and the corresponding increase in Al tolerance now opens up the possibility of enhancing the Al tolerance of crop and pasture species by genetic modification. The isolation of other genes whose products are involved in transport of organic acids across the plasma membrane is likely to provide the molecular tools to further enhance the Al tolerance of plants.

References

Aniol A, Gustafson JP (1984) Chromosome location of genes controlling aluminium tolerance in wheat, rye, and triticale. Can J Genet Cytol 26:701-705

Bartlett RJ, Riego DC (1972) Effect of chelation on the toxicity of aluminum. Plant Soil 37:419-423

Basu U, Basu A, Taylor GJ (1994a) Differential exudation of polypeptides by roots of aluminum-resistant and aluminum-sensitive cultivars of *Triticum aestivum* L. in response to aluminum stress. Plant Physiol 106:151-158

Basu U, Godbold D, Taylor GJ (1994b) Aluminum resistance in *Triticum aestivum* associated with enhanced exudation of malate. J Plant Physiol 144:747-753

Berzonsky WA (1992) The genomic inheritance of aluminum tolerance in "Atlas 66" wheat. Genome 35:689-693

Berzonsky WA, Kimber G (1989) The tolerance to aluminum of *Triticum* N-genome amphiploids. Plant Breeding 103:37-42

Christiansen-Weniger C, Groneman AF, Van Veen JA (1992) Associative N_2 fixation and root exudation of organic acids from wheat cultivars of different aluminium tolerance. Plant Soil 139:167-174

Culvernor RA, Oram RN, Wood JT (1986) Inheritance of aluminium tolerance in *Phalaris aquatica* L. Aust J Agric Res 37:397-408

de Andrade LRM, Ikeda M, Ishizuka J (1997) Excretion and metabolism of malic acid produced by dark fixation in roots in relation to aluminium tolerance of wheat. In: Ando T, Fujita K, Mae T, Mori S, Sekiya J (Eds) Plant nutrition for sustainable food production and environment. Kluwer Academic, Dordrecht, pp 445-446

de la Fuente JM, Ramírez-Rodríguez V, Cabrera-Ponce JL, Herrera-Estrella L (1997) Aluminum tolerance in transgenic plants by alteration of citrate synthesis. Science 276:1566-1568

Delhaize E, Ryan PR (1995) Aluminum toxicity and tolerance in plants. Plant Physiol 107:315-321

Delhaize E, Craig S, Beaton CD, Bennet RJ, Jagadish VC, Randall PJ (1993a) Aluminum tolerance in wheat (*Triticum aestivum* L.) I. Uptake and distribution of aluminum in root apices. Plant Physiol 103:685-693

Delhaize E, Ryan PR, Randall PJ (1993b) Aluminum tolerance in wheat (*Triticum aestivum* L.) II. Aluminum-stimulated excretion of malic acid from root apices. Plant Physiol 103:695-702

Fisher JA, Scott BJ (1987) Response to selection for aluminium tolerance. In: Searle PGE, Davey BG (Eds) Priorities in soil/plant relations research for plant production. University of Sydney, Sydney, pp 135-137

Fisher JA, Scott BJ (1993) Are we justified in breeding wheat for tolerance to acid soils in southern New South Wales? In: Randall P, Delhaize E, Richards R, Munns R (Eds) Genetic aspects of plant mineral nutrition. Kluwer Academic Dordrecht, pp 1-8

Hue NV, Craddock GR, Adams F (1986) Effect of organic acids on aluminum toxicity in subsoils. Soil Sci Am J 50:28-34

Jorge RA, Arruda P (1997) Aluminum-induced organic acids exudation by roots of an aluminum-tolerant tropical maize. Phytochem 45:675-681

Kerridge PC, Kronstad WE (1968) Evidence of genetic resistance to aluminum toxicity in wheat (*Triticum aestivum* Vill., Host). Agronomy J 60:710-711

Kochian LV (1995) Cellular mechanisms of aluminum toxicity and resistance in plants. Annu Rev Plant Physiol Plant Mol Biol 46:237-260

Konishi S, Ferguson IB, Putterill JJ (1988) Effect of acidic polypeptides on aluminium toxicity in tube growth of pollen from tea (*Camellia sinensis* L.). Plant Sci 56:55-59

Ma JF, Hiradate S, Nomoto K, Iwashita T, Matsumoto H (1997a) Internal detoxification mechanism of Al in *Hydrangea macrophylla*. Identification of Al form in the leaves. Plant Physiol 113:1033-1039

Ma JF, Zheng SJ, Matsumoto H (1997b) Secretion of citrate as an aluminium-resistant mechanism in *Cassia tora* L. In: Ando T, Fujita K, Mae T, Mori S, Sekiya J (Eds) Plant nutrition for sustainable food production and environment. Kluwer Academic, Dordrecht, pp 449-450

Ma JF, Zheng SJ, Matsumoto H (1997c) Specific secretion of citric acid induced by Al stress in *Cassia tora* L. Plant Cell Physiol 38:1019-1025

Ma JF, Zheng SJ, Matsumoto H, Hiradate S (1997d) Detoxifying aluminium with buckwheat. Nature 390:569-570

Miyasaka SC, Buta JG, Howell RK, Foy CD (1991) Mechanism of aluminum tolerance in snapbeans: root exudation of citric acid. Plant Physiol 96:737-743

Ojima K, Abe H, Ohira K (1984) Release of citric acid into the medium by aluminum-tolerant carrot cells. Plant Cell Physiol 25:855-858

Ojima K, Koyama H, Suzuki R, Yamaya T (1989) Characterization of two tobacco cell lines selected to grow in the presence of either ionic Al or insoluble Al-phosphate. Soil Sci Plant Nutr 35:545-551

Osawa H, Kojima K, Sasaki S (1997) Excretion of citrate as an aluminium-tolerance mechanism in tropical leguminous trees. In: Ando T, Fujita K, Mae T, Mori S, Sekiya J (Eds) Plant nutrition for sustainable food production and environment. Kluwer Academic, Dordrecht, pp 455-456

Ownby JD, Popham HR (1989) Citrate reverses the inhibition of wheat root growth caused by aluminum. J Plant Physiol 135:588-591

Papernik LA, Kochian LV (1997) Possible involvement of Al-induced electrical signals in Al tolerance in wheat. Plant Physiol 115:657-667

Parker DR, Pedler JF (1998) Probing the "malate hypothesis" of differential aluminum tolerance in wheat using other rhizotoxic ions as proxies for Al. Planta 205:389-396

Pellet DM, Grunes DL, Kochian LV (1995) Organic acid exudation as an aluminum tolerance mechanism in maize (*Zea mays* L.). Planta 196:788-795

Pellet DM, Papernik LA, Kochian LV (1996) Multiple aluminum-resistance mechanisms in wheat. Plant Physiol 112:591-597

Pellet DM, Papernik LA, Jones DL, Darrah PR, Grunes DL, Kochian LV (1997) Involvement of multiple aluminium exclusion mechanisms in aluminium tolerance in wheat. Plant Soil 192:63-68

Polle E, Konzak CF, Kittrick JA (1978) Visual detection of aluminum tolerance in wheat by hematoxylin staining of seedling roots. Crop Sci 18:823-827

Putterill JJ, Gardner RC (1988) Proteins with the potential to protect plants from Al^{3+} toxicity. Biochim Biophys Acta 1988:137-145

Rincón M, Gonzales RA (1992) Aluminum partitioning in intact roots of aluminum-tolerant and aluminum-sensitive wheat (*Triticum aestivum* L.) cultivars. Plant Physiol 99:1021-1028

Ryan PR, DiTomaso JM, Kochian LV (1993) Aluminium toxicity in roots: an investigation of spatial sensitivity and the role of the root cap. J Exp Bot 44:437-446

Ryan PR, Delhaize E, Randall PJ (1995a) Characterization of Al-stimulated malate efflux from the root apices of Al-tolerant genotypes of wheat. Planta 196:103-110

Ryan PR, Delhaize E, Randall PJ (1995b) Malate efflux from root apices and tolerance to aluminium are highly correlated in wheat. Aust J Plant Physiol 22:531-536

Ryan PR, Skerrett M, Findlay GP, Delhaize E, Tyerman SD (1997) Aluminum activates an anion channel in the apical cells of wheat roots. Proc Natl Acad Sci USA 94:6547-6552

Schroeder JI (1995) Anion channels as central mechanisms for signal transduction in guard cells and putative functions in roots for plant-soil interactions. Plant Molec Biol 28:353-361

Tice KR, Parker DR, DeMason DA (1992) Operationally defined apoplastic and symplastic aluminum fractions in root tips of aluminum-intoxicated wheat. Plant Physiol 100:309-318

von Uexküll HR, Mutert E (1995) Global extent, development and economic impact of acid soils. Plant Soil 171:1-15

Part III.
Cell Apoplast in Nutrient Acquisition and Metal Tolerance

The Role of the Outer Surface of the Plasma Membrane in Aluminum Tolerance

Tadao Wagatsuma, Satoru Ishikawa, and Paul Ofei-Manu

Summary. Aluminum (Al) toxicity is the primary stress factor in acid soils. Al tolerance is suggested to be constitutional because of its rapid response. The cell wall of the root-tip acts as the adsorption site for Al ions, but its negativity and Al adsorption capacity cannot explain the differential Al tolerance among crop plant species and cultivars. Among several strategies to tolerate Al toxicity, much evidence has been accumulating on the specific exudation of organic acids (citric, oxalic, and malic acids) from root-tip cells. However, release of organic acids was suggested to be a partial strategy for tolerance to Al ions and not to be a general and satisfactory one. The role of phenolic compounds existing outside and/or inside of root cells under soil acidity stress should be clarified more in detail. Among the apoplastic components (mucilage, cell wall, and the outer surface of the plasma membrane), the plasma membrane is considered to be the determinant of Al tolerance. The structure, function, and genetic aspects of the plasma membrane of root-tip cells that control Al tolerance are future subjects for research. For better growth in acid soils, plants should also possess additional tolerances to low nutrients (phosphorus, calcium and others) under Al stress.

Key words. Al tolerance, CEC of cell wall, Exudation, FDA-PI, Mucilage, Organic acid, Phenolic compounds, Plasma membrane, Protoplast

1. Introduction

Although toxic factors to crop plants in acid soils are generally combined, the common and primary factor is a toxic level of Al ions. The tolerance of crop plants to Al ions is the primary subject of this chapter.

The concept of the apoplast and the word for it were invented by Münch (1930). He identified the apoplast with wet cell walls, vessel contents, and intercellular spaces if they contained water (Canny 1995). However, as the term "apoplast" at present is used in connection with Al tolerance, the area of the apoplast is limited to the outer part of the endodermis, i.e., epidermis and cortex. Apoplast thus means the cell wall in itself. Kochian and Jones (1996), however, treated the cell wall and the outer surface of the plasma membrane (PM) as the apoplastic target for Al. When Al ions enter the cell wall area, they inevitably contact the outer surface of the PM. Therefore, in an actual sense the outer surface of the PM should be included in apoplastic component. Mucilage is considered also as a special form of apoplast though it exists on the outside of the root.

Mucilaginous walls occur between the root cap and the protoderm and also in the peripheral cells of the cap (Esau 1965). It is well established that mucilage arises primarily from export of polysaccharide formed by the golgi apparatus in the peripheral cell layer of the root cap. In maize, mucilage was composed mainly of glucose, galactose, uronic acid, and fucose; the last sugar is apparently not abundant in mucilage of wheat, pea, and sycamore (Puthota et al. 1991).

Reasons for the positive role of mucilage on Al tolerance are as follows: (i) binding of Al ions to carboxylic acid residues in uronic acid, (ii) depression of diffusion of Al ions in bulk solution to the root surface owing to higher viscosity of mucilage, (iii) higher concentration of organic acids released from the root owing to higher viscosity of mucilage. When the mucilage was periodically removed from root tips with a brush, root elongation of cowpea was considerably inhibited by Al. This indicates a protective function of the mucilage against Al injury (Horst et al. 1982). Seedlings of the Al-tolerant wheat cultivar Atlas 66 formed approximately 3 times more mucilage than that formed by the Al-sensitive cultivar Victory. Al ions reduced the formation of mucilage, but a 20-fold higher concentration of Al was required to inhibit mucilage formation in Atlas 66 than in a susceptible cultivar (Puthota et al. 1991). The volume of mucilage droplets produced when plants were grown in a nutrient solution without Al was positively correlated with Al tolerance in 10 out of 11 cultivars of wheat (Henderson and Ownby 1991). Currently, mucilage is considered to contribute positively to Al tolerance. On the contrary, the removal of the root cap of maize had no effect on the Al-induced inhibition of root growth (Ryan et al. 1993). As the primary portion of mucilage production is the root cap, this experiment suggests a lower significance of mucilage in Al tolerance. The role of mucilage in Al tolerance is thus unclear.

Al-tolerant crop plants (Wagatsuma et al. 1995a) and Al-tolerant wheat lines (Delhaize et al. 1993) accumulate less Al in the root-tip cells than Al-sensitive ones. Several mechanisms for this reduced accumulation have been proposed, including metabolism-dependent exclusion of Al (Wagatsuma 1983; Zhang and Taylor 1989) and detoxification of Al by the secretion of organic acids from roots (Delhaize et al. 1993; de la Fuente et al. 1997; Ma et al. 1997a, b). The binding of Al ions to carboxylic groups in pectic substances of cell wall inhibits the apoplastic movement of water and nutrients (Blamey and Dowling 1995) and the metabolism of cell wall polysaccharides (Le van et al. 1994). In addition, Rengel (1992) assumed that Al ions might repress cell elongation induced by acid-mediated wall loosening. However, there is no consensus on the correlation between cation exchange capacity (CEC) of roots and Al tolerance among crop plant species or cultivars (Munn and McCollum 1976; Mugwira and Elgawhary 1979; Wagatsuma 1983; Rengel and Robinson 1989; Allan et al. 1990; Masion and Bertsch 1997). In addition, the primary target of Al ions is the tip portion of the root. Experiments on root CEC have sometimes been carried out without paying much attention both to the strict root portion, i.e., the root-tip, which is deeply concerned with Al tolerance and the purity of the cell wall. The relationship between the Al binding capacity of cell wall and Al tolerance should be clarified in more detail.

The outer surface of the PM is the boundary of the apoplast and symplast, and can bind Al ions with its negative sites, which originate from the phosphate and carboxylic groups of phospholipids and proteins (Obi et al. 1989). The mean value of the zeta potential (electrokinetic potential: in practice,used with the same meaning as the surface potential of a particle) of root protoplasts was generally lower in Al-sensitive plant species, and in the more apical portion of younger roots (Wagatsuma and Akiba 1989; Fig. 1). Wagatsuma et al. (1995b) developed a new technique for the collection of Al-tolerant protoplasts from roots with newly prepared positively charged silica microbeads (PCSM) based on the differences in surface negativity. A mechanism for the isolation of Al-tolerant protoplasts based on DLVO theory (Derjaguin-Landau-Verway-Overbeek theory on repulsive and attractive action operating between particles - the sum of repulsive action by Coulomb's forces in electric double layer and attractive action by London-van der Waals forces) is presented in Fig. 2. When the original population of protoplasts was mixed with PCSM at pH 5.2 and centrifuged on a discontinuous Ficoll gradient composed of successive layers of 3-mL aliquots of 0%, 5%, 7.5%, and 10% Ficoll in 0.7 M mannitol, Al-sensitive protoplasts were dispersed in the upper interface as a consequence of repulsion due to Coulomb's forces, but Al-tolerant protoplasts precipitated as a result of a salting-out effect. However, it remains to be determined whether or not this technique for the isolation of Al-tolerant cells is applicable to other plant species or cultivars. Takabatake and Shimmen (1997) and Jones and Kochian (1997) reported the phospholipids or lipids of PM

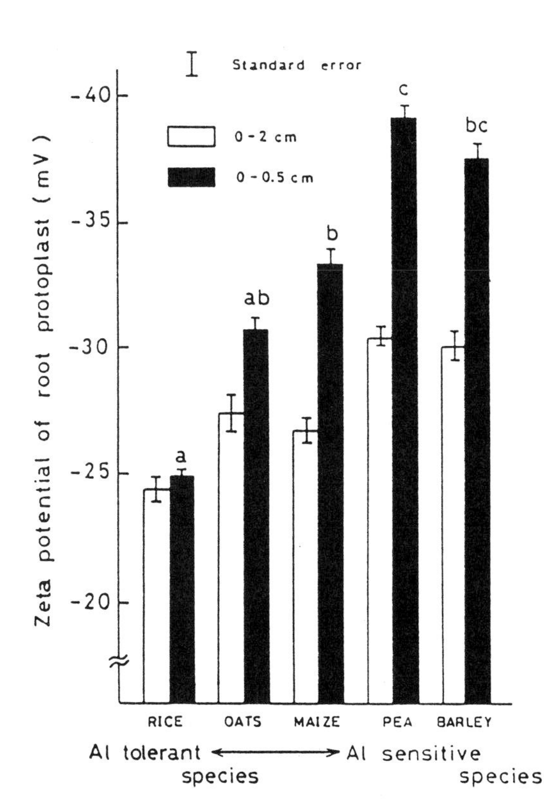

Fig. 1. Zeta potential of root protoplasts from 5 plant species. *Open bars* indicate the values measured on 0–2 cm tip-portions of roots of 5–10 cm length at pH 6; *solid bars* indicate the values measured on 0–0.5 cm tip portion of roots in of 2–3 cm in length at pH 4.7. Different letters on solid bars represent a significant difference using Duncan's multiple range test (P= 0.05)

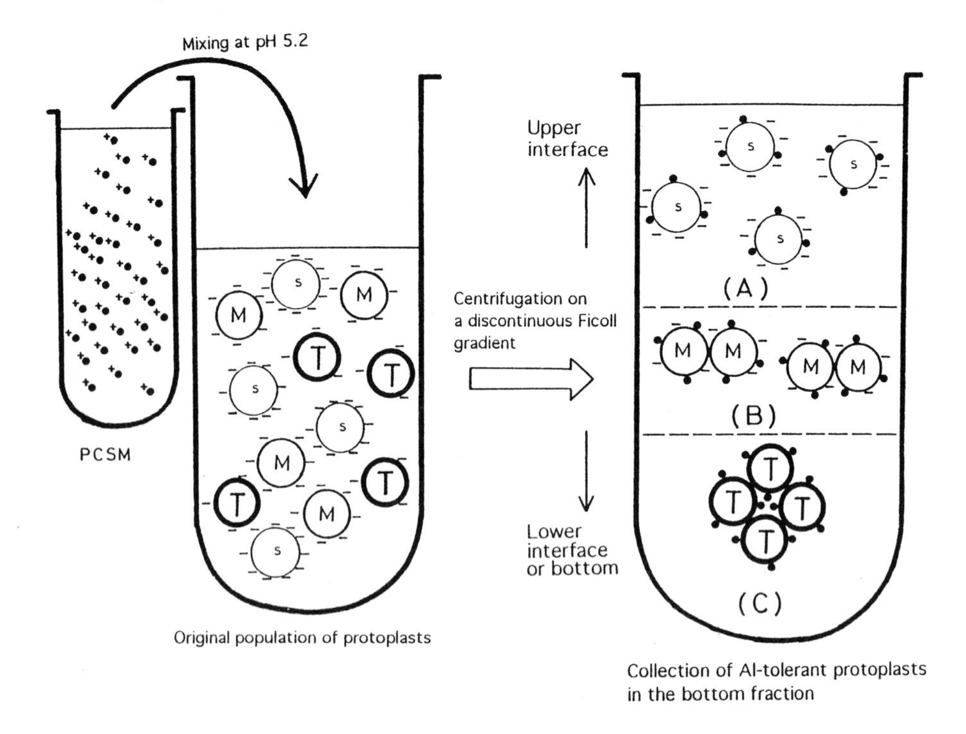

Fig. 2. Schematic representation of the proposed mechanism for the isolation of Al-tolerant protoplasts by mixing with PCSM at pH 5.2 and centrifugation on a discontinuous Ficoll gradient, based on DLVO theory. (**A**), upper interface of Ficoll gradient containing dispersed protoplasts formed as a consequence of repulsion by Coulomb's forces; (**B**), middle interface of Ficoll gradient containing small aggregates formed as a consequence of residual repulsive forces; (**C**), lower interface or bottom of Ficoll gradient containing large aggregates formed as a consequence of absence of repulsive forces. *T*, Al-tolerant protoplasts; *M*, protoplasts with intermediate Al-tolerance; *S*, Al-sensitive protoplasts

to be the primary sites for Al toxicity. Yermiyahu et al. (1997) suggested that surface negativity of PM and Al sorptive capacity probably accounts for some of the sensitivity of wheat cultivar to Al^{3+}. Zhang et al. (1997) on the other hand reported that the ratio of phospholipids to total lipids in the PM from whole roots was lower in Al-sensitive wheat cultivars. Whether or not the PM of root-tip cells is the determinant of Al tolerance should be clarified more clearly.

Most agreement has concentrated on the release of larger amounts of organic acids from the root-tips of Al-tolerant crop plant species and cultivars (Delhaize et al. 1993; Pellet et al. 1995; Ryan et al. 1995; Ma et al. 1997a, b). Of the organic acids, citric acid is the most effective for detoxification of Al ions, and its molar ratio to perfectly detoxify Al ions is known to be 1:1 (citric acid: Al^{3+}). Oxalic acid is less effective and its molar ratio is 1:2 to 3 (Ma et al. 1997b; Ma et al. 1998; Zheng et al. 1998), the ratio of malic acid is 7 to 10, and succinic acid is hardly effective (Delhaize et al. 1993; Ryan et al. 1995; Ma et al. 1997a;). Many of the natural crop plant species and cultivars reported on so far can start to secrete organic acids from the root-tip with a specific signal, i.e., Al ions. Zheng et al. (1998) confirmed the great contribution of the secreted oxalic acid on Al resistance by using phenylglyoxal as an anion-channel inhibitor. Although the secretion of organic acids from the root-tip can be an important strategy to detoxify Al ions, no direct and clear demonstration has been offered yet.

This chapter is the last one that connects this section on root exudates and the next section on the apoplast. Therefore, mechanisms of Al tolerance will be discussed first in connection with the apoplast, next with the PM, and finally with organic acids exuded from roots.

2. The Role of Mucilage in Al Tolerance

The volume of mucilage of two maize and two wheat cultivars was compared by simple observation. Al-tolerant maize cultivar DK789 produced a lower amount of mucilage than an Al-sensitive cultivar, XL61 (Fig. 3 upper). In wheat, the amount of mucilage was very small and was similar in the two cultivars that differed in Al tolerance (Fig.3 lower). In other crops, i.e., rice, pea, and sorghum (all cultivars are listed in Table 1), no differences were detected in the amount of mucilage from the roots of Al-tolerant and Al-sensitive cultivars (data not shown). Both maize cultivars lost their mucilage after a 2-h treatment of 100 μM Al (Fig. 4). Finally, no positive correlation of mucilage volume to Al tolerance was observed.

In order to clarify the role of mucilage in Al tolerance, isolation of intact mucilage from the root-tip is important. Recently, a technique of a 10-min wash in 1 M NH_4Cl was proposed (Archambault et al. 1996). The influence of this technique on PM permeability was therefore tested. The root-tip developed a red fluorescent color in FDA-PI (fluorescein diacetate and propidium iodide) staining after a 10-min wash in 1 M NH_4Cl (Fig.5), and this indicated a remarkable increase in PM permeability. A blue color that developed with Evans blue staining also indicated the same result (data not shown). The concentration of P and especially K in the root-tip decreased considerably under this treatment (Table 2). Owing to the hypertonicity of 1 M NH_4Cl, leakage of cell contents was induced and the PM destroyed. In order to discount the resultant decrease in dry weight owing to the leakage of cell contents, the root-tip portion was first squashed and treatment with 1 M NH_4Cl treatment then carried out. Ca concentration also decreased to one-third of the starting material that already been squashed (data not shown). Consequently, a 10-min wash in 1 M NH_4Cl is established as removing not only mucilage but also cell contents including symplastic Al. Despite expectations, a 10-min wash in 1 M NH_4Cl was too harmful for both the root-tip and the more proximal portion of both wheat and maize roots.

Indirect evidence for less involvement of mucilage in Al tolerance is as follows. After 1-h pretreatment with 20 μM Al (pH 4.9) the roots of two wheat cultivars with a similar accumulation of Al in a 1-cm root-tip portion (Al-tolerant cultivar Kitakami B and Al-sensitive cultivar Muka, Table 1) showed normalities in both appearance and membrane permeability, but re-elongated differently following 10-h treatment with 0.2 mM $CaCl_2$ without Al (Fig. 6). This indicates that even

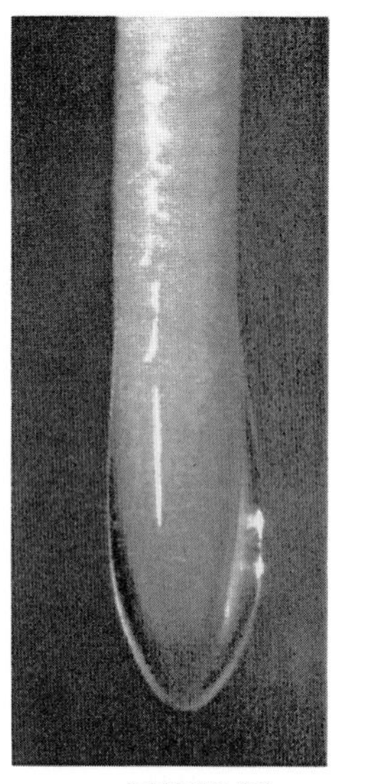

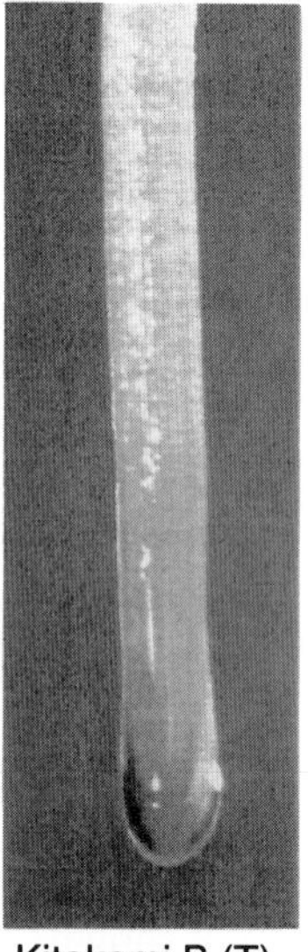

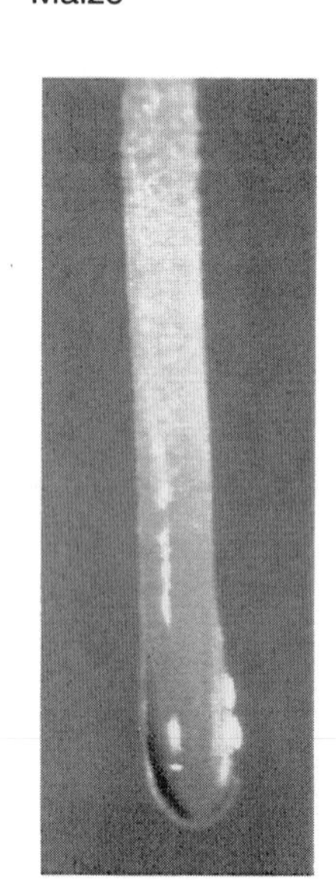

Fig. 3. Mucilage volume of maize cultivars (*upper*) and wheat cultivars (*lower*) grown in 0.2 mM $CaCl_2$. *T*, Al-tolerant cultivar; *S*, Al-sensitive cultivar

though mucilage cannot induce a difference in Al accumulation each cultivar can exhibit a difference in Al tolerance. These results suggest a lower involvement of mucilage in Al tolerance. Many researchers have recently considered the elonga-

Table 1. Al tolerance, capacity of negative charge, and Al accumulation in root-tip portions of Al-tolerant and Al-sensitive cultivars of several crop plant species

Species	Cultivar	Al tolerance[a] T or S	CEC of whole root-tip[b] ($cmol_c\ kg^{-1}$)	CEC of cell wall from root-tip[c] ($cmol_c\ kg^{-1}$)	Al accumulation in root-tip (0.5 cm)[d] ($mg\ kg^{-1}$)
Oryza sativa	Senatore	T	7.0	8.3	321 ± 68
Oryza glaberrima		S	7.9	6.9	267 ± 40
Maize	DK789	T	11	13	215 ± 21
	XL61	S	12	13	239 ± 40
Wheat	Kitakami B	T	8.9	9.4	359 ± 36
	Muka	S	9.4	9.8	386 ± 12
Pea	Harunoka	T	18	53	195 ± 4
	Hyougo	S	16	54	326 ± 20
Sorghum	Energy	T	9.4	12	156 ± 20
	Power	S	8.5	12	151 ± 30

[a] *T*, tolerant; *S*, sensitive. Al tolerance was the ratio of net elongation of the longest root in Al treatment to that in control. Treatments were carried out with 0.2 mM $CaCl_2$ and with or without 5 μM Al for 2 days (pH 5.0)

[b] The dry root powder (48 mesh) of the root-tip portion (1 cm) from young seedlings with 3-5 cm of roots was saturated with H^+ by washing with 0.5 M HCl. The adsorbed H^+ was exchanged with Ca by 0.1 M Ca $(CH_3COO)_2$ (pH 5.0), and finally the Ca was desorbed with 0.05 M HCl

[c] Twenty grams of root-tips (1 cm) were homogenized with a medium consisting of 50 mM Tris-Hepes (pH 7.8), 5 mM EDTA, 10 mM NaF, 2.5 mM DTT, 10 μg ml^{-1} butylated hydroxytoluene, and 250 mM sucrose. The homogenates were filtered, and fully washed with water. This procedure was repeated. The crude cell wall materials were extracted twice in a mixture of chloroform and methanol (1:2). More than 95% of protoplasm was removed from the cell wall materials judging from the ratio of the P concentrations of both materials before and after extraction

[d] The concentrations of Al in the media were: 100 μM (pH 4.5) for rice, 50 μM (pH 4.7) for maize, 20 μM (pH 4.9) for pea and wheat and 10 μM (pH 5.0) for sorghum depending on the differential Al tolerance among crop plant species. The treatment was carried out for 1 h and all media contained 0.2 mM $CaCl_2$. After termination of the treatment, the tip portions of roots (0.5 cm) were harvested, and analyzed

tion zone as the primary target for Al toxicity in roots (Kochian 1995). Therefore, the role of mucilage in Al toxicity may only be limited because the primary site for mucilage production is localized to the tip. In order to clarify the role of mucilage in Al tolerance more directly, the following need to be compared in future under Al stress using cultivars that differ in Al tolerance, i.e., mucilage volume, binding capacity of Al ions, and concentration of organic acids inside of mucilage.

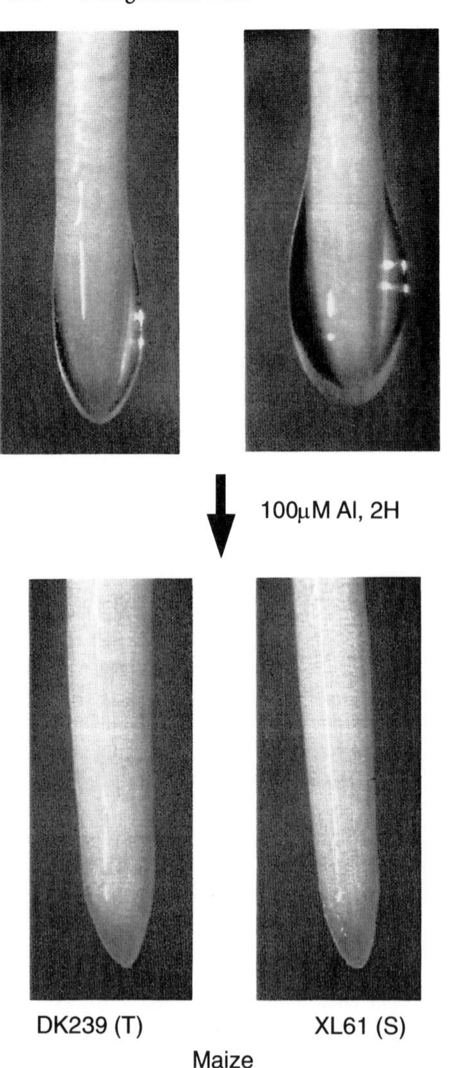

Fig. 4. Change in mucilage volume of two maize cultivars by 2-h treatment with 100 μM Al

3. Al Adsorption Capacity of Cell Wall and Al Tolerance

3.1 Relationship Between the Capacity of Negative Charge, Al Accumulation in the Root-Tip, and Al Tolerance

The cation exchange capacity (CEC) and Al accumulation in the root tips of several crops were investigated. In all crop plants but pea, for which Al concentration in the medium was too high, no significant differences were recognized between Al accumulation in the root-tips of Al-tolerant and Al-sensitive cultivars (Table 1). In this table, the CEC of whole root tip is ascribed not only to the cell wall but also to the PM, organelles, nucleus, and cytoplasmic proteins, because the CEC of all the components of the cell (not only of the cell surface) was studied by measuring the CEC of powdered cell tissue. Al accumulation in the root-tip portion

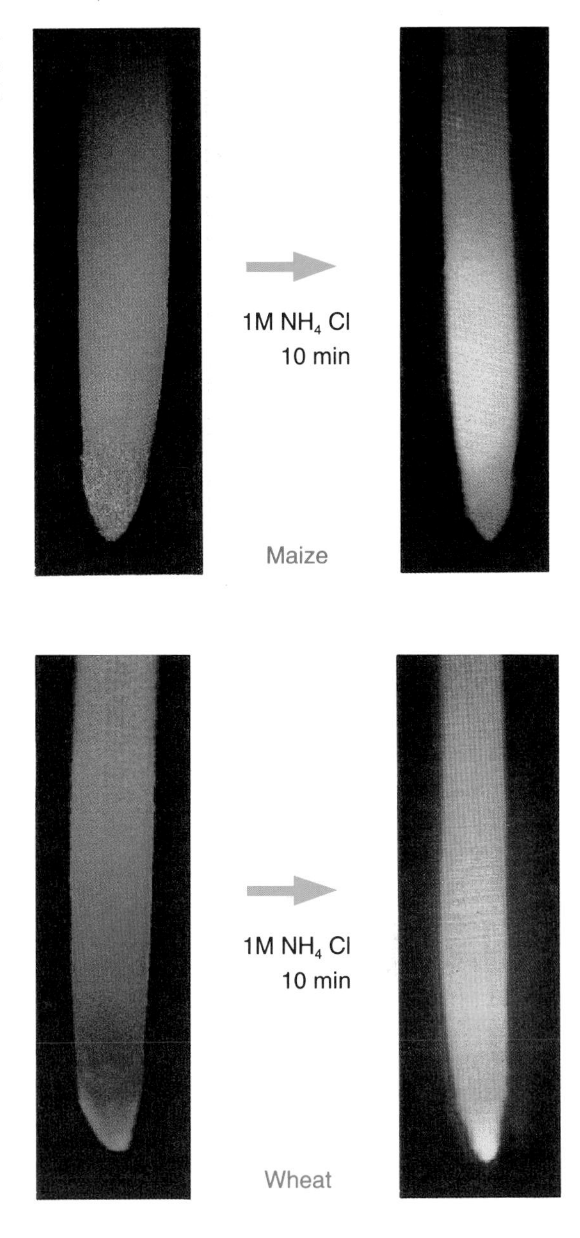

Fig. 5. FDA-PI fluorescence of root-tip portions of maize and wheat before and after 10-min wash in 1M NH_4Cl

was not correlated with the CEC of the whole root-tip. The CEC of the cell wall, on the contrary, is ascribed to pectic substances and proteins, and was not correlated with Al accumulation either.

The relationships between Al tolerance and the CEC of the whole root-tip or cell wall from several crop plants are shown in Table 3. The CECs of root-tips of rice, maize, and barley, which are monocotyledons, were approximately 10 $cmol_c\ kg^{-1}$ dry weight of root, whereas that of pea, a dicotyledon, was approximately 2 times higher than those values. The CECs of the purified cell walls of the monocotyledons

Table 2. Effect of NH_4Cl treatment on the mineral concentration of maize root

Root portion (mm from tip)	NH_4Cl treatment	K	P	Mg	Ca
		(% in dry weight of maize root)			
0-5	—	1.46	1.25	0.136	0.057
	+	0.037	1.04	0.219	0.155
		(0.03)[a]	(0.83)	(1.61)	(2.72)
10-15	—	2.03	0.692	0.148	0.099
	+	0.048	0.193	0.070	0.135
		(0.02)	(0.28)	(0.47)	(1.36)

[a]Ratio of the mineral concentration of the root portion treated with 1 M NH_4Cl for 10 min to that without treatment with 1 M NH_4Cl

Table 3. Relationships between Al tolerance and CEC of whole root-tip or cell wall among several crop plant species

Species	Cultivar	Al tolerance (%)	CEC of whole root-tip ($cmol_c\ kg^{-1}$)	CEC of cell wall from root-tip ($cmol_c\ kg^{-1}$)
Rice	Sasanishiki	80	6.5	6.2
Maize	Pioneer 3352	69	9.8	14
Pea	Kinusaya	60	25	60
Barley	Manriki	4.4	9.8	10

Seedlings were treated with 5 μM Al containing 0.2 mM Ca (pH 5.0) for 3 days. Al tolerance was calculated based on the ratio of the net root elongation of the longest root in the Al treatment to that in the control

were in the range of 6 to 14 $cmol_c\ kg^{-1}$ dry weight of cell wall and that of pea was approximately 6 times higher than this. The correlation coefficients between Al tolerance and CEC of root-tip or cell wall from root-tips among the four crop plants were 0.011 and 0.109 respectively, and each correlation was non-significant.

In other experiments, Al accumulation in the root-tip did not correlate with Al tolerance for a 1-h treatment with a moderately low concentration of Al, but was negatively correlated with Al tolerance under a longer period of Al treatment and higher concentration of Al in the medium, i.e., more than 24 h and more than 50 μM Al (Wagatsuma et al. 1995a; Wagatsuma, et al. 1997; unpublished data). Consequently, it is concluded that a high concentration of Al in the root-tip portion of Al-sensitive crop plant species and cultivars is only the result of the permeation of large amounts of Al into the cytoplasm caused by the destruction of PM, and a difference in Al concentration in the root-tip under the weak stress of Al does not indicate differential Al tolerance among crop plants.

3.2 Al Binding Capacity of the Cell Wall and Al Tolerance

Tables 1 and 3 show no correlation between the CEC of cell walls from the root-tip and Al tolerance in any of the crop plants tested. Further, we compared the Al-binding capacity (amount of Al saturated at an isolated cell wall) and the CEC of the root cell wall. The Al-binding capacity of the cell wall was not correlated with Al tolerance (Table 4). Al-binding capacity was determined at pH 4.0. As CEC was determined at pH 5.0, the negative charge dissociated at pH 4.0 must be lower than that at pH 5.0. Although the Al binding capacity of the cell wall was close to the CEC in general, it slightly oversaturated CEC when all Al ions bound were presumed to be Al^{3+}. The origin of the excess binding sites may have come from proteins, lignin or other compounds in cell wall. From in vitro experiments, Tam and McColl (1990) pointed out that the binding affinities of Al and Ca ions in acidic conditions were different. The cell wall, which contains a larger amount of α-hydroxy carboxylic groups than that of α-carboxy carboxylic groups can be conclusively considered to bind a larger amount of Al than Ca. The surplus amount of Al actually bound over the calculated amount of Al, based on the CEC measured with Ca containing solution and expressed as Ca binding capacity, can partially be explained from the differential amount of these functional groups within the cell wall.

Apart from the point described above, high association constants for aspartic and glutamic acids with Al ions (16.29 and15.12, respectively) were listed by Nordstrom and May (1989). This is a rather high value similar to that of EDTA (16.5) and somewhat higher than that of citric acid (7.98). On the other hand, Orvig (1993) found the stability constants of aspartic and glutamic acids to be rather low, i.e., 2.17 and 2.30, respectively, which are slightly higher than that of acetic acid (1.51). Whether organic ligands are in simple forms or not, the binding behavior of Al ions to organic ligands (e.g., phenolic compounds which can exist not only in the cell wall but also in cytoplasm and even in exudate) is complicated and is a future subject for research.

Table 4. Relationships between Al tolerance and cation exchange capacity (CEC) of cell wall or of Al binding capacity

Species	Cultivar	Al tolerance T or S	CEC of cell wall from root-tip of control plants ($cmol_c\ kg^{-1}$)	CEC of cell wall from root-tip of Al treated plants[a] ($cmol_c\ kg^{-1}$)	Al binding capacity[b] ($mg\ kg^{-1}$)	($cmol_c\ kg^{-1}$)
Maize	DK789	T	13	13	1320	15 (1.15)
	XL61	S	13	13	1290	14 (1.08)
Pea	Harunoka	T	53	51	6270	70 (1.37)
	Hyougo	S	54	49	5520	61 (1.13)

[a] Young seedlings were treated with 0.2 mM Al + 0.2 mM $CaCl_2$ (pH 4.5) for 3 h, and thereafter root-tips were collected. Their cell walls were isolated as before

[b] The isolated cell wall was first treated with 0.5 M HCl, and thereafter incubated with 3 mM Al (maize) or 6 mM Al (pea) at pH 4.0 for 3 h. The unit of $cmol_c\ kg^{-1}$ is the calculated value when all Al ions bound were presumed to be Al^{3+}. Values in parentheses are the equivalent ratio of bound Al^{3+} to CEC

Under Al stress, the root is changed in its morphology and the sugar composition of the cell wall (Le Van et al. 1994). Therefore, the CEC of not only the intact root tips but also of the Al-pre-treated tips was measured and compared with Al tolerance and the Al binding capacity (Table 4). At this stage of Al pre-treatment, roots clearly exhibited Al toxicity symptoms. The CEC of the cell wall of maize roots was not affected by Al pre-treatment; but the CEC of pea slightly decreased with Al pre-treatment. At least in the initial stage of Al toxicity, Al pre-treatment negligibly changed CEC of the cell wall from the root-tip portion, and the differences in the CEC of the cell wall, either intact or after Al pre-treatment, cannot explain the differential Al tolerances between cultivars. Al accumulation in intact root-tip portions (Table 1) was in the range of one-fifth to one-thirty-second of its value of the corresponding cell wall (Table 4, Al binding capacity column). The cell wall possesses the potential to bind Al ions equivalent to CEC, however, the intact root acquires a considerably lower amount of Al ions as compared to that of the cell wall. This is because most of the Al ions initially localize in the cell wall of the epidermis and outer cortex cells (Ishikawa et al. 1996). In conclusion, although Al binding capacity is dependent on the CEC, it cannot control Al tolerance.

Although the purity of the cell wall material was satisfactorily high as judged from the fact that the phosphorus concentration of the cell wall was only 3% to 5% of that of the initial root segments, a milder preparation technique should be applied in order to acquire the compositional and structural intactness of the cell wall material. Binding sites for Al ions other than the carboxylic group from proteins and pectins in the cell wall may be important for plant species such as woody species rich in secondary cell wall materials.

Cosgrove and co-workers (Cosgrove 1996) found two novel apoplastic proteins, the expansins, (29 and 30 kDa) which are involved in wall extension. Expansins act to weaken the non-covalent bonding between cellulose and hemicellulose. Al and copper ions exert their inhibitory effects by binding to the active expansins rather than to pectins or other structural components of the wall (Cosgrove 1989; Cosgrove and Durachko 1994; McQueen-Mason et al. 1992). The amino acid sequence of cucumber expansin was identified (Cosgrove 1996), but the role of expansins in Al tolerance remains to be clarified.

4. Significance of PM of the Root-Tip Cells in Al Tolerance

4.1 Which Controls Al Tolerance, the Cell Wall or the PM of Root-Tip Cells?

In order to clarify the mechanisms of Al tolerance, identification of the primary target for Al toxicity is important. Within the apoplast, the target of Al can be either the cell wall or the outer surface of the PM. The result of our experiments suggested that the PM is the primary target because the increase of the permeability of the PM as a result of Al injury preceded the breakdown of the cell wall. The permeability of the PM of root-tip cells pre-treated with Al for 1 h was determined by a fluorescence staining technique with FDA-PI, but no detectable influences were found (data not shown). However, after these seedlings were re-elongated for 24 h in 0.2 mM $CaCl_2$ without Al, the permeability of the PM for Al-sensitive cultivars in every crop plant species was increased considerably more than that for Al-tolerant cultivars (Fig. 6). When stained with FDA-PI, cells with normal permeability can exclude PI from their PM's; in such cells, FDA passes through the PM and is hydrolized by intracellular esterases to produce fluorescein, and exhibits green fluorescence when excited by UV light. On the other hand, the permeabilized cells exhibit a bright red fluorescence due to the passage of PI through their PM's and intercalation with DNA and RNA. As the bright red color was observed independently of the horizontal cracks in roots but horizontal cracks were always accompanied by the bright red color, it is accepted that the increase in PM permeability precedes the destruction of the cell wall. Ishikawa et al. (1996) also showed the inhibitory effect of Al ions on PM permeability preceding the appearance of the cracks in the cell wall in the root-tip of pea (Fig. 7).

4.2 Relationship Between the Permeability of the PM of Root-Tip Cells and Al Tolerance After Short-Term Exposure to Al Ions

The mineral concentration of wheat root-tip portions (0.5 cm) after 24-h re-elongation in Al-free solution following 1-h pre-treatment with Al is shown in Table 5. During the re-elongation period in Al-free solution, the Al-tolerant cultivar showed

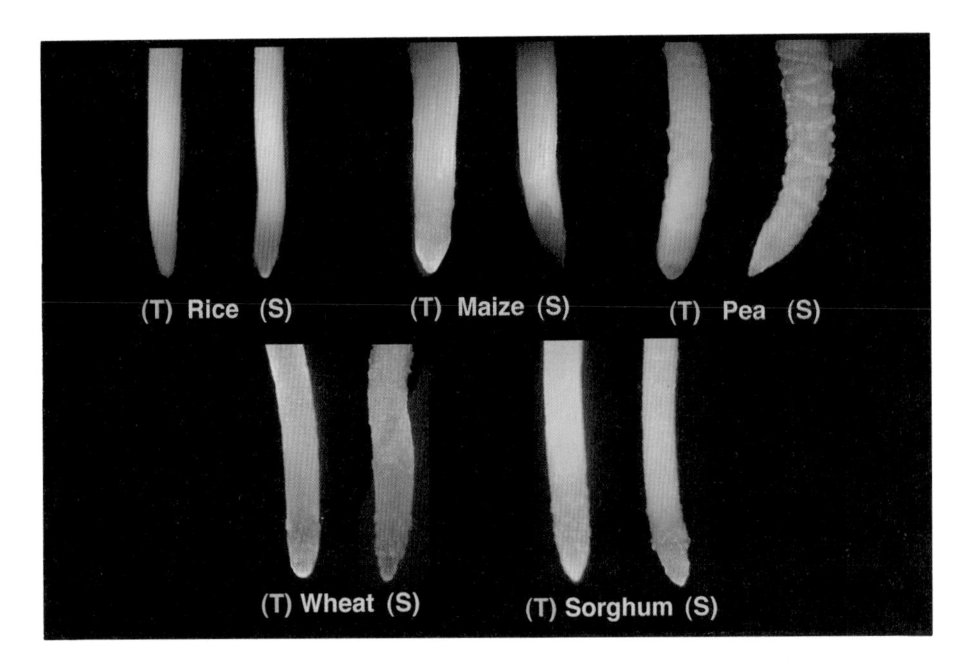

Fig. 6. The permeability of the plasma membrane (PM) of root-tip cells determined by a fluorescence staining technique with FDA-PI. The concentrations of Al in media are indicated in note *d* of Table 1. An increase in permeability was exhibited as non-greenish color. *T*, Al-tolerant cultivar; *S* Al-sensitive cultivar. Photographs were taken at the following durations (h) of re-elongation: rice, 8; maize, 8; pea, 24; wheat, 10; sorghum, 10

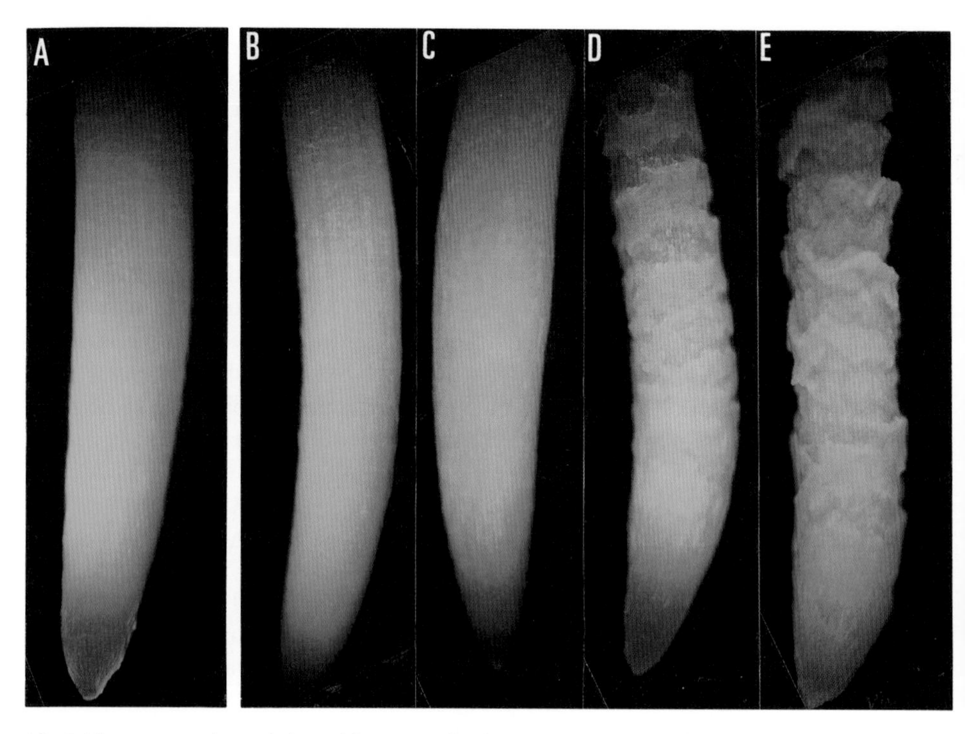

Fig. 7. Fluorescence by staining with FDA-PI for the root-tip portion of peas treated with 20 μM Al^{3+} + 0.2 mM Ca^{2+} at pH 4.9 for 0 h (**A**), 2 h(**B**), 5 h (**C**), 12 h (**D**), and 20 h (**E**)

greater root elongation than the Al-sensitive cultivar. Al concentration of the re-elongated root-tip portion for the Al-tolerant cultivar Kitakami B decreased from the original concentration (cf. Al concentration in Table 1) more than that for the Al-sensitive cultivar Muka. On the contrary, the concentrations of P and K, which are localized in the symplast, decreased more in the Al-sensitive cultivar Muka. A lower concentration of Al in the root-tip portion for the Al-tolerant cultivar can be considered to be the result of greater dilution accompanied by greater elongation. The reason for this is as follows. From the data in Tables 1 and 5, initial Al concentrations before re-elongation for Kitakami B and Muka were 359 and 386, and the final concentrations after re-elongation were 99 and 148 (mg kg^{-1}), respectively. The ratios of the final Al concentrations to the initial Al concentrations were 0.276 and 0.383, respectively, and therefore the latter was 1.39 times higher than the former. On the other hand, the relative re-elongations (%) for the respective cultivars were 99 and 67 (Table 5), and therefore the former was 1.48 times higher than the latter. The relative value of the decrease in Al concentrations (1.39) was close to that of the relative re-elongations (1.48). Finally, other mechanisms, i.e., efflux or dissolution of Al to decrease Al concentration in the root-tip for the Al-tolerant wheat cultivar during the re-elongation period are hardly suggested. Lower concentrations of P and K in the root-tip for the Al-sensitive cultivar can also be considered to be the result of greater leakage accompanied by the necessary elongation of the rigid and less extensible PM of the root-tip.

Table 5. Mineral concentration of wheat root-tip (0.5 cm)[a] after 24-h re-elongation following 1-h pre-treatment of Al

Cultivar	Relative re-elongation (%)	Al	Ca	P (g kg^{-1})	Mg	K
Kitakami B	99	0.099	0.588	10.80	0.851	14.4
			(117)[b]	(104)	(94)	(96)
Muka	67	0.148	0.578	8.96	0.816	11.4
			(123)	(88)	(83)	(86)

[a]Seedlings were treated for 1-h with 0.2 mM Ca and with or without 20 μM Al at pH 4.9, and re-elongated for 24 h in 0.2 mM Ca (pH 4.9). Relative re-elongation was the percentage of re-elongation of the longest root pre-treated with the medium containing Al to that pre-treated with control medium

[b]The ratio of each mineral concentration of the root-tip portion of seedlings pre-treated with the medium containing Al to that pre-treated with control medium

These results suggest that a short-term exposure to Al makes the PM of root-tip cells of the Al-sensitive cultivar more rigid and leaky. Similar results were observed in barley (a highly sensitive crop plant species to Al ions) and by treatment with higher concentrations of Al (Ishikawa and Wagatsuma 1998). However, when pea roots which had been pre-treated with Al for 1 h and were treated with 2 mM citrate (pH4.5) for 0.5 h were re-elongated in Al-free 0.2 mM Ca solution for 8 h, the PM permeability was normal (Fig.8). As the permeation of citric acid through the PM is considerably limited by a large number (eleven) of hydrogen bonding with water molecules (Akeson and Munns 1989), citric acid in apoplast can desorb the Al ions from the outer surface of PM. These results suggest that even though the amount of Al bound to the outer surface of the PM in root-tip cells may be small it can be the cause of permeabilization of the PM and differential Al tolerance among crop plants.

In order to investigate the role of PM in protecting cells from Al, the Al absorption of protoplast with intact PM, or without it, eg., protoplast ghost, was compared. The Al content of protoplasts treated with 100 μM $AlCl_3$ for 10 min was in the order barley > maize > pea > rice (Table 6). Protoplast ghosts took up more Al than the respective protoplasts for all crop plants, but the ratio of Al content of protoplast ghosts to that of the respective protoplasts was in a different order, i.e., rice > maize > pea, barley (Table 6). Protoplast Al is composed of two fractions: Al bound to PM and permeated Al. Microtubule, nucleus and cytoplasmic proteins may be retained in the protoplast ghost (Sonobe and Takahashi 1994), and therefore the permeated Al ions can easily bind to these symplastic contents. Greater Al content in the protoplast ghost relative to protoplast indicates the difficulty of permeation of Al ions into protoplast. Table 6 suggests that the PM of root-tip cells for Al-tolerant crop plant species is less permeable to Al ions. Lower release of K^+ from the protoplast (data not shown) and from root-tip portion (Wagatsuma et al. 1995a) of Al-tolerant crop plant species supports this view.

The presence of Al itself increases the permeability of PM, and it was shown that the degree of increase was higher in Al-sensitive plants than in Al-tolerant

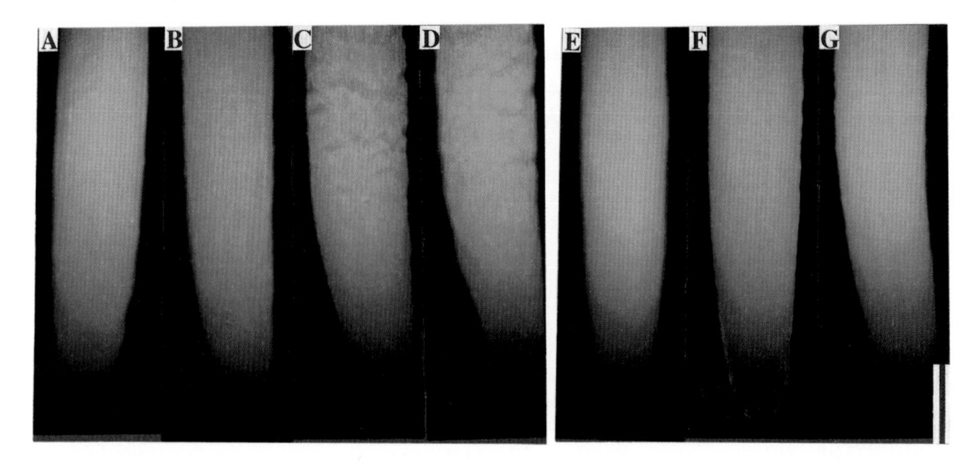

Fig. 8. Effect of post-treatment with citrate on permeability of the PM in root-tip portions when re-elongated in Al-free solution following a short treatment with Al. Pea seedlings were treated with 5, 20, and 100 μM Al for 1 h, post-treated with or without 2 mM citrate (pH 4.5) for 0.5 h thereafter, and subsequently re-elongated in Al-free solution for 8 h. **A**, just after the treatment with 100 μM Al; **B-G**, after the re-elongation of roots following treatment with 5 (**B** and **E**), 20 (**C** and **F**), or 100 (**D** and **G**) μM Al and post-treatment with (**E-G**) or without (**B-D**) 2 mM citrate

Table 6. Al contents of protoplasts and protoplast ghosts (f mol Al protoplast^{-1} or protoplast ghost^{-1})

	Rice	Maize	Pea	Barley
Protoplast (A)[a]	20.9	58.7	48.6	62.3
Protoplast ghost (B)[b]	59.9	104.3	57.9	74.4
(B)/ (A)	2.87	1.78	1.19	1.19

[a]Protoplasts isolated from root-tips (0-1 cm) were treated with isotonic 100 μM Al solution for 10 min

[b]Protoplast ghosts were prepared by the treatment of protoplasts with Al-free solution in the absence of 0.7M mannitol, and subsequently with 100 μM Al solution for 10 min

ones. The permeabilities of the PM for pea protoplasts treated with moderately hypotonic (0.55 M mannitol medium as compared with 0.7 M mannitol of isotonic medium) solutions with or without Al for 30 min are shown in Fig 9. In the solution containing Al, the Al-tolerant pea cultivar Harunoka almost retained normal permeability, but the Al-sensitive pea cultivar Hyougo showed considerably increased permeability. Temporary exposure to Al ions irreversibly induces the PM of the root-tip cells of all Al-sensitive crop plant species and cultivars tested to become more rigid and leaky for the same reason described above. At present, no exceptions have ever been observed.

In every crop plant species tested, i.e., rice, maize, pea, wheat, and sorghum, the Al content of protoplasts isolated from root-tip portions of Al-sensitive cultivars was slightly higher than that of Al-tolerant cultivars (data not shown). In Table 7, Al content of protoplasts from the Al-sensitive maize cultivar (XL61) was 1.19

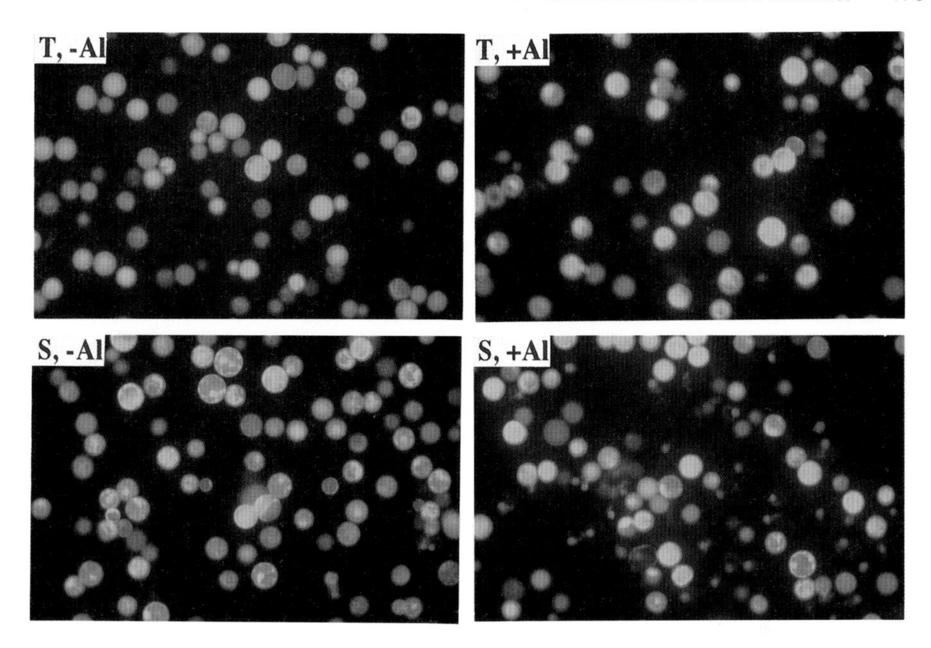

Fig. 9. PM permeability of protoplasts isolated from root-tips of pea cultivars (FDA-PI staining) in moderately hypotonic solutions (0.55 M mannitol) containing 0.2 mM Ca and with or without 100 μM Al for 0.5 h. *T*, Al-tolerant cultivar Harunoka; *S*, Al-sensitive cultivar Hyougo

Table 7. Effect of DNP and hypotonic conditions on Al contents of protoplasts[a] isolated from root-tips of Al-tolerant and Al-sensitive maize cultivars (f mol protoplast^{-1})

Cultivar	50 μM Al in isotonic (0.7 M mannitol) solution	100 μM Al in isotonic (0.7 M mannitol) solution containing DNP	50 μM Al in hypotonic (0.6 M mannitol) solution
DK789	34.3	70.2	42.8
XL61	40.8	97.1	89.7
XL61/DK789	1.19	1.38	2.10

[a]Protoplasts were isolated from root-tips (1 cm) from each maize cultivar and treated for 0.5 h. Each treatment solution contained 0.2 mM Ca and pH was maintained at pH 4.5. Al contents of protoplasts from the control solution (0.2 mM Ca) were 4.9 and 5.8 in DK789 (tolerant cv) and XL61 (sensitive cv), respectively

times higher than that from the Al-tolerant maize cultivar (DK789) in isotonic conditions. DNP (2,4-dinitrophenol, a metabolic inhibitor acting as a protonophore to negate the electrochemical potential difference of H^+ across PM) raised this ratio slightly. In contrast, in hypotonic conditions that induce rupture of the PM, the Al content of protoplasts from the Al-sensitive cultivar was considerably increased and was 2.1 times higher than that from the Al-tolerant cultivar. Therefore, a temporary toxic effect on the PM is suggested to accelerate Al absorption by Al-sensitive root cells more remarkably than that by Al-tolerant root cells.

Al^{3+}, which has a large ionic potential (the ratio of valence to effective ionic radius), preferentially binds covalently to the negative sites of the PM. When hydrated Al^{3+}, i.e., $[Al(OH_2)_6]^{3+}$, binds to the negatively charged phospholipid, seven to eight hydrated water molecules of the phosphate group (Cevc 1982) and six hydrated water molecules of Al^{3+} may be dehydrated together by the Eigen mechanism, i.e., the dissociation of water molecules with complex formation between hydrated metal ions and organic ligands. Dispersed phospholipid molecules in the normal PM form a partial packing area by binding Al ions; the membrane in a liquid crystal state becomes rigid and gel-like as a result of dehydration (Hauser and Phillips 1979; Chen et al. 1991). This dehydration may make the packing area more hydrophobic. Membrane proteins can also join this packing area with their carboxyl groups and Al ions. All these results and other investigations suggest the formation of a more rigid and less extensible, but more permeable PM, of root-tip cells in Al-sensitive crop plant species and cultivars by temporary contact with Al ions. The boundary between the packed area (e.g., phospholipids with higher affinity to Al ions and proteins) and the non-packed area (e.g., sterols and phospholipids with lower affinity to Al ions) may be enlarged by re-elongation or under hypotonic conditions; permeabilization of the PM may be the result of the notch effect of stress concentration in a mechanical sense.

There have been many investigations on the decrease in membrane fluidity and the change in membrane physics induced by in vitro binding of Al ions and other polyvalent cations with artificial membranes, liposomes, membrane vesicles, and PM vesicles. However, there is no information on the change of extensibility and hardness of membranes due to the binding of Al ions (e.g., Hauser and Phillips 1979; Vierstra and Haug 1978). Reaction of Al ions to membrane materials and its influence on membrane physics is very rapid (Deleers et al. 1985, 1986). As the actual composition of the PM in root-tip cells is remarkably complex as compared with these artificial membranes, the reactivity of Al ions and its influence on membrane physics may be too difficult to understand. However, the following characteristics of the PM may be presumed as the cause(s) of induction of a rigid and leaky PM by the binding of Al ions: (i) numerous negatively charged sites originated from phospholipids, or phospholipids and proteins, or boundary phospholipids surrounding proteins, (ii) strong binding of membrane protein in the PM to the cytoskeleton inside the PM, (iii) a high degree of saturation and/or shortness of fatty acid chain, and/or a greater amount of campesterol which makes PM less fluid (Douglas 1985), and/or (iv) other complex chemical compositions of PM lipid which influences the physical properties of PM under Al stress.

The most reliable method to obtain PM at present is an aqueous two-phase polymer partitioning system (Larsson et al. 1987). However, PM acquired by this method is contaminated with other membranes (endoplasmic reticulum (ER), mitochondria, tonoplast, etc.), and its purity is not perfect (Flowers and Yeo 1992; Larsson and Møller 1990). There are no reports on what kind of modification to the partitioning method is needed to acquire PM's of similar purities using different plant cultivars or mutants. Therefore, investigations using the same partitioning method may not be suitable. Further research to acquire PM with higher purity is a subject for future research.

5. The Role of the Release of Organic Acids from the Root-Tip in Al Tolerance, and the Fate of Organic Acids in Acid Soils

Since Delhaize et al. (1993) reported the release of a larger amount of malic acid from the root-tip of an Al-tolerant wheat cultivar, citric acid and oxalic acid have also been found to be secreted from other crop plants and woody species (Osawa et al. 1997) as an important strategy for Al tolerance, as mentioned in other chapters here. This site was presumed to be an anion channel specific to the Al ion, and Ryan et al. (1997) reported that this site was similar to the S-type Cl^- efflux channel. However, the behavior of the hypothetical channel is different from other general channels: several hours lag period is needed for its opening after the start of the Al ion signal, and conversely the channel does not close for several hours even after the removal of Al ion stress (Ma et al. 1997a; Zheng et al. 1998). Details of this hypothetical channel will be clarified hereafter.

We investigated the release of organic acids from the roots of eight plant species ranging in their tolerance to Al. Citric acid was the organic acid released in largest amounts from the roots of these plants. Table 8 shows the release of citric acid from Al-treated crop plant species and their tolerance to Al ions. Malic acid was also released from the roots of rice, maize and tea only in the medium containing Al, but the amount was below 1 μmol g^{-1} dry weight of roots $(12\ h)^{-1}$. No succinic acid was found. All crop plants except *Brachiaria brizanta* released more citric acid in the medium with Al than in the medium without Al. No correlation was found between the amount of citric acid released and Al tolerance among the crop plant species ($r = 0.185$). The concentration of citric acid in the medium with Al was below 1 μM and was considerably lower than that of Al ions, i.e., less

Table 8. Citric acid released from Al-treated roots of several crop plant species and Al tolerance

Plant species	Amount of citric acid released from Al-treated roots[a] (μmol g^{-1}DW of roots $12\ h^{-1}$)	Concentration of citric acid in Al medium (nM)	Al tolerance[b] (%)
Brachiaria brizanta	10.7 (0.91)[c]	243 (0.94)[d]	91
Rice	3.60 (2.77)	170 (2.36)	91
Tea	0.86 (1.62)	294 (1.49)	85
Maize	1.92 (1.92)	241 (1.71)	77
Pea	1.77 (1.84)	208 (1.86)	60
Cassia tora L.	18.4 (5.44)	628 (4.76)	51
Barley	1.20 (4.80)	82 (5.13)	45

[a]Collection and determination of citric acid were identical with the methods described in Table 9
[b]The ratio of the net elongation of the longest root during 24 h in control medium without Al following 1 h of treatment with 50 μM Al to that of treatment without Al
[c]Numbers in parentheses indicate the ratio of the amount of citric acid released from Al-treated roots to that of control roots
[d]Numbers in parentheses indicate the ratio of the concentration of citric acid in Al medium to that in control medium

than one-fortieth. These results suggest that the release of citric or other organic acids is not a common and primary strategy for Al tolerance in these crop plants. Alternative possibilities are as follows: a superior mechanism to the release of organic acids for Al tolerance exists, or the primary tolerance mechanism differs among plant species, or that each crop plant species has multiple tolerance mechanisms of which the components differ in strength.

Cassia tora released a remarkable amount of citric acid, i.e., 18.4 μmol g^{-1} dry weight of roots $(12\ h)^{-1}$ from roots, and this amount is comparable to the data of Ma et al. (1997a). This remarkable amount of citric acid released from roots considerably exceeds that from chickpea and other beans (Ohwaki and Hirata 1992), *Paraserianthes falcataria, Acacia mangium, Leucaena leucocephala* (Osawa et al. 1997), *Sanguisorba minor*, and 17 other species of calcifuge and acidifuge species (Ström et al. 1994). This amount is also comparable to or above the amounts released from rape (Luo et al. 1997; Hoffland et al. 1989) and lupin (Dinkelaker et al. 1989; Luo et al. 1997) which among other plants has been reported to secrete the largest amounts of citric acid under phosphate-starved or Al stress conditions, except for where experimental conditions differed. However, one of the difficulties for the strategy of organic acid secretion against Al stress is that *Cassia tora*, which can secrete a remarkable amount of citric acid under Al stress, is not so tolerant to Al.

Next, we investigated the release of organic acids from the roots of Al-tolerant and Al-sensitive cultivars of 5 crop plant species (Table 9). The data are summarized as follows: (i) except for the Al-sensitive pea cultivar, the presence of Al ions in the medium generally increased any kind of organic acid released from roots. Also with the exception of pea, the ratio of citric acid in Al-containing media to that in control media was greater for Al-sensitive cultivars than for Al-tolerant cultivars, (ii) in maize, wheat and pea, the amount of citric acid released in Al-containing media was greater for Al-tolerant cultivars than for Al-sensitive cultivars. However, the relationship between the difference in Al tolerance and that in the amount of citric acid released within the two cultivars almost agreed only in maize; in wheat, the two-fold greater amount of citric acid released did not correspond to the 11-fold greater tolerance in Kitakami B, and vice versa in two cultivars of pea, and (iii) the concentration of citric acid in media was considerably lower than that of Al ions, i.e., less than one-fortieth. Conclusively, the release of organic acids from roots was suggested to be only a partial strategy for tolerance to Al ions and also not to be a general and satisfactory one. The only explanation for this kind of conflict has been ascribed to the presumed high concentration of citric acid in the apoplastic area (Ryan et al. 1995). However, no direct demonstrations have been offered yet.

Apart from efforts to analyse the real mechanisms of Al tolerance, de la Fuente et al. (1997) were able to make tobacco and papaya plants tolerate high Al by producing transgenic plants that over-express a citrate synthase gene from *Pseudomonas aeruginosa* in their cytoplasm. However, no experiments have been reported on the ability of citric acid to detoxify Al ions under soil conditions. In actual soils, organic acids exuded from roots are partly consumed not only for solubilization of Al ions (Jones et al. 1996a) but also for binding with Fe, Mn and Zn (Dinkelaker et al. 1989) and possibly Cu and other ions directly from soil min-

Table 9. Concentrations of organic acids released from Al-treated roots of Al tolerant and Al-sensitive cultivars of 5 crop plant species (nM)[b]

Plant species	Cultivar	(T or S)[a]	Citric acid[d]	Malic acid	Succinic acid[d]
Oryza sativa[c]	Senatore	(T, 77)	165 (2.6)	0	182 (10.9)
Oryza glaberrima[c]		(S, 54)	211 (3.1)	0	85 (1.3)
Maize	DK789	(T, 71)	472 (3.3)	12	156 (3.7)
	XL61	(S, 28)	212 (3.9)	0	122 (7.2)
Wheat	Kitakami B	(T, 80)	179 (2.8)	228	0
	Hachiman	(S, 7.4)	95 (12.4)	78	0
Pea	Harunoka	(T, 90)	251 (4.8)	0	84 (0.5)
	Hyougo	(S, 72)	35 (0.4)	0	52 (1.7)
Sorghum	Energy	(T, 38)	103 (1.4)	0	80 (1.5)
	Power	(S,19)	141 (2.6)	0	109 (1.5)

[a]*T*, Al-tolerant cv, *S*, Al-sensitive cv; numbers in parentheses are Al tolerance, relative net elongation of the primary root in Al medium to that in control medium for 2 days of treatment in 0.2 mM Ca with or without 2 (sorghum), 5 (rice, wheat and pea) or 10 (maize) μM Al, pH 4.9

[b]Control medium was a solution of 0.2 mM Ca (pH 4.9), and Al medium was composed of 20 μM Al and 0.2 mM Ca (pH 4.9). Treatment was carried out for 12 h in 600 ml volume of medium using 10-20 seedlings each under constant pH and Al concentration (seedling numbers used were equal for each crop species). Preparation of solution medium and the analysis of organic acids were based on the method of Ma et al. (1997a)

[c]These are the different rice species

[d]Numbers in parentheses are the ratio of citric or succinic acid in Al medium to that in the control medium

erals. Finally, release of large amounts of organic acids from roots may increase the concentration of trace elements around roots, accelerate the uptake of trace elements by crop plants, and in some cases induce toxicities of trace elements.

Adsorption of citric acid to soils, especially to the secondary clay minerals, is an additional point that weakens the significance of citric acid in Al tolerance in actual soils (Jones et al. 1996a). In our experiments, a mixed organic acid solution composed of 0.1 mM equal concentration of the representative organic acids, i.e., citric, malic and succinic acids, were incubated in acidic conditions with Choukai Andosol from north-eastern Japan. All organic acids, especially citric acid, were immediately adsorbed by the soil (Fig. 10). Among three secondary minerals, goethite was the specific component for the adsorption of citric acid (Fig. 11). Citric acid or other low molecular weight organic acids are easily available for soil microorganisms (Jones et al. 1996b), and this is a remarkable difference from other substances released from roots under other ionic stresses; mugineic acids are released only at a low temperature in the early morning, which are unsuitable conditions for the activity of microorganisms, and piscidic acid is not so easily attacked by microorganisms because it is a phenolic compound.

Special caution should also be paid to the assumed beneficial role of oxalic acid exuded from roots. Solubility of calcium oxalate is very low: 0.57 mg per 100 ml at 20°C. The maximum concentration of oxalic acid in solution is calculated as 45 μM because soils and nutrient solutions generally contain more than 45 μM

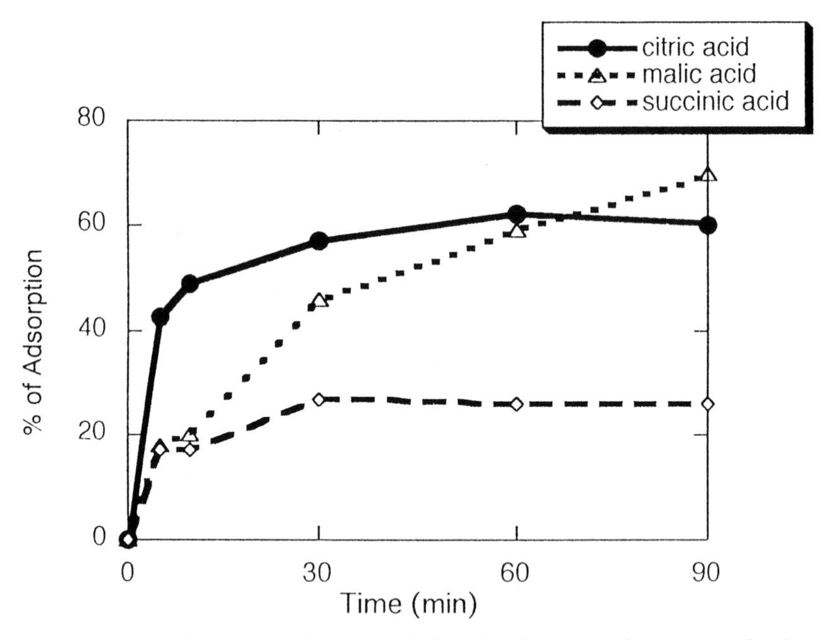

Fig. 10. Adsorption of organic acids by Choukai Andosol in Japan from equimolar (0.1 mM) mixed solution at pH 4.5

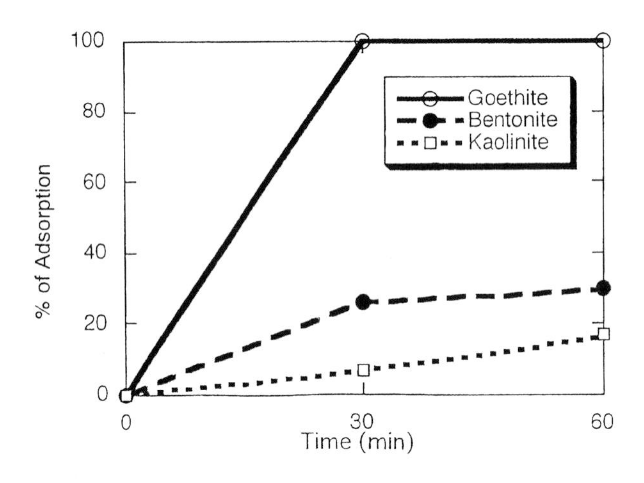

Fig. 11. Adsorption of citric acid from 0.1 mM citrate at pH 4.5 by secondary soil minerals

Ca. If perfect detoxification can be obtained at a 1:3 molar ratio of Al^{3+} to oxalic acid (Ma et al. 1997b) and oxalic acid reacts with Ca in preference to Al^{3+}, the maximum concentration of Al^{3+} to be detoxified in the medium is calculated as at most 15 μM. This point should be clarified further in the future. Moreover, the strategy of oxalic acid release from roots is disadvantageous for Ca nutrition, i.e., a continuous decrease in Ca concentration in the soil rhizosphere solution in acid soils normally deficient in Ca.

The concentration of organic acids in the symplast of roots is higher than the millimolar range and is remarkably higher than the concentration of Al ions in the cell wall area and rhizosphere. A larger amount of organic acids in the symplast can be leaked from the root symplast immediately after the beginning of Al stress because of immediate destruction of the PM. No data have been offered yet on why such Al-stressed or Al-sensitive crop plant species cannot tolerate Al ions or cannot recover from Al toxicity in spite of the leakage of large amounts of organic acids from their roots.

In some cases, especially under deficiencies of some kinds of nutrients, organic substances other than organic acids are produced abundantly, and consequently large amounts of such organic substances can be exuded from the roots. No investigations have been reported yet on the detoxifying effect of such organic substances (for example, phenolics) on Al ions.

To promote better growth in acid soils, research should also be developed on tolerance to low nutrients (phosphorus, calcium and others) under Al stress.

6. Conclusion

In acid soils, crop plants suffer from several growth limiting factors. Among them, Al toxicity is a general and rapid stress for crop plants. Although no common agreement has been accepted yet on whether Al tolerance is an inducible or constitu-

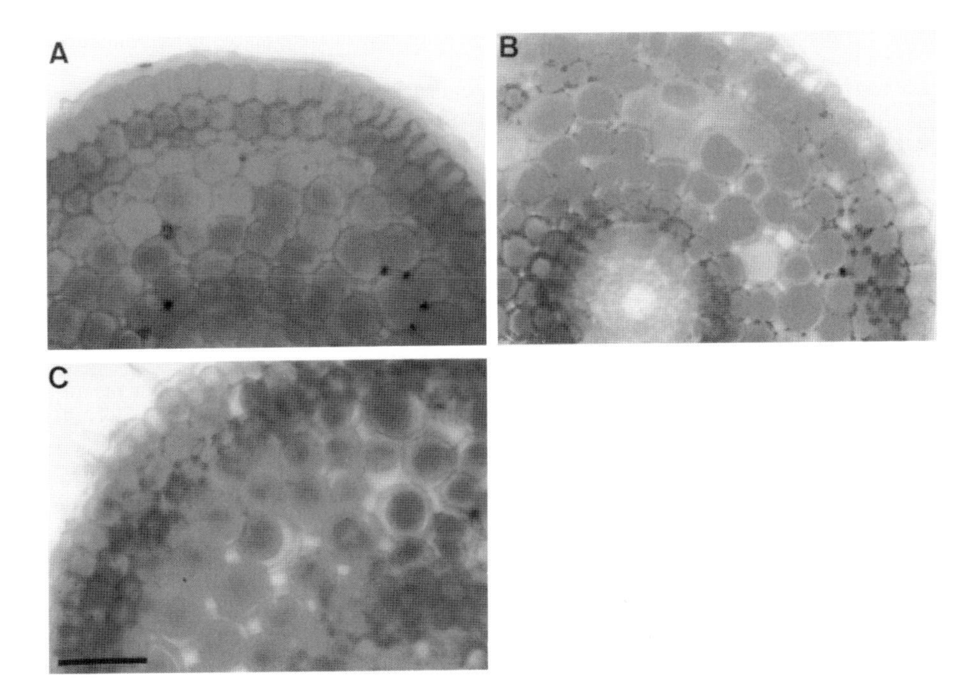

Fig. 12. Plasmolysis of sections of the root-tip (2 mm from the apex) cells of 4-day-old seedlings of rice treated with 0.2 mM Ca (Control) or 20 μM each of trivalent metal ion containing 0.2 mM Ca at pH 4.9 for 24 h (neutral red staining). **A** Control, **B** Al^{3+}, **C** La^{3+}

tional characteristic, the authors are inclined to support the latter characteristic because the response to Al ions or the vast differential Al tolerance among crop plant species or cultivars is developed within the short term of a few tens of minutes of Al stress. As Al ions in medium plasmolyzed the cortex cells (Fig. 12), Al ions are judged to affect the outer surface of the PM as well as the cell wall of the cortex. Among the apoplastic components, although the cell wall has been attracting many researchers for many years as the apparatus for the detoxifier of Al ions with its fixed negative charge, the authors consider the PM to be the controller of Al tolerance because of its rapid response. What kinds of structures and functions of the PM are the causes of differential Al tolerance among crop plant species is a future subject for agronomists, plant nutritionists, or molecular biologists.

References

Akeson MA, Munns DN (1989) Lipid bilayer permeation by neutral aluminum citrate and by three α-hydroxy carboxylic acids. Biochim Biophys Acta 984:200-206

Allan DL, Shann JR, Bertsch PM (1990) Role of root cell walls in iron deficiency of soybean (*Glycine max*) and aluminium toxicity of wheat (*Triticum aestivum*). In: van Beusichem ML (Ed) Plant nutrition – physiology and applications. Kluwer Academic Publishers, Dordrecht, pp 345-349

Archambault DJ, Zhang G, Taylor GJ (1996) Accumulation of Al in root mucilage of an Al resistant and an Al –sensitive cultivar of wheat. Plant Physiol 112:1471-1478

Blamey FPC, Dowling AJ (1995) Antagonism between aluminium and calcium for sorption by calcium pectate. Plant Soil 171:137-140

Canny MJ (1995) Apoplastic water and solute movement: New rules for an old space. Annu Rev Plant Physiol Plant Mol Biol 46:215-236

Cevc G (1982) Water and membranes: The interdependence of their physico-chemical properties in the case of phospholipid bilayers. Studia Biophysica 91:45-52

Chen J, Sucoff EI, Stadelmann EJ (1991) Aluminum and temperature alteration of cell membrane permeability of *Quercus rubra*. Plant Physiol 96:644–649

Cosgrove DJ (1989) Characterization of long-term extension of isolated cell walls from growing cucumber hypocotyls. Planta 177:121-130

Cosgrove DJ (1996) Plant cell enlargement and the action of expansins. BioEssays 18:533-540

Cosgrove DJ, Durachko DM (1994) Autolysis and extension of isolated walls from growing cucumber hypocotyls. J Exp Bot 45:1711-1719

de la Fuente JM, Ramírez-Rodríguez V, Cabrela-Ponce JL, Herrera-Estrella L (1997) Aluminum tolerance in transgenic plants by alteration of citrate synthesis. Science 276:1566-1568

Deleers M, Servais J-P, Wulfert E (1985) Micromolar concentrations of Al^{3+} induce phase separation, aggregation and dye release in phosphatidylserine-containing lipid vesicles. Biochim Biophys Acta 813:195-200

Deleers M, Servais J-P, Wulfert E (1986) Neurotoxic cations induce membrane rigidification and membrane fusion at micromolar concentrations. Biochim Biophys Acta 855:271-276

Delhaize E, Ryan PR, Randall PJ (1993) Aluminum tolerance in wheat (*Triticum aestivum* L.). II. Aluminum-stimulated excretion of malic acid from root apices. Plant Physiol 103:695-702

Dinkelaker B, Römheld V, Marschner H (1989) Citric acid excretion and precipitation of calcium citrate in the rhizosphere of white lupin (*Lupinus albus* L.). Plant Cell Environ 12:285-292

Douglas TJ (1985) NaCl effects on 4-desmethylsterol composition of plasma membrane-enriched preparations from citrus roots. Plant Cell Environ 8:687-692

Esau K (1965) Plant anatomy. John Wiley & Sons, New York, pp 486, 712

Flowers TJ, Yeo AR (1992) Solute transport in plants. Blackie Academic & Professional, London, pp 17-47

Hauser H, Phillips MC (1979) Interactions of the polar groups of phospholipid bilayer membranes. Prog Sur Mem Sci 13:297-404

Henderson M, Ownby JD (1991) The role of root cap mucilage secretion in aluminum tolerance in wheat. Curr Top Plant Biochem Physiol 10:134-141

Hoffland E, Findenegg GR, Nelemans JA (1989) Solubilization of rock phosphate by rape II. Local root exudation of organic acids as a response to P-starvation. Plant Soil 113:161-165

Horst WJ, Wagner A, Marschner H (1982) Mucilage protects root meristems from aluminum injury. Z Pflanzenphysiol 105:435-444

Ishikawa S, Wagatsuma T (1998) Plasma membrane permeability of root-tip cells following temporary exposure to Al ions is a rapid measure of Al tolerance among plant species. Plant Cell Physiol 39:516-525

Ishikawa S, Wagatsuma T, Ikarashi T (1996) Comparative toxicity of Al^{3+}, Yb^{3+}, and La^{3+} to root-tip cells differing in tolerance to high Al^{3+} in terms of ionic potentials of dehydrated trivalent cations. Soil Sci Plant Nutr 42:613-625

Jones DL, Prabowo AM, Kochian LV (1996a) Aluminum-organic acid interactions in acid soils. II. Influence of solid phase sorption on organic acid –Al complexation and Al rhizotoxicity. Plant Soil 182:229-237

Jones DL , Prabowo AM, Kochian LV (1996b) Kinetics of malate transport and decomposition in acid soils and isolated bacterial populations: The effects of microorganisms on root exudation of malate under Al stress. Plant Soil 182:239-247

Jones DL, Kochian LV (1997) Aluminum interaction with plasma membrane lipids and enzyme metal binding sites and its potential role in Al cytotoxicity. FEBS Lett 400:51-57

Kochian LV (1995) Cellular mechanisms of aluminum toxicity and resistance in plants. Annu Rev Plant Physiol Plant Mol Biol 46:237-260

Kochian LV, Jones DL (1996) 4. Aluminum toxicity and resistance in plants. In: Yokel RA, Golub MS (Eds) Research issues in aluminum toxicity. Taylor and Francis, Washington, pp 69-89

Larsson C, Møller IM (1990) The plant plasma membrane. Springer-Verlag, Berlin Heidelberg, pp 16-75

Larsson C, Widell S, Kjellbom P (1987) Preparation of high-purity plasma membranes. Methods Enzymol 148:558-568

Le Van H, Kuraishi S, Sakurai N (1994) Aluminum-induced rapid root inhibition and changes in cell-wall components of squash seedlings. Plant Physiol 106:971-976

Luo HM, Osaki M, Tadano T (1997) Mechanisms of differential tolerance of crop plants to high aluminum and low phosphorus growth conditions. In: Ando T et al. (Eds) Plant nutrition – for sustainable food production and environment. Kluwer Academic Publishers, Dordrecht, pp 473-474

Ma JF, Zheng SJ, Matsumoto H (1997a) Specific secretion of citric acid induced by Al stress in *Cassia tora* L. Plant Cell Physiol 38: 1019-1025

Ma JF, Zheng SJ, Matsumoto H, Hiradate S (1997b) Detoxifying aluminum with buckwheat. Nature 390:569-570

Ma JF, Hiradate S, Matsumoto H (1998) High aluminum resistance in buckwheat. II. Oxalic acid detoxifies aluminum internally. Plant Physiol 117:753-759

Masion A, Bertsch PM (1997) Aluminum speciation in the presence of wheat root cell walls: a wet chemical study. Plant Cell Environ 20:504-512

McQueen-Mason S, Durachko DM, Cosgrove DJ (1992) Two endogenous proteins that induce cell wall extension in plants. Plant Cell 4:1425-1433

Mugwira LM, Elgawhary SM (1979) Aluminum accumulation and tolerance of triticale and wheat in relation to root cation exchange capacity. Soil Sci Soc Am J 43:736-740

Münch E (1930) Die Stoffbewegungen in der Pflanze. p 73 Jena. Fischer 234 pp

Munn DA, McCollum RE (1976) Solution culture evaluation of sweet potato cultivar tolerance to aluminum. Agron J 68:989-991

Nordstrom DK, May HM (1989) 2. Aqueous equilibrium data for mononuclear aluminum species. In: . Sposito G (Ed) The environmental chemistry of aluminum. CRC Press, Boca Raton, pp 29-53

Obi I, Ichikawa Y, Kakutani T, Senda M (1989) Electrophoretic studies on plant protoplasts. I. pH dependence of zeta potentials of protoplasts from various sources. Plant Cell Physiol 30:439-444

Ohwaki Y, Hirata H (1992) Differences in carboxylic acid exudation among P-starved leguminous crops in relation to carboxylic acid contents in plant tissues and phospholipid level in roots. Soil Sci Plant Nutr 38:235-243

Orvig C (1993) The aqueous coordination chemistry of aluminum. In: Robinson GH (Ed) Coordination chemistry of aluminum. VCH Publishers, New York, pp 85-122

Osawa H, Kojima K, Sasaki S (1997) Excretion of citrate as an aluminum-tolerance mechanism in tropical leguminous trees. In: Ando T et al. (Eds) Plant nutrition – for sustainable food production and environment. Kluwer Academic Publishers, Dordrecht, pp 455-456

Pellet DM, Grunes DL, Kochian LV (1995) Organic acid exudation as a mechanism of Al-tolerance in maize (*Zea mays* L.). Planta 197:788-795

Puthota V, Cruz-Ortega R, Johnson J, Ownby J (1991) An ultrastructural study of the inhibition of mucilage secretion in the wheat root cap by aluminum. In: Wright RJ et al. (Eds) Plant-soil interactions at low pH. Kluwer Academic Publishers, Dordrecht, pp 779-789

Rengel Z (1992) Role of calcium in aluminium toxicity. New Phytol 121:499-513

Rengel Z, Robinson DL (1989) Determination of cation exchange capacity of ryegrass roots by summing exchangeable cations. Plant Soil 116:217-222

Ryan PR, Ditomaso JM, Kochian LV (1993) Aluminum toxicity in roots: An investigation of spacial sensitivity and the role of the root cap. J Exp Bot 44: 437-446

Ryan PR, Delhaize E, Randall PJ (1995) Malate efflux from root apices and tolerance to aluminum are highly correlated in wheat. Aust J Plant Physiol 22:531-536

Ryan PR, Skerrett M, Findlay GP, Delhaize E, Tyerman SD (1997) Aluminum activates an anion channel in the apical cells of wheat roots. Proc Natl Acad Sci USA 94:6547-6552

Sonobe S, Takahashi S (1994) Association of microtubules with the plasma membrane of tobacco BY-2 cells *in vitro*. Plant Cell Physiol 35:451-460

Ström L, Olsson T, Tyler G (1994) Differences between calcifuge and acidifuge plants in root exudation of low-molecular organic acids. Plant Soil 167:239-245

Takabatake R, Shimmen T (1997) Inhibition of electrogenesis by aluminum in characean cells. Plant Cell Physiol 38:1264-1271

Tam SC, McColl JG (1990) Aluminum- and calcium-binding affinities of some organic ligands in acidic conditions. J Environ Qual 19:514-520

Vierstra R, Haug A (1978) The effects of Al^{3+} on the physical properties of membrane lipids in *Thermoplasma acidophilum*. Biochim Biophys Res Commun 84:138-144

Wagatsuma T (1983) Characterization of absorption sites for aluminum in the roots. Soil Sci Plant Nutri 29:499-515

Wagatsuma T, Akiba R (1989) Low surface negativity of root protoplasts from aluminum- tolerant plant species. Soil Sci Plant Nutr 35:443-452

Wagatsuma T, Ishikawa S, Obata H, Tawaraya K, Katohda S (1995a) Plasma membrane of younger and outer cells is the primary specific site for aluminum toxicity in roots. Plant Soil 171:105-112

Wagatsuma T, Jujo K, Ishikawa S, Nakashima T (1995b) Aluminum-tolerant protoplast from roots can be collected with positively charged silica microbeads: a method based on differences in surface negativity. Plant Cell Physiol 36:1493-1502

Wagatsuma T, Hitomi H, Ishikawa S, Tawaraya K (1997) Al-binding capacity of plasma membrane of root-tip portion in relation to Al tolerance. In: Ando T et al. (Eds) Plant nutrition - for sustainable food production and environment. Kluwer Academic Publishers, Dordrecht, pp 467-468

Yermiyahu U, Brauer DK, Kinraide TB (1997) Sorption of aluminum to plasma membrane vesicles isolated from roots of Scout 66 and Atlas 66 cultivars of wheat. Plant Physiol 115:1119-1125

Zhang G, Taylor GJ (1989) Kinetics of aluminum uptake by excised roots of aluminum-tolerant and aluminum-sensitive cultivars of *Triticum aestivum* L. Plant Physiol 91:1094-1099

Zhang G, Slaski JJ, Archambault DJ, Taylor GJ (1997) Alteration of plasma membrane lipids in aluminum-resistant and aluminum-sensitive wheat genotypes in response to aluminum. Physiol Plant 99:302-308

Zheng SJ, Ma JF, Matsumoto H (1998) High aluminum resistance in buckwheat. I. Al-induced specific secretion of oxalic acid from root tip. Plant Physiol 117:745-751

Adaptation Mechanisms of Upland Rice Genotypes to Highly Weathered Acid Soils of South American Savannas

Kensuke Okada and Albert J. Fischer

Summary. Tropical sub-humid savannas are generally blessed with climatic conditions such as favorable temperature and sufficient rainfall for agricultural production, but the highly weathered acid soils have hampered crop production in the region due to their high acidity and low nutrient status. Upland rice is an important component of rice-pasture cropping systems which have been developed as both productive and sustainable cropping systems for the savannas. Upland rice has a large genotypic difference in tolerance to acid soils, but the lack of study into the physiological mechanisms behind this genotypic difference has hampered the development of an efficient breeding system. Therefore, the true limiting factor causing the difference between the tolerant and susceptible varieties of upland rice was clarified and the physiological mechanisms for the genotypic difference were investigated.

The severe acidic conditions with high Al saturation and high Al concentration in soil solution were found to be created by the split application of fertilizers (urea and KCl), and were rather limited by time (middle of growing season) and space (topsoil) in upland rice fields in the savannas. The growth reduction of susceptible varieties at a low liming rate was not significantly correlated to the indices of Al toxicity, such as exchangeable Al or Al saturation, nor soil pH, but was significantly correlated with exchangeable Ca. Thus the low Ca was assumed to be the true constraint related to the genotypic difference in tolerance to acid savanna soils. In relation to physiological mechanisms, the root surface of the susceptible variety had a higher apoplastic content of Ca which could be eluted by chelators under non-acidic conditions. However, this fraction of Ca was easily replaced by Al in acidic conditions. The tolerant variety may not depend on the regulating function of cell elongation for this fraction of Ca, and thus root elongation, the earliest and most important phenomenon of acid-soil stress, was less affected by the acidic conditions.

Key words. Acid soil, Aluminum, Aluminum saturation, Apoplast, Calcium, Cation exchange capacity, Chelator, *Oryza sativa* L., Physiological mechanism, Savanna, Tolerance, Upland rice

1. History of the Study of Soil Acidity

Even in the middle of 19th century, the infertility of acid soils was recognized, and the cause of the acidity was ascribed to acid humic substances. Hopkins et al. (1903) first measured the exchangeable acidity by NaCl extraction followed by alkali titration, even before potentiometric methods started to be used for the measurement of soil pH. At the same time, Veitch (1904) found that the NaCl extracts from mineral acid soil contain Al. In Japan, Daikubara and Sakamoto (1910) found that the 1N KCl is the best extractant for the measurement of soil acidity, and concluded that the cause of the soil acidity was Al, as did Veitch (1904).

As soon as scientific studies on tropical agriculture started, the problem of highly weathered tropical soils in the humid tropics was noticed. Extensive studies on the acid tropical soils in Latin America were initiated by various groups such as North Carolina State University (Sanchez 1977), International Center for Tropical Agriculture (CIAT), Brazilian research centers and others. Here, the idea of aluminum saturation was first proposed by Kamprath (1970), which was shown to be useful by many researchers (Kamprath 1984; Farina et al. 1980; Farina and Channon 1991). Because various methods to determine the amount of lime required based on soil pH measurement had been developed and widely used because of their convenience, the method based of Al saturation was a new proposal at that time.

2. Soil Acidity Problem in Highly Weathered Tropical Soils

Of the world's land surface where the agriculture is possible (excluding permanently frozen land and the land covered with ice, rocks etc.), ca. 42% is classified as acid soils (Okagawa 1984) and ca. 74% of the acid soils are classified as "strongly acid" soils. These comprise the Ferralsols (the largest classification among acid soils at 23.3%) and Acrisols (the second largest classification at 18.3%), and others. The cause of the acidity of these two soil types are low saturation of cations due to leaching caused by strong weathering under the tropical humid climate. Most of these two soils exist in the tropics, especially in South America and Africa (43% and 27%, respectively) (Okagawa 1984). In South America, these soils support sub-humid savannas (Cerrados in Brazil, and Llanos in Colombia and Venezuela) and tropical rain forest in the Amazon basin. The degree of utilization of these ecosystems for agriculture is generally very low. The major problems of these soils are the low cation exchange capacity (CEC), low content of cations, the toxicity of exchangeable Al and/or soluble Al, and low level of phosphorus. Because of the long weathering process, silicon availability is also very low.

As was mentioned before, the idea of Al saturation was first proposed for this type of soil and it should be noted that, since the studies began, the inter-relationship between Al and Ca (and other cations) was implied in the idea of Al saturation. Since Ca and other cations have an ameliorative effect against Al toxicity through the cation competition effect, the proportional index such as Al saturation or Al/Ca balance may correlate better with plant growth, even if Al toxicity is the major growth limiting factor.

However, as is shown in Fig. 1, when two soils of the same Al saturation but with different CEC were compared, it was evident that an absolute high concentration of Al may become the major problem in a high CEC soil, while an absolute low concentration of Ca or Ca deficiency (or the deficiency of other cations) may become the major growth limiting factor in low CEC soils.

In some studies on highly weathered acid soils, doubt was cast on Al as the main cause of growth limitation, as in a general case (Zeigler et al. 1995) or specifically for the case of upland rice on Oxisols (Tanaka et al. 1986). In fact, Ca was suggested to be the major limiting factor for growth of some plants on highly weathered soils. For example, Hutton (1985) concluded that the physiological basis for the tolerance of *Centrosema*, a tropical legume forage, on humid savanna soils in Colombia lies in its efficient absorption of Ca but not in tolerance to Al. Blair et al. (1988) mentioned that the main growth limiting factor of *Leucaena* depends on the soil type: on an Oxisol it was Ca (0.17 and 0.59 $cmol_c\ kg^{-1}$ for exchangeable Ca and Al, respectively), but on an Ultisol it was Al (3.78 and 1.00 $cmol_c\ kg^{-1}$ for exchangeable Ca and Al, respectively). Bruce et al. (1988) also concluded that Ca was the main cause for the reduction of soybean yield on various acid soils in Australia.

Therefore, although much effort was initially paid to finding the major growth limiting factor when the problem of soil acidity was first tackled, once Al was established as the main cause of the soil acidity it could be said that less and less effort was paid to elucidate the true limiting factor under various plant growth conditions on different types of soils. The term 'acid-soil syndrome' was coined because of the extreme complexity of the problem, which also may have prompted some to give up on dissecting this syndrome.

In the breeding effort to improve plant genotypes tolerant to acid soils, it may not be necessary to elucidate the major limiting factor as far as the screening is conducted at 'hot spot', i.e. the plants were grown on the target stressful soils and screened based on the performance (usually yield) of each genotype in the field. However, if a laboratory screening method is intended to be developed to improve the precision and efficiency of the screening, or an effort made to identify the genes that relate to the tolerance to the toxicity or deficiency of a certain element is started, it becomes extremely important to elucidate the major limiting factor for the target

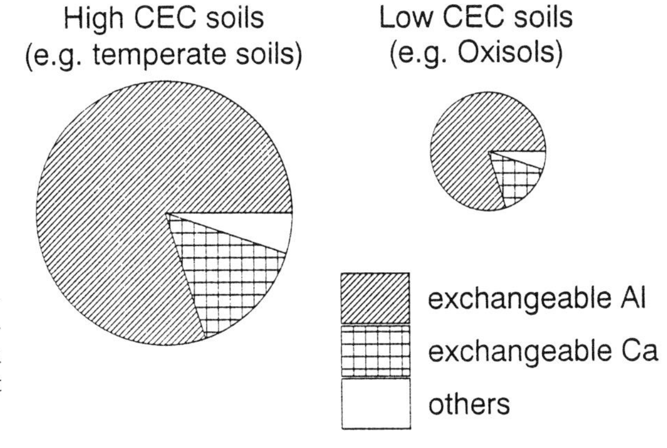

Fig. 1. Amount of exchangeable cations in two contrasting soils with different cation exchange capacity (*CEC*) but same aluminum saturation

combination of plant and soil. This is because the response of plants to the concentration of only one element is measured in laboratory experiments in most cases. Insufficient effort to elucidate the major growth limiting factor before starting laboratory screening might be a reason for the low correlation between the results of laboratory screening and those of field screening in many cases.

In the following part of this chapter, the pursuit of the major limiting factor of the growth of upland rice on the highly weathered acid soils of tropical sub-humid savanna will be presented as an example.

3. Upland Rice: An Important Component of a Rice-Pasture Cropping System on Acid Savanna Soils, and Its Breeding for Acid Soil Tolerance

As was mentioned before, tropical sub-humid savannas are one of the biggest areas of acid soils in the world. Historically, these savannas have been utilized only for extremely extensive cattle raising with native grass pasture. Today, savannas are still used for ranching, or are partly used for crop production such as soybeans after correction of soil pH with high input of lime.

An improved rice-pasture system was developed and introduced into this ecosystem by CIAT and the Brazilian centers (Empresa Brasileira de Pesquisa Agropecuária [EMBRAPA] and Centro Nacional de Pesquisa em Arroz e Feijoao [CNPAF]), and was shown to be successful as a profitable and sustainable cropping system (Toledo et al. 1990). In this system, the native or degraded pasture lands are ploughed and a low rate of fertilizers and dolomitic lime applied at the beginning of the rainy season (the latter not to correct soil pH but to supply Ca and Mg as nutrients) (Fig. 2). Seeds of upland rice are sown by drilling, and the seeds of improved grass and legume are broadcast at the same time. Upland rice is harvested after about 4 months by combine harvester, and is cut at a height higher than normal to avoid excess damage to the pasture which has been growing under the upland rice canopy. By selling the rice, farmers can pay back the initial cost for the establishment such as seeds, fertilizers and machine operation (Toledo et al. 1990). Pastures start to grow rapidly once the rice is harvested, and continue to grow in the dry season. The cattle can then be introduced from the next rainy season. This improved pasture with high productive capacity can be continued for 3 to 5 years after

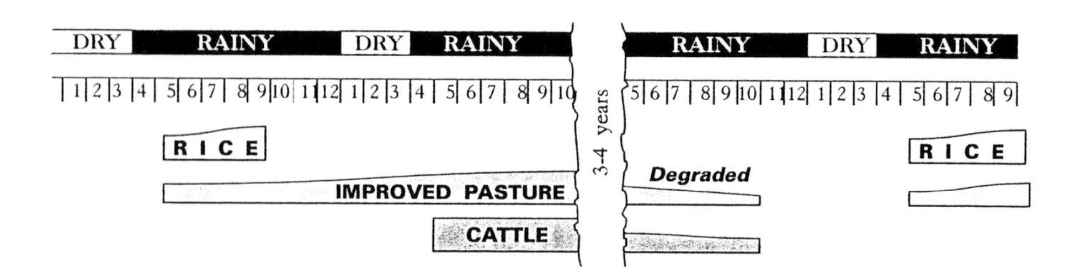

Fig. 2. A schematic time sequence of rice-pasture cropping system

the introduction of the rice-pasture system, depending on the soil characteristics. Once the pastures degrade, the rice-pasture system can be introduced again (Fig. 2). In this way, this rice-pasture system was introduced as an efficient system of renovating native or degraded pasture with early reimbursement of the initial investment by the rice harvest after 4 months (Toledo et al. 1990).

It was only after a new variety of upland rice, "Oryzica Sabana 6", was released in 1991 in Colombia by CIAT and the Colombian Agricultural Institute (ICA) that this system was adopted as a feasible new technology. Oryzica Sabana 6 was selected from a cross of semi-dwarf indicas and tropical japonicas made in 1982, and has the advantages of the both the semi-dwarf indica type (high yield potential and high amylose content) and japonica type (acid-soil tolerance). This new variety has an unusually high yield potential of 3 to 4 t ha^{-1} as an upland rice, has good grain quality with high amylose content and can grow well in Llanos soils with a small dose of lime (300 kg ha^{-1}) to supply Ca. A successor, Oryzica Sabana 10, was also released in 1995. (CIAT 1995).

Breeding is an ever-continuing effort. Whenever new disease-resistance genes or other characters are introduced by crossing there is a risk of introducing susceptibility genes for acid-soil stress. Rice is a relatively tolerant crop to acid soils, but has a wide genotypic difference in the degree of tolerance (Howeler 1991). To establish an efficient screening technique to eliminate lines susceptible to soil acidity from the offspring of the crosses, a clear understanding of the physiological mechanisms behind the genotypic difference in the tolerance to acid soils is indispensable. Therefore, at CIAT we started a project to clarify (1) the true limiting factors, and (2) the physiological mechanisms for acid-soil tolerance, with the purpose of contributing to the improvement of the breeding program.

4. Dynamics of Soil Acidity in Upland Rice Fields in Tropical Savanna Soils of Colombia

Field experiments on the response of upland rice to liming were conducted using both tolerant (Oryzica Sabana 6 and IAC165) and susceptible (Oryzica 1 and Oryzica Llanos 5) genotypes over 4 years at two sites in the Eastern Plains of Colombia (Llanos Orientales).

The soil acidity of the topsoil (Oxisols, typic haplustox, 5.9% organic matter, 77% Al saturation, 0.34 $cmol_c$ kg^{-1} exchangeable Ca) was found to be augmented during crop growth. The soil pH at the beginning of the rainy season was around 4.8, but it decreased to 4.3 during the growing season and then recovered. In accordance with the change of soil pH, the Al concentration in soil solution increased drastically in the middle of the growth period (Fig. 3). At this stage when the soil acidity was at its maximum, severe soil acidity was observed only in the topsoil layer as shown by both a low pH and high Al concentration (Fig. 4). The Ca/Al ratio in the soil solution, which has recently started to be used as the index of soil acidity, also indicated strong acid conditions only at the latter stage of crop growth in the upper soil layer (data not shown). Therefore, the occurrence of soil acidity which might be stressful to upland rice was rather limited in time and space.

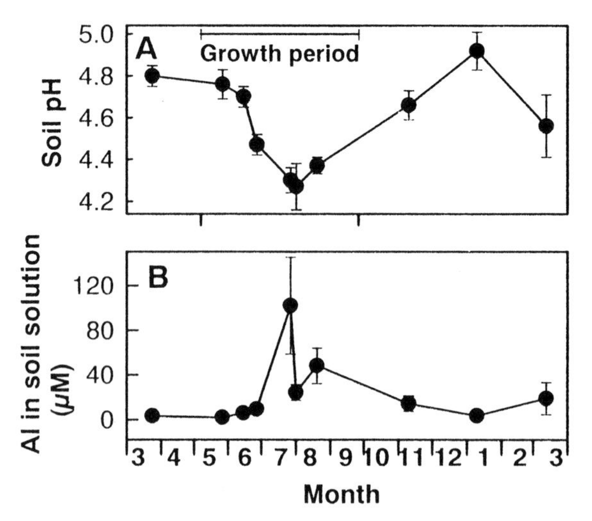

Fig. 3. Change of soil pH (1:1 water) and Al concentration in the soil solution of topsoil (0 to 15 cm) in a field of upland rice in savanna in Colombia, 1995

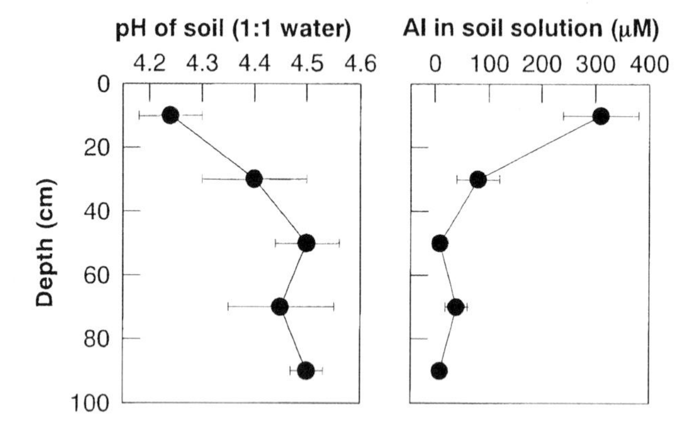

Fig. 4. Distribution of soil pH and the concentration of Al in soil solution, along soil depth at the flowering stage of upland rice

We analyzed the pH of topsoils from 11 sites in the Eastern Plains of Colombia which include various cropping systems (Fig. 5). The occurrence soil pH lower than the critical level (i.e. 4.3) was rather limited, and was only found in upland rice fields with low liming, probably caused by the high rate of fertilizer application. The results of a pot experiment indicated that the increase of soil acidity was caused by the increase of salt concentration in soil solution (which was indicated by high electric conductivity [EC], data not shown) and by a nitrification process, brought about by the accumulation of KCl and urea applied sequentially according to the conventional practice (Fig. 6). It has been well demonstrated that nitrification and the application of other fertilizers enhance soil acidity (Adams 1984). This phenomenon has been observed especially in poorly buffered highly-weathered soils in the tropics (Friesen et al. 1980). These results have shown the sensitive nature of these soils to human intervention and have demonstrated the need to manage this type of soil with due care and knowledge.

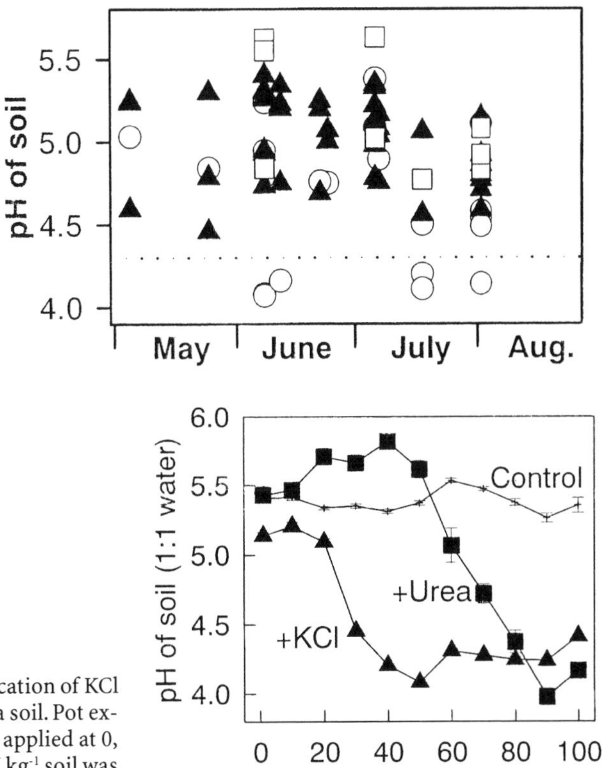

Fig. 5. pH of surface soil at various sites in the Eastern Plains of Colombia in 1995. □, high-liming system with maize or upland rice; ○, low-liming system with upland rice monoculture; ▲, other low-liming systems (native savanna, improved pasture, rice-pasture association, etc.)

Fig. 6. Effect of sequential application of KCl and urea on the pH of a savanna soil. Pot experiment. 25 mg K kg^{-1} soil was applied at 0, 20, 35, and 60 days, and 33 mg N kg^{-1} soil was applied at 15, 35 and 60 days of incubation

5. Response of Upland Rice to Liming on Acid Soils, and the Elucidation of the Major Growth-Limiting Factor

The response of both acid-soil-susceptible and -tolerant varieties of upland rice to the application of lime was investigated in pot experiments conducted in a screenhouse, and also in field experiments.

5.1 Pot Experiment

A pot experiment was carried out using both top- and sub-soil from a native savanna at a farm Matazul in the Eastern Plains of Colombia (Oxisols, typic haplustox). The topsoil (0 to 10 cm depth) had 2.5 ppm P (Bray 2), 84% Al saturation, and exchangeable cations (2.10 Al, 0.18 Ca, 0.13 Mg, 0.09 K; in $cmol_c\ kg^{-1}$), while the subsoil (15 to 30 cm depth) had 1.1 ppm P (Bray 2), 90% Al saturation, and exchangeable cations (1.35 Al, 0.08 Ca, 0.04 Mg, and 0.03 K; in $cmol_c\ kg^{-1}$). An acid-susceptible variety Oryzica 1 and an acid-tolerant variety Oryzica Sabana 6 were used as plant materials.

In the case of the topsoil a varietal difference was observed at a liming rate lower than 25 ppm, which is equivalent to 63 kg ha^{-1} (dolomitic lime; 55% $CaCO_3$ and 35% $MgCO_3$) in the field (Fig. 7). In the case of the subsoil, a genotypic difference in growth appeared from 25 to 100 ppm of application.

The soil pH in the pots at the flowering stage is shown in Fig. 8. Although there was some difference between the varieties, the most striking result is that the pH of non-planted pots was much lower than the planted pots. As a result of the lower pH Al concentration in the soil solution was much higher in non-planted pots than in the planted pots (data not shown). We considered that the more severe acidity in the non-planted pots was due to the applied fertilizers not being absorbed by the growing plant roots but accumulated and increased the EC in the soil solution (Yanai et al. 1995).

The relationship between plant growth and Al concentration in the soil solution is shown in Fig. 9 where total plant growth at higher Al concentrations was shown to be reduced. However, this does not necessarily mean that the difference in these two varieties can be explained by the difference in their tolerance to high Al concentration, since total dry matter at the same Al concentration did not differ between tolerant and susceptible varieties. Rather, we interpreted the results to show that the higher concentration of Al in soil solution was not the cause of the difference but the result was brought about by suppressed plant growth, as discussed above (Figs. 7 and 8).

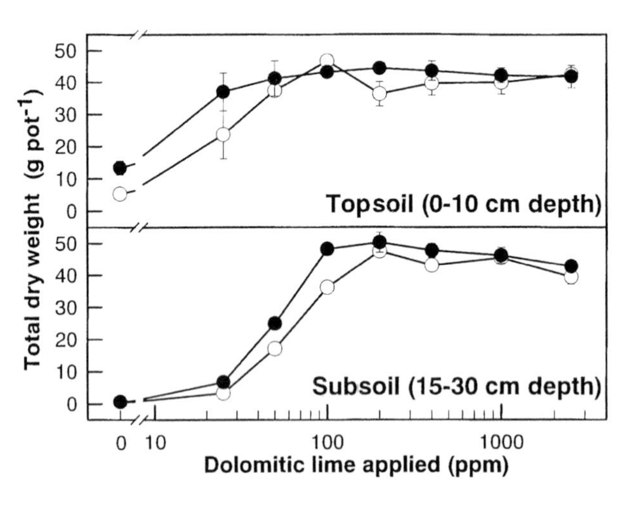

Fig. 7. Response of acid-soil susceptible variety (Oryzica 1, ○) and acid-soil tolerant variety (Oryzica Sabana 6, ●) of upland rice to the application of dolomitic lime on savanna soils (pot experiment)

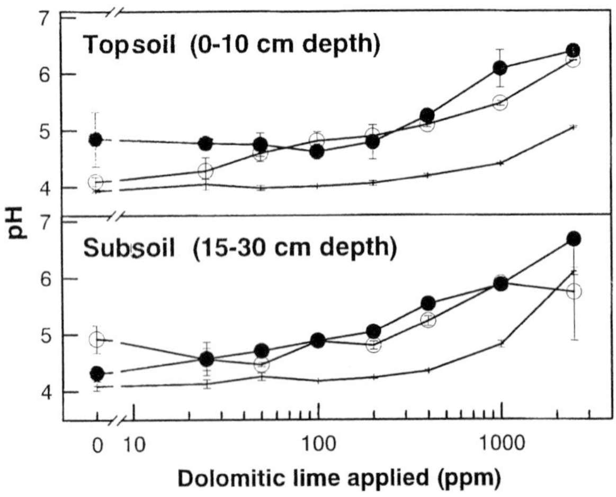

Fig. 8. Effect of liming on the pH of soils in the pots with acid-soil susceptible (Oryzica 1, ○) and acid-soil tolerant (Oryzica Sabana 6, ●) variety of upland rice, or non-planted treatment (+) (pot experiment)

What then was the primary cause that differentiated the growth of the two varieties? Figure 10 shows the relationship between plant growth and the level of exchangeable Ca remaining in the pots at the flowering stage, and it shows that 0.2 $cmol_c\ kg^{-1}$ was the critical level of exchangeable Ca. When growth is plotted against total Ca supply from soil, i.e., the sum of the exchangeable Ca in soils and limed Ca, growth was well correlated even when the results of top- and subsoil were combined (Fig.11). Therefore, since both varieties had the same dry weight under sufficient Ca supply, it was presumed that under the low Ca supply the tolerant variety could continue to grow even when most of the Ca in the pot was absorbed and the Ca level reached a critical level, and thus continued to absorb the added nutrients and maintained the EC in the soil solution low enough. On the other hand, the susceptible variety stopped growing and absorbing nutrients at a smaller plant-growth stage, and so the nutrients in the rhizosphere accumulated and further deteriorated the acid conditions in the pots. Tolerance to low Ca in the soil was the primary cause of the better growth of the tolerant variety, rather than tolerance to high Al. Although the differences between susceptible and tolerant varieties was minimal in this pot experiment in which root volume was limited, the growth difference would be much amplified in fields where enhanced plant growth and root extension could lead to the tapping of more nutrients from the soil.

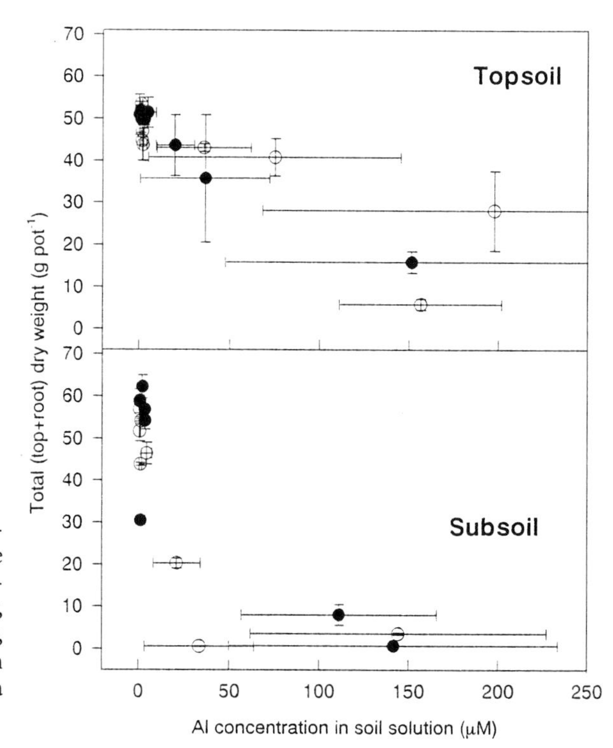

Fig. 9. Effect of Al concentration in soil solution on the growth of upland rice varieties (○: susceptible variety, Oryzica 1; ●: tolerant variety, Oryzica Sabana 6) grown on topsoil and subsoil of savanna (pot experiment)

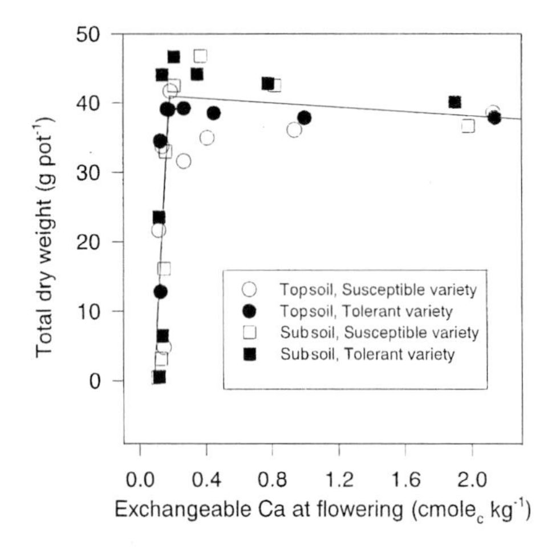

Fig. 10. Relationship between total growth of upland rice at flowering and the remaining exchangeable Ca in the topsoil and subsoil of savanna (pot experiment)

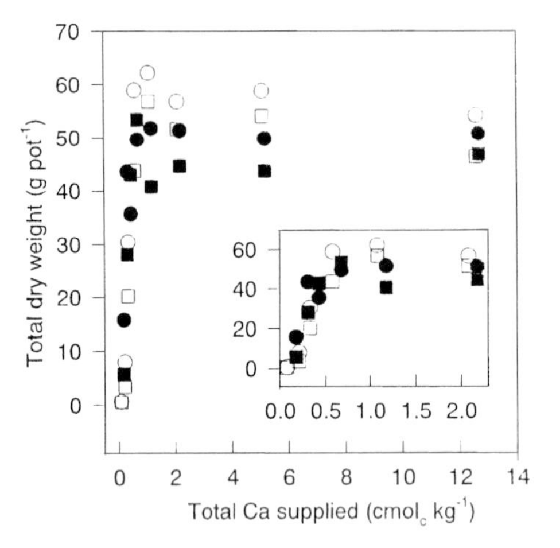

Fig. 11. Effect of total Ca supplied (exchangeable Ca in original soil and Ca supplied with lime) on the growth of upland rice in savanna soils (pot experiment). See **Fig. 10** for explanation of symbols

5.2 Field Experiments

From 1993 to 1996 we conducted field experiments on the response to liming, using two sites (La Libertad and Matazul) in the Eastern Plains of Colombia. The rainfall is higher at La Libertad and the experimental field has been used for cultivation for around 10 years, while at Matazul the native savanna was newly cultivated for the purposes of the experiment. Basically, we compared the growth of both acid-soil susceptible varieties (Oryzica 1 and Oryzica Llanos 5) and tolerant varieties (IAC 165 and Oryzica Sabana 6) with various rates of lime application (0.3 and 3 t ha^{-1} for La Libertad and 0, 0.15, 0.3, 0.6 and 3 t ha^{-1} for Matazul). Except for lime, all other fertilizers were applied sufficiently (80, 40, 75 and 10, for N, P, K and Zn in kg ha^{-1}, respectively). It was expected before the experiment that at the low rate of liming the yield of tolerant varieties would be maintained but that of susceptible ones would decline. However, this yield decline of susceptible varieties was observed only at Matazul in 1995 and 1996 at the very low rate of liming (<300 kg ha^{-1}) as was shown in the relative yield (Table 1).

Table 1. Relative yield of acid-soil susceptible varieries of rice (compared to the yield at 3 t ha^{-1} lime application)

Place		La Libertad		Matazul	
Year		'93	'94	'95	'96
		Oryzica Llanos 5			
Lime Level	0			74 %*	80 %*
(t ha^{-1})	0.15			100 % ns	89 % ns
	0.3	94 % ns	101 % ns	88 %*	89 % ns
		Oryzica 1			
	0			77 %*	
	0.15			87 % ns	
	0.3		110 % ns	96 % ns	

*indicates statistically significant at 5 % level

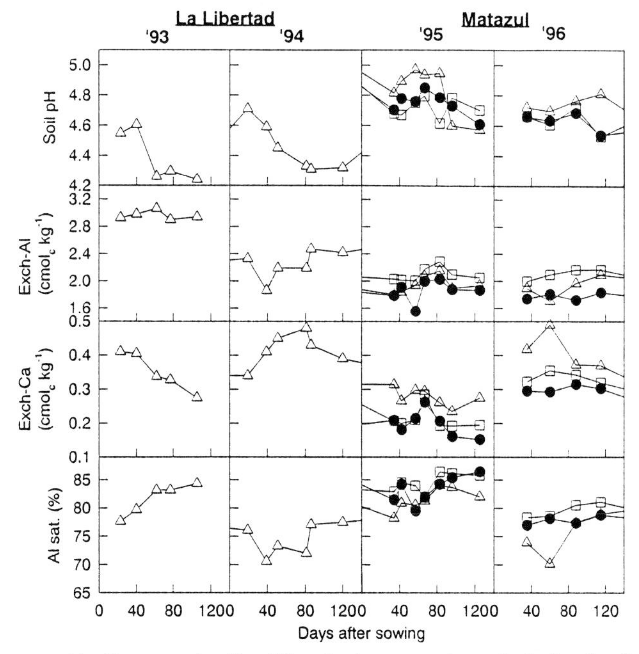

Fig. 12. Changes of indicators of soil acidity during growth period of upland rice in experiments in savannas in Colombia. ● 0, □ 150, △ 300 kg lime ha^{-1} was applied

In this experiment, several indicators of soil acidity were measured through the growing period of upland rice (Fig. 12). Although some changes over time were observed for these indicators, the soil pH was lower and exchangeable Al was higher at La Libertad than at Matazul. These indicators showed an opposite tendency to that which the relative yield of the susceptible varieties indicated. The exchangeable Ca was relatively lower in Matazul, which agreed with the yield results. Al saturation did not differ very much between La Libertad and Matazul.

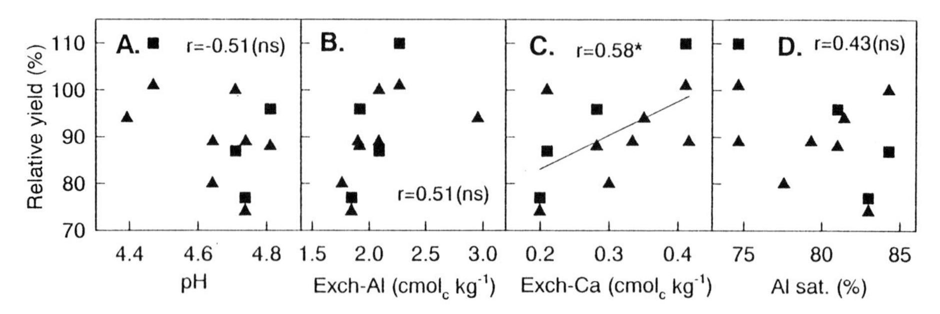

Fig.13. The response of susceptible varieties of upland rice (■, Orizyca 1; ▲, Oryzica Llanos 5) to soil pH (**A**), exchangeable Al (**B**), exchangeable Ca (**C**) and Al saturation (**D**) in the topsoil of savanna (0 to 20 cm). Results of 1993 to 1996 at La Libertad and Matazul fields in the Eastern Plains of Colombia

In summary, the relationship between the relative yield at the low liming rate compared to the yield with 3 t ha^{-1} of liming of susceptible varieties, and the indicators of soil acidity mentioned above averaged over the whole growth period at both sites, are shown in Fig. 13. Here, the relative yield of susceptible varieties was correlated significantly ($P<0.05$) only with the exchangeable Ca of soil (Fig.13c). Other indicators of soil acidity such as soil pH and exchangeable Al had no significant correlation with yield of susceptible varieties, and further, the tendency even seemed contradictory to the indicators of soil acidity (Fig.13a, b). Al saturation, the most widely used indicator for soil acidity, was not correlated significantly, either (Fig.13d). These results suggest that exchangeable Ca should be used as the indicator of soil acidity for upland rice, rather than Al related parameters. This finding has practical implications in choosing stressful sites for on-site field screening for Al tolerance of upland rice: a low-Ca soil rather than a soil of high Al saturation should be chosen. Laboratory screening methods should also put more emphasis on low Ca tolerance.

6. Mechanisms of Genotypic Difference of Upland Rice in Tolerance to Acid Savanna Soils

Aluminum is usually regarded as the major stress factor in acid soils, and therefore, various mechanisms of the tolerance to Al have been proposed and tested. Al tolerance is usually classified into (1) internal tolerance, and (2) exclusion. In the former, Al is absorbed by plant roots, and in many cases it is transported to the shoots. Al is chelated and detoxified by organic acids, as in the case of *Hydrangea macrophylla* (Ma et al. 1997) and *Melastoma malabathricum* (Watanabe et al. 1998), and these plants are usually extremely tolerant to acid soil conditions. On the other hand, many ordinary crops were reported to be excluders.

In the case of the excluders in which Al entrance through the plasma membrane is reduced, we should also take the 'action site of Al' into consideration. If the reaction site is in the symplast, preventing Al entrance through the plasma membrane is enough for the plants to avoid Al toxicity. If, however, the action site is at the membrane/apoplast interface, stopping Al entrance across the plasma membrane is not enough to avoid toxicity. In fact, several hypothesis have been presented referring to the outer surface of the plasma membrane or cell wall as the action site of Al, for example, (1) Al binds with the plasma membrane and affects the membrane integrity (Wagatsuma et al. 1987), (2) Al affects cell wall and reduces its elongation (Blamey et al. 1997), etc. Even in the widely accepted chelate-exudation theory for Al tolerance (see chapter by Delhaize in this volume) by which chelators (organic acids) are thought to bind with and detoxify Al mainly in the apoplast, the action mechanism of Al is not entirely clear.

In the case of rice, even a tolerant japonica variety exuded only small amounts of organic acids (chelators) (Otani et al. 1996), and the contribution of the root exudates may be very low. Therefore, we first investigated the amount of Al which exists in the apoplast held by the negative surface charge or bound by ligands at the root surface. Ca compartmentation was also investigated under acidic conditions, since Ca or Ca/Al status of the soil was seen as playing the key role in the genotypic difference of upland rice as shown in the previous section.

Both acid-soil tolerant (Oryzica Sabana 6) and susceptible (Oryzica 1) varieties of upland rice were grown for 5 days in a simple solution containing only $CaCl_2$ (500 μM) and $AlCl_3$ (0, 30, 100, 300, 600 μM). The pH was adjusted to 4.2 by the daily addition of HCl. The solution was continuously aerated. Compared to the 0 μM Al treatment, root growth was reduced by 50% at the Al concentration of 29 μM for the susceptible variety and 64 μM for tolerant one. The growth difference between the two varieties was greatest at around these concentrations of Al. The root tips (0-1 cm) of the seminal roots were then cut and the content of Al and Ca in the apoplast of these root rips was measured following the method of Tice et al. (1992). Briefly, the root tips were incubated in 10 mL of one of the following eluting solution and were shaken for 30 min. i.e., (1) 5 mM $BaCl_2$, (2) 3.3 mM citric acid, and (3) 5 mM EDTA (pH was adjusted to 4.2 in all solutions). Then each solution was eluted and collected. This process was repeated with another 10 mL of the same solution, and this 10 mL was combined with the former 10 mL. The elements in the combined solution (20 mL) were considered to have existed in the "apoplastic" compartment of the root tips. The concentration of Ca and Al in the eluted solution was measured by an inductively coupled plasma spectrophotometer (ICP).

As is shown in Fig.14a, the Al concentration in the apoplast eluted by $BaCl_2$ increased as the Al concentration in culture solution increased. More Al in the apoplast was held by the susceptible variety than by the tolerant variety at 30 to 100 μM of Al, at which the varietal difference in root elongation was the greatest. This varietal difference in the retention of Al may have reflected the higher cation exchange capacity of the susceptible variety, as was reported for other plant species (see the chapter by Blamey in this volume). When eluted by citric acid or EDTA, no such genotypic difference in the amount of Al was observed (Fig.14b, c).

The Ca content in the apoplast eluted by $BaCl_2$ decreased as the Al concentration in culture solution increased (Fig.14d), which indicates the displacement of Ca held in the apoplast by Al in the surrounding solution. As in the case of Al, more Ca in the

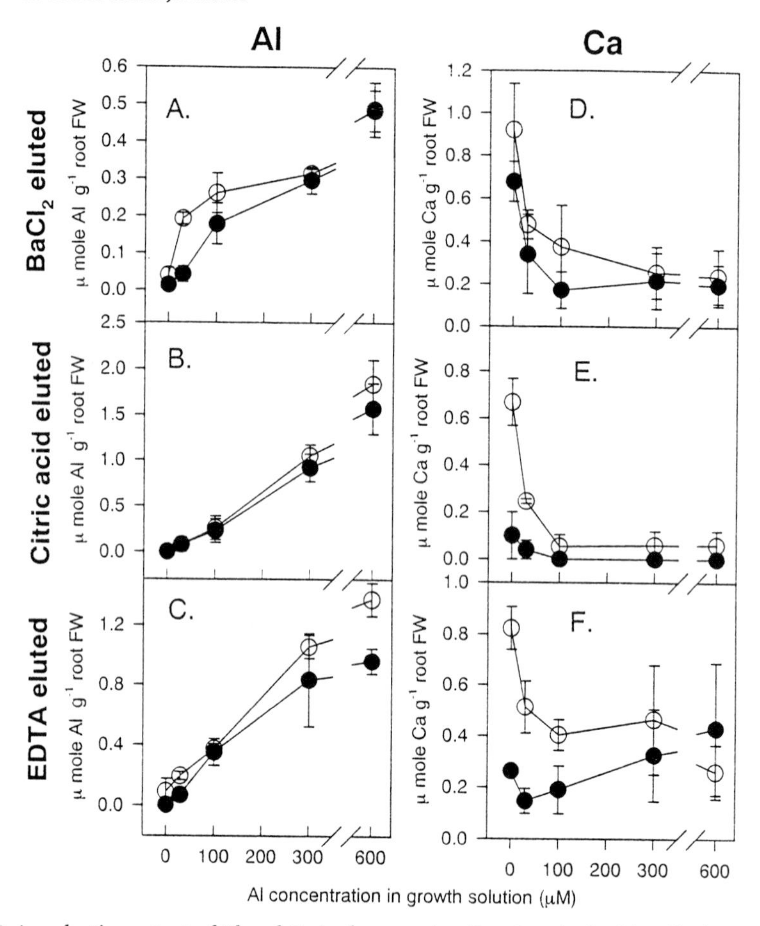

Fig.14. Apoplastic content of Al and Ca in the root tips (0 to 1 cm) of acid-soil tolerant variety (Oryzica Sabana 6: ●) and susceptible variety (Oryzica 1: ○) of upland rice grown with different concentrations of aluminum

apoplast was held by the susceptible variety than by the tolerant variety, probably due to the higher root CEC of the susceptible variety than of the tolerant variety.

The results of the apoplastic content of Ca eluted by the chelators, however, showed a much larger difference between the two varieties (Fig.14e, f). At 0 µM treatment, the susceptible variety had much higher apoplastic content of Ca than the tolerant variety had. The apoplastic Ca content of the susceptible variety, however, decreased by half when the Al concentration in growth solution increased to 30 µM, and then decreased further at the higher Al concentration. On the other hand, the apoplastic Ca content of the tolerant variety, which was very low even at the 0 µM Al treatment, did not decrease much further.

The eluting solution of $BaCl_2$ has a cation displacement ability mainly through its divalent cation Ba, whereas the eluting solution of citric acid or EDTA (both as Na salts) may have a displacing ability through its chelating ligand, because the cation displacement ability of the monovalent cation is much lower than that of the divalent cation.

These results may support the following two mechanisms of genotypic difference in Al tolerance. First, the tolerant genotype has lower root CEC and lower Al retention at the root surface as mentioned above. Second, as Ca held by ligands is assumed to play a key role in regulating the loosening and elongation of the cell wall at elongation site (Hanson 1984), the easy loss of apoplastic Ca in the root tip of the susceptible variety with the presence of Al may cause serious inhibition to normal root elongation, which is the most typical phenomenon of acid soil stress. In the case of the tolerant variety, the original content of ligand-held Ca is lower, and probably Ca held by other mechanisms or some other elements are regulating the cell elongation. Although the Ca displacement theory for Al toxicity was proposed many years ago and was dismissed by several reports, the results of this study suggest the necessity of further study of the same aspect for other crops including rice.

7. Conclusions

Al is widely accepted as the intrinsic growth-limiting factor of acid soils. This view, however, should be verified for each crop in different types of acid soils.

In sub-humid savannas in Latin America, the highly weathered Oxisols are strongly acid soils with pH lower than 5 and Al saturation frequently as high as 70 to 80%. Even in these conditions, the increased Al concentration in the soil solution was a phenomenon rather limited by time and space. The soil chemical stress that differentiated the tolerant and susceptible varieties was more related to the low Ca status rather than the high Al per se. This finding had the following direct implication. (1) Fields with low exchangeable Ca should be chosen for the on-site selection of upland rice for acid-soil tolerance. Choice based on Al saturation, the most common index for acid soils, may mislead in selection. (2) Laboratory screening should also be based on tolerance to low Ca with some level of Al, rather than screening with high levels of Al.

The mechanism for the genotypic difference of upland rice in the tolerance to soil acidity was suggested to be related to the relationship between regulation of cell elongation and ligand-bound Ca at the root apoplast, but further research is needed to clarify this.

Although these results are for upland rice which is relatively tolerant to acid soils, they also suggested, however, that there is room for reconsideration of the focus only on Al in research of acid-soil tolerance of medium tolerant crops such as maize, sorghum, soybean, etc. on Oxisols and Ultisols.

Acknowledgments. This work was carried out at Centro Internacional de Agricultura Tropical (CIAT) in Colombia as a part of the JIRCAS-CIAT collaborative project "Ecophysiological study of upland rice roots in their adaptation to savanna soils in South America (1992-1997)." We especially thank Fredy Perez and other assistants in our laboratory for assisting with the experiments. The help and encouragement of the Rice Program of CIAT is also highly appreciated.

References

Adams F (1984) Crop response to lime in the Southern United States. In: Adams F (Ed) Soil acidity and liming, 2nd edn., ASA, CSSA, SSSA, Madison, Wisconsin, pp 211-265

Blair GJ, Lithgow KB, Orchard PW (1988) The effects of pH and calcium on the growth of *Leucaena leucocephala* in an oxisol and ultisol soil. Plant Soil 106:209-214

Blamey FPC, Edmeades DC, Wheeler DM (1990) Role of root cation-exchange capacity in differential aluminum tolerance of *Lotus* species. J Plant Nutr 13:729-744

Blamey FPC, Ostatek-Boczynski Z, Kerven GL (1997) Ligand effects on aluminium sorption by calcium pectate. Plant Soil 192:269-275

Bruce RC, Warrell LA, Edwards DG, Bell LC (1988) Effects of aluminium and calcium in the soil solution of acid soils on root elongation of *Glycine max* cv. Forrest. Aust J Agric Res 38:319-338

CIAT Rice Program Annual Report (1995) pp 31-55

Daikubara G, Sakamoto Y (1910) Studies on the cause and nature of soil acidity and the distribution of acid soils (in Japanese). Report of Agr Exp Stn 37:1-141

Farina MPW, Sumner ME, Plank CO, Letzsch WS (1980) Exchangeable aluminum and pH as indicators of lime requirement for corn. Soil Sci Soc Am J 44:1036-1041

Farina MPW, Channon P (1991) A field comparison of lime requirement indices for maize. In: Wright RJ et al. (Eds) Plant-soil interactions at low pH. pp 465-473

Friesen DK, Juo ASR, Miller MH (1980) Liming and lime-phosphorus-zinc interactions in two Nigerian Ultisols: I. Interactions in the soil. Soil Sci Soc Am J 44:1221-1226

Hanson JB (1984) The functions of calcium in plant nutrition. In: Tinker PB, Lauchli A, (Eds) Advances in plant nutrition. Vol 1. Praeger Scientific, New York, pp 149-208

Hopkins CG, Knox WH, Pettit JH (1903) A quantitative method for determining the acidity of soils. USDA Bur Chem Bull 73:114-119

Howeler RH (1991) Identifying plants adaptable to low pH conditions. In: Wright RJ et al. (Eds) Plant-soil interactions at low pH. Kluwer Academic, pp 885-904

Hutton EM (1985) *Centrosema* breeding for acid tropical soils, with emphasis on efficient Ca absorption. Trop Agric 62:273-280

Kamprath EJ (1970) Exchangeable aluminum as a criterion for liming leached mineral soils. Soil Sci Soc Am Proc 34:252-254

Kamprath EJ (1984) Crop response to lime on soils in the tropics. In: Adams F (Ed) Soil acidity and liming. Agron Monograph No. 12 (2nd edn), ASA, CSSA, SSSA, pp 349-368

Ma JF, Hiradate S, Nomoto K, Iwashita T, Matsumoto H (1997) Internal detoxification mechanism of Al in *Hydrangea*. Plant Physiol 113:1033-1039

Okagawa N (1984) Global distribution of acid soils and their use (in Japanese). In: Tanaka A (Ed) Acid soils and their agricultural use: Present status and the future in tropics. Hakuyu-sha, Tokyo, pp 21-49

Otani T, Ae N, Tanaka H (1996) Phosphorus (P) uptake mechanisms of crops grown in soils with low P status. II. Significance of organic acids in root exudates of pigeonpea. Soil Sci Plant Nutr (Tokyo) 42:553-560

Sanchez PS (1977) (Ed) A review of soils research in tropical Latin America. Soil Science Department, North Carolina State University at Raleigh, Tech Bull No 219, North Carolina Agr Expt Stn

Tanaka A, Sakuma T, Okagawa N, Imai H, Ito K, Ogata S, Yamaguchi J (1986) Agro-ecological condition of the Oxisol-Ultisol area of the Amazon river system. Report of a Survey of Llanos in Colombia and jungle in Peru, Fac Agric, Hokkaido Univ

Tice KR, Parker DR, DeMason DA (1992) Operationally defined apoplastic and symplastic aluminum fractions in root tips of aluminum-intoxicated wheat. Plant Physiol 100:309-318

Toledo JM, Zeigler RS, Torres F (1990) Sustainable utilization of poor acid soils of tropical America. Proc 24th Intl Symp Trop Agric Res, Kyoto, Aug 14-16, 1990. Trop Agric Res Ser 24:133-145

Veitch FP (1904) Comparison of methods for the estimation of soil acidity. J Amer Chem Soc 637-662

Wagatsuma T, Kaneko M, Hayasaka H (1987) Destruction process of plant root cells by aluminum. Soil Sci Plant Nutr 33:161-175

Watanabe T, Osaki M, Yoshihara T, Tadano T (1998) Distribution and chemical speciation of aluminum in the Al accumulator plant, *Melastoma malabathricum* L. Plant Soil 201:165-173

Yanai J, Araki S, Kyuma K (1995) Effects of plant growth on the dynamics of the soil solution composition in the root zone of maize in four Japanese soils. Soil Sci Plant Nutr 41:195-206

Zeigler RS, Pandey S, Miles J, Gourley LM, Sarkarung S (1995) Advances in the selection and breeding of acid-tolerant plants: Rice, maize, sorghum and tropical forages. In: Date RA et al. (Eds) Plant soil interactions at low pH, Kluwer Academic, Netherlands, pp 391-406

The Role of the Root Cell Wall in Aluminum Toxicity

Frederick Pax C. Blamey

Summary. Aluminum (Al), the most common metal of the earth's crust, is highly toxic to roots of many plant species when present in solution as monomeric cations (e.g., Al^{3+}, $AlOH^{2+}$) at a concentration as low as 10 μM. Despite this long-known effect, there is little consensus on the biochemical basis of Al toxicity, which may be manifest either external to or within the symplasm. There is increasing evidence that the vast majority of Al in roots occurs in the walls of cells at the apex of plant roots; ≥99% of the Al in *Chara corallina* nodal cells occurs in the cell wall, for example. The Al which accumulates at the apex of plant roots is bound both rapidly and strongly by negatively-charged pectic compounds (predominantly α-1,4 D-polygalacturonic acid). It is proposed that the accumulation of Al in the cell wall exerts a detrimental effect on root growth and function in three ways. First, the decrease in apoplastic sorption of basic cations, which have limited ability to displace bound Al, reduces nutrient acquisition per unit root length. Second, the Al sorbed in the cell wall reduces cell expansion, thus reducing root elongation (a major visible effect of soluble Al). This would also reduce nutrient uptake through decreased root proliferation through the soil. Third, sorption of Al in the cell wall reduces the movement of water and solutes through the apoplasm, directly decreasing nutrient acquisition by the root.

Key words. Aluminum, Calcium, Cation-exchange capacity, Cell expansion, Cell wall, pH, Pectin, Root elongation, Soil solution, Sorption

1. Introduction

The father of agriculture gave us all a hard calling, and did not will that the way of cultivation be easy. (Virgil 30 BC)

There are many reasons for the poor growth of plants on acid soils, making the study of acid soil infertility one of the most intriguing and interesting topics in plant nutrition. The epithet, the acid soil infertility complex, is well justified, and studies of root growth and function in acid soils address the important issues of how plants recognize, respond to and, in many instances, modify their environment.

Nutrient acquisition, the topic of this book, is reduced in acid soils due to either the low absolute levels of nutrients or their low availability to plants, especially with respect to phosphorus (P), calcium (Ca), magnesium (Mg) or molybdenum

(Mo). Further, nutrient acquisition is reduced by toxicities induced by soluble aluminum (Al) or manganese (Mn). Specific limitations reduce plant growth in each individual soil, but Al toxicity is probably the most important limitation to plant growth on acid soils worldwide (Foy 1984; Horst 1995). Aluminum is the most common metal of the earth's surface, comprising ca. 8% of the earth's crust and ca. 7% on average in soils (Lindsay 1979). Most of the Al in soils is in aluminosilicate compounds, and has no direct effect on plant growth. In acid soils, however, Al enters solution, and is toxic to plant roots and other biological organisms in concentrations as low as 10 μM when present as monomeric Al (e.g. Al^{3+}, $AlOH^{2+}$). Besides increasing the plants' susceptibility to drought, soluble Al in the root environment decreases nutrient uptake. This is seen by the innumerable instances of nutrient deficiency symptoms in plants growing on acid soils. These symptoms are evident also in nutrient solution culture studies even with adequate concentrations of nutrients.

Despite decades of concerted research, there is no consensus among plant physiologists regarding the biochemical basis of Al toxicity. It is accepted that Al has its primary effect on root growth, specifically at the root apex (Ryan et al. 1993). Possible sites of action include the cell wall, plasmalemma, cytoplasm and nucleus. Aluminum reacts with many compounds in each of these compartments which all occur in close proximity. This, along with the low concentration at which Al is toxic and the rapidity with which Al exerts its toxic effect and reduces root growth, makes studies on Al toxicity difficult. Much research has addressed the reactions of Al with cell wall components (Rengel and Robinson 1989; Taylor 1991; Blamey et al. 1993b; Rengel 1992; Horst 1995). This chapter, with some attention to the chemistry of Al, focuses on three aspects: (1) the accumulation of Al in the cell wall; (2) the strong binding of Al in the cell wall; and (3) the consequent impairment of root growth and function.

The location and timing of symptom development, as with all symptoms of nutritional disorders, are useful in the diagnosis of Al toxicity. The most obvious visible effects of soluble Al are a severe reduction in root elongation (Fig. 1), along with root distortion and damage to the outer cortical and epidermal cells near the root tip (Fig. 2) and reduced root hair development (Hecht-Buchholz et al. 1990; Brady et al. 1993). These effects occur rapidly, certainly within 1 day and in many instances within a few hours of root exposure to soluble Al (Horst et al. 1992; Lazof et al. 1994). Indeed, Rengel (1992) claimed that the effects of Al could be measurable "within minutes or even seconds following an exposure to Al." Any hypothesis of the toxic effect of Al must accommodate all these observed effects.

My interests in the reactions of Al with cell wall components developed from investigations on differences among plant genotypes in sensitivity to soluble Al, a topic reviewed by Carver and Ownby (1995) in wheat (*Triticum aestivum* L.). Before this could be addressed, however, it seemed useful to develop some understanding of the biochemical basis of toxicity. Basic chemistry indicates that Al has a high charge to ionic size ratio, and would react strongly with negative charges specifically associated with pectic compounds in the root cell wall. Also, Al behaves like many heavy metals in that it is not readily translocated from roots to

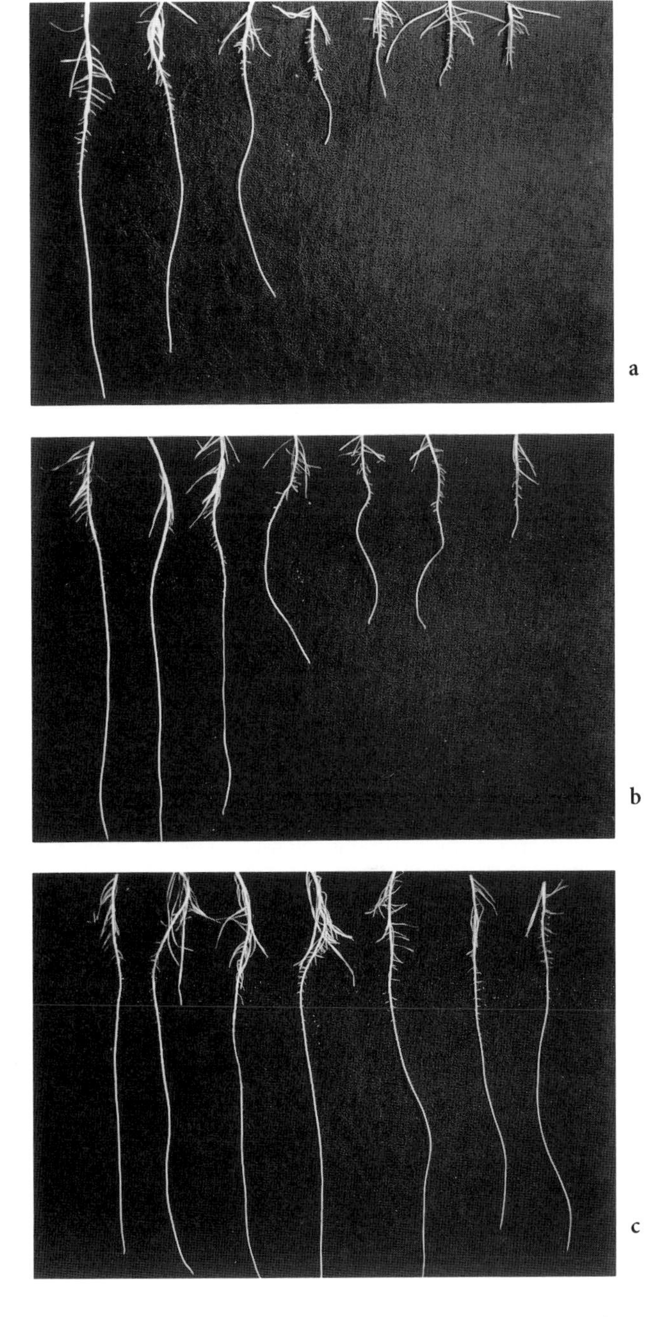

Fig. 1. Aluminum effects on soybean root length grown for 4 days in solution containing (*left to right*) 0, 5, 10, 15, 20 30 and 40 μM Al in solution **a** to which there was no added OH (pH 4.33 ± 0.06), **b** with a 1:1 OH:Al molar ratio (pH 4.47 ± 0.08), and **c** with a 2:1 OH:Al molar ratio (pH 4.70 ± 0.04) (data presented by Blamey et al. 1983)

tops, at least in most species, an exception being tea (*Camellia sinensis* (L.) Kuntze) (Matsumoto et al. 1976). Results of two seminal studies (Vose and Randall 1962; Foy et al. 1967), in which it was shown that Al-sensitive genotypes accumulate more Al in the roots than do Al-tolerant ones, led to further investigations of the relationships between Al and the negative charges in the walls of cells at the root apex.

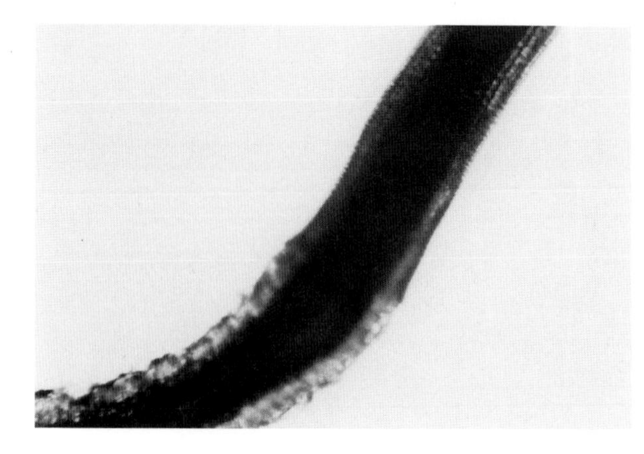

Fig. 2. Aluminum-induced damage to the outer cortical and epidermal cells of a root of *Lotus corniculatus* cv. Maitland 2 days after adding 10 μM Al. The undamaged root towards the top of the photograph developed prior to exposure to soluble Al

2. The Root Cell Wall: Structure and Function

The cell wall is not simply a rigid box, but a responsive mosaic replete with diverse form and function. (Carpita and Gibeaut 1993)

2.1 Cell Wall Composition

It is necessary first to address the structure and function of the cell wall before evaluating any possible effects that Al might have. Further, the effect of Al on root elongation, which occurs near the root apex (Ishikawa and Evans 1995), focuses attention on the primary cell wall. The excellent review of cell walls by Carpita and Gibeaut (1993) has listed the important components of the dicot cell wall to include cellulose, hemicellulose (xyloglucan) and pectin (predominantly α-1,4 D-polygalacturonic acid) along with lignin and numerous enzymes, including extensin. In the grasses (Poaceae), xyloglucan is largely replaced by glucuroarabinoxylan, and the cell walls are low in pectin (Carpita and Gibeaut 1993), some roles of which are assumed by glucuroarabinoxylan.

Cellulose is an unbranched β 1,4 glucan with a high degree of polymerization, and the individual cellulose micro-fibrils form fibres with crystalline and non-crystalline regions (Brett and Waldron 1996). (It is perhaps noteworthy that callose (β 1,3 glucan) forms in response to wounding, the polysaccharide precursors being diverted from the synthesis of cellulose.) Xyloglucan (α-1,4 glucose residues to which are bound xylose residues via α-1,6 bonds) is the predominant hemicellulose in the primary cell wall of dicots, though numerous other forms exist (Brett and Waldron 1996). The hemicellulose is bound to cellulose via strong hydrogen bonding, making extraction difficult. Pectic compounds make up about 30% of the total cell wall mass, are rich in galacturonic acid, and "are some of the most complex polymers known" (Carpita and Gibeaut 1993). Two main types have been recognized, viz. rhamnogalacturan I (RGI) which consists largely of unbranched α-1,4 D-polygalacturonic acid, with interspersed rhamnose residues, and rhamnogalacturan II (RGII), which is highly-branched. The RGI is arranged in an "egg-box" arrangement facilitated by Ca (Brett and Waldron 1996). Importantly, the uronic acid residues of pectin vary in degree of methylation or esterification of the galacturonic acid residues (Fig. 3). Those pectic compounds with highly methylated carboxylic acid residues (i.e., where H is replaced by CH_3) result in roots of low cation-exchange capacity (CEC) since there is no dissociation of the methyl groups.

Most polysaccharides of the cell wall matrix are synthesized in the Golgi apparatus though there is evidence that some synthesis occurs in the endoplasmic reticulum (ER) (Brett and Waldron 1996). Either the raw products or some synthesized polysaccharides move within vesicles from the ER to the Golgi apparatus, where most synthesis occurs. These synthesized products then move within vesicles towards the plasmalemma, and are excreted into the cell wall via exocytosis. As will be evident later, this may have important implications in our understanding of Al toxicity.

2.2 Primary Cell Wall Structure

A knowledge of primary cell wall structure is essential to elucidate any possible effects of Al on root function. Cellulose is important in providing cellular rigidity, the cellulose microfibrils being linked by hemicellulose. For the cell to expand, the hemicellulose must be cleaved by the appropriate hemicellulase enzyme. Pectin provides the gel-like structure between the cellulose and hemicellulose microfibrils (analogous to concrete in a steel-reinforced building), governing (1) the CEC of the plant root; (2) the access of the hemicellulase enzyme to its hemicellulose substrate; and (3) water and solute movement into and out of the symplasm and via the apoplastic pathway.

Because pectin is a weak acid (Fig. 3), plant roots have been considered to behave as a variable-charge compound in the soil (Blamey et al. 1993b), an increase in pH increasing the binding of cations in the cell wall. It is noteworthy that pectic compounds do not occur only in the middle lamella but are present throughout the cell wall (Sasaki and Nagahashi 1989), and are excreted as part of the mucilage at the root tip, reportedly providing protection from the toxic effects of Al (Horst et al. 1982).

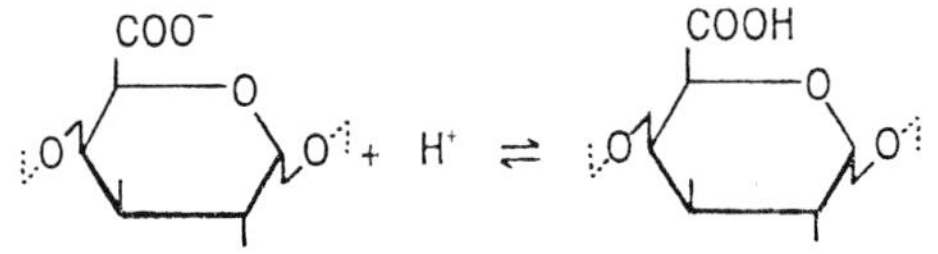

COO^- … O … $+ H^+ \rightleftharpoons$ … COOH … O …

Fig. 3. Non-methylated galacturonic acid residues are present as part of the pectin in cell walls. The carboxylic acid groups of these residues are able to dissociate, the extent of which decreases with a decrease in pH (i.e., increased H^+ activity drives the reaction to the right)

2.3 Cell Wall Functions

As eloquently described by Carpita and Gibeaut (1993), cell walls have numerous functions. These may be summarized as containment of cytoplasm, control of cell elongation (specifically in the elongation zone of the root apex) and control of apoplastic movement of solution. Besides these functions, there are numerous extra-cytosolic enzymes present in the apoplastic space, all of which play a role in root growth and function.

The containment of the cytoplasm, and associated cell turgor and plasmolysis, may be regarded as rather uninteresting. However, there are probably many mechanisms associated with pressure within the cell whereby roots perceive their environment and respond accordingly, e.g., in response to reduced water availability in the rhizosphere (Passioura and Stirzaker 1993) and in relation to Al and signal transduction (Bennet and Breen 1991a).

The role of the cell wall in controlling cell expansion, and the consequent effect on root elongation, is of particular interest because of the clearly-visible effect of soluble Al on root length (Fig. 1). The control of cell expansion has received much attention in the recent past, but there is still no consensus on this topic (Carpita and Gibeaut 1993; Cosgrove 1993). Also, the signals which initiate, control the rate of elongation and finally result in cessation of expansion are little understood. However, one particular aspect of root elongation that has attracted much research is acid mediated cell expansion. This involves the excretion of H^+ ions in response to increased cell turgor, with the H^+ ions replacing Ca^{2+} ions bound by negative charges in the cell wall. The decrease in apoplastic pH results in many changes, including a decrease in the amount of Ca bound by pectic compounds as detailed in Eq. 1, most likely changing the physical and chemical characteristics of these compounds:

$$CaX + 2H^+ \rightleftharpoons 2HX + Ca^{2+} \qquad (1)$$

where X = the negatively-charged pectic compounds.

Specifically, a decrease in pH would result in the displacement of Ca (and other basic cations) from the pectic gel, allowing the individual RGI polymers to move apart.

Carpita and Gibeaut (1993) stated that "artificial gels possess remarkable capacity for reversible volume changes induced by small changes in temperature, type and concentration of salts, pH, and electrical fields." A decrease in apoplastic pH (and consequent decrease in adsorbed Ca) might well increase the access of hemicellulase to its substrate, facilitating the cleavage of hemicellulose and thus allowing cell expansion. This is in addition to the effect of pH on the activity of enzymes which directly catalyze the degradation of pectin.

Again, it is intriguing that it is the pectic compounds in the root cell wall that govern the movement of water and solutes into and out of the cell (Carpita and Gibeaut 1993). The pore size in primary cell walls is of the order of 4 nm. However, not all pores are of the same size, with considerable variation both within and among cells, and variation among genotypes is also likely. Findings that gel volume varies with environmental conditions, as discussed in the previous paragraph, suggest that there are both temporal and spatial variations in pore size also. Larger molecules than those with 4 nm diameter may pass through to the plasma membrane either through larger pores or through reptation (Baron-Epel et al. 1988).

Steudle and Frensch (1996) described a conceptual model of the symplastic and apoplastic movement of solution (Fig. 4a), and further used a mathematical model to evaluate these concepts. Clarkson (1996) presented a conceptual model of the pores through which water, cations and anions move through the cell wall (Fig. 4b). It is important to note that the pectic compounds are highly hydrated and that their gel qualities ensure that the pore sizes are not fixed, but vary in response to external conditions, e.g., Ca concentration (Blamey and Dowling 1995).

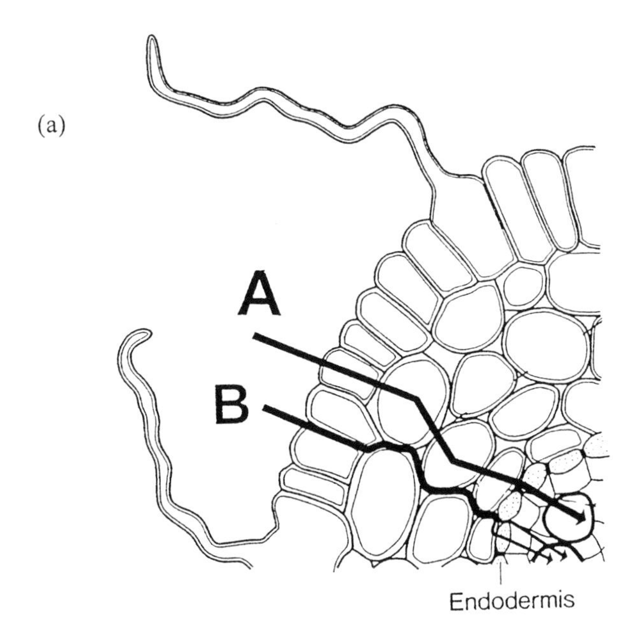

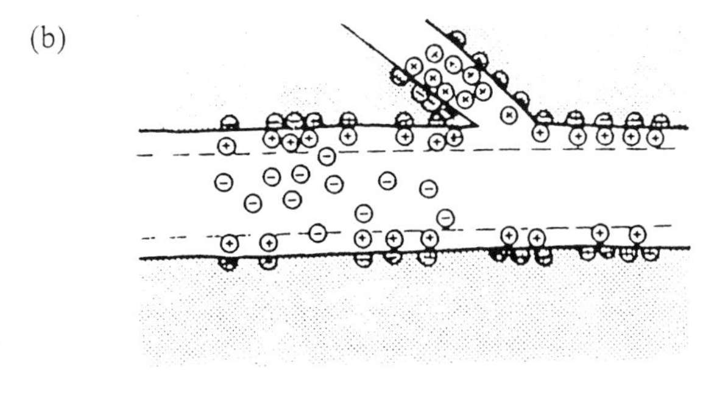

Fig. 4. a A conceptual model of the movement of ions and water through the symplastic (*A*) and apoplastic (*B*) pathways (after Steudle and Frensch 1996) and **b** a conceptual model of the pore sizes within the cell wall. (after Clarkson 1996)

Finally, the composition of the pectic compounds determines root CEC. Knight et al. (1961), and Haynes (1980) determined that increased methylation of uronic acids in root cell walls decreases root CEC. A decrease in root CEC reduces the binding of cations in the cell wall, though this might not affect the overall uptake and translocation of nutrients (Ando 1970).

2.4 Cell Expansion and Root Elongation

Root elongation is a function of, first, cell division at the root meristem and, second, expansion of these cells, the latter having overall the greater influence on root length. There are many specialized zones (Fig. 5) of development in roots (Ishikawa and Evans 1995). Of particular importance to the scope of this chapter are the zones of elongation and of root hair development, given the effects of Al on root length (Fig. 1) and on root hair development (Hecht-Buchholz et al. 1990; Brady et al. 1993).

It is noteworthy that root elongation is determined by forces exerted by the inner layer of the root cortex (Pritchard 1994). Many stresses (e.g., low temperature, soil compaction) affect root elongation, and "in many cases alteration of root growth is consistent with changes in the cell wall properties of the growing cells" (Pritchard 1994).

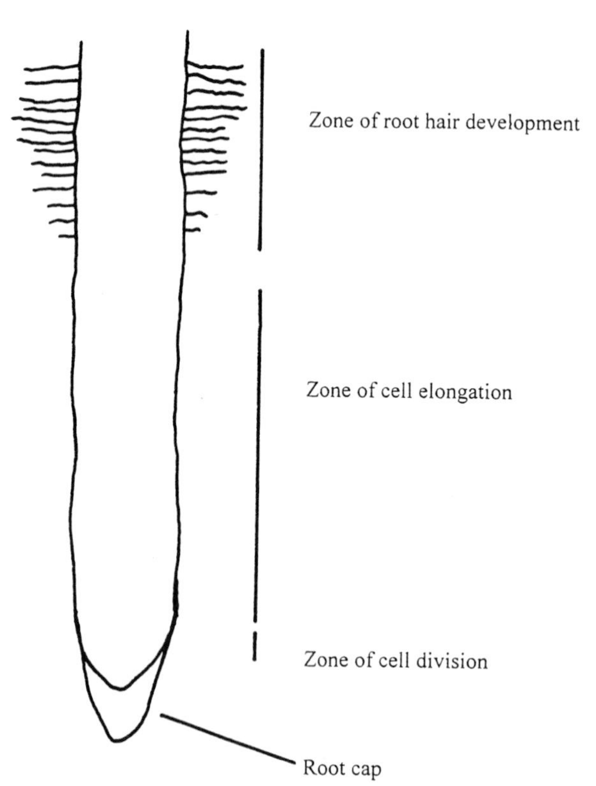

Fig. 5. The root apex illustrating the root cap and the zones of cell division, cell elongation and root hair development. There is some overlap between the zones of cell division and elongation

In contrast with root elongation, root hairs and pollen tubes grow from the tip which is a zone of intense cellular activity. The presence of Ca^{2+} on the external face of the membrane at the tip is necessary for root hair integrity (Jones et al. 1995).

3. Aluminum Chemistry

The chemistry of aluminum in soils and natural waters has emerged as a scientific problem of critical importance because of widespread public interest, (and) environmental chemists...have substantiated the great complexity of aluminum biogeochemistry. (Sposito 1995)

Before addressing the complex chemistry of Al in soils, it is worthwhile considering the position of Al in the Periodic Table (Fig. 6), and to focus on differences between Al and the two major essential basic cations in the root environment, Ca and Mg. These three ions are all Class A (oxygen-seeking) ions, which bind preferentially with F^-, O^{2-}, OH^-, SO_4^{2-}, NO_3^- or C=O-O$^-$ (Nieboer and Richardson 1980). Other Class A metal ions include all elements in the first three columns of the Periodic Table (except H) and the Lanthanide and Actinide series.

Many studies have shown that Al interferes with the uptake and utilization of Ca and Mg. Both Ca and Mg are divalent, with a s^2 outer electron shell; Al, on the other hand, is trivalent at acid pH with a sp^3 outer electron shell. This, along with its relatively small ionic radius (especially *vis a viz* Ca) makes Al highly reactive, and accounts also for its interference with Mg metabolism (Macdonald and Martin 1988; Martin 1994). Indeed, symptoms of Ca or Mg deficiency (Rengel 1990; Rengel and Robinson 1990; Keltjens 1995; Tan et al. 1992; Tan and Keltjens 1995) in plant tops are a common result of high soluble Al in the root environment. The question remains, and will be addressed to some extent below: How does soluble Al reduce the uptake of Ca, Mg and other nutrients?

Following the classic paper by Jackson (1963), excellent reviews have been published in more recent years describing the complex chemistry of Al (e.g., Lindsay 1979; Sposito 1996) and the possible phytotoxic effects of the numerous forms of Al present in soils and in natural waters (Kinraide 1991). Asher et al. (1992) listed the

Table 1. Forms of Al potentially present in solution extracted from acid soils (after Asher et al. 1992)

Form	Examples
Micro-crystals	Kaolinite, Gibbsite
Amorphous precipitates	Al_2SiO_5, $Al(OH)_3$, $AlPO_4$
Inorganic polycations	$Al_2(OH)_2^{4+}$ $Al_3(OH)_4^{5+}$
	$AlO_4Al_{12}(OH)_{24}(H_2O)_{24}^{7+}$
Organic complexes	
Low relative molecular mass	Ca-citrate, Al-oxalate
High relative molecular mass	Al-fulvate, Al-humate
Inorganic complexes	AlF^{2+}, AlF_2^+, $AlSO_4^+$
Inorganic monomers	Al^{3+}, $AlOH^{2+}$, $Al(OH)_2^+$

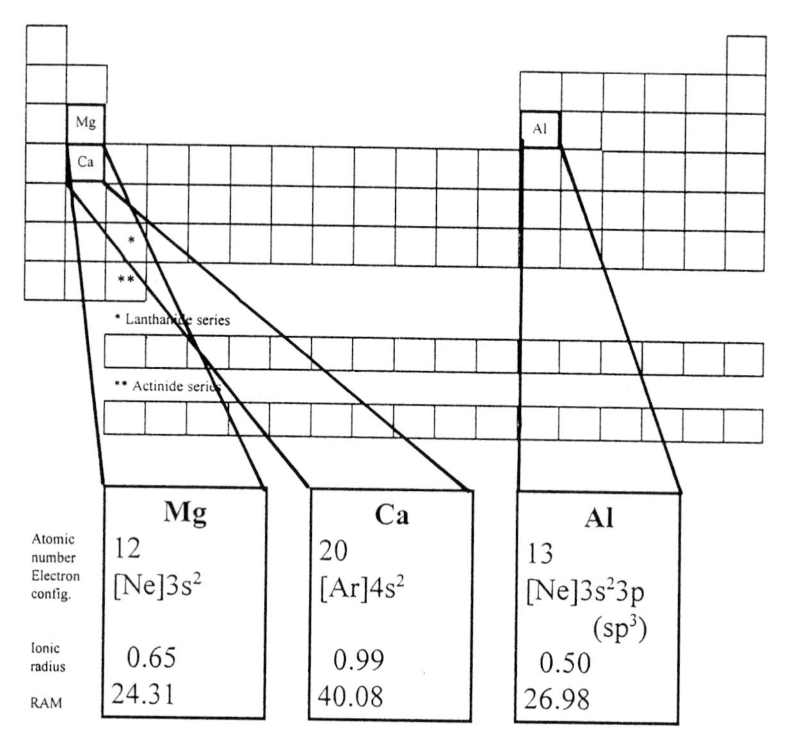

Fig. 6. The positions of magnesium (*Mg*), calcium (*Ca*) and aluminum (*Al*) in the Periodic Table, along with the atomic number, electron configuration, ionic radius, and relative atomic mass (*RAM*) of these elements

forms of Al potentially present in solutions extracted from acid soils (Table 1).

Micro-crystals and amorphous precipitates, both of which are measured with non-discriminatory techniques such as Inductively Coupled Plasma Atomic Emission Spectroscopy (ICPAES), are considered non-toxic. Of the inorganic polymers, the tridecameric polycation, Al_{13}, has been shown to be highly toxic when tested in simple solution culture (Parker et al., 1989) (see below). It is doubtful, however, if this highly-charged cation is present in soil solutions given the inhibition of formation by silicate (SiO_3^{2-}) (Larsen et al. 1995) and sulfate (SO_4^{2-}) (Kerven et al. 1995) in solution. Further, Al_{13}, with its very high charge (7+), is unlikely to be found free in natural systems (e.g., in soil solution). Supporting this conclusion are the data of Menzies (1991) who, in a study of 40 highly-weathered acid soils, found that exchangeable Al occupied ca. 43% of the exchange sites (Table 2). There was less saturation of the exchange sites by Ca and Mg, and substantially less by the monovalent cations, Na and K. The proportion of cations in the soil solution was reversed, however, with very little monomeric Al present. This illustrates the importance of valency in determining the proportion of bound cations to the free concentration in solution (i.e., high valency increases greatly the proportion bound to negatively-charged soil particles) - imagine playing with a tennis ball covered with velcro hooks!

Table 2. Mean quantities and percentages of Ca, Mg, Na, K, and Al present in exchangeable forms and in the soil solutions (filtered to 0.025 μm) of 40 unamended, highly weathered acid soils of tropical and subtropical origin (after Menzies, 1991)

Cation	Exchangeable (cmol(+) kg^{-1})	Soil solution (mM)	Saturation of exch. Sites (%)	Presence in soil solution (%)
Ca	1.24	0.11	25	8
Mg	1.25	0.21	25	15
Na	0.12	0.74	2	52
K	0.19	0.32	4	22
Al	2.13	0.04	43	3

Organic and inorganic ligands vary in their ability to ameliorate Al toxicity, depending on the strength of their complexes with Al. For example, citrate and fulvate greatly reduce Al phytotoxicity, while gallate, phthalate and galacturonate have no ameliorative effect (Cameron et al. 1986; Asher et al. 1992; Ostatek-Boczynski et al. 1995). Of the inorganic ligands, fluoride has been shown to greatly reduce the toxic effect of Al (Cameron et al. 1986) and the $AlSO_4^+$ ion is non-toxic (Kinraide and Parker 1987). In contrast, the inorganic monomers are phytotoxic but the relative severity of the effects of the various monomers (Al^{3+}, $AlOH^{2+}$, $Al(OH)_2^+$) is still not clear (Kinraide 1991, 1997).

Considerable difficulties exist in evaluating the relative toxicity of the various forms of Al, if indeed they have any toxic effect at all. First, measurement of low concentrations of monomeric Al (often < 10 μM) against a background of total Al in soil or in the soil solution is itself a challenge to analysts (Fig. 7). The toxicity of Al may indeed be an order of magnitude lower under some conditions, and especially if Al_{13} is present (Parker et al. 1989). The problems arising from the presence of various forms of Al in solutions extracted from acid soils (Table 1), but with varying rhizotoxicity, can be overcome (or at least minimized) with ultrafiltration to 0.025 μm (Menzies et al. 1991a, b, c). This must be followed by the use of analyses which discriminate among the various forms of Al. Examples include the pyrocatechol violet (PCV) method (Kerven et al. 1988) for determining the concentration of monomeric Al, both in the presence and absence of complexing organic ligands, and the ferron method to determine the concentration of Al_{13}, albeit at a relatively high concentration (Jardine and Zelazny 1986; Parker and Bertsch 1992). Second, it is noteworthy that the concentrations and activities of the various forms of reactive Al in solution can only be estimated, often by using complicated computer programs, e.g., GEOCHEM (Parker et al. 1995). Furthermore, there is the underlying problem that it is not possible to change the concentration or activity of one form of Al without changing those of other forms (Kinraide 1991).

Considerable progress has been made in elucidating the forms of Al that are toxic to plant roots despite these non-trivial problems. Adams and Lund (1966) first related cotton (*Gossypium hirsutum* L.) root growth to the activity of Al^{3+} in soil solution and nutrient solution. Later work by Blamey et al. (1983) and Alva et al. (1986) suggested that soybean (*Glycine max* (L.) Merr.) root growth was most

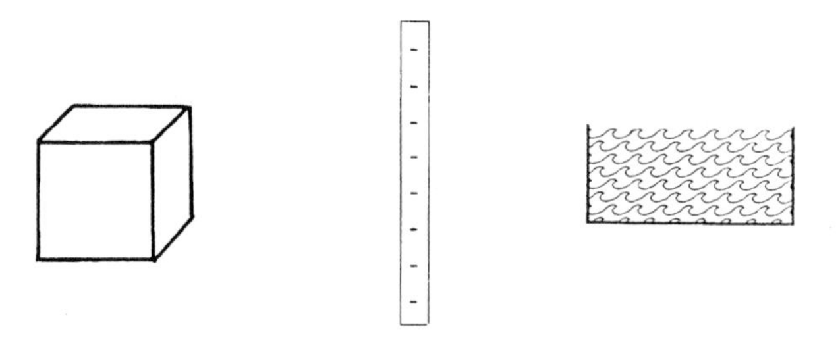

Fig. 7. The concentrations and quantities of total Al present in 1 m^3 of an acid, mineral soil, and those present as exchangeable Al on clays and adsorbing organic matter and in the soil solution that would probably result in toxicity to plants

SOLID	EXCHANGEABLE	MONOMERIC
7 % Al	1.2 cmol(+) Al kg^{-1}	12 μM Al
70 kg Al	100 g Al	0.1 g Al

closely related to the sum of the activities of the monomeric Al forms in solution. In contrast with these findings, Wood and Cooper (1984), working with *Rhizobium trifolii*, and Parker et al. (1989), with soybean, suggested that polymeric forms are toxic. More correctly, these have been identified as polycationic forms, and of these, the tridecameric polycation, Al_{13}, has been shown specifically to be highly phytotoxic (Parker et al. 1989; Bertsch and Parker 1996). Blamey et al. (1983) had earlier concluded that polymeric Al is not toxic to soybean, but the "polymeric Al" in solution was probably micro-particulate material in suspension (see Kerven et al. 1995).

The importance of soil pH on Al chemistry cannot be overemphasized (Fig. 8). First, the solubility of Al decreases with an increase in pH, > 80% of the original 100 μM Al being lost from solution between pH 4.5 and 5.0 (Fig. 8a). This ensures that there is only a narrow range in which to study the toxic effects of Al; too low a pH (probably ≤ pH 3.5) results in H^+ toxicity to plant roots while too high a pH (probably ≥pH 4.2) results in the precipitation of $Al(OH)_3$. Second, the concentration or activity of the various forms of Al change with solution pH (Fig. 8b), Al^{3+} predominating at low pH. With an increase in pH, the various hydrated forms of Al ($AlOH^{2+}$, $Al(OH)_2^+$) increase, there being a concomitant decrease in that of Al^{3+}. At alkaline pH, there is again an increase in solubility, with negatively charged aluminate ions present. The aluminate ion, $Al(OH)_4^-$, has been shown to be nontoxic, however (Kinraide 1990).

Besides the effects on the speciation of monomeric Al, increasing solution pH results in conditions favorable for the formation of polymeric cations (Bertsch and Parker 1996). This has important implications in the study of Al toxicity in solution culture, given the highly toxic nature of Al_{13} (Parker et al. 1989).

Numerous techniques have been used in the study of Al effects on root growth in solution culture. Unfortunately, many of these techniques have been inappropriate, using high pH and concentrations of nutrients, especially P, and of Al that

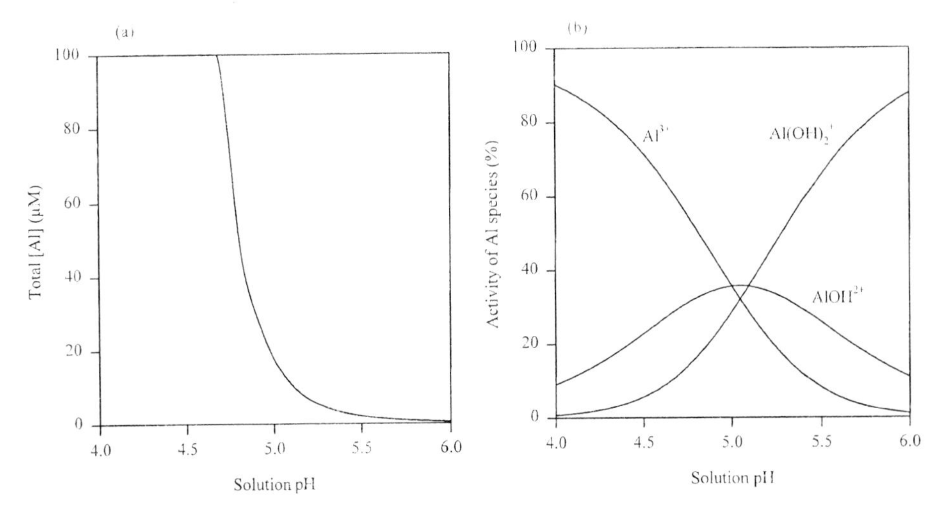

Fig. 8. The total concentration of Al in solution may be calculated using the solubility of amorphous aluminum hydroxide (having a solubility constant = 9.66) in a solution initially containing 1000 µM $CaCl_2$ and 100 µM $AlCl_3$ (used to approximate the soil solution). There is **a** a marked decrease in the total Al concentration with an increase in pH governed by the solubility of $Al(OH)_3$ and **b** a change in the relative activities of inorganic monomeric Al species calculated on the basis of the following equations:

$Al^{3+} + H_2O = AlOH^{2+} + H^+$ pK=5.00

$Al^{3+} + 2H_2O = AlOH^{2+} + 2H^+$ pK=10.10

are orders of magnitude above those found in soil solutions. It is important that care is taken in ensuring that Al remains in solution and that sufficiently large volumes of solution are used (Blamey et al. 1983; Asher et al. 1992). A bare minimum should be the measurement of Al in solution using an appropriate technique, but the results of many studies have been published where even this has not been done. The addition of base (e.g., NaOH, KOH) or phosphate to solutions containing Al results in local volumes in which the solubility of Al is exceeded, and even a small increase in solution pH (which decreases the concentration of monomeric Al in solution) has a large effect on Al toxicity (Fig. 1).

Provided appropriate precautions are taken, however, it is evident that root growth is equally reduced in solutions which contain the same concentration of monomeric Al irrespective of the volume used (Fig. 9). This is, of course, provided that the Al which reacts with compounds in roots does not lower the Al concentration in solution. Further, it is noteworthy that Al-sensitive and Al-tolerant genotypes can be reliably identified whether they are grown in separate containers or communally in the same solution. This indicates that the concentration, rather than the absolute quantity of Al present in the root environment, is important, and suggests that exchange processes may play a role in genetic differences in tolerance to soluble Al.

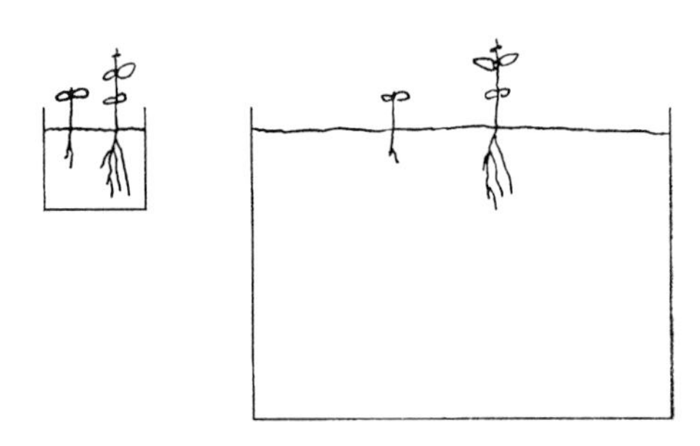

Fig. 9. Schematic description of the difference in root growth of an Al-tolerant (*right*) and an Al-sensitive (*left*) genotype in solutions containing 12 μM soluble Al (a concentration toxic to many plant species) in solutions of 1 or 1000 L. The relative sensitivity of the genotypes is evident irrespective of the volume of solution employed or the total quantity of soluble Al

4. The Root Cell Wall and Al Toxicity

It is surprising how little information is available so far on the physical, physiological and biochemical changes in the cell walls induced by Al. (Horst 1995)

The role of the cell wall in Al toxicity has been addressed in much research. Vose and Randall (1962) were probably the first to suggest that Al-sensitive genotypes accumulate more Al in their roots, and that this is associated with high root CEC. Rengel (1992) emphasized the importance of the cell wall acting as an ion exchanger, and mathematical models have been developed based on this concept (Blamey et al. 1992; Grauer and Horst 1992). Indeed, Grauer and Horst (1992) stated that apoplastic binding of Al is a requirement for the expression of Al toxicity, and Al toxicity is a function of the Al saturation of exchange sites.

Such a concept might be illustrated as two exchange surfaces, clay and root, separated by the soil solution of the rhizosphere (Fig. 10).

There has been lack of consensus on the amount of Al bound in the apoplasm, with a number of studies illustrating considerable quantities of Al in the symplasm of root cells. Tice et al. (1992) grew wheat for 2 days in solutions containing up to 140 μM Al. In all instances, the root tips of the Al-sensitive cultivar (Tyler) contained more Al than those of the Al-tolerant cultivar (Yecora Rojo). Analysis of the fractional distribution of Al indicated that apoplastic Al was consistently only 30 to 40% of the total Al in root tips. Also, Lazof et al. (1994) found there was a rapid accumulation (within 30 min) of Al in root tips of soybean exposed to 38 μM Al, and this Al was mostly present in the outer 30 μm (i.e., the outer two to three cell layers) of the root. (Is it coincidental that root distortion due to soluble Al (Fig. 2) is particularly evident in these layers?) Root elongation was reduced by ca. 50% after 6 h exposure to Al, and Al was measured throughout the root cross section after 18 hours. These studies on the localization of Al led to the conclusion that "a large majority of the Al ... evidently was intracellular" (Lazof et al. 1994).

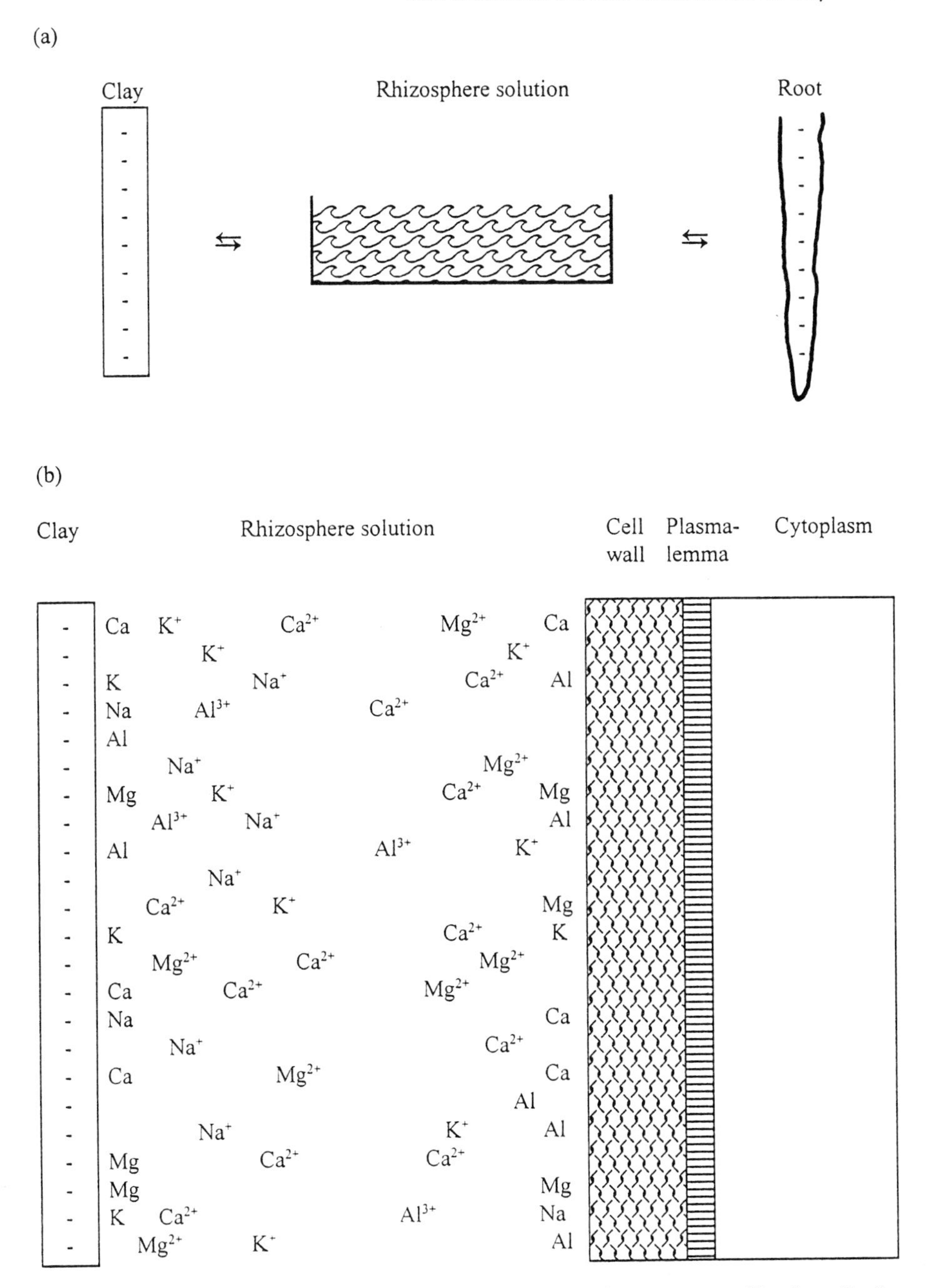

Fig. 10. A conceptual model of two exchange surfaces, clay and root, separated by the soil solution in the rhizosphere showing these compartments at a macro (**a**) and a micro (**b**) scale

In contrast, Clarkson (1967) found that Al is bound strongly by isolated cell wall material. More recent evidence with giant algal cells of *Chara corallina* has shown "that up to 99.99% of the total cellular Al accumulates in the cell wall"

(Rengel and Reid 1997). This has been confirmed (and a value of 99.5% proposed) with the novel technique of ^{26}Al accelerated mass spectrometry in which extremely low Al concentrations were used (G.J. Taylor, personal communication). In vitro studies with Ca pectate have shown very low concentrations ($< 2.5\ \mu M$ Al) in the supernatant solution despite up to 200 μM Al originally present in solution (Blamey et al. 1997). This suggests little free Al available to cross the plasmalemma. However, even very low concentrations of Al, given its reactivity, crossing the plasmalemma would most likely have severe consequences on cell function. In this regard, there may be an analogy with Ca which may be present at concentrations of ≥ 1 mM in the apoplasm but of $\leq 1\ \mu M$ in the symplasm - higher concentrations being highly deleterious.

The Al that reacts with cell wall components is bound both rapidly and strongly, and is difficult to remove with moderate to high concentrations of acids, metal ions or chelates (e.g. Clarkson 1967). Recently, it was found (Blamey et al. 1997) that the presence of organic ligands reduces Al sorption by Ca pectate in accordance with the strength of the Al complex formed by these ligands (Fig. 11). Specifically, the presence of citrate, malate or fluoride substantially reduced the sorption of Al by Ca pectate. In contrast, the presence of galacturonate or sulfate did not decrease Al sorption over that observed in the presence of chloride which does not form a complex with Al. In this context, it is noteworthy that excretion of citrate or malate has been associated with Al-tolerance in wheat (Delhaize et al. 1993) and in a number of other plant species (Kochian 1995).

Illustrative of the relationship between root CEC and Al sorption were the studies conducted on *Lotus pedunculatus* cv. Maku (Al-tolerant) and *L. corniculatus* cv. Maitland (Al-sensitive) (Blamey et al. 1990). (Maku has a lower root CEC than Maitland.) There was an initial, rapid loss of Al from solution on the immersion of roots of both species (Fig. 12), followed by a decreased rate of loss after ca. 10 min. Importantly, the Al-sensitive species (*L. corniculatus*) accumulated the greater quantity of Al in its roots, an effect that persisted for the 40 minutes' duration of the experiment (Fig. 12). Work with wheat by Zhang and Taylor (1989), using a different technique, showed similar effects over a longer time-span. Matsumoto (1994), using a low Al concentration and a short exposure time, also demonstrated

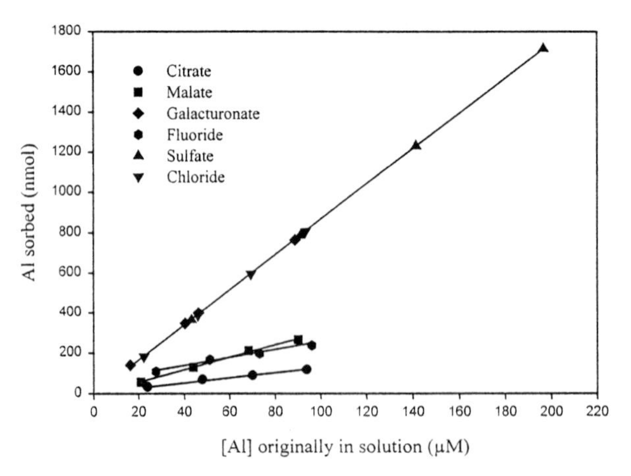

Fig. 11. Effects of three organic ligands, citrate, malate and galacturonate, and three inorganic ligands, fluoride, sulfate and chloride, on the sorption of aluminum by calcium pectate (Blamey et al. 1997)

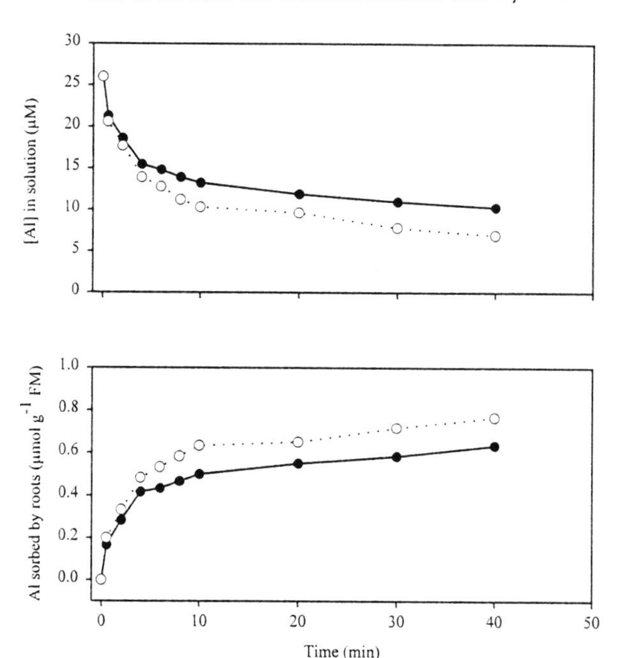

Fig. 12. There was greater sorption of Al by roots of *Lotus corniculatus* cv. Maitland (○) (Al-sensitive) than by those of *L. pedunculatus* cv. Maku (●)(Al-tolerant) from solutions initially containing 26 μM monomeric Al (after Blamey et al. 1990)

the considerably higher concentration of Al in the roots of wheat cv. Scout (Al-sensitive) than of Atlas (Al-tolerant), especially at the root tips.

As mentioned previously, there have been great difficulties in determining which of the monomeric Al species are toxic to plant roots. (Kinraide (1997) has provided an excellent overview of this topic.) In soil, Al toxicity increases with a decrease in pH as a result of increased Al solubility. In contrast, carefully-controlled solution culture studies (in which Al concentration is maintained constant), have shown Al toxicity increases with an increase in pH. For example, Kerridge (1969) reported an increased detrimental effect of 74 μM Al in wheat as solution pH increased from 4.0 to 4.5, and the data of Noble and Sumner (1988), reanalyzed by Kinraide and Parker (1990), indicated increased Al toxicity as pH increased from 4.2 to 4.8. These effects have been ascribed to the greater toxic effect of the hydroxy-aluminum species ($AlOH^{2+}$, $Al(OH)_2^{+}$) than Al^{3+}. An alternative explanation is that the increase in solution pH would result in greater dissociation of the carboxyl groups of the pectic compounds, increasing root CEC and, thus, the binding of Al in the cell wall (Fig. 3).

The rapid and strong binding of Al to cell wall components, specifically pectic compounds, may be related also to Al effects on cell ultrastructure. Importantly, as mentioned above, cell wall polysaccharide synthesis and deposition is closely related to Golgi apparatus function. Ultrastructural studies by Bennet et al. (1985) have shown a disruption of Golgi apparatus activity within 5 h of exposing roots to Al, suggesting this to be a primary site of Al toxicity. Importantly, resumption of root elongation following removal from solution containing Al coincided with secretory activity in root cap cells (Bennet and Breen 1991b).

Both studies with cell wall components and with pectic compounds, therefore, have shown that there is a rapid and strong binding of Al. Furthermore, it would appear that studies indicating high accumulation of Al in the cytoplasm have not considered its high pH, with correspondingly low solubility of Al (Fig. 8), or whether the Al in the cytoplasm is in a reactive form. Having established that a considerable quantity of Al is bound in the cell wall, it is pertinent to ask what effects this Al might have, if any. Three major roles, specifically related to Al bound by pectic compounds in the walls of cells in the elongation zone of the root, may be considered: (1) a decrease in the ability of roots to bind basic metal cations; (2) a decrease in the ability of cell walls to expand, thus reducing root extension; and (3) a decrease in movement of water and solutes via the apoplastic pathway.

4.1 Al Effects on Root CEC

Given the evidence that the cell wall acts as an ion-exchanger, the presence of Al in the root environment would result in desorption of Ca and other basic cations. In vitro evidence with Ca-pectate (Blamey and Dowling 1995; Blamey et al. 1997) indicates that this is so, with 1.5 mole Ca desorbed for each 1.0 mole Al sorbed according to the reaction:

$$3CaX + 2Al^{3+} \rightleftharpoons 2AlX + 3Ca^{2+} \quad (2)$$

where X = the negatively-charged pectic compounds. Once sorbed, Al is bound strongly, and is not readily displaced by divalent or monovalent cations. This would, therefore, decrease root CEC. Again, there is a need for further evidence of such an effect in plant roots, Rengel and Robinson (1989) having found that Al decreased the amounts of Ca and Mg but increased those of K and Na that could be desorbed from ryegrass (*Lolium multiflorum* Lam.) roots. Despite the strength with which Al is bound by pectic compounds, the low activity of monomeric Al in soil solutions relative to that of basic cations (Table 2) may result in only a small proportion of root CEC being occupied by Al.

4.2 Al Effects on Root Elongation

It has long been known that Al greatly reduces root elongation (an effect illustrated in Fig. 1) which is controlled both by the turgor pressure inside the cell and by the resistance to expansion by the cell wall. In this, a fine balance must be maintained: too weak a resistance would result in cellular disintegration; too strong a resistance would restrict cell expansion and, thus, root elongation.

Klimashevskii and Dedov (1980) first suggested the importance of Al reactions with pectic compounds in the cell wall. More recently, there has been added focus on these reactions since pectic compounds carry the bulk of the negative charge within the cell wall. Reactions of Al with pectic compounds are rapid (i.e., complete in < 1 min) (Blamey et al. 1993b), a finding somewhat at odds with studies in roots where a rapid, short-term reaction (complete in 10 to 30 min) is followed by a slow, linear phase (Zhang and Taylor 1989; Blamey et al. 1990). This linear phase was explained by Zhang and Taylor (1989) as Al uptake across the plasmalemma, but an alternative explanation would be the movement of Al through the tortuous apoplastic pathway (Blamey et al. 1993a).

There are two (possibly more) important consequences of the reaction of Al with pectic compounds in the cell wall. First, there would be increased rigidity of the cell wall as originally described by Klimashevski and Dedov (1980), although it is not clear if this results from increased rigidity of the pectic gel per se or decreased cleaving of hemicellulose. Second, the displacement of basic cations, especially Ca, by Al from the exchange sites would reduce the ability of protons to permit cell expansion (acid mediated cell expansion), since Al binds much more strongly to pectin than does Ca. In this regard, Kinraide (1988) found increased H^+ excretion by wheat roots on exposure to Al.

It is intriguing to speculate on the reported beneficial effect of Al on plant growth. In some studies using high concentrations of Al and P (and other nutrients), this may be related to decreased P toxicity. Such an explanation is not correct where appropriate techniques have been used, however. In these studies, low Al concentrations improve plant growth at low pH, and the effect has been ascribed to Al amelioration of H^+ toxicity (Kinraide 1993; 1997). It was concluded that Al^{3+} and H^+ compete for common apoplastic binding sites, with Al^{3+} alleviating H^+ toxicity and vice versa, examples of one cation ameliorating the toxic effects of another. At low pH, Al may act as a Ca analogue, providing some stability to the cell wall by overcoming the poor binding of Ca (Eq. 1).

In a classic study, Ryan et al. (1993) used small agar blocks containing Al to investigate the localization of Al effects on root growth of maize (*Zea mays* L.). Exposure of the meristem was found to be important in the expression of Al toxicity. Further examination of their results, however, suggests that it is cellular expansion that is more rapidly affected by Al. Indeed, Ryan et al. (1993) stated that Al applied to the elongation zone resulted in "considerable visual damage to the epidermal and cortical tissues." Exposure of the root tip to soluble Al certainly resulted in a greater long-term detrimental effect than when the section just behind the root tip was exposed (Fig. 13). In this latter case, however, only those cells currently expanding would have been affected by Al. In contrast, when Al was applied to the root tip, those cells currently expanding and those with the potential to expand would have been exposed to its toxic effects.

Besides the effects of Al on root elongation, exposure of roots results in callose formation at the root tip (Wissemeier and Horst 1995; Horst et al. 1997). It is noteworthy that the intensity of callose is highest at the root tip and is confined to the outer cortical cell layers (Wissemeier et al. 1987) (i.e., those cell layers visibly damaged by Al). Also, callose concentration in the root tip is "closely and positively related to Al-induced inhibition of root elongation" in maize (Horst et al. 1997). However, in contrast with other studies, Al accumulation in the root tip was not related to decreased root elongation. More recent studies (Sivaguru and Horst 1998) have shown that in maize the part of the root apex preparatory to rapid elongation (1-2 mm from the tip) is most sensitive to Al.

Root hair development is particularly sensitive to soluble Al (2 - 3 μM) (Brady et al. 1993; Jones et al. 1995), which has important implications for nutrient acquisition in acid soils high in soluble Al. Furthermore, decreased proliferation of root hairs in

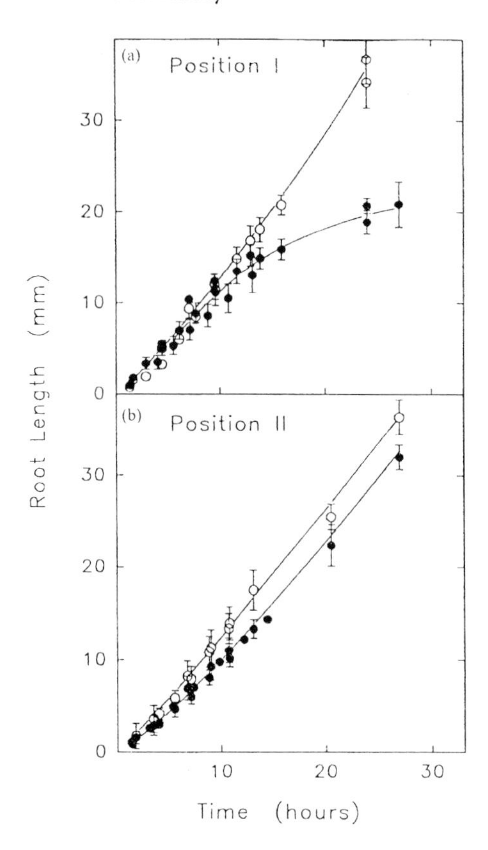

Fig. 13. Agar blocks without (○) and with (●) 2 mM Al were placed on maize roots either at the root tip (Position I), the zone of cell division (**a**), or 2 to 3 mm behind the root tip (Position II), the zone of maximum cell elongation (**b**) (Ryan et al. 1993). Aluminum at the root tip resulted in the greater overall effect on root length, but Al at the elongation zone had the more rapid effect and resulted in considerable visual damage to the epidermal and cortical tissues (a typical effect of soluble Al)

legumes decreases the number of sites for infection by rhizobia, with the consequent detriment to nitrogen fixation. It is noteworthy that the elongation of root hairs was reduced by 2 μM Al, but this did not affect Ca^{2+} fluxes (Jones et al. 1995).

In conclusion, the visible damage to cells of the epidermis and outer cortex (Fig. 2) is a likely result of Al sorption in cell walls since (1) most Al accumulates in the walls of cells in the epidermis and outer cortex (resulting in increased rigidity of these cell layers); and (2) root elongation is governed by the extension of inner cortex.

Because root hairs and pollen tubes grow from the tip, no visible damage is likely; they would just stop elongating on exposure to soluble Al. Such an effect has been noted in soybean at 3 μM Al (Brady et al. 1993), and in *Limnobium stoloniferum* there was swelling of the tip (Jones et al. 1995). Especially noteworthy are recent findings (Wenhao Zhang, personal communication) that, within 10 min of adding Al, La or EGTA (but not the Ca channel-blocker, veripermal), there was bursting of pollen tubes of an Australian native flowering plant (*Chamelaucium uncinatum*); prior to this occurring, however, there was normal pollen tube growth and even cytoplasmic streaming was not affected.

4.3 Al Effects on Apoplastic Flow

Besides the effect of Al on cell expansion, the rapid reaction of Al with pectic compounds may decrease the movement of solution through the apoplasm. This has been demonstrated in vitro in which a synthetic Ca pectate gel imbedded in a cellulose filter paper was used to simulate the plant cell wall (Blamey et al. 1993a). In this study, the addition of 29 μM Al decreased solution movement by > 80% within 2 min of adding Al (Fig. 14). Blamey and Dowling (1995) reported a 20% decrease in the volume of a Ca pectate gel with up to 200 μM Al in solution. It was noteworthy that increased Ca in solution also decreased gel volume, but that 1000 μM Ca was required to have a similar effect, probably due to the increase in ionic strength. The effect of Al, however, was probably due to its high valence and small ionic radius given the low concentration of Al present.

However, such in vitro effects are yet to be demonstrated explicitly in living cells. Some evidence of this was shown by a rapid decrease in water use and nutrient uptake on exposure of cotton roots to Al in solution (Lance and Pearson 1969), and decreased transpiration is a common effect of heavy metals in the rhizosphere (Barcelo and Poschenrieder 1990; Bazzaz et al. 1974). The apoplastic path-

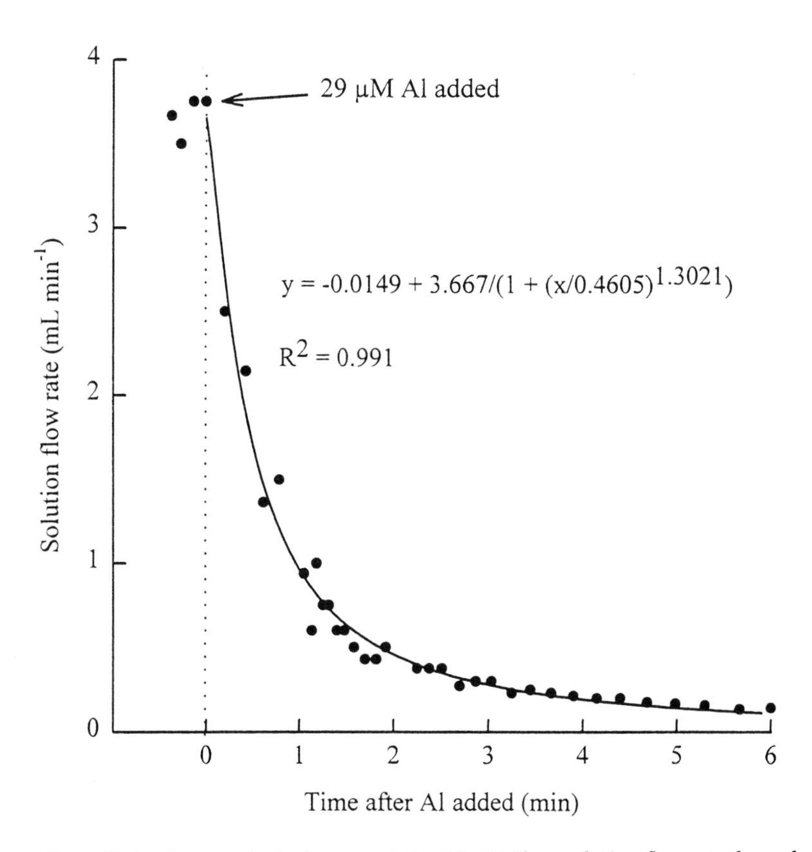

Fig. 14. Effect of injecting 12 ml solution containing 29 μM Al on solution flow rate through a Ca pectate gel embedded in a cellulose filter paper to simulate the plant cell wall (Blamey et al. 1993)

way may contribute substantially to overall hydraulic conductivity of roots (Steudle and Frensch 1996), and any decrease in water movement would detrimentally affect plant water relations and nutrient uptake. Such findings are in keeping with the observed wilting of plants growing in soil high in soluble Al even in the presence of sufficient water (in addition to reduced root proliferation limiting water uptake).

5. Conclusions

Science is inherently open-ended and exploratory, and makes mistakes every day. Indeed that will always be its fate. (Ferris 1988).

The ideas put forward in this chapter have been presented ever mindful of the above statement by Timothy Ferris, who has further stated that "the process (of science) depends on error" and that "theories, however grand, are held hostage to empirical data that have the power to ruin them." The ideas have been presented to review the literature on the topic and hopefully to provide new insights.

It has long been known that soluble Al in the root environment has an adverse effect, often severe, on nutrient acquisition by plants. Reasons for this antagonistic effect are not entirely clear, but may be related both to decreased root elongation and to decreased nutrient uptake per unit root length. The former effect possibly results from the increased rigidity of cell walls in the elongating zone near the root apex, while the latter effect may result from the effect of Al in decreasing root CEC and apoplastic flow of water and solutes.

Aluminum reacts with many compounds in the cell wall, plasma membranes, cytoplasm and nucleus, and these reactions probably play important roles in the long-term demise of cells at the root apex, with the consequent detriment to plant growth. In the short term, however, there is evidence that the toxicity of Al is reduced by any factor, genetic (e.g., excretion of organic ligands, low root CEC) or environmental (e.g., high Ca in solution, low pH), which decreases Al sorption by pectic compounds in the walls of expanding cells at the root apex.

Acknowledgments. I would like to thank Dr. Greg Taylor for information on ^{26}Al studies and Dr. Neal Menzies for his consent to allow the use of unpublished data from his Ph.D. thesis that are summarized in Table 2. Permission from Dr. Wanhao Zhang to mention his recent findings on Al effects on pollen tube growth is much appreciated.

References

Adams F, Lund ZF (1966) Effect of chemical activity of soil solution aluminum on cotton root penetration of acid subsoils. Soil Sci 101:193-198

Alva AK, Blamey FPC, Edwards DG, Asher CJ (1986) An evaluation of aluminum indices to predict aluminum toxicity in plants grown in nutrient solutions. Communications in Soil Science and Plant Analysis 17:1271-1280

Ando T (1970) Cation absorption by excised barley roots in relation to root cation exchange capacity. Journal of the Faculty of Fisheries and Animal Husbandry, Hiroshima University 9:83-121

Asher CJ, Blamey FPC, Kerven GL (1992) Recent research on soil acidity: Identifying toxic aluminium in soils. In: Tovar SJL, Quintero L (Eds) La Investgacion Edafologica en Mexico 1991-1992. Memorias del 15 Congreso Nacional de la Sociedad Mexicana de la Ciencia del Suelo. Sociedad Mexicana de la Ciencia del Suelo, Acapulco, pp 61-68

Barcelo J, Poschenrieder C (1990) Plant water relations as affected by heavy metal stress: A review. Journal of Plant Nutrition 13:1-37

Baron-Epel O, Gharyal PK, Schindler M (1988) Pectins as mediators of wall porosity in soybean cells. Planta 175:389-395

Bazzaz FA, Carlson RW, Rolfe GL (1974) The effect of heavy metals on plants: Part 1. Inhibition of gas exchange in sunflower by Pb, Cd, Ni, and Tl. Environmental Pollution 7:241-252

Bennet RJ, Breen CM (1991a) The aluminium signal: New dimensions to mechanisms of aluminum tolerance. In: Wright RJ, Baligar VC, Murrmann RP (Eds) Plant-Soil Interactions at Low pH. Developments in Plant and Soil Sciences. Kluwer Academic, Dordrecht, pp 703-716

Bennet RJ, Breen CM (1991b) The recovery of roots of *Zea mays* L. from various aluminium treatments: Towards elucidating the regulatory processes that underlie root growth control. Environmental and Experimental Botany 31:153-163

Bennet RJ, Breen CM, Fey MV (1985) The primary site of aluminium injury in the root of *Zea mays* L. South African Journal of Plant and Soil 2:8-17

Bertsch PM, Parker DR (1996) Aqueous polynuclear aluminum species. In: Sposito G (Ed) The environmental chemistry of aluminum. CRC Lewis, Boca Raton, pp 117-168

Blamey FPC, Dowling AJ (1995) Antagonism between aluminium and calcium for sorption by calcium pectate. In: Date RA, Grundon NJ, Rayment GE, Probert ME (Eds) Plant-soil interactions at low pH: Principles and management. Kluwer Academic, Dordrecht, pp 303-306

Blamey FPC, Edwards DG, Asher CJ (1983) Effect of aluminum, OH:Al and P:Al ratios, and ionic strength on soybean root elongation in solution culture. Soil Science 136:197-207

Blamey FPC, Edmeades DC, Wheeler DM (1990) Role of root cation-exchange capacity in differential aluminum tolerance of *Lotus* species. Journal of Plant Nutrition 13:729-744

Blamey FPC, Edmeades DC, Wheeler DM (1992) Empirical models to approximate calcium and magnesium ameliorative effects and genetic differences in aluminium tolerance in wheat. Plant and Soil 144:281-287

Blamey FPC, Asher CJ, Edwards DG, Kerven GL (1993a) In vitro evidence of aluminum effects on solution movement through root cell walls. Journal of Plant Nutrition 16:555-562

Blamey FPC, Asher CJ, Kerven GL, Edwards DG (1993b) Factors affecting aluminium sorption by calcium pectate. Plant and Soil 149:87-94

Blamey FPC, Ostatek-Boczynski Z, Kerven GL (1997) Ligand effects on aluminium sorption by calcium pectate. Plant and Soil 192:269-275

Brady DJ, Edwards DG, Asher CJ, Blamey FPC (1993) Calcium amelioration of aluminium toxicity effects on root hair development in soybean (*Glycine max* (L.) Merr.). New Phytologist 123:531-538

Brett CT, Waldron KW (1996) Physiology and biochemistry of plant cell walls. 2nd ed. Chapman and Hall, London

Cameron RS, Ritchie GSP, Robson AD (1986) Relative toxicities of inorganic aluminum complexes to barley. Soil Science Society of America Journal 50:1231-1236

Carpita NC, Gibeaut DM (1993) Structural models of primary cell walls of flowering plants: consistency of molecular structure with the physical properties of the walls during growth. Plant Journal 3:1-30

Carver BF, Ownby JD (1995) Acid soil tolerance in wheat. Advances in Agronomy 54:117-173

Clarkson DT (1967) Interactions between aluminium and phosphorus on root surfaces and cell wall material. Plant and Soil 27:347-356

Clarkson DT (1996) Root structure and sites of ion uptake. In: Waisel Y, Eshel A, Kafkafi U (Eds) Plant roots: The hidden half. 2nd ed. Marcell Dekker Inc, New York, pp 483-510

Cosgrove DJ (1993) Wall extensibility: its nature, measurement and relationship to plant cell growth. New Phytologist 124:1-23

Delhaize E, Ryan PR, Randall PJ (1993) Aluminum tolerance in wheat (*Triticum aestivum* L.) II. Aluminum-stimulated excretion of malic acid from root apices. Plant Physiology 103:695-702

Ferris T (1988) Coming of age in the milky way. Morrow, New York

Foy CD (1984) Physiological effects of hydrogen, aluminum, and manganese toxicities in acid soil. In: Adams F (Ed) Soil acidity and liming. 2nd ed, Agronomy No. 12, American Society of Agronomy, Madison, pp 57-97

Foy CD, Fleming AL, Burns GE, Armiger WH (1967) Characterization of differential aluminum tolerance among varieties of wheat and barley. Soil Science Society of America Proceedings 31:513-521

Grauer UE, Horst WJ (1992) Modeling cation amelioration of aluminum phytotoxicity. Soil Science Society of America Journal 56:166-172

Haynes RJ (1980) Ion exchange properties of roots and ionic interactions within the root apoplasm: their role in ion accumulation by plants. Botanical Reviews 46:75-99

Hecht-Buchholz C, Brady DJ, Asher CJ, Edwards DG (1990) Effects of low activities of aluminium on soybean (*Glycine max*) II. Root cell structure and root hair development. In: van Beusichem ML (Ed) Plant nutrition - physiology and applications. Kluwer Academic, Haren, pp 335-343

Horst WJ (1995) The role of the apoplast in aluminum toxicity and resistance of higher plants: a review. Zeitschrift für Pflanzenerährung und Bodenkunde 158:419-428

Horst WJ, Wagner A, Marschner H (1982) Mucilage protects root meristems from aluminium injury. Proceedings of the 9th International Plant Nutrition Colloquium, Warwick, 238-243

Horst WJ, Asher CJ, Cakmak I, Szulkiewicz P, Wissemeier AH (1992) Short-term responses of soybean roots to aluminum. Journal of Plant Nutrition 140:174-178

Horst WJ, Puschel AK, Schmohl N (1997) Induction of callose formation is a sensitive marker for genotypic aluminium sensitivity in maize. Plant and Soil 192:23-30

Ishikawa H, Evans ML (1995) Specialized zones of development in roots. Plant Physiology 109:725-727

Jackson ML (1963) Aluminum bonding in soils: A unifying principle in soil science. Soil Science Society of America Proceedings 27:1-10

Jardine PM, Zelazny LW (1986) Mononuclear and polynuclear aluminum speciation through differential kinetic reactions with ferron. Soil Science Society of America Journal 50:895-900

Jones DL, Shaff JE, Kochian LV (1995) Role of calcium and other ions in directing root hair tip growth in *Limnobium stoloniferum*. Planta 197:672-680

Keltjens WG (1995) Magnesium uptake by Al-stressed maize plants with special emphasis on cation interactions at root exchange sites. Plant and Soil 171:141-146

Kerridge PC (1969) Aluminum toxicity in wheat (*Triticum aestivum* Vill. Host) Ph. D. Thesis, Oregon State University (Dissertation Abstracts B29, 3159)

Kerven GL, Edwards DG, Asher CJ, Hallman PS, Kokot S (1988) Aluminium determination in soil solution. II. Short term colorimetric procedures for the measurement of inorganic monomeric aluminum in the presence of organic acid ligands. Australian Journal of Soil Research 27:91-102

Kerven GL, Larsen PL, Blamey FPC (1995) Detrimental sulfate effects on formation of Al-13 tridecameric polycation in synthetic soil solutions. Soil Science Society of America Journal 59:765-771

Kinraide TB (1988) Proton extrusion by wheat roots exhibiting severe Al toxicity symptoms. Plant Physiology 88:418-423

Kinraide TB (1990) Assessing the rhizotoxicity of the aluminate ion, $Al(OH)_4^-$. Plant Physiology 93:1620-1625

Kinraide TB (1991) Identity of the rhizotoxic aluminium species. In: Wright RJ, Baligar VC, Murrmann RP (Eds) Plant-soil Interactions at Low pH. Developments in Plant and Soil Sciences. Kluwer Academic, Dordrecht, pp 717-728

Kinraide TB (1993) Aluminium enhancement of plant growth in acid rooting media. A case of reciprocal alleviation of toxicity by two toxic cations. Physiologia Plantarum 88:619-625

Kinraide TB (1997) Reconsidering the rhizotoxicity of hydroxyl, sulfate, and fluoride complexes of aluminium. Journal of Experimental Botany 48:1115-1124

Kinraide TB, Parker DR (1987) Non-phytotoxicity of the aluminum sulphate ion, $AlSO_4^+$. Physiologia Plantarum 71:207-212

Kinraide TB, Parker DR (1990) Apparent phytotoxicity of mononuclear hydroxy-aluminum to four dicotyledonous species. Physiologia Plantarum 79:283-288

Klimashevskii EL, Dedov VM (1980) Characteristics of an elastic cell wall of the root in relation to genotypic variance of plant resistance to aluminum ions. Izv. Sib. Otd. Akad. Nauk SSSR, Ser. Biol. Nauk 1:108-112 (Chemical Abstracts 93(15):416-417)

Knight AH, Crooke WM, Inkson RHE (1961) Cation-exchange capacities of tissues of higher and lower plants and their related uronic acid contents. Nature 192:142-143

Kochian LV (1995) Cellular mechanisms of aluminum toxicity and resistance in plants. Annual Review of Plant Physiology and Plant Molecular Biology 46:237-260

Lance JC, Pearson RW (1969) Effect of low concentrations of aluminum on growth and water and nutrient uptake by cotton roots. Soil Science Society of America Proceedings 33:95-98

Larsen PL, Kerven GL, Bell LC, Edwards DG (1995) Effects of silicic acid on the chemistry of monomeric and polymeric (Al_{13}) aluminium species in solutions. In: Date RA, Grundon NJ, Rayment GE, Probert ME (Eds) Plant-soil interactions at low pH: Principles and management. Kluwer Academic, Dordrecht, pp 617-621

Lazof DB, Goldsmith JG, Rufty TW, Linton RW (1994) Rapid uptake of aluminum into cells of intact soybean root tips. A microanalytical study using secondary ion mass spectrometry. Plant Physiology 106:1107-1114

Lindsay WL (1979) Chemical equilibria in soils. John Wiley and Sons, New York

Macdonald TL, Martin RB (1988) Aluminum ion in biological systems. Trends in Biochemical Sciences 13:15-19

Martin RB (1994) Aluminum: A neurotoxic product of acid rain. Accounts of Chemical Research. 27:204-210

Matsumoto H (1994) Comparison of the early response to aluminum stress between tolerant and sensitive wheat cultivars: Root growth, aluminum content and efflux of K^+. Journal of Plant Nutrition 17:1257-1288

Matsumoto H, Hirasawa E, Morimura S, Takahashi E (1976) Localization of aluminum in tea leaves. Plant and Cell Physiology 17:627-631

Menzies NW (1991) Solution phase aluminium in highly weathered soils: evaluation and relationship to aluminum toxicity to plants. PhD Thesis. The University of Queensland, Brisbane

Menzies NW, Bell LC, Edwards DG (1991a) A simple positive pressure apparatus for the ultrafiltration of soil solution. Communications in Soil Science and Plant Analysis 22:137-145

Menzies NW, Bell LC, Edwards DG (1991b) Characteristics of membrane filters in relation to aluminium studies in soil solutions and natural waters. Journal of Soil Science 42:585-597

Menzies NW, Bell LC, Edwards DG (1991c) Effects of incubation time and filtration technique on soil solution composition with particular reference to inorganic and organically complexed Al. Australian Journal of Soil Research 29:223-238

Nieboer E, Richardson DHS (1980) The replacement of the nondescript term 'heavy metals' by a biologically and chemically significant classification of metal ions. Environmental Pollution 1:3-26

Noble AD, Sumner ME (1988) Calcium and Al interactions and soybean growth in nutrient solutions. Communications in Soil Science and Plant Analysis 19:1119-1131

Ostatek-Boczynski Z, Kerven GL, Blamey FPC (1995) Aluminium reactions with polygalacturonate and related organic ligands. Plant and Soil 171:41-45

Parker DR, Bertsch PM (1992) Identification and quantification of the "Al13" tridecameric polycation using ferron. Environmental Science and Technology 26:908-914

Parker DR, Kinraide TB, Zelazny LW (1989) On the phytotoxicity of polynuclear hydroxy-aluminum complexes. Soil Science Society of America Journal 53:789-796

Parker DR, Norvell WA, Chaney RL (1995) GEOCHEM-PC: A chemical speciation program for IBM and compatible personal computers. In: Loeppert RH, Schwab AP, Goldberg S (Eds) Soil chemical equilibrium and reaction models. Soil Science Society of America, Inc, Madison, pp 253-269

Passioura JB, Stirzaker RJ (1993) Feedforward responses of plants to physically inhospitable soil. In: Buxton DR, Shibles R, Forsberg RA, Blad BL, Asay KH, Paulsen GM, Wilson RF (Eds) International crop science I. Crop Science Society of America, Inc., Madison, pp 715-719

Pritchard J (1994) The control of cell expansion in roots. New Phytologist 127:2-26

Rengel Z (1990) Competitive Al^{3+} inhibition of net Mg^{2+} uptake by intact *Lolium multiforum* roots. II. Plant age effects. Plant Physiology 93:1261-1267

Rengel Z (1992) Role of calcium in aluminium toxicity. New Phytologist 121:499-513

Rengel Z, Reid RJ (1997) Uptake of Al across the plasma membrane of plant cells. Plant and Soil 192:31-35

Rengel Z, Robinson DL (1989) Aluminum and plant age effects on adsorption of cations in the Donnan free space of ryegrass roots. Plant and Soil 116:223-227

Rengel Z, Robinson DL (1990) Competitive Al^{3+} inhibition of net Mg^{2+} uptake by intact *Lolium multiflorum* roots. I. Kinetics. Plant Physiology 91:1407-1413

Ryan PR, Ditomaso JM, Kochian LV (1993) Aluminium toxicity in roots: An investigation of spatial sensitivity and the role of the root cap. Journal of Experimental Botany 44:437-446

Sasaki K, Nagahashi G (1989) Autolysis-like release of pectic polysaccharides from regions of cell walls other than the middle lamella. Plant and Cell Physiology 30:1159-1169

Sivaguru M, Horst WJ (1998) The distal part of the transition zone is the most aluminum-sensitive apical root zone of maize. Plant Physiology 116:155-163

Sposito G (1995) The environmental chemistry of aluminum. 2nd ed, CRC Lewis, Boca Raton

Steudle E, Frensch J (1996) Water transport in plants: Role of the apoplast. Plant and Soil 187:67-79

Tan K, Keltjens WG (1995) Analysis of acid-soil stress in sorghum genotypes with emphasis on aluminium and magnesium interactions. Plant and Soil 171:147-150

Tan K, Keltjens WG, Findenegg GR (1992) Acid soil damage in sorghum genotypes: Role of magnesium deficiency and root impairment. Plant and Soil 139:149-155

Taylor GJ (1991) Current views of the aluminum stress response; the physiological basis of tolerance. Current Topics in Plant Biochemistry and Physiology 10:57-93

Tice KR, Parker DR, DeMason DA (1992) Operationally defined apoplastic and symplastic aluminum fractions in root tips of aluminum-intoxicated wheat. Plant Physiology 100:309-318

Vose PB, Randall PJ (1962) Resistance to aluminium and manganese toxicities in plants related to variety and cation exchange capacity. Nature 196:85-86

Wissemeier AH, Horst WJ (1995) Effect of calcium supply on aluminum-induced callose formation, its distribution and persistence in roots of soybean (*Glycine max* (L.) Merr.). Journal of Plant Physiology 145:470-476

Wissemeier AH, Klotz F, Horst WJ (1987) Aluminum induced callose synthesis in roots of soybean (*Glycine max* L.). Journal of Plant Physiology 129:487-492

Wood M, Cooper JE (1984) Aluminium toxicity and multiplication of *Rhizobium trifolii* in a defined growth medium. Soil Biology and Biochemistry 16:571-576

Zhang G, Taylor GJ (1989) Kinetics of aluminum uptake by excised roots of aluminum-tolerant and aluminum-sensitive cultivars of *Triticum aestivum* L. Plant Physiology 91:1094-1099

Boron in Plant Nutrition and Cell Wall Development

Toru Matoh

Summary. Intracellular localization of boron (B) was examined and it was found that B both in a water-soluble and water-insoluble form occurs mainly in cell walls. Almost all the cell-wall bound B is in a form of a B-dimeric rhamnogalacturonan II (RG-II) complex, in which B cross-links two RG-II chains through borate-diol ester bonding. The B-RG-II complex ubiquitously occurs in higher plant cell walls. In B-deficient cells, cell walls swelled and monomeric RG-II was detected. The calcium ion is also a native constituent of the B-RG-II complex, and this ion works to strengthen the borate-diol ester bonding. Sequential extraction of Ca^{2+} from cell walls revealed that only a limited portion of the cell-wall Ca^{2+} is responsible for retaining the pectic polysaccharides in cell walls. The Ca^{2+} responsible is that associated with the RG-II region. These results suggest that cross-linking by B and Ca^{2+} at the RG-II region anchors the pectic network into cell walls and arranges the pectic chains to form appropriate coordinate bonding with Ca^{2+} in polygalacturonic acid regions.

Key words. Boron, Cell wall, Calcium, Pectin, Rhamnogalacturonan II

1. Introduction

Boron deficiency of crop plants has been reported in the field in 80 countries and on at least 132 crop species, and it is estimated that about 15 million hectares are annually treated with B-fertilizers (Shorrocks 1997). In B-deficient plants, the shoot tissues first appear as abnormal or there is retarded growth of the apical growing points. The youngest leaves are misshapen, wrinkled and are often thicker than normal. Internodes are shorter, giving the plants a bushy or rosette appearance (for general information on B nutrition of higher plants, see Mengel and Kirkby 1982; Gupta 1993; Marshner 1995; Bell 1997). In the field, sexual reproduction is often more sensitive to low soil B than vegetative growth, and marked seed yield reductions can occur without symptoms being expressed during prior vegetative growth (Dell and Huang 1997). Rerkasem and Jamjod (1997) demonstrated that the expression of B deficiency occurred primarily through male sterility. On the other hand, B-rich soils, to a lesser extent than deficient soils, also occur and considerable genetic variation in response to high B has been identified in a wide range of plant species (Nable et al. 1997).

Warington (1923) conducted convincing experiments that B is an essential element for higher plants. Following her work many growth experiments, such those by Sommer and Lipman (1926), were carried out in order to confirm the essentiality of B for higher plants. Kouchi and Kumazawa (1975) demonstrated that among the six soil-born essential plant nutrient elements, such as Ca^{2+}, Mg^{2+}, Fe^{2+}, Mn^{2+}, Zn^{2+}, and B, deprivation of Ca^{2+} and/or B stopped the elongation of tomato seedling roots within a few hours. This suggests that the need for the supply of B and Ca^{2+} is maintained throughout life. Seventy-five years have passed since the finding of the essentiality of B, however, its physiological function still remains unknown. Loomis and Durst (1992) comprehensively reviewed the postulated roles of B, such as 1) sugar transport, 2) cell wall synthesis, 3) lignification, 4) cell wall structure, 5) carbohydrate metabolism, 6) RNA metabolism, 7) respiration, 8) indole acetic acid (IAA) metabolism, 9) phenol metabolism and 10) membrane function. Almost all these hypotheses were based on observed abnormalities in plants when B was withheld from culture media. Whether the abnormalities are the result of a primary function of the element or merely a secondary consequence of the primary effect has been assessed only by the timing of the noticeable changes in metabolic activities and/or morphology. Therefore, it is most important to detect the earliest manifestation of B deficiency. The timetable of early symptoms of boron deficiency are listed by Hirsch and Torrey (1980). However, it is not possible to judge with confidence which event is the earliest, because the physiological changes due to B deficiency are not always visible and are quite often subtle.

A great abundance of experiments have demonstrated that deprivation of B results in disturbance of membrane function. The immediate stopping of root elongation due to deprivation of B and/or Ca^{2+} (Kouchi and Kumazawa 1975) suggests that the driving force for elongation may be lost because of depletion of elements, that is, B may interact with the membranous components (Blaser-Gril et al. 1989; Schon et al. 1990). However, I hesitate to assume that B works primarily in membranes for the following reasons. 1) the presence of a significant amount of B in microsomal membranes could not be confirmed (Kobayashi et al. 1997b); 2) direct evidence for the involvement of B in isolated and purified membrane systems, not in tissues, has never been presented; and 3) animal cells do not require B for their normal growth. Therefore, there must be some invisible link between B and membranous activity. Recently it was proposed that B may be involved in ascorbic acid production, because deprivation of B results in a sharp decline of ascorbic acid content (Lukaszewski and Blevins 1996). However, direct evidence is necessary to verify this, such as the presence of a B-enzyme or a B-substrate complex, or significant activation of a purified enzyme with B. Recent advances in B study are reviewed by Blevins and Lukaszewski (1998).

The specific function of B might be associated with a certain cellular constituent, and abnormalities due to B deficiency may develop first where B should be localized. Therefore, information on its distribution within the cell would be of value. Boric acid interacts spontaneously with diol groups in the *cis* position. The interconversion of boric acid (HB) and borate anions (B^-), and the interaction of borate with diols (D) are represented by the equilibria (1) to (3) in Fig. 1. These equilibria in aqueous solution are rapidly established and reversible, and the positions of the equilibria depend upon the pH of the solution and the stereochemistry of the diol (Henderson et al. 1973). Accordingly, B in boric acid has been postulated to react and form a complex with sugars and phenolics spontaneously. However, for each of these proposed functions the localization of the element has been neglected and therefore each hypothesis is flawed since they all lack evidence for the presence of the corresponding B-substrate complex. Therefore, I at first concentrated on a study of the intracellular localization of the element in plant cells.

During this study, I found that a B-polysaccharide complex isolated from cell walls contains Ca^{2+}. Therefore, the study was extended to fractionation of Ca^{2+} in cell walls. The essential nature of calcium as a plant nutrient has been recognized for over a hundred years. Calcium has striking characteristics compared to other plant nutrients; the cytosolic concentration is quite low in all eukaryotes in a range of 10^{-8} to 10^{-5} M (Bush 1995). This may reflect a biochemical necessity for maintaining low levels of Ca^{2+} in the phosphate-rich environment of the cytoplasm. However, the low concentration of Ca^{2+} has evolved a complex system for regulating cytosolic metabolic activities through Ca^{2+} concentrations. The concept that Ca^{2+} acts as a second messenger for the signal transduction was first introduced in animal cells and it has now been successfully adopted for plant cells. However, quantitative analysis indicates that Ca^{2+} is primarily an apoplastic nutrient in plants. Therefore, there must be a critical requirement for the element also outside the protoplast, on the exterior surface of the plasma membrane and in the cell wall, even though this has not been established yet. In this review, based on the latest results from the laboratory, I propose that B fulfils its essential function in the cell wall in collaboration with Ca^{2+} by cross-linking pectic chains at the RG-II regions.

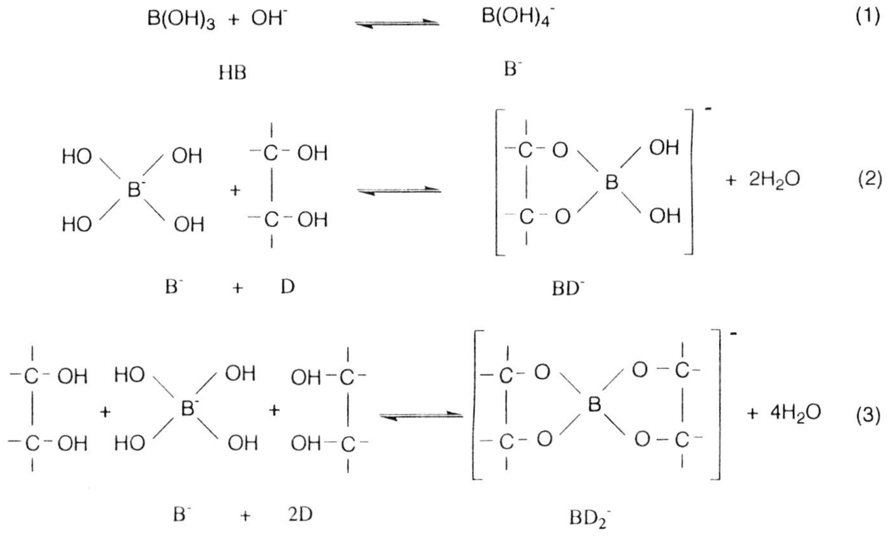

$$\underset{\text{HB}}{B(OH)_3 + OH^-} \rightleftharpoons \underset{\text{B}^-}{B(OH)_4^-} \quad (1)$$

$$\underset{\text{B}^- \; + \; \text{D}}{B^-(OH)_2(OH)_2 + (-C-OH)_2} \rightleftharpoons \underset{\text{BD}^-}{\left[(-C-O)_2B(OH)_2\right]^-} + 2H_2O \quad (2)$$

$$\underset{\text{B}^- \; + \; 2\text{D}}{(-C-OH)_2 + B^-(OH)_2(OH)_2 + (OH-C-)_2} \rightleftharpoons \underset{\text{BD}_2^-}{\left[(-C-O)_2B(O-C-)_2\right]^-} + 4H_2O \quad (3)$$

Fig. 1. Interconversion of boric acid (HB) and borate anions (B^-), and the interaction of borate with diols (D)

2. Boron in Plants Occurs Both in a Water-Soluble and -Insoluble Form

Marsh and Shive (1941) and Marsh (1942) demonstrated that higher plants contain B both in a water-soluble and -insoluble form, and that in broad bean and corn plants, the water-soluble B increases according to increase in B supplied to plants, while the insoluble-B does not. Eaton (1944), Husa and McIlrath (1965), McIlrath (1965), and Parr and Loughman (1983) also reported that both forms of B occur in higher plants. Skok and McIlrath (1958) found that B-deficient sunflower plants contain less water-soluble B, leaving the water-insoluble B unchanged. A decrease in water-soluble B during the development of B deficiency was also noted by Shive and Barnett (1973). I also observed that in Swiss chard plants when the B concentration in culture solutions increased, only the water-soluble B increased while the water-insoluble B remained constant (Fig. 2). In a B-deficient regime, such as less than 0.05 ppm B in the culture media, insoluble B also decreased. These data together with those by Skok and McIlrath (1958) and Shive and Barnett (1973) indicate that the levels of B in plants under adequate to excess B nutrition are ascribed to the water-soluble B and that B in the bound form does not increase even under B-excess conditions.

Where does the water-soluble B occur in plant tissues? In cultured cells, a significant amount of water-soluble B has not been detected in cells after extensive washing (Durst and Loomis 1984; Loomis and Durst 1992; Matoh et al. 1992; Hu and Brown 1994). Protoplasts prepared from cultured tobacco cells (Loomis and Durst 1992; Matoh et al. 1992) and from Swiss chard leaves (Matoh unpublished data) did not contain significant amounts of B. These results suggest that B in a water-soluble form may occur in the cell walls. Boron-11 NMR spectroscopy is certainly one of the most powerful, non-destructive means to study boric acid and borate esters in living tissues. When leaf disks (diameter 4 mm) of the middle leaves of Swiss chard plants grown at a B concentration of 10 mg B L^{-1}, were subjected to ^{11}B-NMR analysis, the majority of B was in a boric acid form (∂ around 19.2, after Oi et al. 1992) with a minor peak at around ∂ 10.6, which is attributable to BD_2^- (Oi et al. 1992) in a concentration of 3.2% of the HB signal (Fig. 3a). The

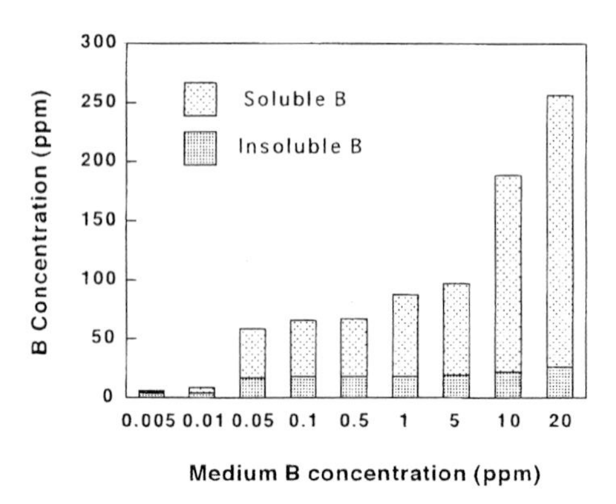

Fig. 2. Water-soluble and water-insoluble B in the young leaves of Swiss chard plants grown under various B concentrations in the culture solutions

signal due to the water-insoluble B that is bound inside the cell walls is likely to be too broad to detect because of decreasing rates of molecular motion within the native cell walls (Roberts 1987). When the leaf discs in a sample tube were frozen in liquid N_2 and thawed at room temperature, the HB signal decreased while that of BD_2^- increased (Fig. 3b). These spectra suggest that almost all the B (>96%) in the leaf discs occurs as boric acid. If boric acid were localized in the cytoplasm in which the ambient pH is in a range of 7 to 8, boric acid would bind to sugar components having *cis* diol, such as ribose, and the significant BD_2^- signal would be detected. Therefore, B must occur in a compartment where the ambient pH is as low as 5 and water-soluble ligands are absent. As the freeze-thaw treatment disrupts membrane integrity, the increase in the BD_2^- signal suggests that boric acid started to bind to water-soluble ligands after disturbance of the intracellular compartmentation. Such a compartment is the vacuole and/or cell wall. As protoplasts contain a negligible amount of B, therefore B may occur mainly in the cell wall. Martini and Thellier (1993) employed neutron capture radiography to trace the intracellular localization of B in leaf parenchyma cells of 3-week old clover plants. The clover leaves contained 43 mg B per kg dried material, and the concentration of B in the cell wall was 82.5 mg B per kg fresh material, while in the cytoplasm it was only 3.78 mg B per kg fresh material, and practically zero in the vacuole. Taking into consideration the ^{11}B-NMR data from intact leaves and protoplast experiments, I speculate that the majority of the water-soluble B may occur as boric acid in the apoplast, cell walls and intercellular spaces, not in the symplast. That is, B both in water-soluble and -insoluble forms occur mainly in cell walls. Figure 2 illustrates that even when the B supply is excessive, only water-soluble B is accumulated in the leaves. Accordingly, some of the soluble B in cell walls may intrude into the cytoplasm through the plasma membrane to form B-polyol complexes there, hence the B may disturb cellular metabolisms, leading to B toxicity.

Many authors have assumed that B is localized in the cytoplasm because the molecular mass of boric acid is small and not ionized under physiological pH, and it was considered to penetrate plasma membrane readily in plant cells (Oertli and Richardson 1970; Raven 1980). However, a survey of the literature reveals

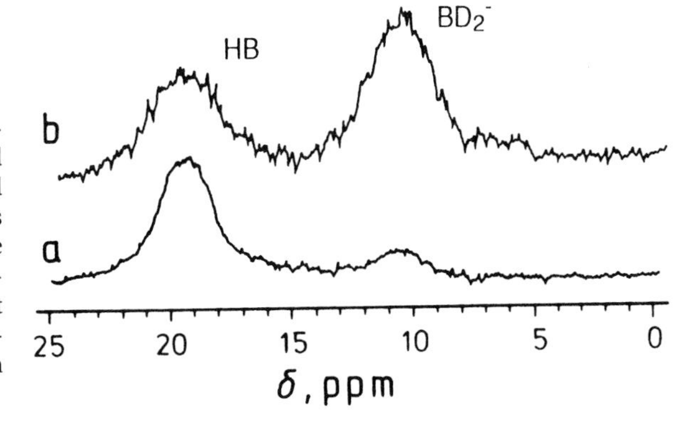

Fig. 3. ^{11}B-NMR spectra of intact leaf disks of Swiss chard plants. (**a**) Spectrum obtained in 45 min. (**b**) The leaf disks were frozen by dipping the disks in the NMR tube in liquid N_2 then were thawed at room temperature. The spectrum was obtained in 30 min (Matoh 1997)

that unequivocal evidence for the localization of this element in the symplast based on experimental results has never been demonstrated. In experiments designed to trace B uptake by plant tissues, time-dependent increase of B in plant tissues has been reported (for a review, see Shelp 1993). However, the increase may not always represent an increase of B in the symplast. Careful desorption of B from the material after the uptake period is critical, as noted by Bowen and Nissen (1976). Bingham et al. (1970) found that nearly all (95%) of the B taken up by excised barley roots was washed out by desorption. Wildes and Neales (1971) also noted in disks of carrot storage roots that the desorption of once absorbed B was rapid, having a half-time of approximately 30 min. Substantial loss of B from leaves through guttation (Oertli 1962) is likely if the majority of leaf B is localized in the cell walls. Oertli and Richardson (1970) postulated a mechanism to explain the apparent immobility of B in plants. According to them, B readily penetrates and is translocated to the phloem in leaves, but at the petiole B re-enters the xylem and then moves back into the leaf via the transpiration stream. I assume, however, that once B penetrates into the cytoplasm it may form a B-diol complex there, such as a B-ribose complex, because the cytoplasmic pH and high concentrations of the substrates in the cytoplasm are favorable for complex formation. Therefore, if B is once taken up into the cytoplasm it may not leak back outside easily. The proposal that B is localized exclusively in cell walls could explain the immobility of B that was once transported to shoots (McIlrath 1965; Oertli and Richardson 1970; Shelp et al. 1995). Then, if B does not penetrate readily through plasma membrane, redistribution of soluble B from mature tissues to immature ones through phloem would also be difficult. Of course, this result does not rule out a possible involvement of B in metabolic reactions in the symplast, and the possibility of a phloem transport of B is postulated (Shelp et al. 1995; Brown and Hu 1996). The mechanism for the retranslocation of B between organs should be examined further. Recently, Hu et al. (1997) isolated a borate-mannitol complex from celery plants, which is a candidate for the B transport complex in phloem tissues. Localization of the water-soluble B in intact plant tissues, as well as the membrane transport mechanism for this element, should be examined further.

3. Cell-Wall Bound B

3.1 Characterization of the B-Polysaccharide Complex

Schmucker (1933) suggested a close relationship between B and pectic substances in growing pollen tubes. Later, Smith (1944) demonstrated in squash leaves that water-insoluble B is cell wall-bound and is released with diluted HCl. Analogous to Ca^{2+}, he proposed that pectic polysaccharides are a candidate for fixing B. Torssell (1956) speculated that B may be involved in gel formation of polysaccharides which were secreted into cell walls. Yamanouchi (1971) was the first to demonstrate that the amount of water-insoluble but 0.5 M HCl-soluble B is highly correlated with that of the water- and oxalate-insoluble but 0.05 M HCl-soluble pectic polysaccharides in the cell walls of 52 species of higher plants (Fig. 4). Hu et al. (1996)

obtained similar results. Yamauchi et al. (1986) presented the possibility that cell-wall polysaccharides which contain B could be solubilized by hydrolysis of cell walls with pectinase enzymes.

Matoh et al. (1993) first isolated and purified a particular B-polysaccharide complex from radish root cell walls. In the complex, B was present as a tetravalent 1:2 borate-diol complex. Because of the ability of B to form complexes spontaneously with substrates having *cis* diol, it seems possible that B may bind to polysaccharides in cell walls at random. However, nearly 80% of the cell-wall B bound solely to this particular polysaccharide. Sugar analysis suggested that the sugar component of the complex is rhamnogalacturonan II (RG-II) (Kobayashi et al. 1995). Since the B content of the complex was 0.23% and the molecular weight was 9900, 1 mol of B-RG-II complex contains 2.1 mol of B. Acid hydrolysis (0.1 M HCl for 15 min at 25°C) of the B-RG-II complex released B and decreased its molecular weight by one-half with only a slight increase (14%) in the proportion of reducing end groups. Fragments generated by this mild hydrolysis were identical in terms of size and charge, indicating that B cross-links two identical RG-II through B-diol ester bonds to form a B-dimeric RG-II complex (Kobayashi et al. 1996). This finding was subsequently confirmed for the RG-II isolated from cultured sycamore cells and pea shoots (O'Neill et al. 1996), red wine (Pellerin et al. 1996), beet taproots (Ishii and Matsunaga 1996) and bamboo shoots (Kaneko et al. 1997). Occurrence of the B-RG-II complex in growing pollen tubes, which are the most dramatic manifestation of B requirement (Schmucker 1933), was confirmed by sugar analysis of the B-polysaccharide complex (Table 1).

Pectic polysaccharides are considered to be comprised of at least three fundamental constituents, namely, homogalacturonan, rhamnogalacturonan I (RG-I), and RG-II (Carpita and Gibeaut 1993). Homogalacturonans are made up of α 1,4-linked chains of galacturonic acid, which may be partly methyl esterified. RG-I contains a backbone of α 1,4-linked galacturonic acid and α 1,2-linked rham-

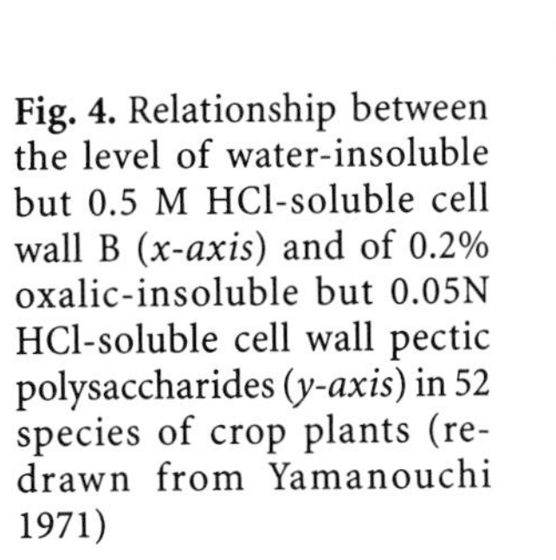

Fig. 4. Relationship between the level of water-insoluble but 0.5 M HCl-soluble cell wall B (*x-axis*) and of 0.2% oxalic-insoluble but 0.05N HCl-soluble cell wall pectic polysaccharides (*y-axis*) in 52 species of crop plants (redrawn from Yamanouchi 1971)

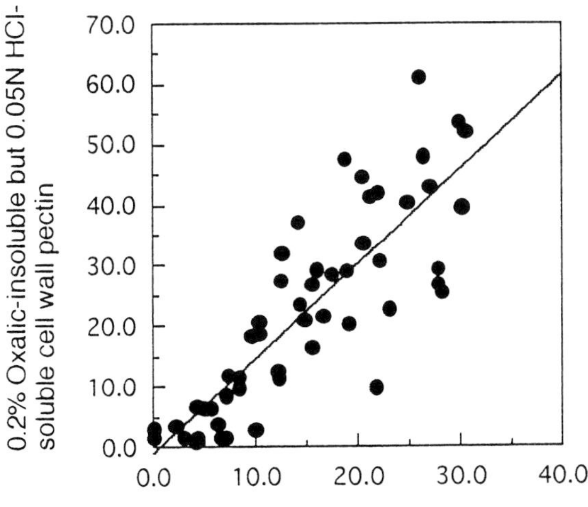

nose, probably alternating sequence. Side chains rich in neutral sugar residues, such as arabinose and galactose, attach to the *O*-4 of many of the rhamnosyl residues along the RG-I region. RG-II was first isolated from the primary cell walls of cultured sycamore cells, having the richest diversity of sugars and linkage structures, by Darvill et al. (1978). As presented in Fig. 5, the backbone of RG-II is composed of at least seven α 1,4-linked galacturonic acid residues (Thomas et al. 1989), and a variety of oligosaccharide side chains are attached to the backbone. The side chain sugar residues are characterized by rare sugars such as aceric acid, apiose, 2-keto-3-deoxysugars, *O*-methy fucose, and *O*-methyl xylose. As RG-II is released through an endopolygalacturonase enzyme, RG-II is considered to be attached to homogalacturonan and RG-I in vivo and the isolated RG-II complex are comprised of two chains of RG-II cross-linked with boric acid, two pectic chains are cross-linked in the RG-II region with two molecules of boric acid to form a supramolecular network. In the RG-II sugar moiety, O'Neill et al. (1996) revealed that apiosyl residue is responsible for fixing B to the polysaccharide chains.

Occurrence of the B-RG-II complex was surveyed in the cell walls of 24 species of higher plants (Matoh et al. 1996). As presented in Fig. 6, amounts of B which were associated with high-molecular weight polysaccharides were positively correlated (r^2= 0.778) with those of 2-keto-3-deoxysugars, which is the diagnostic monosaccharide for RG-II. The monomeric RG-II with 30 glycosyl residues contains two residues of 2-keto-3-deoxysugars (Puvanesarajah et al. 1991) and a unit B-RG-II complex contained 2.1 mol of B. Therefore, the molar ratio of B to 2-keto-3-deoxysugars in a unit B-RG-II complex should be 0.525. This value is close to the slope of the regression line, 0.535, which represents the molar ratio. The re-

Table 1. Glycosyl-residue composition of the B-RG-II complexes of lily pollen, radish roots, tobacco and sycamore cells

Residue	Lily pollen[a]	Radish root[b]	Tobacco cells[c]	Sycamore cells[d]
		(Mol %)		
Rhamnose	13.1	8.3	18.4	12.4
Fucose	2.5	1.9	3.0	2.8
2-O-Methylfucose	3.2	3.2	2.9	3.5
Arabinose	11.8	6.9	10.0	10.0
Xylose	1.2	—	0.8	—
2-O-Methylxylose	3.1	3.8	3.3	4.8
Apiose	6.1	11.4	8.3	12.2
Galactose	8.2	7.3	8.4	9.0
Mannose	2.0	—	0.2	—
Aceric acid	1.5	1.1	2.0	3.5
2-Keto-3-deoxysugars	4.8	5.4	4.1	7.0
Uronic acids	42.5	50.4	38.4	31.5

[a]Kobayashi et al. 1997a
[b]Kobayashi et al. 1996
[c]Kobayashi et al. 1997b
[d]Stevenson et al. 1988

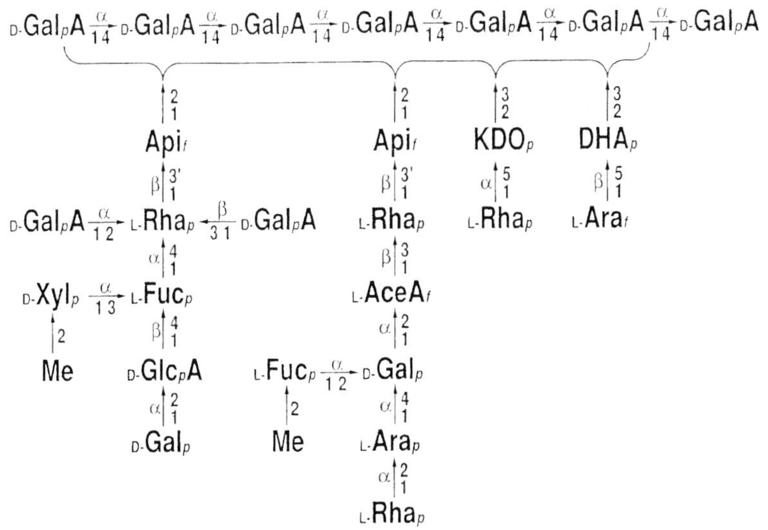

D-Gal$_p$A $\xrightarrow[14]{\alpha}$ D-Gal$_p$A $\xrightarrow[14]{\alpha}$ D-Gal$_p$A $\xrightarrow[14]{\alpha}$ D-Gal$_p$A $\xrightarrow[14]{\alpha}$ D-Gal$_p$A $\xrightarrow[14]{\alpha}$ D-Gal$_p$A $\xrightarrow[14]{\alpha}$ D-Gal$_p$A

2 1 Api$_f$ — 2 1 Api$_f$ — 3 2 KDO$_p$ — 3 2 DHA$_p$

β 3' 1 — β 3' 1 — α 5 1 — β 5 1

D-Gal$_p$A $\xrightarrow[12]{\alpha}$ L-Rha$_p$ $\xleftarrow[31]{\beta}$ D-Gal$_p$A — L-Rha$_p$ — L-Rha$_p$ — L-Ara$_f$

α 4 1 — β 3 1

D-Xyl$_p$ $\xrightarrow[13]{\alpha}$ L-Fuc$_p$ — L-AceA$_f$

2 — β 4 1 — α 2 1

Me — D-Glc$_p$A — L-Fuc$_p$ $\xrightarrow[12]{\alpha}$ D-Gal$_p$

α 2 1 — 2 — α 4 1

D-Gal$_p$ — Me — L-Ara$_p$

α 2 1

L-Rha$_p$

Fig. 5. Proposed model for RG-II (Puvanesarajah et al. 1991)

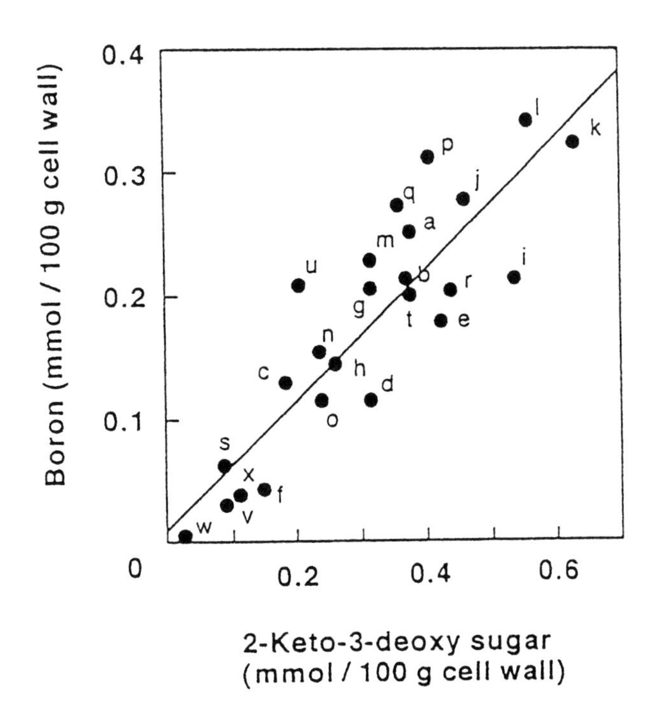

Fig. 6. Quantitative relationship of boron and 2-keto-3-deoxysugars in the high-molecular weight fragments of cell walls solubilized by Driselase from 24 species of higher plants. 2-Keto-3-deoxysugars are the diagnostic monosaccharide for RG-II. The regression line is Y= 0.0100 + 0.535X and the regression coefficient $r^2 = 0.778$. *a* cabbage, *b* Brussels sprout, *c* cucumber, *d* gourd, *e* towel gourd, *f* soybean, *g* pea, *h* *Vicia sepium*, *i* cowpea, *j* carrot, *k* parsley, *l* table beet, *m* spinach, *n* sweet pepper, *o* tomato, *p* Garland chrysanthemum, *q* lettuce, *r* lily, *s* duckweed, *t* leek, *u* Welsh onion, *v* Italian ryegrass, *w* rice, *x* corn. (Matoh et al. 1996)

gression line crossed the Y axis close to the origin, suggesting that almost all the B which is associated with high-molecular weight polysaccharides exists as a 2:2 complex with RG-II, and moreover that no RG-II exists without B and vice versa. Therefore it is likely that the B which is associated with high-molecular weight polysaccharides occurs as a B-RG-II complex, with RG-II being the exclusive carrier of B in cell walls. During this experiment, the levels of B associated with high-molecular weight polysaccharides were found to be nearly equal to the B level in the water-insoluble form shown in Fig. 2, except for some species, such as those in the Cucurbitaceae. Therefore, it is likely that the amount of B which is needed by plants may be determined by the amount of RG-II in cell walls. It is also likely that higher the RG-II contents the higher the B requirement. This is compatible with the notion that gramineous plants require less B than dicotyledonous ones, because the former have different structures in cell walls with less content of pectic polysaccharides. Duckweed is exceptional and polysaccharides other than RG-II are involved in fixing B in their cell walls (Matoh et al. 1996). Selected cells of cultured tobacco BY-2 which had been acclimated to lower doses of B (1 to 10 μg B L^{-1} medium whereas the control cells were grown under 1 mg B L^{-1} medium) were characterized as having swollen cell walls (Fig. 7; Matoh et al. 1992). The oven-dried cells of the selected (10 μg B L^{-1} medium) and wild cells (1,000 μg B L^{-1} medium) contained B at 1.88 mg and 7.74 mg kg^{-1} dried cells, respectively (Takasaki, 1998). Cell walls of both selected and control cells were digested with Driselase, and the digest was chromatographed on DEAE-Sepharose. As shown in Fig. 8a, 2-

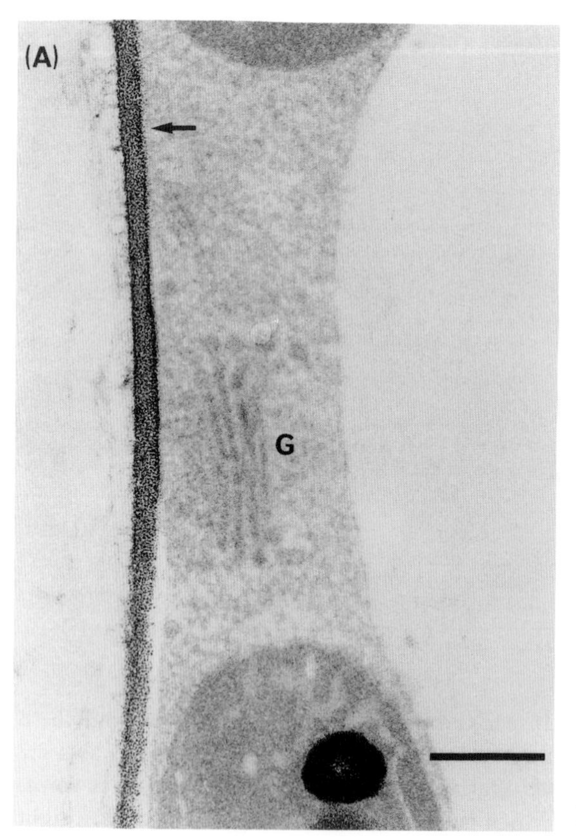

Fig. 7. Ultrastructure of the control cells (**A**) and of the cells that tolerated the reduced concentration of B in the medium (**B**). The cell wall (*arrow*) is thicker and seemed to be not tightly packed in the selected cells. The Golgi apparatus (*G*) was accompanied by more strongly staining secretory vesicles. *Bar* 0.5 μm. (Matoh et al. 1992)

keto-3-deoxysugars, the diagnostic monosaccharide for RG-II, were eluted as a single peak at the same position as that of the B-dimeric RG-II complex (Takasaki 1998), while two peaks for 2-keto-3-deoxysugars (the former ascribed to the monomeric RG-II and the latter to the B-dimeric RG-II) were detected from the digest of the cell walls of the selected cells (Fig. 8b). These results illustrate that the amount of the dimeric B-RG-II complex in cell walls is determined by the supply of B to cell walls, not by the secreted amounts of the RG-II polysaccharides. Furthermore, the amount of RG-II excreted into cell walls may be irrespective of the B supply. Limited supply of B resulted in lower amounts of the B-RG-II complex in cell walls and yielded swollen cell walls (Fig. 7). This will be discussed further in the following section.

Intracellular localization of the B-RG-II complex was examined using a polyclonal antibody against the B-RG-II complex raised in rabbits (Matoh et al. 1998). The antibody recognizes not only the B-RG-II complex but also the monomeric RG-II, suggesting that the epitope is on the sugar component of the complex. However, all the RG-II in cell walls may occur as the B-RG-II complex as described above (Fig. 8a), therefore, localization of the epitope may also indicate

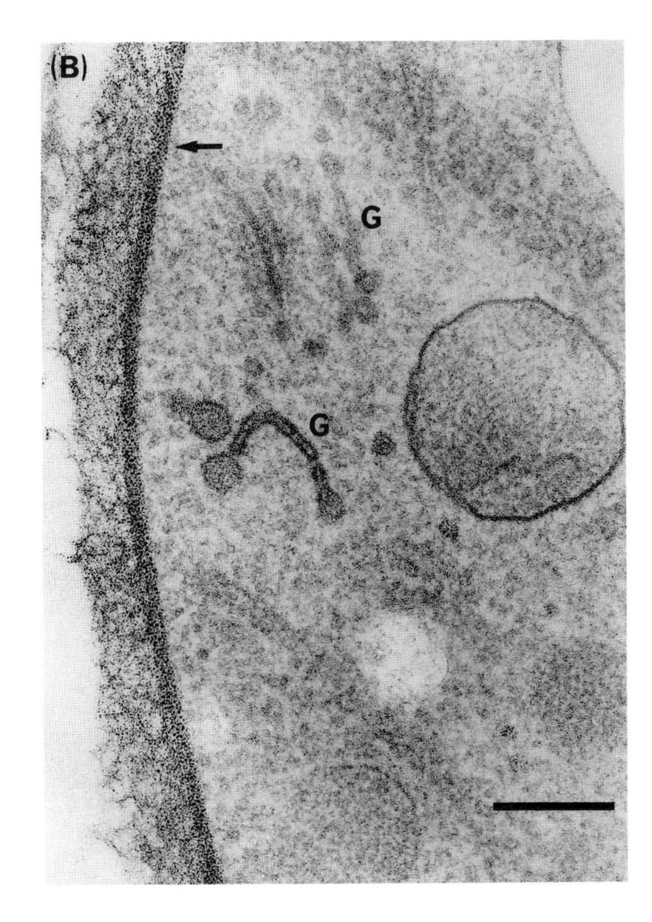

Fig. 7. *continued*

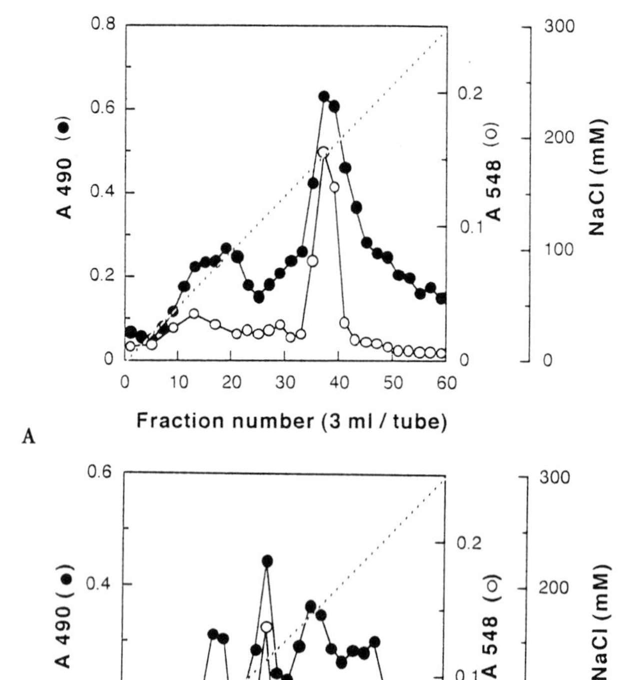

Fig. 8. Chromatogram of Driselase digests of cell walls of the control (**A**) and selected cells (**B**) on DEAE-Sepharose. Cells were cultivated under a lower dose of B (10 μg B L^{-1} medium) and a standard dose (1,000 μg B L^{-1} medium). Sugars were determined at OD 490 nm (*solid circles*) and 2-keto-3-deoxysugars at OD 548 nm (*open circle*). (Takasaki 1998)

the site of B. The epitope was ubiquitous in cell walls of all the cells in radish and rice roots, cultured tobacco cells and red clover root nodules (Fig. 9). Omnipresence of the B-RG-II complex in cell walls may substantiate the essential requirement of B by higher plants. The label was denser proximal to the plasma membrane (Fig. 10), suggesting that B may cross-link newly secreted pectic polysaccharides at the membrane-cell wall interface. Our results confirm the observation by Williams et al. (1996) that the RG-II epitope is localized proximal to the plasma membrane. The epitope of RG-I and of deesterified pectin are localized in these regions and also in the middle lamella and in the intercellular spaces (Moore et al. 1986; Moore and Staehelin 1988; Knox et al. 1990; Liners and Von Cutsem 1992). On the other hand, the RG-II epitope was not detected in the middle lamella or in the intercellular spaces (Fig. 10) suggesting that pectic polysaccharides in the primary cell walls and those in the middle lamella may be different in their constituents.

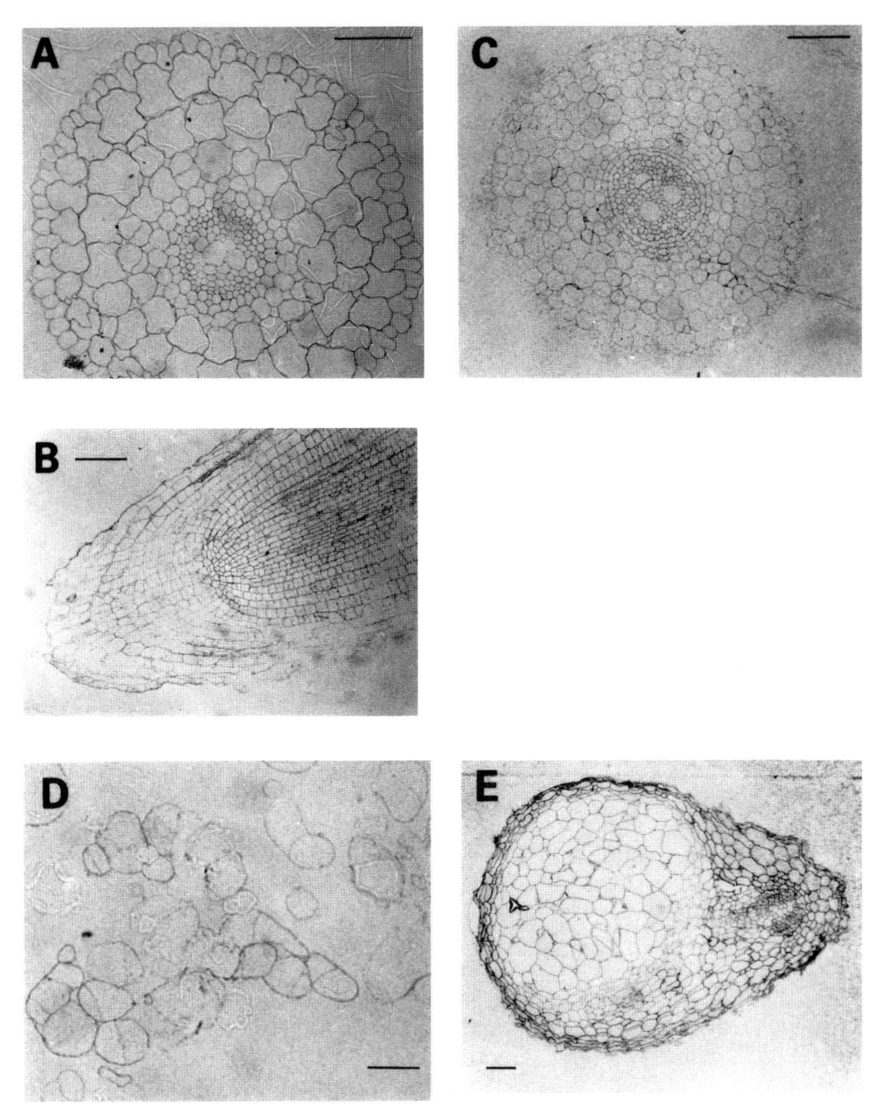

Fig. 9. Immunogold labeling with the RG-II antisera of transverse section of radish root tip (**A**), longitudinal section (**B**), transverse section of rice root tip (**C**), cultured tobacco cells (**D**), and root nodules of red clover (**E**). The gold particles were visualized by enhancement with silver. *Bars* 20 μm. (Matoh et al. 1998)

4. Function of Ca^{2+} in the B-RG-II Complex

O'Neill et al. (1996) reported that for the native B-RG-II complexes of sycamore cells and pea stems the complexes contain divalent cations other than B, such as Mg^{2+}, Ca^{2+}, Sr^{2+}, Ba^{2+} and Pb^{2+}. Occurrence of these cations is reported for the native B-RG-II complexes of beet and bamboo (Matsunaga et al. 1997) and for the

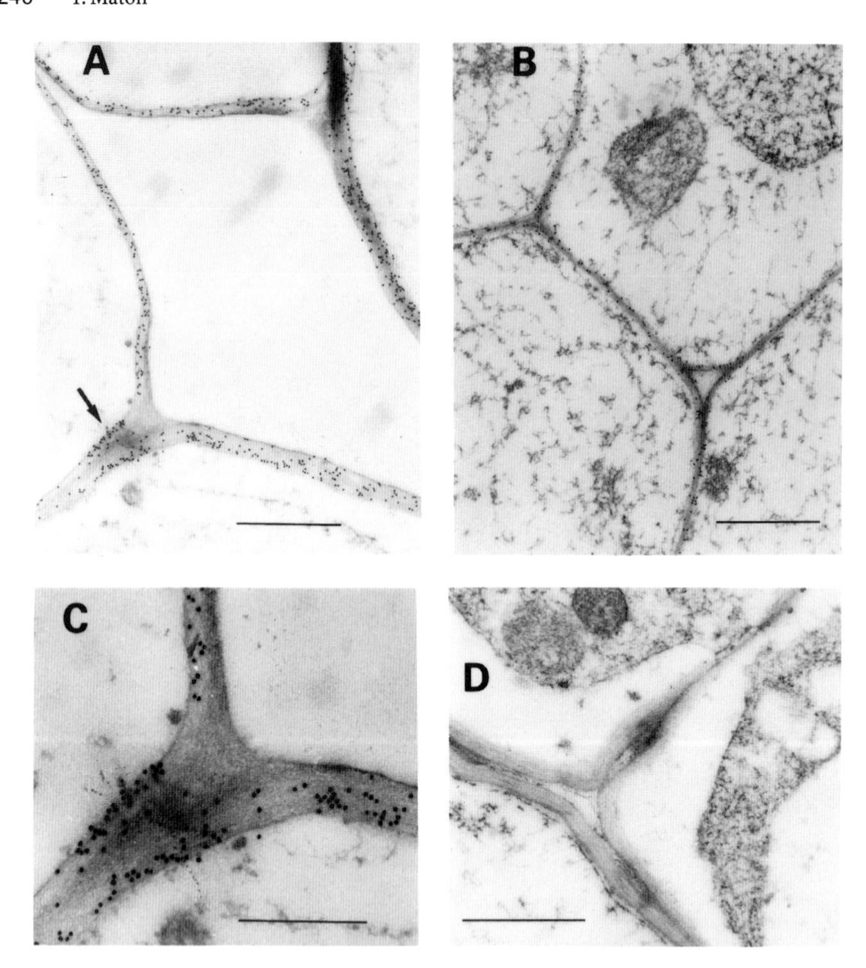

Fig. 10. Immunogold labeling with the RG-II antisera of radish root tips. **A** Mature cells in the cortex tissues. **B** Immature cells close to cambium cells in the stele tissues. **C** Enlargement of part designated by the *arrow* in **A**. **D** Preimmune sera was used. *Bars* in **A**, **B** and **D**, 1 μm; in **C**, 0.5 μm. (Matoh et al. 1998)

RG-II related polysaccharides of red wine (Pellerin et al. 1996). The radish B-RG-II complex also contained Ca^{2+}, Sr^{2+}, Ba^{2+} and Pb^{2+} other than B at molar ratios to B of 1.2, 0.1, 0.01 and 0.04, respectively (Kobayashi et al. 1999). Removal of these divalent cations from the native B-RG-II complex with CDTA induced release of B and hence decay of the complex (Fig. 11; Kobayashi et al. 1999).

O'Neill et al. (1996) showed that the dimeric B-RG-II complex is reconstituted spontaneously only by mixing the monomeric RG-II with B, and we confirmed their results with tobacco and radish RG-II (Kobayashi et al. 1997b). However, required concentrations of the monomeric RG-II and B for rapid reconstitution were very high, and the reconstituted dimer decomposed gradually. O'Neill et al. (1996) demonstrated that supplement of Sr^{2+}, Pb^{2+} and Ba^{2+} to the reconstitution medium stimulated dimer formation. In the reconstitution solution, the ratio of the dimeric B-RG-II complex to the monomeric one was under equilibrium, there-

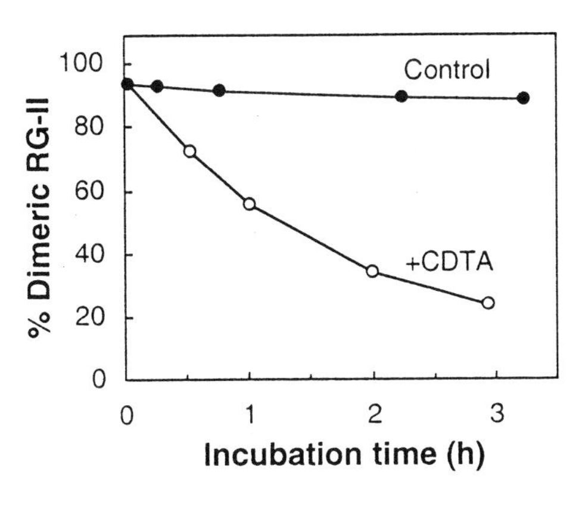

Fig. 11. Decomposition of the native B-RG-II complex with CDTA. Radish B-RG-II complex (1 mM) was incubated with 50 mM CDTA in 50 mM sodium acetate (pH 6.5, ○) or 10 mM MES-NaOH (pH 6.5, ●). Numbers on the *vertical axis* indicate the mol percent of RG-II converted into the dimeric form. (Kobayashi et al. 1999)

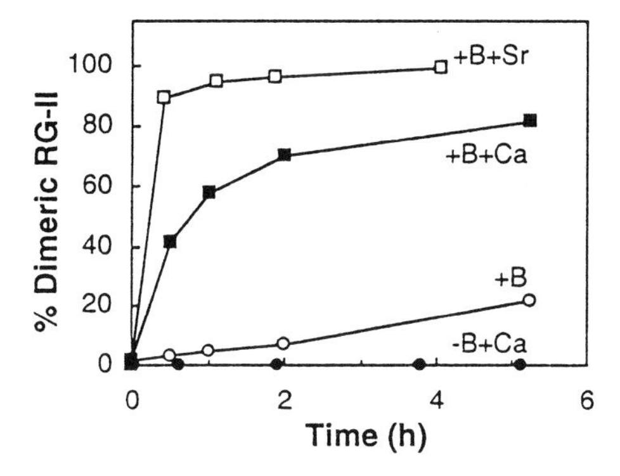

Fig. 12. Time-dependent reconstitution of the B-RG-II complex. Monomeric RG-II (1 mM) was incubated in 50 mM sodium acetate buffer (pH 4.0) with 10 mM H_3BO_3 (○), 10 mM $CaCl_2$ (●), 10 mM each of H_3BO_3 plus $CaCl_2$ (■), and 10 mM each of H_3BO_3 plus $SrCl_2$ (□). Numbers on the *vertical axis* indicate the mol percent of RG-II converted into the dimeric form. (Kobayashi et al. 1999)

fore, the divalent cations may stabilize the formed dimeric B-RG-II complex probably through coordinating bonding and hence prevent concomitant break down of the dimeric complex. They suggested that the divalent cations having an ionic radius larger than 1.1 Å may form coordination complexes with the B-RG-II complex. Haug and Smidsrød (1970) also noted a higher selectivity of pectic acid for Sr^{2+} than Ca^{2+}. The smaller cations, such as Ca^{2+}, may not interact strongly enough with the ligands on the two chains in order to stabilize the complex (O'Neill et al. 1996). However, higher doses of Ca^{2+} may compensate for the lower affinity. In accordance with this suggestion, we found that Ca^{2+} is also effective for the reconstitution (Fig. 12; Kobayashi et al. 1999). Divalent cations alone did not bring about the reconstitution. Even though the effect of Sr^{2+} on the stabilization of the RG-II was higher than Ca^{2+}, the content of Sr^{2+} in angiosperms is in the range of three orders lower than that of Ca^{2+} (Bowen 1966) and Ca^{2+} is far more abundant than Sr^{2+} in the soil and in cell walls. Furthermore, the nutritional requirement of Sr^{2+}, Ba^{2+} and Pb^{2+} has not been reported unequivocally for higher plants (Bowen 1966), except for the substitution effect of Sr^{2+} under Ca^{2+} deficient conditions (Da Silva 1962). From these discussions, together with the results of the gradual decomposition of the native complex with CDTA (Fig. 11) and the stimulative effect of Ca^{2+} on the reconstitution (Fig. 12), I am inclined to conclude that Ca^{2+} is a native constituent of the B-RG-II complex.

In the reconstitution, germanic acid could to some extent substitute for B in the dimer formation (Fig. 13; Kobayashi et al. 1997b). Skok (1957) found that germanic acid can substitute for B to some extent in the growth of B-deprived sunflower plants. In tomato plants, germanic acid fulfils a similar growth-promoting effect by stimulating the upward transport of B (Brown and Jones 1972). Loomis and Durst (1992) confirmed the effect of germanic acid on B deficiency by establishing cell lines which could grow using germanic acid without boric acid and Cakmak et al. (1995) demonstrated that germanic acid prevents solute leakage from B-deficient sunflower leaves. The ability of germanic acid to form a dimeric RG-II complex in vitro may explain the substitution effect of Ge when B is deprived from culture media.

5. Fractionation of Cell Wall Ca^{2+}

To evaluate the physiological significance of the Ca^{2+} in the B-RG-II complex the distribution of Ca^{2+} in cell walls was examined by sequential extraction. By incubating radish root cell walls in a 1.5% (w/v) sodium dodesyl sulfate (SDS) solution (pH 6.5), 96% of the cell-wall Ca^{2+} was released probably due to an ion-exchange action of Na^+, as the SDS solution contains *ca.* 50 mM Na^+, while only 6% of the pectic polysaccharides and 13% of the cell-wall B were released (Table 2; Kobayashi et al. 1999). Using the SDS treatment, the cell wall swelled two-fold. Subsequent extraction with 50 mM CDTA (pH 6.5) of the SDS-washed cell walls removed all the remaining Ca^{2+} and 80% of B, as well as 61% of pectic polysaccharides. The released B was in the form of boric acid.

The SDS-washed cell walls, which retain 4, 87 and 94% of Ca^{2+}, B and pectic polysaccharides of the original cell walls, respectively, were digested with Driselase and the digest was subjected to ion-exchange chromatography on DEAE-Sepharose. As shown in Fig. 14 (Kobayashi et al. 1999), Ca^{2+} co-migrated with B and 2-keto-3-deoxysugars, and significant peaks for Ca^{2+} were not detected other than that for RG-II. This chromatogram indicates that almost all the Ca^{2+} retained in the SDS-washed cell walls occurs in the RG-II region. As the CDTA treatment eliminates not only Ca^{2+} but also B from the native RG-II complex which then

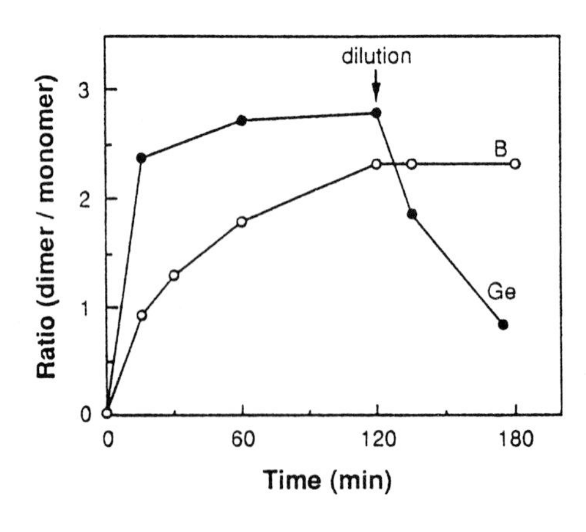

Fig. 13. Formation of the dimeric RG-II with boric and germanic acid. The monomeric RG-II and boric or germanic acid were incubated in 50 mM sodium acetate buffer (pH 4.0) at a final concentration of 5 mM each. At 120 min (*arrow*) the mixture was diluted 5-fold with 50 mM sodium acetate buffer (pH 4.0). (Kobayashi et al. 1999)

Table 2. Sequential extraction of Ca^{2+}, B and pectic polysaccharides from radish root cell walls. (Kobayashi et al. 1999)

Sample	Ca^{2+}	B	Pectic polysaccharides[a]
Crude cell walls	100 (6.7)[b]	100 (0.26)	100 (28.8)
SDS-treated cell walls	4	87	94
SDS- and CDTA-treated cell walls	0	7	33
SDS-, CDTA- and Na_2CO_3-treated cell walls	—[c]	—	15

[a]Determined by a *m*-hydroxydiphenyl method using galacturonic acid as a standard
[b]Values in the parentheses are μ mol Ca^{2+} and B g^{-1} crude cell walls, and mg uronic acid g^{-1} crude cell walls
[c]—, Not determined

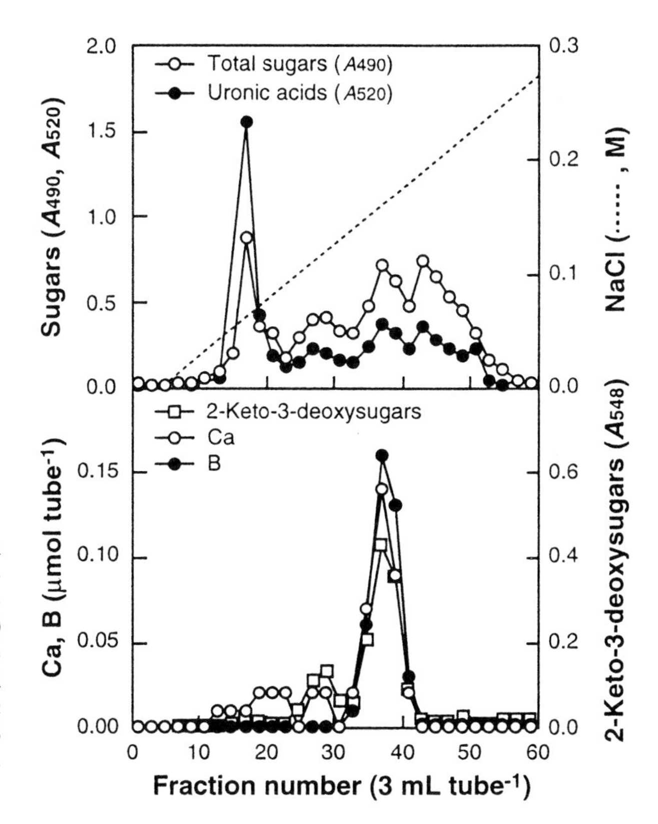

Fig. 14. Chromatogram of a Driselase digest of the SDS-washed cell walls on a DEAE-Sepharose column (1 × 20 cm). *Top panel* shows distribution of neutral sugars and uronic acid, and the *bottom panel* shows that of Ca^{2+}, B and 2-keto-3-deoxysugars. (Kobayashi et al. 1999)

results in decomposition of the B-RG-II complex, it is likely that the pectic polysaccharides which are not released with the SDS but with the CDTA are retained in cell walls through the cross-linking by Ca^{2+} and B in the RG-II regions.

In cell walls, pectic polysaccharides are cross-linked in various ways such as by Ca^{2+} bridges, phenolic coupling, galacturonate ester bonds (Fry 1986), and borate-diester bonds (Kobayashi et al. 1996). Jarvis (1982) demonstrated that an application of CDTA was successful in releasing Ca^{2+}-bound pectic polysaccharides, and they have been termed as chelator-soluble pectic polysaccharides. Our results indicate that the SDS solution released almost all (>96%) the cell-wall bound Ca^{2+}, however, more than 90% of the pectic polysaccharide in cell walls remained insoluble. Furthermore, the remaining Ca^{2+} is associated with the RG-II polysaccharide. This suggests that the Ca^{2+} associated with the RG-II region is responsible for anchoring the chelator-soluble pectic polysaccharides in cell walls. Or, Ca^{2+} which is sensitive to SDS washing does not contribute significantly to anchor the pectic network into cell walls. As the SDS washing increased the packed-cell wall volume twice, the SDS-sensitive Ca^{2+} may function in packing the cell wall tight. At present, I cannot explain the different sensitivities of the cell-wall Ca^{2+} to the SDS washing between those associated with the RG-II region and those in the rest of cell walls. The RG-II region may provide a kind of cavity for Ca^{2+}.

After the SDS- and CDTA-treatment, cell walls still contained 33% of pectic polysaccharides. According to Fry (1986), alkaline CDTA-sensitive pectic polysaccharides are retained by ester bonding. As the alkaline CDTA-liberated pectic polysaccharides, which amount to 15% of the cell wall pectic polysaccharides (Table 2), also contain the RG-II region (Matoh et al. 1998), these pectic polysaccharides are probably retained by both ester bonding and Ca^{2+}-B bridges in cell walls.

6. Function of B and Ca^{2+} in the Pectic Network

Shea et al. (1989) demonstrated that pectic polysaccharides are secreted into cell walls in advance of cellulose microfibrils during the regeneration of cell walls of carrot protoplasts. Shedletzky et al. (1990) isolated a cell line of tomato, of which cell walls appear to consist solely of a cross-linked meshwork of pectin. A pectic network is no longer compatible with the former notion that they are simply in a gel form, but is another cell wall network (Roberts 1990). Physiological roles of pectic polysaccharides are summarized as follows; 1) limit porosity of cell walls, 2) cement cells together at middle lamella, 3) provide an alternate co-extensive network, and 4) transduce environmental signals to cells with their fragments (McCann and Roberts 1991; McNeil et al. 1984). Plasma membranes of root cells do not directly perceive the ionic conditions in the soil solution but are in contact with a modified medium, the composition of which is affected by the interactions between the outer solution and the cell walls. Therefore, cell walls, especially pectic polysaccharides, are responsible in accommodating the symplast with the ionic and aquatic environment. According to Grignon and Sentenac (1991), pectic polysaccharides in cell walls may affect the cell wall physical properties as an ion exchanger, a diffusion barrier, a buffering system for excreted H^+, and as an ion reservoir. As cell walls carry non-diffusible negative charges due to uronic residues along pectic polysaccharides they behave as a Donnan system. Therefore, the total concentration of the mobile cations has to be larger in the wall than it is in the bathing medium (Grignon and Sentenac 1991). However, when Ca^{2+} is the major cation in the bathing medium, the carboxyl groups behave as if they are little ionized, because the interaction of Ca^{2+} with uronic residues is especially strong (Sentenac and Grignon 1981; Demarty et al. 1984). This is because Ca^{2+} can interact with pectic polysaccharides through coordination bonding with the oxygen atoms of the carboxyl and hydroxyl groups of uronic residues, not through electrostatic (ionic) interaction alone (Demarty et al. 1984). Thus, the strong interaction of Ca^{2+} with uronic residues may mask the fixed negative charges in cell walls. Under the current experimental conditions, a portion of 96% of the cell wall Ca^{2+} was released by SDS-washing at pH 6.5. Therefore, almost all the Ca^{2+} and H^+ were replaced by Na^+, and as a result they absorbed water osmotically. That is, Ca^{2+} is responsible for wall shrinkage. The selected cells which tolerated a reduced dose of B in the culture medium were also characterized by a thickened cell walls (Fig. 7), and the cell wall thickening may be a result of osmotic water absorption. The B contents differed significantly between the cell lines, however, the Ca^{2+} contents did not differ (data not presented). I assume that in the selected cells Ca^{2+} may not be properly distributed and arranged along the pectic chains to form an "egg-box" like conformation (Grant et al. 1973). As a result the cell wall may have swollen.

Skok (1958) discussed the possibility that if B is required for some other more basic physiological function than that of cell wall synthesis, it is reasonable to expect B requirement to be more widespread among living organisms. Loomis and Durst (1992) concluded that the rapid response of pollen tube elongation to withdrawal of B from the culture medium indicates that neither nucleic acid metabolism nor protein synthesis is involved in the function of B. They add that it is most likely that B's role is in the assembly of cell walls. Hu and Brown (1994) provided evidence that B is involved in the structural role in cell wall by binding to pectic polysaccharides. As has been shown here, it is likely that B and Ca^{2+} associated with the RG-II region are indispensable in anchoring the pectic network into cell walls. Calcium ions alone have the potential to cross-link pectic chains through ionic bonding at the polygalacturonic acid region (Jarvis 1984), however, the binding sites are arbitrary. On the other hand, as the site for borate-diester bonding is confined to apiosyl residues of the RG-II region, cells could control the site and frequency of the borate-diester linkage by regulating the frequency of RG-II insertion into pectic polysaccharide chains. This may be the merit of the borate bridges over the Ca^{2+} bridges. However, borate-diester bonding is rather unstable, and the bonding has to be strengthened with Ca^{2+}. The RG-II region provides a site for cross-linking of pectic chains through the B-diester and Ca-coordinate bonding to form a supramolecular pectic network, and the knot anchors the pectic network into cell walls. The rest of Ca^{2+} in cell walls may contribute to modify the physical properties of pectic gel in cell walls.

Teasdale and Richards (1990) found that in the culture medium for pine cells the B concentration that gives a maximum growth rate varies considerably according to the concentrations of coexisting Ca^{2+} and Mg^{2+} in the medium. From these interactions, the authors drew the conclusion that a critical acceptor molecule is activated only by binding both Ca^{2+} and B at separate sites, and therefore B has a primary role in forming carbohydrate gel surrounding the cellulose fibrils. Mühling et al. (1998) revealed that B deficiency reduces plasma membrane-bound Ca^{2+} before visible symptoms of B deficiency were noticed, suggesting a close relationship between B and Ca^{2+}. Blevins's group demonstrated that aluminum stress is prevented with supplemental B supply (LeNoble et al. 1996). As an increase in Ca^{2+} supply also alleviates Al stress to plants, the effect of B may become apparent through facilitating Ca^{2+} arrangement in the pectic network, or aluminum ions themselves may inhibit the formation of the B-Ca^{2+}-RG-II bridge.

7. Future Study

What is clear at present is that B cross-links two pectic polysaccharide chains at the RG-II region and Ca^{2+} strengthens the bonding, but this does not directly elucidate the mechanism underlying the abrupt cease of elongation of roots (Kouchi and Kumazawa 1975) and pollen tubes (Schmucker 1933) and the immediate death of cultured cells (Matoh et al. 1992) due to withdrawal of B. At cell division, cell plates are produced in the equatorial plane. The end of the plates fuses with cell surface from inside and then new cell walls and plasma membranes of the daughter cells are formed. Materials for the new cell walls and plasma membranes are successively secreted. At this moment, if B is absent from the apoplastic solute, newly secreted pectic polysaccharides may not be cross-linked, and then the pectic network will not be retained in the cell walls. Without Ca^{2+} in the apoplastic solute, B alone may not be able to form a stable RG-II cross-linking and this may result in the same consequence. If some systems perceive this then the signal may trigger off various disorders. One end of a pectic strand which has the RG-II region may bind covalently to a membrane protein and the other newly secreted pectic polysaccharide may be anchored to the membrane-bound pectic strand through the B and Ca^{2+} bridge, or the negative charge due to the B-diol bonding in the RG-II region may be attracted to a positive charge on a membrane protein, an electrostatic bonding of polysacharide chains with membrane proteins. Anyway, there must be a link between the integrity of a pectic network and the viability of cells, which is invisible at present. It is valuable and important to find out the missing link between the structure and function of the B-Ca^{2+}-RG-II complex.

Acknowledgments. Experiments on B and Ca^{2+} presented here have been carried out in the Laboratory of Plant Nutrition, Faculty of Agriculture, Kyoto University since 1986. I wish to express my profound gratitude to the laboratory members, Instructor Mr. Masaru Kobayashi, and my students, and to my colleagues Dr. Keiji Takabe, Dr. D. Ohta, and Dr. M. Mizutani. I appreciate the steadfast encouragement of Professor Emeritus Eiichi Takahashi and Professor Jiro Sekiya.

References

Bell RW (1997) Diagnosis and prediction of boron deficiency for plant production. In: Dell B, Brown PH, Bell RW (Eds) Boron in soils and plants: reviews. Kluwer Academic, Dordrecht, pp 149-168

Bingham FT, Elseewi, A and Oertli JJ (1970) Characteristics of boron absorption by excised barley roots. Soil Sci Soc Amer Proc 34:613-617

Blaser-Grill J, Knoppik D, Amberger A, Goldbach H (1989) Influence of boron on the membrane potential in *Elodea densa* and *Helianthus annuus* roots and H^+ extrusion of suspension cultured *Daucus carota* cells. Plant Physiol 90:280-284

Blevins DG, Lukaszewski KM (1998) Boron in plant structure and function. Annu Rev Plant Physiol Plant Mol Biol 49:481-500

Bowen HJM (1966) Trace elements in biochemistry. Academic, London, pp 61-84

Bowen JE, Nissen P (1976) Boron uptake by excised barley roots. 1. Uptake into the free space. Plant Physiol 57:353-357

Brown JC, Jones WE (1972) Effect of germanium on utilization of boron in tomato (*Lycopersicon esculentum* Mill.). Plant Physiol 49:651-653

Brown P, Hu H (1996) Phloem mobility of boron in species dependent: Evidence for phloem mobility in sorbitol-rich species. Ann Bot 77:497-505

Bush DS (1995) Calcium regulation in plant cells and its role in signaling. Ann Rev Plant Physiol Plant Mol Biol 46:95-122

Cakmak I, Kurz H, Marshner H (1995) Short-term effects of boron, germanium and high light intensity on membrane permeability in boron deficient leaves of sunflower. Physiol Plant 95:11-18

Carpita NC, Gibeaut DM (1993) Structural models of primary cell walls in flowering plants: consistency of molecular structure with the physical properties of the walls during growth. Plant J 3:1-30

Darvill AG, McNeil M, Albersheim P (1978) Structure of plant cell walls. VIII. A new pectic polysaccharide. Plant Physiol 62:418-422

Da Silva PGP (1962) On the possibility of substitution of strontium for calcium in maize plants. Agronomia Lusitana 24:133-164

Dell B, Huang L (1997) Physiological response of plants to low boron. In: Dell B, Brown PH, Bell RW (Eds) Boron in Soils and Plants: Reviews. Kluwer Academic, Dordrecht, pp 103-120

Demarty M, Morvan C, Thellier M (1984) Calcium and the cell wall. Plant Cell Environ 7: 441-448

Durst RW, Loomis WD (1984) Substitution of germanium for boron in suspension-cultured carrot cells. Plant Physiol 75S:14

Eaton FM (1944) Deficiency, toxicity, and accumulation of boron in plants. J Agric Res 69:237-277

Fry SC (1986) Cross-linking of matrix polymers in the growing cell walls of angiosperms. Ann Rev Plant Physiol 37:165-186

Grant GT, Morris, ER, Rees DA, Smith, PJC, Thom D (1973) Biological interactions between polysaccharides and divalent cations: The egg-box model. FEBS lett 32:195-198

Grignon C, Sentenac H (1991) pH and ionic conditions in the apoplast. Ann Rev Plant Physiol Plant Mol Biol 42:103-128

Gupta UC (1993) Deficiency and toxicity symptoms of boron in plants. In: Gupta UC (Ed) Boron and its role in crop production. CRC, Boca Raton, pp 147-155

Haug A, Smidsrød O (1970) Selectivity of some anionic polymers for divalent metal ions. Acta Chem Scand 24:843-854

Henderson WG, How MJ, Kennedy GR, Mooney EF (1973) The interconversion of aqueous boron species and the interaction of borate with diols: A 11B N.M.R. study. Carbohydr Res 28:1-12

Hirsch AM, Torrey JG (1980) Ultrastructural changes in sunflower root cells in relation to boron deficiency and added auxin. Can J Bot 58:856-866

Hu H, Brown PH (1994) Localization of boron in cell walls of squash and tobacco and its association with pectin. Evidence for a structural role of boron in the cell wall. Plant Physiol 105:681-689

Hu H, Brown PH, Labavitch JM (1996) Species variability in boron requirement is correlated with cell wall pectin. J Exp Bot 47:227-232

Hu H, Penn SG, Lebrilla CB, Brown PH (1997) Isolation and characterization of soluble boron complexes in higher plants. The mechanism of phloem mobility of boron. Plant Physiol 113:649-655

Husa, JG, McIlrath WJ (1965) Absorption and translocation of boron by sunflower plants. Botan Gaz 126: 186-194

Ishii T, Matsunaga T (1996) Isolation and characterization of a boron-rhamnogalacturonan-II complex from cell walls of sugar beet pulp. Carbohydr Res 284:1-9

Jarvis MC (1982) The proportion of calcium-bound pectin in plant cell walls. Planta 154: 344-346

Jarvis MC (1984) Structure and properties of pectin gels in plant cell walls. Plant Cell Environ 7:153-164

Kaneko S, Ishii T, Matsunaga T (1997) A boron-rhamnogalacturonan-II complex from bamboo shoot cell walls. Phytochemistry 44:243-248

Knox JP, Linstead PJ, King J, Cooper C, Roberts K (1990) Pectin esterification is spatially regulated both within cell walls and between developing tissues of root spices. Planta 181:512-521

Kobayashi M, Matoh T, Azuma J (1995) Structure and glycosyl composition of the boron-polysaccharide complex of radish roots. Plant Cell Physiol 36S:139

Kobayashi M., Matoh T, Azuma J (1996) Two chains of rhamnogalacturonan II are cross-linked by borate-diol ester bonds in higher plant cell walls. Plant Physiol 110:1017-1020

Kobayashi M, Kawaguchi S, Takasaki M, Miyagawa I, Takabe K, Matoh T (1997a) A borate-rhamnogalacturonan II complex in germinating pollen tubes of lily (*Lilium longiflorum*). In: Ando T et al (Eds) Plant Nutrition-For Sustainable Food Production and Environment. Kluwer Academic, Dordrecht, pp 89-90

Kobayashi M, Ohno K, Matoh T (1997b) Boron nutrition of cultured tobacco BY-2 cells. II. Characterization of the boron-polysaccharide complex. Plant Cell Physiol 38:676-683

Kobayashi M, Nakagawa H, Asaka T, Matoh T (1999) Borate-rhamno-galacturonan II bonding reinforced by Ca^{2+} retains pectic polysaccharides in higher-plant cell walls. Plant Physiol 119:199-203

Kouchi H, Kumazawa K (1975) Anatomical responses of root tips to boron deficiency I. Effects of boron deficiency on elongation of root tips and their morphological characteristics. Soil Sci Plant Nutr 21:21-28

Linears F, von Cutsem P (1992) Distribution of pectic polysaccharides throughout walls of suspension-cultured carrot cells. An immunocytochemical study. Protoplasma 170:10-21

LeNoble ME, Blevins DG, Sharp RE, Cumbie BG (1996) Prevention of aluminum toxicity with supplemental boron I. Maintenance of root elongation and cellular structure. Plant Cell Environ 19:1132-1142

Loomis WD, Durst RW (1992) Chemistry and biology of boron. BioFactors 3:229-239

Lukaszewski KM, Blevins DG (1996) Root growth inhibition in boron-deficient or aluminum-stressed squash may be a result of impaired ascorbate metabolism. Plant Physiol 112:1135-1140

Marschner H (1995) Mineral nutrition of higher plants, 2nd edn., pp 379-396

Marsh RP (1942) Comparative study of the calcium-boron metabolism of representative dicots and monocots. Soil Sci 53:75-78

Marsh RP, Shive JW (1941) Boron as a factor in the calcium metabolism of the corn plant. Soil Sci 51:141-151

Martini F, Thellier M (1993) Boron distribution in parenchyma cells of clover leaves. Plant Physiol Biochem 31:777-786

Matoh T, Ishigaki K, Mizutani M, Matsunaga W, Takabe K (1992) Boron nutrition of cultured tobacco BY-2 cells. I. Requirement for and intracellular localization of boron and selection of cells that tolerate low levels of boron. Plant Cell Physiol 33:1135-1141

Matoh T, Ishigaki K, Ohno K, Azuma J (1993) Isolation and characterization of a boron-polysaccharide complex from radish roots. Plant Cell Physiol 34:639-642

Matoh T, Kawaguchi S, Kobayashi M (1996) Ubiquity of a borate-rhamnogalacturonan II complex in the cell walls of higher plants. Plant Cell Physiol 37:636-640

Matoh T (1997) Boron in plant cell walls. In: Dell B, Brown PH, Bell RW (Eds) Boron in soils and plants: reviews. Kluwer Academic, Dordrecht, pp 59-70

Matoh T, Takasaki M, Kawaguchi S, Kobayashi M (1998) Immunocytochemistry of rhamnogalacturonan II in cell walls of higher plant. Plant Cell Physiol 39:483-491

Matsunaga T, Ishii T, Watanabe-Oda H (1997) HPLC/ICP-MS study of metals bound to borate-rhamnogalacturonan-II from plant cell walls. In: Ando T et al (Eds) Plant Nutrition-For Sustainable Food Production and Environment, Kluwer Academic, Dordrecht, pp 81-82

McCann MC, Roberts K (1991) Architecture of the primary cell wall. In: Lloyd CW (Ed) The cytoskeletal basis of plant growth and form. Academic, London, pp 109-129

McIlrath WJ (1965) Mobility of boron in several dicotyledonous species. Botan Gaz 126: 27-30

McNeil M, Darvill AG, Fry SC, Albersheim P (1984) Structure and function of the primary cell walls of plants. Ann Rev Biochem 53:625-663

Mengel K, Kirkby EA (1982) Principles of plant nutrition, 3rd edn., International Potash Instutute, Bern, pp 533-543

Moore PJ, Darvill AG, Albersheim P, Staehelin LA (1986) Immunogolod localization of xyloglucan and rhamnogalacturonan I in the cell walls of suspension-cultured sycamore cells. Plant Physiol 82:787-794

Moore PJ, Staehelin LA (1988) Immnogold localization of the cell-wall matrix polysaccharides rhamnogalacturonan I and xyloglucan during cell expansion and cytokinesis in *Trifolium pratense* L.; implication for secterory pathways. Planta 174:433-445

Mühling KH, Wimmer M, Goldbach HE (1998) Apoplastic and membrane-associated Ca^{2+} in leaves and roots as affected by boron deficiency. Physiol Plant 102: 179-184

Nable RO, Bañuelos GS, Paull JG (1997) Boron toxicity. In: Dell B, Brown PH, Bell RW (Eds) Boron in soils and plants: reviews, Kluwer Academic, Dordrecht, pp 181-198

Oertli JJ (1962) Loss of boron from plants through guttation. Soil Sci 94:214-219

Oertli JJ, Richardson WF (1970) The mechanism of boron immobility in plants. Physiol Plant 23:108-116

Oi T, Takeda T, Kakihana H (1992) Esterification of boric acid with 1,2-propanediol, 3-amino-1,2-propanediol, and (±)-3-dimethylamino-1,2-propanediol as studied by 11B NMR spectroscopy. Bull Chem Soc Jpn 65:1903-1909

O'Neill MA, Warrenfeltz D, Kates K, Pellerin P, Doco T, Darvill AG, Albersheim P (1996) Rhamnogalacturonan-II, a pectic polysaccharide in the walls of growing plant cells, forms a dimer that is covalently cross-linked by a borate ester. J Biol Chem 271:22923-22930

Parr AJ, Loughman BC (1983) Boron and membrane function in plants. In: Robb DA, Pierpoint WS (Eds) Metals and micronutrients. Uptake and utilization by plants. Academic, New York, pp 87-107

Pellerin P, Doco T, Vidal S, Williams P, Brillouet JM, O'Neill MA (1996) Structural characterization of red wine rhamnogalacturonan II. Carbohydr Res 290:183-197

Puvanesarajah V, Darvill AG, Albersheim P (1991) Structural characterization of two oligosaccharide fragments formed by the selective cleavage of rhamnogalacturonan II: evidence for the anomeric configuration and attachment sites of apiose and 3-deoxy-2-heptulosaric acid. Carbohydr Res 218:211-222

Raven JA (1980) Short- and long-distance transport of boric acid in plants. New Phytol 84:231-249

Rerkasem B, Jamjod S (1997) Genotypic variation in plant response to low boron and implications for plant breeding. In: Dell B, Brown PH, Bell RW (Eds) Boron in soils and plants: reviews, Kluwer Academic, Dordrecht, pp 169-180

Roberts JKM (1987) NMR in plant biochemistry. In: Davis DD (Ed) The biochemistry of plants, vol 13. Academic, London, pp 181-227

Roberts K (1990) Structures at the plant cell surface. Curr Opinion Cell Biol 2:920-928

Schmucker T (1933) Zur Blütenbiologie Tropischer Nymphaea-Arten. II. (Bor als entscheidender faktor.). Planta 18:641-650

Schon MK, Novacky A, Blevins DG (1990) Boron induces hyperpolarization of sunflower root cell membranes and increases membrane permeability to K^+. Plant Physiol 93:566-571

Sentenac H, Grignon C (1981) A model for predicting ionic equilibrium concentrations in cell walls. Plant Physiol 68:415-419

Shea EM, Gilbeaut DM, Carpita NC (1989) Structural analysis of the cell walls regenerated by carrot protoplasts. Planta 179:293-308

Shedletzky E, Shmuel M, Delmer DP, Lamport DTA (1990) Adaptation and growth of tomato cells on the herbicide 2,6-dichlorobenzonitrile leads to production of unique cell walls virtually lacking a cellulose-xyloglucan network. Plant Physiol 94:980-987

Shelp BJ (1993) Physiology and biochemistry of boron in plants. In: Gupta UC (Ed) Boron and its role in crop production. CRC, Boca Raton, Florida, pp 53-85

Shelp BJ, Marentes E, Kitheka AM, Vivekanandan P (1995) Boron mobility in plants. Physiol Plant 94:356-361

Shive JB Jr, Barnett NM (1973) Boron deficiency effects on peroxidase, hydroxyproline, and boron in cell walls and cytoplasm of *Helianthus annuus* L. hypocotyls. Plant Cell Physiol 14:573-583

Shorrocks VM (1997) The occurrence and correction of boron deficiency. In: Dell B, Brown PH, Bell RW (Eds) Boron in soils and plants: reviews, Kluwer Academic, Dordrecht, pp 121-148

Skok J (1957) The substitution of complexing substances for boron in plant growth. Plant Physiol 32:308-312

Skok J (1958) The role of boron in the plant cell. In: Lamb CA, Bentley OG, Beattie JM (Eds) Trace elements. Academic, London, pp 227-243

Skok J, McIlrath WJ (1958) Distribution of boron in cells of dicotyledonous plants in relation to growth. Plant Physiol 33:428-431

Smith ME (1944) The role of boron in plant metabolism. 1. Boron in relation to the absorption and solubility of calcium. Aust J Exp Biol Med Sci 22:257-263

Sommer AL, Lipman CB (1926) Evidence on the indispensable nature of zinc and boron for higher green plants. Plant Physiol 1:231-249

Stevenson TT, Darvill AG, Albersheim P (1988) 3-Deoxy-D-lyxo-2-heptulosaric acid, a component of the plant cell-wall polysaccharide rhamnogalacturonan-II. Carbohydr Res 179:269-288

Takasaki M (1998) Immunocytochemistry of the B-RG-II complex in higher plant cell walls. Master thesis, Kyoto University, Graduate School of Agriculture

Teasdale RD, Richards DK (1990) Boron deficiency in cultured pine cells. Quantitative studies of the interaction with Ca and Mg. Plant Physiol 93:1071-1077

Thomas JR, Darvill AG, Albersheim P (1989) Isolation and structural characterization of the pectic polysaccharide rhamnogalacturonan II from walls of suspension-cultured rice cells. Carbohydr Res 185:261-277

Torssell K (1956) Chemistry of arylboric acids. VI. Effects of arylboric acids on wheat roots and the role of boron in plants. Physiol Plant 9:652-664

Warington K (1923) The effect of boric acid and borax on the broad bean and certain other plants. Ann Bot 37:629-672

Wildes RA and Neales TF (1971) The absorption of boron by disks of plant storage tissues. Aust J Biol Sci 24:873-884

Williams MNV, Freshour G, Darvill AG, Albersheim P, Hahn MG (1996) An antibody Fab selected from a recombinant phage display library detects deesterified pectic polysaccharide rhamnogalacturonan II in plant cells. Plant Cell 8:673-685

Yamanouchi M (1971) The role of boron in higher plants I. The relations between boron and calcium or the pectic substance in plants. J Soil Sci Manure Jpn 42:207-213

Yamauchi T, Hara T, Sonoda Y (1986) Distribution of calcium and boron in the pectin fraction of tomato leaf cell wall. Plant Cell Physiol 27:729-732

The Significance of the Root Cell Wall in Phosphorus Uptake

Noriharu Ae, Renfang Shen, and Takashi Otani

Summary. Groundnut grew well on an extremely phosphorus (P) deficient soil in which soybean and sorghum did not survive. This ability could not be attributed to differences in root development, C_{min}, or root exudates that are capable of solubilizing sparingly soluble iron- and aluminum-bound P in soils. Since plants acquire P by diffusion, poor survival of soybean and sorghum indicated that P diffusion to their roots was negligible. Because P solubilization must occur in the contact zone between root and soil, the "contact reaction" between root epidermal cell walls and P-fixing minerals was examined in detail. Upon treatment with naturally-occurring P minerals such as strengite and variscite, root cell walls of groundnut solubilized twice as much P than those of sorghum and soybean. Such activity partly explains the superior growth of groundnut on P-deficient soils. The ability of pigeonpea to take up P from a low-P soil was associated with its production of root exudates. However, since exudation is high mainly during reproductive growth, its ability to take up P from P-deficient soils in earlier growth stages is worth examining. Tests on pigeonpea root cell walls indicated that its P solubilization activity was as high as that of groundnut. The findings from both crops suggest that "contact reaction" may contribute to P uptake by plants.

Key words. Groundnut, P acquisition, P solubilizing active substances, Root cell wall, Shedding of root surface, Cropping system

1. Introduction

The main driving force for transport of phosphorus (P) to plant roots is diffusion, as P is strongly adsorbed on surface of minerals even in soils supplied with large doses of P fertilizer (Barber and Ernani 1990; Barber and Cushman 1991). Moreover, the P diffusion coefficient is very low in most soils (Jungk 1991). For these reasons, plants surviving on soils with low P fertility develop strategies such as profuse root surface development and/or an increased solubilizing ability of sparingly soluble P in soils. Root surface development through root elongation and/or root hair formation can help increase soil volume from which P is taken up by roots by diffusion. Solubilizing ability is related to production of root exudates with which P-fixed minerals react in the rhizosphere.

Soil P is generally classified into inorganic and organic P. There are three forms of both inorganic and organic P, namely calcium-bound (Ca-P), aluminum-bound (Al-P) and iron-bound (Fe-P). Organic Fe-P, for example, may exist as Fe-associated phytin. From the viewpoint of solubility, all forms of organic P behave exactly in a similar manner to those of inorganic P. Ca-P, whether organic or inorganic, is more soluble than Al-P and Fe-P. Ca-P is soluble by acidification but Al-P and Fe-P are not. The latter forms become soluble, however, with chelating agents like citric acid. Further, Fe-P may become soluble by reduction of Fe^{3+} to Fe^{2+}, especially on the root surface of dicots (Marschner 1995). Phosphatases produced by soil microbes and/or roots (Tarafdar and Claassen 1988; Helal and Dressler 1989; Tadano and Sasaki 1991) can solubilize organic P that is not associated with Al or Fe, as the enzymes do not act on organic Al-P and Fe-P. Therefore, de-coupling of minerals—Ca, Al or Fe—is the most important step for plant utilization of soil P.

There are 3 main processes—acidification, reduction, and chelating with Al^{3+} and Fe^{3+}—to remove minerals from P-fixed minerals in a soil. The better growth of chickpea on calcareous Vertisols in India has been associated with secretion of large amounts of organic acids, especially citric and malic acids, from roots (Ohwaki and Hirata 1992). Likewise, citric acid in proteoid root exudates of white lupin is effective in P mobilization in acid soils (Gardner et al. 1983). By using ^{32}P-labelled hydroxyapatite (Ca-P), Moghimi et al. (1978) reported that an increased P uptake by wheat was due to 2-ketogluconic acid, a major organic acid component of rhizosphere. Citric, malic and 2-ketogluconic acids act as readily available sources of hydrogen ion (H^+) for dissolution of Ca-P. Groundnut roots exude formate under P-deficient conditions and the reducing power of formate increases P solubility through conversion of Fe^{3+} phosphate to Fe^{2+} phosphate (Tanaka et al.1995). The ability of pigeonpea to take up P from Alfisols and Andosols with low P availability was attributed to root exudation of piscidic, malonic and oxalic acids (Ae et al. 1990; Otani and Ae 1997a). Citric, malonic and piscidic acids have a chelating effect with Al-P and Fe-P to release P in Alfisols and Andosols.

The above reports show how root exudates contribute to P acquisition from low-P soils. However, a few important questions still remain unanswered (Rovira 1969). These include: (a) How much organic material is exuded from roots into soil to solubilize sparingly soluble P? (b) How far from roots do these compounds diffuse? (c) What concentration gradients exist? (d) What effects do soil properties have upon exudation? In addition, microbial degradation of root exudates in the rhizosphere must be considered. From these points of view, the relationship between the amount of root exudates and net P solubilization is unclear. Based on a series of experiments in which we did not detect significant production of chelating root exudates that can solubilize sparingly soluble Al-P and Fe-P in a low P soil, we propose a new P uptake mechanism in groundnut.

2. P Acquisition by Groundnut in Soils with Low P Fertility

2.1 Groundnut Cultivation in Japan

Andosols (volcanic ash soil) in Japan are well known for their extremely high P-fixing capacity. In an effort to increase food production after World War II, Japanese farmers started cultivating many virgin Andosols, which are difficult to irrigate, for growing paddy rice. They tried to introduce many crops but failed to meet consumption demands, especially because of low chemical fertilizer production due to the collapse of factories during the war. Of all crops, only groundnut and upland rice gave reasonable yields in these soils. Sanchez and Salinas (1981) reported that upland rice and groundnut are the world's most important basic food crops having a considerable degree of acid soil tolerance. The term "acid soil tolerance" in their report covers tolerance to many individual adverse soil factors and their interactions. Quantitative assessments of tolerance to acid soil stress include tolerance to high levels of Al or manganese, and to deficiencies of Ca, P, magnesium and certain micronutrients, principally zinc and copper. Although groundnut has a relatively high Ca requirement, only small amounts of lime can provide enough Ca for maximum yield in Oxisols and Ultisols of the Venezuelan Llanos (Sanchez 1977).

2.2 Growth Response of Groundnut on Andosols

To confirm "acid tolerance", groundnut was cultivated on an Aldosol field (P0) that never received P fertilizer and with P availability of only 1.3 mg P kg^{-1} as per the Bray 2 method. Another Andosol field (P1), which received NPK fertilizers for 15 years at conventional rates and with P availability of 6.4 mg P kg^{-1}, was also used (Table 1). Seed yield and P uptake by groundnut were 2690 kg ha^{-1} and 8.9 kg ha^{-1} in the P0 field, and 2360 kg ha^{-1} and 13.6 kg ha^{-1} in the P1 field, respectively (Table 2). In contrast, buckwheat yielded little in the P0 field with P uptake of only 0.1 kg ha^{-1}. In the P1 field, P uptake was 5.2 kg ha^{-1}. These findings suggest that groundnut can grow efficiently on low-P soils. Besides Al-P, Fe-P was a major P fraction in Andosols (Table 1). Although P availability was low in this field, the amount of P available to plants was not very low because of the high soil volume occupied by the rooting system under field conditions. Further, the amounts of Al-P and Fe-P were relatively high. It was thus hard to determine which P fraction was preferentially taken up by groundnut, Al-P or Fe-P.

Table 1. Phosphorus (P) status of Andosols used in field experiment (Tsukuba 1993)

		Total P (mg P kg^{-1})	Available P (mg P kg^{-1})		Inorganic P (mg P kg^{-1})		
Soil fertilization	pH (H_2O)		Truog	Bray 2	Ca-P	AI-P	Fe-P
No-P fertilization (P0)	6.1	988	0.5	1.3	0.0	255	134
Conventional fertilization (P1)	6.1	1472	7.3	6.4	2.5	563	261

Two well-weathered acid soils that never received any fertilizer and with low P availability were collected from Ishigaki (Acrisol), Okinawa Prefecture, and Nishinasuno (Andosol), Tochigi Prefecture, for pot experiments to examine P uptake in restricted soil volume conditions. Inorganic P in these soils, as determined by Truog's and Bray's methods (Table 3), was as low as or the same as the Andosol (Table 1), despite an apparently adequate total P in the Nishinasuno soil. Total P was 109 and 829 mg P kg^{-1} in the Ishigaki and Nishinasuno soils respectively. The Nishinasuno soil was rich in organic matter (C = 3.0%) and the discrepancy between available and total P can be attributed to a high amount of organic P that is not readily available to plants. Fe-P was predominant in the Ishigaki soil, whereas both Al-P and Fe-P were the main forms in Nishinasuno soil. Ca-P was negligible in both soils (Otani and Ae 1997b).

2.3 P Uptake Response of Groundnut in Pot Culture

Groundnut (*Arachis hypogaea* L. cv. Chibahandachi), maize (*Zea mays* L. cv. DK250), soybean (*Glycine max* (L.) Mirrill cv. Tachinagaha), and sorghum (*Sorghum bicolor* L. Moench cv. Hybrid-sorgo) were sown in plastic pots filled with 3.0 kg soil, grown in a glasshouse and supplied with P-free nutrient solution. In the Ishigaki soil, sorghum, maize and soybean died within 2 months after sowing due to P deficiency while groundnut survived and set seeds. A plant has in common its own optimum pH condition for optimum growth. To determine whether the poor growth of soybean, maize and sorghum in the Ishigaki soil was be due to low pH, the effect of pH on P uptake in the Nishinasuno soil was examined by raising soil pH from 5.1 to 6.5 by addition of $CaCO_3$. In the Nishinasuno soil too, groundnut took up more P than maize, sorghum and soybean at all soil pH levels

Table 2. P uptake and growth of groundnut and buckwheat in field experiment (Tsukuba 1993)

	Groundnut (cv. Chibahandachi)				Buckwheat (cv. Akisoba)		
	DW	Grain yield	P uptake	Root length	DW	P uptake	Root length
Soil fertilization	(kg ha^{-1})	(kg ha^{-1})	(kg P ha^{-1})	(m/m^{-2})	(kg ha^{-1})	(g P/m^{-2})	(m/m^{-2})
No-P (P0)	6590	2690	8.9	300	240	0.1	484
Conventional fertilization (P1)	8260	2360	13.6	424	4100	5.2	588

Table 3. P status of soils used in the pot experiment

		Total P (mg P kg^{-1})	Available P (mg P kg^{-1})		Inorganic P (mg P kg^{-1})			Organic P (mg P kg^{-1})		
Soil	pH (H_2O)		Truog	Bray 2	Ca-P	AI-P	Fe-P	Ca-P	AI-P	Fe-P
Ishigaki	4.5	109	0.9	0.9	0	20	58	0	4	7
Nishinasuno	5.1	829	0.2	1.6	2	90	75	1	111	439

(Table 4). These results show that superior growth of groundnut in these low P soils is not due simply to its acid tolerance.

The better growth and P uptake by groundnut than soybean in pots with restricted root development showed that groundnut took up Fe-P in Ishigaki and Al-P in Nishinasuno soils (Ae et al. 1996). Hartzog and Adama (1988) pointed out that high-yielding, low-input crops should receive high priority for cultivation in developing countries. From the standpoint of fertilizer input, therefore, groundnut must be considered as one among the more desirable crops. Walker et al. (1979, 1982) reported that groundnut generally showed few field responses to P and K fertilization. Likewise, wild *Arachis* (*A. pintoi*), which has high forage potential, is known to adapt better than grasses such as *Brachiaria dictyoneura* in low P and acid soils (Rao and Kerridge 1994). These results have raised our interest further to examine characteristics of P uptake by *Arachis* spp. in soils with Al-P and Fe-P.

3. Factors Affecting P Uptake by Groundnut

At least five factors are known to affect the uptake of sparingly soluble P in soils. These are:

1 Low P uptake parameter (C_{min})
2 Root elongation (root length including root hairs and root surface area)
3 Vesicular-Arbuscular mycorrhizae (VAM) association
4 Reducing power from Fe^{3+} to Fe^{2+}, and
5 Root exudation for chelation with Al and/or Fe.

3.1 Low P Uptake Parameter (C_{min})

C_{min} is an important factor in ion uptake, because it is the lowest concentration at which roots can extract an ion from soil solution. In other words, a plant with lower C_{min} is more advantageous than a plant with higher C_{min} for cultivation in low P conditions. C_{min}s differ, widely among plant species (Asher 1978). Measurements of C_{min}s at 6 and 8 weeks after sowing of five crops including groundnut

Table 4. P uptake by groundnut, soybean, maize, and sorghum in low-P soils (Pot experiment: P-free nutrient solution was applied every 3 days)

		P uptake (mg P pot^{-1})			
Soil	pH (H_2O)	Groundnut	Soybean	Sorghum	Maize
Ishigaki	4.5	2.0 ± 0.22	0.0 ± 0.05 [a]	0.0 ± 0.01 [a]	0.0 ± 0.01 [a]
Nishinasuno	5.1	2.1 ± 0.48	0.1 ± 0.13	1.2 ± 0.01	0.1 ± 0.15
	5.7 [b]	3.4 ± 0.48	0.7 ± 0.16	1.1 ± 0.23	0.1 ± 0.14
	6.5 [b]	3.8 ± 0.57	0.8 ± 0.20	0.4 ± 0.08	0.0 ± 0.04

[a]Died within 2 months after sowing
[b]pH in Nishinasuno soil was increased by adding $CaCO_3$

showed that C_{min} values in rice (1.97μM) and sorghum (1.70 μM) were largely similar but higher than in three legumes. Among legumes, C_{min} in groundnut was not significantly different from that in pigeonpea and soybean (Table 5).

3.2 Root Elongation

Root length is considered important in P uptake, but it is not known if a crop with long roots is always efficient in absorbing P from soils with low available P. The relationships among root length, soil volume and P uptake were, therefore, examined in buckwheat, groundnut, and sorghum using pots of varying size filled with low P Andosol (P0) (see Table 1). P uptake, root length, root density, P uptake per unit soil weight, and P uptake per unit root length, were determined (Table 6). P uptake was higher in groundnut than in sorghum or buckwheat. P uptake by all plants increased with an increasing soil volume. Sorghum did not survive in the smallest pot due to severe P deficiency.

Table 5. P absorption characteristics on the basis of Eadie-Hofstee plots in the P depletion experiment

	Km (μM P)		*Cmin* (μM P)	
Crop	6WAS	8WAS	6WAS	8WAS
Rice	14.4	19.5	1.95	—
Sorghum	13.7	21.3	1.70	1.85
Pigeonpea	4.4	6.9	1.03	0.82
Groundnut	2.5	8.2	1.06	0.64
Soybean	2.1	6.3	0.76	0.75
LSD (5%)	6.7	4.2	0.54	0.75

WAS, weeks after sowing

Table 6. Effect of soil volume on P uptake and root length on low-P Andosol (see Table 1)

Crops	Pot size	Soil (kg pot^{-1})	P uptake[a] (mg P pot^{-1})	Root length (mg pot^{-1})	P uptake per unit soil weight (mg P kg^{-1} soil)	P uptake per unit root length (μg P/m)
Buckwheat	Small	0.8	0.36	28.6	0.45	12.8
	Medium	2.0	0.77	41.4	0.39	18.2
	Large	7.5	3.98	61.7	0.53	67.4
Groundnut	Small	0.8	2.77	93.5	3.46	30.4
	Medium	2.0	15.61	194.0	7.81	80.9
	Large	7.5	43.53	334.5	5.80	130.7
Sorghum	Small	0.8	0.02	10.6	0.03	1.4
	Medium	2.0	2.47	148.0	1.23	17.0
	Large	7.5	11.73	453.5	1.56	25.7

[a]Net P uptake: seed P was subtracted

Root length in sorghum was maximum when grown in large pots. P uptake was, however, maximum in groundnut irrespective of soil volume variations. P uptake per unit soil weight was also more in groundnut than in sorghum and buckwheat. This observation implies that superior P uptake by groundnut cannot be explained by root elongation. Groundnut roots normally lack root hairs except for a small number of rosette-type root hairs around the lateral root primordia and short root hairs near the lateral root tips (Meisner and Karnox 1991). The groundnut genotype (Chibahandachi) used in this study also lacked root hairs. This observation suggests that groundnut employed some other mechanisms of P uptake beyond root length. Recently, Wissuwa and Ae (1999) examined genotypic variation for uptake of Fe-P and reported that direct P uptake through fruiting organs (facilitated by the presence of root hair like outgrowths on pegs) was possible in groundnut genotypes such as Wasedairyu. However, Chibahandachi had neither root hairs nor root hair like outgrowths on pegs.

3.3 VAM Association

Mycorrhizal fungus (*Glomus fasciculatum*) enhanced groundnut growth, dry matter and P uptake (Krishna and Bagyaraj 1984). However, P uptake in both VAM and non-mycorrhizal plants is normally from the same labile pool (Thomas et al. 1982; Bolan, 1991). Further, mycorrhizal and non-mycorrhizal plants take up P with similar specific activity (Sanders and Tinker 1971; Hyman and Mosse 1972; Tinker 1975). On an extremely low P Alfisol, mycorrhizal infection enhanced growth of pigeonpea. In contrast, sorghum died due to P deficiency despite mycorrhizal infection. The results indicate that mycorrhizal infection does not always necessarily give a host plant superior ability to solubilize sparingly soluble P in a low fertility soil and that a host plant itself must be endowed with superior P uptake ability (Ae et al. 1990).

3.4 Iron Reduction

Plants can be classified into two categories depending on their response to Fe deficiency (Strategy I and Strategy II). Strategy I is usually occurs in dicots and non-graminaceous monocots, and is characterized by distinct components of response such as an increased Fe-reducing capacity and enhanced net excretion of protons. Strategy II is confined to graminaceous plants and is characterized by Fe deficiency-induced release of phytosiderophore (Marschner 1995). A comparison of three crops showed that Fe-reducing activity in groundnut was largely similar to that of soybean. Fe-deficiency treatment enhanced Fe-reducing capacity in all crops. However, P-deficiency decreased Fe-reducing activity (Fig. 1). These findings support that Fe-reducing power does not significantly contribute to P uptake, as total P requirement is much higher than Fe requirement. Also this Fe-reduction mechanism does not account for the superior ability of groundnut to absorb Al-P in Andosols.

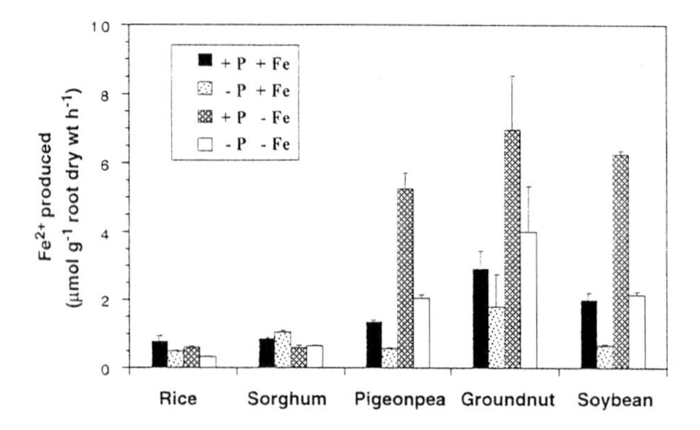

Fig. 1. Effect of phosphorus (P) and/or iron starvation on iron-reduction activity. *Bar*, SE ($n = 4$)

Table 7. Organic acids released by roots of various crops in 22 h

Sampling		Organic acids (μmol g^{-1} root dry weight)					
Time	Crops	Oxalic	Malonic	Malic	Tartaric	Citric	Piscidic
4 weeks[a]							
	Rice	Tr.	Tr.	Tr.	0.02	Tr.	ND
	Sorghum	Tr.	Tr.	Tr.	0.01	Tr.	ND
	Pegeonpea	4.30	16.54	Tr.	0.03	0.05	1.83
	Groundnut	Tr.	Tr.	Tr.	0.02	Tr.	ND
	Soybean	Tr.	Tr.	Tr.	0.0μ1	Tr.	ND
5 weeks							
	Rice	Tr.	Tr.	Tr.	0.02	Tr.	ND
	Sorghum	Tr.	0.02	Tr.	Tr.	0.02	ND
	Pegeonpea	3.87	8.85	0.01	0.01	0.04	2.10
	Groundnut	Tr.	Tr.	Tr.	Tr.	Tr.	ND
	Soybean	Tr.	0.02	Tr.	0.02	Tr.	ND

Tr., trace; ND, not detected
[a]Period in solution culture

3.5 Root Exudation

Since the early 1950s, several investigators demonstrated the ability of organic acids in vitro to increase P availability from Al-P and Fe-P. Acetic, aconitic, citric, fumaric, glycolic, malic, malonic, oxalic and succinic acids have all been identified in root exudates of a variety of crop species and trees. In our tests on root exudates of five crops, no distinct solubilizing ability of $FePO_4$ and $AlPO_4$ was detected in groundnut (Otani et al. 1996). A considerable amount of root exudates with solubilizing ability of $FePO_4$ and $AlPO_4$ was produced in pigeonpea only (Ae et al. 1990, 1993; Otani et al. 1996; Subbarao et al. 1997a). The contents of organic acids in root exudates are presented in Table 7.

While discussing the role of root exudates in nutrient uptake, at least two factors must be considered. First, root exudates act not only upon P-fixed minerals but also on other ion components having no P. Secondly, soil microbes degrade exudates to some extent. We measured Fe and Al contents of the Al-P fraction in the Nishinasuno soil, where groundnut might take up the P from this Al-P fraction that could be extracted with 1 M NH_4F (at pH 7.0). (This chemical extraction method can allow solubilization of most of the Al-P and some part of the Fe-P fraction) Total contents of the P ([P]), Al ([Al]) and Fe ([Fe]) in the Al-P fraction (including the organic Al-P fraction) of the Nishinasuno soil were 6.5, 113.0 and 42.5 μmol kg^{-1} respectively, and the molar ratio of the total of [Al] and [Fe] to [P] was 24. On the other hand, the same molar ratio in the Tsukuba soil was about 10. If we presume that groundnut might take up P from the Al-P fraction in these two Andosols through root exudates only, at least 10-24 times the moles of chelating agents must be secreted as root exudates to solubilize one mole of P. Further, if we consider decomposition of exudates by soil microbes, we may be overestimating the total amount of P solubilizing compounds in root exudates to account for the entire P solubilizing effect. Tests on soils in which 3 phenolic acids were incorporated revealed that most phenolic acid was destroyed by soil microbes within 3-4 days in non-sterilized soil but not in sterilized soil (Table 8).

An enhanced root exudation of organic acids in dicots in general, and legumes in particular is often observed under P deficiency conditions. Rapeseed, for example, secreted citric and malic acids much more in P-deficient than in P-sufficient conditions (Hoffland et al. 1989), thereby suggesting that P uptake might be partly controlled by P status in plants. However, our results show that the $FePO_4$ solubilizing ability of pigeonpea root exudates increased with increasing P application (Otani and Ae 1997a). Likewise, citrate exudation from P-stressed alfalfa (*Medicago sativa*) was 80% of that of plants receiving a complete nutrient solution. The results may, therefore, depend not only on P status, but also on growing stage, varieties, light intensity, humidity, and so on. Definite factors in regulating quantitative and qualitative root exudation are still unclear.

Taken together, all the above five factors fail to account for the superior ability of P uptake by groundnut in low-P soils. It is therefore important to investigate other mechanisms.

Table 8. Decomposition of phenolic acids in soils at field capacity

	Total phenolics extracted (mg kg^{-1} soil as ferulic acid)					
Treatment	0	2	4	7	11	20 (days)
No-amendment	8	...	...	...	...	6
Phenolics-added soil[a] (soil: not sterilized)	732	682	586	23	7	4
Phenolics-added soil[b] (soil: sterilized)	732	691	637	542	300	...

[a] A mixture of phenolic compounds (vanillic acid, vanilline, and ferulic acid) was added
[b] Phenolic mixture was added to a sterilized soil but incubation was not under aseptic conditions

4. Roles of Root Epidermal Cell Walls in P Solubilizing Ability

As mentioned earlier, P is supplied to plant roots by diffusion. Poor survival of sorghum, soybean and maize on these two extremely P deficient soils (Table 4) indicates that diffusion of P to their root surfaces was negligible. This was because of very low P concentration in the soil solution of the rhizosphere, which might be less than C_{min} of these crops (see Table 5). Otherwise, soybean, sorghum and maize should have taken the same amount of P as P uptake by groundnut. If superior growth of groundnut in P-deficient soils is due to P solubilization, it can occur only at the interface between root surface and soil particle. We, therefore, expected that root epidermal cell walls may play an important role in such a reaction.

4.1 Preparation of Plant Root Cell Walls and P-Fixed Minerals

Groundnut, soybean, maize and sorghum were grown on sand and supplied with Arnon's nutrient solution. Thirty grams of fresh roots collected from 60-day old plants were cut into 5-mm long pieces and ground with 250 mL distilled water in a blender (Polytron PT 3000) at 27 000 rpm for 5 min. The suspension was placed onto Whatman No.6 filter paper and washed with 250 mL of 0.5 *M* HCl to remove minerals. The homogenized roots were then thoroughly washed twice with water, rinsed with 200 mL ethanol, and dried at 50°C to obtain a "crude cell wall" preparation. Goethite (Atkinson et al. 1967), Fe_2O_3-removed allophane (Jackson et al. 1986), silica-alumina (Si/Al) gel (Nanzyo and Watanabe 1981) and hematite were used for preparing P-fixing minerals. P as NaH_2PO_4 was adsorbed onto these 4 minerals and washed with water. In addition, variscite and strengite (naturally occurring P-containing rock minerals) were used as P-fixing minerals besides commercially available chemicals of $FePO_4$ and $AlPO_4$.

4.2 Measurement of P Solubilizing Ability by Root Cell Walls

P solubilizing activity was determined after adding 40 mg "crude cell wall" to a centrifuge tube containing 1.5 mL of either 1/7 M barbital-acetate buffer (pH 3.8 to 6.7) or 2-M acetate buffer (pH 5.0) and 1.5 mL mineral-suspended solution (10 mg P-fixed mineral mL^{-1}). After thorough shaking and centrifugation, the suspension was passed through a 0.20 μm membrane filter and the P concentration of the filtrate measured (Murphy and Riley 1962). To estimate the P solubilizing effect, P concentration was determined prior to adding cell walls and this value was subtracted from the P concentration of the filtrate. The reaction between P-fixed minerals (including $FePO_4$ and $AlPO_4$) and cell walls was tested at pH 5.0. This pH was chosen because root surfaces of groundnut and soybean were acidic due to secretion of H^+ (Raven et al. 1991). Moreover, rhizosphere pH is known to decrease under P-deficient conditions (Grinsted et al. 1982).

4.3 P Solubilization by "Crude Root Cell Wall" from P-Fixed Minerals

The crude root cell walls of groundnut solubilized (a) more P than those of sorghum and soybean from all minerals except variscite, (b) relatively more P from hematite than other minerals indicated by P solubility, and (c) more P from minerals with Fe-P than from minerals with Al-P (Table 9). Goethite and hematite are common in red-yellow colored soil, and allophane is a major clay form of not only Nishinasuno soil, but also volcanic ash soils found all over Japan. When fertilizers are applied onto these soils, allophane-bound, goethite-bound and hematite-bound P forms can actually develop (see Table 9). Variscite and strengite are usually present in many acid soils (Tan 1933). Strengite, an amorphous end product of reaction between Fe and P, is especially very stable due to its low solubility (10^{-32}) product (Sample et al. 1980). Even so, the root cell wall of groundnut solubilized P from strengite more than did that of soybean or sorghum. The findings further support previous contentions on groundnut response to P fertilizers (Hartzog and Adama 1988; Walker et al.1979; Walker et al. 1982) and our own observations of its better growth on soils of low P fertility.

5. Characteristics of Active Site Related to Fe-P Solubilizing Ability

5.1 Cation Exchange Capacity of Roots and Fe-P Solubilizing Ability

Plant species differ considerably in cation exchange capacity (CEC) of roots, that is, in the number of cation-exchange sites located on the cell wall. CEC is derived from carboxylic groups in polygalacturonic acids of pectic substances (Knight et al. 1961). In general, the CEC of dicots is much higher than that of monocots.

Table 9. The effects of "crude root cell wall" from various crops on P solubility of minerals containing sparingly soluble P

P form	P content	Released P by root cell wall (μg P g^{-1} cell wall)			P solubility (%)[a]		
Minerals	mg P g^{-1}	Sorghum	Soybean	Groundnut	Sorghum	Soybean	Groundnut
Fe-P							
$FePO_4$	164.630	273.0	247.1	382.2	0.36	0.33	0.50
Strengite	32.290	46.7	47.1	81.6	0.28	0.29	0.50
Goethite-P	2.950	26.4	32.8	58.1	1.79	2.22	3.94
Hematite-P	0.290	2.6	7.8	12.1	1.78	5.35	8.33
AI-P							
$AI PO_4$	241.09	103.4	123.5	201.1	0.12	0.14	0.23
Variscite	49.81	112.1	140.2	135.6	0.45	0.56	0.54
Si/AI Gel-P	160.76	123.6	126.4	229.9	0.15	0.16	0.29
Allophane-P	87.81	94.8	117.8	232.7	0.22	0.27	0.53

[a]P solubility % is calculated from the ratio of solubilized P to added mineral P

Jenny and Overstreet (1939) proposed a theory, which assumes cation exchange between H^+ on plant roots and K^+, Mg^{2+}, Ca^{2+} and other cations on soil colloids. However, Marschner (1995) reported that exchange adsorption in the apparent free space was not an essential step for ion uptake or transport through the plasma membrane into the cytoplasm. "Mucilage" secreted from plant roots consists of partially methylated polygalacturonic acid similar to pectic substances. Such mucilage was reported to form a complex with Fe^{3+} resulting in desorption of P from the surface of iron-oxide minerals (Nagarajah et al. 1970). Also, for many years, differences among species and cultivars in the CEC of root apoplast have been discussed in relation to aluminum tolerance (Wagatsuma 1983; Blamey et al. 1990). Moreover, the polygalacturonate network bound to Fe^{3+} plays an important role in stabilizing Fe^{3+} reducible species (Gessa et al. 1997). These results led us to consider the possibility that the arrangement of carboxylic groups in pectic substances may lead to formation of complex with Fe^{3+} and Al^{3+} through the 3-dimensional matrix of the root cell wall. In order to determine the relationships between the CEC of the cell wall and the solubilizing reaction with $FePO_4$ and/or $AlPO_4$ instead of P-fixing minerals, solubilizing activity by cell walls was examined at various pH levels (Fig. 2).

Groundnut root cell walls showed a higher $FePO_4$ solubilizing ability than soybean, sorghum and maize irrespective of variations in pH. Both legumes showed a higher $AlPO_4$ solubilizing ability than cereals. CEC, which is derived from carboxylic groups on cell walls, increased with an increasing pH. In contrast, P solubilizing ability decreased with an increasing pH in all crops. Root cell walls of soybean had a higher CEC than those of groundnut but groundnut showed higher solubilizing ability for $FePO_4$ and $AlPO_4$ than did soybean. The results suggest that the P solubilizing ability of the cell wall was not related to anions derived from pectic substances.

To investigate the relationship between P solubilizing ability and pectic substances further, we prepared "pure cell wall" of roots. The homogenized root suspension pre-washed with 0.5 M HCl was placed on Whatman No. 6 filter paper and washed three times with water. The washed roots were re-suspended in 0.1% sodium deoxycholate (Masuda et al. 1989) to remove intracellular components and plasma membrane. The suspension was again washed with water, rinsed with ethanol and dried. Purified root cell walls from all tested species showed nearly similar activity to that of crude cell walls.

5.2 Stability of P Solubilizing Active Site to Enzyme Degradation and Acid Hydrolysis

Active sites for P solubilization in purified root cell walls were further characterized for stability (Table 10). About 98% of activity was retained even after boiling at 100°C for 10 min. This result suggests that the active site is considerably stable against heating. Similarly, the filtered cell wall fraction after enzymatic digestion with a mixture of cellulase and pectinase (this mixture is commercially available as dricellase or macerozyme, these are commonly used in protoplast preparation) for 20 h showed largely similar solubilizing activity, despite considerable

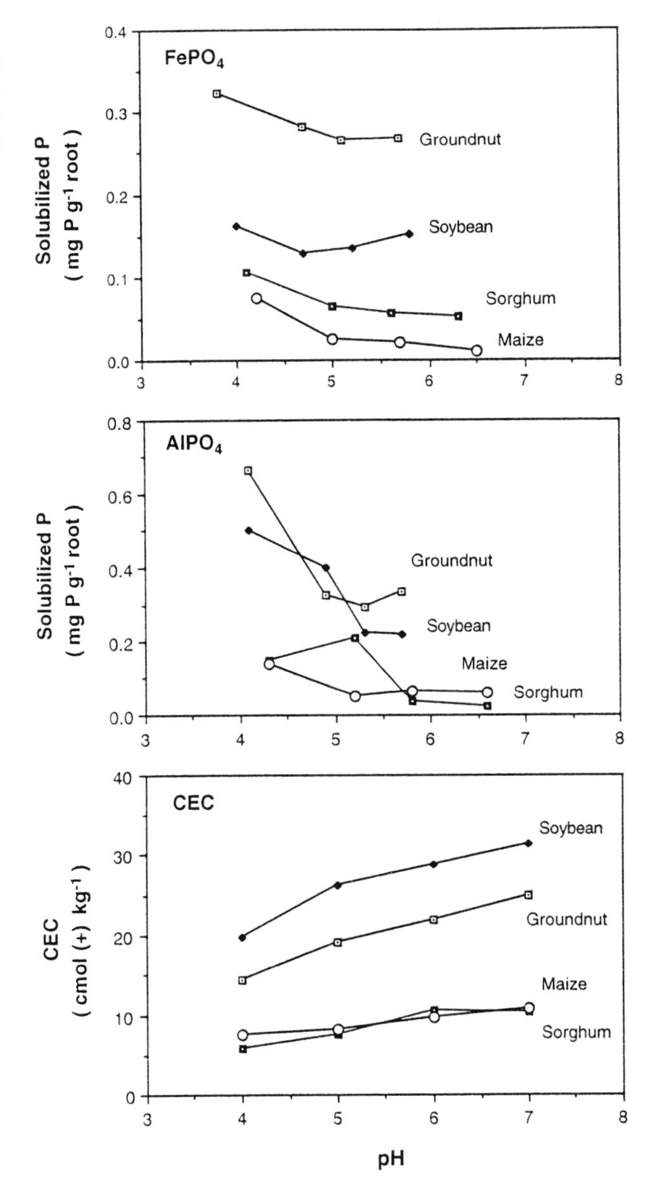

Fig. 2. Effect of pH on P solubilizing ability of plant root cell walls from $FePO_4$ or $AlPO_4$, and on cation exchange capacity (CEC) of roots in different crop species

reduction in the weight of cell wall. Digestion with 0.7 M HCl at boiling temperature, however, led to a substantial decrease in P solubilizing activity, perhaps due to degradation of essential chemical substances in the cell wall that may be related to chelation with Fe^{3+} and/or Al^{3+}.

We then compared $FePO_4$ solubilizing ability of purified groundnut root cell wall with that of commercially available pure pectin and cellulose (Sigma Chemical., USA; Table 11). "Purified root cell wall" showed a much higher activity than cellulose and pectins. In other words, the results confirmed that P solubilizing ability was not entirely from pectic substances (Ae et al. 1996).

Table 10. P solubilizing activity of "pure root cell wall" of groundnut from $FePO_4$ as affected by various treatments

Treatment	Weight of cell wall After treatment (mg)[a]	P solubilizing activity (mg P g^{-1} cell wall)
Control	30.0	297.5 ± 9.0[b] (100%)
Heating at 100°C for 10 min.	30.0	292.8 ± 13.7 (98%)
Digestion with enzyme mixture[c] for 20h	21.3	277.5 ± 17.3 (93%)
Digestion with boiling HCI		
0.7 *M* HCI for 12 min.	23.8	193.5 ± 10.9 (65%)
0.7 *M* HCI for 40 min.	20.4	96.7 ± 10.8 (39%)

[a]Initial weight of "pure root cell wall" was 30 mg
[b]Mean ± SE ($n = 3$)
[c]Cellulase and pectinase

Table 11. P solubilizing activity by pectin, cellulose "pure cell wall" from $FePO_4$

Materials	P solubilized (μg P g^{-1} CW)
Purified cell wall of groundnut roots	264 ± 5.8[a]
Cellulose (fibrous: Sigma Chemical)	0 ± 0.0
Pectin (from citrus fruits: Sigma Chemical)[b]	95 ± 5.1
Pectin (from citrus fruits: Sigma Chemical)[c]	23 ± 6.0

[a]Mean ± SE ($n = 3$)
[b]Galacturonic acid content 84%, methoxy content 10%
[c]Galacturonic acid content 74%, methoxy content 7.8%

5.3 Chemical Characteristics of the Active Site of Fe-P Solubilizing Ability

The effect of various cations on solubilizing activity of "purified root cell wall" was then tested. After soaking the cell wall in H_2O, NaCl, KCl, $CaCl_2$, $MgCl_2$, $AlCl_3$, $FeCl_3$, or $Ga(NO_3)_3$ solutions, it was then washed with water and P solubilizing ability from $FePO_4$ was measured. Solubilizing activity of cell wall treated with mono-valent and di-valent cations remained almost the same as that of the control (H_2O). However, tri-valent cations like Al^{3+}, Fe^{3+} or Ga^{3+} decreased solubilizing activity drastically resulting in negative P solubilization (Fig. 3). When solubilized P concentration in the mixture of treated cell wall with tri-valent cations and $FePO_4$ suspension was less than that of the control (in which only $FePO_4$ suspension without cell wall was used), a negative solubilization was reported to occur. If CWs (cell walls) have a reaction site with tri-valent cation, but not mono- or di-valent cation, pre-treated CWs with $FeCl_3$ must be CW-bound Fe. When CW-bound Fe is put into Fe-P in a buffer, Fe in CW-bound Fe reacts with partially soluble P in Fe-P suspension to make insoluble Fe-P. P concentration in a reaction mixture of CW-bound P and Fe-P suspension thus will be lower than that in untreated CWs and Fe-P suspension. This result suggests that the active site may be able to recognize only tri-cations by chelation with Al^{3+}, Fe^{3+} or Ga^{3+} but not di- and mono-cations (Ae and Otani 1997).

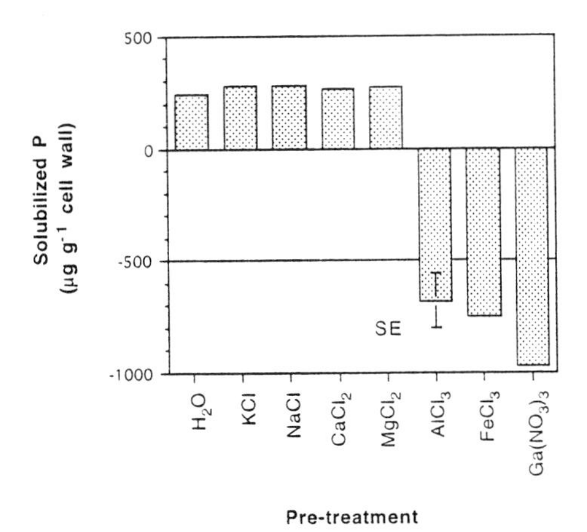

Fig. 3. Effects of pre-treatment with various cations on P solubilizing activity of root cell walls. P concentration of the control containing only $FePO_4$ in buffer without cell wall was referred to as "Zero solubilization". Standard errors for all treatments except $AlCl_3$ were negligible

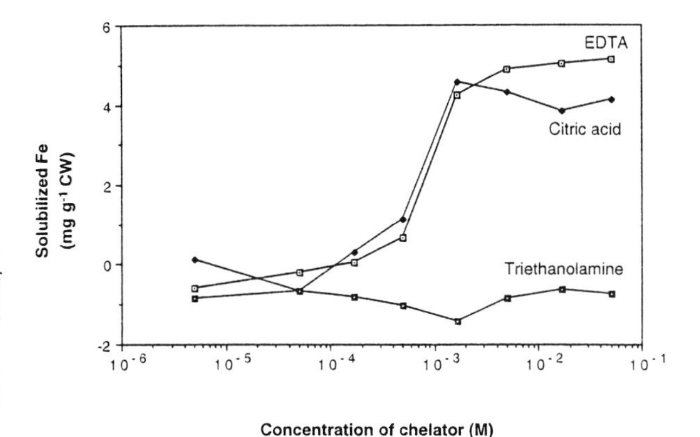

Fig. 4. Effect of chelating compounds on solubilization of iron (III) from Fe-saturated root cell wall (CW) of groundnut. This reaction was conducted under pH 5.0 of 1.0 M acetate buffer

5.4 The Relative Strength of Chelating Ability of Root Cell Walls

The results so far indicate that the reaction of P solubilizing active components in groundnut root cell walls may be related to a chelating effect. In order to confirm this, we tested the chelating ability of cell walls by treating the active site of purified root cell wall with $FeCl_3$ solution. Fe-saturated cell walls were thus prepared and put into acetate buffer (pH 5.0) solution containing various concentrations of chelating compounds (EDTA, citric acid and triethanolamine). After thoroughly shaking the suspended solution with Fe-saturated cell walls for 1 h, Fe^{3+} concentration liberated from cell walls was determined (Fig. 4).

Solubilized Fe content released from Fe-bound cell walls increased with an increasing concentration of citric acid and EDTA, especially when their concentration was above 10^{-3} M. Triethanolamine did not affect Fe liberation from Fe-saturated cell wall. Stability constants (K) of citric acid and EDTA with Fe^{3+} were similar (pKa = 24) but that of triethanolamine was very low (pKa = 2). These observations suggest that the active site of P solubilizing ability can perform as a strong chelating reaction with Fe^{3+} and/or Al^{3+} as EDTA and citric acid. We, therefore, propose a new P uptake mechanism called "contact reaction" hypothesis.

6. Location of P Solubilizing Activity

6.1 Localization of P Solubilizing Active Sites

It is obvious that P solubilization activity must be located at the surface of roots. It means that reaction sites for P solubilization are on the surface of root epidermal cells, which are in close contact with soil particles containing fixed P. To test this assumption, fresh roots of groundnut grown for 2 months in sand-filled pots were collected and soaked in 0.5 M NaOH for varying periods. Upon treatment with NaOH, roots were washed in water, and plants separated into 2 equal groups. One group of plants was placed in water for monitoring their growth while the other group was used for preparing root cell walls and determining P solubilizing ability. Short exposure to NaOH is expected to affect only the outer layers of roots whereas long exposure should affect entire cells. Even a 30-s exposure of roots to 0.5 M NaOH decreased P solubilizing activity by 30% (Fig. 5). Soaking roots in NaOH for more than 5 min decreased P solubilizing activity by an additional 30%. In this case, the NaOH solution turned yellow and plants wilted upon transfer to water. All wilted plants, however, recovered to a normal state the next day. Plants treated with NaOH for more than 20 min did not recover and died, implying that exposure to 0.5 M NaOH destroyed not only the epidermal cell wall but also cytoplasm thereby leading to death of root tissues. These results suggest that at least 33% of total P solubilizing activity of root cell walls was located on the root surface and that a "contact reaction" between the active site on the epidermal cell wall and clay minerals containing P is a likely process in the rhizosphere (Ae and Otani 1997).

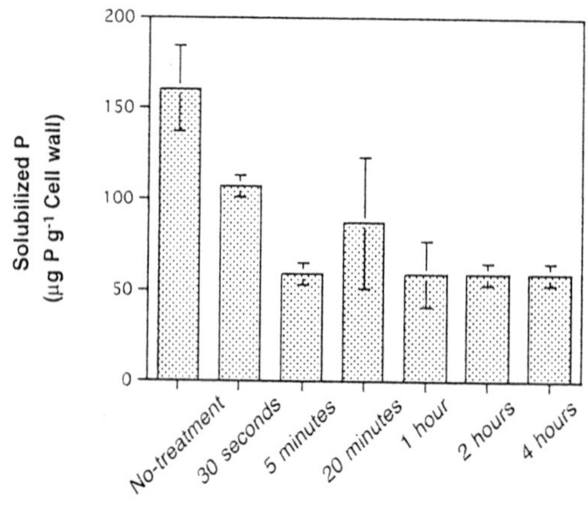

Fig. 5. P solubilization as affected by treatment of root with 0.5 M NaOH

6.2 P Solubilizing Ability at Various Growth Stages

Whether or not the "contact reaction" theory can account for entire P uptake by groundnut from a low-P soil is, however, unclear. To answer this question, it is important to monitor root growth and P solubilizing ability at various growth stages. P solubilizing ability, which was moderately high at 19 days after sowing (DAS), gradually increased with an increasing plant age (Fig. 6). Total P solubilizing capacity at 109 DAS corresponded to 1.33 mg P per plant. Total solubilizing capacity was calculated from root growth and P solubilizing ability of roots. This total P solubilizing capacity at 109 DAS corresponded to 1.33 mg P per plant. Because, as discussed in section 3.5, at least 10 times molar Fe^{3+} in the soil, which is more than equivalent one molar content for corresponding P requirement, must be solubilized from Nishinasuno soil (see Table 3). Thus, solubilizing capacity of groundnut roots only accounted for uptake of 0.133 mg P from the Nishinasuno soil. So the "contact reaction" theory may not be satisfactory to account for the entire P uptake by groundnut from low fertility soil.

6.3 Shedding of Root Epidermal Cell Wall

Yarbrough (1949) observed that surface layers of groundnut roots shed continuously. He pointed out that, at least for absorption of mineral ions, cells of the outermost zone provided a continuous renewal substitute for root hairs. This mode of absorption of nutrients is unique. Uheda et al. (1997) reported that epidermal cells and cells originating in the outer cortex form the surface layers of groundnut roots and the outermost surface layer which separates and sheds. Shedding takes place continuously over the whole surface of roots including lateral ones. Shedding involves modification of cell walls and separation of intact cells. Wall breakdown and cell expansion associated with it may facilitate separation of intact cells. The shedding phenomenon partly supports the "contact reaction" theory as the genotype (Chibahandachi) used in our study has no root hairs. When all active sites on the epidermal cell wall reacted with Fe^{3+} and Al^{3+}, active sites satu-

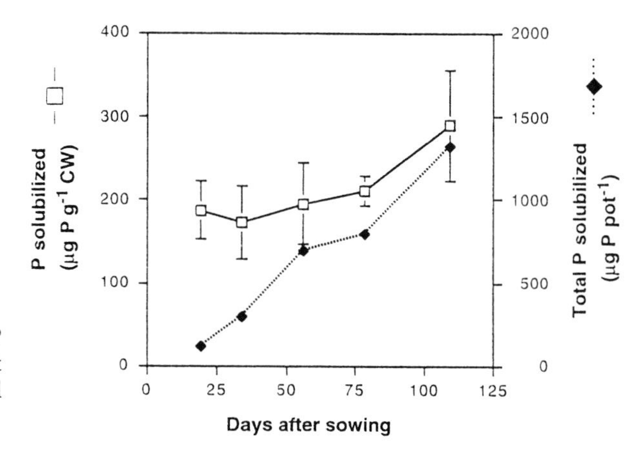

Fig. 6. Effect of growth on P solubilizing activity by root cell wall and estimated total P solubilization

rated with Fe^{3+} and Al^{3+} are expected to lose P solubilizing ability. However, by shedding surface layers, fresh epidermal root surface with P solubilizing activity appears. This phenomenon makes the "contact reaction" hypothesis more convincing in that total P solubilizing capability of root cell walls could be derived by multiplying activity with the number of shedding times in a growing season. Continuous shedding of root epidermis may, therefore, play an important role in P solubilization.

7. "Contact Reaction" Hypothesis and its Relevance in P Uptake by Other Crops

7.1 What is "Contact"?

As discussed earlier, locating the site of solubilizing activity in root cell walls is a necessary condition for accepting the theory of "contact reaction". Another necessary condition is the existence of tight contact between the root surface and clay minerals. Czaenes et al. (1998) investigated suction values on root/soil adhesion energy in maize seedlings and compared energy values associated with root/soil adhesion and root/water adhesion. Though the values vary with soil physical properties including water pressure, root/soil adhesion energy is always much more than root/water adhesion energy. This result shows evidence for tight contact between soil and root surface.

Many higher plants, except for aquatic plants, secrete mucilage, which is a high molecular weight gelatinous material consisting mainly of polysaccharides and polygalacturonic acids. Fibrillar outgrowths of polysaccharide chains of mucilage have been shown to project into the soil and establish minute bridges to the root surface (Leppard and Ramamoorthy 1975). Root surfaces, particularly in the apical zone, are covered by mucilage as cells of the root cap and epidermis secrete mucilage (Vermeer and McCully 1981). The presence of mucilage on the root epidermal surface of groundnut can be seen in a photograph by Uheda et al. (1997). Mucilage is released from the epidermal surface when cuticle is mechanically ruptured or chemically destroyed by soil microbes. It is derived from the primary cell wall layer, and its production is positively correlated with root growth rate (Trolldenier and Hecht-Buchholz 1984). In soil-grown plants, mucilage is usually invaded by microorganisms, and both organic and inorganic soil particles are embedded in the primary wall (Foster 1981). Taken together, the reports indicate that mucilage from roots helps soil particles to closely adhere to root surface and also prevents the released inorganic P to re-adsorb onto soil particles.

7.2 P Solubilizing Ability of Pigeonpea Root Cell Walls

The ability of pigeonpea to take up P from low P soils was attributed to its root exudates (Ae et al. 1990; Otani et al. 1996). In order to determine how P solubilizing agents in root exudates are secreted, pigeonpea was grown on sand with varying P levels. The dry weight and P concentration of plant tissue increased and plants grew faster with an increasing P level. P-solubilizing capacity by root exudates was highest in the highest P treatment (P sufficient condition). The exudation of active organic acids by roots sharply increased during the flowering stage at 9-11 and 11-13 weeks after sowing in high and low P application treatments respectively (Otani and Ae 1997a). Subbarao et al. (1997a, b) demonstrated significant genotypic variation in P uptake from Fe-P in pigeonpea. They also indicated that Fe-P solubilizing activity expressed in root exudates of pigeonpea was possibly regulated by internal P status and to some extent by internal P demand, which is determined by the growth stage. These results led us to pose another question as to how pigeonpea genotypes take up P from low-P soils, especially in the earlier growing stages.

Except for the role of root exudation from pigeonpea in P uptake in a low-P soil, the solubilizing activity of pigeonpea and groundnut root cell wall was also tested. The P-solubilizing activity of the root cell wall of pigeonpea was as high as that of groundnut and greater than that of maize and soybean (Table 12). The active site of pigeonpea root cell walls showed a chelating reaction with Al^{3+} and Fe^{3+}, but not with Ca^{2+} and Mg^{2+} and this reaction pattern was similar to that of groundnut (see Fig. 7). Furthermore, although soybean, maize and sorghum showed less P solubilizing ability, the chelating reaction of their cell walls was similar to that of pigeonpea and groundnut in that chelation occurred with only Al^{3+} and Fe^{3+}, but not with di- and mono-cations.

7.3 Extraction of Active Components of P Solubility

Root cell walls of groundnut, maize, pigeonpea and soybean were used for extraction of active component(s) and to analyze chemical structure. Because active sites might be bound with the pectic chain in cell walls, a suitable extraction procedure needs to be developed first. The "purified cell walls" from 2 month-old plants grown in sand were treated with water (heating), acid (1.0M HCl) and alkaline solution (0.5M NaOH) at 80°C. Cell wall solutions were then passed through a membrane filter with pore size of 0.20 μm, and P solubilizing activity of filtrate and residue was measured. Acid hydrolysis did not adversely affect P solubilizing

Table 12. P solubilizing ability of "pure cell wall" (CW) of several crops following digestion with HCl and NaOH

Digestion	Groundnut	Soybean	Maize	Pigeonpea
	Solubilized P (μg P g^{-1} CW)			
"Pure cell wall"	135 ± 20.0	78 ± 8.7	29 ± 4.6	117 ± 8.2
0.5 *M* –NaOH				
Residue	55 ± 2.9	18 ± 7.5	- nil -	30 ± 0.79
Filtrate	500 ± 109.0	629 ± 17.3	281 ± 48.0	580 ± 6.36
1.0 *M* –HCI				
Residue	122 ± 6.81	61 ± 8.2	15 ± 3.7	69 ± 1.0
Filtrate	- nil -	- nil -	- nil -	- nil -

activity of residue. Alkaline treatment with 0.5M NaOH, however, led to transfer of solubilizing activity from cell walls to the filtrate. Treatment of cell walls with 0.5M NaOH at 80°C for 24 h was thus found to be appropriate to extract the "active site" (Table 12).

P solubilizing activity of both "pure cell wall" and alkaline-extracted cell wall components decreased sharply with the addition of tri-cations like Fe^{3+}. In contrast, di-cations like Ca^{2+} and Mg^{2+} had no effect. The similarity in chemical characteristics of the cell wall and the extracted cell wall components suggests that active sites in cell wall components extracted by NaOH were from cell walls (Fig. 7, data not shown for pigeonpea and soybean). P solubilizing activity in the filtrate following NaOH treatment was much higher than the activity of untreated original cell walls in all four crops. The results suggest that active sites are located not only on root epidermal cell wall, but also on the inner parts of the cell wall. The fact that P solubilizing activity can be extracted not only from the root cell wall of groundnut but also from that of other crops like soybean, pigeonpea and maize suggests that the active components are common in the cell walls of many species.

The extracted active components following NaOH treatment were applied to a LH-20 column chromatography and the eluted samples were monitored for P solu-

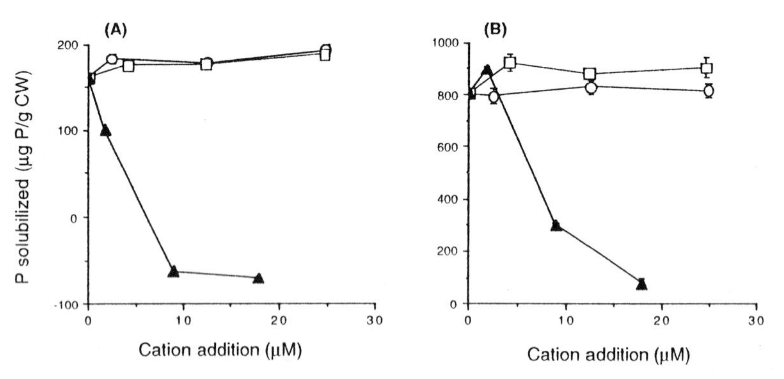

Fig. 7. P solubilizing activity of cell wall (CW) and cell wall extracted components as affected by Ca^{2+}, Mg^{2+}, and Fe^{3+}. **A** Groundnut root cell wall; **B** Extracted component from groundnut root cell wall. *Circles*, Ca; *squares*, Mg; *triangles*, Fe

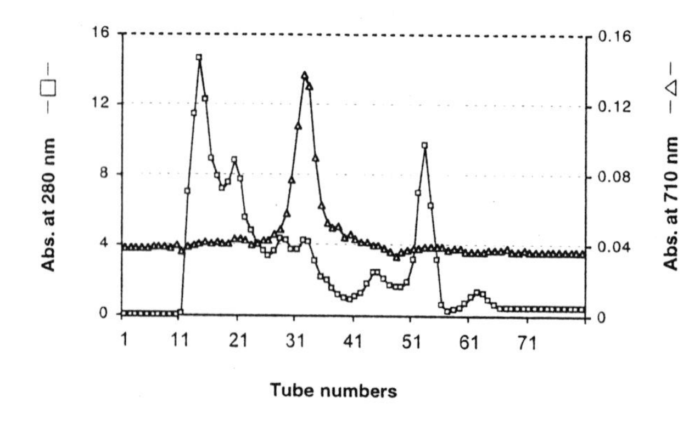

Fig. 8. Column chromatography by LH-20 extracted groundnut root cell wall components with NaOH. *Right vertical axis* represents the P solubilizing activity as expressed by absorbance (*Abs.*) at 710 nm when doing the P measurement. *Left vertical axis* represents the UV absorbance at 280 nm

bilizing ability at 280 nm (Fig. 8). Fractions with higher P solubilizing activity were applied to a DEAE A-25 column. After DEAE column chromatography was completed, active components were acidified to pH2.0. The acidified sample was fractionated with ethyl acetate but the water layer was kept for the further purification, as no activity was found in ethyl acetate fraction. The water layer was dried, solubilized with H_2O and applied to the column with BioGel P2 (2.5 cm × 50 cm). HPLC analysis (TSK gel ODS 80T) showed that the active components might be two substances that are normally present in the root cell walls of all crops, although the amount could vary among crops (Shen and Ae 2001). Caffeic acid (Inanaga et al. 1988; Schmutz et al. 1993) was excluded from the candidate substances although it showed the ability to chelate with Fe^{3+}. Further studies on identifying the chemical structure are continuing.

8. Practical Implications for Cropping Systems

In a long-term field experiment conducted for 6 years to investigate the effects of fertilizer P on groundnut and wheat yields in a groundnut-wheat rotation (Aulakh et al. 1991), only small and inconsistent responses by groundnut to direct, residual, and cumulative P application were observed. Although the initial soil test value of Olsen's extractable-P was low, little differences occurred among direct (applied to groundnut), residual (applied to preceding wheat), and cumulative-P (applied to both groundnut and wheat) treatments. On the other hand, the responses of wheat were substantial and consistent in all years (Table 13). Direct application of fertilizer P to wheat produced significantly higher yields than its application to previous groundnut crop, and yields were as good as for cumulative-P application. The

Table 13. Six-year average yields of groundnut and wheat in plots receiving different rates and frequency of fertilizer P

	P frequency		
P rate ($kg\ P\ ha^{-1}$)	Applied to peanut only	Applied to wheat only	Applied to both peanut and wheat
Pod yield of peanut ($kg\ ha^{-1}$)			
13	1950	1870	1850
26	1930	2020	1980
39	2000	1950	2000
Control	1870		
LSD (0.05): P rate=320 and P frequency=270			
Grain yield of wheat ($kg\ ha^{-1}$)			
13	3510	4070	4150
26	4270	4530	4600
39	4160	4660	4530
Control	3020		
LSD (0.05): P rate=500 and P frequency=290			

From Aulakh et al. 1991

results suggest that groundnut may have better nutrient absorption from soil-derived P. Thus, growing a crop that requires large amounts of fertilizers (such as corn) immediately before groundnut is a widely recommended practice (Chapman and Carter 1976).

In a rice-groundnut cropping system in India, P was applied at 30 and 60 kg ha^{-1} as super phosphate, rock phosphate and as a mixture in a ratio of 1:2. At both levels, application of P as rock phosphate singly or in combination with super phosphate produced better yields than application of super phosphate alone (Krishnappa et al. 1991). These results suggest that residual effect of rock phosphate was significant in the second crop of groundnut. Both the above experiments lead us to draw the conclusion that fertilizer management of groundnut is unusual or peculiar (Chapman and Carter 1976). Unlike other crop species such as maize and cotton, groundnut can use fertilizers applied to crops preceding it in a rotation and can take advantage of P that is not normally available to other crops (Hallock 1962).

An FAO-sponsored workshop "Expert Consultation on Fertilizer Use Under Multiple Cropping Systems" held at New Delhi in 1982 made the following fertilizer scheduling recommendations (FAO 1983). In a rice-groundnut sequential system, N, P, and other nutrients as required, to be applied to rice crop, and only 20 kg P_2O_5 ha^{-1} to be applied to the sequential legume crop if moisture conditions are favorable. Also, in a sorghum + pigeonpea, and sorghum + groundnut intercropping system, N, P, K, S and Zn to be applied to sorghum only. From the viewpoint of sustainable agriculture and shortage of rock phosphate resources in the future, saving of P fertilizer application could be possible by knowing the characteristics of the growth responses to fertilizer.

We still have the question on why groundnut can take up residual P and/or already Al- or Fe-fixed P preferentially even under the existence of soluble P fertilizer as single super phosphate. If groundnut can take up P by the expedient "contact reaction" hypothesis, its preferential uptake of residual and soil derived P can easily be understood. Assuming that the release of free phosphate ions due to reaction of Fe^{3+} of Fe-P with the "active site" on the root surface is followed by transfer of P into the cytoplasm through the phosphate transporter, the mechanism of P uptake by groundnut can be readily explained. The relationship between P solubilization and P transfer to cytoplasm needs further examination, however. Appropriate "soil fertility regimes" for crops such as groundnut and pigeonpea with unique P uptake mechanisms must be developed for agronomic reasons.

References

Ae N, Arihara J, Okada K, Yoshihara T, Johansen C (1990) Phosphorus uptake by pigeonpea and its role in cropping systems of the Indian subcontinent. Science 248:477-480

Ae N, Arihara J, Okada K, Yoshihara T, Otani T, Johansen C (1993) The role of piscidic acid secreted by pigeonpea roots grown in an Alfisol with low-P fertility. In: Randall PJ et al. (Eds) Genetic aspects of plant mineral nutrition. Kluwer Academic, Dordrecht. pp279-288

Ae N, Otani T, Makino T, Tazawa J (1996) Role of cell wall of groundnut roots in solubilizing sparingly soluble phosphorus in soil. Plant Soil 186:197-204

Ae N, Otani T (1997) The role of cell wall component from groundnut roots in solubilizing sparingly soluble phosphorus in low fertility soils. Plant Soil 196:265-270

Asher CJ (1978) Natural and synthetic culture media for Spermatophytes. CRC Handb Ser Nutr Food Sect G3:575-609

Atkinson RJ, Posner AM, Quirk JP(1967) Adsorption of potential —determining ions at the ferric oxide-aqueous electrolyte interface. J Phys Chem 71:550-558

Aulakh MS, Pasricha NS, Baddaesa HS, Bahl GS (1991) Long-term effects of rate and frequency of applied P on crop yields, plant available P, and recovery of fertilizer P in a peanut-wheat rotation. Soil Sci 151:317-316

Barber SA, Ernani PR (1990) Use of a mechanistic uptake model to evaluate phosphate fertilizer. In: Koshino M et al. (Eds) Transactions, 14th International Congress of Soil Science, International Society of Soil Science, Kyoto, Japan, Vol II pp136-140

Barber SA, Cushman JH (1991) Nitrogen uptake model for agronomic crops. In: Iskandar IK (Ed) Modeling wastewater renovation - land treatment. Wiley Interscience, New York, pp382-409

Blamey FPC, Edmeades DC, Wheeler DM (1990) Role of root cation-exchange capacity in differential aluminum tolerance of lotus species. J Plant Nutr 13:729-744

Bolan NS (1991) A critical review on the role of mycorrhizal fungi in the uptake of phosphorus by plants. Plant Soil 134:189-207

Chapman SR, Carter LP (1976) Crop production: principles and practices. WH Freeman, San Francisco, pp 359-370

Czaenes S, Hiller S, Dexter A, Bartoli F (1998) Root/soil adhesion in the maize rhizosphere : the rheological approach. In: 16th World Congr Soil Sci, Symposium 43, [CD]

FAO (1983) Fertilizer use under multiple cropping system. Report of an expert consultation held in New Delhi, February 3-6, 1982. FAO UN, Rome, pp 1-8

Foster RC (1981) The ultrastructure and histochemistry of the rhizosphere. New Phytol 89:263-273

Gardner WK, Barber DA, Parbery DG (1983) The aquisition of phosphorus by *Lupinus albus* L. III. The probable mechanism by which phosphorus movement in the soil/root interface is enhanced. Plant Soil 70:107-114

Gessa C, Deiana S, Premoli A, Ciurli A (1997) Redox activity of caffeic acid towards iron (III) complexed in a polygalacturonate network. Plant Soil 190:289-299

Grinsted MJ, Hedley MJ, White RE, Nye PH (1982) Plant-induced changes in the rhizosphere of rape (*Brassica napus* var. Emerald) seedlings. New Phytol 91:19-29

Hallock DL (1962) Effect of time and rate of fertilizer application on yield and seed size of jumbo runner peanut. Agron J 54:428-430

Hartzog DL, Adams JF (1988) Relationship between soil-test P and K and yield response of runner peanut to fertilizer. Commun Soil Sci Plant Anal 19:1645-1653

Helal HM, Dressler A (1989) Mobilization and turnover of soil phosphorus in the rhizosphere. Z. Pflanzenernšhr Bodenk 152:175-180

Hoffland E, Findenegg GR, Nelemans JA (1989) Solubilization of rock phosphate by rape. II. Local root exudation of organic acids as response to P-starvation. Plant Soil 113:161-165

Hyman DS, Mosse B (1972) Plant growth responses to vasicular-arbuscular mycorrhiza. III. Increased uptake of labile P from soil. New Phytol 71:41-47

Inanaga S, Ishimoto Y, Nishihara T (1988) Ca-binding components in cell wall of peanut shell. Soil Sci Plant Nutr 34:319-326

Jackson ML, Lim CH, Zelanzny LW (1986) Oxide, hydroxides, and aluminosilicates. In: Klute A (Ed) Methods of soil analysis. Part 1. Physical and mineralogical methods. Am Soc Agron, Madison, Wisconsin, pp101-150

Jenny H, Overstreet R (1939) Cation interchange between plant roots and soil colloids. Soil Sci. 47:257-272

Jungk AO (1991) Dynamics of nutrient movement at the soil-root interface. In: Waisel Y, Eshel A, Kafkafi U (Eds) Plant roots: The hidden half. Marcel Dekker Inc, New York, pp 455-481

Knight AH, Crook WM, Inkson RHE (1961) Cation-exchange capacities of tissues of higher and lower plants and their related uronic acid contents. Nature 192:142-143

Krishna KR, Bagyaraj DJ (1984) Growth and nutrient uptake of peanut inoculated with the mycorrhizal fungus *Glomus fasciculatum* compared with non-inoculated ones. Plant Soil 77:405-408

Krishnappa KM, Panchaksharaih SK, Sharma KMS, Vageesh TS (1991) Efficiency of rock phosphate in rice-groundnut cropping system in acid soils of coastal Karnataka, India. In: Wright RJ et al. (Eds) Plant-soil interactions at low pH. Kluwer Academic Press, Dordrecht. pp 533-538

Lepppard GG, Ramamoorthy S (1975) The aggregation of wheat rhizoplane fibrils and the accumulation of soil bound cations. Can J Bot 53:1729-1735

Marschner H (1995) Mineral nutrition in higher plants, 2nd edition. Academic, London

Masuda H, Komiyama S, Sugawara S (1989) Cell wall proteins from sugar beet cells in suspension culture. Plant Physiol 89:712-716

Meisner CA, Karnox KJ (1991) Root hair occurrence and variation with environment. Agron J 83:814-818

Moghimi A, Tate ME, Oades JM (1978) Characterization of rhizosphere products especially 2-ketogluconic acid. Soil Biol Biochem 10:283-287

Murphy J, Riley JP (1962) A modified single solution method for the determination of phosphate in natural waters. Anal Chemi Acta 27:31-36

Nagarajah S, Posner AM, Quirk JP (1970) Competitive adsorptions of phosphate with polygalacturonate and other organic anions on kaoline and oxide surfaces. Nature 228:83-84

Nanzyo M, Watanabe Y (1981) Material balance associated with phosphate sorption by amorphous clays, silica-alumina gel and synthetic goethite. Soil Sci Plant Nutr 27:329-337

Ohwaki Y, Hirata H (1992) Differences in carboxylic acid exudation among P-starved leguminous crops in relation to carboxylic acid contents in plant tissues and phospholipid level in roots. Soil Sci Plant Nutr 38:235-243

Otani T, Ae N (1997a) The exudation of organic acids by pigeonpea roots for solubilizing iron- and aluminum-bound phosphorus. In: Ando T et al. (Eds) Plant nutrition - for sustainable food and environment. Kluwer Academic Publishers, Dordrecht, pp 325-326

Otani T, Ae N (1997b) The status of inorganic and organic phosphorus in some soils in relation to plant availability. Soil Sci Plant Nutr 43:419-429

Otani T, Ae N, Tanaka H (1996) Phosphorus (P) uptake mechanisms of crops grown in soils with low P status. II. Significance of organic acids in root exudates of pigeonpea. Soil Sci Plant Nutr 42:533-560

Rao IM, Kerridge PC (1994) Mineral nutrition of forage Arachis. In: Biology and agronomy of forage Arachis. CIAT, Cali, Colombia, pp 71-83

Ravan JA, Rothemund C, Wollenweber B (1991) Acid -base regulation by *Azolla* spp. with N_2 as sole N source and with supplementation by NH^{4+} or NO^{3-}. Acta Bot 104:132-138

Rovira AD (1969) Plant root exudates. Bot Rev 35:35-57

Sample EC, Soper RJ, Racz GJ (1980) Reactions of phosphate fertilizers in soils. In: Khasawneh FE, Sample EC, Kamprath EJ (Eds) The role of phosphorus in agriculture. Am Soc Soil Chem, Madison, Wisconsin, USA, pp 263-310

Sanchez C (1977) Encalamiento de ultisoles de Sabana. Univ del Oriente, Jusepin, Venezuela

Sanchez PA, Salinas JG (1981) Low-input technology for managing oxisols and ultisols in tropical America. Adv Agron 34:280-406

Sanders FE, Tinker PB (1971) Mechanism of absorption of phosphate from soil by Endogone mycorrhizas. Nature 23, 278-279

Schumutz A, Jenny T, Amrhein N, Ryser U (1993) Caffeic acid and glycerol are constituents of the suberin layers in green cotton fibres. Planta 189:453-460

Shen R, Ae N (2001) Extraction of P solubilizing active substances from the cell wall of groundnut roots. Plant Soil 228:243-252

Subbarao GV, Ae N, Otani T (1997a) Genetic variation in iron-, and aluminum-phosphate solubilizing activity of pigeonpea root exudates under P deficient conditions. Soil Sci Plant Nutr 43:295-305

Subbarao GV, Ae N, Otani T (1997b) Genetic variation in acquisition, and utilization of phosphorus from iron-bound phosphorus in pigeonpea. Soil Sci Plant Nutr 43:511-519

Tadano T, Sasaki H (1991) Secretion of acid phosphatase by roots of several crop species under phosphorus-deficient conditions. Soil Sci Plant Nutr 37:129-140

Tan KH (1933) Principles of soil chemistry. Marcel Dekker Inc, New York

Tanaka H, Adachi N, Isoi T (1995) Secretion of formate in phosphorus-deficient nutrient solution from peanut seedlings (*Arachis hypogaea* L.). Soil Sci Plant Nutr 41:389-392

Tarafdar JC, Claassen N (1988) Organic phosphorus compounds as phosphorus source for higher plants through the activity of phosphatase produced by plant root and microorganisms. Biol Fertil Soils 5:308-312

Thomas GW, Clarke CA, Mosse B, Jackson RM (1982) Source of phosphate taken up from two soils by mycorrhizal (*Telephora terrestris*) and non-mycorrhizal *Picea sitchensis* seedlings. Soil Biol Biochem 14:73-75

Tinker PHB (1975) Effects of vasicular-arbuscular mycorrhizas on higher plants. Symp Soc Exp Biol 29:325-349

Trolldenier, Hecht-Buchholz (1984) Effect of aeration status of nutrient solution on microorganisms, mucilage and ultrastructure of wheat roots. Plant Soil 80:381-390

Uheda E, Akasaka Y, Daimon H (1997) Morphological aspects of the shedding of surface layers from peanut roots. Can J Bot 75:607-611

Vermeer, McCully (1981) Fucose in the surface deposits of axenic and field grown roots of *Zea mays* L., Protoplasma 109:233-248

Wagatsuma T (1983) Characterization of absorption site for aluminum in the roots. Soil Sci Plant Nutr 29:499-515

Walker ME, Flowers RA, Henning RJ, Keisling TC, Millinix BG (1979) Response of early bunch peanuts to calcium and potassium fertilization. Peanut Sci 6:119-123

Walker ME, Gaines TP, Henning RJ (1982) Foliar fertilization effects on yield, quality, nutrient uptake, and vegetative methods for soil testing. Peanut Sci 9:53-57

Wissuwa M, Ae N (1999) Genotypic variation for phosphorus uptake from hardly soluble iron-phosphate in groundnut (*Arachis hypogaea* L.). Plant Soil 206:163-171

Yarbrough JA (1949) *Arachis hypogaea*. The seedling, its cotyledons, hypocotyl and roots. Am J Bot 36:758-772

Structure of Plant Cell Walls and Implications for Nutrient Acquisition

Yoji Kato

Summary. The structure and function of plant cell walls are characterized by the chemical structures of their constituents, cellulosic polysaccharides, matrix polysaccharides, glycoproteins and lignin. In this chapter, the restriction of wall extensibility in growing cells and the aggregation of cells in rice callus tissues by the coupling of feruloyl residues on matrix polysaccharides, the permeability of the plant cell wall, and a possible role for cell wall components, phenolic acids, in nutrient acquisition are described in addition to the chemical structures of the major components of plant cell walls.

Key words. Cell wall, Pectic substances, Hemicelluloses, Phenolic acids, Aggregation, Permeability, Cross-linking

1. Introduction

The cell wall of higher plants is very complex. It has many functions including control of the expansion and rigidity of the plant cell and interaction between plant hosts and their pathogens (McNeil et al. 1984). The cell wall is made up of a thin primary wall and a thicker secondary wall. It is now established that the primary wall is a metabolically active component of the cell. The cell wall of higher plants is constructed with polymers such as cellulosic polysaccharides, matrix polysaccharides, glycoproteins, and lignin (Darvill et al. 1980; Carpita and Gibeaut 1993). Some polysaccharides contain a small amount of hydroxycinnamic acid derivatives, such as *p*-coumaric acid and ferulic acid (Hartley and Ford 1989). In addition, the cell wall contains numerous enzymes (Fry 1985). The structure of the cell wall is characterized by the structure of its constituents. Recent works on mineral nutrient acquisition by plants suggest the significance of additional studies on cell wall formation and function (Ae and Otani 1997). This chapter deals with the structure of cell wall components and its implications for nutrient acquisition.

2. Major Components of Plant Cell Walls

2.1 Polysaccharides

The cell wall polysaccharides are roughly classified into pectic substances, hemicellulose, and cellulose. The approximate ratio of pectic substances, hemicellulose, and cellulose are 35:30:30 for dicot primary cell walls, and 5:65:30 for monocots (Ishii 1997b). Pectic substances are polygalacturonan, rhamnogalacturonan I, ramnogalacturonan II, araban, arabinogalactan and (1→4)-β-D-galactan. (1→3)-β -D-Glucan, (1→3),(1→4)-β-D-glucan, xyloglucan, arabino(glucurono)xylan, glucomannan and others belong to hemicellulose (Darvill et al. 1980). Pectic substances are obtained by extraction with hot water, EDTA, dilute acid or ammonium oxalate, and hemicelluloses are obtained by the subsequent extraction with alkali. However, acid or alkali extraction results in partial cleavage of several types of bonds. This bond disruption is hydrolysis of some glycosidic linkages, particularly the α-L-arabinofuranosidic linkage, hydrolysis of the ester linkage of sugar acetyl groups, hydrolysis of methyl esters of uronic acids, hydrolysis of ester linkages either between polysaccharides or between carbohydrates and noncarbohydrate components, and catalysis of the transelimination of uronic acids. But, the selective dissociation of isolated cell walls by enzymic hydrolysis has the potential to release components with greater retention of structural integrity than fragments derived from treatment with chemical reagents (Talmadge et al. 1973; Kato and Nevins 1984a). Structure of cell wall polysaccharides is determined using samples prepared by extraction with chemical reagents and by treatment with carbohydrolases. Structural formulas proposed for the major polysaccharides from structural analyses are as follows.

2.1.1 Pectic Substances

Pectic polysaccharides are polygalacturonic acid (homogalacturonan), rhamnogalacturonan I (RG-I), and RG-II (McNeil et al. 1984).

Polygalacturonic acid is a homopolymer of (1→4)-α-D-galacturonic acid (Fig. 1). The main chains are normally modified, often to a very considerable extent, by methyl esterification of carboxyl groups. RG-I is a heteropolymer of repeating (1→2)- α-L-rhamnosyl-(1→4)-α-D-galacturonic acid disaccharide units (Fig. 2). RG-II has an extremely complex glycosyl residue (2-*O*-methylfucose, rhamnose, fucose, 2-*O*-methylxylose, arabinose, apiose, galactose, galacturonic acid, glucuronic acid, aceric acid, 3-deoxy-D-manno-2-octulosonic acid, and 3-deoxy-D-lyxo-2-heptulosaric acid) and glycosyl-linkage composition (O'Neill et al. 1990).

Arabans, galactans, and highly branched arabinogalactans (type I) of various configurations and sizes (Fig. 3) are attached to the *O*-4 of the rhamnosyl residues of RG-I. It is known that these side chains contain phenolic acid as described later. In addition, acetylation occurs in polygalacturonic acid and RG-I (O'Neill et al. 1990; Ishii 1997a).

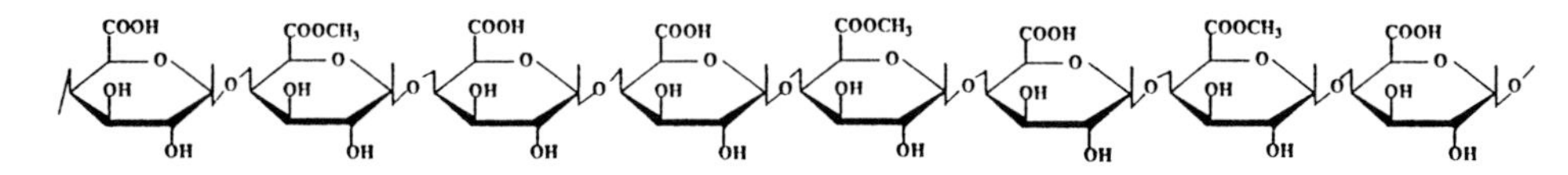

Fig. 1. Partial structure of homogalacturonan. The main chain is modified by methyl esterification of carboxyl groups

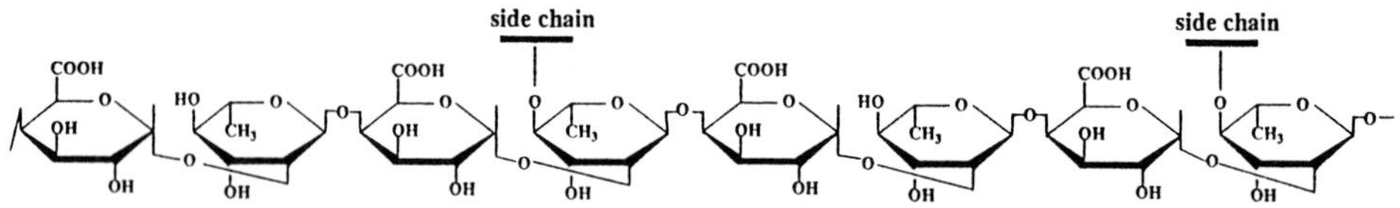

side chain

side chain

Fig. 2. Partial structure of rhamnogalacturonan I (RG-I). Arabans, galactans, and highly branched arabinogalactan are attached to the *O*-4 of the rhamnosyl residues of the backbone as side chains (see Fig. 3) (McNeil et al. 1984). With permission, from the Annual Review of Biochemistry, Volume 53, ©1984, by Annual Reviews www.AnualReviews.org.

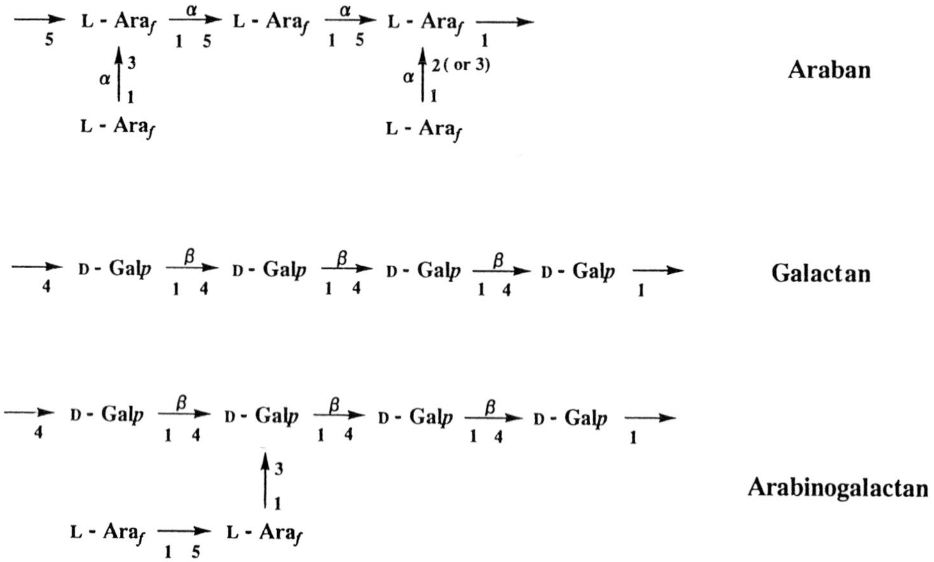

$\xrightarrow[5]{}$ L-Ara$_f$ $\xrightarrow[1\ 5]{\alpha}$ L-Ara$_f$ $\xrightarrow[1\ 5]{\alpha}$ L-Ara$_f$ $\xrightarrow[1]{}$

α (1→3) L-Ara$_f$ α (1→2 (or 3)) L-Ara$_f$

Araban

$\xrightarrow[4]{}$ D-Gal$p$ $\xrightarrow[1\ 4]{\beta}$ D-Gal$p$ $\xrightarrow[1\ 4]{\beta}$ D-Gal$p$ $\xrightarrow[1\ 4]{\beta}$ D-Gal$p$ $\xrightarrow[1]{}$

Galactan

$\xrightarrow[4]{}$ D-Gal$p$ $\xrightarrow[1\ 4]{\beta}$ D-Gal$p$ $\xrightarrow[1\ 4]{\beta}$ D-Gal$p$ $\xrightarrow[1\ 4]{\beta}$ D-Gal$p$ $\xrightarrow[1]{}$

(1→3) L-Ara$_f$ $\xrightarrow[1\ 5]{}$ L-Ara$_f$

Arabinogalactan

Fig. 3. Partial structure of araban, galactan, and arabinogalactan. L-Ara$_f$ and D-Gal$_p$ denote L-arabinofuranosyl and D-galactopyranosyl residues, respectively

2.1.2 Hemicelluloses

In monocot primary walls, the major hemicelluloses are acidic arabinoxylans and neutral (1→3),(1→4)-β-D-glucan, and in dicots they are neutral xyloglucans and acidic arabinoxylan.

Arabinoxylan has a linear backbone chain of β-(1→4)-D-xylopyranosidic residues, some of which are substituted at the *O*-2 or *O*-3 position with arabinose, glucuronic acid, and other substituents (Carpita 1996). Some arabinoxylans contain a small amount of *p*-coumaric acid and ferulic acid. In *Zea* shoot cell walls ferulic acid bonds to arabinose residue. Figure 4 shows a proposed structure for the feruloylated glucuronoarabinoxylan from *Zea* shoot cell walls (Kato and Nevins 1984b).

(1→3),(1→4)-β-D-glucan consists mainly of cellotriosyl and cellotetraosyl residues joined by single (1→3)-linkages (Carpita 1996). Certain other glucosyl sequences are also present in the polysaccharide from *Zea* shoot cell walls including (a) two, three, or four contiguous (1→3)-linkages; (b) blocks of more than four (1→4)-linked glucose residues; (c) regions having alternating (1→3)- and (1→4)-linkages (Kato and Nevins 1986).

Xyloglucan is thought to be one of the key components of the primary cell wall and a regulatory factor in cell enlargement (Hayashi 1989). Xyloglucan consists of a linear backbone chain of (1→4)-linked β-D-glucopyranosyl units with D-xylosyl side chains a-linked to *O*-6 of glucosyl units. Some of the xylosyl side chains are extended by the addition of other mono- or di-saccharides to *O*-2 of xylosyl units. Figure 5 shows a proposed structure for the xyloglucan from mung bean hypocotyl cell walls (Kato and Matsuda 1994; Konishi et al. 1997a). In general, the main chain of Gramineae xyloglucan is less substituted than that of dicot xyloglucan, and the side chain of monocot xyloglucan is shorter than that of dicot xyloglucan. To investigate the arrangement of the glucose residues which are not xylosylated at *O*-6 in the glucan backbone of some xyloglucans, a purified *Penicillium* sp. M451 xyloglucan specific endo-1,4-β-glucanase (xyloglucanase) was used. The enzyme can hydrolyze the glucosidic bonds between unbranched glucose residues and *O*-4 of 4,6-di-substituted glucose residues, and can not hydrolyze Avicel SF, phosphoric acid swollen cellulose, carboxymethyl cellulose, hydroxyethyl cellulose and cello-oligosaccharides (cellobiose to cellohexaose). Xyloglucan preparations obtained from some plant cell walls were hydrolyzed with M451 xyloglucanase to give xyloglucan oligosaccharide subunits corresponding to the respective xyloglucan polymer. The xyloglucan oligosaccharides were further treated with *Eupenicillium* sp. M9 isoprimeverose-producing oligoxyloglucan hydrolase. The xyloglucan oligosaccharides before and after treatment with *Eupenicillium* enzyme were analyzed by high-performance anion exchange chromatography (on Dionex CarboPac PA1 in 0.1 M NaOH with a gradient of NaOAc) and matrix-assisted laser desorption/ionization time-of-flight mass spectrometry. Figure 6 shows the structures of xyloglucan oligosaccharide units of dicots, Solanaceae, and Gramineae xyloglucans (Konishi et al. 1997b, 1998; Kato and Mitsuishi 1999). Xyloglucan molecules are enzymatically modified during cell growth or fruit ripening (Fry 1995). This may be confirmed by investigating changes in the ratio of xyloglucan oligosaccharide subunits of xyloglucan polysaccharides before and after cell growth or fruit ripening. Xyloglucans extracted from cell walls are detected by their characteristic blue staining with iodine reagent (Kooiman 1960), but the polysaccharides pre-treated with cellulase or xyloglucanase do not form any colored complex with iodine reagent (Kato and Matsuda 1977).

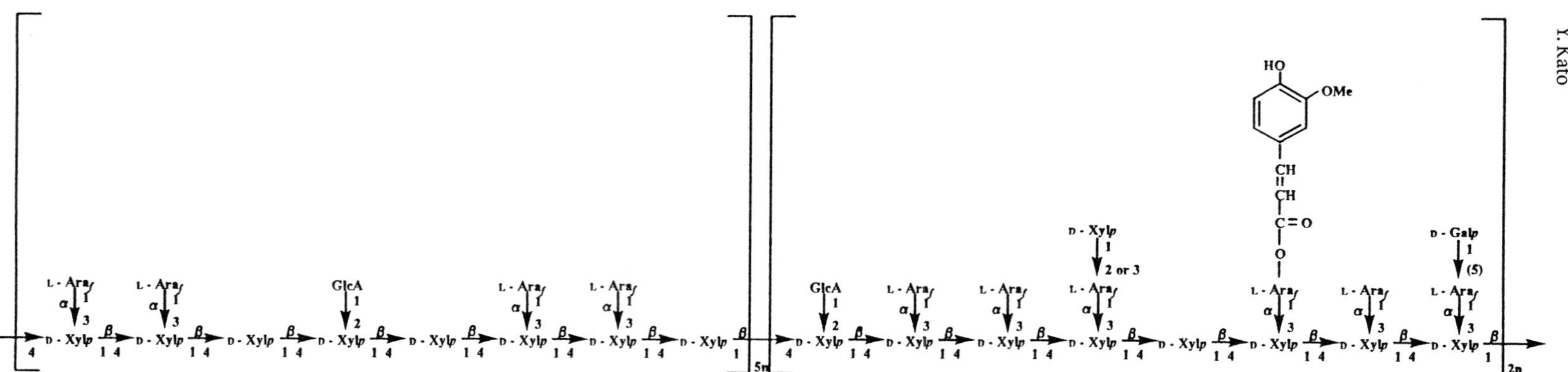

Fig. 4. Proposed structure of feruloylated glucuronoarabinoxylan from the cell walls of *Zea* shoots (Kato and Nevins 1984b)

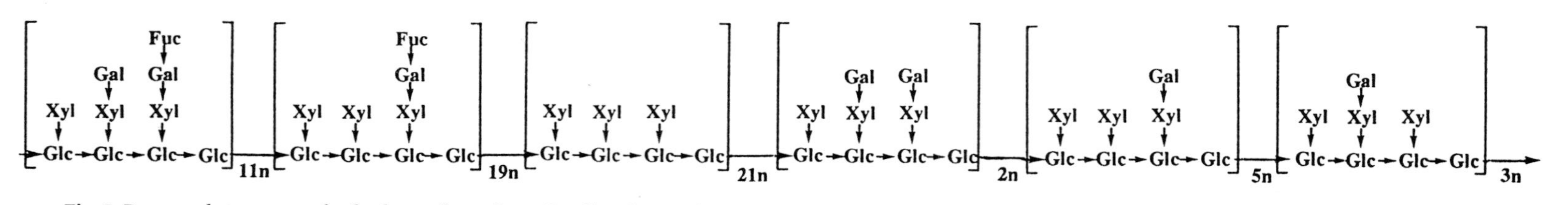

Fig. 5. Proposed structure of xyloglucan from the cell walls of mung bean hypocotyls (Konishi et al. 1997a). *Fuc*, α1,2-linked L-fucopyranose; *Gal*, β1,2-linked D-galactopyranose; *Xyl*, α1,6-linked D-xylopyranose; *Glc*, β1,4-linked D-glucopyranose

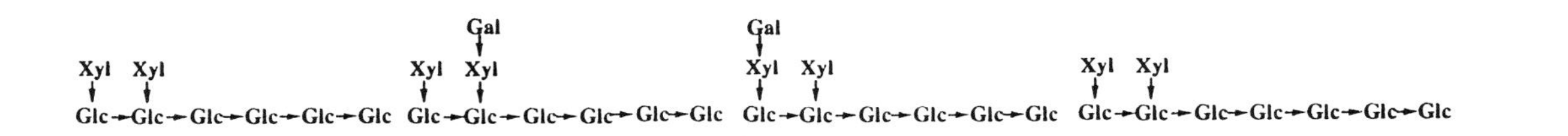

Fig. 6. Structure of xyloglucan oligosaccharides obtained from xyloglucans of *dicotyledons* (mung bean, soybean, cabbage, Chinese cabbage, spinach, chingentsuai, lettuce, turnip, Japanese radish, edible burdock, carrot, East Indian lotus) (Konishi et al. 1997a, 1998), *Solanaceae* (sweet pepper, eggplant, tomato and red pepper) (Konishi et al. 1997b), and *Graminineae* (barley) (Kato and Mitsuishi 1999). Sugar residues and their linkages are the same as those described in the legend for Fig. 5

Fig. 7. Structure of cinnamic acid derivatives, ferulic acid (**a**) and *p*-coumaric acid (**b**)

2.1.3 Cellulose

Cellulose is a simple polymer composed of glucose residues joined together by (1→4)-β-linkages, and which is deposited in partially crystalline microfibrils in plant cell walls. Crystalline microfibrils make up an important part of the framework of cell walls (Darvill et al. 1980).

2.1.4 Phenolic Acids

The cell wall polysaccharides, arabinoxylan, xyloglucan, and pectic (arabino)galactan contain a small amount of phenolic acid (ferulic acid and *p*-coumaric acid, Fig. 7), which is present as ester-linked side chains. The location of phenolic acid-containing polysaccharides in plant tissues can be visualized by the autofluorescence of tissue (cell walls) excited with UV light (Smith and O'Brien 1979). The linkage of phenolic acid to wall polysaccharides has been investigated by determining the structure of wall fragment released from walls by treatment with a mixture of carbohydrate hydrolases free from esterase activities (Ishii 1997b). Structures of some fragments obtained from cell walls by hydrolysis with commercial enzyme preparation are shown in Fig. 8. It is now confirmed that monocot arabinoxylans have *p*-coumaric acid and ferulic acid, and monocot xyloglucans have ferulic acid. In dicots a feruloyl group is attached to araban and (1→4)-galactan of pectic polysaccharides and to the arabinose residues of arabinoxylan. Furthermore, ferulic acid is known to be implicated in cross-linking grass cell wall polysaccharides with lignin; arabinoxylans feruloylated at *O*-5 of an arabinofuranosyl residue attach to lignins during wall development (Ralph et al. 1995).

2.2 Extensin

Extensin is an insoluble, hydroxyproline-rich glycoprotein, which is very tightly associated with the primary cell wall by covalent linkages (McNeil et al. 1984). The amino acid composition of dicot extensin is approximately 41 mol% hydroxyproline(Hyp), 12 mol% serine, 10 mol% lysine, 8 mol% tyrosine and 0-11 mol% histidine. Most of the hydroxyproline residues of dicot extensin contain arabinofuranosyl residues as β-L-Ara$_f$(1→2)-β-L-Ara$_f$(1→2)-β-L-Ara$_f$(1→4)Hyp and α-L-Ara$_f$(1→3)-β-L-Ara$_f$(1→2)-β-L-Ara$_f$(1→2)-β-L-Ara$_f$(1→4)Hyp, and serine residues contain single galactosyl residues (Akiyama et al. 1980).

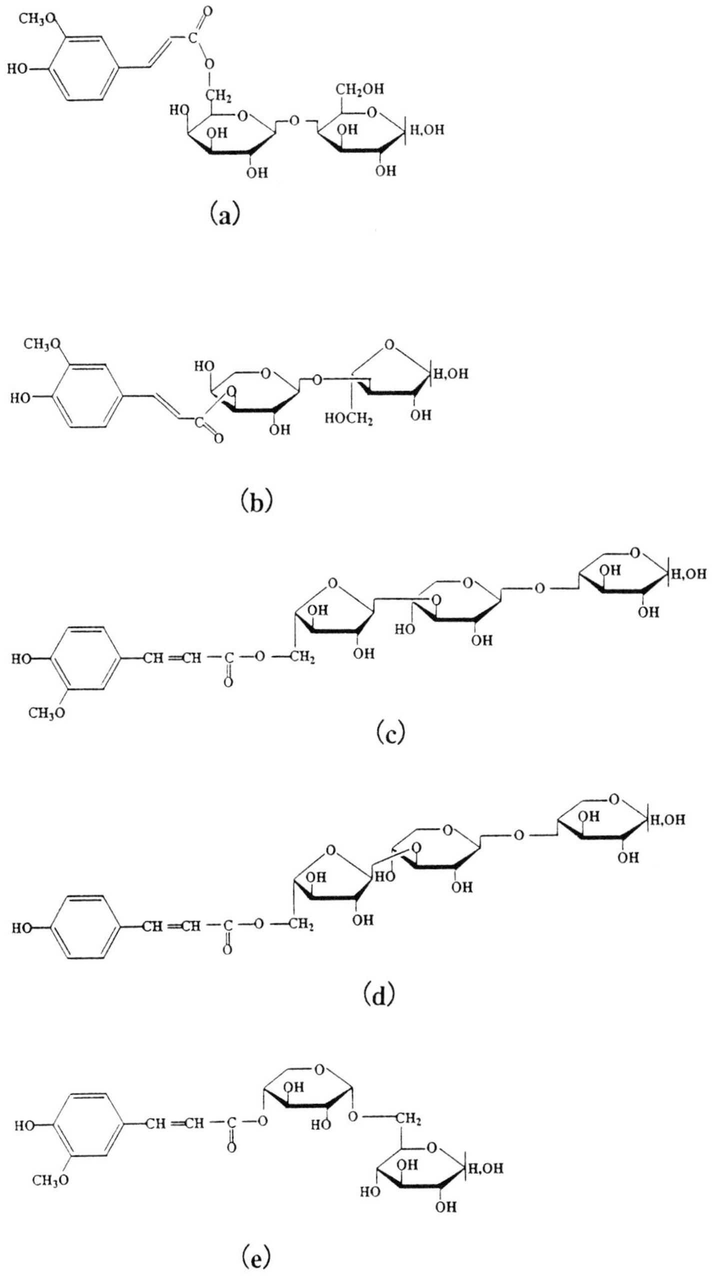

CH_3O HO O C O CH_2 HO O O CH_2OH O OH OH H,OH OH OH

(a)

CH_3O HO HO O O C O O O H,OH HOCH$_2$ OH OH

(b)

HO CH =CH — C — O — CH_2 O OH OH O HO OH O O OH H,OH OH CH_3O

(c)

HO CH =CH — C — O — CH_2 O OH OH O HO O OH O O OH H,OH OH

(d)

HO CH =CH — C — O O OH HO O — CH_2 O OH H,OH HO OH CH_3O

(e)

Fig. 8. Structure of major feruloylated and *p*-coumaroylated oligosaccharides obtained from cell walls. **a** feruloyl-6-D-Gal$_p$(1→4)-D-Gal$_p$ and **b** feruloyl-3-L-Ara$_p$(1→3)-L-Ara$_f$ were obtained from cell walls of suspension-cultured spinach cells (Fry 1983); **c** feruloyl-5-L-Ara$_f$(1→3)-D-Xyl$_p$(1→4)-D-Xyl$_p$ was obtained from cell walls of *Zea* shoots (Kato and Nevins 1985). **d** *p*-coumaroyl-5-L-Ara$_f$(1→3)-D-Xyl$_p$(1→4)-D-Xyl$_p$ and **e** feruloyl-4-D-Xyl$_p$(1→6)-D-Glc$_p$ were obtained from cell walls of bamboo shoots (Ishii and Hori 1990)

2.3 Cell Wall Enzymes

Plant cell walls contain numerous enzymes (Fry 1985, 1995). Some are extracted with salts and are ionically bound to wall components, and others are not extracted with salts and are covalently bound. Included in the wall enzymes are peroxidase (EC 1.11.1.7), malate dehydrogenase (EC 1.1.1.37), endo-(1→4)-β-D-glucanase (EC 3.2.1.4), pectinesterase (pectin methylesterase)(EC 3.1.1.11), polygalacturonase (EC 3.2.1.15), β-D-fructofuranosidase (EC 3.2.1.26), exo-*O*-glycosylhydrolases (for example, β-D-glucosidase, EC 3.2.1.21), xyloglucan endo-transglycosylase and so on. They modify the structural components of the wall of living cells. Figure 9 shows the hydrolysis mechanism of pectinesterase and polygalacturonase.

3. Cross-Linking of Matrix Polymers

3.1 Cross-Linkings of Wall Polymers

It is now accepted that various types of cross linkings (H-bonds, Ca-bridge, other ionic bonds, ester bonds) including coupled phenols play a role in building cell-walls as shown in Fig. 10 (Fry 1986). In addition, it was found that boron cross-links two RG-II chains through borate-diolester bonding. The role of boron cross-linking in cell walls is described in detail in the chapter by Matoh T.

The linear β-(1→4)-glucan chains of cellulose are bound together by hydrogen bonds to form microfibrils. Xyloglucan, which has a cellulose-like β-(1→4)-glucan backbone with several different side chains attached at the *O*-6 position of the glucosyl residues, binds tightly to the surface of individual cellulose microfibrils and functions as cross-links between cellulose microfibrils (Hayashi 1989). This polysaccharide was obtained from cell walls by treatment with 24% potassium hydroxide, not 4% potassium hydroxide (Kato and Matsuda 1976).

Certain pectin molecules cross-link in the presence of Ca^{2+} by forming Ca bridges (Fig. 11). Pectins having a number of side chains, such as RG-I (see Figs. 2

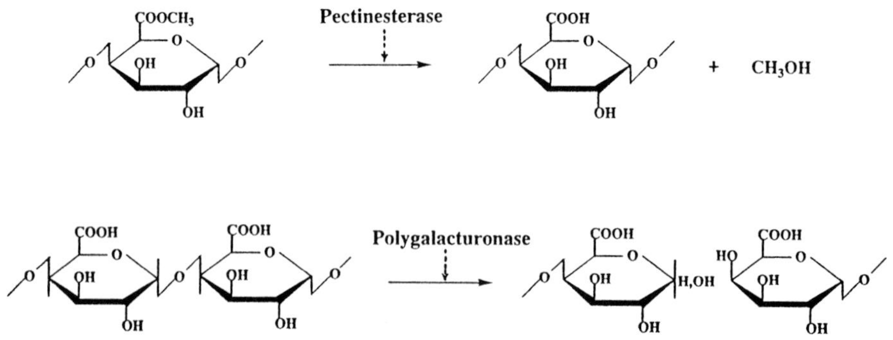

Fig. 9. Action mechanism of pectinesterase and polygalacturonase

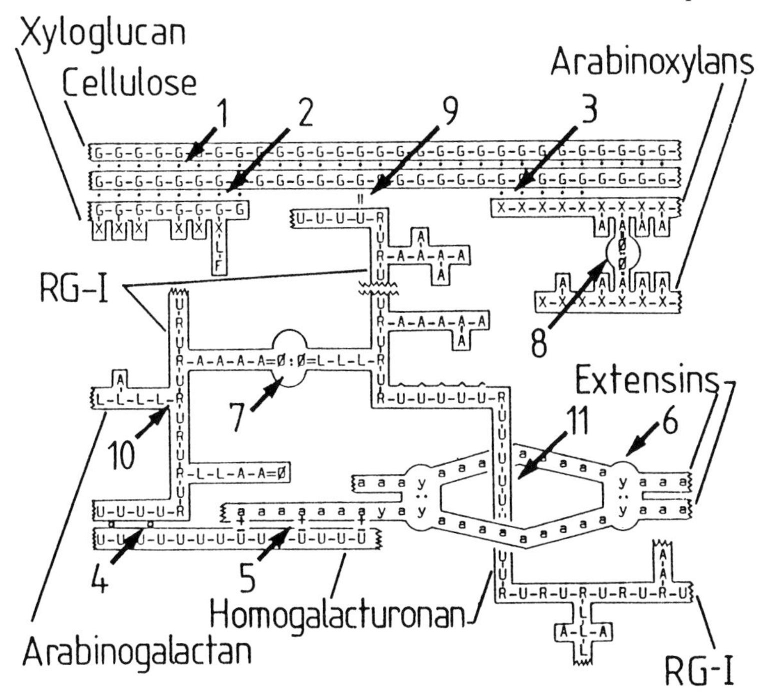

Fig. 10. Possible cross-links of wall polymers. (Fry 1986) The diagram illustrates:

(•) Hydrogen-bonds	*1.* cellulose-cellulose,
	2. xyloglucan-cellulose
	3. xylan-cellulose
(o) Calcium bridges	*4.* homogalacturonan-homogalacturonan
(±) Other ionic bonds	*5.* extensin-pectin
(:) Coupled phenols	*6.* extensin-extensin
	7. pectin-pectin
	8. arabinoxylan-arabinoxylan
(=) Ester bonds	*9.* pectin-cellulose
(-) Glycosidic bonds	*10.* arabinogalactan-rhamnogalacturonan
(⏀) Entanglement	*11.* pectin-in-extensin

A, arabinose; *F*, fucose; *G*, glucose; *L*, galactose; *R*, rhamnose; *U*, galacturonic acid; *Û*, galacturonic acid methyl ester; *a*, amino acid other than tyrosine; *y*, tyrosine; *y:y*, isodityrosine; ∅, ferulic acid; ∅ : ∅ , diferulic acid

With permission, from the Annual Review of Plant Physiology, Volume 37, ©1986, by Annual Reviews www.AnnualReviews.org

and 3), and highly methylesterified homogalacturonan (see Fig. 1), can not form Ca bridges.

Arabinoxylans, when substituted with phenolic acid such as ferulic acid and *p*-coumalic acid, cross-link via phenolic substituents (Fig. 12).

Extensins are rich in the phenolic amino acid tyrosine as described above. These glycoproteins are not extracted from cell walls with any of the conventional protein solvents. Therefore, the extensins are covalently bound through dimerization of their tyrosine residues to form isodityrosine (Fig. 13; McNeil et al. 1984).

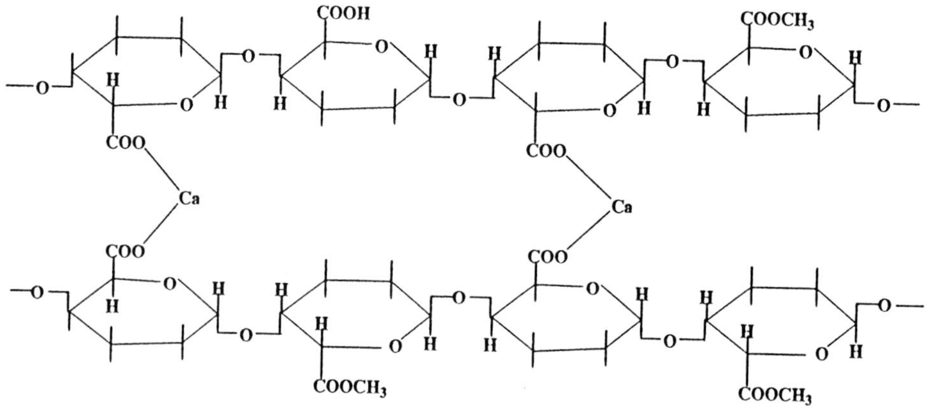

Fig. 11. Ca bridges between homogalacturonan molecules

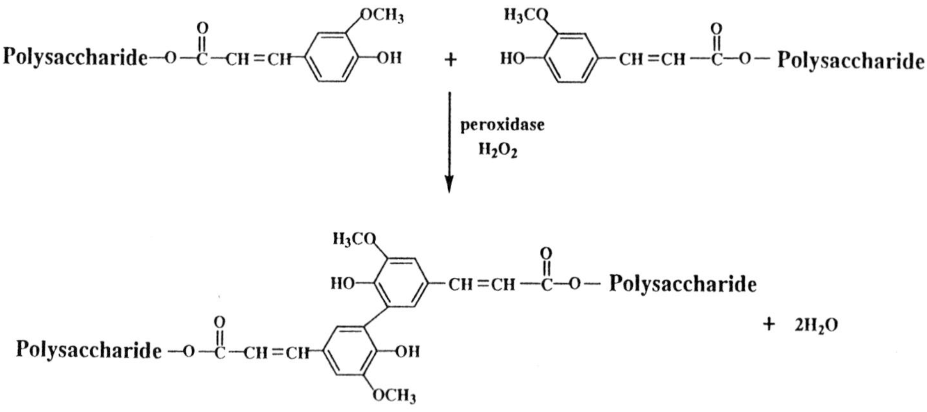

Fig. 12. Coupling of feruloyl esters on oxidation with H_2O_2 and peroxidase to form diferulate-polysaccharide cross-links. (McNeil et al. 1984) With permission, from the Annual Review of Plant Physiology, Volume 37, ©1986, by Annual Reviews www.AnnualReviews.org

3.2 Restriction of Wall Extensibility by Coupling of the Feruloyl Residues on Arabinoxylan

Geissmann and Neukom (1971) found that water-soluble feruloylated arabinoxylans from wheat flour formed gels upon oxidation with H_2O_2 plus peroxidase. The gel-forming mechanism was considered to be due to oxidative coupling of the feruloyl residues on arabinoxylan to give diferulic acid (dehydrodiferulates) (Fig. 12). This was one of several possible polysaccharide cross-linking mechanisms in the growing cells. Such oxidative coupling of wall matrix polysaccharides was considered to drastically affect the wall extensibility and its growth rate (Fry 1986). Ishii (1991) isolated a diferuloyl arabinoxylan hexasaccharide from bamboo shoots to give firm evidence for the existence of the diferuloyl diester cross-link in the cell walls (Fig. 14).

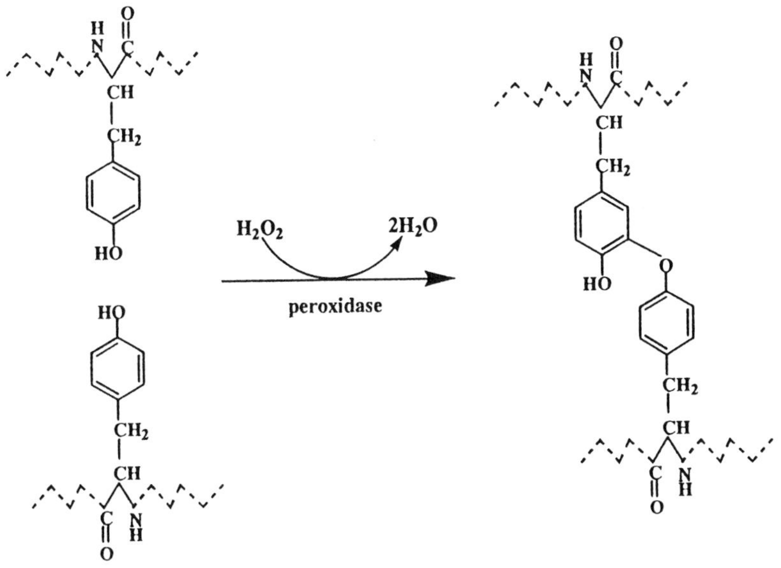

Fig. 13. Oxidative coupling of tyrosine residues to produce dityrosine. (Fry 1986)
With permission, from the Annual Review of Plant Physiology, Volume 37, ©1986, by Annual Reviews www.AnnualReviews.org

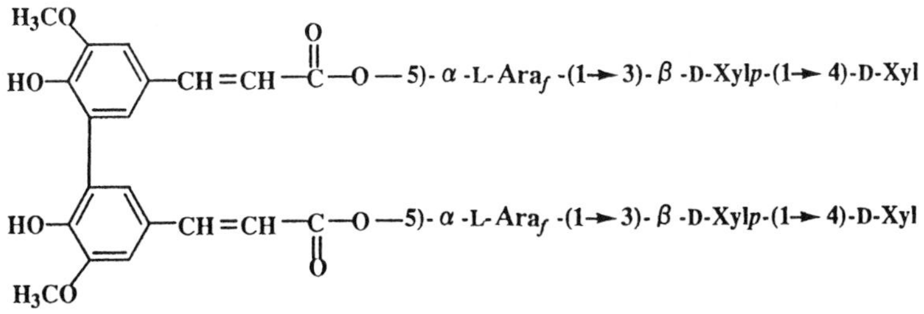

Fig. 14. Structure of a diferuloyl arabinoxylan hexasaccharide (Ishii 1991)

Kamisaka et al. (1990) indicated that in growing cells, ferulic acid, ester-linked to cell wall polysaccharides, is oxidized to give diferulic acid, which makes the cell wall mechanically rigid by cross-linking matrix polysaccharides and results in limited cell extension growth.

3.3 Feruloyl Polysaccharides Involved in the Aggregation of Cells in Rice Callus Tissues

Suspension-cultured plant cells secrete matrix polysaccharides and proteins into culture medium during cultivation (Talmadge et al. 1973). Rice (*Oryza sativa* L) cells are usually obtained as the aggregated cells on cultivation in Gamborg's B5 medium. Toriyama and Hinata (1985) found that amino acid medium, which is composed of aspartic acid, arginine, glycine, and glutamine as a nitrogen source, evoked a finely dispersed cell suspension in suspension-cultured rice cells. The effects of amino acid medium on the dissociation of rice callus tissues was examined (Hayashi et al. 1994). It was found that arginine is metabolized in rice cells to form urea, which is then secreted and solubilizes a part of the wall matrix polysaccharides, arabinoxylan and (arabino)galactan between cells. In the course of the study, differences in chemical properties between cell walls of aggregated and non-aggregated cells were investigated (Kato et al. 1994).

Rice cells were cultured in amino acid medium (AA medium) and Gamborg's B5 medium (B5 medium), respectively. Scanning electron microscopy showed that the intercellular layer of rice cells cultured in B5 medium was more deeply embedded than that in AA medium (Fig. 15). Fluorescence microscopy of thin sections from rice callus tissues showed that all of the walls fluoresced blue in water (pH 5.8). The fluorescence of the walls increased in intensity and changed colour to green when treated with ammonium hydroxide solution (pH 10.0). These phenomena are characteristics of feruloyl esters. Such fluorescence in the walls of cells cultured in B5 medium was much stronger than that in AA medium (Figs. 16-A and -B). Laser scanning microscopy showed that the level of fluorescence was higher in the intercellular layer, especially at corner junctions between cells (Fig. 16C). The hydroxycinnamic esters of wall preparations and extracellular polysaccharide fractions were then analyzed by HPLC. The walls of cells cultured in AA medium mainly contained feruloyl esters and those of cells cultured in B5 medium contained similar proportions of feruloyl and *p*-coumaroyl

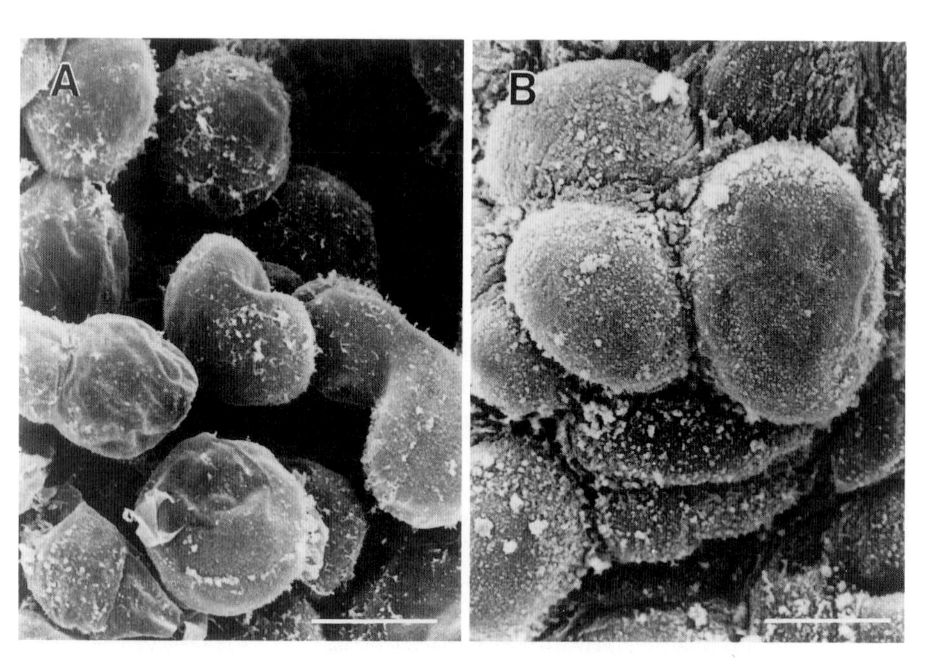

Fig. 15. Scanning electron microscopy of rice cells cultured in AA (**A**) and B5 (**B**) media. *Bar* 20 μm (Hayashi et al. 1994)

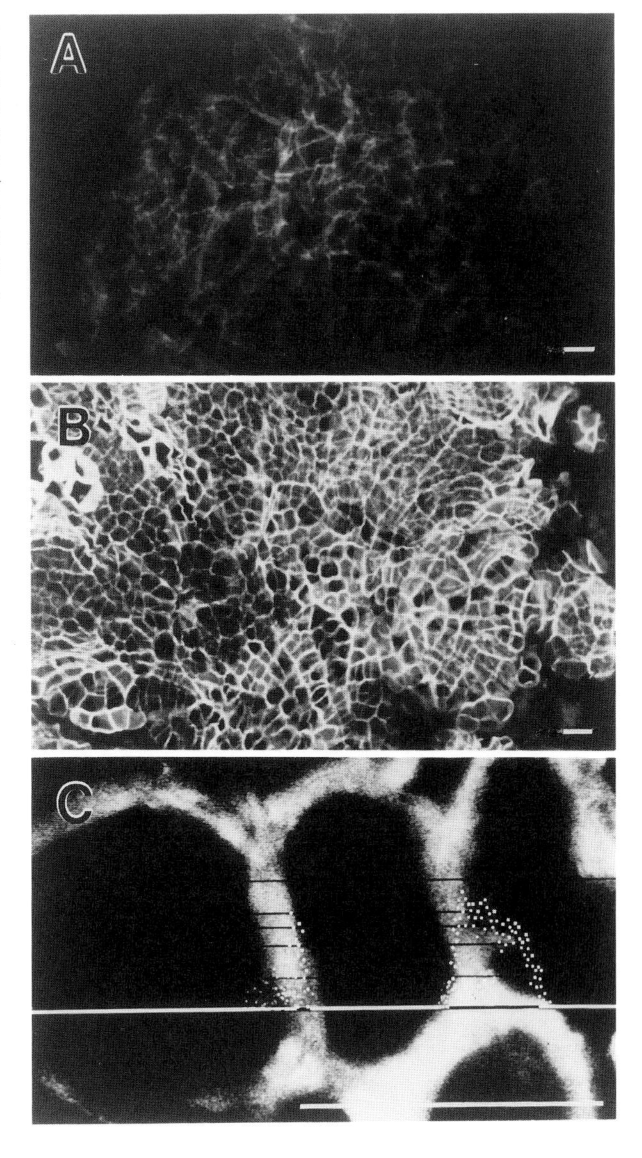

Fig. 16. Fluorescence microscopy of feruloyl esters in the wall of rice cells. Fluorescence of walls obtained from rice cells cultured in AA medium (**A**) and B5 medium (**B**), and a thin section (**C**) scanned by laser scanning microscopy. The sections of the walls were treated with ammonia (pH 10.0). *Bar* 20 μm (Kato et al. 1994)

esters (Fig. 17). Each also contained a small amount of diferuloyl ester. Levels of feruloyl and diferuloyl esters in the walls increased during cultivation in AA medium. On the other hand, the walls of cells cultured in B5 medium maintained constant levels and proportions of ferulic acid, diferulic acid and *p*-coumaric acid during cultivation. The level of the hydroxycinnamic esters in extracellular polysaccharides was much higher in the cultured AA medium than in the cultured B5 medium (Fig. 18). This indicates that the secreted polysaccharides were highly feruloylated. That is, feruloyl esters accumulated mainly with extracellular polysaccharides in AA medium but within the wall in B5 medium. These findings indicated that feruloyl (and coumaroyl) polysaccharides are involved in the aggregation of cells in rice callus tissues. The function of diferuloyl

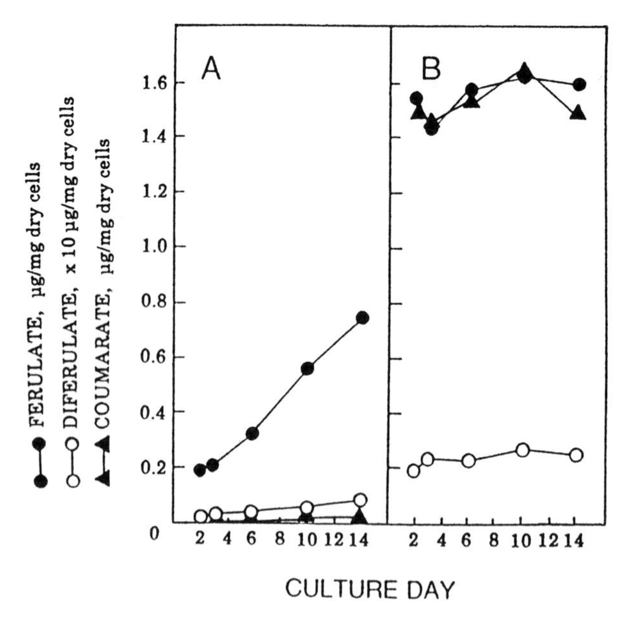

Fig. 17. Changes in the amount of phenolic esters bound to walls. **A** Cells cultured in AA medium; **B** cells in B5 medium (Kato et al. 1994)

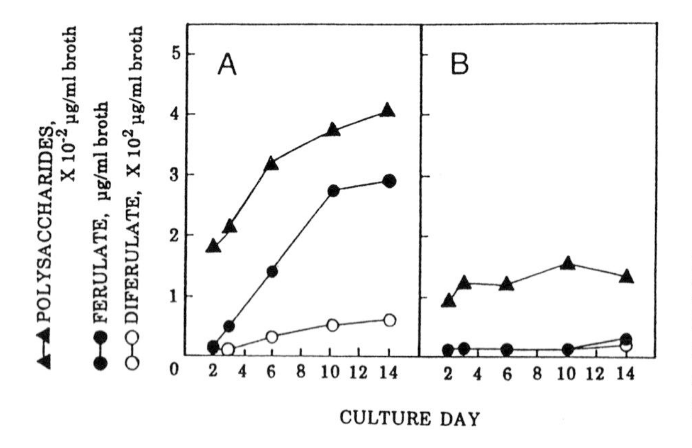

Fig. 18. Accumulation of feruloyl and diferuloyl extracellular polysaccharides in culture medium during cultivation. **A** AA medium; **B** B5 medium (Kato et al. 1994)

ester cross-links is considered most likely to be restriction of wall extensibility (Fry 1986; Kamisaka et al. 1990). Our work indicated a new function of feruloyl ester in cell aggregation.

4. Permeability of Plant Cell Walls

The permeability of plant cell walls determines the ability of enzymes and extracellular polymer to penetrate and alter the cell walls. Tepfer and Taylor (1981) reported that proteins of molecular weight up to 60,000 could permeate a substantial portion of the cell walls isolated from the hypocotyls of bean seedlings.

As described in section 3.1, various kinds of cross-linkings of polymers are present in cell walls. This means that the wall matrix has pores of various sizes, which play the role of a molecular sieve. We investigated the permeability of water-insoluble cell walls to authentic dextrans and maltodextrins by gel-filtration chromatography (Kato and Akiyama 1993; Kato 1993). Water-insoluble cell wall materials were prepared from Japanese radish. The cell wall materials consist of pectic substances : hemicelluloses : cellulose in a ratio of 51:11:38. Major polysaccharides of pectic substances and hemicelluloses of Japanese radish are homogalacturonan and/or rhamnogalacturonan, araban, galactan, arabinogalactan, xylan, xyloglucan, and cellulose. The cell walls suspended in 50 mM Na-acetate buffer, pH 5.5 containing 0.1 M NaCl were packed into a glass column (0.5 × 46 cm) and equilibrated with the same buffer containing 0.1 M NaCl. Figure 19a shows elution profile of a mixture of Blue Dextran (MW = 20×10^5)and glucose (MW = 180). The cell wall materials were found to resolve Blue Dextran and glucose. Figures 19b and c show the elution profiles of Dextran T-10 (MW = 1×10^4), a mixture of maltodextrins (MW = 180 ~ 2,000), respectively. These carbohydrates are eluted between the elution position of glucose and that of Blue Dextran. Figures 20a-e show the elution profiles of Blue Dextran, Dextran T-500 (MW = 50×10^4), Dextran T-110 (MW = 10×10^4), Dextran T-10, maltodextrins and glucose on a column of water-insoluble cell walls from Japanese radish. The cell wall materials were found to resolve dextrans and maltodextrins of molecular weight 2.0×10^3 ~ 5.0×10^5, although the resolving power was rather low. Maltodextrins of molecular weight less than about 2.0×10^3 completely permeated into the pores of the wall matrix. Next, we investigated the permeability of ammonium oxalate-insoluble cell wall materials (pectin-removed fraction) and 24% potassium hydroxide-insoluble cell wall materials (pectin and hemicellulose-removed fraction), respectively, to Blue

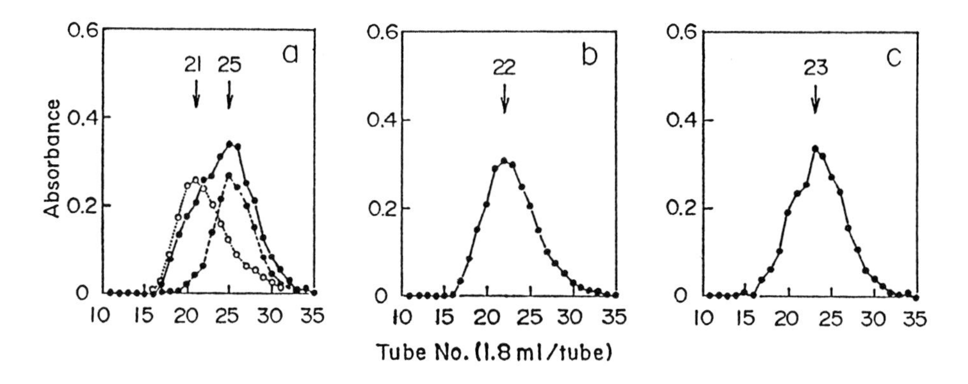

Fig. 19. Chromatography of Blue Dextran, Dextran T-10, maltodextrins, and glucose on a column of water-insoluble cell walls from Japanese radish. A mixture of Blue Dextran and glucose (2 mg each/0.2 ml 50 mM Na-acetate buffer, pH 5.5) **(a)**, Dextran T-10 (10 mg/0.1 ml 50 mM Na-acetate buffer) **(b)**, and a mixture of maltodextrins (10 mg/0.1 ml 50 mM Na-acetate buffer) **(c)** were subjected separately to chromatography. The column (1.0 × 46 cm) was chromatographed in 50 mM Na-acetate buffer, pH 5.5 containing 0.1 M NaCl. Carbohydrates were determined by the phenol-sulfuric acid method (absorbance at 490 nm, -•-). Reducing power (glucose) was measured by the Nelson-Somogyi method (absorbance at 500 nm, --•--). Absorbance at 360 nm was read for Blue Dextran (--∘--). *Arrows* in figures indicate the elution positions of individual samples (Kato and Akiyama, 1993)

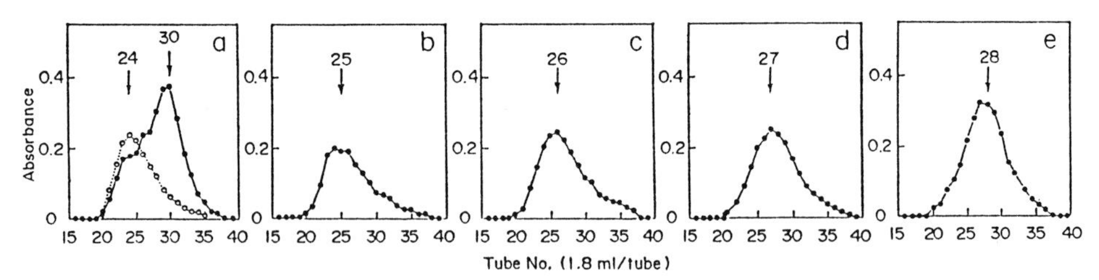

Fig. 20. Chromatography of Blue Dextran, Dextran T-500, T-110 and T-10, maltodextrins and glucose on a column of water-insoluble cell walls from Japanese radish. A mixture of Blue Dextran and glucose (10 mg each/0.1 ml water) **(a)**, Dextran T-500 (10 mg/0.1 ml water) **(b)**, Dextran T-110 (10 mg/0.1 ml water) **(c)**, Dextran T-10 (10 mg/0.1 ml water) **(d)**, and a mixture of maltodextrins (10 mg/0.1 ml water) **(e)** were subjected separately to chromatography. The column was chromatographed in water. Carbohydrates were determined by the phenol-sulfuric acid method (absorbance at 490 nm, -•-). Absorbance at 360 nm was read for Blue Dextran (-- ∘ --) (Kato and Akiyama, 1993)

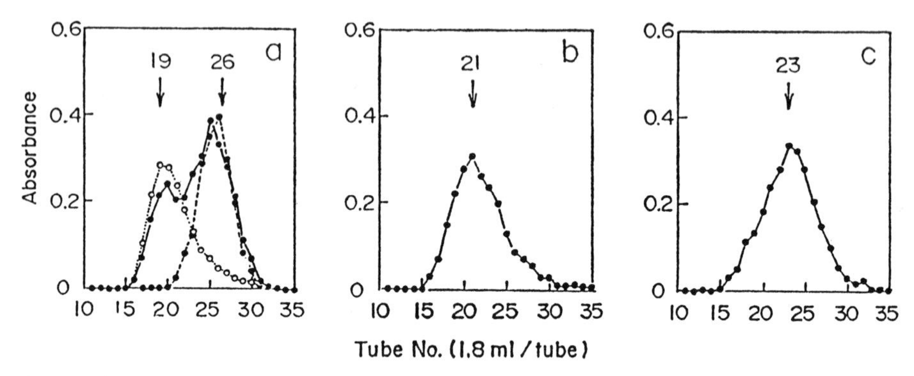

Fig. 21. Chromatography of Blue Dextran, Dextran T-10, maltodextrins, and glucose on a column of ammonium oxalate-insoluble cell walls from Japanese radish. A mixture of Blue Dextran and glucose **(a)**, Dextran T-10 **(b)** and maltodextrins **(c)** were subjected separately to chromatography in the same manner as described in the legend of Fig. 19 (Kato and Akiyama, 1993)

Dextran, Dextran T-10, maltodextrins and glucose by gel-filtration chromatography (Figs. 21 and 22). These cell wall materials were found to resolve the authentic carbohydrates used, although the resolving power was rather low. This indicates firm evidence that pectic substances, hemicelluloses and cellulose contribute to ability of molecular sieve. In addition, we obtained similar results in the case of cell wall materials prepared from bamboo shoots, the constituent polysaccharides of which are different from those of Japanese radish. The ratio of pectic substances, hemicelluloses and cellulose of bamboo shoot cell walls was 12:41:47. The hemicellulosic polysaccharides are mainly composed of (1→3),(1→4)-β-D-glucan, feruloylated glucuronoarabinoxylan, xyloglucan. These findings suggest that the ability of the molecular sieve of cell walls is common in plant cell walls without reference to their constituted polysaccharides.

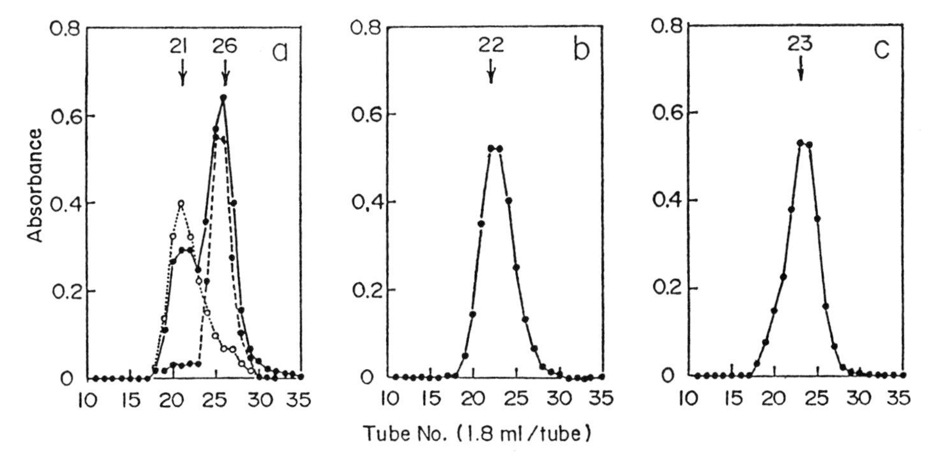

Fig. 22. Chromatography of Blue Dextran, Dextran T-10, maltodextrins, and glucose on a 24% potassium hydroxide-insoluble cell walls from Japanese radish. A mixture of Blue Dextran and glucose (**a**), Dextran T-10 (**b**) and maltodextrins (**c**) were subjected separately to chromatography in the same manner as described in the legend of Fig. 19 (Kato and Akiyama, 1993)

5. The Role of Cell Wall Components in Nutrient Acquisition

Production of root exudates and extensive root development are important factors affecting mineral nutrient acquisition by plants. The former includes organic acids, for example, mugineic acid on Fe acquisition by rice (Takagi 1976), citric acid on P acquisition by lupin (Gardner et al. 1983), and piscidic acid on P acquisition by pigeon pea (Ae et al. 1993). The latter is related to extent of surface of root epidermal cells. Generally, cation exchange capacities (CEC) which derived from carboxylic groups in polygalacturonic acids of pectic substances (Knight et al. 1961), is discussed in relation to acquisition of mineral nutrient. In addition, it was reported that mucigel secreted from plant roots forms a complex with Fe^{3+} resulting in dissociation of P from the surface of iron-oxide minerals (Nagarajah et al. 1970). It is known that polygalacturonic acid is secreted as highly methylesterified polymers, and the pectin esterase (pectin methylesterase, Fig. 9) which are present in the cell walls, cleave some of the methyl groups. Then, Ca^{2+} can bind to non-methylated galacturonic acid residues of pectin molecules (Carpita and Gibeaut 1993). Recently, Ishii (1997a) reported that pectins of potato tuber cell walls are acetylated, and the physical properties of the pectin of the walls of plant cells are affected by the presence of acetyl substituents. Pectin molecules are considered to be modified during cell growth to result in change of CEC of cell wall materials.

Otani and Ae (1996a, b) reported that groundnut had a superior ability to take up P from a soil with low P fertility compared with sorghum and soy bean, and that such an ability was not related to root development or production of root exudates capable of solubilizing iron- and aluminum-bound P. In a subsequent study, Ae and Otani (1997) found that root cell walls prepared from groundnut showed a higher P-solubilizing activity [P solubilized from inorganic P (Al-P, Fe-P)] than those from soybean or sorghum. Interestingly, they reported that P solubility is not related to anions (galacturonic acids) derived from pectic substances. These results suggested that the active component of the walls, except the pectic substances, was located on the root epidermal cell surfaces. Among the cell wall components, phenolic acids, some structural protein and/or lignin are considered to be candidate chemicals able to chelate with Fe^{3+} and/or Al^{3+}.

It was shown that ferulic acid exists in wheat root walls with lesser amounts of other components including diferulic and *p*-coumaric acid (Smith and O'Brien 1979). The presence of phenolic residues in the epidermal cell walls of all dicotyledons was suggested using UV-fluorescence microscopy (Fry and Miller 1989). But, there is no detailed information on this. We carried out a preliminary characterization of the cell walls of groundnut root cell walls. Sugar composition and sugar linkage composition analyses of the water-insoluble fraction of groundnut root cell walls suggested that the fraction consisted mainly of rhamnogalacturonan, arabinogalactan, arabinoxylan, xyloglucan and cellulose. Enzymic hydrolysis of the insoluble fraction followed by analysis of the fragments indicated that some kinds of phenolic acids including *p*-coumaric acid are bonded to wall polysaccharides by an ester linkage (Kato and Ae unpublished data). It is known that phenolic hydroxyl groups form chelate with ferric ion (Fe^{3+}). Therefore, a candidate chemical able to chelate Fe^{3+} could be material containing phenolic acid (not only the monomer, but also dimer and higher oligomers), which is located in cell walls to build material of three-dimensional structures. Another candidate is structural protein. However, the possibility of acidic polysaccharide taking this role is not negligible. In any case, to identify the active component present in groundnut root cell walls, it is necessary to isolate the component containing Fe^{3+} together with greater retention of structural integrity after treatment of groundnut cell walls with $FePO_4$. In this case, the dissociation of $FePO_4$-treated cell walls by enzymic hydrolysis has the potential to release objective components, as already mentioned. Such a study will provide a new perspective for studies on mineral nutrient acquisition by plant cell walls and for studies on additional functions of plant cell walls.

References

Ae N, Otani T (1997) The role of cell wall components from groundnut roots in solubilizing sparingly soluble phosphorus in low fertility soils. In: Ando T et al. (Eds) Plant nutrition for sustainable food production and environment. Kluwer Academic, Dordrecht, Netherlands, pp 309-314

Ae N, Arihara J, Okada K, Yoshihara J, Ohtani T, Johansen C (1993) The role of piscidic acid secreted by pigeonpea roots grown in an Alfisol with low-P fertility. In: Randall PJ (Ed) Genetic aspects of plant mineral nutrition. Kluwer Academic, Dordrecht, Netherlands, pp 279-288

Akiyama Y, Mori M, Kato K (1980) ^{13}C-NMR analysis of hydroxyproline arabinosides from *Nicotiana tabacum*. Agric Biol Chem 44:2487-2489

Carpita NC (1996) Structure and biogenesis of the cell walls of grasses. Annu Rev Plant Physiol Plant Mol Biol 47:445-476

Carpita NC, Gibeaut DM (1993) Structural models of primary cell walls in flowering plants: consistency of molecular structure with the physical properties of the walls during growth. Plant J 3:1-30

Darvill A, McNeil M, Albersheim P, Delmer DP (1980) The primary cell walls of flowering plants. In: Tolbert NE (Ed) The biochemistry of plants Vol 1. Academic, New York, pp 91-162

Fry SC (1983) Feruloyled pectins from the primary cell wall: their structure and possible functions. Planta 157:111-123

Fry SC (1985) Primary cell wall metabolism. In: Miflin BJ (Ed) Oxford surveys of plant molecular and cell biology 2. Oxford, Clarendon, pp 1-42

Fry SC (1986) Cross-linking of matrix polymers in the growing cell walls of angiosperms. Ann Rev Plant Physiol 37:165-186

Fry SC (1995) Polysaccharide-modifying enzymes in the plant cell wall. Ann Rev Plant Physiol Plant Mol Biol 46:497-520

Fry SC, Miller JG (1989) Toward a working model of the growing plant cell wall: phenolic cross-linking reaction in the primary plant cell walls of dicotyledons. In: Lewis NG, Paice MG (Eds) Plant cell wall polymers: biogenesis and biodegradation, ACS Symp Ser 399, Am Chem Soc, Washington DC, pp 33-46

Gardner WK, Barber DA, Parbery DG (1983) The acquisition of phosphorus by *Lupinus albus* L. III. The probable mechanism by which phosphorus movement in the soil/root interface is enhanced. Plant Soil 70:107-124

Geissmann T, Neukom H (1971) Vernetzung von Phnolcarbonsaureestern von Polysacchariden durch oxydative phenolishe Kupplung. Helv Chim Acta 54:1108-1112

Hartley RD, Ford CW (1989) Phenolic constituents of plant cell walls and wall biodegradability. In: Lewis NG and Paice MG (Eds) Plant cell wall polymers: biogenesis and biodegradation. ACS Symp Ser 399, Am Chem Soc, Washington DC, pp 137-149

Hayashi T (1989) Xyloglucans in the primary wall. Annu Rev Plant Physiol Plant Mol Biol 40:139-168

Hayashi T, Ohsumi C, Kato Y, Yamanouchi H, Toriyama K, Hinata K (1994) Effects of amino acid medium on cell aggregation in suspension-cultured rice cells. Biosci Biotech Biochem 58:256-260

Ishii T (1991) Isolation and characterization of a diferuloyl arabinoxylan hexasaccharide from bamboo shoot cell walls. Carbohydr Res 219:15-22

Ishii T (1997a) *O*-Acetylated oligosaccahrides from pectins of potato tuber cell walls. Plant Physiol 113:1265-1272

Ishii T (1997b) Structure and functions of feruloylated polysaccharides. Plant Sci 127:111-127

Ishii T, Hori T (1990) Linkage of phenolic acids to cell-wall polysaccharides of bamboo shoot. Carbohydr Res 206:297-310

Kamisaka S, Takeda S, Takahashi K, Shibata K (1990) Diferulic and ferulic acid in the cell wall of *Avena* coleoptiles-their relationships to mechanical properties of the cell wall. Physiol Plant 78:1-7

Kato Y (1993) Influence of water-insoluble dietary fiber on *in vitro* glucose diffusion speed (in Japanese). J Jpn Soc Nutr Food Sci 46:351-355

Kato Y, Akiyama M (1993) Gel-filtration chromatography of dextrans and maltodextrins on a water-insoluble dietary fiber column (in Japanese). J Jpn Soc Nutr Food Sci 46:161-166

Kato Y, Matsuda K (1976) Presence of a xyloglucan in the cell wall of *Phaseolus aureus* hypocotyls. Plant Cell Physiol 17:1185-1198

Kato Y, Matsuda K (1977) Distribution of xyloglucan in *Phaseolus aureus* seeds. Plant Cell Physiol 18:1089-1098

Kato Y, Matsuda K (1994) Examination of the fine structures of xyloglucans using endo-(1→4)-β-D-glucanases. Methods Carbohydr Chem 10:207-216

Kato Y, Mitsuishi Y (1999) Structural investigation of xyloglucan fragments obtained from cell walls of immature barley plants II (in Japanese). Nippon Nogeikagaku Kaishi 73:10

Kato Y, Nevins DJ (1984a) Enzymic dissociation of *Zea* shoot cell wall polysaccharides. I. Preliminary characterization of the water-insoluble fraction of *Zea* shoot cell walls. Plant Physiol 75:740-744

Kato Y, Nevins DJ (1984b) Enzymic dissociation of *Zea* shoot cell wall polysaccharides. IV. Dissociation of glucuronoarabinoxylan by purified endo-(1→4)-β-xylanase from *Bacillus subtilis*. Plant Physiol 75:759-765

Kato Y, Nevins DJ (1985) Isolation and identification of *O*-(5-*O*-feruloyl-α-L-arabinofuranosyl)-(1→3)-*O*-β-D-xylopyranosyl-(1→4)-D-xylopyranose as a component of *Zea* shoot cell-walls. Carbohydr Res 137:139-150

Kato Y, Nevins DJ (1986) Fine structure of (1→3),(1→4)-β-D-glucan from *Zea* shoot cell-walls. Carbohydr Res 147:69-85

Kato Y, Yamanouchi H, Hinata K, Ohsumi C, Hayashi T (1994) Involvement of phenolic esters in cell aggregation of suspension-cultured rice cells. Plant Physiol 104:147-152

Knight AH, Crooke WM, Inkson RHE (1961) Cation-exchange capacities of tissues of higher and lower plants and their related uronic acid contents. Nature 192:142-143

Konishi T, Mitsuishi Y, Kato Y (1997a) Analysis of the oligosaccharide units of xyloglucans by digestion with isoprimeverose-producing oligoxyloglucan hydrolase followed by anion-exchange chromatography. Biosci Biotechnol Biochem 62:2421-2424

Konishi T, Mitsuishi Y, Kato Y (1997b) Structural analysis of the Solanaceae xyloglucans using xyloglucanase hydrolysis method. Plant Physiol 114:83 (suppl)

Konishi T, Mitsuishi Y, Kato Y (1998) Compositional analysis of the oligosaccharide units of xyloglucans isolated from the cell walls of leaf and root vegetables. J Appl Glycosci 45:401-405

Kooiman P (1960) A method for the determination of amyloid in plant seeds. Recl Trav Chim Pays-Bas 79:675-678

McNeil M, Darvill AG, Fry SC, Albersheim P (1984) Structure and function of the primary cell walls of plants. Ann Rev Biochem 53:625-663

Nagarajah S, Posner AM, Quirk JP (1970) Competitive adsorptions of phosphate with polygalacturonate and other organic anions on kaolinite and oxide surfaces. Nature 228:83-84

O'Neill M, Albersheim P, Darvill A (1990) Pectic polysaccharides. Methods Carbohydr Chem 1:478-512

Otani T, Ae N (1996a) Sensitivity of phosphorus uptake to changes in root length and soil volume. Agron J 88:371-375

Otani T, Ae N (1996b) Phosphorus (P) uptake mechanisms of crops grown in soils with low P status. I. Screening crops for efficient P uptake. Soil Sci Plant Nutr 42:155-163

Ralph J, Grabber JH, Hatfield RD (1995) Lignin-ferulate cross-links in grasses: active incorporation of ferulate polysaccharide esters into ryegrass lignins. Carbohydr Res 275:167-178

Smith MM, O'Brien TP (1979) Distribution of autofluorescence and esterase and peroxidase activities in the epidermis of wheat roots. Aust J Plant Physiol 6:201-219

Takagi S (1976) Naturally occurring iron-chelating compounds in oat- and rice-root washing. I. Soil Sci Plant Nutri 22:423-433

Talmadge KW, Keegstra K, Bauer WD, Albersheim P (1973) The structure of plant cell walls. I. The macromolecular components of the walls of suspension-cultured sycamore cells with a detailed analysis of the pectic polysaccharides. Plant Physiol 51:158-173

Tepfer M, Taylor IEP (1981) The permeability of plant cell walls as measured by gel filtration. Science 213:761-763

Toriyama K, Hinata K (1985) Cell suspension and protoplast culture in rice. Plant Sci 41:179-183

Part IV.
Contribution of Soil Microorganisms and Soil Fauna

Phosphate-Solubilizing Microorganisms and Their Use

Mary Leggett, Sanford Gleddie, and Greg Holloway

Summary. The importance of phosphate (P) in plant nutrition cannot be underestimated. Its role in energy transfer is critical to plant reproduction and productivity. Many soil microorganisms make P more available to plants by solubilizing otherwise inaccessible forms of bound P. Although hundreds of P solubilizers have been studied, only two have been developed into commercial phosphate inoculants: *Bacillus megatherium* var. *phosphaticum* (Phosphobacterin) and *Penicillium bilaii* (Provide).

The lack of available phosphate inoculants is a result of the many biological, agronomic, commercial, and marketplace criteria a P solubilizer must meet to become commercially successful. The biological reality is that the organism, plant, and environment all interact to produce an effect on plant growth and yield. The P solubilizer must effectively colonize the rhizosphere and function in a broad range of environments. The agronomic reality is that the P solubilizer must perform under the full range of conditions seen by the farmer and must demonstrate a contribution to increased yield. This requires multi-year, multi-site field trials, an evaluation of the types of P solubilized, and a study of the range of crops for which an organism is effective. The commercial reality is that the P solubilizer must be amenable to a cost-effective production process, genetically stable, viable in formulation, and efficacious at a reasonable application rate and cost. The marketplace reality is that the organisms must be compatible with current farming practices (e.g. other inoculants and chemicals) and of proven economic benefit to the farmer. A well planned research program addressing all these areas is necessary to develop a commercial phosphate inoculant.

Key words. Commercial, Inoculant, *Penicillium*, Phosphate, Plant growth promotion, Rhizobium, Solubilizer

1. Introduction

Phosphate (P) is an essential nutrient for plant growth. It functions as an agent of energy transfer, and a deficiency of available P is more likely to limit reproduction and productivity than any other material, except water (Stewart and McKercher 1982). Application of P fertilizer has been considered essential for crop production since the mid 1800s, and the need for readily available P is common to most plants.

In the developed world, chemically produced water-soluble P fertilizers are routinely applied to crops. The P in these fertilizers is initially available for plant use, then rapidly reacts with soil and becomes progressively less available for plant uptake (Kucey et al. 1989). The use efficiency of applied P fertilizer in the year of application ranges from 10% to 25% (Spinks and Barber 1947). In the developing world, where chemical fertilizers are not available or are too expensive, ground rock P offers a less costly option (Nahas 1996). This form of P is much less available to plants than standard fertilizers, except in acidic soils.

Many soil microorganisms are able to solubilize otherwise unavailable forms of bound P. These organisms are found in most soils but the numbers and types will vary with the soil. The ability of microorganisms to solubilize P in soil, and make it available for plant use was first demonstrated by Gerretsen (1948). His experiments showed that organisms in the rhizosphere affected the uptake of P by plants. In the 50 years since this discovery, researchers world-wide have worked to isolate P-solubilizing organisms, study their characteristics, and use them in inoculants to make P more readily available in commercial agriculture.

The first step is usually to perform a qualitative screen. The organisms are placed on agar plates containing insoluble P and a cleared zone around the colony is used to indicate P-solubilizing activity. The number of organisms which give positive results using this type of screen ranges from 0.5%–1% (Kucey et al. 1989) to 26%–39% (Sperber 1958a) of the total population. This normally will still leave the researcher with too many organisms to test in greenhouse or field studies and a quantitative screen may then be used to rank the organisms. A quantitative screen may be performed by growth in liquid culture in the presence of insoluble P subsequently, quantify the solubilized P in the supernatant by ICP or HPLC. While these tests may help to indicate potential useful inoculant organisms, they may not correlate with field response. The organism must be able to grow and remain active in the rhizosphere. Kucey found that in his qualitative screen, bacteria out numbered fungi but many bacteria lost this ability after continuous culture and the quantitative screen showed that fungi had greater P-solubilizing activity than bacteria (Kucey 1983).

Despite the attention that has been paid to the study of P-solubilizing microorganisms, few have been developed as commercial inoculants. In the 1950's, Russian workers developed a product containing *Bacillus megatherium* var. *phosphaticum* known as Phosphobacterin. Approximately 10 million hectares were treated with the product in Russia in 1958 (Brown 1974). Reports indicated that approximately 50%–70% of the field crops inoculated with Phosphobacterin benefited from the treatment. Yield increases from 10 to 70 percent were reported (Smith et al. 1961). The best results were seen on vegetable crops, but good responses were also obtained on potatoes and cereals (Smith et al. 1961). Researchers in other countries tried Phosphobacterin with mixed results. The original data from the USSR were questioned as the experiments were not statistically designed or analyzed (Smith et al. 1961). The use of Phosphobacterin in North America was essentially terminated following these reports (Kucey et al. 1989). Phosphobacterin was eventually found to solubilize organic P in soil (Smith et al.

1961). This solubilization of organic instead or inorganic P could be viewed as detrimental to agriculture as the organic matter serves an important role as a soil structural component (Kucey et al. 1989).

The work on P-solubilizing organisms continued, but no commercial inoculants were produced until 1990 when Philom Bios brought Provide® to the marketplace in Western Canada. Provide contains spores of the fungus *Penicillium bilaii* (ATCC 20851, formerly reported as *P. bilaii*). The organism was first isolated and tested at the Agriculture and Agri-Food Canada (AAFC) research station in Lethbridge in 1983 (Kucey 1983). Provide is now registered under the Canadian Feed and Fertilizers act for use on various agricultural and horticultural crops.

Research continues as scientists in different areas of the world search for organisms which suit their conditions. Jones et al. (1991) looked for isolates capable of solubilizing the iron and aluminum P minerals found in Scottish soils. Australian workers screened for organisms that would be effective in the acidic soils of the wheat-belt of New South Wales (Whitelaw et al. 1997; Bender 1998). African researchers look for organisms that fit the agricultural practices in the local area (Mba 1994)

The abundance of research has not produced an abundance of commercial inoculants. In 35 papers dealing with organisms other than *P. bilaii*, more than 60 different microbe species have been reported to solubilize P. Thirteen of the reports only described laboratory experiments, only eight studies tested organisms on plants in sterile soil or sand, eight more studies described the use of organisms on plants in non-sterile soil, and only six papers described field trials. It is difficult to tell from the literature whether the organisms were abandoned at this point, or whether greenhouse and field tests were done and the results were not reported due to negative or ambiguous results.

It may not be unrealistic to expect to screen a large number of organisms in order to produce one or two inoculants. The chemical industry screens hundreds of compounds to identify one commercial product. While this is true, we can still make gains in efficiency by understanding the organism, its behavior, and the factors that affect its use. At the very least, we must increase the number of organisms tested and evaluated in field trials for their commercial capability.

Laboratory studies may indicate an organism's potential as a commercial inoculant, but they should only be viewed as simple screens. Laboratory studies can show that the microorganism is capable of solubilizing P, but not that it will function effectively with plants in the soil. Greenhouse trials in sterile soil or sand include the plant but ignore other microorganisms that exist in the soil and rhizosphere. Even greenhouse tests in non-sterile soil may not accurately predict field response. The roots explore more of the available soil volume in pots than they could in the field. Consequently, if soil P is solubilized by an inoculated plant the response in pots will be greater than in the field (Kucey et al. 1989). Small plot field trials done at one location are done under realistic conditions, but they may still not be enough to be used as the final means of selecting an organism that can be used in a commercial inoculant. Multiple-site, multiple-year, and multiple-crop trials are necessary to truly determine the efficacy of a P-solubilizing organism in the field.

Studies on P-solubilizing organisms often fail to result in a commercial inoculant because the authors ignore the biological, agronomic, commercial or marketplace realities involved in inoculant production and use. In this paper we use examples from published literature and our experiences in bringing Provide to market to identify some of the reasons the large amount of research has not resulted in several widely used P solubilizing inoculants. As the topic is broad, we will restrict the discussion to non-symbiotic organisms that actively solubilize soil P. The discussion will identify some of the measures that can be taken to identify good potential organisms for future development as well as the limitations of potential inoculants and some of the practical research which must be done to successfully commercialize and market an inoculant.

2. Biological Realities

The biology of the inoculant organism and its reaction with the environment is complex. The organism, the plant, and the environment all interact to produce an effect on plant growth and crop yield. Lack of a full understanding of this system does not necessarily prevent commercialization, but the chances of success will be improved by increased knowledge of the organism, and of the factors that affect its ability to improve plant growth and development.

2.1 Bulk Soil Versus Rhizosphere Soil

Organisms intended for use as P-solubilizing inoculants must be present in sufficient numbers in the rhizosphere of crop plants to exert their effect. It is the rhizosphere, not the bulk soil, where their activity is most likely to improve plant growth (Brown 1974). The effect of the introduced organism is usually related to the size of the population (Kundu and Guar 1980). To ensure that enough organisms are distributed over the root, the inoculant organism must colonize the root surface and function in the rhizosphere. Mixing the inoculant in with the bulk soil often results in failure to observe a significant effect on plant growth as the bulk soil cannot support large numbers of organisms. Lee and Bagyaraj (1986) found that the population of bacteria added to soil at a rate of 1.21×10^9 cells/g dry soil, declined rapidly. They proposed that selection of organisms that could survive in the bulk soil would solve this problem, however soil cannot be expected to sustain a level of organisms higher than exists in nature. Normally bulk soil contains only 10^5 to 10^7 culturable organisms/g dry soil. Soil is a nutrient-poor medium in which competition for limited carbon substrates is intense (Carlile and Watkinson 1994). On the other hand, the rhizosphere will support larger populations due to a continual supply of readily assimilable organic substances from the root (Lynch and Whipps 1991). The inoculant organism should be introduced in a way that will increase the probability of the organism reaching and colonizing the plant root. As organisms can only move a few centimeters in soil, inoculants should either be applied directly on seed where they will colonize the root as it grows, or on granules placed so that the growing root will intercept the added organisms.

2.2 Colonization of the Rhizosphere

Once the organism is established in the rhizosphere it must be able to continue to colonize the growing root. The ability of organisms to colonize roots is affected by the biology of the organism, its competitiveness with other microbes, the characteristics of the plant root growth, and the physical characteristics of the soil and the environment.

The success of a rhizosphere microorganism is enhanced by a rapid growth rate on a wide range of common exudates, the production of antibiotics or a tolerance to the products of other microorganisms (Bowen 1991). It is difficult to predict the rhizosphere competence of an organism based on taxonomic information. Good rhizosphere colonizers may come from different taxonomic groups, and their competitive ability can differ between strains (Milus and Rothrock 1993). Different species of bacteria and fungi have differing abilities to colonize the root surface. Hozore and Alexander (1991) found that bacteria isolated from the rhizosphere could move further down the root from the point of inoculation than bacteria isolated from the bulk soil. Fungi, with larger nutrient reserves, can move to the root and along the root over larger distances (Bowen 1991), but not all fungi are able to effectively colonize the rhizosphere (Baker 1991). Baker's experiments with a mutant of a *Trichoderma* sp. demonstrated that this ability was related to cellulose production.

The tolerance of organisms to varying levels of water activity affects the ability of the inoculant organism to colonize roots. Water availability affects the activity of both fungi and bacteria by affecting gas exchange, affecting dissolution of solutes, influencing the motility of the organisms and, affecting the availability of water for growth and metabolism (Cook and Baker 1984). The sensitivity to variation in moisture varies with species. Generally bacteria need higher moisture levels than fungi. Within each group different genera have different moisture tolerances. Gram negative bacteria such as Pseudomonads have higher water potential thresholds of slightly below –0.7 MPa. Gram positive bacteria such as *Bacillus* spp can survive water potentials of –2.5 MPa (Cook and Baker 1984), while fungi such as *Aspergillus* spp. and *Penicillium* spp. are osmotolerant and can grow at water potentials between –30 to –69 MPa (Carlile and Watkinson 1994).

The soil, the environment, and the plant also influence the ability of inoculant organisms to colonize roots. Bacteria move along the root surface by increasing their population size and by movement with or through water. The movement of bacteria along the root is improved by the downward flow of water, and irrigation may therefore help their spread (Bowen 1991). Fungi do not require water movement for transport but are able to grow along the root from the point of inoculation. The growth of the roots of the inoculated plant also determine the extent of root colonization by the inoculant organism. Bacteria and fungal spores can be carried passively by the root tip. The relative growth rate of the root and bacteria will determine the extent of the root colonization. A slow-growing organism combined with a fast-growing root will result in low population on the root due to a dilution effect (Bowen 1991).

The rhizosphere offers a more nutrient rich environment for the growth of the inoculant than bulk soil, however it also contains higher populations of other microorganisms that compete with the inoculant for space and nutrients. A good inoculant must be able to colonize roots in non-sterile soil and laboratory experiments may not accurately assess the competitiveness of an inoculant in the rhizosphere. *Penicillium radicum* (Whitelaw et al. 1997) and *P. bilaii* (Levesque et al. 1997) are able to colonize roots in non-sterile soil, and both have improved plant growth in field trials. Because of a lack of competition from indigenous microorganisms inoculated organisms will almost always reach higher populations in sterile compared to non-sterile soil (Kundu and Guar 1980). Studies comparing P-solubilization is sterile versus non-sterile soil have produced mixed results. In some cases the inoculant was more effective in sterile soils (Kundu and Guar 1980) while in others the ability of the inoculant to solubilize P was greater in non-sterile compared to sterile soil. (Banik and Dey 1982). Young (1990) found different responses to sterilization in different soils. Soils may contain large indigenous populations of P-solubilizing microorganisms that could mask the effect of the introduced organism in non-sterile soil (Thomas et al. 1985). Other microbes in the soil may supply the inoculant organisms with important growth factors (Banik and Dey 1982).

Sterilization of soil will also eliminate the mycorrhizal fungi in soil. Because of the importance of mycorrhizal fungi to P nutrition many researchers add a mycorrhizal inoculant to sterilized soil to study the interaction of the two organisms. The effect of the non-mycorrhizal P-solubilizing inoculant can be more effective (Barber et al. 1976; Kucey 1987; Raj et al. 1981; Young 1990), have no effect (Asea et al. 1988), or be less effective (Leyval and Bertelin 1989) when used in combination with endo or ectomycorrhiza. Toro et al. (1996) found that the effect of mycorrhiza depended on both the species of the P-solubilizer and the species of the mycorrhizal fungus. Kucey (1987) proposed that vesicular arbuscular mycorrhizae (VAM) serve to increase the ability of the plant root to absorb the P that has been solubilized by *P. bilaii*.

Soil sterilization can also influence the results of P-solubilization experiments through affects on the physiology of the host plant. Soil sterilization alters the nature of the root exudates and therefore rhizosphere colonization. Carbon loss from roots of plants grown in sterile soil is less than the carbon loss from plants grown in non-sterile soil (Tinker 1984). These changes in the host affect the ability of the inoculant organism to colonize the rhizosphere.

2.3 Function in the Rhizosphere

Once the organism establishes itself in the rhizosphere, it must be able to function effectively by continually solubilizing P, making it available for the plant. Different organisms have different abilities to solubilize P (Thomas et al. 1985; Banik and Dey 1982), and different organisms may function differently under different conditions. The most efficient P-solubilizers in Egyptians soils were aerobic spore formers, with the most abundant being *B. megatherium* (Taha et al. 1969). Scottish results using isolates from maize showed that the most active bacterial P-solubilizers were motile, short, gram-negative rods (Duff et al. 1969). P-solubilizing fungi are usually either *Aspergillus* spp. or *Penicillium* spp. (Kucey 1983; Thomas et al. 1985) although a wide range of fungal genera have solubilizing activity (Thomas et al. 1985; Jones et al. 1991).

Most of the organisms chosen for inoculant studies have been bacteria, although fungi are often more effective P solubilizers (Nahas 1996). In liquid culture studies Kucey (1983) found that fungi solubilized an average of ten times more P than the bacteria. Thomas et al. (1985) found that *Aspergillus* spp. solubilized more P than *Penicillium* spp., while experiments by Salih et al. (1989) found that a *Penicillium* sp. was more efficient at solubilizing rock P than the two *Aspergillus* spp tested. These seemingly conflicting results are not surprising. Organisms have adapted over the centuries to survive, grow, and function in their local conditions. Taha et al. (1969) found that a local Egyptian strain of *B. megatherium* was more effective at increasing plant growth and P uptake in Egyptian soils than the strain from the USSR. *Penicillium bilaii* was isolated from soil in Western Canada and was selected, in part for its ability to grow and solubilize P at low temperatures. These traits may contribute to its success as an inoculant in Western Canada where soils are cool at the time of seeding.

Plant root exudates exert a strong selective effect on rhizosphere organisms, as organisms in the rhizosphere are more physiologically active than organisms in the bulk soil. Again we find that laboratory experiments do not necessarily evaluate the organisms under conditions that occur in nature. Solutions of apatite used in the laboratory experiments usually contain high levels of carbohydrates. High levels of carbohydrates are more likely to occur in the rhizosphere than in soil (Sperber 1958a), but it is difficult to know how closely the type and concentrations of carbohydrates used in laboratory experiments mimic the variety of nutrients that may exist in the root zone. Illmer and Schinner (1992) found that P-solubilization in liquid culture was higher in medium low in C. They observed that this is not unexpected as suboptimal conditions for growth and biomass production may be optimal for the production of metabolites which be responsible for P-solubilization (Illmer and Schinner 1992). However, minimum threshold levels of C must be present to fuel the process of P-solubilization. Goldstein (1995) proposed that, at least in gram negative bacteria, a supply of carbon is necessary as the bacteria use a direct oxidation pathway to essentially trade reduced carbon for P. Root exudates should provide a large enough nutrient supply for this action.

Organisms to be used as inoculants must function effectively in a broad range of environments. To be commercially acceptable there must be a broad enough market to justify the amount of research involved. Different organisms have different optimal conditions for their activity. Asea et al. (1988) compared the effect of ammonium ions on the P-solubilizing activity of two *Penicillium* species in liquid culture. Greater amounts of P were solubilized in solution culture when ammonium was added to the medium. *Penicillium bilaii* lowered the pH in the medium whether or not ammonium was added but *P. fuscum* only lowered pH in the presence of ammonium. Phosphate-solubilizing activity was then related to acid production. Very little oxalic acid was produced under nitrogen limited conditions while citric acid was produced in both carbon limited and nitrogen limited conditions

(Cunningham and Kuiack 1992). Again we must remember that studies done in liquid culture do not necessarily reflect the solubilization that occurs on roots and in soil (Cunningham and Kuiack 1992). Bajpai and Sundara Rao (1971) found that each of the four organisms they studied, (*Bacillus megatherium* var. *phosphaticum* (Phosphobacterin), *B. megatherium* (Indian Strain), *B. circulans*, and *Escherichia freundi*) each had a different optimal pH for maximum solubilization. The soils that gave the best result from the use of Phosphobacterin are those that were neutral to alkaline and had a high level of organic matter (Smith et al. 1961). In most cases the greatest effects of P-solubilization are seen in calcareous soils, but Young (1990) found that their P-solubilizing bacteria were most effective in acid soils. Kundu and Guar (1980) found that the addition of organic manure provided an improved environment for proliferation of introduced bacteria by improving the chemical and physical properties of the soil. Latour et al. (1996) discovered that the populations of different introduced fluorescent pseudomonads strains differed with soil type and with temperature.

3. Agronomic Realities

A potential inoculant organism, which has been shown to colonize the rhizosphere and solubilize P in the laboratory and greenhouse must be tested under the full range of conditions that will be experienced by farmers. This research must be proactive and conducted with an eye towards changes in agronomic practice five to ten years in the future (Leggett and Gleddie 1995). Field trials must be established across a broad range of soil and environmental conditions, and must be conducted within the context of current and/or future farming practices.

Interpretation of the results of P-solubilization field research requires careful attention. Yield increases after inoculation with an organism that solubilizes P in liquid culture are not necessarily caused by P-solubilization in soil. Microorganisms capable of producing acids, such as citric or gluconic acid are capable of dissolving relatively insoluble P-compounds (Goldstein 1995). Many researchers think that P-solubilization around the root is likely to be too small to have a significant effect on the plant (Tinker and Sanders 1975). Tinker and Sanders (1975) feel it is unrealistic to expect a relatively small number of microbes to alter the chemistry of the bulk soil. They feel that the acid production and subsequent P-solubilization around the root will likely to be too small to have a significant effect on the plant. The solubilization may, however, occur as a micro-site effect at the root surface. Some researchers believe that yield increases after inoculation with P-solubilizing organisms may in reality be increasing plant growth by other mechanisms (Brown 1974). Some organisms may affect root growth by producing plant hormones. However, others have ruled out this mechanism by testing their organisms in the laboratory to see if they produced known hormones (Raj et al. 1981). Microorganisms may increase the absorption of P of seedlings by producing substances that are capable of stimulating absorptive processes (Barber et al. 1976). Organisms inoculated into the rhizosphere may modify the nature or amount of root exudates and, as these exudates can affect P uptake, the inoculant may improve P-nutrition indirectly (Laheurte et al. 1990). Given the complexity of natural systems we should not necessarily expect an organism to affect plant growth by only one mechanism. Multiple modes of action may be the rule rather than the exception. Chabot et al. (1993) attribute yield increases in maize and lettuce in their field studies to a combination of P-solubilization, siderophore and auxin production.

The contribution of P-solubilization to yield increases can be shown by measuring plant P-uptake using ^{32}P labeling to trace the transfer of soil P into the plant and by studying the effects of the inoculant compared to applications of fertilizer P. Experiments with *P. bilaii* have used all three methods to prove that P. solubilization is the major factor contributing to yield increase. Whitelaw et al. (1997) found that *P. radicum* solubilized P in solution culture, however, in field trials *P. radicum* improved plant growth regardless of P level, indicating that some other mechanism of growth promotion may have been acting. Understanding the mode or modes of action for the organism can help predict the success or failure of an inoculant in any given situation.

3.1 Soil Factors

To establish the effectiveness of an organism in a range of soil types across the normal range of environmental conditions it is essential to conduct multiple-site, multiple year field trials. Locations must be chosen to represent the range of environments in which the product will be marketed (Leggett and Gleddie 1995). We have tested Provide in soils across the Canadian prairies and have found it to be effective in all of the major soil groups of Western Canada. We do not recommend its use in extremely sandy soils (> 85% sand), soils with extremely high organic matter (> 14% organic matter) or fields that have been heavily manured over the last several years. Most of the P in these soils is in organic forms, and organisms that solubilize inorganic forms of P such as *P. bilaii* will have little or no effect. We do not recommend its use in soils that are high in available P. Increasing P availability in these soils will have little or no effect on plant P uptake and growth.

3.2 Type and Availability of Phosphate

The availability of P in soils in addition to the types of insoluble P in soil influence the effectiveness of P-solubilizing organisms. Nahas (1996) and Jones et al. (1991) found that fungi that actively solubilized iron phosphates were ineffective in solubilizing calcium phosphates. Banik and Dey (1982) found that solubilization of P by their organisms was highest for $Ca_3(PO_4)$ followed by $AlPO_4$ and $FePO_4$. In most cases the organisms they tested solubilized aluminum and iron phosphates to the same extent, except for the *Aspergillus* spp. which solubilized aluminum phosphates to a greater extent than the iron phosphates. Agnihotri (1970) tested 18 different fungal isolates in liquid culture and found that 14 solubilized tricalcium phosphate, 11 solubilized fluorapatite, 7 solubilized hydroxyapatite and 6 were able to solubilize all 3 types of P minerals. Whitelaw et al. (1997) found that *P. radicum* solubilized significant amounts of calcium and colloidal aluminum phosphate in vitro. Our experiments with *P. bilaii* in the laboratory have shown solubilization of various P minerals. Five different phosphate minerals have been tested: Idaho Rock P, Varasite, Vivianite, Lazulite, and Augulaite. *Penicillium bilaii* was grown in flasks containing nutrient medium with and without added P. The flasks were incubated at 30° C for 5 days, and then the culture filtrate was assayed for phosphorus by ICP and HPLC. In the first of two experiments only the Idaho Rock P and the Vivianite were solubilized, while in the second experiment Varasite was solubilized also. These experiments need to be repeated before any conclusions can be made. We also need to conduct extensive field trials in areas where iron and aluminum forms of P predominate before we can recommend the use of *P. bilaii* in these locations. We also need to verify that the laboratory assays are good screens for the ability *P. bilaii* to solubilize aluminum and iron P-minerals in soil.

Organisms that act by solubilizing inorganic P and making it available to plants cannot be expected to increase plant P uptake or growth if the level of available P in the soil is already at the maximum for plant growth. Many greenhouse and field studies do not take this into account when evaluating a lack of response to the addition of the inoculant. In soils with high available P, lack of increased plant P uptake or yield due to the addition of the inoculant should not be viewed as an indication of the failure of the inoculant (Table 1). Goos et al. (1994) found that *P. bilaii* did not increase yield of wheat in three of four sites. However these sites did not respond to P fertilization. In the one site where the yield data showed a response to P fertilization, a yield response to *P. bilaii* was also observed. Goos et al. (1994) proposed that the lack of response to inoculation with *P. bilaii* might be due to inadequate initial inoculum levels, competition with the native soil microflora, soil type or level of soil ammonium.

Work in Australia with *P. radicum* has revealed that the effect of the inoculant is not linked to application of P fertilizer (Fig. 1), and the authors speculate that another mode of action may be involved (Bender, 1998). However the interaction between P availability and the response to inoculation with *P. radicum* cannot be evaluated as results from only one field trial have been reported.

3.3 Crop

To be commercially viable the P-solubilizing organism should be able to grow and function on the roots of different crop species. Crucifers exude high concentrations of compounds that can inhibit the growth of microorganisms, and this is known to prevent VAM infection of crops such as canola (Gleddie 1993; Kucey and Leggett 1989). *Penicillium bilaii* effectively colonizes roots of both canola and mustard (Table 2) and increases P uptake and yield of these crops in the field (Kucey and Leggett 1989). Populations of fluorescent pseudomonads can differ in their ability to use various organic compounds, suggesting that plant species support specific populations of these bacteria (Latour et al. 1996). Chanway and Nelson (1991) found that the relationship between the plant and the bacterial population was extremely specific, as the effects of an introduced plant growth promoting rhizobacteria (PGPR) were restricted to specific wheat varieties. Organisms with this degree of specificity do not make good inoculants. We have shown that *P.*

Table 1. Effect of soil available phosphate and phosphate fertilizer on the effectiveness of *Penicillium bilaii* on wheat and canola in Western Canada

Phosphate fertilizer applied (Kg /ha P_2O_5)	Yield difference between treatments inoculated with *P. bilaii* and the P. fertilizer control (kg ha^{-1})*			
	Wheat[a]		Canola[b]	
	Low to medium phosphate soils *<20 lbs available P / acre*	High phosphate soils *>20 lbs available P / acre*	P responsive soils[c]	P non-responsive soils
0	65*	(2) ns	136*	(15) ns
10	66*	18 ns	124*	49 ns
20	43*	(15) ns	77*	39 ns
30	(16) ns	30 ns	N/T	N/T

ns, non-significant difference; N/T, not tested.
*Significant difference at $P<0.05$
[a]Adapted from Gleddie et al. (1991)
[b]Adapted from Gleddie et al. (1993)
[c]P responsive is defined by a response to applied P fertilizer

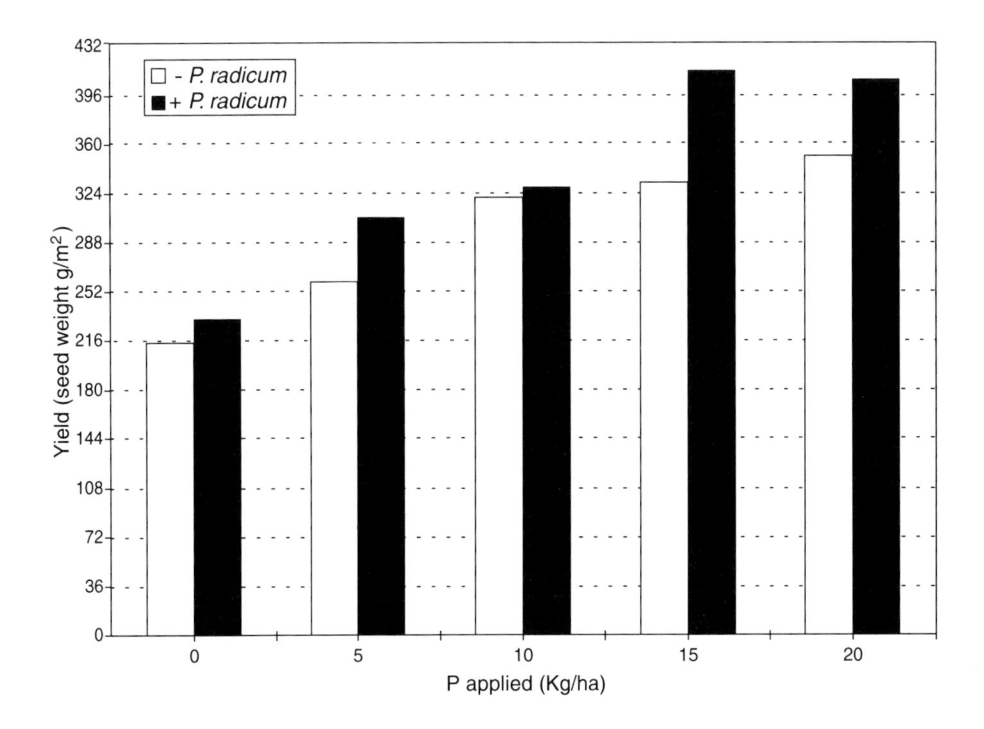

Fig. 1. Effect of presence (+) or absence (-) of *Penicillium radicum* inoculation on yield and protein levels (g/m²) in Dollarbird wheat under field conditions (adapted from Whitelaw et al. 1997). Horizontal grids spaced at LSD (P=0.05) of 36 for *P. radicum*

bilaii can actively colonize the roots of plants from many different genera. In house assays, where sections of root from plants grown in sterile soil are placed on an agar containing insoluble phosphate demonstrated that *P. bilaii* can colonize root in sterile soil. A positive response was scored if *P. bilaii* was observed growing from the section and if there was a clearing of the agar. These results on some of the crops in both sterile and non sterile soil using a DNA probe developed at Agriculture and Agri-Food Canada in British Columbia (Levesque et al. 1997; Table 2). Greenhouse tests and field tests showed that the organism was effective in increasing P uptake of many crops (Table 3).

4. Commercial Realities

An organism that is effective in increasing yield or P uptake in the field may still not result in a commercial inoculant. Often organisms fail to be commercialized for want of a cost effective production process. An organism that will be used in a commercial inoculant must be able to grow rapidly in commonly available inexpensive media, be genetically stable, retain viability in a commercial formulation, and it must be efficacious at an application rate that can be produced, packaged, and sold at a reasonable cost. Many researchers work for years on an inappropriate organism because they ignore or are unaware of these commercial realities.

4.1 Genetic Stability

Organisms that are very effective immediately after isolation from soil or roots often lose this characteristic in continuous culture (Ilmer and Schinner 1992; Kucey 1983; Sperber 1958b). This could be due to a loss in P solubilizing function or a reduction in root colonization ability (Iswandi et al. 1987). In Sperber's (1958a) studies of soil and root bacteria, thirty to forty percent of rhizosphere organisms and fifty percent of those isolated from soil failed to solubilize P after only the

Table 2. Crop plants colonized by *Penicillium bilaii* in sterile and non-sterile soil as determined by growth on agar and DNA probe assays

Agar plate tests (sterile soil)[a]	DNA probe (sterile)	DNA probe non-terile[b]
Wheat Pea	Wheat	Wheat
Lenti Canola	Pea	Pea
Mustard Barley	Lentil	Lentil
Alfalfa Corn	Canola	Canola
Flax Jack pine	Alfalfa	Alfalfa
Douglas fir	Barley	Mustard
White spruce		
Black spruce Soybean		
Barley Rye		

[a]E. Prusinkiewicz and B. Willie, unpublished results
[b]Levesque et al. (1997)

Table 3. Greenhouse and field results of experiments with Provide on different crops

Crop	Type of	Number of field experiment	Response to inoculation with trial locations	Reference *P. bilaii*
Wheat	Greenhouse		Enhanced P uptake and increased dry matter at early stages of growth	Chambers and Yeomans (1990)
Wheat	Greenhouse		Increased dry matter and P uptake	Asea et al. (1988)
Wheat	Greenhouse		Increased dry matter and P uptake	Kucey (1987)
Wheat	Greenhouse		No yield effects Radioisotope studies showed increased soil P uptake	Chambers and Yeomans (1991)
Wheat	Field	4 Sites North Dakota	Increased grain yield in the one site that showed a response to P fertilization No response at the other three sites	Goos et al. (1994)
Wheat	Field	3 Sites in Manitoba	Dry matter yield increased	Chambers and Yeomans (1990)
Wheat	Field	55 Sites Western Canada	Increased grain yield	Gleddie et al. (1991)
Bean	Greenhouse		Increased dry matter and P uptake	Kucey (1987)
Canola	Greenhouse		Increased P uptake, but not dry matter	Kucey and Leggett (1989)
Canola	Field		Increased grain yield with Rock P in field	
Canola	Field	2 Sites Western Canada	Increased grain yield, dry matter, and P uptake	Kucey and Leggett (1989)
Canola	Field	22 Sites Western Canada	Increased grain yield and P uptake	Gleddie et al. (1993)
Pea	Greenhouse		Increased dry matter, nitrogen uptake, and P uptake	Gleddie (1992)
Pea / Lentil	Field	19 Sites Western Canada	Increase in growth, P uptake, N uptake, and grain yield	Gleddie (1993)
Alfalfa	Field	10 Sites Western Canada	Increased dry matter and P uptake at three trifoliate leaf stage Increased forage yields	Schlechte et al. (1996)

second subculturing. This loss of solubilizing ability is a serious concern for an inoculant, but not every organism will exhibit this effect. Kucey (1983) found that 67 of 68 fungi and only 23 of 44 bacteria retained their ability to solubilize P after subculturing. We have also studied the genetic stability of *P. bilaii*, monitoring its growth and acid production over 10 generations. The fungus was inoculated onto agar plates containing Potato Dextrose Agar (PDA) and Alizarin Red as the acid indicator. Fifty-five replicate plates per generation were assessed for colony diameter and diameter of the acidified zone at 96 h. *Penicillium bilaii* was found to be stable over ten generations as measured by growth and acid production (Table 4).

4.2 Production

Most of the researchers reporting on the effectiveness of P solubilizers do not study the growth characteristics of the organisms, or evaluate the feasibility of producing organisms in a commercially acceptable large scale process. Inoculant companies must find an economical way to produce a large amount of inoculum with minimal contamination (Leggett and Gleddie 1995). Often researchers assume that production should be easy because the organisms can be grown on readily available and inexpensive materials in shake flasks. This is not necessarily true, as scale-up of production processes is neither usually straightforward nor easy.

Some researchers have proposed that as production of inoculum in the laboratory is relatively simple, they should be able to establish a cottage industry to produce inoculants. This type of production would make inoculants readily available at the local level in developing countries where farmers cannot afford higher priced inoculants (Mba 1994). However, many pitfalls could prevent the efficient production of effective inoculum and the plan should be carefully evaluated before local production is established. The government or other parties interested in developing the system must implement measures to ensure that the quality is maintained. The culture must be kept pure, the titre must be maintained at an acceptable level and the contamination level must be kept low. This becomes more difficult as more people in diverse areas are involved in production. In the industrialized world, where farmers want easy to use packaged products, production must be done on a large scale. The formulation must be economical to produce, stable in storage, safe, easy to use, and effective.

The organism must be present in sufficient numbers in the formulation to be effective under most field conditions. The effectiveness of an inoculant is affected

Table 4. Stability of *Penicillium bilaii* over 10 generations as measured by fungus growth and acid production[a]

Generation	1	2	3	4	5	6	7	8	9	10
Colony size (mm)	1.38	1.38	1.39	1.41	1.38	1.39	1.38	1.40	1.38	1.38
Zone size (incl. colony) (mm)	3.23	3.15	3.01	3.22	3.21	3.21	3.49	3.38	3.32	3.33

[a]Chris Lindsay, unpublished results

by the rate of application. Organisms must be added to the seed or soil in high enough numbers to ensure that they can colonize plant roots even under less than ideal conditions. Inoculation should be managed so that the survival of the inoculant on the seed is maximized during the period between its application to soil and the development of a rhizosphere which it can colonize (Brockwell et al. 1995). We would amend Brockwell's diagram to include survival on seed prior to seeding. The required population on seed can be achieved by increasing the numbers of organisms applied or by increasing the survival of the organism on seed (Fig. 1).

As increasing the number of organisms is not always economical, inoculant companies try to increase the stability on seed though formulation changes or application recommendations. The survival of any inoculant organism is affected by the formulation, the method of application, storage conditions, or other materials co-applied on the seed.

4.3 Formulation

The effectiveness of the organisms must be tested using the formulation that will be used in the marketplace. In the original research on *P. bilaii*, a bran-based inoculum was developed and added to soil (Kucey 1987). This material could not be produced on a large scale, and was extremely dusty and unpleasant to use. Continued formulation research has led to two seed-applied formulations for agricultural use, and a granular material for horticultural use. These formulations are economical to manufacture and are relatively easy to use. We continue to invest in research to improve the ease-of-use and handling characteristics of our inoculants.

The connection between the production process and the performance of the organism must be recognized. Any major changes in the production process may result in physiological changes in the organism, which may translate into altered shelf life and a change in efficacy in the field. Several researchers working with potential biocontrol organisms have noted links between growth conditions and efficacy (Jackson et al. 1992). Growth conditions that are optimal for yield of the organism are not always optimal for efficacy.

5. Marketplace Realities

Prior to marketing the P-solubilizing microbial inoculant the benefit to the farmer must be clearly demonstrated and the conditions for the use of the organism must be clearly defined. An inoculant that is applied as a seed treatment must be compatible with other organisms (such as rhizobia) and compounds (such as seed applied pesticides) that are added to the seed. Several researchers have evaluated combining P-solubilizing with other inoculant organisms. Kundu and Guar (1980, 1984) obtained the best results by using combinations of nitrogen-fixing and P-solubilizing organisms. Downey and Van Kessel (1990) predicted that the acid producing *P. bilaii* would not be compatible with *Rhizobium* spp. Gleddie (1992)

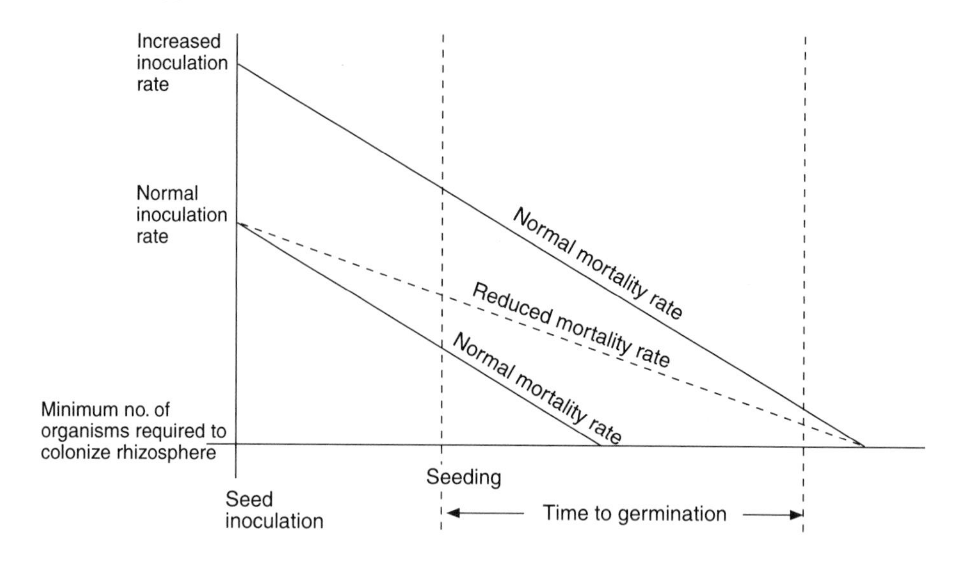

Fig. 2. Relationship of inoculation rate, mortality rate, and time to seeding and germination on the likelihood of root colonization (adapted from Brockwell et al. 1995)

demonstrated that this was not the case. In fact, nodulation and N-fixation was increased when *P. bilaii* was added along with rhizobial inoculants. The two organisms may even be grown together and produced as a dual inoculant (Rice et al. 1994). Philom Bios has commercialized such a product in Western Canada.

Most of the studies reported on inoculants do not list any tests of compatibility with seed- applied chemicals commonly used by farmers. Brown (1974) pointed out that all the tests performed on Phosphobacterin were done on clean seed free of fungicide. We have tested *P. bilaii* with all of the major chemicals used in Western Canada on wheat, canola, pea, lentil, and alfalfa and have established a time allowed in suspension and a planting window for each seed treated with our product and for each pesticide. Many farmers in Western Canada use air-seeding equipment. Philom Bios has worked with Flexicoil, one of the major manufacturers, to test the survival in a unique "seed treatment unit" designed to apply the inoculant while seeding. Organisms used in the unit must survive for at least 6 h in suspension, as this is the maximum length of time required for seeding with the Flexicoil air seeder. Farmers in Canada buy canola seed pre-treated with fungicides, and are reluctant to apply an inoculant on farm. We have also worked within seed processors in Western Canada to certify them to inoculate seed with *P. bilaii* for farmers. Processors use their own equipment and standard operating procedures. Inoculated seed is stored in the processor's warehouses and monitored at regular intervals by Philom Bios to ensure that there are enough viable spores on the seed (Prusinkiewicz et al. 1998).

Above all, to be commercially successful inoculants must have a proven benefit to farmers. The benefits to crop productivity must be reproducibly demonstrated if the use of the inoculant is to prove successful in practice. Yield increases, reduced input costs, and improved quality, which result in increased profits are very important to farmers while quantification of phosphorus uptake is of minor interest. These trials, which must be done to generate this data are different than those done by most researchers in small plot trials. Conducting a large number of split-field trials with farmers to evaluate the economic benefits of the inoculant is essential. Small plot field experiments supply valuable basic information on efficacy and mechanism of action, but larger scale split-field demonstration trials are needed to prove the economic benefits of the inoculant to farmers. We have successfully demonstrated a significant economic benefit to the use of *P. bilaii* in Western Canada (Table 5). This represents a greater than three to one return on the price of *P. bilaii* inoculant.

6. Conclusions

Microorganisms can solubilize P. If we are to make use of these organisms in inoculants we must carry out careful studies that take into account the biological, commercial, and agronomic realities. The organism must colonize and function on the roots of different crop in diverse environments. The inoculant must be produced, formulated and packaged so that it is stable, effective, and easy to use without disrupting normal farming practices, and can be made available to farmers at a reasonable price. Farmers must see an economic benefit to using the inoculant. Our experiences with *P. bilaii* in Western Canada, have shown that an effective P solubilizing inoculant can be developed, if the research from production to application to efficacy is carefully planned and executed.

Acknowledgments. The authors thank Beth Campbell, Chris Lindsay, Elizabeth Prusinkiewicz, and Bonnie Willie for allowing us to use their unpublished data. We also thank John Cross and Beth Campbell for editing the paper.

Table 5. Summary of economic response to *Penicillium bilaii* as determined by split-field demonstration trials done by farmers across Western Canada (since 1988)

Crop	Number of split-field trials	Average net return[a] (Cdn. dollars) / ha
Wheat	77	20.45
Canola	56	27.91
Pea / Lentil	39	29.59
Alfalfa	4	31.00
All Crops	176	25.10

[a]Net returns were calculated using 5-year average commodity prices for crops and fertilizer and the 1997 suggested retail price for Provide. The average net return of $10.16 acre across all crops has a 95% confidence limit of $2.50/acre

References

Agnihotri VP (1970) Solubilization of insoluble phosphates by some soil fungi isolated from nursery seed-beds. Can J Microbiol 16:877-880

Asea PEA, Kucey RMN, Stewart JWB (1988) Inorganic phosphate solubilization by two *Penicillium* species in solution culture and in soil. Soil Biol Biochem 20:459-464

Bajpai PD, Sundra Rao WVB (1971) Phosphate-solubilizing bacteria. Part 2. Extracellular production of organic acids by selected bacteria solubilizing insoluble phosphate. Soil Sci Plant Nutri 17:44-45

Baker R (1991) Induction of rhizosphere competence in the biocontrol fungus *Trichoderma*. In: Keister DL, Cregan PB (Eds) The rhizosphere and plant growth. Kluwer Academic, Boston, pp 221-228

Banik S, Dey BK (1982) Available phosphate content of an alluvial soil as influenced by inoculation of some isolated phosphate solubilizing microorganisms. Plant Soil 69:353-364

Barber DA, Bowen GD, Rovira AD (1976) Effects of microorganisms on absorption and distribution of phosphate in barley. Austr J Plant Physiol 3:801-808

Bender G (1998) Friendly fungus lifts wheat yields. Austr Grain Dec 1997-Jan 1998:3-4

Bowen (1991) Microbial dynamics in the Rhizosphere. Possible strategies in managing rhizosphere populations. In: Keister DL, Cregan PB (Eds) The rhizosphere and plant growth. Kluwer Academic, Boston, pp 22-32

Brockwell J, Bottomley PJ, Thies JE (1995) Manipulation of rhizobia microflora for improving legume productivity and soil fertility: A critical assessment. Plant Soil. 174:143-180

Brown ME (1974) Seed and root bacterization. Ann Rev Phytopathol 12:181-197

Carlile MJ, Watkinson SC (1994) The fungi. Academic, London, 482 pp

Chabot R, Antoun H, Cecas P (1993) Stimulation de la croissance du mais et de la laitue romaine par des microorganismes dissolvant le phosphore inorganique. Can J Microbiol 39:941-947

Chambers JW, Yeomans JC (1990) The effect of PB50 (*Penicillium bilaii* inoculant) on yield and phosphorus uptake by wheat. In: Proceedings of the Annual Society of Soil Science Meeting 33:283-293

Chambers JW, Yeomans JC (1991) The influence of PB50 on crop availability of phosphorus from soil and fertilizer as determined by ^{32}P dilution. In: Proc Annu Soc Soil Sci Meeting 34:123-125

Chanway CP, Nelson LM (1991) Characterization of cultivar specific growth promotion of spring wheat by *Bacillus* sp. In: Keister DL, Cregan PB (Eds) The rhizosphere and plant growth. Kluwer Academic, Boston, p 365

Cook RJ, Baker KF (1984) The nature and practice of biological control of plant pathogens. APS, St Paul, 539 pp

Cunningham JE, Kuiack C (1992) Production of citric and oxalic acids and solubilization of calcium phosphate by *Penicillium bilaii*. ApplEnvironl Microbiol 58:1451-1458

Datta M, Banik S, Gupta RK (1982) Studies on the efficacy of a phytohormone producing phosphate solubilizing *Bacillus firmus* in augmenting paddy yield in acid soils of Nagaland. Plant Soil 69:365-373

Downey J, van Kessel C (1990) Dual inoculation of *Pisum sativum* with R. *leguminosarum* and *Penicillium bilaii*. Biol Fertil Soils 10:194-196

Duff RB, Webley DM, Scott RO (1969) Solubilization of minerals and related materials by 2-ketogluconic acid-producing bacteria. Soil Sci 95:105-114

Gerretsen FC (1948) The influence of microorganisms on the phosphate intake by the plant. Plant Soil 1:51-81

Gleddie SC (1992) Responses of pea and lentil to inoculation with the phosphate-solubilizing fungus *Penicillium bilaii*. MSc Thesis, University of Saskatchewan, Saskatoon

Gleddie SC (1993) Response of pea and Lentil to inoculation with the phosphate- solubilizing fungus *Penicillium bilaii* (PROVIDE). In: Proc Soils Crops Workshop, Saskatoon, Saskatchewan, 25-26 Feb 1993 pp 47-52

Gleddie SC, Hnatowich GL, Polonenko DR (1991) A summary of spring wheat response to *Penicillium bilaii*, a phosphate inoculant. In: Proc Western Phosphate/Sulfur Workshop Conf. Colorado State University, Fort Collins, Colorado, March 21-22 1991

Gleddie SC, Schlechte D, Turnbull G (1993) Effect of inoculation with *Penicillium bilaii* (Provide) on phosphate uptake and yield of Canola in Western Canada. In: Proc Alberta Soil Sci Workshop, Edmonton, Alberta, 19-20 Feb 1993, pp 155-160

Goldstein AH (1995) Recent progress in understanding the molecular genetics and biochemistry of calcium phosphate solubilization by gram negative bacteria. Biol AgricHortic 12:185-193

Goos RJ, Johnson BE, Stack RW (1994) *Penicillium bilaii* and phosphorus fertilization effects on the growth, development, yield and common root rot severity of spring wheat. Fertilizer Res 39:97-103

Hozore E, Alexander M (1991) Bacterial characteristics important to rhizosphere competence. Soil Biol Biochem 23:717-723

Illmer P, Schinner F (1992) Solubilization of inorganic phosphates by microorganism isolated from forest soils. Soil Biol Biochem 24:389-395

Iswandi A, Bossier P, Vandenabeele J, Verstraete W (1987) Effect of seed inoculation with the rhizopseudomonad strain 7NSK2 on the root microbiota of maize (*Zea Mays*) and barley (*Hordeum vulgare*). Biol Fertil Soil 31: 153-158

Jackson MA, Schisler DA, Slininger PJ, Boyette CD, Silman RW, Bothast RJ (1992) Fermentation strategies for improving the fitness of a bioherbicide. Weed Technol 10:645-650

Jones D, Smith BFL, Wilson MJ, Goodman BA (1991) Phosphate solubilization in Scottish upland soil. Mycolog Res 95:1090-1093

Kucey RMN (1983) Phosphate-solubilizing bacteria and fungi in various cultivated and virgin Alberta soils. Can J Soil Sci 63:671-678

Kucey RMN (1987) Increased phosphorus uptake by wheat and field beans inoculated with a phosphorus solubilizing *Penicillium bilaii* strain and with vesicular arbuscular mycorrhizal fungi. Appl Environ Microbiol 53:2699-2703

Kucey RMN, Leggett ME (1989) Increased yields and phosphorus uptake by Westar canola (*Brassica napus* L) inoculated with a phosphate-solubilizing isolate of *Penicillium bilaii*. Can J Soil Sci 69:425-432

Kucey RMN, Janzen HH, Leggett ME (1989) Microbially mediated increases in plant available phosphorus. Adv Agron 42:199-228

Kundu BS, Guar AC (1980) Establishment of nitrogen fixing and phosphate solubilizing bacteria in the rhizosphere and their effect on the nutrient uptake of the wheat crop. Plant Soil 57:223-230

Kundu BS, Guar AC (1984) Rice responses to inoculation with N_2 fixing and P-solubilizing microorganisms. Plant Soil 79:227-234

Laheurte F, Leyval C, Berthelin J (1990) Root exudates of maize pine and beech seedlings as influenced by mycorrhizal and bacterial inoculations. Symbiosis 9:111-116

Latour X, Corberland T, Laguerre G, Allard F, Lemanceau P (1996) The composition of fluorescent pseudomonad populations associated with roots is influenced by plant and soil type. Appl Environ Microbiol 62:2449-2456

Lee A, Bagyaraj AJ (1986) Effect of soil inoculation with vescicuar-arbuscular mycorrhizal fungi and either phosphate rock dissolving bacteria or thiobacilli on dry matter production and uptake of phosphorus by tomato plants. N Z J Agric Res 29:525-531

Leggett ME, Gleddie SC (1995) Developing biofertilizer and biocontrol agents that meet farmer's expectations. Adv Plant Pathol 11:59-74

Levesque CA, O'Gorman DT, Peterson SW, Leggett ME (1997) Molecular techniques to identify and detect *Penicillium bilaiae*. Abst Mycolog Soc America

Leyval C, Berthelin J (1989) Interaction between *Laccaria laccata*, *Agrobacterium radiobacter* and beech roots: Influence on P, K, Mg and Fe mobilization from minerals and plant growth. Plant Soil 117:103-110

Leyval C, Berthelin J (1991) Weathering of a mica by roots and rhizospheric microorganisms of pine. Soil Sci Socf America J 55:1009-1016

Lynch JM, Whipps JM (1991) Substrate flow in the rhizosphere. In: Keister DL, Cregan PB (Eds) The rhizosphere and plant growth Kluwer Academic, Boston, pp15-24

Mba CC (1994) Field studies on two rock phosphate-solubilizing actinomycete isolates as biofertilizer sources. Environ Manage 18:263-269

Milus EA, Rothrock CS (1993) Rhizosphere colonization of wheat by selected soil bacteria over diverse environments. Can J Microbiol 39:335-341

Nahas E (1996) Factors determining rock phosphate solubilization by microorganisms isolated from soil. World J Microbiol Biotechnol 12:567-572

Prusinkiewicz E, Klutz M, Audette S (1998) Development of a certification program which ensures proper commercial inoculation of seed with the fungal inoculant Provide. Can J Plant Pathol (Abstr in press)

Raj J, Bagyaraj DJ, Manjunath A (1981) Influence of soil inoculation with Vesicular Arbuscular Mycorrhiza and a phosphate dissolving bacterium on plant growth and P uptake. Soil Biol Biochem 13:105-108

Rice WA, Olsen PE, Leggett ME (1994) Development of a multi organism inoculant. In: Germida J (Ed) Proc BIOREM 16-18

Salih HM, Yahya AI, Abdul-Rahem AM, Munam BH (1989) Availability of phosphorus in a calcareous soil treated with rock phosphate or superphosphate as affected by phosphate-dissolving fungi. Plant Soil 120:181-185

Schlechte D, Beckie H, Gleddie SC (1996) Response of alfalfa in the establishment year to inoculation with the phosphate-solubilizing fungus *Penicillium bilaii* (PROVIDE). In: Proc Soils Crops Workshop, Saskatoon, Saskatchewan, 22-23 Feb 1996, pp 309-317

Smith JH, Allison FE, Soulides DA (1961) Evaluation of phosphobacterin as a soil inoculant. Soil Sci Soc Proc 25:109–111

Sperber JI (1958a) The incidence of apatite solubilizing organisms in the rhizosphere. Austr J Agric Res 9:778-781

Sperber JI (1958b) Solution of apatite by soil microorganisms producing organic acids. Austr J Agric Res 9:782-788

Spinks JWT, Barber SA (1947) Study of fertilizer uptake using radioactive phosphorus. Sci Agron 27:145-155

Stewart JWB, McKercher RB (1982) Phosphorus cycle. In: Burns RG, Slater JH (Eds) Experimental microbial ecology. Blackwell, Oxford

Taha SM, Mahmoud SAZ, Halim El-Damaty A, Abd El-Hafez AM (1969) Activity of phosphate-dissolving bacteria in Egyptian soils. Plant Soil 31:149-160

Thomas FV, Shantaran MV, Saraswathy N (1985) Occurrence and activity of phosphate-solubilizing fungi from coconut plantation soils. Plant Soil 87:357-364

Tinker PB (1984) The role of microorganisms in mediating and facilitating the uptake of plant nutrients from soil. Plant Soil 76:77-91

Tinker PBH, Sanders FF (1975) Rhizosphere microorganisms and plant nutrition. Soil Sci 119:363-368

Toro M, Azcon R, Herrera R (1996) Effects on yield and nutrition of mycorrhizal and modulated *Pueraria phaseoloides* exerted by P-solubilizing bacterial. Biol Fertil Soils 21:23-29

Whitelaw MA, Harden TJ, Bender GL (1997) Plant growth promotion of wheat inoculated with *Penicillium radicum sp nov.* Austr J Soil Res 35:291-300

Young C (1990) Effects of phosphorus solubilizing bacteria and Vesicular Arbuscular Mycorrhizal fungi on the growth of tree species in subtropical soils. Soil Sci Plant Nutri 36:225-231

Phosphorus Nutrition in Cropping Systems Through Arbuscular Mycorrhizal Management

Joji Arihara and Toshihiko Karasawa

Summary. Improvement of phosphorus (P) uptake and plant growth by arbuscular mycorrhizae (AM), one of the most common symbiotic systems, is well known in many crops. Because AM fungi require living roots of the host to complete their life cycle, the number of AM fungal spores in the soil and subsequent AM colonization of crop roots were enhanced with the cultivation of mycorrhizal crops. This enhancement is affected by the degree of mycorrhizal association of crops grown in previous seasons. Thus, preceding crops influence growth and P uptake of succeeding mycorrhizal crops. Enhancements of the growth and P uptake of succeeding mycorrhizal crops are influenced by the degree of AM symbiosis of preceding crops. Because of the high cost of AM inoculum and the usual predominance of indigenous AM fungi over inoculated AM fungi, effective utilization of the indigenous AM fungi is desirable. This improving effect is also dependent on the degree of AM symbiosis of succeeding crops. The improving effect is especially clear on crops grown under dry weather conditions because the efficiency of AM colonization of crop roots and the diffusion of phosphorus in soil are decreased with the decrease of soil moisture. The improving effect of AM colonization on crop growth is also clear when crops are grown in soils of high P-fixing capacity such as Andosol. To establish cropping systems efficient in P utilization, arranging crop sequence from the point of view of AM symbiosis is important for environmentally sound and sustainable crop production. AM symbiosis, however, does not always improve the growth of mycorrhizal crops even though they are grown on soils of low P availability and high P-fixing capacity. Factors involved in this scarce enhancement of growth have to be well understood to further extend P-efficient cropping systems over wider areas. Effects of soil properties, agrochemicals, and agronomic practices on AM symbiosis also are not completely clear. Additional research is necessary to explore interactions among environmental factors, AM, and host plants, and their benefits in different cropping systems.

Key words. Phosphorus nutrition, Arbuscular mycorrhizae (AM), Mycorrizal association, Cropping systems, AM inoculation, Soil moisture, Soil phosphorus, Soil type, Agronomic practice

1. Introduction

The importance of rotations to obtain stable and high crop yields has long been recognized by farmers worldwide. Prior crops are known to affect growth and yield of succeeding crops through several means such as altering nutrients and water availability of soils, soil physical conditions, soil pH, biodiversity, disease and pest incidence, and through allelopathic effects etc. Some effects of previous crops on following crops, however, cannot be fully explained by such mechanisms. For example, the reasons for reduced growth and yield of crops such as soybean or sweet corn following sugar beet, which is often observed in Hokkaido, Japan (Nishiiri et al. 1981), are not entirely clear. Similarly, the enhanced growth and yield of maize following soybean even in conditions of adequate nitrogen (N) supply cannot be explained by residual effects of N fixed by soybean. Further analysis of mechanisms whereby previous crops affect the following crops is, therefore, necessary to select the most appropriate crops in rotations suited to a particular environment.

Volcanic ash (Andosols) and acidic (podsolic Molisols) soils, which are widely distributed in Hokkaido, have very high phosphorus (P) fixing capacity. As P deficiency has been a serious constraint for crop production in these soils, application of heavy doses of P fertilizer was common. Although excessive P fertilization is still practiced in many areas, it is becoming increasingly controversial and difficult due to environmental concerns associated with pollution of watersheds by soil sediments containing high amounts of P, and the rising cost of P fertilizer. Increasing affinity of crops for P by improving efficacy to P uptake from soils would be the most effective way to reduce dosage of P fertilizer to crops.

Improvement of P uptake and plant growth by arbuscular mycorrhizae (AM), one of the most common symbiotic systems, is well known in many crops (Mosse 1973). AM are found in roots of a diverse range of plants but they are not formed in plants belonging to families such as Cruciferae (rape seed, cabbage, radish, mustard, etc.), Chenopodiaceae (sugar beet, spinach, etc.), Polygonaceae (buckwheat), and Amaranthaceae (amaranthus). Crops like lupin (Leguminosae) and perennial ryegrass (Gramineae) are also not mycorrhizal. Because AM fungi require living roots of the host to complete their life cycle, growing mycorrhizal crops may enhance AM populations in the soil, while cultivating non-mycorrhizal crops may decrease them. Prior crops can affect AM formation in succeeding crops due to their direct effects on soil AM population (Black and Tinker 1977), but there is very little direct evidence to suggest that crop rotation affects P relationships (Bullock 1992).

Inoculating AM fungi can improve AM associations of crops but it is not yet practical for widespread adoption in field crop production. Utilization of indigenous AM fungi is, therefore, more appropriate to enhance the growth and P uptake of crops. In this chatper, we consider the importance of optimizing AM associations by cropping systems to improve P nutrition in crop production systems in Hokkaido practiced on soils containing low or sparingly soluble P. We then examine the role of soil, climate, and factors such as agrochemicals and tillage on associations between crops and AM fungi.

2. Effects of Preceding Crops

In a field experiment conducted on a volcanic ash soil for two consecutive years, growth and yield of maize (*Zea mays* L. cv. Okahomare) were significantly influenced by previous crops (Arihara and Karasawa 2000). Shoot weight and grain yield of maize following sunflower (*Helianthus annuus* L. cv. DO-707), maize, soybean (*Glycine max* Merr. cv. Kitahomare), potato (*Solanum tuberosum* L. cv. Danshku-imo) and spring wheat (*Triticum aestivum* L. cv. Haruyutaka) were much higher than after rape (*Brassica napus* L.) or cabbage (*Brassica oleracea* L.), fallow and sugar beet (*Beta vulgaris* L. cv Monohomare). The variation in grain yield was, however, much higher than in shoot weight (data not shown). Additional P (87 kg ha^{-1}) application could not offset the adverse effects of cultivating rape and sugar beet or fallowing on maize yield (Fig. 1). The wide variations in P uptake by succeeding maize (from 2.7 to 38.5 kg ha^{-1}) showed that preceding crops strongly

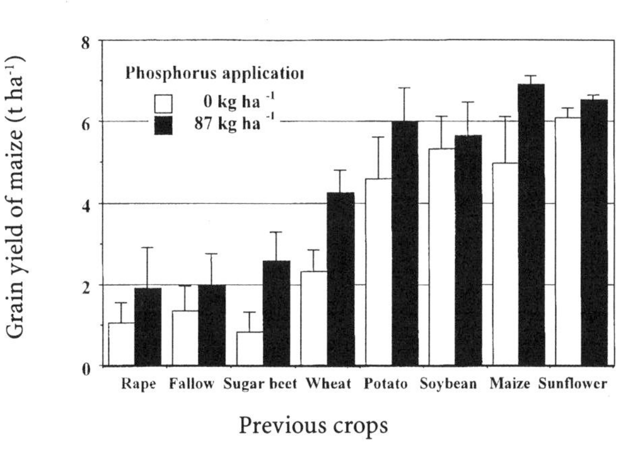

Fig. 1. Effect of cultivation of crops in 1990 on grain yield of succeeding maize in 1991 (Arihara and Karasawa 2000)

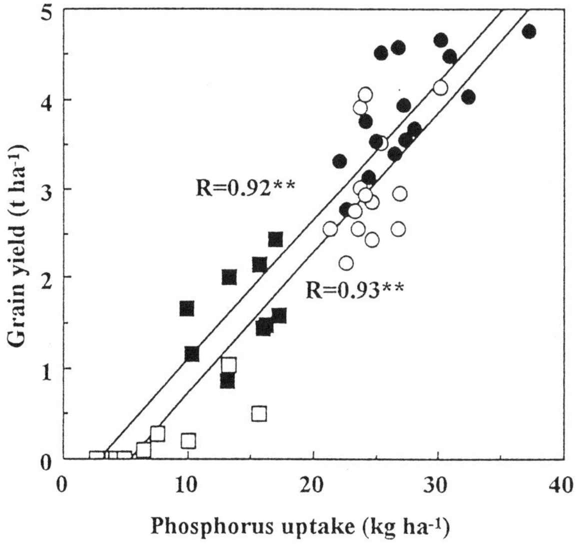

Fig. 2. Relationship between phosphorus uptake at harvesting time and grain yield of succeeding maize in 1991. *Circles* represent after mycorrhizal crops and *squares* represent after non-mycorrhizal crops and fallow. *Open* and *shaded symbols* represent 0 and 87 kg ha^{-1} of phosphorus applications, respectively. (Arihara and Karasawa 2000)

affected the amount of P taken up by succeeding crops. A strong correlation (Fig. 2) between grain yield of succeeding maize and P uptake (r=0.93 for 0 kg P ha^{-1} and r=0.92 for 87 kg P ha^{-1}) further suggested that the influence of previous crops was perhaps mediated through effects on P availability and uptake.

Notwithstanding the considerable variation in P uptake by preceding crops, there was no relationship between P uptake by them and the growth of succeeding maize. Available soil P status at the time of sowing of succeeding maize was also unaffected by treatments in the previous season (53.0 ±2.0 mg P kg^{-1} soil). AM colonization of maize, which varied from 2% to 100%, was greatly affected by the type of preceding crop but not P application (Fig. 3). Because crops differ in mycorrhizal dependency and their association with AM fungi (Plenchette et al. 1983), it is logical to expect that crop rotations alter AM status in the soil.

Sunflower, maize, wheat, soybean, potato are mycorrhizal while rape and sugar beet are non-mycorrhizal (Plenchette et al. 1983; Tester et al. 1987; Becard and Piche 1990). The higher shoot weight and grain yield of maize after mycorrhizal than non-mycorrhizal crops, and a strong positive correlation between shoot weight and AM colonization (r=0.80 to 0.84) suggest that previous crops might play an important role in determining AM status in soils in the following season. An increased AM colonization during cultivation of mycorrhizal crops perhaps improved the efficiency of AM colonization of succeeding maize, thereby promoting AM formation. The increased AM formation may in turn stimulate P uptake and maize growth. Black and Tinker (1979) also reported that spore number of AM fungi and subsequent colonization of barley roots were higher following barley (mycorrhizal) than after kale (non-mycorrhizal) and fallow treatments.

In another experiment, cultivation of mycorrhizal crops indeed increased AM population in the following season (Karasawa et al. 2000a). The soils after cultivation of mycorrhizal crops such as sunflower, maize, adzuki bean (*Vigna angularis*

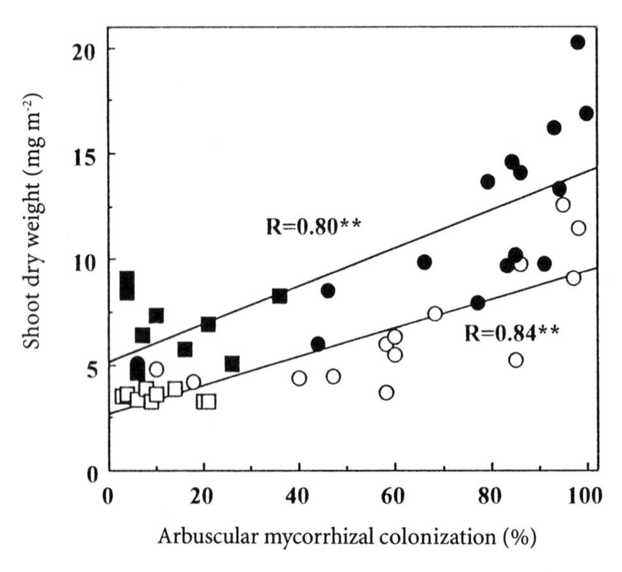

Fig. 3. Relationship between arbuscular mycorrhizal formation (71 days after sowing) and shoot dry weight of succeeding maize in 1992. *Circles* and *squares* indicate after mycorrhizal crops and after non-mycorrhizal crops and fallow, respectively. *Open* and *shaded symbols* indicate 0 and 87 kg ha^{-1} of phosphorus applications, respectively. (Karasawa et al. 2000b)

Table 1. Number of arbuscular mycorrhizae(AM) spores in soil cultivated with mycorrhizal and non-mycorrhizal plants or left fallow in the previous season

Previous crops	Number of spores
Sunflower	1575
Maize	1650
Soybean	1500
Kidney bean	1850
Adzuki bean	1675
Potato	600
Buckwheat	800
Mustard	800
Japanese white radish	775
Fallow	900

(Karasawa et al. 2000a)

(wild) Ohwi et Ohashi), kidney bean (*Phaseolus vulgaris* L.) or soybean contained more AM spores than those after cultivation of non-mycorrhizal crops such as buckwheat (*Fagopyrum esculentum* Moench), mustard (*Brassica juncea* (L.,) Czern et Coss), potato, Japanese radish (*Raphanus sativus* L.) or fallow treatment (Table 1).

3. Effects of Combinations of Preceding and Succeeding Crops

Combination of preceding and succeeding crops are thought as the minimum unit of cropping systems. In order to design cropping systems efficient in P utilization by enhancing optimum AM associations, information on combinations of crops differing in mycorrhizal association and dependency is useful. Though mycorrhizal crops depend for their P uptake on mycorrhizal association, there are variations in dependency of P uptake on mycorrhizal association. Combination of preceding and succeeding crops even among mycorrhizal crops, therefore, may produce differences in productivity of cropping systems.

A series of experiments was, conducted from 1991 to 1996 with a range of crops displaying wide variation in mycorrhizal associations. Highly mycorrhizal crops (Sunflower, maize, soybean, adzuki bean and kidney bean), moderately mycorrhizal crops (potato and spring wheat), and non-mycorrhizal crops (buckwheat, Japanese white radish, sugar beet and mustard) were included in the study. The effects of previous crops varied widely with the degree of mycorrhizal association of succeeding crops (Table 2).

In highly mycorrhizal crops, the stronger the association of previous crops with AM, the better was the yield of the following crop. Moderately mycorrhizal crops were also affected by preceding crops but the magnitude of effect was less than in highly mycorrhizal crops. By contrast, yields of non-mycorrhizal crops were only slightly affected. The yields of non-mycorrhizal crops following mycorrhizal crops or after fallow were higher, however, than those after non-mycorrhizal crops. Fac-

Table 2. Effect of cultivation of previous crops on yield of crops with different mycorrhizal association

	Previous crops											
Following crops	SF	MZ	SB	AB	FB	PT	SW	BW	RD	SBT	MTD	Fallow
Highly mycorrhizal crops												
Sunflower (SF)	74.0	97.8	91.0	82.3	76.0	87.7	77.5	66.6	66.7	74.2	62.2	64.8
Maize (MZ)	100.0	96.2	94.0			91.8	75.1		85.7	86.5		75.6
Soybean (SB)	100.0	83.1	71.6			75.1	66.2		72.8	75.9		43.6
Adzuki bean (AB)	92.3	99.2	75.9	97.8	93.8	83.6	94.6	76.1	65.9	73.0	79.1	63.5
Field bean (FB)	87.3	100.0	62.8			75.0			65.3			63.6
Moderately mycorrhizal crop												
Potato (PT)	96.7	88.1	92.8	91.7	100.0	88.5	95.3	91.4	89.4	87.1	71.0	83.6
Spring wheat (SW)	97.6	100.0	91.1	99.1		95.1		73.6	94.7			93.3
Non-mycorrhizal crops												
Buckwheat (BW)	96.8	91.9	89.6	82.6	82.2	85.6	90.8	85.0	88.1	83.8	87.2	90.9
Radish (RDS)	90.3	97.3	97.6			94.1	40.5		84.8	92.3		91.8
Sugar beet (SBT)	86.4	76.6	74.4			84.1	87.2		74.0	76.0		95.9
Mustard (MTD)	80.1	86.0	100.0			78.5			69.7			88.6

Figures in the table are average yield of following crops obtained from a series of experiments conducted in the same fields for five years.
Yields of each following crop with a different prior crop combination are expressed as percentages of the maximum yield plot in each year. Number of years for experimentation is different for each crop combination, as the combination of prior and following crops was changed each year
Field experiments were conducted with three replications for each year

tors other than AM association may have been involved in such effects.

Based on the above results, the ideal combination of crops in a rotation from the point of optimum AM association can be outlined. Highly mycorrhizal crops may be preferred as previous crops for cultivating any crop in the following season, irrespective of its mycorrhizal association. However, in case of highly mycorrhizal crops, cultivation after highly mycorrhizal crops gives best results. Moderately mycorrhizal crops grown in the previous season may have beneficial effects on highly to moderately mycorrhizal crops, but the effects may not be clear in case of non-mycorrhizal crops. Non-mycorrhizal preceding crops are not suitable for most crops irrespective of their mycorrhizal association, though negative effects are generally for non-mycorrhizal crops. Fallow treatment is detrimental to mycorrhizal crops, but it has some beneficial effects on non-mycorrhizal crops. It must be noted that the above crop sequences are based solely on the AM association. To design a productive rotation, other factors affecting crop productivity such as nematodes, diseases and pests, must also be considered.

4. Interactions with Climatic Conditions

4.1 Precipitation

A field experiment conducted in 1993, a cool wet year, showed different results from the previous two years. When no P was applied, grain yield of succeeding maize was less in plots where soybean and maize were grown as preceding crops, and more when spring wheat, potato, radish and sugar beet were grown or when the field was kept fallow in the previous season. When P fertilizer was applied, grain yield of maize was slightly more in plots with radish, potato, and fallow treatments than other plots but the differences were marginal (Fig. 4).

AM colonization on maize roots, however, exceeded 70% in all plots irrespective of the previous crop. As all agronomic management practices had been similar in all years, cool and wet weather conditions especially during first two months after sowing in 1993 were considered to account for such differences. Because precipitation in June 1993 was 78mm, as against 5 mm 1991 and 26.5 mm in 1992, high soil moisture status during early vegetative growth of maize was suspected to have enhanced AM colonization. A pot experiment — in which soil moisture status was varied — showed that the effect of previous crop was more obvious in dry than moist or wet soils (Fig. 5), thereby raising the possibility that P uptake by maize was stimulated more by AM fungi when soil moisture was low. Because available P in soils after various treatments was not markedly different, and because P uptake was positively correlated with AM colonization (Fig.6), the varia-

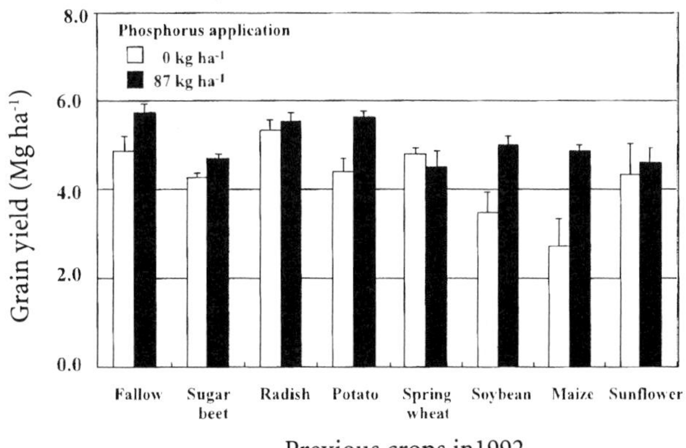

Fig. 4. Effect of previous crops on grain yield of succeeding maize grown in the cool and wet year of 1993. *Bars* indicate SE (Arihara et al., unpublished data)

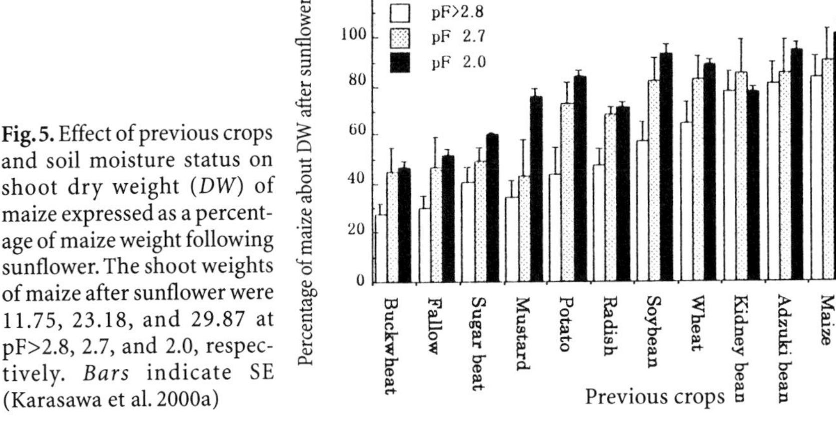

Fig. 5. Effect of previous crops and soil moisture status on shoot dry weight (*DW*) of maize expressed as a percentage of maize weight following sunflower. The shoot weights of maize after sunflower were 11.75, 23.18, and 29.87 at pF>2.8, 2.7, and 2.0, respectively. *Bars* indicate SE (Karasawa et al. 2000a)

tion in maize P uptake in different treatments was considered to be due to differences in AM stimulated P uptake. The effect of soil moisture on AM colonization varied with the population density of AM spores in the soil. When AM spore population was high, the average AM colonization of maize roots was 73.0% in dry soil (pF > 2.8) and 82.2% in wet soil (pF 2.0). When AM spore population was low, AM colonization was very low (29.2 %) in dry soil but it increased to 61.6% in wet soil (pF 2.0) (Fig. 7).

The results suggest that an increased soil moisture status improved the efficiency of AM colonization in maize roots thus promoting AM formation, which in turn might stimulate P uptake and increase plant growth, thereby minimizing the influence of cropping history. Reid and Bowen (1979) reported that an increase in water content up to 0.22 kg kg^{-1} soil enhanced AM colonization of pri-

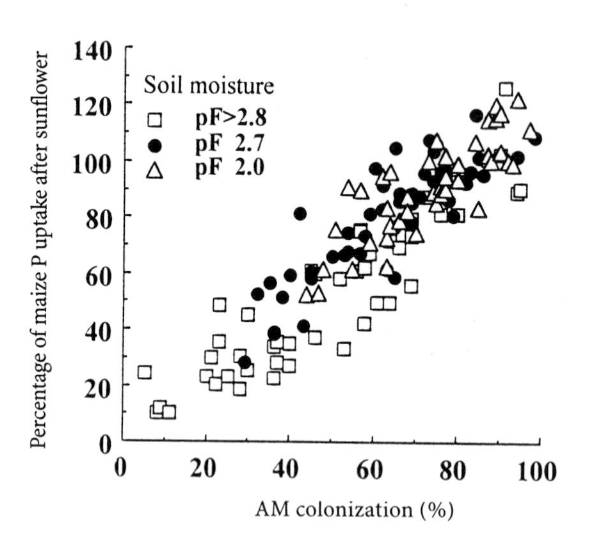

Fig. 6. Relationship between arbuscular mycorrhizae (AM) colonization and P uptake by maize in various treatments expressed as a percentage of P uptake by maize following sunflower (Karasawa et al. 2000a)

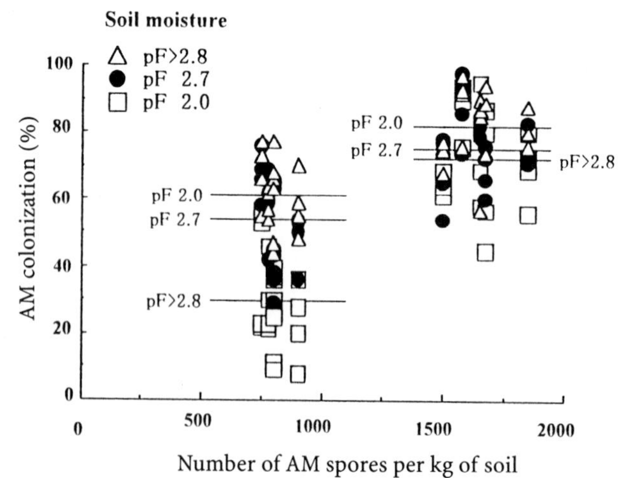

Fig. 7. Relationship between number of AM spores in soil and percentage colonization of maize seminal roots. *Horizontal lines* denote means of AM colonization at pF 2.0, 2.7, >2.8 when AM spores populations were low or high (Karasawa et al. 2000a)

mary and lateral roots of barrel medic (*Medicago tribuloides* Desv.). Our findings corroborate their results, although the optimal soil moisture level for AM formation in maize was more than in barrel medic. Jasper et al. (1993) reported that AM spores were drought resistant but the hyphae did not survive very long in dry soil as they lost the vigor of colonization within six weeks. This suggests that a higher soil moisture status might improve the vitality of hyphae grown from spores, and enhance the efficiency of AM colonization even in soils with a few spores following cultivation of non-mycorrhizal crops in the previous season. Read et al. (1976) reported that AM penetration and colonization in *Cladium* spp. was low and/or absent in marsh plants, as infection occurred only during a dry period. However, soil moisture in a marsh would be too excessive and cannot be comparable to a crop field.

Soil moisture also directly affects root growth and P uptake by crops, as low moisture content reduces P diffusion through the soil to the root surface (Olsen et al. 1965). Karasawa et al. (2000b) showed that increasing effects of AM inoculation on growth of maize was evident in dry soil, but not evident in moist and wet soil (Fig. 8). On the other hand, P uptake of maize was increased with the amount of AM inoculum irrespective of soil moisture conditions (Fig. 9). Weak negative correlations between P uptake and shoot dry weight of maize suggest that growth of maize plants was not regulated by factors other than P uptake. In contrast to the results of their experiment, the grain yield of maize was improved with an application of P as seen in the field experiment conducted under wet soil conditions (Fig. 4). However, improvement of grain yield of maize with an application of P indicates that under field conditions maize can not uptake a sufficient amount of P even under the wet soil conditions in 1993 (Fig. 4). The effect of preceding crops was thus less distinct in wet soils than in dry soils due to a higher efficiency of AM colonization and increased P availability in the former, which is strongly associated with higher soil moisture content. AM management thorough cropping systems would assume more significance in rain fed cropping systems than in irrigated cropping systems under drier climate conditions.

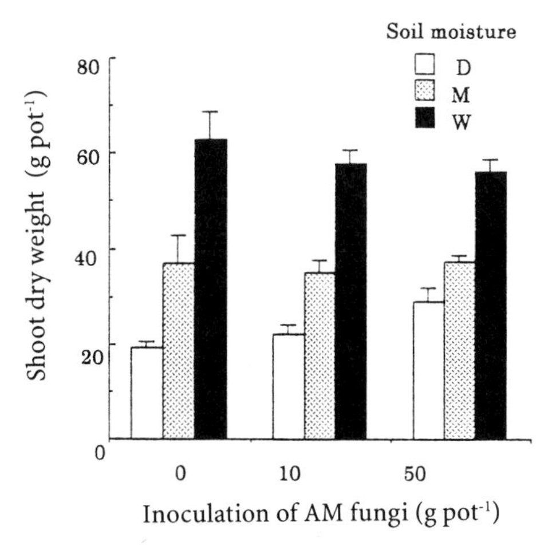

Fig. 8. Effects of AM inoculation and soil moisture status on shoot dry weight of pot-grown maize. Soil water potential was adjusted from 11 to 75 days after sowing around −10 kPa (field capacity: wet, W), −50 kPa (moisture of rupture of capillary bound: moist, M), and <−63 kPa (drier than moisture of rupture of capillary bound: dry, D). Bars indicate SE. (Karasawa et al. 2000b)

4.2 Temperature

The effects of temperature on the rate and extent of AM colonization are complex, as responses vary with the host plant and the fungus. AM colonization usually increases up to about 30°C, but some plant-fungus combinations develop normally up to 35°C or more (Bowen 1987). Such variations may represent adaptations to different climates. While rapid germination and colonization would be an advantage in the humid tropics, subtler interactions between soil moisture and optimum temperature for germination and colonization may have evolved in more seasonal environments. For example, Bowen (1987) found significant variation between species in spore germination. Minimum temperature for germination of *Glomus* sp. and *Acaulospora* spores varied between 10 and 15°C, and maximum temperature around 30°C but the optimal temperature was usually between 20 to 25°C. As a group, species of *Gigaspora* appear to have a higher temperature range and higher optima.

AM inoculation experiments are often conducted at temperatures above 15°C, but many plants, both wild and cultivated, grow and develop mycorrhizae at lower soil temperatures especially in temperate regions. Allen et al. (1989) found high (40%–60%) colonization in *Agropyron* by both field (*Glomus* spp.) and pot-grown (*Gigaspora margarita*) inoculum at 12°C, while Daft et al. (1980) reported that colonization in English bluebells increased rapidly in the winter months when soil temperatures were about 5°C. An assessment of the effects of four soil temperatures on growth of cotton and root colonization by three AM fungi (Smith and Roncadori 1986) showed that shoot and root growth were not improved by mycorrhizae at 18°C and total root length was in fact suppressed. The three AM fungi generally stimulated plant growth equally well at 24°C and 30°C, but *G. ambisporum* was slightly more effective as a symbiont at 36°C than either *G. intraradices* or *G. margarita*. Further work directed at understanding the biology of propagule survival, and germination and colonization of roots in different habitats is, however, necessary.

5. Interactions of AM with Soil Factors

5.1 Soil Phosphorus

Soil P availability strongly affects percent AM colonization. While very low P availability inhibits AM colonization, as in subterranean clover (Bolan et al. 1984), addition of small amounts of P increases percent colonization. However, further P addition reduces both proportion of root length infected and dry weight of mycorrhizal roots. The sensitivity of response to P supply also varies with host plant and its variety. Whereas roots of wheat, barley and rye grown in soil with 5 mg kg^{-1} bicarbonate extractable P were colonized by AM fungi up to 40% of their length, P addition at 5 mg kg^{-1} markedly reduced AM colonization. Further addition of 30 or 60 mg P kg^{-1} eliminated the fungus (Baon et al. 1992). In *Trifolium subterraneum*, however, application of P necessary to achieve maximum growth reduced AM colonization by only 21% (Oliver et al. 1983). Even within a crop, cultivars of barley were colonized to a varying extent by *Glomus etunicatum*, and they showed different sensitivities to P addition in terms of AM colonization (Baon et al. 1993).

In our field experiments, cultivation of mycorrhizal crops in the previous season enhanced AM association even when 87 kg P ha^{-1} was applied (Fig. 1). Such application was considered equivalent to 72.5 mg kg^{-1} of soil, assuming that fertilizer was mixed into top 15 cm soil with a specific weight of 0.8, a typical value for volcanic ash soils in Japan. The results indicated that AM colonization was not reduced even when moderately high amounts of P were applied to soils of very high P fixing capabilities.

The response of AM colonization to applied P were compared among crops that differed in mycorrhizal associations in 1994. AM colonization of sunflower - a crop known for high mycorrhizal dependency - grown after mycorrhizal crops was 67.0%, 62.6%, and 63.6% respectively when 0, 44, and 132 kg P ha^{-1} was applied, as against 73.8%, 75.8%, and 84.9% for adzuki bean. In potato, however, AM colonization decreased from 44.3% at 44 kg P ha^{-1} to 28.0% at 132 kg P ha^{-1}. A similar trend was seen in crops grown after non-mycorrhizal crops, but colonization rates were much less than those after mycorrhizal crops. Thus, an increase in P application of up to 132 kg P ha^{-1} did not decrease AM colonization of highly mycorrhizal crops such as sunflower and adzuki bean, but decreased those of moderately mycorrhizal crops such as potato. Mean AM percent colonization values in crops grown after mycorrhizal and non-mycorrhizal hosts respectively were 64.4 and 51.7 (sunflower), 78.2 and 71.9 (adzuki bean), and 44.7 and 27.5 (potato).

When no P was applied, average yield advantages from cultivating sunflower, adzuki bean and potato after mycorrhizal crops as compared with those after non-mycorrhizal crops were 28.8%, 15.4%, and 36.7% respectively. The yield advantage values were respectively 31.7%, 12.4%, and 20.2%, and 26.1%, 7.0% and 6.9%, when 44 and 132 kg ha^{-1} of P were applied. No yield advantage was evident for buckwheat, a non-mycorrhizal crop. The results clearly demonstrate that choice of rotations to enhance AM associations is important in cultivation of mycorrhizal crops on soils of high P fixing capability even when high rates of P are applied. For cultivation of non-mycorrhizal crops, however, considerations on rotation are less important from the viewpoint of P nutrition.

The role of soil P availability in relation to AM associations was examined in another trial, in which spring wheat was grown after buckwheat, potato, adzuki bean and sunflower, and the amount of P application was varied. Grain yield of wheat differed with the type of preceding crop at all levels of soil P availability. Wheat yield following sunflower was more than that after buckwheat when Truog P level was less than 110 mg kg^{-1} (Fig. 9). The influence of preceding crops was especially evident at low Truog P level, with the effect declining with an increase in soil available P status. AM colonization of wheat roots also depended on the preceding crops, with a higher AM percent colonization following sunflower than after buckwheat (Fig. 10).

With an increase in Truog P level, however, AM colonization of wheat following sunflower decreased. Thompson et al. (1986) reported that formation of penetration points, colonization, length of external hyphae, and spore numbers were highest at moderate P levels and that they were reduced at very low and very high P levels. Our results largely corroborate these observations, although a reduction in AM colonization was seen only at a higher soil P level than previously reported. While the adverse effects of cultivating non-mycorrhizal crops on succeeding crops can be offset partly by heavy P fertilization, the spore population of effective AM fungi in soils might decrease due to high soil P availability. In order to optimize the populations of effective AM fungi and reduce application of P fertilizer simultaneously, improvement of P uptake by succeeding crops through an effective management of indigenous AM fungi with cropping systems is important especially in soils with high P fixing capacity.

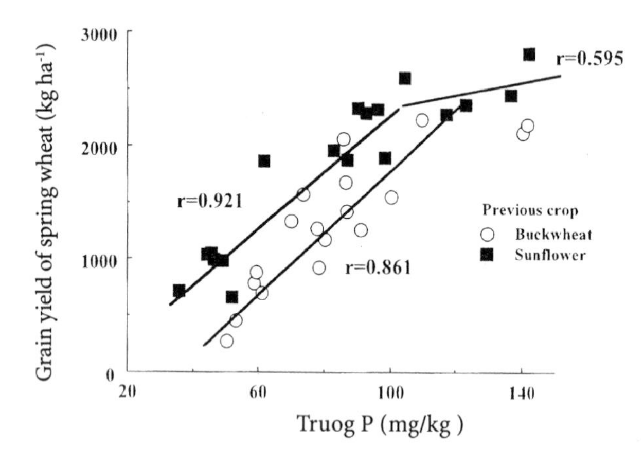

Fig. 9. Relationship between Truog P and grain yield of spring wheat after cultivation of buckwheat, potato, adzuki bean, and sunflower (Karasawa et al. unpublished data)

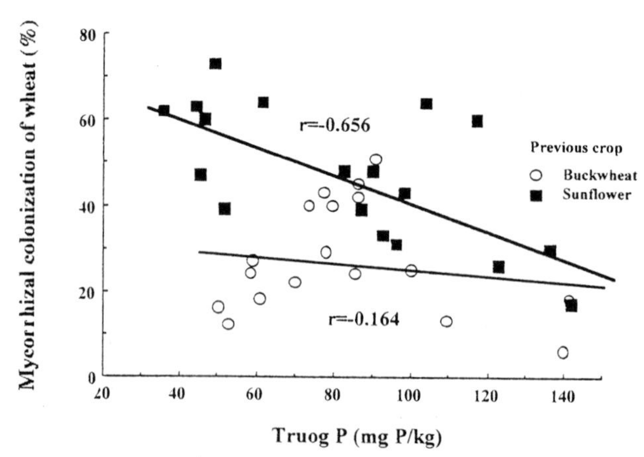

Fig. 10. Relationship between Truog P and AM colonization of spring wheat roots after cultivation of buckwheat, potato, adzuki bean, and sunflower (Karasawa et al. unpublished data)

5.2 Soil Type

Cultivation of mycorrhizal crops does not always improve the growth of succeeding mycorrhizal crops however. For example, the effect of mycorrhizal crops in Hokkaido varies with location without any distinct relationship to soil P availability. An experiment, therefore, was conducted to study the effects of preceding crops (sunflower or mustard) on AM formation and growth of succeeding maize in 17 soil types (Table 3).

Influence of the preceding crops on maize growth was distinct in 10 soil types. AM colonization in maize after sunflower was significantly higher than that after mustard in all volcanic ash soils (A1 to A11) (Table 4). Shoot dry weight of maize following sunflower was greater than that following mustard (non-mycorrhizal host) in all volcanic ash soils except A1, A6, A8, and A9. The enhanced growth of maize following sunflower in those volcanic ash soils irrespective of differences in soil P status led us to postulate that sunflower improved AM formation of succeeding maize, which might in turn enhance P uptake and growth of succeeding

Table 3. Classification, land use, and selected chemical properties of soils used in experiments

Soil	Soil site	Soil Group	Preceding crop[a] (Land use)	pH (H_2O)	EC (m mho cm^{-3})	Available P (Truog P) (mg kg^{-1})	Exchangeable cations (c mol(+) kg^{-1}) Ca	Mg	K
Andisols									
A1	Chitose	Udivitrands	maize	6.09	0.045	116.2	6.35	1.04	0.281
A2	Memuro	Hapludands	oat	5.81	0.037	16.3	3.35	0.57	0.608
A3	Shihoro	Hapludands	wheat	6.50	0.102	81.5	9.94	1.86	0.839
A4	Shikaoi	Hapludands	wheat	5.90	0.103	132.9	14.67	2.90	0.838
A5	Sapporo	Melanudands	oat	5.42	0.036	51.5	8.46	0.74	0.575
A6	Sapporo	Melanudands	grass[b]	5.78	0.027	6.6	15.07	1.45	0.117
A7	Sapporo	Melanudands	maize	5.56	0.026	15.3	7.88	1.25	0.375
A8	Shihoro	Hapludands	wheat	5.66	0.078	118.3	12.79	2.23	0.352
A9	Makubetsu	Melanaquands	wheat	5.60	0.061	40.8	8.72	1.12	0.522
A10	Shihoro	Hapludands	—[c]	6.09	0.107	197.0	14.29	4.90	2.455
A11	Miyagi	Melanudands	wheat	5.70	0.068	159.7	10.62	2.52	0.863
Other Soils									
B1	Sapporo	Dystrochrepts	forest	5.88	0.035	0.0	7.43	1.34	0.802
B2	Furano	Eutrochrepts	wheat	5.97	0.020	144.0	2.85	0.99-	0.553
B3	Ebetsu	Hydraquents	—[c]	5.35	0.058	82.5	6.77	0.69	0.740
B4	Makubetsu	Eutrochrepts	bean[d]	5.81	0.080	192.4	10.90	2.56	0.979
B5	Shihoro	Dystrochrepts	wheat	5.91	0.094	166.1	9.80	2.33	0.938
B6	Toyokoro	Medihemists	grass[b]	5.30	0.085	94.2	11.20	2.15	0.496
B7	Niki	Eutrochrepts	grass			100.0			
B8	Yoichi	Dystrochrepts	grass			117.0			
B9	Niki	Dystrochrepts	grass			341.0			

[a] A6 and B6 soils had been used as grassland and the B1 soil was sampled from forest. Other soils were sampled from an upland field

[b] Used as grassland

[c] Uncertain

[d] Adzuki bean

Table 4. Comparison of prior crop effect on AM colonization and shoot dry weight of maize grown in a pot filled with different soils

	AM colonization (%) of Maize after		Short dry weight of maize (g pot^{-1}) after	
Soils	Mustard	Sunflower	Mustard	Sunflower
Volcanic ash soil				
A1	45.3	67.3**	5.24	6.88
A2	18.0	75.0***	1. 18	4.07**
A3	25.7	64.3*	7.12	12.95**
A4	4.3	62.0***	6.24	12.69**
A5	11.0	80.7***	1.06	5.32*
A6	21.3	55.3*	0.96	1.03
A7	28.7	79.7*	1.59	5.14*
A8	26.3	64.7*	10.50	13.88
A9	26.7	83.7***	4.91	7.16
A10	23.3	71.0***	9.70	15.46**
A11	63.0	73.7*	7.22	11.35*
Other soils				
B1	19.7	21.7	0.88	0.72
B2	12.3	24.0	8.58	7.13
B3	27.3	81.7**	1.97	6.02**
B4	26.7	18.7	9.73	13.73
B5	37.3	71.7**	9.78	13.59*
B6	39.3	68.0	7.64	15.77**

The values bearing symbols for Mustard and Sunflower are significantly different
*$P<0.05$, **$P<0.01$, ***$P<0.001$
Symbols for soils are the same as in Table 3

maize. An increased AM colonization of maize following sunflower, however, did not increase shoot weight of maize in soil samples A1, A6, A8, and A9 (Table 5). In the case of A1 and A8 soils, considerably high available P levels might have reduced the improving effects of AM on P uptake of maize. Although it is not statistically significant, 45% greater shoot weight of maize after sunflower than after mustard indicates that improved AM colonization might be effective in increasing P uptake in A9 soil. However, it is difficult to explain why the significant increase in AM colonization was not at all effective in increasing the shoot weight of maize in A6 soil, where available P was especially low. This anomaly might be associated with differences in soil traits related to P availability and in the effectiveness of native AM populations to improve P uptake by the host plant. Further studies are necessary to clarify the anomaly. In soils other than volcanic ash soil, the effect of preceding crops on AM colonization and growth of succeeding maize were not evident in B1, B2, and B4. Low AM colonization of maize in B2 and B4 would be due to high P status of those soils. In case of B1, however, soil factors other than P availability might account for the effect of preceding crops on growth and AM colonization of succeeding maize in B1.

Table 5. Effect of mycorrhizal inoculation on shoot dry weight of maize grown on five types of soils fertilized with different rate of P

		Applied P (g pot^{-1})				
Soils	Inoculation	0.00	0.22	0.44	1.31	2.18
A5	-AM	0.41	1.10	1.45	3.79	5.62
	+AM	1.29***	1.56**	1.82	4.18	5.74
A6	-AM	0.60	1.06	2.07	6.80	10.86
	+AM	0.59	1.10	1.71	6.34	10.00
B7	-AM	3.17	5.94	5.76	5.68	6.90
	+AM	3.28	5.78	5.54	5.18	7.32
B8	-AM	3.88	5.80	5.84	7.81	6.71
	+AM	5.33**	6.62	5.31	7.09	7.16
B9	-AM	6.30	5.98	7.19	7.27	9.80
	+AM	8.23	6.25	7.64	8.15	8.98

Values bearing symbols are significantly different between -AM and +AM
**$P<0.01$, ***$P<0.001$
Symbols for soils are the same as those of Table 3

5.3 AM Inoculation

Maize was grown in five different soils (A5, A6, B7, B8 and B9 as in Table 3) with or without AM inoculum and supplemented with 0, 0.22, 0.44, 1.31 and 2.18 g P pot^{-1} to study interactions among soil types, AM inoculation and P application. Available P level was increased proportionally to the application of P in each soils. When P was not applied, shoot weight of maize inoculated with AM fungi was more than that of the controls in A5 and B8 (Table 5). However, shoot weight of maize was not increased with AM inoculation in A6, B7and B9 soils. AM inoculation had no effect on shoot weight in all soils when they were supplemented with 0.44 g P pot^{-1} or more. The results obtained with A5 and B8 soils corroborate those of Thomson et al. (1986), who reported that the growth promoting effect of AM inoculation was evident only in low P soils and not in high P soils. Thus the lack of response to AM inoculum in the B9 soil even when no P was applied might be due to its high available P status (332 mg P kg^{-1}).

In A6 and B7 soils, where the effect of AM inoculum was not apparent, AM colonization was similar to that in A5 and B8 soils, where the effect was evident. This raises the possibility that efficiency of P uptake by AM fungi colonizing maize roots varied with soil type. However, reasons for lack of response in A6 and B7 soils with low available P are not entirely clear. The results thus again indicate that AM formation is dependent on factors besides available P status.

5.4 Plant-Fungus Combinations

Because both infection and colonization by AM fungi are affected by a range of environmental factors, the effect of preceding crops might vary with indigenous AM fungal species and other soil properties. Many different mycorrhizal fungi colonize a single plant root system, but not all of them are equally effective in improving nutrient uptake. Notwithstanding the lack of clear evidence yet on absolute specificity between taxa of AM fungi and taxa of potential host plants, large variation in AM colonization and nutrient uptake was observed with different plant-fungus combinations. For example, *Plantago lanceolata, Zea mays* and *Sorghum sudanense* are extensively colonized by a wide variety of AM fungi (Smith and Read 1997). Studies on compatibility between hosts and AM fungi, and on influence of host species on competition among fungi revealed that the most suitable plant species for proliferation (spread of infection and sporulation) of an individual fungus was different among different fungi. For instance, inoculation of marigold, soybean and asparagus with one (non-competitive) or three (competitive) species of AM fungi [*Glomus etunicatum* (Ge), *G. mosseae* (Gm) and *Gigaspora* sp. (C1)] revealed that root infection and growth of all plants was promoted by inoculation of Ge and Gm but not C1.

The growth responses of host plants to AM inoculation also differ among the plant-fungus combinations because of differences in colonization. For example, growth responses to *Glomus epigaeus* were significant in soybean, asparagus, tomato and Sudan grass but not in citrus. And growth responses to *G. mosseae* occurred on tomato, citrus and cotton; to *G. fasiculatus* on citrus only; to *G. epigaeus* on tomato, citrus, maize, Sudan grass and asparagus; and to *G. constrictus* on maize only. Root-infection was evident, however, in all AM-inoculated plants except *G. epigaeus* inoculated citrus (Daniels and Menge 1981). Growth promoting effect may thus depend on plant-fungus combination even if AM fungi can colonize the roots. Other soil factors such as soil P status affect the optimum plant-fungus association. For instance, colonization of an indigenous fungus in subterranean clover was optimum when no P was applied but it was maximum for an introduced fungus when 0.5 g P per pot was applied (Bolan et al. 1984).

5.5 Soil pH

Several studies showed that soil pH affected the activity of indigenous AM fungi (Mosse 1972; Graw 1979), which in turn can mediate the effects of preceding crops on succeeding crops. Soil pH may indeed be as important as soil P status in determining both the success of colonization and outcome of interactions between fungi and crops. Many endophytes are known to infect roots well at a wide range of pH, but they do not enhance plant growth. Furthermore, pH optima for colonization and plant growth may be different depending on species. Assessment of nine AM fungi for infectivity and ability to enhance the growth of alpine strawberry in sterilized, low P soils which were limed to different pH, showed that only *Glomus clarum* greatly stimulated growth at pH 4 while the others had no effect (Hayman and Tavares 1985). The most effective endophytes at pH 5 were E3 and *G. clarum*. Plant growth was best at pH 7, but inoculations with *Glomus clarum, Acaulospora laevis, Gigaspora heterogama* and *G. margarita* were ineffective at this pH. Percent colonization also varied with soil pH.

Soil pH also affects P solubility and the response of a particular crop-fungus combination to applied P. For example, *Tagetes minuta* and *Guizotia abyssinica* reacted differently to changes in soil pH and to fertilization with various phosphates despite inoculation with the same strain of *Glomus macrocarpus* (Graw 1979). At pH 4.3, *Guizotia* was unable to absorb P from all compounds leading to a severe growth inhibition even after inoculation, whereas *Tagetes* grew well at this pH in the presence of $Ca(H_2PO_4)_2 \cdot 5H_2O$, $Ca_5(PO_4)_3OH$ and $AlPO_4$. P uptake and growth of mycorrhizal *Guizotia* were improved at pH 5.6 and its growth surpassed that of non-mycorrhizal plants at pH 6.6. By contrast, growth of mycorrhizal *Tagetes* was reduced at pH 5.6 in the presence of $Ca(H_2PO_4)_2 \cdot 5H_2O$ and $Ca_5(PO_4)_3OH$, but it was improved at pH 6.6 in all treatments with the exception of $Ca_5(PO_4)_3OH$ as a P source. The results demonstrated that both soil pH and the type of P compound strongly influenced the effects of AM.

6. Effects of Agrochemicals and Tillage

An understanding of the effects of agrochemicals on AM fungi is important to preserve and effectively use AM associations in cropping systems. Heavy N fertilization can adversely affect penetration and colonization of mycorrhizae due to its direct effects on the fungus rather than simply the result of changes in root growth. Ammonium ions, added as ammonium sulfate, decrease AM colonization more than nitrate ions, added as sodium nitrate (Chambers et al. 1980).

The effects of pesticides on mycorrhiza vary with the type of chemical. Soil fumigants such as methyl bromide are toxic (Kleinschmidt and Gerdemann 1972), whereas herbicides have little adverse effect on most AM fungi. Paraquat and simazine, however, showed an adverse effect on *Glomus etunicatum* (Nemec and Tucker 1983). Insecticides also differ in their effects on AM. For example, parathion decreased the formation of AM but carbaryl and endosulphan at 5 mg L^{-1} had negligible effect except for a delay in AM infection (Parvathi et al. 1985). Fungicides such as chloroneb, metalaxyl, and captan (at low concentrations) had no adverse effects on mycorrhizal infection of sour orange (Nemec 1980). On the other hand, copper, maneb and chlorothalonil sharply reduced sporulation at medium and high levels of application, while benomyl, drazoxolon and triamefon inhibited mycorrhizal infection.

Soil disturbance (tillage) reduced AM infection and P absorption in maize and wheat (Evans and Miller 1990). The decreased P absorption following tillage was attributed to changes in root distribution in soil where stratification of plant-available P occurred. By contrast, tillage showed no effects on P uptake by two non-mycorrhizal crops (Evans and Miller 1988).

7. Conclusions and Future Needs

Preceding crops influenced growth and P uptake of succeeding crops by affecting soil AM populations. Number of AM fungal spores, subsequent colonization and P uptake by crops following mycorrhizal crops were higher than after non-mycorrhizal crops. Because of the high cost of AM inoculum and the usual predominance of indigenous AM fungi over inoculated AM fungi, effective utilization of the indigenous AM fungi is important to enhance growth and P uptake of succeeding crops. Soil pH, moisture status and temperature mediate the effects of AM inoculum or preceding crops. Nutrient benefits in a cropping system ultimately depend on the effectiveness of plant-fungus combination. Although high P availability resulted in reduction of AM colonization, the effect of preceding crops differed among various soils even when soil P availability was low. This raised the possibility of differences in soil traits (especially those related to P availability) and in the effectiveness of native AM populations to improve P uptake by the host plant. Additional studies to determine the role of environmental factors and host plants on dominance of AM fungi and their growth-promoting effect in different cropping systems are therefore necessary. Because the ability of colonized AM also differs among soils, studies to identify specific soil factors will be important for designing sustainable and highly productive cropping systems.

References

Allen MF, Allen EB, Friese CF (1989) Responses of the non-mycotrophic plant *Salsola kali* to invasion by vesicular-arbuscular mycorrhizal fungi. New Phytol 111:45-49

Arihara J, Karasawa T (2000) Effect of previous crops on arbuscular mycorrhizal formation and growth of succeeding maize. Soil Sci Plant Nutr 46:43-51

Baon JB, Smith SE, Alston AM, Wheeler RD (1992) Phosphorus efficiency of three cereals as related to indigenous mycorrhizal infection. Aust J Agric Res 43:479-491

Baon JB, Smith SE, Alston AM (1993) Mycorrhizal responses of barley cultivars differing in P efficiency. Plant Soil 157:97-105

Becard G, Piche Y (1990) Physiological factors determining vesicular-arbuscular mycorrhizal formation in host and nonhost Ri T-DNA transformed roots. Can J Bot 68:1260-1264

Black RLB, Tinker PB (1977) Interaction between effects of vesicular-arbuscular mycorrhiza and fertilizer phosphorus on yields of potatoes in the field. Nature 267:510-511

Black RLB, Tinker PB (1979) The development of endomycorrhizal root systems. II. Effect of agronomic factors and soil conditions on the development of vesicular-arbuscular mycorrhizal infection in barley and on the endophyte spore density. New Phytol 83:401-413

Bolan NS, Robson AD, Barrow NJ (1984) Increasing phosphorus supply can increase the infection of plant roots by vesicular-arbuscular mycorrhizal fungi. Soil Biol Biochem 16:419-420

Bowen GD (1987) The biology and physiology of infection and its development. In: Sifir GR (Ed) Ecophysiology of VA mycorrhizal plants. CRC, Boca Raton, Florida, USA, pp 27-70

Bullock DG (1992) Crop rotation. Crit Rev Plant Sci 11:309-326

Chambers CA, Smith SE, Smith FA (1980) Effects of ammonium and nitrate ions on mycorrhizal infection, nodulation and growth of *Trifolium subterraneum*. New Phytol 85:47-62

Daft MJ, Chilvers MT, Nicolson TH (1980) Mycorrhizas of the Liliiflorae. I. Morphogenesis of Endymion non-scriptus (L) Garke and its mycorrhizas in nature. New Phytol 85:181-189

Daniels BA, Menge JA (1981) Evaluation of the commercial potential of the vesicular-arbuscular mycorrhizal fungus, *Glomus epigaeus*. New Phytol 87:345-354

Evans DG, Miller MH (1988) Vesicular-arbuscular mycorrhizas and the soil-disturbance-induced reduction of nutrient absorption in maize. I. Causal relations. New Phytol 110:67-74

Evans DG, Miller MH (1990) The role of the external mycelial network in the effect of soil disturbance upon vesicular-arbuscular mycorrhizal colonization of maize. New Phytol 114:65-71

Ezawa T, Kuwahara S, Yoshida T (1995) Compatibility between host and arbuscular mycorrhizal fungi, and influence of host plant species on the competition among the fungi. Soil Microorg 45:9-19

Graw D (1979) The influence of soil pH on the efficiency of vesicular-arbuscular mycorrhiza. New Phytol 82:687-695

Hayman DS, Tavares M (1985) Plant growth responses to vesicular-arbuscular mycorrhiza. XV. Influence of soil pH on the symbiotic efficiency of different endophytes. New Phytol 100:367-377

Jasper DA, Abbott LK, Robson AD (1993) The survival of infective hyphae of vesicular-arbuscular mycorrhizal fungi in dry soil: an interaction with sporulation. New Phytol 124:473-479

Karasawa T, Arihara J, Kasahara Y (2000a) Effects of previous crops on arbuscular mycorrhizal formation and growth of maize under various soil moisture conditions. Soil Sci Plant Nutr 46:53-60

Karasawa T, Takebe M, Kasahara Y (2000b) Arbuscular mycorrhizal (AM) effects on maize growth and AM colonization of roots under various soil moisture conditions. Soil Sci Plant Nutr 46: 61-67

Karlen DL, Sharpley AN (1994) Management strategies for sustainable soil fertility. In: Hatfield JL, Karlen DL (Eds) Sustainable agriculture systems. Lewis Publ, CRC, Boca Raton, Florida. pp 47-108

Kleinschmidt GD, Gerdemann JW (1972) Stunting of citrus seedlings in the fumigated nursery soil related to the absence of endomycorrhizae. Phytopathol 62:1447-1452

Lioi L, Giovanetti M (1987) Variable effectivity of three vesicular-arbuscular mycorrhizal endophytes in *Hedysarum coronarium* and *Medicago sativa*. Biol Fertil Soils 4:193-197

Mosse B (1972) Effect of different Endogone strains on the growth of *Paspalum notatum*. Nature 239:221-223

Mosse B (1973) Advances in the study of vesicular-arbuscular mycorrhiza. Ann Rev Phytopathol 11:171-196

Nemec S (1980) Effects of 11 fungicides on endomycorrhizal development in sour orange. Can J Bot 58:522-526

Nemec S, Tucker D (1983) Effects of herbicides on endomycorrhizal fungi in Florida USA citrus (*Citrus* spp.) soils. Weed Sci 31:427-431

Nishiiri K, Matsui S, Watanabe Y (1981) Studies on a crop sequence in Hokkaido with special emphasis on behavior of agronomical characteristics. I. Growth and yield of soybeans following afferent crops. Res Bull Hokkaido Natl Agric Expt Stn 130:113-122

Oliver AJ, Smith SE, Nicholas DJD, Wallace W, Smith FA (1983) Activity of nitrate reductase in *Trifolium subterraneum*: effects of mycorrhizal infection and phosphate nutrition. New Phytol 94:63-79

Olsen SR, Kemper WD, van Schaik JC (1965) Self diffusion coefficients of phosphorus in soil measured by transient and steady-state methods. Soil Sci Soc. Am Proc 29:154-158

Parvathi K, Venkateswarlu K, Rao AS (1985) Effects of pesticides on development of *Glomus mosseae* in groundnut. Trans Br Mycol Soc 84:29-33

Plenchette C, Fortin JA, Furlan V (1983) Growth responses of several plant species to mycorrhizae in a soil of moderate P-fertility. I. Mycorrhizal dependency under field conditions. Plant Soil 70:199-209

Reid CPP, Bowen GD (1979) Effects of soil moisture on V/A mycorrhiza formation and root development in Medicago. In: Harley JL, Russell RS (Eds) The soil-root interface. Academic, London, pp 211-219

Read DJ, Koucheki HK, Hodgson J (1976) Vesicular-arbuscular mycorrhiza in natural vegetation systems. I. The occurrence of infection. New Phytol 77:641-653

Smith GS, Roncadori RW (1986) Responses of three vesicular-arbuscular mycorrhizal fungi at four soil temperatures and their effects on cotton growth. New Phytol 104:89-95

Smith SE, Read DJ (1997) The symbionts forming VA mycorrhizas. In: Smith SE, Read DJ (Eds) Mycorrhizal symbiosis. 2nd Edn. Academic, London, pp 11-32

Tawaraya K, Kinebuchi T, Watanabe S, Wagatsuma T, Suzuki M (1996) Effect of arbuscular mycorrhizal fungi *Glomus mosseae*, *Glomus fasciculatum* and *Glomus caledonium* on phosphorus uptake and growth of welsh onion (*Allium fistulosum* L.) in Andosols. (in Japanese) Jpn J Soil Sci Plant Nutr 67:294-298

Tester M, Smith SE, Smith FA (1987) The phenomenon of "nonmycorrhizal" plants. Can J Bot 65:419-431

Thomson BD, Robson AD, Abbott LK (1986) Effects of phosphorus on the formation of mycorrhizas by *Gigaspora calospora* and *Glomus fasciculatum* in relation to root carbohydrates. New Phytol 103:751-765

A New Control Method of Soybean Cyst Nematode Using Animal Feces

Kazuyuki Matsuo

Summary. We observed strong attachment of soybean cyst nematodes (SCN, *Heterodera glycines* Ichinohe) to soybean roots when seeds were sown soon after application of dried cattle feces (DCF) to the soil. The increase in the attachment of SCN seemed not to be related to changes in soil condition or in crop growth. Similar accelerating effects were found in other animal fecal treatments in spite of an interaction with inorganic nitrogen contained in animal feces. Moreover, increase in the number of second-stage juveniles was observed regardless of soybean plant presence or absence. Although the active substance has not been identified, hydrophobic extracts from DCF showed superior effects to intact DCF. Delaying the time of sowing of soybean seed until after DCF application reduced the number of female adults on soybean roots because of starvation. In addition, DCF application combined with a delayed sowing time of more than 30 days reduced the density of cysts remaining in the soil by 80% of that of chemical fertilizer treated plots. Although the yield reduction caused by nematode injury was alleviated with DCF application, the application of animal feces in the integrated management of SCN should be carried out together with crop rotation and resistant cultivars, especially in soil severely infested by SCN.

Key words. Soybean cyst nematode (SCN), Animal feces, Soybean, Hatching agent, Acceleration of hatching, Nematode control, Reducing nematode injury

1. Introduction

A recent report (Tsuiki and Harada 1997) showed that the by-products from livestock production in Japan could supply as much as 150 kg N /ha/year for arable lands. However, a large part of these products is wasted without being restored to fields and causes underground water pollution through nitrate leaching (Kumazawa 1999). This problem, which threatens the existence of livestock production itself, stems from the extreme dependence of animal feed in Japan on overseas suppliers. Needless to say, a solution to this problem can occur only with fundamental changes to the regime of the animal feed supply. At the same time, farmers tend to neglect restoration of organic matter to their fields through their dependence on chemical fertilizers. Such management practices lead to degradation of soil physical and biological properties. Therefore, developing new cropping systems in which animal by-products are utilized more highly is an urgent problem. Re-evaluation of animal by-products as plant nutrients is still an important subject. Research on the direct absorption of organic nutrients by plants may offer a new explanation for the discrepancy between nutrient uptakes of the crop and the amount of nutrients released from organic materials. Research also may clarify the effects of organic matter on the growth or yield of crops in years of cold weather.

The use of animal by-products that qualify as plant nutrients alone will not be attractive to farmers, however, because of the poor handling properties and slow effects on crop growth of these materials. We need, therefore, to find additional functions of animal by-products, such as suppressive effects on pests or diseases, that cannot be replaced by chemicals.

Originally, we did not have a particular interest in the latter function of organic materials, but a strange phenomenon we observed enticed us to begin the work reported here. While evaluating the role of dried cattle feces (DCF) in soybean nutrition, we noticed strong attachment of soybean cyst nematode (SCN) on soybean roots in the DCF-treated plots. Such a phenomenon had not been reported before and was completely inexplicable in common-sense terms. It had been reported that organic matter was effective in suppressing nematode populations or in reducing their damage to crops (Chikaoka et al. 1981; Ueda and Kasai 1993; Toida et al.1993). Although the mechanism of this suppression has not been clearly understood, the mode of action was roughly classified into five categories (Nakasono 1990): (1) being an obstacle to movement of nematodes, (2) increasing numbers of natural enemies, (3) providing toxic effects of substances derived from organic materials, (4) enhancing nematode-resistance properties of crops and (5) enhancing crop growth and damage compensation. Nevertheless, the observed phenomenon seemed to be due to a promoting effect rather than a suppressive effect of the DCF, and such an effect differed from the above five categories.

The soybean cyst nematode (*Heterodera glycines* Ichinohe) is a very pathogenic species of cyst nematode and attacks mainly Fabaceae plants and its distribution is wide, including Japan, China, the former USSR, Korea, Colombia, Brazil, and Java in Indonesia (Riggs and Niblack 1993). For soybean production, the soybean cyst nematode is of major economic importance in many parts of the world and is recognized as the main cause of yield decrease in continuous cropping in Japan.

Strategies that have been used to manage cyst nematodes include chemicals, cultivar resistance, rotations, cultural practices, and biological control (Riggs and Schuster 1998). Application of chemicals such as D-D (1,2-dichloropropanes-1,3-dichloropropenes) or nematicides such as Aldicarb can reduce the number of nematodes (Carter 1943; Nishizawa et al. 1972; Smith et al. 1991) but chemical control is rarely practiced in commercial soybean production for economical or environmental reasons. Cultivar resistance is the most economical practice for managing soybean cyst nematode. Accordingly, breeding programs are carried out systematically in major soybean-producing countries to develop resistant cultivars with commercial value and with high yielding potential. However, as there are several races of nematode with differing abilities to reproduce on resistant cultivars (Golden et al.1970), continuous planting of resistant cultivars often results in the appearance of resistance-breaking nematode races. Because of the narrow host range, cropping practices such as fallow and crop rotations including non-host, resistant cultivar and susceptible cultivar may be most effective (Slack et al. 1981; Shimizu et al. 1989). However, the effectiveness and the length of fallow and rotation depend on several factors because of the longevity of the eggs (Riggs and Schuster 1998). Cyst formation is a major mechanism for the survival of *Heterodera* spp. Eggs of these species can resist many stresses such as severe desiccation and attack by soil microorganisms.

It is therefore useful to identify any cultural practice that can accelerate hatching of eggs in the cyst so that the nematode population may be effectively controlled. Many chemicals are known to accelerate hatching in vitro (Clark and Shepherd 1966; Okada 1977; Tefft and Bone 1984; Masamune et al. 1982; Fukuzawa et al. 1985) but the effective application of such stimulants has never been reported in the field. This is probably due to the instability of the chemicals associated with absorption by the soil or intense soil microbial activity.

Thus our findings of strong attachment of SCN on soybean roots in the soil are in contrast to the above knowledge. In addition, such a phenomenon seems highly suggestive of a new active substance against nematodes. Accordingly, we carried out experiments to clarify the basis of the increase in the number of female adults on soybean roots while investigating the possibility of controlling SCN using animal feces.

2. Effects of Animal Feces on Soybean Cyst Nematode

DCF application to the soil changes the nutrient condition of the soil and spontaneously affects plant growth. For example, we reported that DCF application to the soil caused a reduction of the inorganic nitrogen level through an increase of microbial activity, and this state continued for a considerably long period of time (Matsuo et al. 1993a). As inhibition of SCN hatching by NH_4^+ or NO_3^- was observed in vitro by Lehman et al. (1971), a reduction of the inorganic nitrogen level with application of DCF may affect hatching of SCN. Also, changes in root development might have an influence on the opportunity for SCN to attach to the roots. The association between the increase in the number of female adults and changes in environmental condition or in the growth of soybeans was then investigated. The number of female adults on soybean roots was significantly higher in the DCF treatment than in the chemical fertilizer treatment or the control when soybean seeds were sown on the day of application (Table 1). The number of female adults also increased with an increasing amount of DCF. Figure 1 shows the relationship between the amount of DCF and the number of female adults. Although DMW and soybean root length were not affected by DCF treatment, both the nitrogen content of soybean and the inorganic nitrogen content of the soil were reduced. Similar changes of nitrogen content in the soil and the plant were observed in sucrose treated plots; however, there was no associated increase in female adults. Therefore, increase in the number of female adults seemed not to be related to a reduction of inorganic nitrogen.

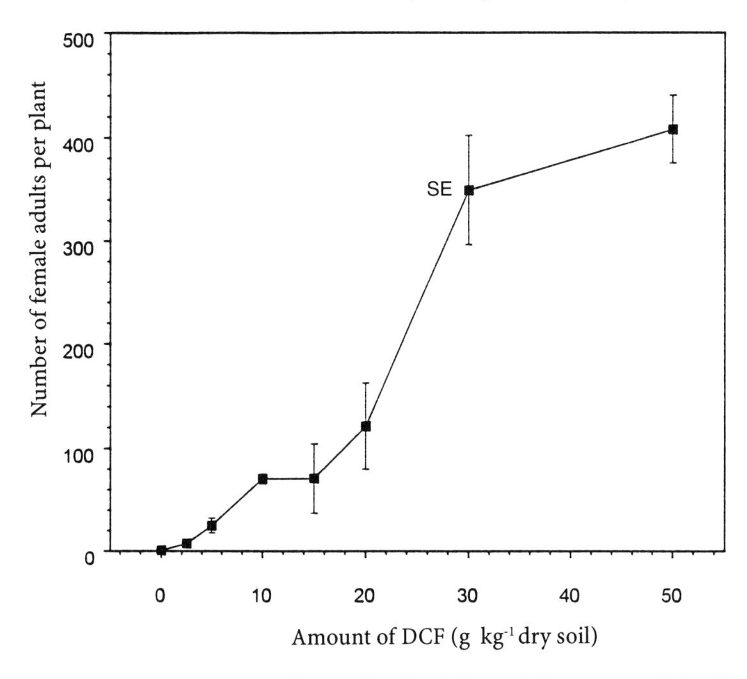

Fig. 1. Relationship between the number of female adults and the amount of dried cattle feces (*DCF*) applied to the soil. The number of female adults on soybean roots was counted by the naked eye 18 days after seed sowing

Table 1. Effects of application of dried cattle feces (DCF), sucrose and chemical fertilizer (NPK) on the number of female adults of the soybean cyst nematode, dry matter weight, root length, and N content in soybean and inorganic N content in the soil

Treatment	Number of female adults per plant[a]	Dry matter (mg plant^{-1})	Root length (m plant^{-1})	N content (mg g^{-1})	Soil inorganic N (mg kg^{-1})
Control	7.7 c	471.5 ab	20.1 ab	22 b	22 b
DCF 10g	278.3 ab	602.2 ab	24.4 a	16 a	11 a
DCF 20g	431.0 a	567.7 ab	19.1 b	15 a	9 a
Sucrose 30g	12.3 c	429.7 b	18.0 b	17 a	4 a
NPK 0.88g	8.0 c	643.6 a	19.6 ab	29 c	143 d
DCF 20g + NPK 0.88g	183.7 b	648.5 a	22.1 ab	27 c	95 c

[a]Number of female adults was counted by the naked eye 21days after seed sowing
Values followed by the same letter are not significantly different (P=0.05)

In our experiments, we commonly used non-nodulating soybean genotype (T201) as a test plant to make counting the number of SCN easier. So such a possibility remained that this phenomenon was unique to this genotype. In addition, it has been reported that the susceptibility of soybean plants to SCN is affected by environmental conditions and the number of female adults on soybean markedly increase with rising air temperature (Shimizu 1986). The number of female adults on soybean roots was then compared among five different soybean cultivars (Lee 68, Peking, Picket 71, PI88788 and PI90763; these soybean cultivars are commonly used for determining the race of SCN populations in the soil) with and without DCF, and with chemical fertilizer. There was no attachment of SCN to PI90763 regardless of treatments (Table 2). In the other four cultivars, the number of female adults was higher in DCF-treated plots than in chemical fertilizer treated plots or the control. But the female indices (Golden et al. 1970) of each cultivar, derived from the formula 100 (number of female adults on test soybean/number on cultivar Lee 68) were not so different among treatments. These results suggested that DCF treatment increased only the number of female adults but did not affect the relative resistance among soybean cultivars to SCN. Therefore, there was a slight possibility that DCF induced the increase in number of female adults through a change in the tolerance of soybeans, as in the case of higher temperatures. Undoubtedly the increase of female adults on roots caused by DCF was not necessarily unique to non-nodulating soybean genotypes.

The above results were obtained from use of single dried cattle feces produced by the Livestock Feeding & Forage Laboratory of NARC (National Agriculture Research Center, Japan). Needless to say, the contents of animal feces differed not only from each kind of animal but also from the feeding conditions of animals (Table 3). Therefore, the effect on increase in female adults was compared among six animal feces types (CF1, CF2, CF3, CF4, LH and PF, see Table 3). There was a large difference in the root length of soybeans among the treatments at 18 days after seeds were sown. Both the numbers of female adults per plant or per unit root length were significantly higher in the animal feces plots than in the chemical fertilizer plot or the control, except for the PF plot. In the PF-treated plot, the number of female adults was higher than the control but not significantly different (Table 4). A negative association was found between the inorganic nitrogen

Table 2. Difference in the number of female adults on soybean roots among 5 soybean cultivars with three different treatments

Cultivar	Treatment					
	DCF		Chemical fertilizer		Control	
	Number of female adults[a]	Female index[b]	Number of female adults	Female index	Number of female adults	Female index
Lee 68	283.5 ± 13.6	100	6.0 ± 1.2	100	21.8 ± 3.8	100
Pickett 71	31.7 ± 6.2	13.3	0.5 ± 0.3	8.3	1.8 ± 0.4	8.0
Peking	18.0 ± 2.7	7.5	0.8 ± 0.2	12.5	2.3 ± 1.1	10.3
PI88788	91.8 ± 10.3	38.5	1.8 ± 0.4	29.2	8.5 ± 1.2	39.1
PI90763	0.0 ± 0.0	0.0	0.0 ± 0.0	0.0	0.0 ± 0.0	0

[a]The number of female adults was counted by the naked eye 18 days after seed sowing
[b]Female index was derived from the formula: 100 (Number of female adults on test plant/Number of female adults on cultivar Lee 68)
Numbers represent means ±standard error

Table 3. Carbon and nitrogen analyses of animal feces used in this study

Code	Animal	Producer	Total C ($g\ kg^{-1}$)	Total N ($g\ kg^{-1}$)	NO_3-N ($mg\ kg^{-1}$)	NH_4-N ($mg\ kg^{-1}$)	C/N ratio
CF1	Lactating cow	NARC[a]	404	24	660	211	16.8
CF2	Lactating cow	NRIAI[b]	434	23	614	472	18.9
CF3	Heifer cow	NRIAI	436	20	380	325	21.8
CF4	Heifer beef	NRIAI	420	20	364	287	21.0
LH	Laying hen	NRIAI	282	24	739	628	11.8
PF	Pregnant sow	NRIAI	444	33	987	950	13.4

[a]National Agriculture Research Center, MAFF, Japan
[b]National Research Institute of Animal Production

Table 4. Effects of application of various types of animal feces on the number of female adult Soybean cyst nematodes, root length of soybean, and inorganic N content in the soil

Treatment	Number of female adults per plant[a]	Number of female adults per unit root length (m^{-1})	Root length of soybean ($m\ plant^{-1}$)	Inorganic N in the soil ($mg\ kg^{-1}$)
CF1	668 a	33 b	20.2 a	8.2 d
CF2	315 c	23 bc	13.8 d	2.5 e
CF3	684 a	50 a	14.3 cd	3.3 e
CF4	515 ab	47 a	12.1 d	4.1 e
LF	349 bc	18 c	19.6 ab	15.3 c
PF	239 cd	18 cd	13.4 d	48.4 b
Chemical fertilizer	28 d	2 d	13.6 d	103.2 a
Control	94 d	5 d	17.0 bc	13.4 c

[a]The number of female adults was counted by the naked eye 18 days after seed sowing.
Values followed by the same letter are not significantly different (P=0.05)

contents of animal feces and the number of female adults. This association might be similar to the decrease in the number of female adults caused by the addition of chemical fertilizer to DCF shown in Table 1. If we consider the suppressive effects of inorganic nitrogen (discussed later in more detail), swine feces (PF) may have had a similar effect to the other animal feces. At the least, these results suggested that cattle feces and hen feces had similar effects on SCN regardless of the difference in feeding conditions or in kind of animal.

3. Mechanism for the Increase of Soybean Cyst Nematode Attachment

The juveniles of SCN generally do not emerge in large numbers until stimulants that are usually found in secretions from roots of host plants are present. Previous researchers (Schmitt and Riggs 1991; Tefft and Bone 1985; Hill and Schmitt 1989; Tutsumi and Sakurai 1966) pointed out that (a) secretions from legume roots accelerated the hatching of SCN, (b) there was positive association between the amount of secretion and root dry matter, and (c) maximum hatching occurred at flowering of soybeans. In our experiments, however, the number of 2nd-stage juveniles and hatching rate was higher in DCF treatments with and without soybean plants (Table 5). These results indicated that DCF accelerated the hatching of eggs and increased the number of 2nd-stage juveniles in the soil regardless of host plant presence or absence.

In order to clarify the interaction between animal feces and inorganic nitrogen that was observed in the above experiments, the effect of nitrogen fertilizer treatment on the number of 2nd-stage juveniles in the soil was investigated under conditions with or without application of DCF. Potassium nitrate (KNO_3) application at a level below 80 mg N kg^{-1} (dry soil) did not reduce the effects of DCF but did do so at rates above this (Fig. 2). Meanwhile ammonium sulfate ($(NH_4)_2SO_4$) reduced the effects of DCF more drastically than potassium nitrate (Fig. 3). It was evident that inorganic nitrogen had an inhibitory action on SCN emergence; however, the number of 2nd-stage juveniles was constantly higher in the DCF treatment than in the control even if the amount of nitrogen fertilizer was considerably increased. Although immobilization of inorganic nitrogen after DCF application might slightly reduce the inhibitory action of inorganic nitrogen on the emergence of 2nd-stage juveniles that existed naturally in the soil, such an effect was not the true nature of the accelerating effect of DCF on SCN emergence. These results may support the hypothesis that animal feces, including that from cattle, hens, and swine, commonly had an accelerating effect on the hatching of SCN.

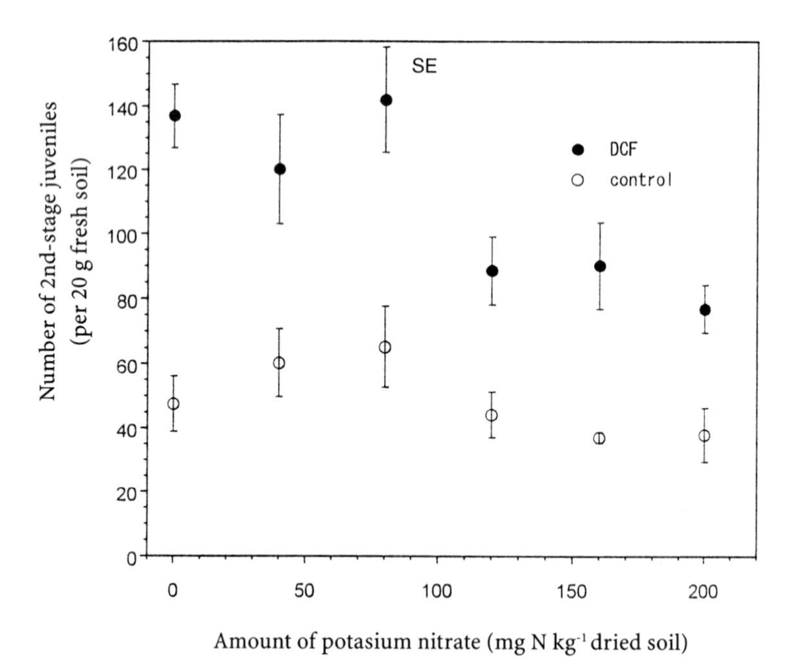

Fig. 2. Effects of potassium nitrate (KNO_3) supplement on the emergence of 2nd-stage juveniles with and without DCF application. Second-stage juveniles were separated from the soil by the Baermann funnel method 10 days after DCF application

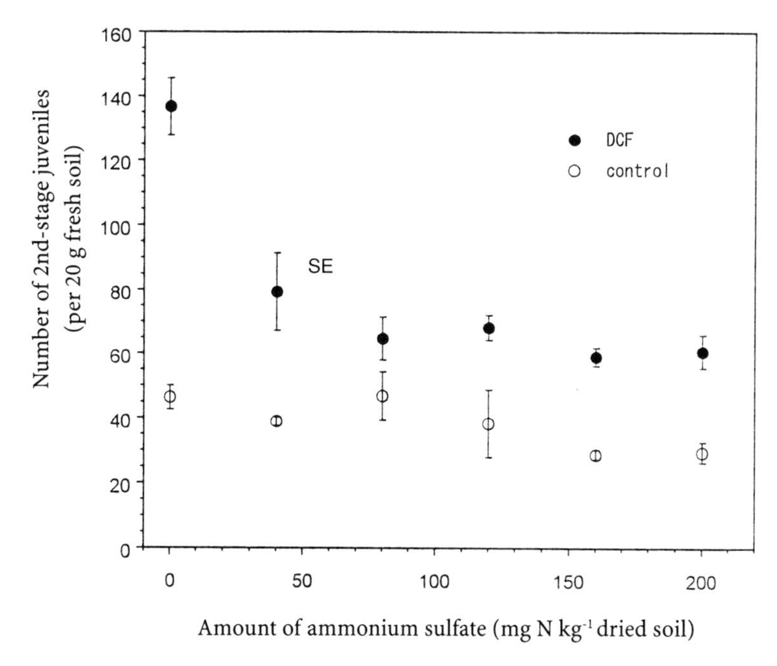

Fig. 3. Effects of ammonium sulfate ($(NH_4)_2SO_4$) supplement on the emergence of 2nd-stage juveniles with and without DCF application. Second-stage juveniles were separated by the Baermann funnel method 10 days after DCF treatment

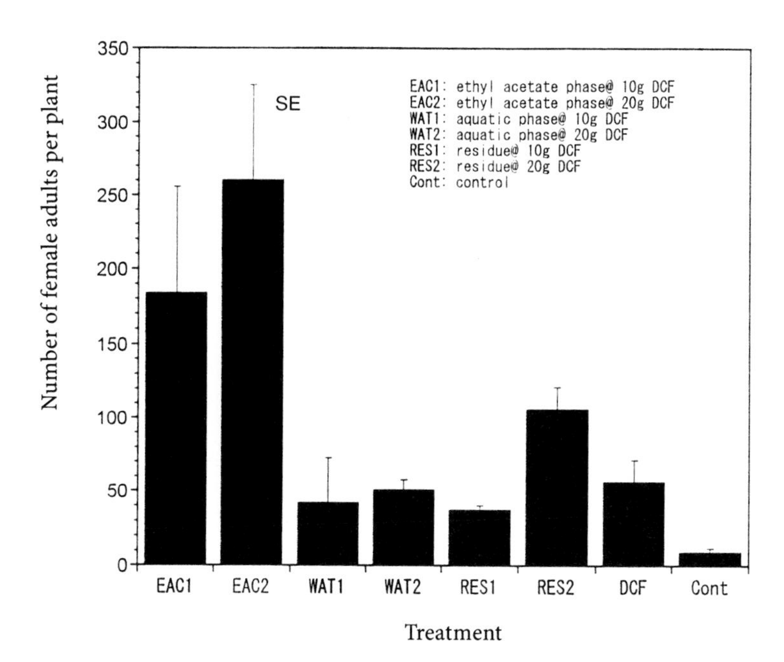

Fig. 4. Effects of application of extracts from DCF on the number of female adults on soybean (T201) roots. The number of female adults was counted by the naked eye 18 days after seed sowing. The dose of treated extracts was in terms of DCF

The activity of several fractions (WAT, aquatic phase; EAC, ethyl acetate phase; RES, the residue of extraction) from DCF on the emergence of 2nd-stage juveniles was compared. Application of EAC markedly increased the number of 2nd-stage juveniles compared to intact DCF or the control (Fig. 4). The effects of WAT and RES were inferior to that of EAC. Additionally, the effect was compared among three fractions from EAC (F1, strong acidity fraction; F2, weak acidity fraction; F3, neutral fraction) on the attachment of female adults on soybean roots (Fig. 5) and on the emergence of 2nd-stage juveniles in the soil (Fig. 6). In both bioassays, the neutral fraction showed the highest activity even though this fraction was insoluble in aquatic solution regardless of the pH.

Because we are now continuing with the separation of the active substance, its structure and the properties are not yet clear. However, this substance might be hydrophobic and tolerant to heat, given that sterilization of DCF with autoclave did not reduce its activity (Matsuo et al. 1992).

Although zinc salts, such as $ZnCl_2$ and $ZnSO_4$, could accelerate the hatching of SCN eggs, recent test proved that zinc salts applied to SCN infested soil did not induce hatching of eggs (Behm et al. 1995). A similar result was obtained for glycinoeclepin A (Tylka 1994), extracted from kidney-bean roots (Masamune et al. 1982), a well-known stimulant of SCN eggs. Therefore the behavior of extracts of DCF in the soil is a very interesting method for seeking useful hatching agents to control SCN.

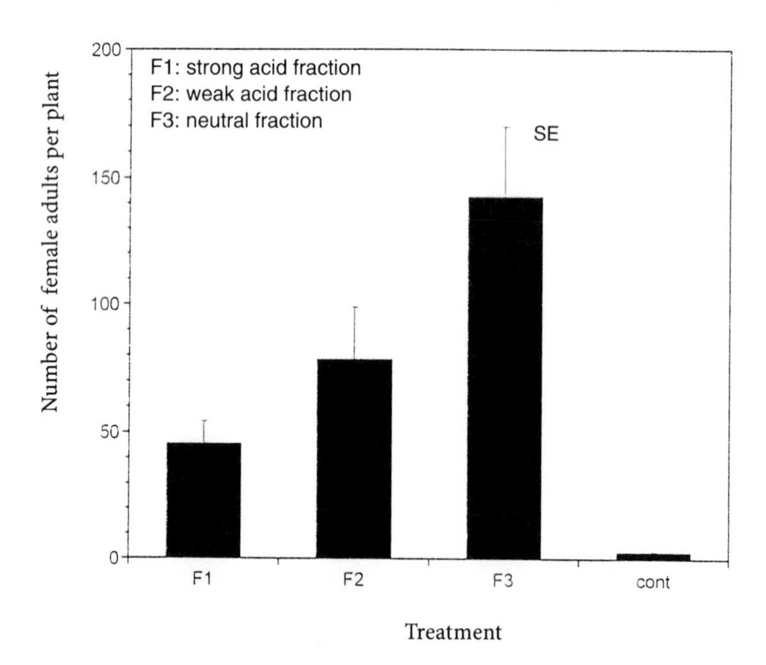

Fig. 5. Effects of three fractions from the ethyl-acetate phase on the attachment of soybean cyst nematodes (SCN). The number of female adults was counted by the naked eye 18 days after seed sowing of soybean (T201). The dose of treated extracts was equivalent to 10 g DCF kg^{-1} dried soil

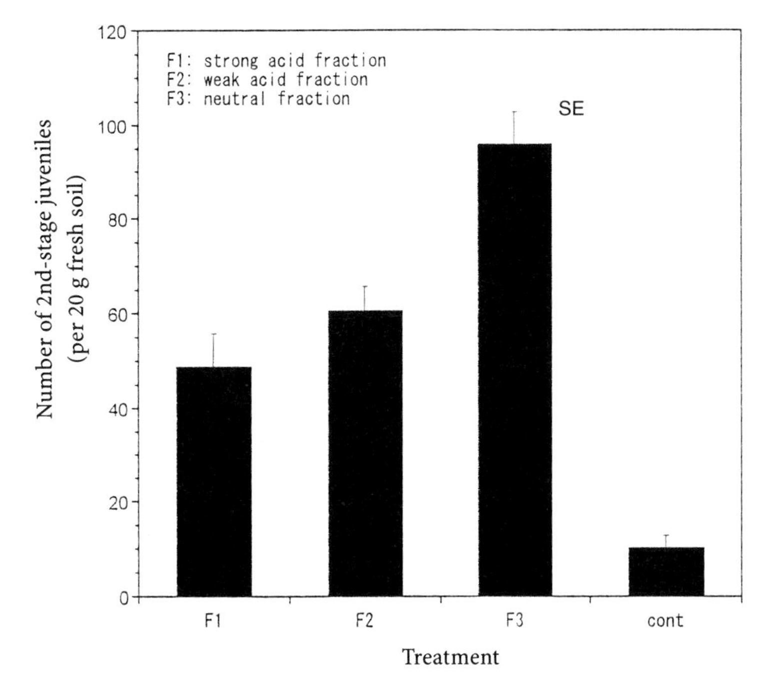

Fig. 6. Effects of three fractions from the ethyl-acetate phase on the emergence of 2nd-stage juveniles. Second-stage juveniles were separated from the soil by the Baermann funnel method 10 days after treatment. The dose of treatments was equivalent to 10 g DCF kg^{-1} dried soil

4. Soybean Cyst Nematode Control with Animal Feces

Because soybean cyst nematode absolutely depends on its host plant for its nutrition, 2nd-stage juveniles emerged from cysts cannot live for long without a host plant. Moreover, the host range of SCN is very narrow (Ichinohe 1955) compared with other nematodes such as *Pratylencus* spp. and *Meloidogyne* spp. Consequently, SCN has the ability to resist severe environmental stresses in the cyst form during host plant absence and then hatching synchronizes the life cycle with the root development of the host plant. Accordingly, DCF application in the absence of the host plant appears to be a useful method for practical management of SCN.

In order to clarify the effects of delaying seed sowing under DCF-treated conditions, the number of female adults was examined 18 days after each seed sowing. The number of female adults was found to have decreased by nearly 75% at 25°C when sowing was delayed for 10 days after DCF application. The reduction in the number of female adults was more rapid at 25°C than at 20°C, but there was scarcely any attachment of SCN at either temperature when sowing was delayed until 60 days after application (Fig. 7). These results suggested that DCF application had contrasting effects. When soybean seed was sown close to the time of DCF application SCN attack on soybean plants was severe, but when sowing was delayed for a certain period SCN would lose their ability to attack soybean plants, probably due to starvation. The effects of DCF were somewhat of a double-edged sword and might explain the fluctuation in the effects of animal by-products in practical usage (Nakasono 1990; Toida et al. 1993; Ueda and Kasai 1993). Although the period of time for starvation will depend on the temperature, we may expect suppressive effects to occur after more than 30 days following application of DCF at the temperatures that occur during the soybean cropping season.

In the evaluation of the SCN control methods, not only the effects on the reduction of crop injury but also the effects on the number of cysts remaining after soybean cropping have to be considered. The effects of DCF application and sowing time of soybean were investigated in the field. The number of cysts following a soybean cropping decreased with the delay in the dates of seed sowing because of the shortening of soybean growing duration in both treatments. When seeds were sown on the same day as treatment application, there was no significant difference in the number of cysts between treatments in spite of the intense attachment of female adults on soybean roots in the DCF treated plots. A theoretical model (Koenning and Sipes 1998) indicated that the reproductive factor derived from the formula Pf/Pi (Pf, final nematode population density; Pi, initial population density) for the host plant might be less than one in those situations where the host plant was severely damaged by high level of the parasite. This may be due, therefore, to a reduction in cyst formation through intraspecific competition by female adults. Nevertheless, the number of cysts in DCF-treated plots decreased by nearly 80% compared to the control when seeds were sown 30 days

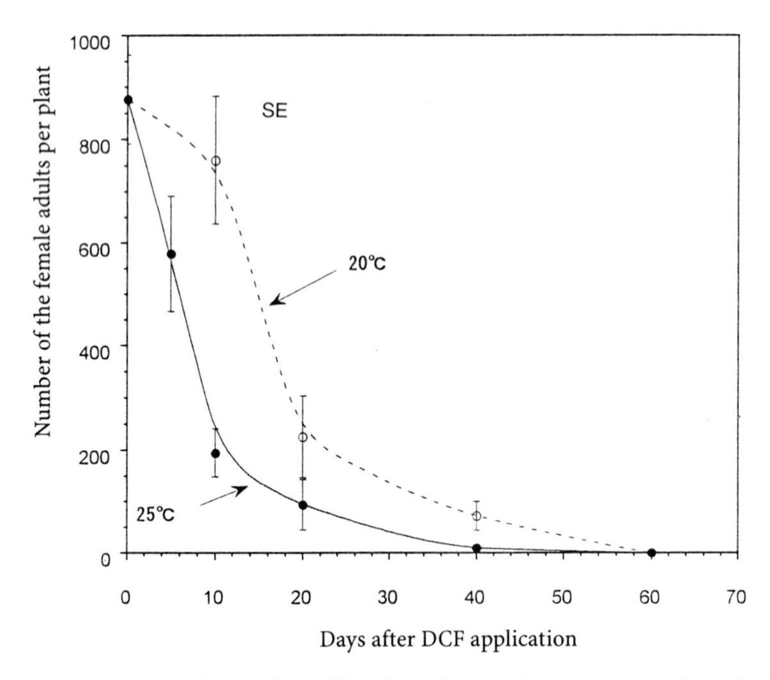

Fig. 7. Relationship between the number of female adults per plant and sowing time after DCF application. The number of female adults on the soybean roots was counted 18 days after each sowing time.

after application (Fig. 8). In the following year in a similar experiment, the number of eggs remaining in the soil after soybean cropping was found to be significantly lower in DCF-treated plots than in the control when seeds were sown 30 days after application (Fig. 9). Therefore, if the DCF application is carried out 30 or more days before seed sowing, not only will nematode attachment be reduced but we may also expect a decrease in cyst density.

In the field experiment, growth of soybean was more vigorous in DCF applied plots than in chemical fertilizer applied plots except for two plots in which the initial population density of cysts was much higher than in other plots. The relationship between grain yield of soybean and initial density of cysts in the previous autumn is shown in Fig. 10. Although there was a tendency that grain yield was reduced with increasing initial density of cysts in all treatments, yield reduction with increasing initial density of cyst was drastic in chemical fertilizer applied plots but only moderate in DCF applied plots. These results suggested that DCF had the effect of reducing injury caused by SCN. However, the effects of DCF differed with the times of application. Application in the previous autumn showed less effect than in early summer. This might be due to seasonal changes in the sensitivity of SCN eggs to hatching stimulants. Although dormancy of SCN was not remarkable in subtropical or tropical regions, in the temperate zone dormancy characterized by low hatching rates was observed even when temperatures were suitable for hatching (Ross 1963; Okada 1972).

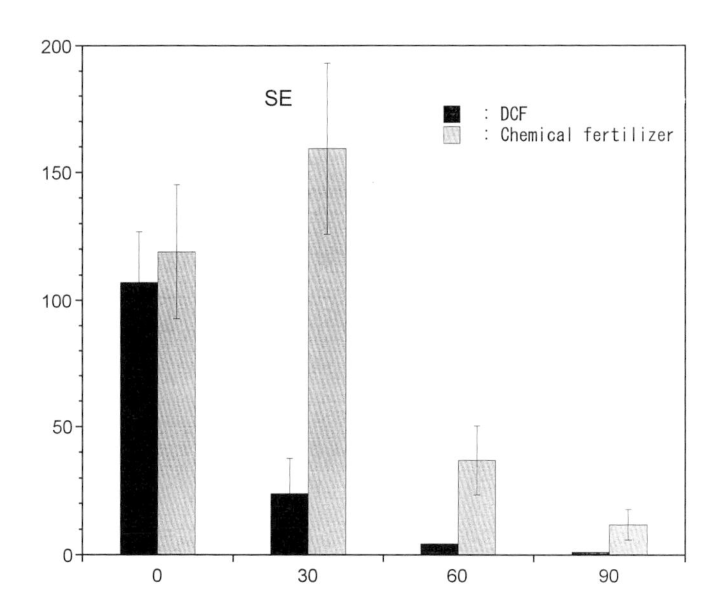

Fig. 8. Effects of DCF application at different sowing times of seeds on the number of remaining cysts after soybean (cv. Tachinagaha) cropping in 1994. Cysts were separated from the soil with sieving and empty cysts were excluded from the number of cysts

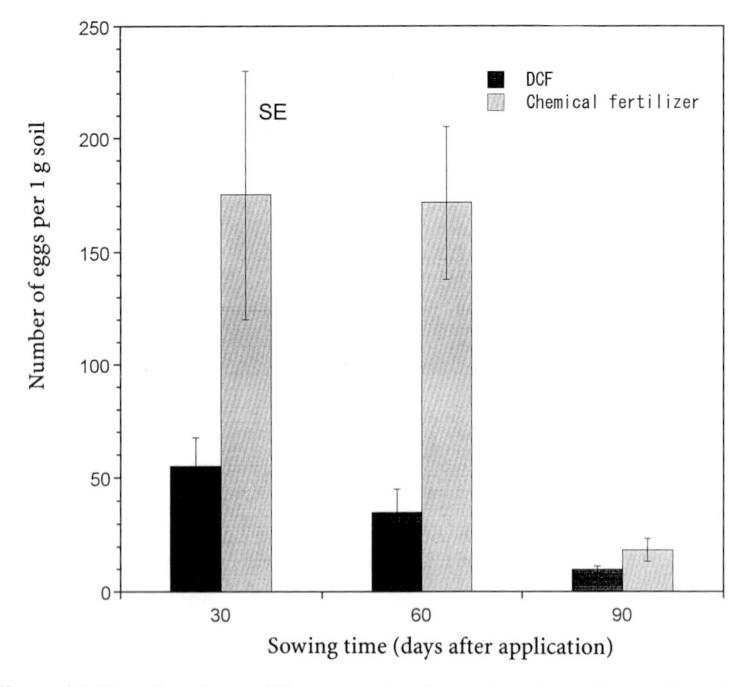

Fig. 9. Effects of DCF application at different sowing times of seeds on the number of remaining eggs after soybean (cv. Tachinagaha) cropping in 1995. Cysts were separated from the soil by sieving and eggs were obtained by homogenizing cysts

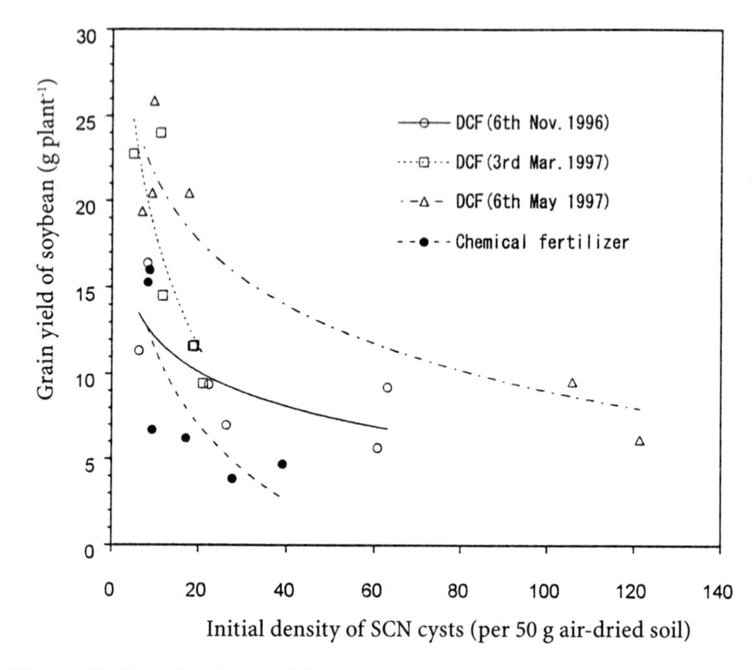

Fig. 10. Effects of DCF application at different times on the grain yield of soybean (cv. Enrei)

5. Concluding Remarks

The effects of DCF in regulating SCN population density will produce good results if a proper balance among DCF, SCN, and the soybean plant is achieved. Because the effects of DCF are based on acceleration of hatching, we can relieve the yield damage of SCN and decrease the population density of SCN effectively if a sufficient period of time from DCF application to seed sowing is ensured. Conversely, DCF will aggravate the injury of SCN when the sowing time is missed; therefore, the relationship among DCF, SCN, and soybean plants in the practical use of DCF as a hatching agent must be considered carefully.

There exists a limitation in reducing yield damage by SCN with DCF application when the population density of cysts is remarkably high. In such a case, there is no choice but to take measures such as an appropriate crop rotation for decreasing the population density of SCN. We previously reported that adequate yields of corn and barley in a double-cropping system could be obtained with DCF application at 400-500 kg N ha^{-1} without additional chemical fertilizer application in an Andosol (Matsuo et al. 1993b). Because an effect of DCF on reduction of SCN density can be expected in the absence of host plants, crop rotation, including the above double-cropping system, and use of DCF will contribute to SCN control and alternation of fertilization. In addition, the term of rotation to decrease the population density will be shortened. Such a multipurpose use must eventually enhance the utilization of by-products from livestock production.

References

Behm JE, Tylka GL, Niblack TL, Wiebold WJ, Donald PA (1995) Effect of zinc fertilization of corn on hatching of *Heterodera glycines* in soil. J Nematol 27:164-171

Carter W (1943) A promising new soil amendment and disinfectant. Science 97:383-384

Clarke AJ, Shepherd AM (1966) Inorganic ions and the hatching of *Heterodera* spp. Ann Appl Biol 58:497-508

Chikaoka I, Fujiwara S, Takezawa H (1981) Effect of livestock excrements against soil nematodes (in Japanese). Proc Kanto Plant Prot Soc 28:140-141

Fukuzawa A, Furusaki A, Ikura M, Masamune T (1985) Glycinoeclepin A, a natural hatching stimulus for the soybean cyst nematode. J Chem Soc 4:222-224

Golden AM, Epps JM, Riggs RD, Duclos LA, Fox JA, Bernard RL (1970) Terminology and identity of intraspecific forms of the soybean cyst nematode (*Heterodera glycines*). Plant Dis Reporter 54:544-546

Hill NS, Schmitt DP (1989) Influence of temperature and soybean phenology on dormancy induction of *Heterodera glycines.* J Nematol 21:361-369

Ichinohe M (1955) Studies on the morphology and ecology of the soybean nematode *Heterodera glycines* in Japan. Res Bull Hokkaido National Agricultural Experimental Station 48:1-64

Koenning SR, Sipes BS (1998) Biology. In: Sharma SB (Ed) The cyst nematodes. Chapman & Hall, London, pp156-190

Kumazawa K (1999) Present state of nitrate pollution in groundwater. Jpn J Soil Sci 70:207-213

Lehman PS, Baker KR, Huisingh D (1971) Effects of pH and inorganic ions on emergence of *Heterodera glycines.* Nematologica 17:467-473

Masamune T, Anetai M, Takasugi M, Katsui N (1982) Isolation of a natural hatching stimulus glycinoeclepin A, for the soybean cyst nematode. Nature 297:495-496

Matsuo K (1992) Effects of application of dried cattle feces on population dynamics of the soybean cyst nematode (*Heterodera glycines* Ichinohe) 1. Number of the second-stage juveniles in the soil and female adults on the soybean roots. Jpn J Nematol 24:69-74

Matsuo K, Tsuji H, Yamamoto H, Matsumoto D (1993a) Initial growth of upland crops under dried feces applied condition (in Japanese). Jpn J Crop Sci, 62 (extra issue 1):60-61
Matsuo K, Tsuji H, Yamamoto H, Matsumoto D (1993b) Crop growth and nitrogen dynamics under dried feces applied condition (in Japanese). Jpn J Crop Sci, 62 (extra issue 1):58-59
Nakasono K (1990) Organic amendments and crop injuries by plant parasitic nematodes (in Japanese). Plant Protection (Tokyo) 44:535-538
Nishizawa T, Shimizu K, Nagashima T (1972) Chemical and cultural control of the rice cyst nematode, *Heterodera oozy* Luc *et* Berdon Brizuela, and hatching responses of larvae to some root extracts (in Japanese with English summary). Jpn J Nematol 2:27-32
Okada T (1972) Hatching inhibitory factor in the cyst contents of the soybean cyst nematode *Heterodera glycines*. Appl Entomol Zool 9:49-51
Okada T (1977) Studies on the hatching factors of the soybean cyst nematode, *Heterodera glycines* Ichinohe (in Japanese with English summary). Jpn J of Nematol 6:1-49
Riggs RD, Niblack TL (1993) Nematode pests of oilseed crops and grain legumes. In: Evans K, Trudgill DL, Webster JM (Eds) Plant parasitic nematodes in temperate agriculture. CAB International, Wallingford, pp 209-258
Riggs RD, Schuster RP (1998) Management. In: Sharma SB (Ed) The cyst nematodes. Chapman & Hall, London, pp 388-416
Ross JP (1963) Seasonal variation of larvae emergence from cyst of the soybean cyst nematode, *Heterodera glycines*. Phytopathol 53:608-609
Schmitt DP, Riggs RD (1991) Influence of selected plant species on hatching of eggs and development of larvae of *Heterodera glycines*. J Nematol 23:1-6
Shimizu K (1986) Effect of temperature on the resistance of soybean to *H. glycines* Ichinohe (in Japanese with English summary). Jpn J Nematol 16:32-34
Shimizu K, Aiba S, Mitsui Y (1989) Soybean cyst nematode of soybean. Ext Bull No.302, Food and Fertilizer Technology Center, ASPAC, pp 7-9
Slack DA, Riggs RD, Hamblen ML (1981) Nematode control in soybeans: Rotation and population dynamics of soybean cyst and other nematodes. Arkansas AES Rep Ser 263:1-36
Smith GS, Niblack TL, Minor HC (1991) Response of soybean cultivars to aldicarb in *Heterodera glycines*-infested soils in Missouri. J Nematol (Suppl) 23:693-698
Tefft PM, Bone LW (1984) Zinc mediated hatching of eggs of soybean cyst nematode, *Heterodera glycines*. J Chemist Ecol 16:361-372
Toida Y, Keetrrwan S, Tangjitsomkid N (1993) Effects of addition of organic materials on population density of *Meloidogyne javanica* attacking mungbean in Thailand. Jpn J Nematol 23:90-93
Tsuiki M, Harada Y (1997) Quantitative estimation of nutrient flow in dairy farms (1) Nitrogen flow. Jpn J Agric Syst Sci 12:113-117
Tsutsumi M, Sakurai K (1966) Influence of root diffusates of several host and non-host plants on the hatching of the soybean cyst nematode, *Heterodera glycines* Ichinohe (in Japanese with English Summary). Jpn J App Entomol Zool 10:129-137
Tylka G (1994) A biorational approach to soybean cyst nematode management in Iowa. Leopold Center Progress Reports 3:41-44
Ueda Y, Kasai Y (1993) Efficacy of mixing organism in soybean nematode-inhabited soil (in Japanese). Proc Kanto Plant Protection Soc 40:301-302

Part V.
Direct Incorporation of Soil Micro and Macro Organic Molecules

The Effectiveness of Nitrogen Derived from Organic Matter: Results from Long-Term Experiments

Paul R. Poulton

Summary. All agricultural systems make use of nitrogen derived from organic matter. Long-term experiments like those at Rothamsted have been used to quantify the effectiveness with which that N is used.

On a sandy soil at Woburn Experimental Farm, soil organic matter (SOM) was increased by regular additions of animal manure or alternating periods of grass or grass/legume leys with arable crops. Residues from crops grown with inorganic fertilizers alone were not enough to prevent SOM levels declining unlike on the heavier silty clay loam at Rothamsted. At Woburn yields of cereals and root crops were greater, and less fertilizer N was needed to achieve maximum yield, on soils with higher organic matter inputs even though any increase in SOM was modest. However, losses of N by leaching were also greater.

At Rothamsted the additional benefits of SOM have become more apparent with higher yielding cultivars. On soils with less SOM the same yield could not be matched with fertilizer N. The mineralization of N at times during the season and its distribution in the soil profile is one reason for increased crop yield. However, this is also one reason why the risk of loss by leaching or denitrification is greater, i.e., the lack of synchrony between mineralization and N uptake by the crop.

The long-term experiments are also being used to improve our understanding of the complex processes and interactions which govern mineralization, nitrification, immobilization, uptake and loss. Labeled ^{15}N has been used to follow the fate of crop residues and more recently to measure gross rates of mineralization and nitrification. Better understanding of how nitrogen is cycled within the soil should allow us to make better use of a valuable resource.

Key words. Long-term experiments, Soil organic matter, Gross mineralization, Leaching, Denitrification

1. Introduction

All plants take up nitrogen derived from organic matter to a greater or lesser extent. Some agricultural systems depend solely on N mineralized from organic matter, others rely largely, but not exclusively, on fixed N. In many systems mineralized N will be supplemented by inorganic fertilizer N to achieve a greater yield. Understanding the role played by "organic" N, i.e., its source, rate of mineralization, nitrification and immobilization, uptake and loss, is essential if the farming community is to make the best use of the resources available to it. Better understanding will also help minimize the risk of environmental pollution by nitrate leaching or gaseous loss. It is important, therefore, to quantify, as far as possible, the amount of N being made available for crop uptake or loss. It is also important to note that N derived from biological N fixation has to go through the mineralization process before becoming available to other, non-fixing, crops.

Recent papers by Schulten and Schnitzer (1998), Jarvis et al. (1996), and Mengel (1996) have reviewed current knowledge on the chemistry of soil organic nitrogen, the processes involved in its assimilation and mineralization and its availability to plants and/or loss. These papers illustrate the complexity of the subject, the difficulty in measuring many of the processes involved and the interactions between those processes. The processes, mediated by the microbial population, are subject to considerable variability. They are influenced by many factors including moisture, temperature, soil physical characteristics and conditions, interactions with growing plants and the amount of inorganic N present in the soil. The material which is being acted upon will also vary greatly in e.g., amount, C:N ratio, the degree of lignification or the proportion of N present in protein. It will include animal manures (farmyard manure or slurry), grass or legume leys, recent crop residues ranging from cereal straw to green leafy material. It will also include root exudates, dead microbial biomass and older organic matter present in the soil.

Quantifying the amount of N made available for uptake or loss in the field is not easy. Crop uptake can be measured in the absence of fertilizer N but there is in some areas a significant, confounding, input of N by atmospheric deposition. The crop will reflect to a certain extent the net effects of mineralization, nitrification, and immobilization. Losses of N, either gaseously or by leaching are much more difficult to measure. Both uptake and loss will vary greatly from season to season.

This paper gives examples of how long-term experiments may be used to further our understanding of the subject. Because of the inherent variability mentioned above, and the fact that some of the interactions may not become apparent for some time, it is important to draw conclusions only when several years data are available. There are many long-term experiments around the world (Paustian et al. 1996; Körschens 1996; Steiner 1995). These cover a wide range of soil types, climatic conditions and management regimes. In this paper examples are taken from experiments run by Rothamsted. These have a well-documented history. Manure and fertilizer inputs are recorded; crop offtakes are measured. Soils, some of them dating back > 100 years, have been regularly sampled and analyzed for total N and organic C. The sites are at Rothamsted itself on a silty clay loam (Chromic Luvisol; FAO) containing about 20%–25% clay (Avery and Catt 1995) and on a sandy loam (Cambic Arenosol; FAO) at Woburn which contains about 8%–14% clay (Catt et al. 1980). Both sites are in southeast England, a cool temperate region. Average annual rainfall is 600-700 mm. Nitrogen is not the only nutrient made available during the mineralization of SOM. Organic P and S will also be mineralized and limited amounts made available for crop growth. In addition animal manures contain substantial amounts of P and K together with micronutrients. In the experiments referred to in this chapter where the effects of SOM on crop yield are being evaluated every effort has been made to ensure that yield is not limited by lack of plant available P, K or other nutrients, soil pH, pests, or disease unless otherwise stated.

2. Discussion

2.1 Effect of Management on Soil Organic Matter

The Broadbalk Wheat Experiment is the best known of the long-term experiments at Rothamsted (Johnston 1994). First sown in autumn 1843, winter wheat has been grown on all or part of the experiment every year since. Treatments include a control which receives no inorganic fertilizer or organic manure. Other plots have received P, K and Mg fertilizer together with different rates of inorganic N. The effect of these treatments on crop yield and soil organic matter is compared with plots receiving 35 t ha^{-1} of farmyard manure (FYM) each year, supplying ca. 220-240 kg N ha^{-1} $year^{-1}$. Figure 1 shows the %N in the topsoil (0–23 cm) of some treatments of the Broadbalk experiment. The effect of the FYM applications can be clearly seen. On this soil type total N and organic C increased rapidly in the years immediately after manure applications started as 30%–50% of the N and C was immobilized (Johnston 1969). The decline in the 1920s is because at this time the experiment was divided into five sections, one section was fallowed each year to control weeds and FYM was not applied in the fallow years. On the control plot, soil organic matter has been in equilibrium for more than 100 years, it contains c. 0.10% N. Plots which have received inorganic N have grown larger crops, thus returns to the soil of stubble, roots and root exudates have been greater (straw is removed) and the total N content is higher, e.g. 0.12% N where 144 kg N ha^{-1} has been applied each year since 1852. These soils have also been in equilibrium for many years, taking perhaps 20–30 years to reach a new equilibrium value (Glendining et al. 1996). Compared with FYM the effects of long-continued N fertilizer applications are smaller and more subtle. Some of these effects are discussed later.

On Broadbalk the soil had been in arable cultivation for several hundred years before the experiment began and the total N and organic C content of the soil was relatively low (for this soil type) compared to other fields on the Rothamsted farm (Johnston et al. 1981). In contrast, where soils have been in grass for > 300 years, total N in the 0-23 cm layer is about 0.3% N. Soils plowed out of permanent grass into arable cropping will lose N rapidly at first as the organic matter is mineralized. Some of this mineralized N will be available for crop uptake but much will be lost gaseously or by leaching. The rate of decline in SOM on a particular soil type will depend on the management system adopted. Thus, a rotation which included two root crops every three years lost N and C more rapidly than one in which root crops were grown in only two out of every six years (Fig. 2; Johnston 1986).

The amount of organic matter in the soil is very dependent on soil type, particularly clay content. Hence similar management systems on different soils will

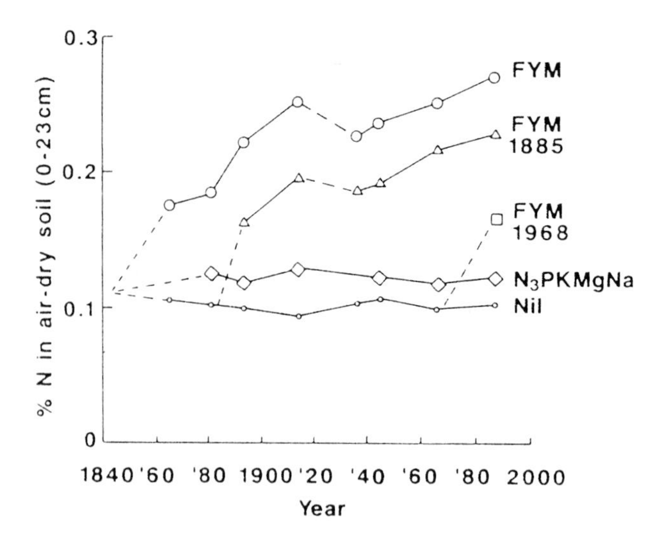

Fig. 1. Broadbalk Wheat experiment, Rothamsted. Total nitrogen in topsoil (0–23 cm). *Smaller circles*, Nil, unmanured since 1843; *diamonds*, PKMgNa fertilizer + 144 kg N ha^{-1} since 1852; *larger circles*, farmyard manure (FYM), 35 t ha^{-1} since 1843; *triangles*, FYM, 35 t ha^{-1} since 1885; *squares*, FYM, 35 t ha^{-1} since 1968. (Adapted from Johnston 1969)

give rise to different amounts of SOM. In contrast to the silty clay loam at Rothamsted which contains 20%–25% clay, the sandy soil at Woburn Experimental Farm contains 8%–16% clay. Experiments started at Woburn in 1876. At that time the topsoil contained about 1.6% C. Soils which have been in continuous arable cropping since then now contain c. 0.8% C (Fig. 3) and have not yet reached equilibrium (Mattingly et al. 1975).

2.2 Effect of Soil Organic Matter on Yield

The value of maintaining or increasing SOM became apparent much earlier on the lighter soil at Woburn than at Rothamsted. Several experiments were started which looked at the benefits of organic inputs. These included experiments on green manures (Dyke et al. 1977) and various sewage sludges and composts (Johnston et al. 1989).

One experiment, the Woburn Ley-Arable, started in 1938. The soil contained 10%–16% clay and had an initial organic carbon content of 0.98%. A 5-year rotation consisted of a 3-year treatment phase followed by two arable test crops to assess the effects of organic matter. Managements during the treatment phase include: i) continuous arable crops, ii) all grass ley + fertilizer N, iii) a grass/clover ley. Data from the first 30 years of the experiment were given by Johnston (1973).

Soils in continuous arable cropping declined in organic carbon from 0.98% C in 1938 to 0.81% C by the mid-1980s. Soils with 3-year all grass or grass/clover leys followed by two arable test crops have increased to c. 1.10%–1.15% C. The test crops both tested four amounts of fertilizer N on subplots (Poulton and Johnston 1996). Table 1 shows the mean yield, over 10 years, of winter wheat and spring barley. Without fertilizer N, wheat following the all grass or grass/clover

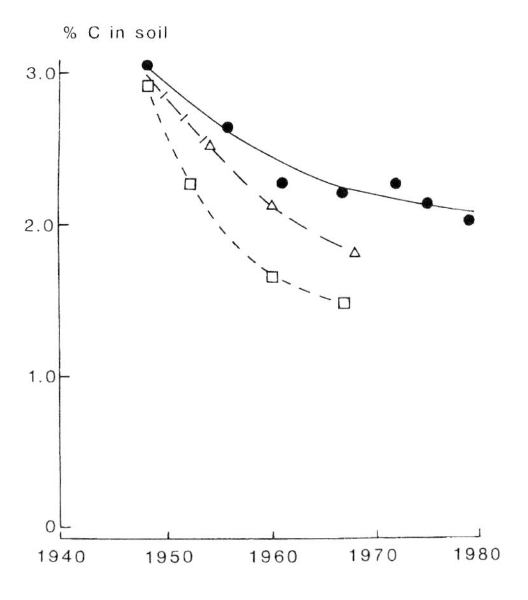

Fig. 2. Rothamsted. Effect of rotation on organic carbon in soil (0–23 cm) plowed out of old grassland. *Circles*, six-course rotation, 2 roots, 3 cereals, 1-year ley; *triangles*, three-course rotation; 2 roots, 1 cereal; *squares*, permanent fallow. (From Johnston 1986)

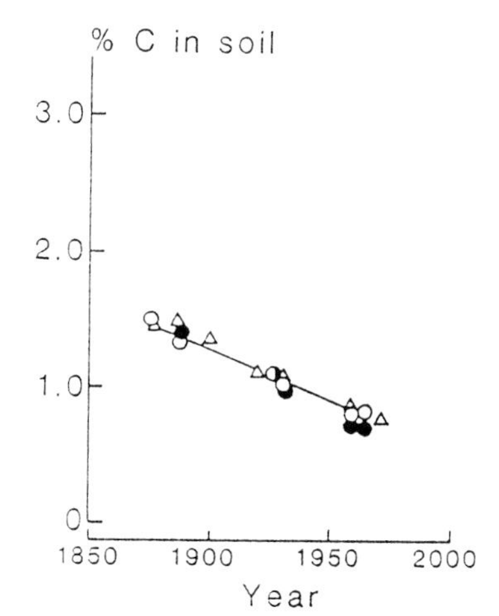

Fig. 3. Woburn. Organic carbon in soil (0–23 cm). Cereals each year, given inorganic fertilizer or occasional organic manure. (From Mattingly et al. 1975)

leys yielded 0.9 and 2.3 t ha^{-1}, respectively, more grain than wheat in the continuous arable rotation, reflecting the different amounts of N made available from the breakdown of the different residues. For maximum yield 210 kg N ha^{-1} was required in the continuous arable rotation and after the grass ley but only 140 kg N ha^{-1} was needed when wheat followed the grass/clover ley. However, maximum yield was almost achieved with 140 kg N ha^{-1} after the all grass ley and 70 kg N ha^{-1} after grass/clover. The same maximum yield was achieved in the continuous arable rotation and following the all grass leys, 7.54 and 7.58 t ha^{-1} respectively but was greater following grass/clover.

Yields of spring barley in the second year following the plowing of the all grass and grass/clover leys were the same at each rate of N tested. This suggests that the extra N made available to the first test crop of winter wheat from the leguminous residue was not available to a second, spring sown, cereal. Any additional N produced by mineralization of the leguminous residue in autumn may have been leached during the intervening winter.

Table 2 shows that winter wheat and spring barley both took up more N when following leys than in the continuous arable rotation. Furthermore, in the first year, winter wheat acquired 23 kg ha^{-1} more N following the grass/clover ley compared to the all grass ley. The large amount of available N from the leguminous residues was not unexpected. In the second year spring barley acquired the same amount of extra N from both leys.

Spring barley in all rotations gave its maximum yield with 120 kg N ha^{-1} and lower yields with 180 kg N ha^{-1}, suggesting that N was not limiting yield. Thus, the larger yields following the leys (Table 1) suggests that there was a benefit from the extra organic matter accumulated in soil where leys had been grown for three in every five years since 1938. Spring-applied fertilizer N did not achieve the same maximum yield; similar effects have been reported for potatoes (Johnston et al. 1994). This is discussed later. This benefit of extra soil organic matter was not so apparent in the yields of winter wheat. In the rotation with the grass/clover ley maximum yield, 8.1 t ha^{-1}, was given by 140 kg N ha^{-1} and yield was less with 210 kg N ha^{-1}, again suggesting that the amount of N was not limiting yield. This maximum yield of 8.1 t ha^{-1} following grass/clover suggests, as with the barley, that winter wheat benefited from the extra soil organic matter accumulated under this rotation. That the maximum yield of wheat following the all grass ley was not better than in the continuous arable rotation needs to be explained. The lack of benefit from the extra soil organic matter may be related to the amount

Table 1. Woburn Ley-Arable. Yield of winter wheat (1981–90) and spring barley (1982–91) following different rotations (from Poulton and Johnston 1996)

Fertilizer N kg ha^{-1}	Rotation: Continuous arable	3 year grass + N	3 year grass/clover
		Wheat grain, t ha^{-1} at 85% DM	
0	3.26	4.18	5.54
70	6.00	6.73	7.70
140	7.07	7.46	8.10
210	7.54	7.58	7.56
		Barley grain, t ha^{-1} at 85% DM	
0	2.29	4.14	4.16
60	4.57	5.70	5.64
120	5.41	5.93	5.85
180	5.14	5.60	5.40

and timing of the nitrate released by mineralization and nitrification of the ley residues which were less in total amount following the all grass ley compared to the grass/clover ley. The amount of extra N, compared to the continuous arable rotation, varied greatly from year to year. In some years uptake following the grass leys was less than that on the continuous arable treatments, presumably as N was temporarily immobilized as grass residues having a wide C-to-N ratio decomposed.

The extra N taken up by the test crops (Table 2) represents only part of the N mineralized from SOM and/or readily mineralizable residues. Much N will be mineralized in early autumn when there is little or no crop uptake. It will therefore be at risk to loss by leaching or denitrification (see later).

In 1964 the Organic Manuring experiment was started at Woburn. This was seen as a logical extension to the three experiments mentioned above. An initial treatment phase (1965-71) included leys, FYM, green manures and straw additions as well as plots which received inorganic fertilizers only. This treatment phase was followed by two 4-course rotations of arable crops (Mattingly 1974).

Table 3 shows the increases in organic C and total N in the topsoil for selected treatments during the initial treatment phase (see Mattingly et al. 1974 for further details). With organic treatments both organic C and total N were increased. Where FYM was applied between 1965–1971 about 50% of the carbon and 37% of the nitrogen remained in the top 23 cm in 1971. In contrast, where straw was incorporated, only 13% of the carbon but much more of the N (48%) remained in the soil. Where grass/clover leys were grown, about 8-10 t ha^{-1} of organic matter (c. 5 t C ha^{-1}) was accumulated which contained about 170 kg N ha^{-1}. Plots receiving only inorganic fertilizers declined in C and N. The annual rate of decline (0.008% C) was similar to the average longer-term decline measured in this soil type.

Table 2. Woburn Ley-Arable. Mean annual N uptake, kg ha^{-1}, by grain on plots given no fertilizer N (from Poulton and Johnston 1996)

Rotation	Wheat 1981-90	Barley 1982-91	Total in 2 years	"Extra" N from leys
Continuous arable	48	32	80	—
3-year grass + N	66	57	123	43
3-year grass/clover	89	60	149	69

Table 3. Woburn Organic Manuring. Organic C and total N in topsoil (0–23 cm) in 1971 and increase between 1964–1971 (from Mattingly 1974)

Treatment 1965-71	Organic C, %		Total N, %	
	1971	Increase	1971	Increase
Straw	0.923	0.094	0.094	0.004
FYM	1.034	0.304	0.097	0.017
Grass/clover	0.918	0.153	0.091	0.006
PK fertilizer	0.696	-0.048	0.078	-0.004

The effect of these treatments on the yield of four subsequent test crops is shown in Table 4 (Mattingly 1977). Eight amounts of N were tested on each crop. Table 4 shows data averaged over the four smallest and the four largest rates of N. Yields of all four crops were always greater on soils with more organic matter. Yields were least, at all levels of applied N, on soils with least organic matter. After some modifications a further cycle of treatments started in 1979 followed by test cropping which began in 1987. In 1997 winter rye was grown with no added fertilizer N. There were still significant differences in yield depending on the previous organic matter treatment (Table 5, unpublished data). Yields were least on the fertilizer only treatment and largest on treatments which had grown grass/clover leys or received FYM in the two treatment phases, i.e., those treatments on which SOM was increased most.

2.2.1 Separating and Quantifying the Effects of Soil Organic Matter

The long-lasting benefits of the extra organic matter is not necessarily due to any extra N which may be mineralized and made available to the crop. It can be difficult to disentangle the physical effects of increased SOM, e.g. improved soil structure and therefore better root growth, greater water holding capacity, etc. from the effect of extra N. Dyke et al. (1983) described a curve fitting technique which attempts to separate and quantify these effects. We used the same techniques with data from an experiment on the sandy soil at Woburn (Johnston et al. 1994). Test crops of winter wheat and potatoes were grown following grass/clover leys of different age. Each test crop tested six amounts of fertilizer N. The response curves for the first and second test crops, wheat and potatoes, are shown in Fig. 4a,c. The six curves can be brought into coincidence by a series of diagonal shifts (Fig. 4b,d). The diagonal shift consists of a vertical and a horizontal component. The horizontal component can be attributed to an effective difference in spring fertilizer N addition. The vertical component can be attributed to factors other than spring applied N. These factors would include soil structure, water holding capacity, and others. They may also include nitrogen which is mineralized from SOM at times during the growing season and distributed through the soil profile in a way which is not mimicked by surface-applied spring N. In this experiment there was a significant benefit of the extra OM in the first year after plowing up the grass/clover leys which could be equated to fertilizer N. Compared to the one year old ley this benefit ranged from 6 to 126 kg N ha^{-1} for the 2- to 6-year leys. The vertical shift was worth up to 0.95 t ha^{-1}, an 11% increase in the maximum yield. For the second test crop, potatoes, the horizontal shift was small but the vertical shift ranged from 1.6 to 8.1 t ha^{-1} tubers, an increase of up to 13% in the maximum yield that could be achieved. For winter wheat grown in the third year after the leys were plowed (data not shown), the horizontal component was again small but the vertical shift was worth up to 1.5 t ha^{-1} of grain. In this year summer rainfall was below average and mean yields were less than in the first test year; the effect of the extra SOM was proportionately greater resulting in a 26% increase in maximum yield (Johnston et al. 1994). Thus, on this soil type, some of the ben-

Table 4. Woburn Organic Manuring. Yields, t ha^{-1}, of potatoes, winter wheat, sugar (from sugar beet) and spring barley, 1972–76 (from Mattingly 1977)

Crop	Years	N rate	Treatment 1965-71			
			Straw	FYM	Grass/ clover	PK fertilizer
Potato tubers	1972–73	Low[1]	35.4	38.3	43.8	32.0
		High	47.4	50.8	49.7	43.8
Wheat grain	1973–74	Low	3.60	3.85	5.14	3.22
		High	6.10	5.73	5.66	5.42
Sugar from sugar beet	1974–75	Low	2.40	2.75	2.83	2.11
		High	3.08	3.38	3.25	2.90
Barley grain	1975–76	Low	3.45	3.32	3.56	2.96
		High	4.14	3.91	4.12	3.81

[1] Means of four lowest and four highest rates of N tested on each crop (see text)

Table 5. Woburn Organic Manuring. Organic C in 1995 and yield, t ha^{-1}, of winter rye in 1997

Treatment 1965–71 & 1979-86	Organic C, % 1995	Grain yield t ha^{-1} 1997
Straw	0.79	1.99
FYM	0.96	2.24
Grass/clover	0.82	2.30
PK fertilizer	0.68	1.70

efits associated with increased SOM could not be equaled simply by adding more fertilizer N. Similar effects have also been seen in tropical climates. For example Nambiar (1994) reporting data from long-term experiments in India showed that yields could be maintained or increased by using organic manures and inorganic fertilizers together but that yields declined when fertilizers alone were used even though the amount of N P K supplied was greater in the latter instance. Nambiar attributed these benefits to the correction of marginal deficiencies in secondary nutrients and micronutrients during decomposition of the organic manure and its beneficial effect on the physical and biological properties of the soil.

Although the benefits of maintaining or increasing SOM can be demonstrated in the experiment described above, some of the disadvantages are also apparent. Losses of NO_{3-} N after the grass/clover leys were plowed were large. The leys were cultivated in July 1986 and sown to wheat in September 1986. Between October and the following April the minimum net loss (decrease in soil inorganic N minus uptake by crop) from the crop:soil system was 115 kg N ha^{-1} following the three to six year leys; most is likely to have been lost by leaching (Johnston et al. 1994). Losses of N are discussed in greater detail later.

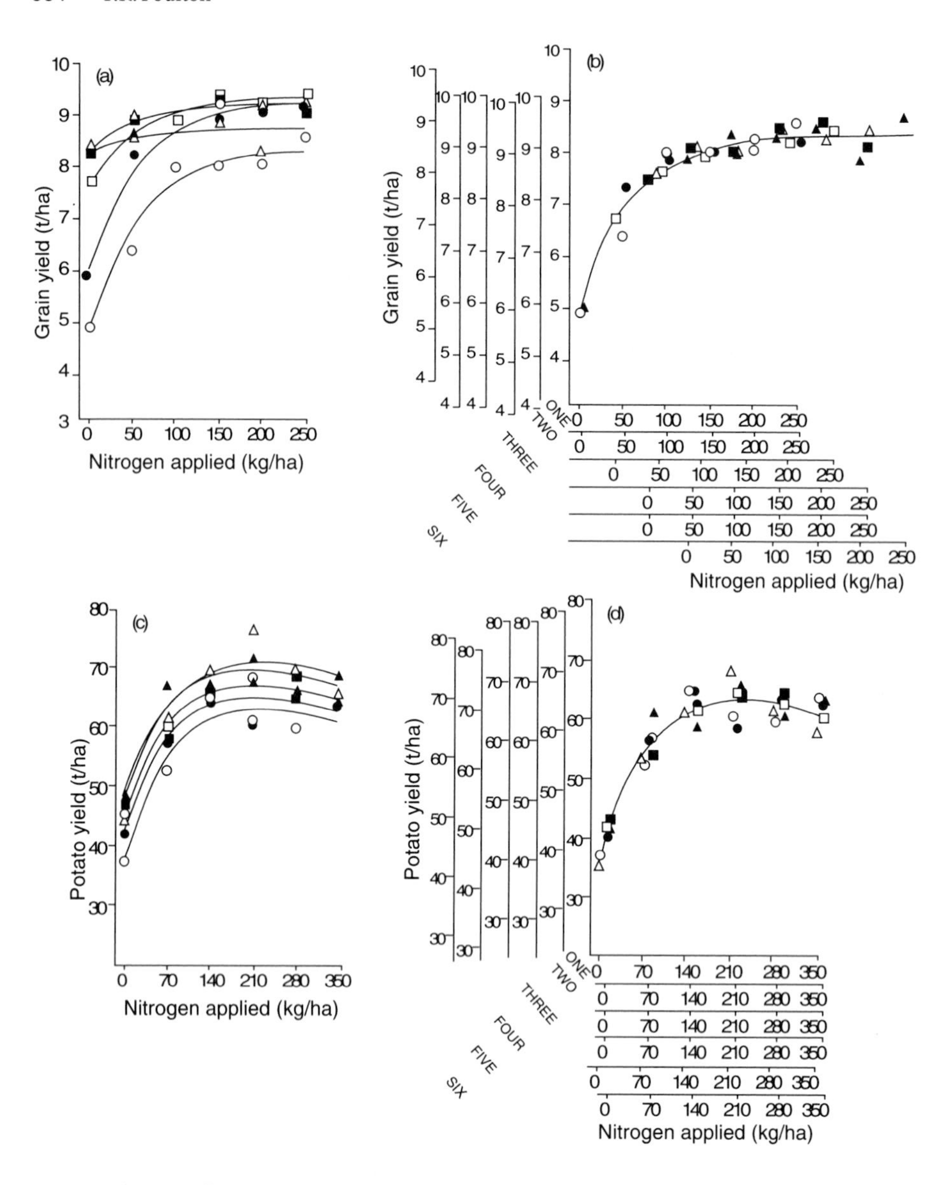

Fig. 4. Woburn. Effect of ley age (years) on response to fertilizer N of two arable crops. **(a)** winter wheat **(c)** potatoes grown in succeeding years following plowing the leys. The individual N response curves were then brought into coincidence by appropriate vertical and horizontal shifts: **(b)** winter wheat **(d)** potatoes. Ley age in years: 1, *open circles*; 2, *closed circles*; 3, *open squares*; 4, *closed squares*; 5, *open triangles*; 6, *closed triangles*. (From Johnston et al. 1994)

2.3 Exploiting the Potential of Modern Cultivars

The benefits of maintaining SOM on the sandy soil at Woburn have been clearly seen over many years. On the heavier textured soil at Rothamsted any additional benefits were not appreciated until the 1960s. It then became clear that root crops grown on soils with more SOM did benefit from that extra organic matter and that yields could be greater than when inorganic fertilizers only were used (Johnston 1994). Similarly with both winter wheat and spring barley there was apparently little or no additional benefit to be gained in terms of increasing crop yield by increasing SOM. The same yield could be obtained on a low organic matter soil simply by adding sufficient fertilizer N (Johnston 1986). In recent years however where cultivars with a higher yield potential have been grown and where that yield potential has been protected by the proper use of pesticides, this is no longer true. The Hoosfield Spring Barley Experiment at Rothamsted started in 1852. Like Broadbalk it has plots which have received 35 t ha^{-1} FYM or inorganic fertilizer only since the start of the experiment. These high and low organic matter soils now contain c. 3.2% and 1.0% organic C (0.30% and 0.10% N) respectively (Jenkinson and Johnston 1977). Since 1970, four rates of N (0, 48, 96, 144 kg N ha^{-1}) have been tested on successive cultivars of barley (Johnston 1994). Figure 5 shows that for cultivars Julia and Georgie yields on the high organic matter soils could be matched by those on the lower organic matter soils when sufficient fertilizer N was given. In the next period when cultivar Triumph was grown (and subsequently) yields on the low organic matter soils could not match those where FYM has been applied for many years. The figure shows that the difference was not caused by any decline in average yield on the inorganic fertilizer only treatments but rather by higher yields on the FYM-treated plots. The effect was particularly clear in 1984 and 1985 (two very good growing seasons in the UK) when the yield of barley on the FYM plots given no additional N were 2.0 and 1.0 t ha^{-1} greater than the best yields on the low organic matter plots given fertilizer N.

The introduction of improved cultivars has been a feature of the long-term experiments, particularly since the 1960s. The cultivars chosen were from those recommended by the National Institute of Agricultural Botany for use by farmers. Improvements would include better disease resistance and less susceptibility to lodging, and have invariably led to increases in potential grain yield. Silvey (1986) found that between 1947 and 1985, varietal improvement accounted for 49% of the increase in the yield of spring barley. In a series of comparative trials Riggs et al. (1981) found that the cultivars Julia, Georgie, and Triumph yielded 6.20, 6.30 and 6.68 t ha^{-1} of grain respectively, demonstrating the increasing yield potential of newer varieties. In the same trials cultivar Chevalier, grown on Hoosfield between 1852 and 1880, yielded 4.61 t ha^{-1}. The influence of pests and disease on potential yield is clearly seen on the Broadbalk experiment. In addition to comparing inorganic fertilizers and organic manures, discussed earlier, the yields of wheat grown as the first crop after a 2-year break (thus minimizing the effects of soil-borne pests and disease, particularly *Gaeumannomyces graminis*; Gutteridge et al. 1993) can be compared with wheat grown continuously, either with or without fungicides. On plots receiving P K fertilizers mean maximum yields between 1991–1995 were 8.52 t ha^{-1} for the first wheat and 7.93 t ha^{-1} for continuous wheat where 192 kg N ha^{-1} and 288 kg N ha^{-1} respectively was applied. For continuous wheat without fungicides the mean maximum yield was much reduced, averaging 6.13 t ha^{-1} when 192 kg N ha^{-1} was applied (unpublished data).

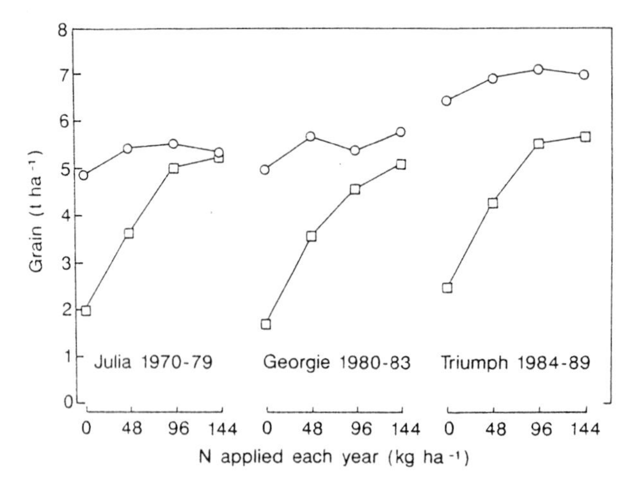

Fig. 5. Hoosfield Spring Barley experiment, Rothamsted. Yields of spring barley grown on soils with different amounts of soil organic matter. *Circles*, FYM, 35 t ha^{-1} since 1852. *Squares*, inorganic fertilizers only since 1852. (From Johnston 1994)

2.4 Using Labeled Nitrogen to Investigate the Uptake of N by the Crop

Casual observation of the response curves in Fig. 5 would suggest that fertilizer N was not used efficiently on the FYM-treated soil or at the highest rate of applied N on the PK treatment. To gain a better understanding of how fertilizer N was used in this experiment, double-labeled $^{15}NH_4{}^{15}NO_3$ was used as a tracer. This allowed us to measure the recovery of labeled N by the crop, the amount and form (organic or inorganic) of N remaining in the soil and by difference, the amount lost from the crop:soil system. It also allows us to calculate the uptake of unlabeled N by the crop. Although a small proportion of this will originate from atmospheric deposition, most will be derived from the mineralization of SOM. Thus, differences in unlabeled N uptake are likely to reflect differences in the amount of N made available.

In 1986 microplots were established within the PK and FYM plots on Hoosfield. These received labeled N at the appropriate rate. Separate, labeled plots were established in 1987. Mean grain yields on these microplots for these two years were 2.51, 4.34, 5.31, 5.34 t ha^{-1} where 0, 48, 96 or 144 kg N ha^{-1} was applied to the PK treatment; 5.69, 6.54, 6.73, 6.12 t ha^{-1} where the same rates of N were applied to the FYM treated soil. These yield responses were similar to those seen over the longer term (Fig. 5), i.e., there was little yield response to added N on the FYM plots but a big response to 48 or 96 kg N ha^{-1} on the PK treatment. Figure 6 shows the total amount of both labeled and unlabeled N taken up by barley given labeled fertilizer N. It shows, perhaps surprisingly, that the labeled fertilizer N was taken up with similar efficiency (c. 51% of the amount applied) at all rates of N on both the high and low organic matter soil (Glendining et al. 1997). This was despite the fact that, in both years, at the time of N application (GS 11; 8 weeks after drilling) the soil (0-50 cm) on the FYM plot contained more than twice as much inorganic N as on the PK plots; 67 v 33 and 97 v 40 kg N ha^{-1} in 1986 and 1987 respectively. At harvest, barley grown on the FYM treatment had taken up more than twice as much unlabeled N, i.e. N derived from the mineralization of SOM, as crops grown on the lower organic matter soil. At the highest rate of fertilizer N, particularly on the FYM treatment uptake of unlabeled N was depressed (Fig. 6), presumably as the crop's capacity was exceeded; evidence of a negative real ANI (Added Nitrogen Interaction, Jenkinson et al. 1985). In both years yields on the FYM plots given no fertilizer N were the same or greater than maximum yield on the PK treatment (given 96 kg N ha^{-1}). In these instances, total unlabeled N uptake on the FYM plot was the same or greater than labeled plus unlabeled N uptake on the PK treatment given N (Table 6). Although fertilizer N was taken up with the same efficiency at all rates of N on both high and low organic matter soils, it was not necessarily used effectively, particularly on the FYM plot, to increase grain yield. Given that the total N supply, and other nutrients, on the FYM treatment receiving additional N was more than adequate to give a greater yield than was actually achieved, other, physiological, factors which govern final grain yield must be considered. These factors are numerous, the interactions are complex and our understanding of them is limited (see Evans 1993, for an appraisal of the physiological aspects of crop improvement). Some factors will affect growth and uptake of N early in the season, others relate to the translocation of N within the plant during the grain filling period. All of these will be subject to external influences such as management and climate, particularly rainfall, drought, temperature and radiation. In 1986 and 1987 for example maximum grain yield was 7.11 t ha^{-1} and 6.34 t ha^{-1} respectively. However, the yield of straw, chaff and stubble totaled 5.07 t ha^{-1} and 7.95 t ha^{-1} (at 85% dry matter) respectively. In 1986, 78% of the total N uptake was in the grain; in 1987 only 61% was found in the grain. Labeled N was similarly distributed (Glendining et al. 1997). In 1987 far more ear-bearing tillers survived than in 1986, c. 1100 cf. c. 600 but the weight of individual grains was much less, 26.6 mg cf. 39.0 mg. We cannot be certain of the reasons for these differences. However, the three months before harvest in 1987 were cooler, wetter and had less hours of sunshine than in 1986. In addition, following heavy rain, the crop became lodged about one month before harvest.

2.5 Losses of N

As with the experiments at Woburn the benefits of extra organic matter are now clearly seen on the heavier soil at Rothamsted. But, like Woburn, there are problems associated with higher organic matter inputs. In the Hoosfield Spring Barley experiment some of the N added in the annual FYM dressing will have been immobilized in the soil. Jenkinson and Johnston (1977) calculated that, over the course of the experiment, about 15% of the N would have been immobilized in the topsoil (0-23 cm). Initially this proportion would have been greater. A further 5% is in the 23–46 cm layer. They also calculated that about 25% of the N had been removed in the crop. About 55% has therefore been lost from the system; losses being shared between NH_3 volatilization (likely to be greatest soon after manure application), denitrification and leaching. Measurements of inorganic N in the soil profile (0-110 cm) during the autumn/winter period show that there is always much more inorganic N under the organic treatment than under plots receiving inorganic fertilizer only (Fig. 7). The figure shows a small decrease in the amount of inorganic N in the FYM plot before plowing. This decrease, as the soils returned to field capacity, was mainly in the lower soil horizons as NO_3-N

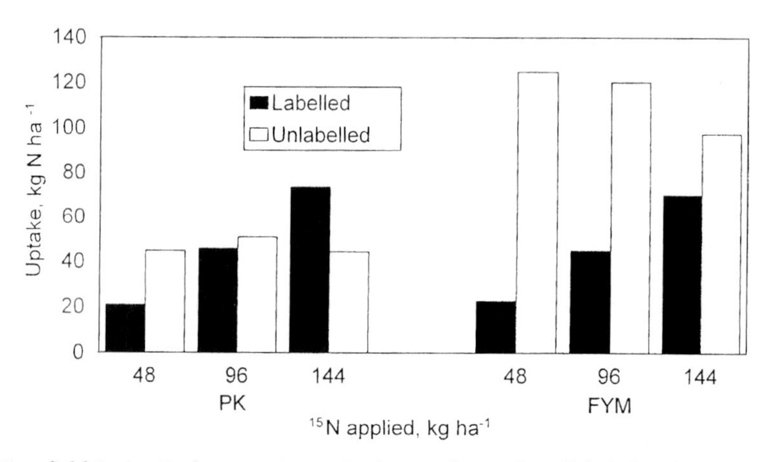

Fig. 6. Hoosfield Spring Barley experiment, Rothamsted. Uptake of labeled and unlabeled N by spring barley given ^{15}N-labeled fertilizer at different rates and grown on soils with different amounts of soil organic matter; 1986 and 1987. *FYM*, 35 t ha^{-1} since 1852. *PK*, inorganic fertilizer only since 1852. (From Glendining et al. 1997)

Table 6. Hoosfield Spring Barley. Grain yield and uptake of labeled and unlabeled N, 1986 and 1987 (from Glendining et al. 1997)

Year	Treatment	N applied kg ha^{-1}	Yield t ha^{-1}	Uptake, kg N ha^{-1}		
				Labeled	Unlabeled	Total
1986	PK	96	4.12	49	51	100
	FYM	0	4.85	—	108	108
1987	PK	96	4.90	43	51	94
	FYM	0	4.82	—	122	122

was leached. The subsequent increases in inorganic N, particularly where FYM had been plowed in, initially occurred in the topsoil (0-23 cm). However, by April all horizons down to 70 cm showed an increase in inorganic N (Powlson et al. 1989). Thus, during the autumn/winter as the soils come back to field capacity, and where there is little or no crop uptake, there is a risk of greater leaching losses under the FYM treated soils. A model which simulates mineralization and leaching (Addiscott and Whitmore 1987) was used to estimate leaching from the two treatments. Cumulative leaching loss during winter 1986/87 was estimated to be 25 and 124 kg NO_3-N ha^{-1} on the PK and FYM treatment respectively.

Losses by denitrification will occur when the soil is both warm and wet, giving rise to anaerobic conditions. The process, mediated by the soil microbial biomass, also requires a carbon source. Webster and Goulding (1989) showed that losses were greater on the higher organic matter soil than on the soil with a lower organic matter content. They measured denitrification in the autumn period and found that rates were highest after the soil wetted up in early October. Over the next few weeks losses totaled 4.5 kg N ha^{-1} on the low organic matter soil but on the FYM soil losses totaled 29 kg N ha^{-1}, six times greater.

Applications of 35 t ha^{-1} of FYM over long periods, as in the Rothamsted experiments, is atypical; FYM applications in agricultural practices are usually much lower than this. However there are many farmers in Europe with intensive livestock enterprises who need to dispose of large quantities of organic manure and slurry on limited areas of land. Consequently application rates, particularly of slurry, can be very high.

In common with the plowing of grass or legume leys, the potential for large losses of N following the incorporation of animal manures is caused by a lack of synchrony between the mineralization of N from those residues and crop requirement. For example FYM may be incorporated during October of November prior to drilling winter wheat. Although the crop will take up N slowly during winter months maximum demand for N will be during April-June, i.e., between the start of stem extension and anthesis. Following anthesis there will be little further N uptake (Gregory et al. 1979). Mineralization of the predominantly organic N present in the manure will proceed at a rate which will depend on the composition of that manure and on soil and climatic conditions. In temperate regions mineralization will continue for much of the year although it will be slowed during the colder winter months. Higher rates tend to occur during autumn (September–October) when the soil is still warm and starting to wet up (Johnston and Jenkinson 1989) Thus, N will be mineralized at times when there is little or no N uptake by the crop. If soil and climatic conditions favor leaching or denitrification, much of that N is at risk.

2.6 Effect of Inorganic Fertilizer N on Soil Organic N

In contrast to the large effect that organic manures can have (on some soils) on soil organic matter, mineralizable N, crop uptake and loss, the effects of inorganic N fertilizer are smaller but still important. On the Broadbalk experiment some treatments have received the same rate of fertilizer N since 1852, others have been

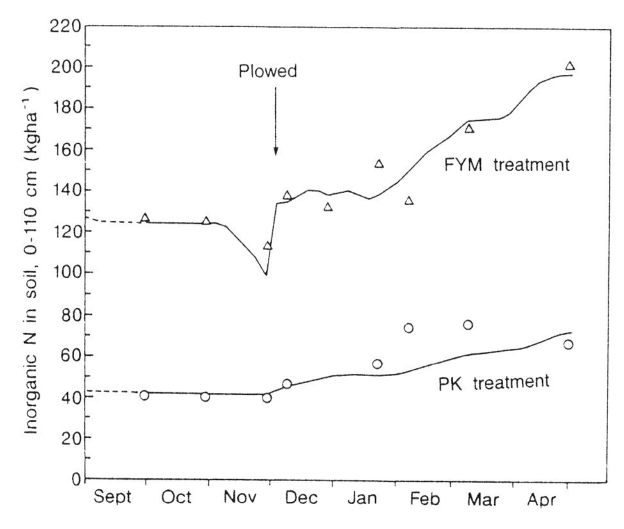

Fig. 7. Hoosfield Spring Barley experiment, Rothamsted. Inorganic N (ammonium + nitrate) in soil to a depth of 110 cm on plots with different amounts of soil organic matter as measured (*data points*) or simulated (*lines*). *FYM*, 35 t ha^{-1} since 1852. *PK*, inorganic fertilizer only since 1852. (From Powlson et al. 1989)

imposed more recently. Figure 1 shows that soils receiving no fertilizer N have been in equilibrium for more than 100 years, at about 1.0% C, 0.10% N. Where fertilizer N has been applied, larger crops have been grown and there has been a greater return of organic matter in stubble, roots and root exudates. Consequently the total N content of the soil has been raised to a higher equilibrium level, e.g., where 144 kg N ha^{-1} has been applied since 1852 the soil now contains 0.12% N. This represents an increase of about 570 kg N ha^{-1}, slightly more than 4 kg N ha^{-1} y^{-1}: about 3% of the N added. Glendining et al. (1996) looked at the effect of these long-continued fertilizer dressings on the mineralization capacity of the soil. The usual fertilizer N dressing was withheld from a small area (2.5 × 6 m) of selected treatments for up to four years. Results from laboratory incubations highlighted the importance of recent residues on N mineralization. In short-term incubations, with soil sampled at harvest (soon after root and stubble from the previous crop had entered the soil), net mineralization was 60% greater on soils normally receiving 144 kg N ha^{-1} compared to soil which has never received N. In soils without recent residues, differences in net mineralization between long-term treatments were less marked. N uptake by the crop in the absence of fertilizer N was 12–30 kg N ha^{-1} greater on plots which had received 144 kg N ha^{-1} since 1852 or 192 kg N ha^{-1} since 1968, respectively, than the c. 30 kg N ha^{-1} taken up from the soil which has never received N. These differences in N uptake were similar in amount to the extra unlabeled N (compared to the control plot) taken up by plots given labeled fertilizer N at the usual rate (Powlson et al. 1986).

2.7 The Fate of N in Crop Residues

Labeled N can also be used to follow the fate of crop residues. In the early 1980s a series of trials were established in which ^{15}N-labeled fertilizer at different rates was applied to microplots within three long-term experiments growing winter wheat (Powlson et al. 1986; Powlson et al. 1992). The experiments were on the sandy loam at Woburn, the silty clay loam at Rothamsted and a sandy clay loam (Eutric Gleysol) at Saxmundham. Following the application of ^{15}N in spring, between 18 and 39 kg N ha^{-1} remained in the soil at the end of the first harvest, predominantly as organic N in the plow layer but with some in stubble and roots. The fate of this residue was monitored over the next few years (Hart et al. 1993). Much of the labeled N became stabilized in the soil with only small, and ever decreasing, amounts of N being mineralized and either recovered by subsequent crops or lost. Averaged over all sites, years and fertilizer applications, 6.6% of the labeled residue was taken up by the following wheat crop. A further 3.5% was recovered in the second "residual" crop, 2.2% in the third, and 2.2% in the fourth. Losses of residual N were more rapid on the lighter sandy soil at Woburn than on the two heavier textured soils at Rothamsted and Saxmundham. After four residual crops on the Broadbalk experiment at Rothamsted, 55% of the labeled residue still remained in the soil, 16% had been taken up by subsequent crops, and 29% had been lost (Hart et al. 1993).

2.8 Modeling the Turnover of Soil Organic Matter

Much of the data generated by the long-term experiments at Rothamsted and Woburn has been used to construct and validate the Rothamsted Carbon Model (ROTH-C) which simulates the turnover of soil organic matter (Jenkinson 1990). Similarly, SUNDIAL (Smith et al. 1996) a model which simulates nitrogen turnover and can be used to predict fertilizer N requirements, has relied heavily on data from long-term experiments.

Fertilizer N requirement can be defined as the difference between likely crop N uptake (which will depend to a large extent on expected yield) and the supply of N through the mineralization of SOM. Better prediction of the amount, timing and fate of mineralized N is required to improve fertilizer recommendations. Generally, measurements of N mineralization have been determined from the change in the amount of inorganic N in the soil over time. This only identifies the net difference between two opposing processes, ammonification and immobilization. Measurements of net N mineralization are usually conducted over incubation periods of several weeks to obtain a measurable change in the size of the inorganic N pool (Raison et al. 1987).

One difficulty with the prediction of net mineralization using empirically based models is that gross rates of ammonification and immobilization are not known. The development of process-based (i.e. mechanistic) models which contain gross N transformation routines will improve the accuracy with which N mineralization is predicted, especially when soil and environmental conditions change. However, at present there is a lack of data on gross measurements for model development and validation.

2.8.1 Measuring Gross N Mineralization

Gross N mineralization can be determined independently of immobilization using an ^{15}N isotopic dilution technique (Kirkham and Bartholomew 1954; Barraclough 1991; Murphy et al. 1997, 1999a). The technique entails the labeling of the soil ammonium pool with ^{15}N and subsequent short-term incubation. Gross N mineralization is then calculated from the decline in ^{15}N enrichment and change in the size of the ammonium pool as ammonium, at natural abundance, is released from the mineralization of organic matter. Net N mineralization is determined from the change in the size of the inorganic (ammonium plus nitrate) pool. A negative rate of net mineralization indicates that immobilization was greater than mineralization during the incubation period. We are using this technique to examine gross N transformations with respect to land use, plant inputs, organic amendments, soil type and climate. Initial results indicate large differences between net and gross rates of mineralization. On the Ley-Arable experiment at Woburn gross mineralization under a grass ley was more than twice that on a continuous arable soil (D.V. Murphy, personal communication) while on the Broadbalk experiment gross mineralization on the soil which has received 144 kg N ha^{-1} since 1852 was almost double that on the soil which has never received N (J.A. Wakefield, pers. comm.). Measurements of gross N transformations are being used in model development and validation.

2.9 Soluble Organic Nitrogen

One area which has received little attention with respect to agricultural soils is soluble organic nitrogen (SON). Although its importance in natural ecosystems, especially arctic tundra and forests, has been recognized for some time (see chapter by K. Kielland, this volume; Qualls and Haines 1991; Atkin 1996), it has not been considered to be of significance in the cycling of N within cropped systems (Harper 1984). The development of simple, rapid techniques to measure total soluble nitrogen (TSN) has made it easier to estimate SON. (Persulphate oxidation of the same soil extract used to determine inorganic N gives a value for TSN. SON = TSN minus inorganic N).

Measurements of SON on the Broadbalk experiment indicate that the pool of SON may be similar in size to the inorganic N pool and represents perhaps 20%–50% of the TSN. Similarly in the Ley-Arable experiment at Woburn SON represented 30%–60% of the total soluble N present in the soil profile under both continuous arable and grass ley treatments.

Although these measurements indicate a significant pool of N, its relevance in agricultural soils is not clear. There are still many unanswered questions. For instance, to what extent is SON readily mineralized and thus available for plant uptake, immobilization or loss? Can it be immobilized directly by the soil microbial biomass? Recent work by Barraclough (1997) gives evidence of some direct uptake by the biomass. Are arable crops able to take up significant amounts of SON? To what extent is the pool of SON at risk to loss by leaching? Some of the questions are considered by Murphy et al. (1999b).

Meanwhile we hope to make further studies on our long-term sites to look at the effects of inorganic nitrogen, organic manures, soil type, crop rotation, etc. on soluble organic nitrogen.

3. Conclusions

Nitrogen derived from organic matter which is available for crop uptake or loss is the end result of many complex, interacting processes. Data from long-term experiments can be used to evaluate the effectiveness with which that N is used in agricultural systems. By measuring crop yield, N uptake, changes in soil nitrogen and N losses we can at least quantify the net effect of these processes. Because of the considerable variability in the processes involved and the factors which affect them it is important to assess these overall effects over several years.

It is clear that there are distinct benefits to be gained from maintaining or increasing organic matter in the soil by the addition of organic manures or the incorporation of leys or other crop residues. The proportion of added organic N retained in, or lost from, the soil will depend greatly on the current SOM content, management, soil type and climate.

On poorer soils, low in organic matter, even modest increases in the total N content of the soil will result in significantly higher crop yields. Where fertilizer N is applied maximum yield is often higher on soils with more organic matter and less fertilizer N is required to achieve that yield. On better structured soils the benefits of increased SOM may not be so clear. However, since modern, higher yielding cultivars have been introduced (and where that yield potential has been protected) higher yields have been obtained on soils with more SOM. It has not been possible to match these yields on soils with less SOM simply by adding more fertilizer N.

Much of the benefit can be ascribed to N being mineralized from organic matter throughout the soil profile and at times during the growing season which are not imitated by fertilizer N. Some of the benefit will be due to improved water holding characteristics of soil with a higher organic matter content or its superior physical structure that permits more extensive root growth.

Along with the benefits discussed above, the possible disadvantages must also be considered. Losses of N to the wider environment are potentially greater on soils with higher organic matter inputs. N may be lost by leaching as nitrate or gaseously as a result of denitrification or ammonia volatilization. This potentially greater loss is the result of a lack of synchrony between the mineralization of N and crop N requirement. It is important that the environmental implications of such losses are taken into account. However, spatial heterogeneity in the soil means that estimates of both leaching and denitrification are subject to large errors.

The increased use of labeled ^{15}N in the last 20 years has provided us with a powerful means of studying the cycling of N in the soil. In particular, isotopic dilution techniques now allow us to measure gross rates of mineralization, nitrification, and immobilization. Being able to make these measurements on well-characterized long-term sites should give us a greater insight into the complex processes and interactions governing N transformations in the soil.

Acknowledgments. I thank AE Johnston, DS Powlson and DV Murphy for useful discussion. IACR-Rothamsted receives grant-aided support from the Biological and Biotechnological Sciences Research Council.

References

Addiscott TM, Whitmore AP (1987) Computer simulation of changes in soil mineral nitrogen and crop nitrogen during autumn, winter and spring. J Ag Sci Cambridge 109:141–157

Atkin OK (1996) Reassessing the nitrogen relations of arctic plants: a mini review. Plant Cell Environ 19:695–704

Avery BW, Catt JA (1995) The soil at Rothamsted. Lawes Agricultural Trust, Harpenden, UK

Barraclough D (1991) The use of mean pool abundances to interpret ^{15}N tracer experiments. I. Theory. Plant Soil 131:89–96

Barraclough D (1997) The direct or MIT route for nitrogen immobilization: a ^{15}N mirror image study with leucine and glycine. Soil Biol Biochem 29:101–108

Catt JA, Weir AH, Norrish RE, Rayner JH, King DW, Hall DGM, Murphy CP (1980) The soils of Woburn Experimental Farm. III. Stackyard. Rothamsted Experimental Station Report for 1979, Part 2, pp 5–39

Dyke GV, Patterson HD, Barnes TW (1977) The Woburn long-term experiment on green manuring, 1936-67; results with barley. Rothamsted Experimental Station Report for 1976, Part 2, pp 119–152

Dyke GV, George BJ, Johnston AE, Poulton PR, Todd AD (1983) The Broadbalk Wheat Experiment 1968-78: yields and plant nutrients in crops grown continuously and in rotation. Rothamsted Experimental Station Report for 1982, Part 2, pp 5–44

Evans LT (1993) Crop evolution, adaption and yield. Cambridge University Press

Glendining MJ, Powlson DS, Poulton PR, Bradbury NJ, Palazzo D, Li X (1996) The effects of long-term applications of inorganic nitrogen fertilizer on soil nitrogen in the Broadbalk Wheat Experiment. J Agric Sci Cambridge 127:347-363

Glendining MJ, Poulton PR, Powlson DS, Jenkinson DS (1997) Fate of ^{15}N-labelled fertilizer applied to spring barley grown on soils of contrasting nutrient status. Plant and Soil 195:83–98

Gregory PJ, Crawford DV, McGowan M (1979) Nutrient relations of winter wheat. I. Accumulation and distribution of Na, K, Ca, Mg, P, S and N. J Agric Sci Cambridge 93:485–494

Gutteridge RJ, Hornby D, Hollins TW, Prew RD (1993) Take-all in autumn-sown wheat, barley, triticale and rye grown with high and low inputs. Plant Path 42:425–431

Harper JE (1984) Uptake of organic nitrogen forms by roots and leaves. In: Hauck RD (Ed) Nitrogen in crop production. ASA, CSSA, SSSA. Madison, WI, pp 165–170

Hart PBS, Powlson DS, Poulton PR, Johnston AE, Jenkinson DS (1993) The availability of the nitrogen in the crop residues of winter wheat to subsequent crops. J Agric Sci Cambridge 121:355–362

Jarvis SC, Stockdale EA, Shepherd MA, Powlson DS (1996) Nitrogen mineralization in temperate agricultural soils: processes and measurement. Adv Agron 57:187–235

Jenkinson DS (1990) The turnover of organic carbon and nitrogen in soil. Philos Trans Roy Soc, Series B, 329:361–368

Jenkinson DS, Johnston AE (1977) Soil organic matter in the Hoosfield Continuous Barley Experiment. Rothamsted Experimental Station Report for 1976, Part 2, pp 87–101

Jenkinson DS, Fox RH, Rayner JH (1985) Interactions between fertilizer nitrogen and soil nitrogen — the so-called "priming" effect. J Soil Sci 36:425–444

Johnston AE (1969) Plant nutrients in Broadbalk soils. Rothamsted Experimental Station Report for 1968, Part 2, pp 93–115

Johnston AE (1973) The effects of ley and arable cropping systems on the amounts of soil organic matter in the Rothamsted and Woburn Ley Arable experiments. Rothamsted Experimental Station Report for 1972, Part 2, pp 131–159

Johnston AE (1986) Soil organic matter, effects on soils and crops. Soil Use Manage 2:97-105

Johnston AE (1994) The Rothamsted Classical Experiments. In: Leigh RA, Johnston AE (Eds) Long-term experiments in agriculture and ecological sciences. CABI, Wallingford, UK, pp 9-37

Johnston AE, Jenkinson DS (1989) The nitrogen cycle in UK arable agriculture. Proceedings No. 286 of The Fertilizer Society, pp 1-24

Johnston AE, Poulton PR, McEwen J (1981) The soils of Rothamsted Farm. The carbon and nitrogen content of the soils and the effect of changes in crop rotation and manuring on soil pH, P, K and Mg. Rothamsted Experimental Station Report for 1980, Part 2, pp 5-20

Johnston AE, McGrath SP, Poulton PR, Lane PW (1989) Accumulation and loss of nitrogen from manure, sludge and compost: long-term experiments at Rothamsted and Woburn. In: Hansen JAA, Henriksen K (Eds) Nitrogen in organic wastes applied to soils. Academic, pp 126–139

Johnston AE, McEwen J, Lane PW, Hewitt MV, Poulton PR, Yeoman DP (1994) Effects of one to six-year-old ryegrass-clover leys on soil nitrogen requirements of the arable sequence winter wheat, potatoes, winter wheat, winter beans (*Vicia faba*) grown on a sandy loam soil. J Agric Sci Cambridge 122:73–89

Kirkham D, Bartholomew WV (1954) Equations for following nutrient transformations in soil, utilizing tracer data. Soil Sci Soc Am Proc 18:33–34

Körschens M (1996) Long-term data sets from Germany and Eastern Europe. In: Powlson DS, Smith P, Smith JU (Eds) Evaluation of soil organic matter models. NATO ASI Series, Vol 138. Springer, Berlin, pp 69–80

Mattingly GEG (1974) The Woburn Organic Manuring Experiment I. Design, crop yields and nutrient balance, 1964–72. Rothamsted Experimental Report for 1973, Part 2, pp 98–133

Mattingly GEG (1977) Grass ley and FYM: Organic Manuring Experiment. In: Woburn Experimental Station Centenary. Guide to demonstrations. Rothamsted Experimental Station, Harpenden, 56 pp

Mattingly GEG, Chater M, Poulton PR (1974) The Woburn Organic Manuring Experiment II. Soil analyses, 1964-72, with special reference to changes in carbon and nitrogen. Rothamsted Experimental Station Report for 1973, Part 2, pp 134–151

Mattingly GEG, Chater M, Johnston AE (1975) Experiments made on Stackyard Field, Woburn 1876–1974 III. Effects of NPK fertilizers and farmyard manure on soil carbon, nitrogen and organic phosphorus. Rothamsted Experimental Station Report for 1974, Part 2, pp 61–77

Mengel K (1996) Turnover of organic nitrogen in soils and its availability to crops. Plant and Soil 181:83–93

Murphy DV, Fillery IRP, Sparling GP (1997) A method to label soil cores with $^{15}NH_3$ gas as a prerequisite for ^{15}N isotopic dilution and measurement of gross N mineralization. Soil Biol Bioc 29:1731-1741

Murphy DV, Bhogal A, Shepherd MA, Goulding KWT, Jarvis SC, Barraclough D, Gaunt JL (1999a) Comparison of ^{15}N labelling methods to measure gross mineralization. Soil Biol Bioc 31:2015-2024

Murphy DV, Fortune S, Wakefield JA, Stockdale EA, Poulton PR, Webster CP, Wilmer WS, Goulding KWT, Gaunt JL (1999b) Assessing the importance of soluble organic nitrogen in agricultural soils. In: Wilson WS, Ball AS, Hinton RH (Eds) Managing risks of nitrates to humans and the environment. Proc R Soc Chem pp 65–86

Nambiar KKM (1994) Soil fertility and crop productivity under long-term fertilizer use in India. India Council of Agricultural Research, New Delhi

Paustian K, Elliot ET, Paul EA, Collins HP, Cole V, Frey SD (1996) The North American site network. In: Powlson DS, Smith P, Smith JU (Eds) Evaluation of soil organic matter models. NATO ASI Series Vol 138. Springer, Berlin, pp 37–54

Poulton PR, Johnston AE (1996) The long-term effect of ley-arable cropping on soil organic matter and yield. Transactions of the 9th Nitrogen Workshop, Braunschweig, September 1996, pp 437–440

Powlson DS, Pruden G, Johnston AE, Jenkinson DS (1986) The nitrogen cycle in the Broadbalk Wheat Experiment — recovery and losses of ^{15}N-labelled fertilizer applied in spring and inputs of nitrogen from the atmosphere. J Agric Sci Cambridge 107:591–609

Powlson DS, Poulton PR, Addiscott TM, McCann DS (1989) Leaching of nitrate from soils receiving organic or inorganic fertilizers continuously for 135 years. In: Hansen JAA, Henriksen K (Eds) Nitrogen in organic wastes applied to soils. Academic, pp 334–345

Powlson DS, Hart PBS, Poulton PR, Johnston AE, Jenkinson DS (1992) Influence of soil type, crop management and weather on the recovery of ^{15}N-labelled fertilizer applied to winter wheat in spring. J Agric Sci Cambridge 118:83–100

Qualls RG, Haines BL (1991) Geochemistry of dissolved organic nutrients in water percolating through a forest ecosystem. Soil Sci Soc Am J 55: 1112–1123

Raison RJ, Connel MJ, Khanna PK (1987) Methodology for studying fluxes of soil mineral N in situ. Soil Biol Bioc 19: 521–530

Riggs TJ, Hanson PR, Start ND, Miles DM, Morgan CL, Ford MA (1981) Comparison of spring barley varieties grown in England and Wales between 1890 and 1980. J Agric Sci Cambridge 97:599–610

Schulten H-R, Schnitzer M (1998) The chemistry of soil organic nitrogen: a review. Biol Fert Soils 26:1–15

Silvey V (1986) The contribution of new varieties to cereal yields in England and Wales between 1947 and 1983. J Natl Inst Agric Bot 17:155–168

Smith JU, Bradbury NJ, Addiscott TM (1996) SUNDIAL: a PC-based system for simulating nitrogen dynamics in arable land. Agron J 88:38–43

Steiner RA (1995) Long-term experiments and their choice for the research study. In: Barnett V, Payne R, Steiner R (Eds) Agricultural sustainability: economic, environmental and statistical considerations. Wiley, Chichester, pp 15–22

Webster CP, Goulding KWT (1989) Influence of soil carbon content on denitrification from fallow land during autumn. J Sci Food Agric 49:131-142

Short-Circuiting the Nitrogen Cycle: Ecophysiological Strategies of Nitrogen Uptake in Plants from Marginal Environments

Knut Kielland

Summary. Physical and biological characteristics of arctic soils can vary substantially over small spatial scales resulting in large gradients in soil moisture, pH, and organic matter content. Plant species of different growth forms show inherent differences in their physiological response to the environment, and thus are differentially distributed along such environmental gradients. Notwithstanding these physical and biological contrasts, arctic ecosystems are uniformly characterized by short growing seasons, low primary productivity, slow rates of decomposition, and thus low annual rates of soil organic matter turnover. These conditions contribute to low mineral nutrient availability, a factor that exerts a severe limiting effect on net primary production.

The traditional concept, however, that (non N-fixing) plant species are dependent on their nitrogen supply exclusively through microbially mediated nitrogen mineralization has recently been brought into question. Biological and biochemical evidence, both at the organismal and ecosystem level, strongly suggest that soil-plant relationships in arctic systems are more dynamic than previously believed.

There is increasing evidence for the hypothesis that organic forms of nitrogen, particularly free amino acids, are direct sources of nitrogen for plants in nitrogen-limited soils across a wide variety of ecosystems. This hypothesis is based primarily on work in the Arctic, but the idea has been proposed in studies of temperate ecosystems as well. Moreover, analyses of nitrogen budgets in both arctic and subarctic ecosystems show that the flux of inorganic nitrogen cannot meet the annual requirements for nitrogen by plants. In the Arctic this observation, coupled with findings demonstrating high uptake capacity for amino acids by tundra plants, both in the laboratory and the field, strongly suggests that organic forms of nitrogen may represent a major source of nitrogen to arctic species that hitherto has been unaccounted for. These findings suggest that we need to seriously reevaluate our understanding of nitrogen cycling in arctic ecosystems, and perhaps in temperate and tropical ecosystems as well. These conceptual revisions may have serious ramifications for agricultural practices in marginal soils, as well as forming an essential component of modeling studies of carbon and nitrogen cycles in the context of global change.

Key words. Amino acid uptake, Arctic ecosystems, Nutrient availability, Nutrient cycling, Soil amino acids, Tundra soils

1. Introduction

Nitrogen availability in Alaskan taiga and tundra ecosystems is considered to be low because of slow rates of decomposition and nitrogen mineralization (Chapin et al. 1988; Giblin et al. 1991; Klingensmith and Van Cleve 1993). However, the total soil nitrogen pools are substantial, due to large accumulation of soil organic matter with high concentrations of nitrogen. Nitrogen exists predominantly in organic form in arctic and subarctic ecosystems both in terrestrial (Walker 1989; Shaver et al. 1990; Kielland 1995) and aquatic environments (Hershey et al. 1996). In soils, the organic fraction typically constitutes 99% of total soil nitrogen (Chapin et al. 1988). Dissolved organic nitrogen (DON) is typically an order-of-magnitude greater in concentration than ammonium and nitrate (Shaver et al. 1990; Kielland 1995, 1997). Recently, organic nitrogen, both the total dissolved fraction (DON) and total free amino acids (TFAA), has attracted attention as soil nitrogen pools that modulate nitrogen availability and play a direct role in the nitrogen nutrition of plants (Kielland 1994, 1995, 1997; Northup et al. 1995; Raab et al. 1996; Näsholm et al. 1998).

The absolute dependence of plants on mineral nitrogen for uptake and subsequent metabolism has been the traditional perspective of plant-nitrogen relations in terrestrial ecosystems, although plants' capacity to absorb and utilize organic sources of nitrogen has been known for a long time (Melin 1953; Melin and Nilsson 1953; Miettinen 1959). In some temperate forests, nitrogen uptake by the vegetation appears to be balanced by nitrogen mineralization and other major nitrogen fluxes (Nadelhoffer et al. 1984). Moreover, in most agricultural systems worldwide, inorganic nitrogen supplied in fertilizer and released from soil organic matter via microbial action is by far the major flux of nitrogen. Until recently, this biochemical scenario has been applied to natural ecosystems as well. However, a challenge to the generality of this view is recently coming into focus from a variety of research focusing on marginal ecosystems (including arctic tundra, taiga forests, alpine, and temperate heath communities). The fundamental argument involves the finding that the annual plant requirements for nitrogen greatly exceeds the annual (inorganic) nitrogen supply. This observation has been made both in arctic (Shaver et al. 1990; Giblin et al. 1991), alpine (Labroue and Carle 1977; Fisk and Schmidt 1995) and taiga ecosystems (Ruess et al. 1996), as well as determined experimentally in ^{15}N-labeling experiments (Stribley and Read 1980; Read 1991). Although it is possible that some of the discrepancies between nitrogen supply and plant nitrogen demand can be explained by inadequate methodology, it seems evident that species from these ecosystems obtain nitrogen from sources other than ammonium and nitrate.

One mechanism by which plant species can respond to low (inorganic) nitrogen supply is by absorbing organic nitrogen, thus circumventing N mineralization (Read 1991; Kielland 1994). The importance of soluble organic nitrogen in terrestrial ecosystem nitrogen cycling and plant nutrition is only beginning to be appreciated (Abuarghub and Read 1988; Read 1991; Chapin et al. 1993; Chapin 1995; Kielland 1995; Northup et al. 1995). Recent studies show that a host of species from arctic tundra (Kielland 1994, 1997), boreal forests (Näsholm et al. 1998) as well a temperate alpine species (Raab et al. 1996), readily absorb organic nitrogen in the form of amino acids. There is a wide diversity of plant growth forms that exhibit this capacity, including graminoids, deciduous and evergreen shrubs, lichens, and mosses. Calculations based on physiological measurements and soil chemistry show that amino acid uptake may contribute up to 80% of the annual nitrogen budget in these species. This form of nitrogen acquisition has altered many long-held concepts of plant nitrogen nutrition and nitrogen cycling (Chapin 1995). Because species differ significantly in their capacity to absorb this nitrogen form, partially due to different levels of mycorrhizal infection, differences among species in their potential to absorb amino acids provide ample basis for niche differentiation among species in competing for what was previously considered a single resource (Kielland 1994). One potential consequence of such physiological niche differentiation is that it allows more species to coexist because they tap different nitrogen sources. Moreover, by directly absorbing organic (i.e. non-mineralized) nitrogen, plants in effect short-circuit the process that hitherto has been considered the bottle-neck in the nitrogen cycle (Van Cleve and Alexander 1981; Chapin 1983).

2. Characteristics of Arctic Ecosystems

2.1 Climate

The Arctic is a largely treeless zone, the southern boundary of which approximately coincides with the southern boundary of the arctic front or the July 10°C isotherm (Barry et al. 1981). The low radiation input results in a negative mean annual temperature producing permafrost (permanently frozen ground) and a short growing season. Air temperature during the growing season is low, averaging less than 5°C in coastal areas such as Barrow, Alaska. The length of the growing season (the time the plants are snow-free and thawed) is approximately 10-14 weeks. However, the 24-h photoperiod between early June until early August somewhat compensates for the low radiation intensity, so that the total radiation is approximately similar ($\approx$20 MJ m^{-2} d^{-1}) in arctic and temperate regions during this time period (Barry et al. 1981). Nevertheless, by the time the plant root systems are thawed and growth can be supported by current nutrient uptake, the radiation regime is on its downhill slide. Moreover, the Arctic receives very little precipitation, on the order of 50-250 mm y^{-1}. This annual amount is comparable to temperate deserts, but in the Arctic half of this precipitation falls as snow during the winter. Despite the low rainfall during the summer months, arctic soils are typically wet (>40% volumetric water content) because permafrost prevents drainage and the low air temperature curtails evaporative losses. The often water-logged conditions in sedge meadow tundras have profound impacts on the energy balance of these ecosystems. For example, up to 70% of the net radiation in summer is dissipated in evaporating water, leaving about 18%-32% to warm the air, and only 2%-9% to warm the soil (Weller and Holmgren 1974).

2.2 Soil Properties

Arctic tundra soils have a diverse history, but are dominated by Inceptisols (tussock tundra), Histosols (wet meadow tundra), and Mollisols (lichen heaths and deciduous shrub tundra) (Rieger 1983). High soil organic matter content and permafrost dominate the soils over much of the arctic landscape (Tedrow 1977). Nutrient availability is generally low in arctic soils because external inputs from the atmosphere in the form of deposition or N-fixation are limited by low precipitation and low temperatures, respectively. Moreover, low soil temperatures severely curtail weathering of the parent material, as well as reducing inorganic nutrient release from soil organic matter (Babb and Whitfield 1977; Nadelhoffer et al. 1991). High organic matter content and high soil acidity ($pH < 4.0$) limits the rates of nitrification, which has important consequences for soil nitrogen composition (see Section 3.1). The concentrations of soil nutrients and oxygen decrease sharply with depth so that approximately 90% of the biological activity, including root and microbial biomasses, are concentrated in the top 5-10 cm of soil (Barel and Barsdate 1978; Chapin et al. 1979).

2.3 Decomposition

Temperature, moisture, and the chemical makeup of the decomposing substrate (substrate quality) are prominent controls over decomposition in ecosystems from the tropics to arctic tundra. For example the strong positive relationship between decomposition and actual evapotranspiration (AET) on a global scale (Meentemeyer 1978) suggest that decomposition is controlled largely by the same set of factors. Cold soils and a short growing season contribute to the generally inverse relationship between latitude and annual rates of decomposition. The combined effects of the primary controls over decomposition and the short growing season at high latitudes account for the large accumulation of organic matter in arctic soils. This process conserves nutrient stocks in the ecosystem, but also severely curtails the availability of inorganic nutrients to plants.

The rate of litter turnover, k, as defined by production divided by the standing crop (Jenny 1980), vary over two orders-of-magnitude between tundra and tropical forest. Thus, the turnover time for litter can vary from over 30 years in arctic tundra to less than three months in the tropics (Table 1). Variations in local conditions also produce significant differences in decomposition potentials whether it be among different vegetation types of arctic tundra, or broader comparisons such as among tundra, taiga forests, and agricultural fields in Alaska (Table 2).

3. Distribution and Abundance of Nitrogen Fractions in Arctic Soils

Low rates of decomposition in arctic ecosystems produce soils with large organic matter pools. Because of the long turnover times of soil carbon, rates of nutrient release is slow, in particular for environmentally sensitive processes such as nitrogen mineralization and nitrification (Marion and Miller 1982; Nadelhoffer et

Table 1. Productivity, litter standing crop, and decomposition across five major biomes. Decomposition is defined in terms of litter turnover time (l/k). (Data calculated from Swift et al. 1979)

Biome	NPP ($t\ ha^{-1}\ year^{-1}$)	Litter biomass ($t\ ha^{-1}$)	Decomposition (years)
Tundra	1.5	44	33.3
Boreal forest	7.5	35	4.8
Temperate deciduous forest	11.5	15	1.3
Savannah	9.5	3	0.3
Tropical forest	30.0	5	0.2

Table 2. Cellulose decomposition potentials in arctic, taiga, and agricultural field soils in Alaska

Soil origin	Horizon	Annual mass loss (%)
Tundra[1]		Litter layer
Dry heath		4
Wet meadow		8
Tussock tundra		14
Shrub tundra		25
Taiga[2]	Litter layer	
Willow stage		80
Alder stage		72
Field[3]	Litter layer	
Fallow barley		62

[1]From Kielland 1990
[2]From Kielland et al. 1997
[3]From Sparrow et al. 1992

al. 1991). Thus, arctic soils have small pools and slow fluxes of inorganic nitrogen, despite having large pools of total soil nitrogen.

3.1 Soil Nitrogen Composition

Arctic soils are dominated by organic nitrogen forms. The pools of bound organic nitrogen (total Kjeldahl nitrogen - [extractable inorganic plus extractable organic nitrogen]) are typically orders-of-magnitude greater than soluble organic nitrogen, which again are orders-of-magnitude greater than dissolved inorganic nitrogen (Table 3). A sizable fraction of the exchangeable nitrogen is in the form of free amino acids (Kielland 1995). Fifteen principal amino acids have been identified in arctic tundra soils, including neutral, acidic, and basic species (Table 4). The concentration of water-extractable free amino acids varies substantially over the season, but on the average the pool of free amino acids is nearly 10-fold greater than ammonium (Table 5), the principal form of inorganic nitrogen (Everett et al.1981). In a similar fashion, the KCl-extractable (2M) amino acid concentration in moor

Table 3. Composition of soil nitrogen in four major tundra vegetation types in arctic Alaska. Values are mean pool sizes (g N m^{-2}). (Data from Kielland 1990)

	Dry heath	Wet meadow	Tussock tundra	Shrub tundra
Bound organic N	90	245	140	180
Soluble organic N	0.50	0.72	1.35	0.89
Dissolved inorganic N	0.004	0.01	0.03	0.008

Table 4. Range of monthly mean concentrations of individual free amino acids from arctic tundra soils (Data from Kielland 1995)

Amino acid	Ng/g dry soil
Acidic	
Aspartic acid	50 - 830
Asparagine	<10 - 80
Glutamic acid	20 - 670
Glutamine	<10 - 120
Basic	
Arginine	200 - 1800
Lysine	10 - 100
Neutral	
Glycine	50 - 730
Alanine	<10 - 50
Leucine	50 - 150
Phenylalanine	<10 - 30
Serine	30 - 2860
Threonine	10 - 900
Ornithine	50 - 150
Methionine	<10 - 100
Valine	<10 - 30

Table 5. Mean seasonal concentrations of water-extractable total free amino acids and ammonium in four major arctic tundra communities. Coefficients of variation in parentheses. (Data from Kielland 1995)

	Nitrogen concentration (µg/g dry soil)			
	Dry heath	Wet meadow	Tussock tundra	Shrub tundra
Amino acids	2.19	1.57	8.29	2.88
	(82%)	(29%)	(89%)	(80%)
Ammonium	0.34	0.38	0.84	0.64
	(59%)	(4%)	(4%)	(58%)

soils from a boreal forest site in northern Sweden were approximately three-fold greater than that of inorganic nitrogen (Näsholm et al. 1998), attesting to the prevalence of amino acid-nitrogen in a variety of northern soils.

3.2 Nitrogen Turnover

Net nitrogen mineralization can vary over 50-fold across latitudes from a maximum 1.7 g N m^{-2} y^{-1} in the Arctic (Kielland 1990) to 90 g m^{-2} y^{-1} in tropical forests (Chandler 1985; Vitousek and Denslow 1986), which roughly reflects the 80-fold range in net primary productivity aboveground between tropical forests and arctic tundra (Nadelhoffer et al. 1992). Maximum annual nitrogen mineralization decreases in an exponential fashion with increasing latitude (Fig. 1). Even though the specific capacity for nitrogen mineralization (mg N g^{-1} d^{-1}) in arctic soils can be similar to that observed in temperate (Wedin and Tilman 1990) and tropical (Ruess and Seagle 1994) soils, the much lower annual flux of inorganic nitrogen in the Arctic is compounded by the much shorter growing season.

Even within the Arctic, the turnover in the soil of a given nitrogen form varies substantially among community types as a function of differences in soil physicochemical attributes. Using net annual nitrogen mineralization as the flux to estimate the turnover time of soil nitrogen fractions in arctic tundra, we find that the net turnover for total soil N is on the order of hundreds of years. For soluble organic N the net turnover time is 0.7-14 years, whereas dissolved inorganic nitrogen turns over in a matter of a few days (Table 6). The available data on the proteolytic activity in tundra soils (Linkins and Neal 1982; Chapin et al. 1988) suggest that the potential of free amino acid generation is far greater than the rate of ammonification (Table 7), which may in part explain the high concentrations

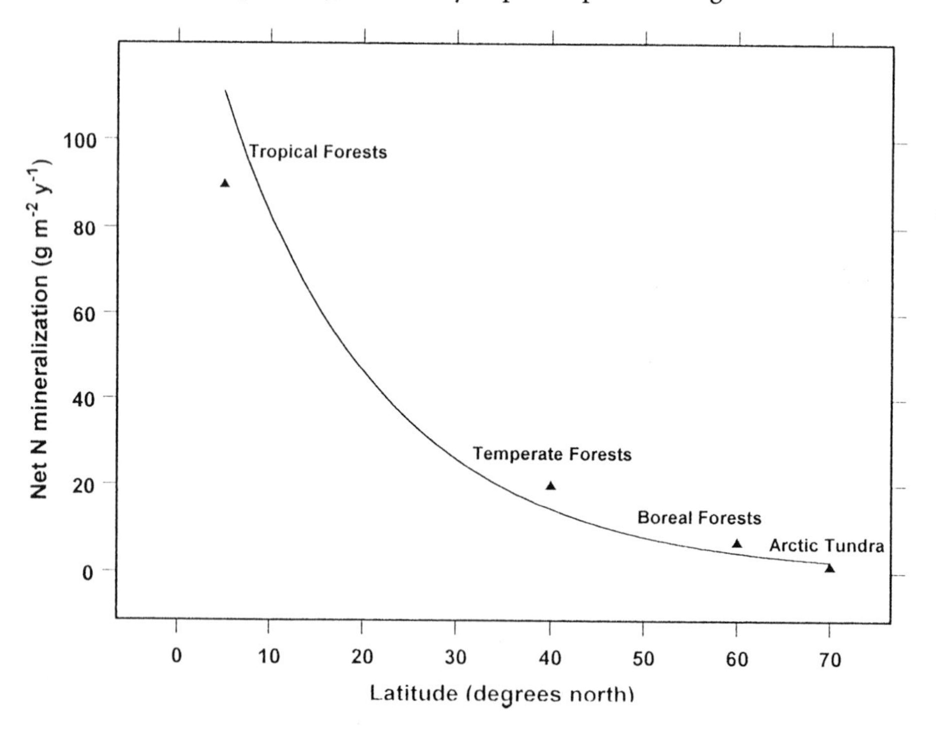

Fig. 1. Comparative rates of maximum net nitrogen mineralization across latitudes from tropical forests to arctic tundra. (Data from Nadelhoffer et al. 1992 and Kielland 1990)

of free amino acids relative to inorganic nitrogen. Moreover, the turnover time of individual amino acids (Fig. 2) is much shorter than that of inorganic nitrogen. Thus, the flux of organic nitrogen, both expressed as a vector of ecosystem element cycling and as a supplier of plant-available nitrogen, appears to be substantially greater than that of ammonium and nitrate.

4. Nitrogen Uptake by Arctic Plants

From the evolutionary perspective of the individual plant, resource use is modulated by resource supply. Therefore it is not surprising that variations in resource supply might result in variations in strategies of resource acquisition as reflected by inter-specific differences in physiological capacities, growth patterns, or spatial distributions. The observation that the annual flux of inorganic nitrogen in taiga and tundra is insufficient to meet the annual requirements of the major plant species occurring there (Table 8), strongly suggests that these plants obtain additional nitrogen through "unconventional" mechanisms. One such mechanism that has been elucidated, is the uptake of simple organic nitrogen forms, such as free amino acids. The physiological capacity to absorb free amino acids is found in all the major growth forms occurring in arctic tundra (Table 9). The species exhibiting this capacity span the range of relative growth rate potential for tundra species, and are found across the range of soil fertility extant in arctic soils. Though mycorrhizal infection appears to enhance the capacity of amino acid absorption, this physiological trait is also found in non-mycorrhizal species.

Table 6. Turnover rates (% of pool $year^{-1}$) of the major nitrogen fractions in four arctic tundra soils from northern Alaska. (Data from Kielland 1990)

	Dry heath	Wet meadow	Tussock tundra	Shrub tundra
Bound organic N	0.04	0.06	1.3	0.6
Soluble organic N	7	20	143	122
Dissolved inorganic N	910	1428	5882	14285

Table 7. Protease activity and net nitrogen mineralization potentials in tussock tundra soil measured under laboratory conditions at 35°C

Soil horizon (organic soil)	Protease activity[1] (μmol g^{-1} day^{-1})	N mineralization[2,3] (μmol g^{-1} day^{-1})
Fibric	11.76 (1.92)	1.05
Hemic	7.92 (1.44)	1.60
Sapric	0.72 (0.48)	1.20
Mineral	0.24 (0.00)	0.002

[1]Data from Chapin et al. 1988
[2]Calculated from Marion and Miller 1982
[3]Variance not available

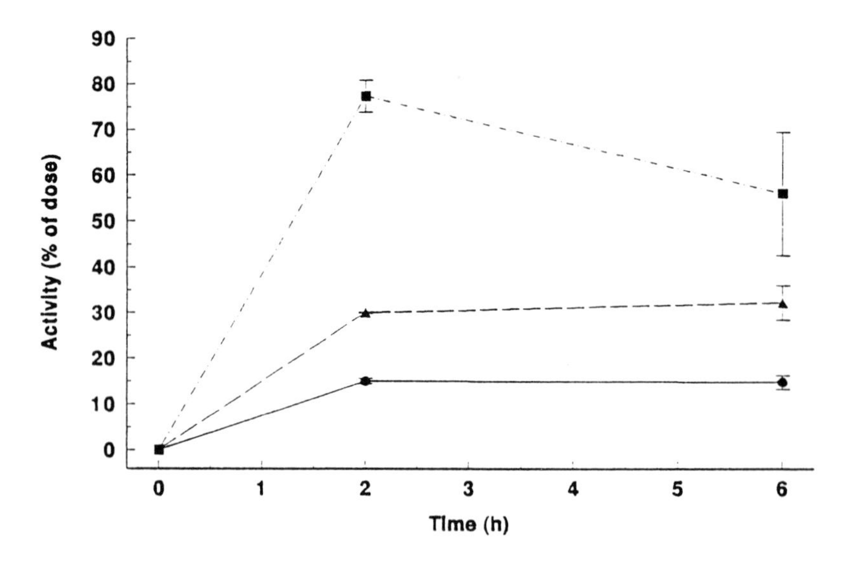

Fig. 2. Activity of ^{14}C-labeled amino acids as respired CO_2 in arctic tundra heath soil, incubated under aerobic conditions in the laboratory at 20 °C. The amino acids are glycine (●), aspartic acid(■), and glutamic acid(▲). (Data from Kielland 1995)

Table 8. Nitrogen budgets for tundra and taiga ecosystems showing contributions of inorganic nitrogen fluxes and annual nitrogen requirement of the vegetation. Units are in g m^{-2} $year^{-1}$

	Taiga forests					
	Floodplain			Upland		
	Alder-balsam poplar	Balsam poplar	White spruce	Black spruce	Birch-aspen	White spruce
Supply[1]	4.34	2.40	1.47	0.38	2.45	1.93
Requirement[1]	7.46	2.23	1.35	0.92	5.25	1.69
S/R[2]	58	1.0	109	41	47	114

	Tundra communities			
	Dry heath	Wet meadow	Tussock tundra	Shrub tundra
Supply[3]	0.03	0.14	1.76	1.08
Requirement[4]	0.09	0.84	1.80	2.34
S/R[2]	33	17	100	46

[1]Data from Ruess et al. 1996
[2]Proportion of annual requirement supplied by inorganic N (%)
[3]Data from Kielland 1990
[4]Data from Shaver and Chapin 1991

Table 9. Characteristics of arctic species showing physiological uptake capacity of free amino acids. (Data from Kielland 1994)

Species	Growth form	Mycorrhizal association
Carex aquatilis	Graminoid	Largely non-mycorrhizal
Carex bigelowii	Graminoid	Largely non-mycorrhizal
Eriophorum angustifolium	Graminoid	Largely non-mycorrhizal
Eriophorum vaginatum	Graminoid	Largely non-mycorrhizal
Betula nana	Deciduous shrub	Ectomycorrhizal (e.g. Hebeloma)
Salix pulchra	Deciduous shrub	Ectomycorrhizal (e.g. Hebeloma)
Ledum palustre	Evergreen shrub	Ericoid endomycorrhizal
Cetraria richardsonii	Lichen	Non-mycorrhizal (?)
Sphagnum rubellum	Moss	Non-mycorrhizal (?)

4.1 Kinetics of Amino Acid Absorption

Maximum uptake capacities for methylamine (ammonium) in seven vascular species from northern Alaska appear in general to be greater than maximum absorption capacity for any individual amino acid so far examined (Fig. 3). The differences in rates of ammonium vs. amino acid uptake are greatest at the highest solution concentration. At this concentration (500 μM), ammonium uptake ranges from 4-30 μmol g^{-1} h^{-1} among all species, as compared to 4-6 μmol g^{-1} h^{-1} for glycine and aspartate uptake, and <1 mmol g^{-1} h^{-1} for absorption of glutamate. Among the amino acids examined, glycine is absorbed at a higher rate than aspartic acid (Fig. 3d), which in turn is taken up at a higher rate than glutamic acid (Fig. 3c). However, at lower, more ecologically realistic solution concentrations, the sharp differences in absorption rate among nitrogen forms are less pronounced. For example, at naturally occurring solution concentrations glycine uptake can be as high or higher than ammonium uptake.

Across the vascular species examined, different growth forms show differences in their amino acid preference. The deciduous shrubs have the greatest uptake capacity for glycine, whereas non-mycorrhizal graminoids have the lowest capacity (Fig. 4). By contrast, growth form differences for V_{max} of aspartate and glutamate uptake potential are largely insignificant (Fig. 4).

The highest absolute uptake capacity for organic N found across all growth forms examined are in the non-vascular species (cryptogams) (Fig. 5). This may be a reflection of the combined effects of the order-of-magnitude higher concentration of soluble organic nitrogen compared to inorganic soil nitrogen in the soil they occupy (Kielland 1997), and an evolutionary physiological adjustment to high temporal variability of the nutrient supply (Kielland and Chapin 1994). Moreover, the available data from cryptogams show substantially higher uptake capacity for the organic form of nitrogen (glycine) than the inorganic form (am-

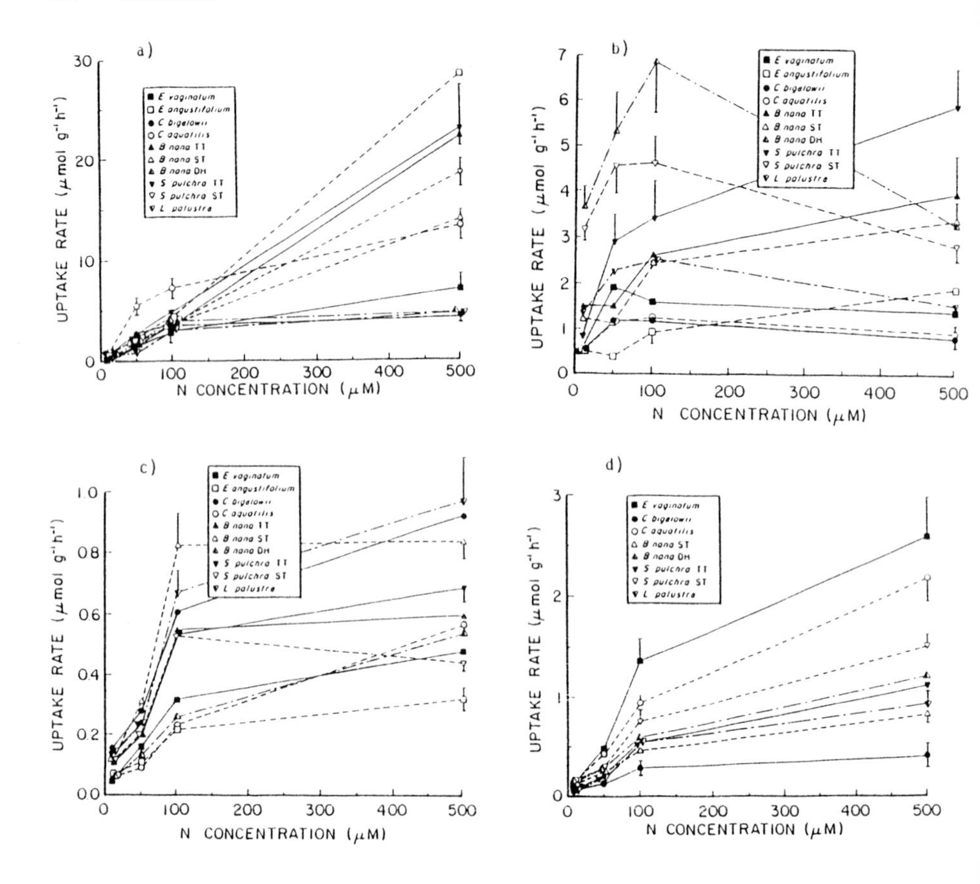

Fig. 3. Uptake kinetics of (**a**) ammonium, (**b**) glycine, (**c**) glutamic acid, and (**d**) aspartic acid by dominant vascular plants in arctic tundra. Data from species occurring in more than one community type are identified. DH, dry heath; ST, shrub tundra; TT, tussock tundra. (Data from Kielland 1994)

monium) (Fig. 5). The relationship between organic vs. inorganic nitrogen uptake potential is well illustrated when expressed as the "preference ratio"; the rate of absorption of glycine/ammonium at the average seasonal concentration of glycine and ammonium. Thus expressed, in a comparison of 12 species by site combinations, 75% showed equal or greater preference for the organic form of nitrogen (Fig. 6). The lowest preference ratio (0.09) is found in a wet meadow sedge species (*Eriophorum angustifolium*), and the highest uptake ratio (6.7) was found in a tussock tundra moss species (*Sphagnum rubellum*).

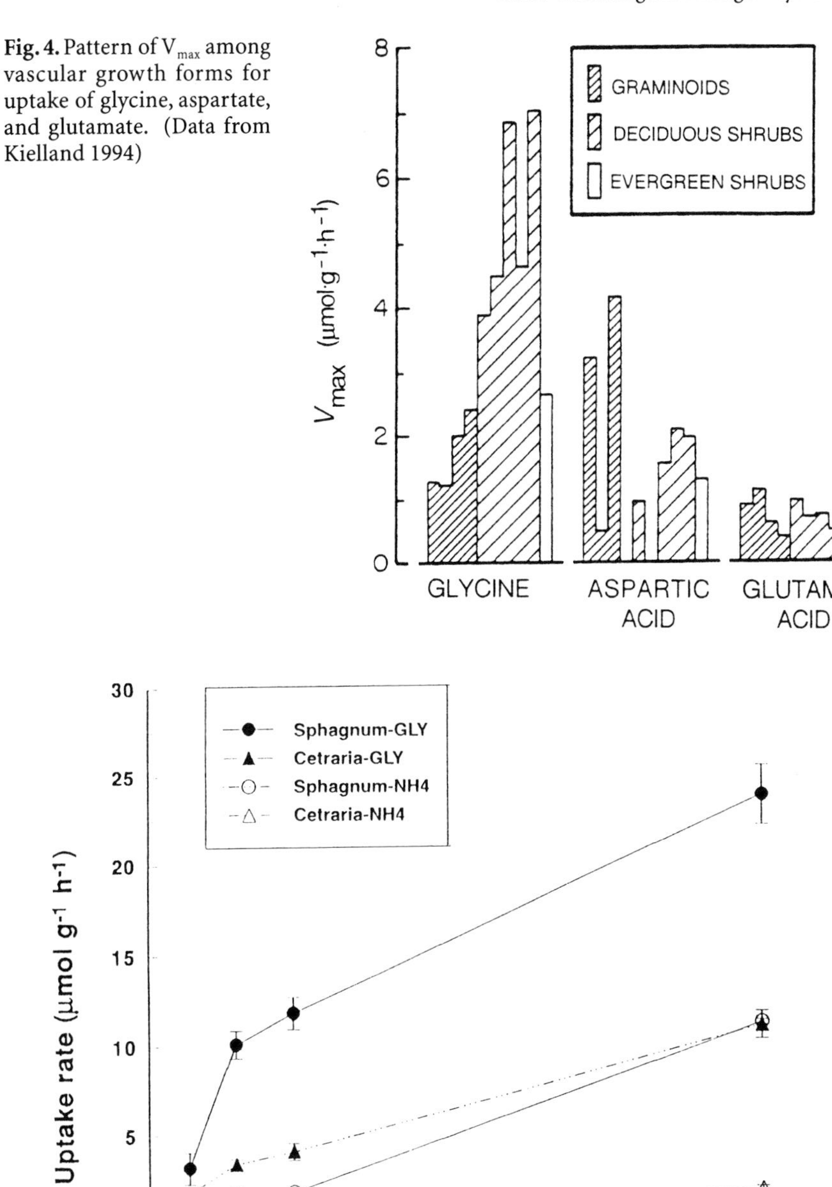

Fig. 4. Pattern of V_{max} among vascular growth forms for uptake of glycine, aspartate, and glutamate. (Data from Kielland 1994)

Fig. 5. Uptake kinetics of ammonium and glycine absorption in two arctic cryptogams, *Sphagnum rubellum* and *Cetraria richardsonii*. (Data from Kielland 1997)

The response of uptake to solution concentration also differs among amino acids: most species show high affinity (low K_m) for glycine, a low affinity for aspartate, and an intermediate affinity to glutamate (Table 10). The half-saturation constant (K_m) for amino acid absorption is in general lower than 200 mM, whereas the corresponding value for ammonium uptake is greater than 200 mM (Table 11). Likewise, V_{max} of ammonium absorption are consistently higher than V_{max} of amino acid absorption, in part as a consequence of the shape of the kinetic curves.

Variance in the capacity to absorb different nitrogen forms has potentially important ramifications for how natural communities are structured, as controlled by differential growth responses. Experimental evidence shows that plants adapted to soil of different nitrogen compositions show divergent growth responses when supplied with inorganic vs. organic nitrogen (Chapin et al. 1993). For example, biomass accumulation of *Eriophorum vaginatum*, a ubiquitous arctic sedge, was greatest when grown in a mixture of amino acids, rather than inorganic nitrogen (Fig. 7a). This growth response is consistent with an adaptation to organic nitrogen, the form that prevails in arctic soils (cf. Table 5) By contrast, *Hordeum vulgare*, an agricultural species adapted to well-fertilized mineral soils, grew best when supplied with

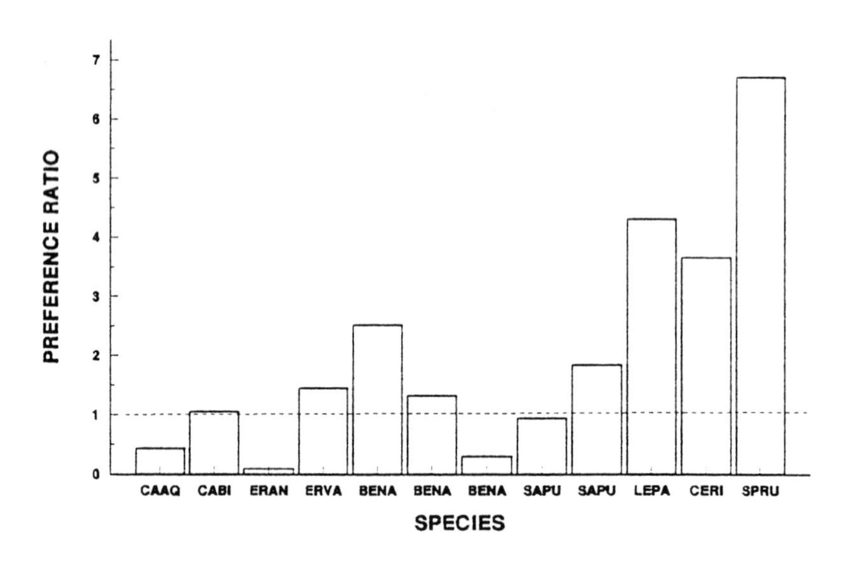

Fig. 6. Ratio of glycine:ammonium uptake ("Preference ratio") calculated from kinetic parameters and the average seasonal concentration of soil glycine and ammonium from the respective communities in which the species occur. Species designations are: CAAQ, *Carex aquatilis*; CABI, *Carex bigelowii*; ERVA , *Eriophorum vaginatum*; BENA, *Betula nana* (from respectively left to right; shrub tundra, tussock tundra, and dry heath); SAPU, *Salix pulchra* (from respectively shrub tundra and tussock tundra); LEPA, *Ledum palustre*; CERI, *Cetraria richardsonii*; SPRU, *Sphagnum rubellum*. (Data from Kielland 1994, 1995, 1997)

Table 10. Kinetic parameters for amino acid uptake by vascular and non-vascular species arctic tundra species. (Data from Kielland 1994)

Community	Species	Amino acid	K_m (μmol/L)	V_{max} (μmol · g^{-1} · h^{-1})
DH	*Betula nana*	Gly	13	6.8
		Asp	152	1.5
		Glu	176	0.7
WM	*Carex aquatilis*	Gly	9	1.3
		Asp	258	3.2
		Glu	296	0.9
	Eriophorum angustifolium	Gly	143	2.4
		Asp	n.d.	n.d.
		Glu	86	0.4
TT	*Betula nana*	Gly	64	4.5
		Asp	n.d.	n.d.
		Glu	70	0.7
	Carex bigelowii	Gly	7	1.2
		Asp	81	0.5
		Glu	100	1.1
	Eriophorum vaginatum	Gly	12	2.0
		Asp	293	4.1
		Glu	118	0.6
	Ledum palustre	Gly	9	2.6
		Asp	202	1.3
		Glu	123	1.2
	Salix pulchra	Gly	87	7.0
		Asp	377	1.9
		Glu	30	0.6
	Sphagnum rubellum	Gly	99.8	28.1
		Asp	n.d.	n.d.
		Glu	n.d.	n.d.
	Cetraria richardsonii	Gly	142.3	13.5
		Asp	n.d.	n.d.
		Glu	n.d.	n.d.
ST	*Betula nana*	Gly	96	3.9
		Asp	107	0.9
		Glu	69	0.9
	Salix pulchra	Gly	19	4.6
		Asp	192	2.0
		Glu	32	0.5

DH, dry heath; WM, wet meadow; TT, tussock tundra; ST, shrub tundra; Gly, glycine; Asp, aspartic acid; Glu, glutamic acid

inorganic nitrogen (Fig. 7a). The pattern of nitrogen uptake between these species mirrored the differential pattern of growth (Fig. 7b). However, nitrogen use efficiency, i.e., the amount of carbon fixed per unit of nitrogen absorbed, was over two-fold greater in the tundra sedge. Thus, physiological diversity regarding absorption potential for different nitrogen forms can be manifested by differential growth, and

Table 11. Kinetic parameters for methylamine (ammonium) uptake by vascular and non-vascular arctic tundra species. (Data from Kielland 1994)

Community	Species	K_m (μmol/L)	V_{max} (μmol • g^{-1} • h^{-1})
DH	*Betula nana*	67	5.6
WM	*Carex aquatilis*	153	17.8
	Eriophorum angustifolium	935	17.8
TT	*Betula nana*	3821	142.8
	Carex bigelowii	78	5.4
	Eriophorum vaginatum	242	13.7
	Ledum palustre	256	7.6
	Salix pulchra	1197	75.9
	Sphagnum rubellum	1000.6	31.1
	Cetraria richardsonii	290.0	2.6
ST	*Betula nana*	663	31.7
	Salix pulchra	6717	272.8

Community abbreviations are in Table 7

thus dominance of one species over another. This is one mechanism by which small changes in resource conditions induced by climate change, could result in large differences in community composition over time, with concomitant changes in primary productivity and other ecological parameters.

Though there is considerable variation among arctic tundra species in their capacity to absorb organic nitrogen, it is clear that arctic plants potentially derive a large proportion (up to 80%) of their annual nitrogen requirement directly from organic nitrogen sources. This appears to be the case irrespective of mycorrhizal infection, and applies equally to vascular and non-vascular plants. Thus, the capacity of absorbing nitrogen in organic form appears to be a general adaptive syndrome of arctic species (Kielland 1997).

4.2 Evolution of Absorptive Potentials

The evolution and maintenance of organic nitrogen uptake potentials in arctic plants are likely a function of several environmental drivers, including as discussed previously, high concentrations and rapid flux rates of organic N compared with inorganic N. However, in addition to these factors, plants also respond to temporal variations in resource supply. For example, species experiencing large temporal variation over the growing season in light conditions (e.g., sun-fleck plants) or soil moisture (e.g., desert plants) exhibit high capacity for, respectively, photosynthesis (Chazdon and Pearcy 1991) and water uptake (Mooney 1980). Thus, low annual flux of resource supply may not necessarily dictate low physiological capacity. Likewise, large temporal variation in soil nutrient availability can have profound effects on nutrient uptake by plants (Campbell and Grime 1989). Under constant, "steady-state", nutrient availability, species from infertile soils generally exhibit lower capacity for nutrient absorption than do

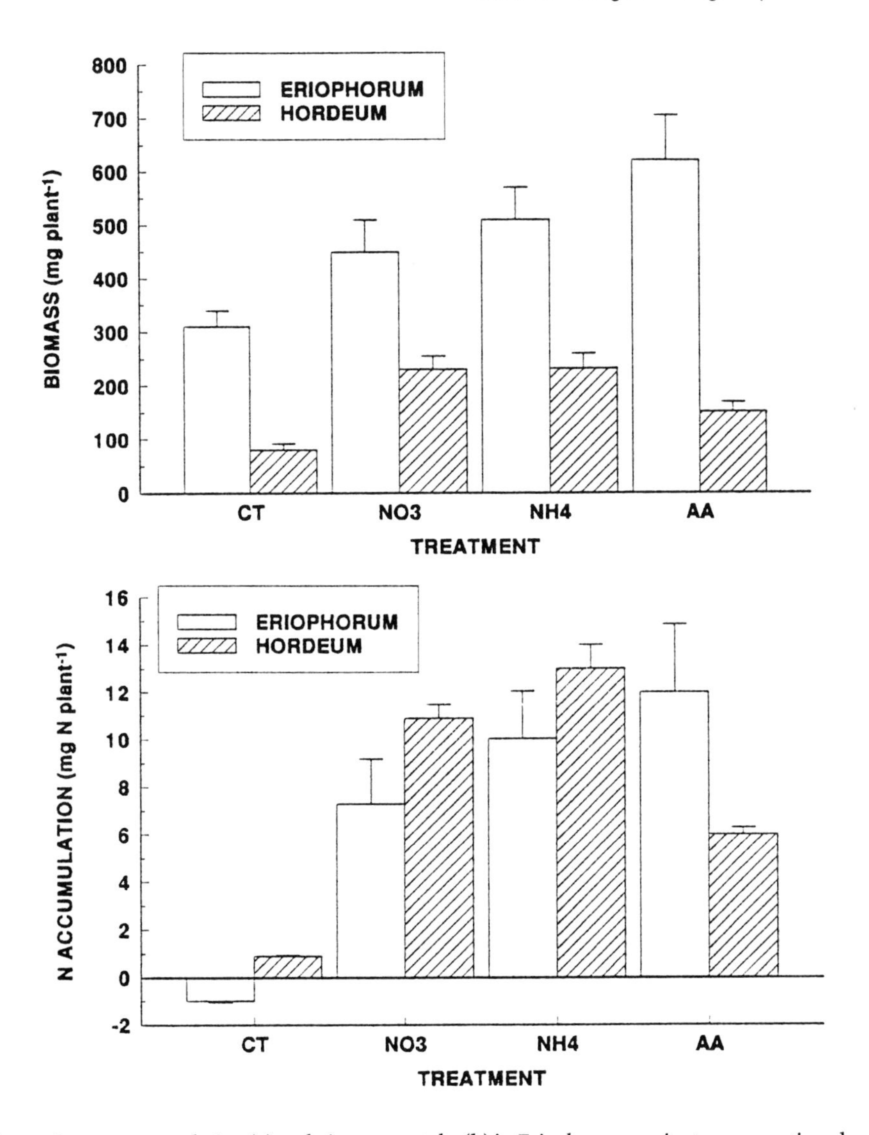

Fig. 7. Biomass accumulation (**a**) and nitrogen uptake (**b**) in *Eriophorum vaginatum*, an arctic sedge, and *Hordeum vulgare*, an agricultural cereal, supplied with different nitrogen sources in solution culture in the laboratory. CT, no nitrogen; NO3, nitrate-N; NH4, ammonium-N; AA, equimolar concentrations of alanine, arginine, aspartate, glutamate, and glycine. (Data from Chapin et al. 1993)

species from more fertile soils (Chapin 1988). By contrast, when nutrient supply fluctuates greatly over the growing season ("pulsed"), in an otherwise infertile soil, it should be advantageous to have the capacity to exploit the resource during the short time it is available (Fig. 8). Consequently, two populations from different sites with the same annual nutrient flux (thus classified similarly as either high or low nutrient sites), but with different modes of nutrient supply ("steady-state" vs. "pulsed"), would be subject to different selection pressures regarding

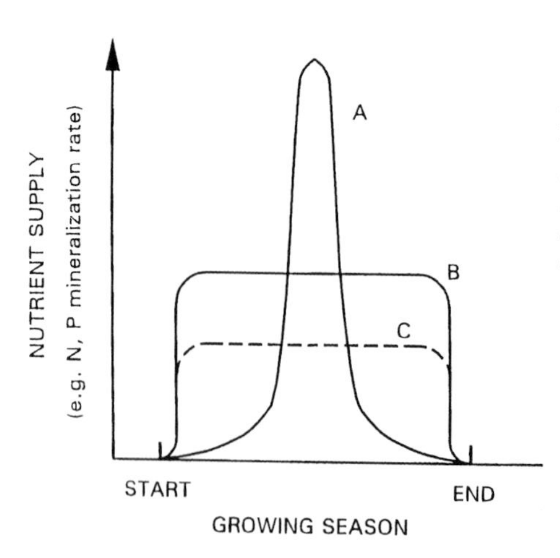

Fig. 8. Idealized diagram of three soil nutrient regimes conferring differential selective pressures on plant nutrient absorption potential, characterized by either pulsed (**A**) or steady-state (**B**, fertile; **C**, infertile) nutrient supply. Area under the curves equal the net annual nutrient supply. (Data from Kielland and Chapin 1994)

nutrient absorption capacity. It would seem that species from a highly "pulsed" soil could evolve towards higher nutrient absorption potentials than species experiencing "steady-state" nutrient supply, even if the latter have greater annual nutrient flux (Fig. 8). Thus, the magnitude and duration of nutrient pulses can exert significant control over plant nutrient acquisition (Campbell and Grime 1989). The extreme temporal variation in nutrient supply that characterizes arctic soils (Barel and Barsdate 1978; Giblin et al. 1991) results in conditions that may favor high nutrient uptake potentials, despite the overall low nutrient supply of these ecosystems. This may explain why bryophytes and lichens, which receive the majority of their nutrient during brief events like runoff and rainfall, often exhibit high nutrient uptake capacities.

4.3 Organic N Uptake In Situ

The importance of organic nitrogen in the nitrogen economy of plants ultimately must be assessed by in situ measurements. Schimel and Chapin (1996) used double-labeled (^{13}C, ^{15}N) glycine in a field experiment with *Eriophorum vaginatum* in arctic tundra. They found that glycine-derived nitrogen was taken up as rapidly as ammonium, but could not demonstrate that glycine was not mineralized prior to uptake. In a recent, elegant study, Näsholm et al. (1998) reported that boreal forest plants take up glycine in situ, thus lending strong support to the hypothesis that boreal plants are not exclusively dependent on microbially mediated nitrogen mineralization for their nitrogen supply. In a series of field experiments using double, universally (U) labeled glycine (U-$^{13}C_2$, ^{15}N), Näsholm and his coworkers showed that glycine was taken up as an intact molecule in ectomycorrhizal trees (*Pinus sylvestris* and *Picea abies*), an ericoid-mycorrhizal shrub (*Vaccinium myrtillus*), and an arbuscular-mycorrhizal grass (*Deschampsia flexuosa*) (Fig. 9). Though the proportion of intact glycine absorbed varied amongst species, these experiments demonstrate that direct uptake of organic nitrogen can occur under field conditions. Uptake responses of crops supplied

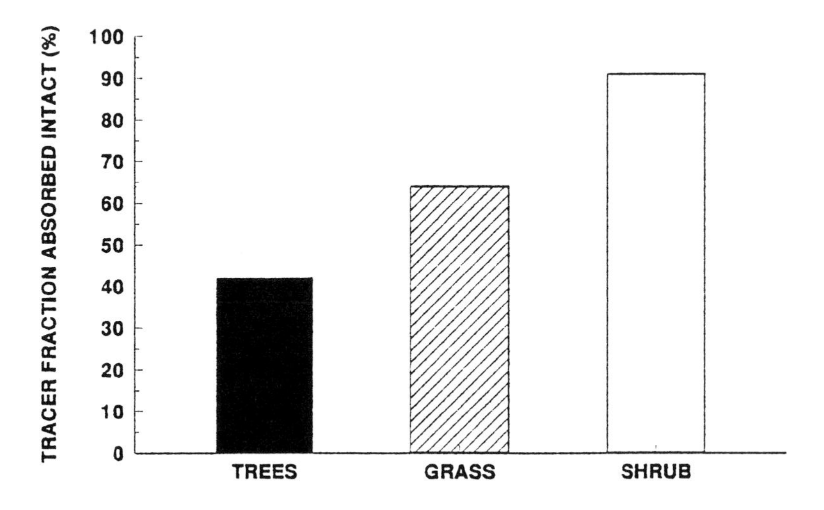

Fig. 9. Proportion of organic N-tracer (glycine) absorbed intact under field conditions in three forest growth forms from northern Sweden. (Data from Näsholm et al. 1998)

with organic or inorganic N also suggest that organic N is absorbed intact in a range of agricultural species (Yamagata and Ae 1996). These findings have major implications for our understanding of crop production, intensified forestry, and nitrogen deposition.

4.4 Organic N Uptake and Species-Specific Variation in $\delta^{15}N$ Signatures

In addition to direct physiological studies of organic and inorganic nitrogen absorption, evidence for niche differentiation regarding the acquisition of different nitrogen forms is suggested by the $\delta^{15}N$ signatures of tundra and taiga species (Schulze et al. 1994; Michelsen et al. 1996; Nadelhoffer et al. 1996; Kielland 1997; Kielland et al.1998). Arctic species of different growth forms show pronounced variation in leaf $\delta^{15}N$ values, which strongly suggests differences in the source, and possibly composition, of absorbed soil nitrogen (Michelsen et al. 1996; Kielland 1997). Graminoids that primarily rely on inorganic nitrogen and exploit deeper soil horizons are enriched in $\delta^{15}N$, whereas deciduous and evergreen shrubs that rely more heavily on organic nitrogen are depleted in the heavy isotope (Fig. 10). Recent work in taiga ecosystems show similar results (Schulze et al. 1994; Kielland et al. 1998), suggesting that similar physiological traits may exist among species in taiga forest as in the Arctic. In particular, the very depleted isotope signature of black spruce (*Picea mariana*) is contrary to what is expected of species growing in century-old peat, where soil nitrogen has had a long history of processing, and thus would tend to be enriched. The observed range of interspecific isotope values are approximately similar in taiga (Fig. 11) as in arctic tundra, which is consistent with the hypothesis that within each respective ecosystem plant species have evolved a diversity of strategies for acquiring soil nitrogen.

The consequences of these findings have important ramifications for how we understand how nitrogen cycles through northern ecosystems, and it poses a whole suite of new questions of how these ecosystems may respond to climate warming and other aspects of global change.

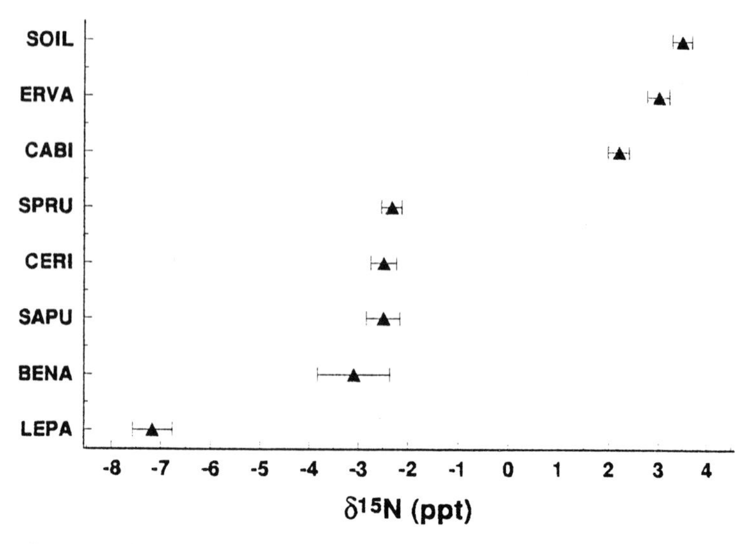

Fig. 10. Bulk soil and foliar $\delta^{15}N$ signatures of arctic tundra species across major growth forms. Species designations as in Fig. 6. (Data from Kielland 1997, and unpublished information)

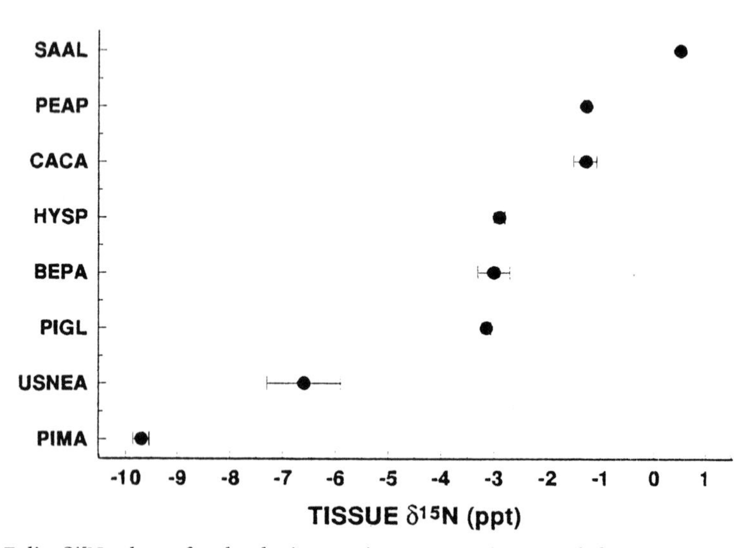

Fig. 11. Foliar $\delta^{15}N$ values of upland taiga species across major growth forms. Species designations are SAAL, *Salix alaxensis*; PEAP, *Peltigera apthosa*; CACA, *Calamagrostis canadensis*; HYSP, *Hylocomnium splendens*; BEPA, *Betula papyrifera*; PIGL, *Picea glauca*; USNEA, *Usnea* sp.; PIMA, *Picea mariana*. (Data from Kielland 1997, and unpublished information)

5. Conclusions

The capacity to absorb free amino acids both in the laboratory and in the field of more than a dozen species from a variety of ecosystems ranging from arctic tundra to temperate heathlands, demonstrate that the conventional representation of the terrestrial nitrogen cycle is inadequate. Moreover, evidence from agricultural systems (see the chapters by Yamagata M et al. and Nishizawa NK and Mori S) suggest that the uptake of organic nitrogen is not restricted to plants adapted to low-nitrogen soils. Several agricultural species show growth responses and chemical characteristics that imply direct uptake of organic nitrogen both in solution culture and in the field (the chapter by Yamagata M et al.). In the context of low-input sustainable agriculture, these findings are extremely important (Yamagata and Ae 1996). A reevaluation of the nitrogen cycle (Fig. 12), and the controls thereof, needs to incorporate at least the five following points (sensu Kielland 1994):

1) Nitrogen mineralization, which hitherto has been considered the bottleneck in the nitrogen cycle of natural ecosystems, may underestimate perhaps by as much as 50%, the rate of nitrogen supply to plants.
2) The pool sizes of ammonium and nitrate are insufficient measures of instantaneous plant-available soil nitrogen.

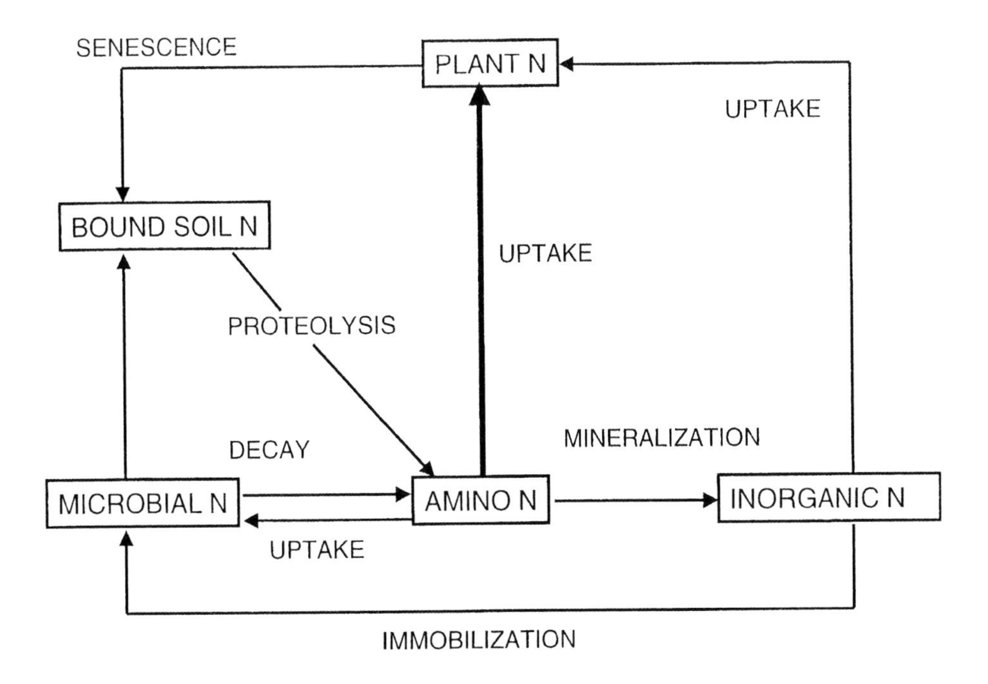

Fig. 12. Schematic diagram of the nitrogen cycle in tundra ecosystems showing the hypothesized role of organic nitrogen. (From Kielland 1994)

3) Differences in absorption potential of organic nitrogen at the level of growth forms and species provide ample basis for niche differentiation among species in competing for what previously was considered a single resource.
4) By absorbing soil nitrogen in organic form, plants in effect "short-circuit" the nitrogen cycle, demonstrating that plants are not solely passive recipients of microbially mediated nitrogen mineralization.
5) By circumventing nitrogen mineralization plants may accelerate nitrogen turnover and by doing so shift a greater part of the control of nitrogen cycling from soil microbes to the plants themselves.

The presence of amino acid uptake under as different environmental settings and in as diverse species as lichens on the arctic tundra of Alaska to upland rice in the paddies of Hokkaido suggests that this physiological strategy is far more prevalent both in natural and agricultural ecosystems than considered previously. These findings warrant a considerable reassessment of plant-nitrogen relationships both at the level of the individual plant and at the ecosystem level, which has important ramifications for both global change research on natural ecosystems and applied agricultural practices.

References

Abuarghub SM, Read DJ (1988) The biology of mycorrhiza in the Ericaceae. XII. Quantitative analysis of individual 'free' amino acids in relation to time and depth in the soil profile. New Phytologist 108:433-441

Babb TA, Whitfield DWA (1977) Mineral nutrient cycling and limitation of plant growth in the Truelove Lowland ecosystem. In: Bliss LC (Ed) Truelove Lowland, Devon Island, Canada: A high arctic ecosystem. University of Alberta Press, Edmonton, Canada, pp 589-606

Barèl D, Barsdate RJ (1978) Phosphorus dynamics of wet coastal tundra soils near Barrow, Alaska. In: Adriano DC, Brisbin IL (Eds) Environmental chemistry and cycling processes. DOE Symposium Series, Conf-760429, Washington DC, pp 516-537

Barry RG, Courtin GM, Labine C (1981) Tundra climates. In: Bliss LC, Heal OW, Moore JJ (Eds) Tundra ecosystems: A comparative analysis. Cambridge University Press, Cambridge, pp 81-114

Campbell BD, Grime JP (1989) A new method of exposing developing root systems to controlled patchiness in mineral nutrient supply. Ann Bot 63:396-400

Chandler G (1985) Mineralization and nitrification in three Malaysian forest soils. Soil Biol Biochem 17:347-353

Chapin FS III (1983) Direct and indirect effects of temperature on arctic plants. Polar Biol 2:47-52

Chapin FS III (1988) Ecological aspects of plant mineral nutrition. Adv Miner Nutr 3:161-191

Chapin FS III (1995) New cog in the nitrogen cycle. Nature 377:199-200

Chapin FS III, Van Cleve K, Chapin M (1979) Soil temperature and nutrient cycling in the tussock growth form of *Eriophorum vaginatum*. J Ecol 67:169-189

Chapin FS III, Fetcher N, Kielland K, Everett K, Linkins AE (1988) Productivity and nutrient cycling of Alaskan tundra: enhancement by flowing soil water. Ecology 69:693-702

Chapin FS III, Moilanen L, Kielland K (1993) Preferential use of organic nitrogen for growth by a non-mycorrhizal arctic sedge. Nature 631:150-153

Chazdon RL, Pearcy RW (1991) The importance of sunflecks for forest understory plants. Bioscience 41:760-766

Everett KR, Vassiljevskaya VD, Brown J, Walker BD (1981) Tundra and analogous soils. In:Bliss LC, Heal OW, Moore JJ (Eds) Tundra Ecosystems: A comparative Analysis. Cambridge University Press, Cambridge. pp 139-179

Fisk MC, Schmidt SK (1995) Nitrogen mineralization and microbial biomass N dynamics in three alpine tundra communities. Soil Sci Soc Am J 59:1036-1043

Giblin AE, Nadelhoffer KJ, Shaver GR, Laundre JA, McKerrow AJ (1991) Biogeochemical diversity along a riverside toposequence in arctic Alaska. Ecology 61:415-435

Hersey AE, Bowden WB, Deegan LA, Hobbie JE, Peterson BJ, Kipphut GW, Kling GW, Lock MA, Merritt RW, Miller MC, Vestal JR, Schuldt JA (1996) The Kuparuk River: A long-term study of biological and chemical processes in an arctic river. In: Milner AM, Oswood MW (Eds) Freshwater of Alaska, ecological syntheses. Springer, New York. Ecological Studies 119:107-130

Jenny H (1980) The soil resource. Springer, New York

Kielland K (1990) Processes controlling nitrogen release and turnover in arctic tundra. Ph.D. dissertation. University of Alaska Fairbanks, p177

Kielland K (1994) Amino acid absorption by arctic plants: implications for plant nutrition and nitrogen cycling. Ecology 75:2373-2383

Kielland K (1995) Landscape patterns of free amino acids in arctic tundra soils. Biogeochemistry 32:1-14

Kielland K (1997) Role of free amino acids in the nitrogen economy of arctic cryptogams. Ecoscience 4:75-79

Kielland K, Bryant JB, Ruess R (1997) Mammalian herbivory and carbon turnover in early successional stands in interior Alaska. Oikos 80:25-30

Kielland K, Chapin FS III (1992) Nutrient absorption and accumulation in arctic plants. In: Chapin FS III, Jefferies RL, Reynolds JF, Shaver GR, Svoboda J (Eds) Arctic ecosystems in a changing climate: an ecophysiological perspective. Academic, San Diego, pp 321-335

Kielland K, Chapin FS III (1994) Phosphate uptake in arctic plants in relation to phosphate supply: the role of spatial and temporal variability. Oikos 70:443-448

Kielland K, Barnett B, Schell D (1998) Intraseasonal variation in the $\delta^{15}N$ signatures of taiga trees and shrubs. Can J For Res 28:485-488

Klingensmith KM, Van Cleve K (1993) Patterns of nitrogen mineralization and nitrification in floodplain successional soil along the Tanana River, interior Alaska. Can J For Res 23:964-969

Labroue L, Carle J (1977) Le cycle de l'azote dans les solsalpins du Pic du Midi de Bigorre (Hautes-Pyrennees). Oecologia Plantarum 12:55-77

Linkins AE, Neal JL (1982) Soil cellulase, chitinase, and protease activity in *Eriophorum vaginatum* tussock tundra at Eagle Summit, Alaska. Holarctic Ecology 5:135-138

Marion GM, Miller PC (1982) Nitrogen mineralization in a tussock tundra soil. Arctic and Alpine Res. 14:287-293

Meentemeyer V (1978) Macroclimate and lignin control over litter decomposition rates. Ecology 59:465-472

Melin E (1953) Physiology of mycorrhizal relations in plants. Ann Rev Plant Physiol 4:325-346

Melin E, Nilsson H (1953) Transfer of labelled nitrogen from glutamic acid to pine seedlings through the mycelium of *Boletus variegas* (S.W.) Fr. Nature 171:434

Michelsen A, Schmidt IK, Jonasson S, Quarmby C, Sleep D (1996) Leaf ^{15}N abundance of subarctic plants provide field evidence that ericoid, ectomycorrhizal and non- and arbuscular mycorrhizal species access different sources of soil nitrogen. Oecologia 105:53-63

Miettinen KJ (1959) Assimilation of amino acids in higher plants. In: Symposium 13 Society of Experimental Biology. Academic, New York, pp 210-230

Mooney HA (1980) Seasonality and gradients in the study of stress adaptation. In: Turner NC, Kramer PJ (Eds) Adaptation of plants to high water and temperature stress. Wiley, New York, pp 279-295

Nadelhoffer KJ, Aber JD, Melillo JM (1984) Seasonal patterns of ammonium and nitrate uptake in nine temperate forest ecosystems. Plant Soil 80:321-335

Nadelhoffer KJ, Giblin AE, Shaver GR, Laundre JL (1991) Effects of temperature and substrate quality on element mineralization in six arctic soils. Ecology 72:242-253

Nadelhoffer KJ, Giblin AE, Shaver GR, Linkins AE (1992) Microbial processes and plant nutrient availability in arctic soils. In: Chapin FS III, Jefferies RL, Reynolds JF, Shaver GR, Svoboda J (Eds) Arctic ecosystems in a changing climate: an ecophysiological perspective. Academic, San Diego, pp 281-300

Nadelhoffer KJ, Shaver GR, Fry B, Giblin A, Johnson L, McKane R (1996) ^{15}N natural abundances and N use by tundra plants. Oecologia 107:386-394

Northup RR, Yu Z, Dahlgren RA, Vogt KA (1995) Polyphenol control of nitrogen release from pine litter. Nature 377:227-229

Näsholm T, Ekblad A, Nordin A, Giesler R, Högberg M, Högberg P (1998) Boreal forest plants take up organic nitrogen. Nature 392:914-916

Raab TK, Lipson DA, Monson RK (1996) Non-mycorrhizal uptake of amino acids by roots of the alpine sedge *Kobresia myosuroides*: implications for the alpine nitrogen cycle. Oecologia 108:488-494

Read DJ (1991) Mycorrhizas in ecosystems. Experientia 47:376-391

Read DJ, Bajwa R (1985) Some nutritional aspects of the biology of ericaceous mycorrhizas. Proc Royal Soc Edinburgh 85B:317-332

Rieger S (1983) The genesis and classification of cold soils. Academic, London

Ruess RW, Seagle SW (1994) Landscape patterns in soil microbial processes in the Serengeti National Park, Tanzania. Ecology 75:892-904

Ruess RW, Van Cleve K, Yarie J, Viereck LA (1996) Contributions of fine root production and turnover to the carbon and nitrogen cycling in taiga forests of the Alaskan interior. Can J For Res 26:1326-1336

Schimel JP, Chapin FS III (1996) Tundra plant uptake of amino acid and NH_4^+ nitrogen in situ: plants compete well for amino acid N. Ecology 77:2142-2147

Schulze E-D, Chapin FS III, Gebauer G (1994) Nitrogen nutrition and isotope differences among life forms at the northern treeline of Alaska. Oecologia 100:406-412

Shaver GR, Chapin FS III (1991) Production : biomass relationships and elements cycling in contrasting arctic vegetation types. Ecol Monogr 61:1-31

Shaver GR, Nadelhoffer KJ, Giblin AE (1990) Biochemical diversity and element transport in a heterogeneous landscape, the North Slope of Alaska. In: Turner, MG, Gardner RH (Eds) Quantitative methods in landscape ecology. Springer, New York, pp 105-125

Sparrow SD, Sparrow E, Cochran VL (1992) Decomposition in forest and fallow subarctic soil. Biol Fertil Soill 14:253-259

Stribley DP, Read DJ (1980) The biology of mycorrhiza in the Ericaceae. VII. The relationship between mycorrhizal infection and the capacity to utilize simple and complex organic nitrogen sources. New Phytol 86:365-371

Swift M, Heal OH, Anderson J (1979) Decomposition in Terrestrial Ecosystems. Blackwell Scientific Publishers, London

Tedrow JCF (1977) Soils of polar landscapes. Rutgers University Press, New Brunswick, New Jersey

Van Cleve K, Alexander V (1981) Nitrogen cycling in tundra and boreal ecosystems. In: Clark FE, Rosswall T (Eds) Terrestrial nitrogen cycles. Ecol Bull 33:375-404

Vitousek PM, Denslow JS (1986) Nitrogen and phosphorus availability in treefall gaps of a lowland rainforest succession in Costa Rica, Central America. Oecologia 61:99-104

Walker LR (1989) Soil nitrogen changes during primary succession on a floodplain in Alaska. Arct Alp Res 21:341-349

Wedin DA, Tilman D (1990) Species effects on nitrogen cycling: a test with perennial grasses. Oecologia 84:433-441

Weller G, Holmgren B (1974) The microclimate of the arctic tundra. J Appl Meteorol 13:854-862

Yamagata M, Ae N 1996 Nitrogen uptake response of crops to organic nitrogen. Soil Sci Plant Nutr 42:389-394

Possibility of Direct Acquisition of Organic Nitrogen by Crops

Makoto Yamagata, Shingo Matsumoto, and Noriharu Ae

Summary. We found unusual growth responses of some crop species to organic nitrogen (N). Upland rice could take up more N in soil amended with organic materials than with chemical fertilizer, which had a higher inorganic N level. On the other hand, maize, sugar beet and potato grew better in the soil with chemical fertilizer than with organic materials. We proposed two possibilities for these results: (1) Significant N mineralization occurs in the rhizosphere of upland rice. Upland rice may enhance the N mineralization in the rhizosphere by secreting enzymes such as protease. (2) Upland rice can take up organic N directly.

Protease activity in the rhizosphere soil of upland rice was lower than that of other crops, which indicates that upland rice did not enhance the N mineralization in the rhizosphere. Amino acid and protein contents in the rhizosphere soil of upland rice were lower than the levels in the rhizosphere soil of maize and non-rhizosphere soils. Furthermore, when ^{15}N-labelled rice bran was applied as an organic material, the ^{15}N content of upland rice was higher than that of maize. These results suggest that upland rice might take up organic N preferentially compared to maize.

We examined the N uptake response of vegetable crops such as *chingensai* (a kind of Chinese cabbage), pimento, carrot, and leaf lettuce, whose growing duration is much shorter than that of upland rice. For this, two plots were employed; one plot received rapeseed cake as organic N, and the other received ammonium sulfate. Growth and N uptake by carrot and *chingensai* was stimulated more on the organic N plot than on the plot with ammonium sulfate, but in the case of pimento and leaf lettuce, the reverse was true. These results suggest that, like upland rice, carrot and *chingensai* have an ability to take up organic N preferentially.

In order to obtain evidence that *chingensai* may take up organic N from organic N amended soil, we investigated xylem sap of *chingensai* grown with rapeseed cake using high-pressure liquid chromatography (HPLC) analysis. We detected the same peak in the xylem sap of *chingensai* grown in a soil with rapeseed cake as in the phosphate buffer extract from soil with organic N. However, this peak was not detected in xylem sap of pimento, which showed poor growth with rapeseed cake. The compound found in the soil extract was estimated to be a protein-like nitrogenous compound, which is likely to be produced through bacterial biomass when organic materials are supplied to a soil.

In conclusion, upland rice and *chingensai*, both of which responded well to organic N, might have an ability to take up directly this particular organic N produced in the soil. The presence or absence of this ability could be a source of N uptake differences among crops when organic matter is applied.

Key words. Organic nitrogen, Upland rice, *Chingensai*, Carrot, Mineralization, Protein, Protease, Rhizosphere, Phosphate buffer

1. Introduction

Nitrogen (N) is the most crucial element for crop production because it is a limiting factor for the growth of almost all crops. Organic matter has long been known as one of the most important N sources in agriculture. Organic N in a soil amended with organic matter is decomposed by soil microorganisms, changed to protein and then to amino acids and ultimately to inorganic N such as ammonium and nitrate, which are the main N sources for crops. Since these inorganic N sources are also incorporated into soil microorganisms, crops meet severe competition in N uptake with soil microorganisms.

There are several methods for assessing soil N fertility. One method proposed by Stanford et al. (1965) evaluates N fertility by measuring the amount of inorganic N mineralized in soils after several weeks of incubation at the proper temperature in a laboratory. This method is valid under the hypothesis that crops mainly take up mineralized inorganic N. The N fertility measured by this method is, in fact, well correlated with the N uptake by crops (Grunes et al. 1963; Harmsen and Van Schreven 1955; Saito and Ishii 1987; Stanford et al. 1965). Various extraction methods have also been developed for evaluating soil N fertility, for example, hot water, alkaline and acid extractions. The N contents in these extracts are then measured (Numan et al. 1998; Saito 1988). These extraction methods showed good agreement with the incubation method. Yet another approach uses ^{15}N tracer in paddy fields (Yamamuro 1986). ^{15}N-labelled fertilizer was added to soil weekly, and then mineralization and immobilization were analyzed by determining the ^{15}N. However, all these methods are based on the concept that plants mainly take up inorganic N.

According to this concept, N uptake by crops should correspond to the amount of soil inorganic N under similar conditions of growing period and rooting soil volume among crop species. Matsumoto et al. (1999b) confirmed that growth and N uptake by spinach was greater in soil with organic materials compared to soil with chemical fertilizer, even though the inorganic N level was lower in the soil amended with organic matter than in that amended with chemical fertilizer. In fact, it is widely known that organic matter is especially effective in some crops. For example, farmers commonly apply organic matter to vegetables such as spinach and carrot for higher production and better quality.

In order to assess N uptake by crops grown with organic matter, we must know the N status in the soil amended with organic matter. A phosphate buffer extraction method (Higuchi 1981,1982) has been developed recently for evaluating easily mineralized organic N. This extraction procedure is very useful for determining the N status in a soil.

2. N Status of Organic Materials Applied to Soils

2.1 Relationship Between N Mineralized and the Amount of Protein-Like N Compounds

Understanding the N availability in soil is important for optimizing application rate of N fertilizer. An incubation method in which inorganic N released is determined during incubation at 30 °C under field moisture conditions has been commonly used (Keeney 1982). The major source of inorganic N in this method is organic N compounds in the soil (Stevenson 1958; Stevenson and Braids 1968; Wagner and Mutatkar 1968). Ideally, organic N that releases inorganic N during incubation should be the same as the organic N extracted by the extraction method.

Higuchi (1981, 1982) reported that the organic N extracted by 1/15 M phosphate buffer ($Na_2HPO_4 \cdot 12H_2O$, 14.6 g l^{-1}; KH_2PO_4, 3.5 g l^{-1} ; pH 7.0) was in the form of amino acids and proteins. This organic N correlated with inorganic N produced by the incubation method. Using 20 soil samples collected from various soil types (Table 1), we also confirmed that this "phosphate buffer-extractable organic N" (PEON) correlated strongly (r=0.92) with the inorganic N mineralized by the incubation method (Table 1). The phosphate buffer extract certainly included not only PEON but also inorganic N. The N of PEON was determined by subtracting inorganic N from total N in the phosphate buffer extract. Higuchi and the other authors detected amino acid formation by hydrolysis of this PEON (Higuchi 1982). Also, this PEON reacted positively with reagent Coomassie brilliant blue for protein measurement. This implies that PEON contains protein-like N compounds in a soil.

2.2 HPLC Analysis of the Soil Extract with Phosphate Buffer

The question remains what was the source of the protein-like N compound. Analysis using high-pressure liquid chromatography (HPLC) was done to identify the actual substances found in the PEON fraction. Size-exclusion and ion-exchange chromatograms of soil extracts with 1/15 M phosphate buffer were obtained from different types of soil distributed in Shimane prefecture, Japan. Regardless of soil type, only one major peak with a retention time of 8.4 min could be detected by the size-exclusion method (Fig. 1a). The corresponding molecular weight was estimated to be approximately 8 000 – 9 000 based on the comparison with retention times of standard molecular weight compounds. The ultraviolet (UV) detector at 280 nm in the ion-exchange HPLC also showed only one major peak with a retention time of 2.8 min in the extract by phosphate buffer (Fig. 1b). This major peak area on the size-exclusion chromatogram highly correlated with PEON (Fig. 2a, r = 0.95), and the major peak on ion-exchange HPLC also correlated with PEON (r = 0.93, not shown). Likewise, a strong correlation was observed between these peak areas and protein-N concentration in PEON (Fig. 2b) measured by the Bradford method (Bradford 1976). This implies that this major peak at 280 nm was derived from the protein-like N compound formed in the soil, but protein accounted for only approximately 30% of N of PEON among the tested soils (Table 1). Similar results

Table 1. Characteristics of soils collected from 20 sites in Shimane Prefecture, Japan. The soils were incubated at 30°C; soil moisture was maintained at 60% of maximum waterholding capacity. Mineralized N was extracted by 2 M KCl. Phosphate buffer extractable organic N (PEON) was calculated by subtracting inorganic N from the total N in the extract with 1/15 M phosphate buffer. Protein-N in PEON was measured by the Bradford method (Bradford 1976)

No.	Soil type[a]	Textural class	Total-C ($g\ kg^{-1}$)	Total-N ($g\ kg^{-1}$)	pH (H_2O)	Mineralized -N ($mg\ kg^{-1}$)	Phosphate buffer extractable organic-N (PEON) ($mg\ kg^{-1}$)	Protein-N in-PEON ($mg\ kg^{-1}$)
1	Andosol	Loam	50.8	4.6	6.0	351	224	52.9
2	Andosol	Loam	30.3	2.6	5.0	327	182	50.8
3	Andosol	Clay loam	53.7	4.4	5.7	215	189	42.7
4	Andosol	Clay loam	58.5	5.3	5.9	225	210	44.7
5	Cambisol	Loam	58.0	5.8	7.4	363	250	60.0
6	Cambisol	Loam	15.1	1.4	6.6	178	90	32.9
7	Cambisol	Loam	20.8	1.6	6.8	220	136	33.9
8	Cambisol	Loam	46.0	3.8	6.2	118	102	26.9
9	Cambisol	Loam	22.3	2.0	6.5	180	137	37.9
10	Cambisol	Loam	32.1	2.6	6.7	240	130	36.9
11	Fluvisol	Loam	9.8	0.9	5.7	68	70	25.8
12	Fluvisol	Loam	50.6	4.5	5.7	170	105	34.0
13	Gleysol	Light clay	12.3	1.1	5.3	253	207	51.7
14	Gleysol	Light clay	15.9	1.3	5.7	203	115	40.7
15	Gleysol	Light clay	31.6	3.4	4.8	251	174	38.7
16	Gleysol	Clay loam	19.7	1.8	5.5	283	215	53.0
17	Gleysol	Clay loam	12.2	1.0	6.2	135	75	20.3
18	Regosol	Sandy loam	4.1	0.4	6.4	16	27	12.2
19	Regosol	Sandy loam	1.2	0.1	6.2	13	20	9.0
20	Regosol	Sandy loam	5.5	0.5	5.8	71	90	19.6

All properties except pH represent content in dry soil
[a] The soils were classified according to the recent FAO (Food and Agriculture Organization of the United Nations) soil classification system

were reported that hydrolyzable organic N composed of amino acids accounted for only 23% to 53% of the organic N extracted by electro-ultrafiltration and $CaCl_2$ solution (Nemeth et al. 1988; Appel and Mengel 1992, 1993; Groot and Houba 1995).

Then what are the compounds included in the 280 nm peak other than proteins? PEON contains other N compounds such as amino sugars (Higuchi 1981), and Marumoto et al. (1982) also found glucosamins and some sugars in the easily available organic nitrogen fraction. Humic substances show an absorbance at UV-280 nm, and phosphate buffer can extract fulvic acid (Honda 1980). Highly polymerized humic substances are not dissolved in a neutral phosphate solution, whereas humic substances with a low polymerization can be extracted with this solution. Therefore, some humic substances may be included in the PEON. Berden and Berggren (1990) reported that one major peak in a soil solution was detected by size-exclusion HPLC, and this peak might originate from humic substances based

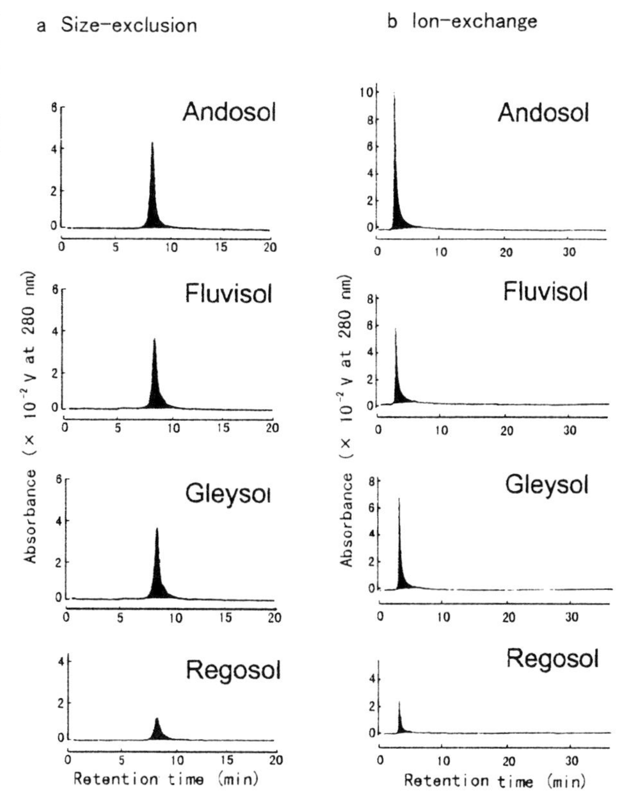

Fig. 1. Size-exclusion and ion-exchange high-pressure liquid chromatography (HPLC) chromatograms of soil extracts with 1/15 M phosphate buffer from 4 different types of soil

on its characteristics of UV absorbance. In our study, although the phosphate buffer extracts seemed to include both protein and humic substances, one major peak was detected on the chromatograms in two types of HPLC analysis. Therefore, the peak at 280 nm may not correspond to a pure protein but to a complex of protein, humic substances, and amino sugar. This complex, the so-called "protein-like N compound," in the PEON has a molecular weight of 8 000 to 9 000 daltons.

Summary of the components related to organic N is as follows: PEON is an organic part of N extracted by phosphate buffer from soils, and contains HPLC-measurable protein-like N, which is considered to be a complex of protein, humic substances and amino sugar.

2.3 Origin of Protein-Like N Compound in a Soil

Mineralized N measured by the incubation method highly correlated with PEON having uniform molecular weight. Our next question was how this PEON was produced in a soil. For this purpose, three kinds of organic materials were added to each pot filled with the No. 3 Andosol shown in Table 1. A mixture of ammonium sulfate and glucose (the carbon/nitrogen (C/N) ratio was adjusted to 20), a mixture of rice bran and rice straw (RBS) (the C/N ratio was 18.9), and egg albumin were used. These soils were incubated at 30°C at 60% of maximum water holding capacity and extracted

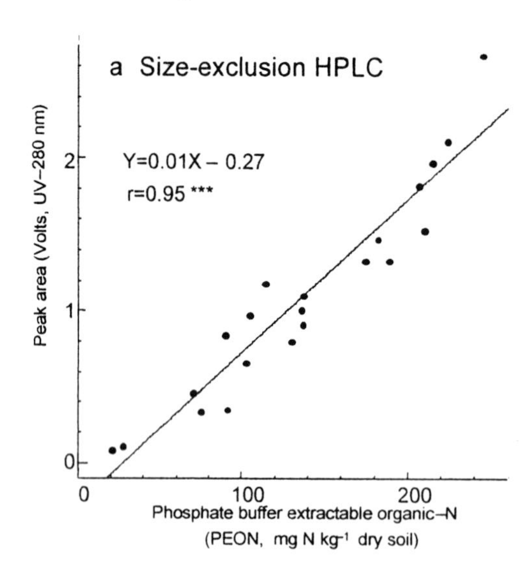

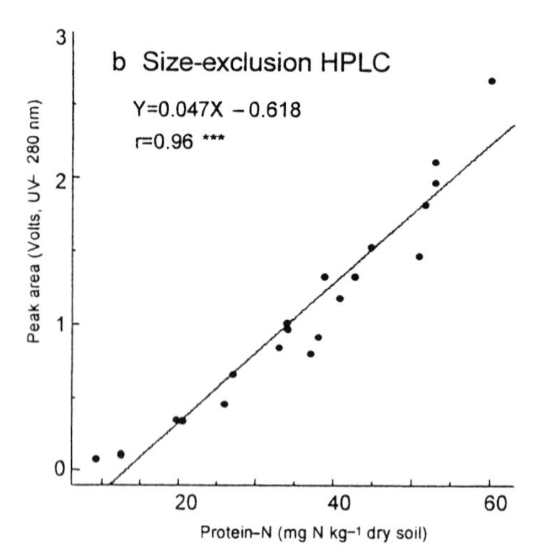

Fig. 2. a Relationship between phosphate buffer extractable organic N (PEON) and peak area measured by size-exclusion HPLC in the 20 different soils in Table 1. PEON was calculated by subtracting inorganic N from total N in the phosphate buffer extract. **b** Relationship between protein N in PEON and the peak area determined by size-exclusion HPLC in these same soils

with 1/15 M phosphate buffer. The extracts were then analyzed with size-exclusion HPLC (Fig. 3), which showed several peaks of organic substances in the case of RBS, as shown in Fig. 3. These peaks appeared after addition of the mixture and disappeared rapidly after 14 days of incubation. One major peak with a retention time of 8.4 min appeared gradually after 14 days of incubation. This trend was similar irrespective of whether the organic material added to the soil was glucose with ammonium sulfate, albumin, or RBS. The retention time (8.4 min) of the major peak was identical to that of extract with phosphate buffer from non-amended soil, as shown in Fig. 1. These results suggest that organic materials added to the soil were decomposed rapidly and seemed to form a protein-like N compound with a UV absorption at 280 nm and a uniform molecular weight (Matsumoto et al. 2000a).

Bremner and van Kessel (1990) showed that ^{14}C-labelled glucose added to soil was incorporated into microbial biomass within 48 h. Urquiga et al. (1998) also

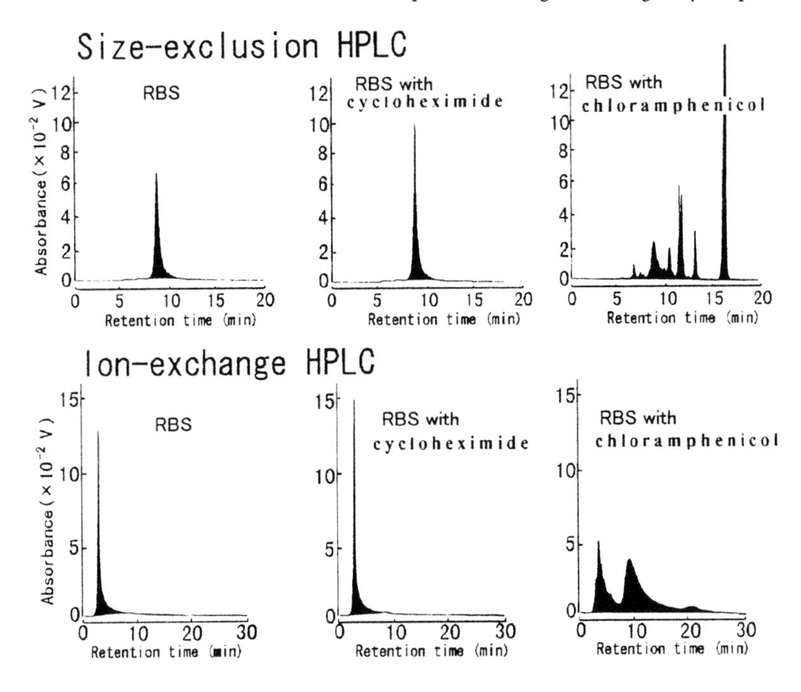

Fig. 3. Chromatograms of size-exclusion HPLC of soil extracts with 1/15 M phosphate buffer applied with 3 different organic matters

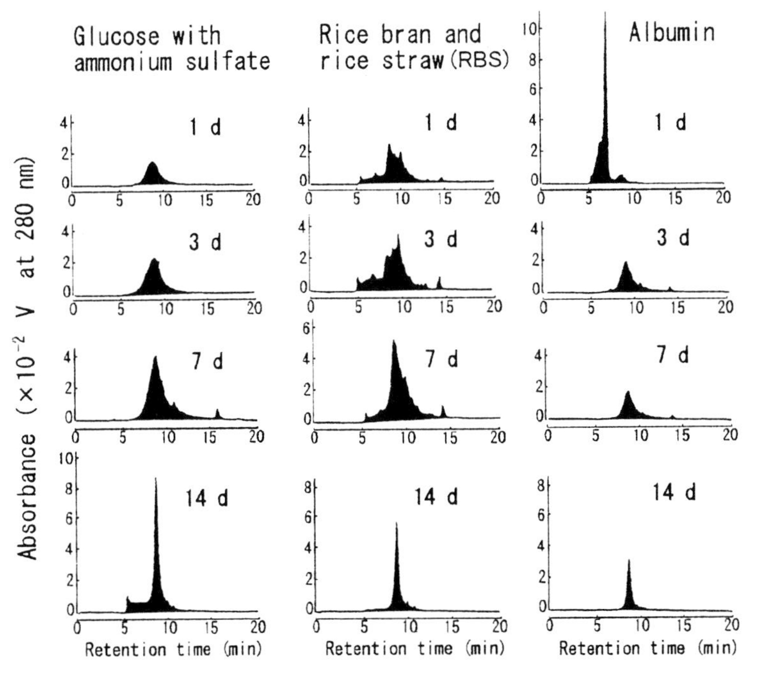

Fig. 4. Size-exclusion and ion-exchange HPLC chromatograms of soil extracts (1/15 M phosphate buffer) after the addition of RBS (a mixture of rice bran and rice straw) and antibiotics. Chloramphenicol and cycloheximide were used

reported that grass roots in a soil were decomposed by soil microbes, and CO_2 was produced. This CO_2 evolution process reached a peak after 5 days of addition of the organic matter. These results support our observation that the protein-like N compound seen on the HPLC chromatogram was produced by soil microorganisms within a few days after the addition of organic matter.

To confirm the microbial process in relation to the production of PEON, we added antibiotics to the RBS amended soil and performed HPLC analysis of the phosphate buffer extract. Two antibiotics, chloramphenicol, an anti-bacterial agent, and cycloheximide, an anti-fungal agent, were employed. The peak with a retention time of 8.4 min was suppressed by the addition of chloramphenicol, which inhibits bacterial growth and eventually stimulates fungal growth (Fig. 4).

In contrast, the formation of the peak was stimulated with the addition of cycloheximide, whose anti-fungal agent eventually enhances bacterial growth. These results strongly indicate that the protein-like N compound in PEON can be derived from soil bacteria. Further, it has been reported that easily decomposable organic substances accumulating in soil were derived from the remains of soil microorganisms (Chantigny et al. 1997; Degens 1998; Marumoto et al. 1982). These researchers also detected some amino sugars in easily decomposable organic substances. This may mean that PEON is derived from bacterial cell wall components but not from fungal cell walls (Matsumoto et al. 2000b).

Table 2. Amino acid composition of soil organic matter (data from Marumoto et al. 1974) and soil extracts (data from Higuchi 1982 and Ogiuchi et al. 2000) with phosphate buffer. Amino acid in organic matter in an alluvial soil and soil extracts of an alluvial soil and an Andosol were obtained by hydrolysis of the soil with 6 M HCl

	Amino acid composition of the hydrolysates (%)		
Amino acid	Soil organic matter	Soil extract from an alluvial soil	Soil extract from an Andosol
Gly	14.3	13.4	11.0
Ala	10.8	11.6	9.7
Val	6.0	6.8	10.7
Leu	6.2	10.0	5.3
Ile	3.7	3.4	3.8
Ser	5.9	8.6	5.7
Thr	6.3	9.1	8.1
Phe	3.8	2.1	3.0
Tyr	1.4	3.7	2.7
Pro	4.6	5.0	5.1
Asp	12.2	11.8	13.9
Glu	12.4	8.0	11.9
His	1.4	1.0	1.1
Lys	10.8	3.4	5.3
Arg	0.3	2.1	2.7
Total	100.0	100.0	100.0

Although we have discussed already that PEON contained a protein-like N compound, Higuchi (1982) showed that PEON had a constant amino acid composition and C/N ratio irrespective of soil type, and Ogiuchi et al. (2000) also showed similarities in amino acid composition in different soil extracts. Marumoto et al. (1974) determined the amino acid composition of easily decomposable organic matter after the addition of Italian ryegrass into a soil. These are summarized in Table 2, which shows amino acid composition of soil organic matter and PEON in soils that received glucose and ammonium sulfate followed by incubation for 14 days. These amino acid compositions were quite similar, irrespective of soil types and added organic matters (Ogiuchi et al. 2000). Thus we can confirm that the mineralizable N of the soil was uniform not only in molecular weight but also in amino acid composition.

To summarize, when we incorporate organic matter into a soil, the microbial biomass increases and reaches a maximum within a few days. Enriched soil bacteria may then be decomposed through self-digestion and other microbial attack. Because the cell wall component of soil bacteria resists self-digestion or decomposition, a protein-like N compound in PEON is released and accumulated. The PEON or pro-

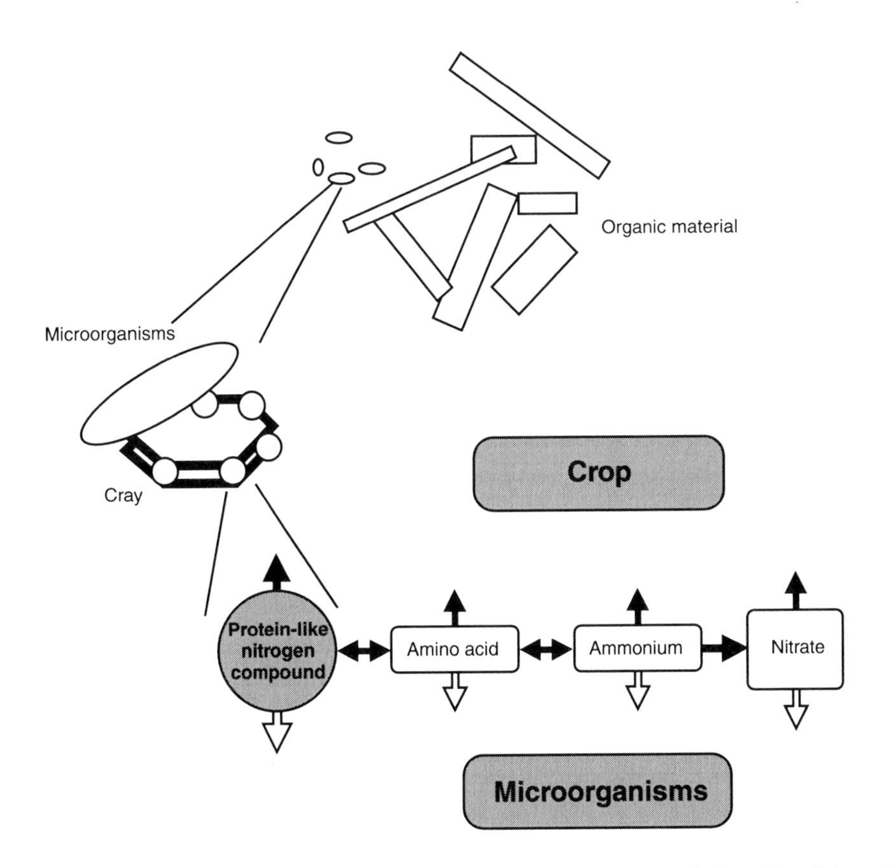

Fig. 5. Schematic presentation of the N mineralization–immobilization cycle in soils and the utilization of various forms of N by crops and microorganisms in the presence of organic materials

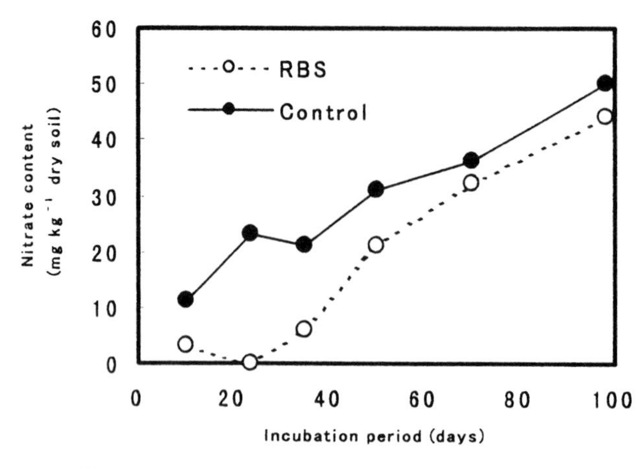

Fig. 6. Changes in the amount of nitrate released from soils after the addition of RBS. Soil ammonium levels were constantly very low so that only nitrate was given here

tein-like N is adsorbed on clay surfaces in the soil, probably through linkage with aluminum and iron, which stabilizes it against microbial attack; however, this protein-like N will be gradually decomposed to ammonium and to nitrate (see Fig. 5).

3. N Uptake Response of Crops to Organic N

3.1 Field Trials for Response of Crops to Organic Matter

To examine crop growth response to organic matter, a mixture of rice bran and rice straw (RBS; 22.3 g N kg^{-1}, C/N 19.9) was applied as organic N to an Andosol field (Tsukuba; 41.7 g C kg^{-1}; 3.4 g N kg^{-1} dry soil, C/N ratio 12.3). Andosol is characterized by its high organic material content. Figure 6 shows the N mineralization pattern from this organic matter amended soil compared to the control soil without any N under incubation (30°C, at 60% of maximum water holding capacity).

Nitrate levels during the first 60 days after the addition of RBS were significantly lower than in the control soil due to N starvation caused by the relatively high C/N ratio of RBS. At 22 days, no nitrate was found in the soil with RBS. After 60 days, the nitrate level with RBS was almost the same as in the control soil.

This RBS was applied to an Andosol field at 80 kg N ha^{-1} with ammonium sulfate at rates of 0, 30, 60, 120 kg N ha^{-1}. Phosphorus (200 kg P_2O_5 ha^{-1}) and potassium (100 kg K_2O ha^{-1}) were added to all field plots as basal nutrients. Maize (*Zea mays* L. var. DK250), upland rice (*Oryza sativa* L. var. Toyaohatamochi), soybean (*Glycine max* Merr. Var. Tachinagaha), potato (*Solanum tuberosum* L. var. Toyoshiro), and sugar beet (*Beta vulgaris* L. Beetmonobar) were grown for 100 days after sowing. Table 3 shows the N uptake response to RBS.

N uptake by upland rice, soybean, and potato was higher in the soil amended with RBS than in the control soil with chemical fertilizer. When RBS was supplied to upland rice, its N uptake doubled. Since the root zone of upland rice is shallow compared to that of maize, which showed better N uptake in the control soil than in the RBS plot, the total amount of available N from rooting soil of upland rice was considered to be smaller than that of maize. In the cases of maize and sugar beet, presumably due to their wide and deep root zone, the amount of N uptake was higher in the control soil than the amount in the case of upland rice, and no increase occurred in N uptake as a result of the addition of RBS. On the other

Table 3. Nitrogen uptake of 5 plant species after addition of organic N (RBS, 80 kg N ha^{-1}) in comparison to a control without RBS in a field experiment. Ammonium sulfate was applied to both fields with and without RBS at the rates of 0, 30, 60, 120 kg N ha^{-1}

Crops	Nitrogen uptake (kg ha^{-1})		Response to organic nitrogen (%)
	RBS	Control	
Maize	107	122	88
Upland rice	105[a]	50	210
Soybean	179[a]	145	123
Potato	103[a]	81	127
Sugar beet	109	104	105

[a] Significant at 5% level

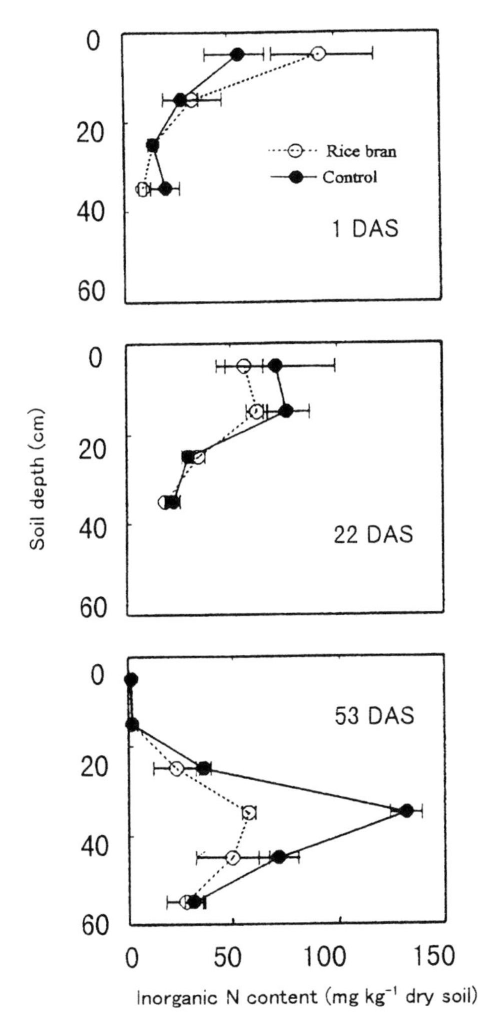

Fig. 7. Changes in inorganic nitrogen content at each depth of fallow soil in a field to which RBS was applied. Ammonium sulfate was evenly applied to both the control and organic matter-amended plots. *Bars*, SE (n=3)

hand, the inorganic N level in the fallow soil was lower in the RBS amended soil than in the control soil, although ammonium sulfate was added to the RBS plot at the same rate as to the control plot (Fig. 7). This result shows that N uptake of upland rice did not respond to inorganic N in a soil (Yamagata and Ae 1996).

3.2 Pot Trials for N Uptake by Plants in Limited Root Zones

To eliminate the effect of root size, we conducted a pot experiment (pot size 3.8 l) that was similar to the field trial with RBS. RBS was applied as N to a soil at rates of 0, 43, 87, and 130 mg N kg^{-1}. N uptake by upland rice increased with increasing RBS application, while no increase was found in maize. However, maize took up N prominently with ammonium sulfate in contrast to upland rice and soybean. These results suggest that upland rice could take up N derived from RBS preferentially compared to maize.

Based on both the field and pot experiments, we could confirm different N uptake response among crop species to organic matter amended soil. For example, maize responded well to inorganic N level, but upland rice grew well even when the soil inorganic N level was lower.

4. Factors for N Uptake Differences Among Crops

We proposed two hypotheses regarding the high response to organic N by upland rice. Compared to other crops such as maize, upland rice (1) accelerates N mineralization from organic N by secreting some enzymes or stimulating microorganisms in its rhizosphere (Clarholm 1985; Hart et al. 1979; Moister et al. 1987; Rovira and Mcdougall 1967), or (2) is able to take up organic N as soil protein (including

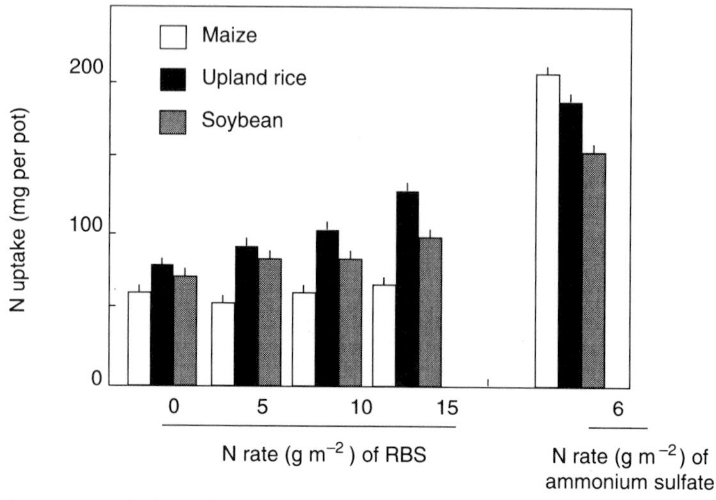

Fig. 8. Nitrogen uptake by upland rice, maize, and soybean (non-nodulated line) grown with RBS and ammonium sulfate in a pot experiment. *Bars*, SE (n=3)

PEON) and amino acid (Ghosh and Burris 1956; Wright 1962) which were produced more in RBS amended soil than in the control soil.

4.1 Differences in Protease Activity in Rhizosphere Among Crops

The possible organic N status in upland rice rhizosphere is considered to be PEON. If mineralization is stimulated by enriched soil microorganisms in upland rice rhizosphere, upland rice will take advantage to absorb ammonium N much more in RBS amended soil than other crops. To assess this mineralization process, we compared the protease activities of rhizosphere among upland rice, maize, and soybean (T201, non-nodulated line).

Table 4 shows the protease activities in the whole soil of the pots. The protease activities of the soils were measured using 2 substrates, casein or Benzyloxycarbonyl phenylalanyl leucine (Z-phe-leu) (Hayano 1993; Ladd and Butler 1972), to ensure a wide range of protease in the soil. The activities in maize were greater than those in upland rice except for the case of Z-phe-leu at 26 days after sowing. Thus, the "rhizosphere effect" of N mineralization by enhancement of protease activity failed to explain the higher response to organic N by upland rice compared to maize (Yamagata et al. 1997a).

According to Kanazawa and Chino (1997), protease activity was not very high in wheat rhizosphere soil, while phosphatase, β-acetylglucosaminidase, and denitrifying activities were higher in the soil within the range of 1 mm from the root plane than in the soil over 1 mm from the root plane. This indicates that there was no secretion of protease from wheat root and no stimulation by soil microorganisms in relation to organic N decomposition. Thus, the occurrence of the "rhizosphere effect" seems to vary not only with plant species but also with the enzymes in the soil.

4.2 N Uptake Differences Among Crops Derived from ^{15}N-Labelled Rice Bran

To confirm that upland rice takes up organic N preferentially, we conducted a pot experiment using ^{15}N-labelled RBS (^{15}N 4.36 atom%), a 4:1 mixture of ^{15}N-labelled rice bran and rice straw. The control plot received ^{15}N-labelled ammonium chlo-

Table 4. Protease activities in pot soils planted with upland rice, maize, and soybean (non-nodulated line) with RBS in comparison with fallow soil

	Protease activities (μmol h^{-1} g^{-1} dry soil)			
	Caseine		Z-phe-leu	
Crop	26 days	34 days	26 days	34 days
Upland rice	1.62 ±0.17	1.64 ±0.14	1.57 ±0.25	1.78 ± 0.08
Maize	2.67 ±0.17	2.11 ±0.06	1.56 ±0.08	1.96 ± 0.08
Soybean	1.52 ±0.15	1.78 ±0.19	1.71 ±0.07	1.90 ± 0.03
Fallow	2.19 ±0.32	1.31 ±0.04	1.48 ±0.03	1.67 ± 0.05

Mean ±SE (n=4)

ride (^{15}N 13.3 atom%, 43 mg N kg^{-1} dry soil). The same three crops as used in the protease assay were grown in pots for 82 days. Table 5 shows the ^{15}N concentration (atom%) of N taken up by the crops.

The ^{15}N concentration in upland rice with RBS was higher compared with concentrations in maize and soybean (non-nodulated line, T201), while the differences in ^{15}N concentration especially between upland rice and maize were smaller with ammonium chloride. This result suggests that upland rice may have taken up N forms other than nitrate. If plants take up mainly nitrate N, ^{15}N concentration of the plants becomes much lower due to the dilution of ^{15}N concentration by inorganic N derived from soil. These results indicate that upland rice would have an ability to absorb an organic form of N derived from RBS (Yamagata et al. 1997b).

4.3 Amino Acid Absorption by Upland Rice in Solution Culture

According to Noguchi (1992), the major amino acids in rice bran protein are aspartate, glutamate, and arginine. We tested N uptake response to these three amino acids by crops, since they are possibly released in soils supplied with RBS.

As N sources, 1 M solution of each amino acid, ammonium, and nitrate was prepared to observe the growth response of upland rice and maize. This solution also contained all mineral nutrients except N, and ampicilline (antibiotic), and the pH of the solution was adjusted to 6.8. Plants of 30-day-old seedlings grown without nutrient were transplanted to this nutrient solution, and the solution was daily replaced by fresh solution twice a day to minimize the formation of organic compounds by microbial contamination. The plants were grown for 4 days and harvested to measure N uptake (Fig. 9). Upland rice took up more aspartate, arginine, and ammonium compared to maize. This result suggests that upland rice may take up amino acid and ammonium more efficiently than maize. If organic matter is applied to a soil and the amino acid content in the soil is increased, upland rice will have an advantage over maize in growth potential.

When crops can take up amino acid, less energy is required for its incorporation because of its higher energy state in chemical linkage in a molecule compared to nitrate (Mori 1981; Mori et al. 1985). If amino acid is the predominant form of ni-

Table 5. The ^{15}N concentration (atom%) of nitrogen taken up by upland rice, maize, and soybean (non-nodulated line, T201) planted with ^{15}N-labelled RBS and ^{15}N-labelled ammonium chloride

^{15}N-labelled source		Days after sowing		
		56	69	82
RBS	Upland rice	1.44 ±0.04	1.53 ±0.01	1.54 ±0.02
	Maize	1.39 ±0.06	1.36 ±0.01	1.38 ±0.03
	Soybean	1.26 ±0.03	1.33 ±0.03	1.38 ±0.01
Ammonium chloride	Upland rice	3.11 ±0.01	3.05 ±0.02	3.02 ±0.04
	Maize	3.06 ±0.06	3.09 ±0.03	2.90 ±0.04
	Soybean	2.80 ±0.06	2.90 ±0.02	2.89 ±0.02

Mean ±SE (n=3)

trogenous compounds in the soil, the plant can adopt to absorb it principally instead of inorganic N. In fact, sedge (*Eriophorum vaginatum* L.) grass preferentially took up amino acid compared to barley (*Hordeum vulgare* L., cv. Stepto) in a tundra soil where very little N mineralization, especially nitrification, occurred due to low temperature and extremely low pH (Chapin et al. 1993). Birch (*Betula pendula* L.) took up ammonium and amino acid predominantly rather than nitrate when organic N such as amino acid existed abundantly (Arnebrant and Chalot 1995).

4.4 Content of Protein-Like N Compound in Soils Planted with Crops

We determined the concentration of protein-like N compound in PEON in rhizosphere soils to learn how much protein-like N compound derived from RBS was present in soils depending on crop species. Maize and upland rice were grown on small pots holding 125 ml of an Andosol with RBS. Plants were cultivated for 35 days, and rhizosphere soil was taken. Protein-like N compounds in rhizosphere and unplanted soil (fallow soil: non-rhizosphere soil) was extracted with 1/15 M phosphate buffer at pH 7.0. The concentration of this buffer-extractable organic compound, that is, PEON, determined by the Lowry method (Rej 1974) was lower in upland rice than in maize and fallow soil at all sampling dates over 35 days after sowing (Fig. 10) (Yamagata and Ae 1997). As described before, the protein-like N compound in PEON is assumed to be a complex of protein, humic substances, and amino sugar with 8 000 to 9 000 daltons of molecular weight. Lower content of this compound in the rhizosphere soil of upland rice than that of maize and unplanted soil suggests that upland rice may have taken up the protein-like N compound through a direct and/or indirect process. Because the protease activity of the rhizosphere of upland rice was lower than that of maize, the low level of protein-like N in the rhizosphere of upland rice could not be responsible for protease but for direct uptake.

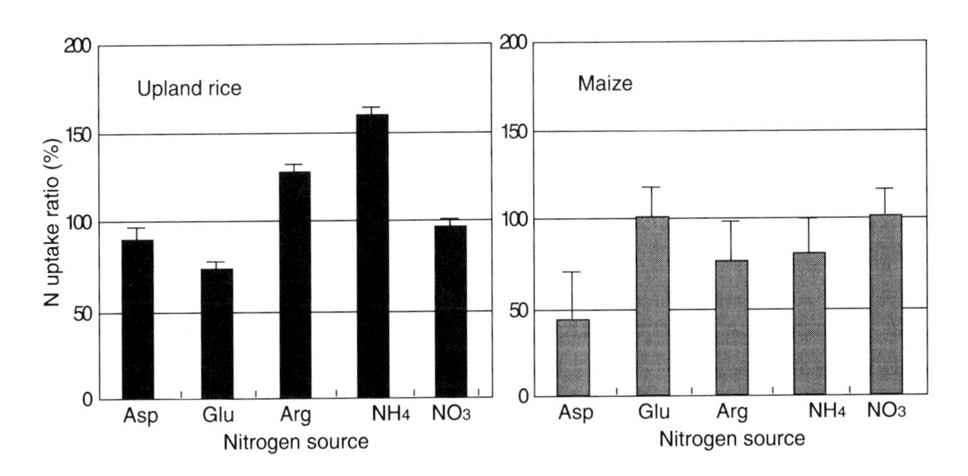

Fig. 9. Nitrogen uptake efficiency of ammonium, aspartate, glutamate, and arginine in upland rice and maize in comparison with nitrate treatment in a solution culture. *Bars*, SE (n=4)

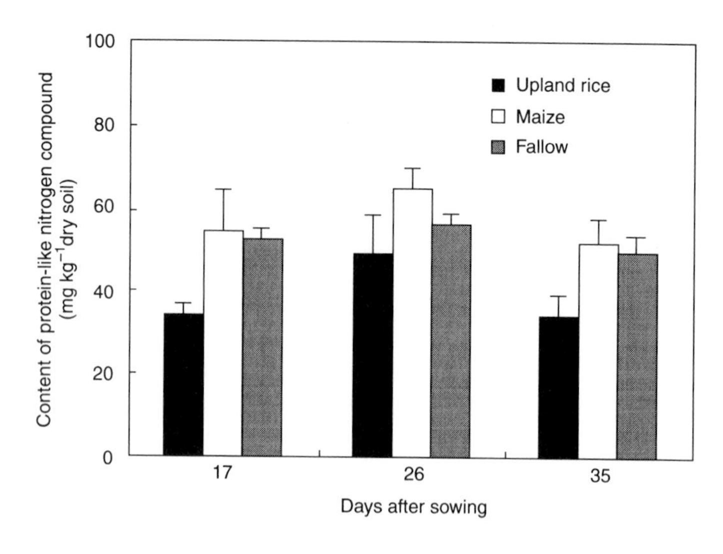

Fig. 10. Content of protein-like N compound in the rhizosphere soil planted with upland rice and maize amended with RBS in a pot trial in comparison with unplanted soil (fallow soil). Protein-like N compound was determined by the Lowry method (Rej 1974). Plants were grown in cubic pots holding 125 ml of soil for 35 days after sowing. *Bars*, SE (n=3)

Mori and Nishizawa (1979) reported that upland rice and barley absorbed large molecular proteins such as albumin and hemoglobin under sterilized solution culture. Naked barley (*Hordeum vulgare* L. emend. Lam) also grew better in a solution containing RNA as the sole N source (Mori 1986). A mechanism of "heterophagy" has been observed in relation to direct protein uptake in upland rice (Nishizawa and Mori 1980). Hemoglobin molecules were taken up by root cell membranes through invagination of plasma lemma, and the molecules then moved to vacuoles, where digestion of hemoglobin for N nutrition took place.

5. N Uptake by Vegetable Crops with Organic Materials

5.1 N Uptake Response of Vegetables to Organic Matter (Rapeseed Cake)

We carried out further experiments to clarify whether crops other than upland rice show similar responses to organic materials. Instead of RBS, a pot experiment using rapeseed cake (RC) (50.0 g N kg^{-1}, C/N ratio 7.0) was tested to examine the N uptake response of four vegetables: pimento (*Capsicum annuum* L. var.*angulosum* Miller), *chingensai* (Chinese cabbage, *Brassica campestris* L. var. *chinensis* (L)), carrot (*Daucus carota* L.var. *sativa* DC.), and leaf lettuce (*Lactuca sativa* L. var. *capitata* L.). As the control, ammonium sulfate was put into a pot filled with an Andosol. These vegetables have a much shorter growing duration than upland rice, and quick response to organic N would be expected. N at a rate of 100 mg N kg^{-1} dry soil was applied to a pot filled with an Andosol in the form of either RC or ammonium sulfate. Inorganic N levels in unplanted fallow soil with RC were higher than in the control soil without any N source but lower than in the soil with ammonium sulfate during 28 days of the growth period (Table 6). In contrast, the content of organic N such as amino acid and protein was higher in RC amended soil than in the control soil and the ammonium sulfate amended soil.

The N uptake by *chingensai* and carrot was higher in soil amended with RC than in soil amended with ammonium sulfate, while that by pimento and leaf lettuce was lower in RC amended soil (Fig. 11). These facts imply that *chingensai* and carrot preferentially took up organic N such as amino acid and protein. However, the content of amino acid in an unplanted plot with RC soil was only 0.6 mg N kg^{-1} dry soil at the most. This amount of amino acid is not sufficient for *chingensai* and carrot to increase N uptake by RC application. Since the amino acid content was much less compared to the protein content, the effective factor in N uptake may have been protein N rather than amino acid (Matsumoto et al. 1999b).

Table 6. Inorganic nitrogen (inorganic-N), amino acids (amino acids-N), and protein nitrogen (protein-N) in an Andosol applied with ammonium sulfate (100 mg N kg^{-1} dry soil), rapeseed cake (50.0 g N kg^{-1}, C/N ratio 7.0; 100 mg N kg^{-1} dry soil), and no nitrogen source over a period of 28 days. Inorganic-N was extracted with 2 M KCl. Amino acids were extracted with 0.2 M H_2SO_4 and measured by the ninhydrin method (Smith and Stockel 1954). Protein was extracted with 1/15 M phosphate buffer and determined by the Lowry method

Nitrogen form	Without N	Ammonium sulfate	Rapeseed cake
Inorganic N (mg kg^{-1})	27.5 – 50.0	90.5 – 133.0	41.0 – 82.5
Amino acid N (mg kg^{-1})	0.1 – 0.2	0.3 – 0.4	0.4 – 0.6
Protein-like N (mg kg^{-1})	18.7 – 34.7	18.9 – 31.0	34.6 – 55.9

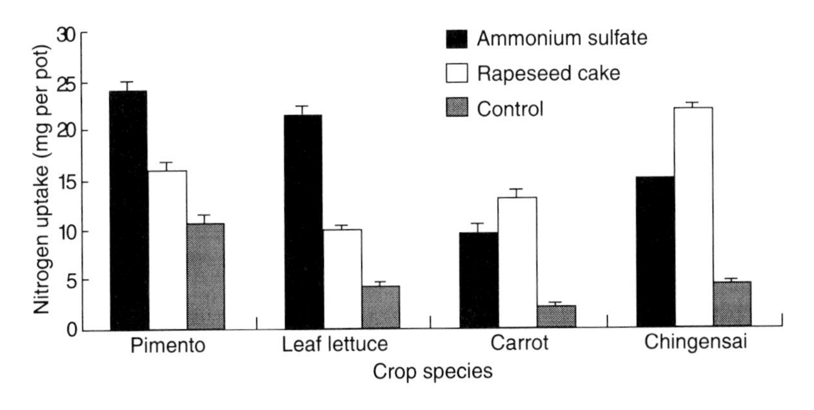

Fig. 11. N uptake by pimento, leaf lettuce, carrot, and *chingen-sai* supplied with ammonium sulfate and rapeseed cake and with no N source at 28 days after planting. *Bars*, indicate SE (n=3)

5.2 N Compounds in the Xylem Sap of Crops

One possible explanation for the superior N uptake by *chingensai* and carrot in the soil amended with RC could be direct uptake of organic N, especially of a protein-like N compound with a uniform MW of 8 000 to 9 000 daltons, accumulated in the soil to which RC and RBS had been applied. In order to test the possibility of direct uptake of protein-like N compound in a soil, two typical vegetables were chosen: *chingensai*, which responds better to organic N, and pimento, which responds better to inorganic N. *Chingensai* and leaf lettuce were grown in Andosol pots amended with RC and on solution culture without any organic nitrogen for 35 days. The plants were cut at 10 mm above ground level for collection of xylem sap. The xylem sap was analyzed by size-exclusion HPLC at 280 nm (Fig. 12).

Xylem sap collected from *chingensai* had a peak with the same retention time (8.4 min) as the soil extract on the chromatogram of size-exclusion HPLC (Fig. 12a). This

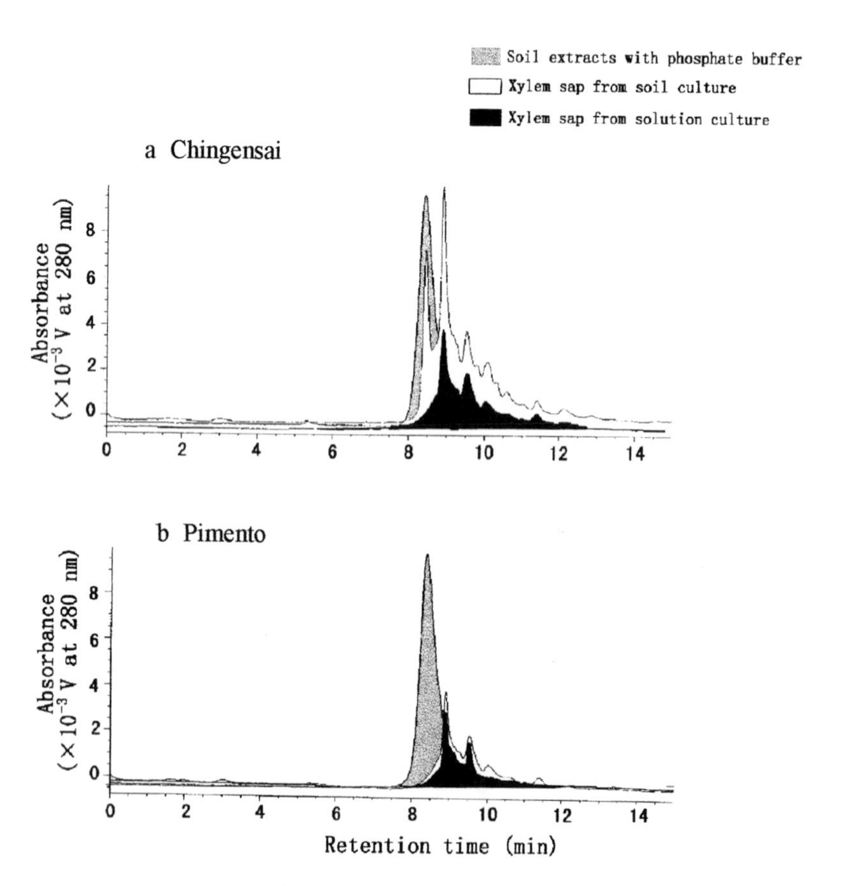

Fig. 12. Size-exclusion HPLC chromatograms of unplanted soil extract with 1/15 M phosphate buffer, xylem sap collected from *chingensai* and pimento grown in an Andosol with rapeseed cake, and xylem sap from the crops grown in a Hoagland solution culture (Aiello and Graves 1997) with no organic N at 28 days after transplanting

peak was absent from the chromatogram of xylem sap from *chingensai* grown in a Hoagland solution containing only inorganic nutrients, while common peaks at 8.9, 9.5, 10.1, 10.6, 11.4, and 12.2 min were present. The common peaks might represent peptides that form complexes with polyvalent heavy metal cations during xylem transport (Catald et al. 1988). We concluded that the peak with a retention time of 8.4 min was not produced by *chingensai* itself because there was no such peak at 8.4 min in the xylem sap from *chingensai* grown on solution culture without any organic N. However, when pimento was grown in the RC treated soil or in solution culture, no such peak was detected in the xylem sap (Fig. 12b). Thus, this peak seems to be related to the higher N uptake of *chingensai* under RC application compared to chemical fertilizer application (Matsumoto et al. 2000c).

6. Possibility of Protein-Like Compound Uptake by Crops

We have shown the possibility that plants such as upland rice and *chingensai* take up a protein-like N compound with a molecular weight of 8 000-9 000 daltons. This protein-like N compound was extracted by phosphate buffer. The function of phosphate is to react with iron and aluminum, which fixed the N compound on a clay surface (unpublished data). If the same mechanism works in the rhizosphere by secretion of chelating substances such as organic acids, the N compound is released from the clay and might become available to particular crops such as upland rice, *chingensai*, and carrot, which had high responses to organic matter in our experiments. A chromatogram of size-exclusion HPLC of xylem sap collected from upland rice grown in an Andosol with RBC was similar to that of *chingensai* and carrot (data not shown), that is, upland rice had a peak at 8.49 min of retention time on the chromatogram similar to the peak found in the soil extract by phosphate buffer to which organic materials had been added. Accordingly, the chelating mechanism allowed a high response of N uptake by upland rice to organic materials.

Minerals in soil solution are taken into the root through the surface of the root epidermis. One of the major routes for transfer of materials between organs is the vascular bundle, which is composed of xylem and phloem. The xylem mainly consists of xylem vessels, which form a kind of apoplastic space, in which the xylem sap flows from the roots to the shoots. The vital activities of organs depend on the supply of inorganic and organic compounds from the xylem sap, which are produced and blended in the root and transported via xylem vessels (Nooden and Mauk 1987). Satoh et al. (1992) showed that several proteins, i.e., those with molecular weights of 75 000, 40 000, 32 000, 19 000, and 14 000 daltons, were transported into xylem sap in squash (*Cucurbita maxima* Duchens x *C. moschesne* Duchene) seedlings. The proteins might be secreted into the cell wall or apoplastic space of the cells in the central cylinder according to their signal sequences and then apoplastically transported to the xylem vessels with the movement of water. Sakuta et al. (1998) reported that proteins estimated to have a molecular weight of 30 000 daltons were transported through the xylem from the root to aboveground organs such as leaves and shoots in cucumber (*Cucumis sativus* L.). Cleve et al. (1991) reported that a poplar (*Populus deltoides* Bartr. ex Marsh) storage protein (32 000 daltons) could be detected in xylem sap during the dormant period and especially during bud break. A long distance transport of not only sugars and amino acids but also of protein molecules may therefore be possible.

7. Future Research and Practical Implications

"Inorganic N uptake theory" has long been studied, while "organic N uptake theory" has been overlooked since the research of Liebig became known. Liebig's conclusion was correct, and crop production in agriculturally advanced nations has been extraordinarily enhanced by the application of chemical fertilizer based on this theory. As a result of intensive application of fertilizer, modern agricultural production can nourish the huge number of people on earth. However, chemical fertilizers are produced from limited natural resources such as petroleum, phosphate rock, and potassium salts. On the other hand, organic materials are increasing because of their occurrence as by-products in various industries, and sometimes these materials cause serious environmental problems. Thus, the intensive use of organic materials must be inevitably promoted to save natural resources and protect the environment.

We have shown that some crops could possibly take up high molecular weight protein-like N compound from soils supplied with certain organic materials and grow more efficiently than they can with the application of chemical fertilizers. The ability to take up organic N appears to vary with plant species, however, so we need to further study the mechanism enabling plants to take up organic N. This effort could allow us to improve crop production by satisfying both environmental and food production requirements by efficient use of both chemical fertilizer and organic materials.

References

Aiello AS, Graves WR (1997) Two leguminous tree species differ in growth and ion uptake in Hoagland solution. J Plant Nutr19:1061-1073

Appel T, Mengel K (1992) Nitrogen uptake of cereals on sandy soils as related to nitrogen fertilizer application and soil nitrogen fraction obtained by electroultrafiltration (EUF) and $CaCl_2$ extraction. Eur J Agronomy 1:1-9

Appel T, Mengel K (1993) Nitrogen fractions in sandy soils in relation to plant uptake and organic matter incorporation. Soil Biol Biochem 25:685-691

Arnebrant K, Chalot M (1995) Uptake of ammonium and alanine by intact birch mycorrhizal systems. Dynamics of physiological processes in woody roots. In: Dynamics of physiological processes in woody roots. Booklet of Abstracts pp 64

Berden M, Berggren D (1990) Gel filtration chromatography of humic substances in soil solutions using HPLC-determination of molecular weight distribution. J Soil Sci 41:61-72

Bradford MM (1976) A rapid and sensitive method for the quantitation of microgram quantities of protein utilizing the principle of protein-dye binding. Anal Biochem 72:248-254

Bremner E, van Kessel C (1990) Extractability of microbial ^{14}C and ^{15}N following addition of variable rates of labelled glucose and $(NH_4)_2SO_4$ to soil. Soil Biol Biochem 22:707-713

Catald A, Mcfadden K, Garland T, et al (1988) Organic constituents and complexation of nickel (II), iron (III), cadmium (II), and plutonium (IV) in soybean xylem exudates. Plant Physiol 86:734-739

Chantigny MH, Anger DA, Prevost D, et al (1997) Soil aggregation and fungal and bacterial biomass under annual and perennial cropping system. Soil Sci Soc Am J 61:262-267

Chapin FS, Moilanen L, Kielland K (1993) Preferential use of organic nitrogen for growth by a non-mycorrhizal arctic sedge. Nature 361:150-153

Clarholm M (1985) Interactions of bacteria, protozoa and plants leading to mineralization of soil nitrogen. Soil Biol Biochem17:181-187

Cleve B, Just J, Suter J (1991) Poplar storage protein in xylem sap. J Plant Physiol 137:763-764.

Degens BP (1998) Microbial functional diversity can be influenced by the addition of simple organic substrates to soil. Soil Biol Biochem 30:1981-1988

FAO (1989) Soil map of the world. Tech Paper 20 ISRIC Wageningen, Netherlands

Ghosh BP, Burris RH (1956) Utilization of nitrogenous compounds by plants. Soil Sci 70:187-203

Groot JJR, Houba VJG (1995) A comparison of different indices for nitrogen mineralization. Biol Fertility Soils 19:1-9

Grunes DL, Haise HR, Turner F, et al (1963) Relationship between yield response to applied fertilizers and laboratory measures of nitrogen and phosphorus availability. Soil Sci Soc Am Proc 27:675-679

Harmsen GW, Van Schreven DA (1955) Determination of the mineralization of nitrogen in soil. Adv Agron 7:361

Hart PBS, Goh KM, Ludecke TE (1979) Nitrogen mineralization in fallow and wheat soils under field and laboratory conditions. NZ J Agric Res 22:115-125

Hayano K (1993) Protease activity in a paddy field soil: origin and some properties. Soil Sci Plant Nutr 39:539-546

Higuchi M (1981) Characterization of immobilized nitrogen and organic nitrogen extracted with various buffer solution (in Japanese). Jpn J Soil Sci Plant Nutr 52:481-489

Higuchi M (1982) Characterization of soil organic nitrogen extracted with phosphate buffer (in Japanese.) Jpn J Soil Sci Plant Nutr 53:1-5

Honda T (1980) Characterization of soil organic matters extracted with acid solution (in Japanese). Jpn J Soil Sci Plant Nutr 51:183-192

Kanazawa S, Chino M (1997) Distribution of denitrifying and enzyme activities across the rhizosphere of wheat. In: Ando T, Fujita K, Mae T, et al (Eds) Plant nutrition for sustainable food production and environment. Kluwer Academic, Dordrecht, pp 517-518

Keeney DR (1982) Nitrogen-availability indices. In: Page AL, Miller RH, Keeney DR (Eds) Methods of soil analysis. Part 2. Chemical and microbiological properties. Agronomy 9 (2nd edn) Madison pp 711-733

Ladd JN, Butler JH (1972) Short-term assays of soil proteolytic enzyme activity using proteins and dipeptide derivatives as substrates. Soil Biol Biochem 4:19-30

Marumoto T, Hurukawa K, Yoshida T, et al (1974) Contribution of microorganisms and its cell wall to easily decomposable soil organic matter (in Japanese). Jpn J Soil Sci Plant Nutr 45:23-28

Marumoto T, Anderson JPE, Domsch KH (1982) Decomposition of ^{14}C- and ^{15}N-labelled microbial cells in soil. Soil Biol Biochem 14:461-467

Matsumoto S, Ae N, Yamagata M (1999a) Influence of organic fertilizers on the growth and contents of nitrate, oxalic acid and ascorbic acid in spinach (in Japanese with English summary). Jpn Soil Sci Plant Nutr 40:31-38

Matsumoto S, Ae N, Yamagata M (1999b) Nitrogen uptake response of vegetable crops to soil incorporated organic nitrogen. Soil Sci Plant Nutr 45:269-278

Matsumoto S, Ae N, Yamagata M (2000a) The status and origin of available nitrogen in soils. Soil Sci Plant Nutr 46:139-149

Matsumoto S, Ae N, Yamagata M (2000b) Extraction of mineralizable organic nitrogen from soils by a neutral phosphate buffer solution. Soil Biol Biochem 32:1293-1299

Matsumoto S, Ae N, Yamagata M (2000c) Possible direct uptake of organic nitrogen from soil by *chingensai* (*Brassica campestris* L.) and carrot (*Daucus carota* L.) Soil Biol Biochem 32:1301-1310

Moister A, Haider K, Heinemeyer O (1987) Effects of plants on organic carbon and nitrogen turnover in soil. Agronomy Abstracts, American Society of Agronomy, Madison, pp 188

Mori S (1981) Primary assimilation process of triply (^{15}N,^{14}C and ^{3}H) labeled arginine in the roots of arginine-fed barley. Soil Sci Plant Nutr 27:29-43

Mori S (1986) High response of naked barley growth to RNA (in Japanese). Jpn J Soil Sci Plant Nutr 57:171-178

Mori S, Nishizawa N (1979) Nitrogen absorption by plant root from the culture medium where organic and inorganic nitrogen coexist. II. Which nitrogen is preferentially absorbed among [U-^{14}C]GluNH$_2$,[2,3-^{3}H]Arg and Na$^{15}NO_3$? Soil Sci Plant Nutr 25:51-58

Mori S, Uchino H, Sago F, et al (1985) Alleviation effect of arginine on artificially reduced grain yield of NH_4^+- or NO_3^--FED rice. Soil Sci Plant Nutr 31:55-67

Nemeth K, Bartels M, Vogel M, et al (1988) Organic nitrogen compounds extracted from arable and forest soils by electro-ultrafiltration and recovery rates of amino acids. Biol Fertility Soils 5:271-275

Nishizawa N, Mori S (1980) Vacuole formation as a result of intracellular digestion. Soil Sci Plant Nutr 26:525-540

Noguchi K (1992) Composition of organic fertilizer and decomposition and change in soil microflora. Katakura Chikkarin, Tsukuba Research Institute Report 1:14-41

Nooden LD, Mauk CS (1987) Changes in the mineral composition of soybean xylem sap during monocarpic senescence and alternations by depodding. Physiol Plant 70:735-742

Numan N, Morgan MA, Herlihy M (1998) Ultraviolet absorbance (280 nm) of compounds release from soil during chloroform fumigation as an estimate of the microbial biomass. Soil Biol Biochem 30:1599-1603

Ogiuchi K, Nakajima N, Ae N, et al (2000) Amino acid status of organic nitrogen in soil extract with phosphate buffer (in Japanese). Jpn J Soil Sci Plant Nutr (in press)

Rej R (1974) Interference by tris buffer in the estimation of protein by the Lowry procedure. Anal Biochem 62:240-247

Rovira AD, Mcdougall B (1967) In: Maclaren, Peterson GH (Eds) Soil biochemistry. Marcel Dekker, New York, pp 417-463

Saito M (1988) Estimation of nitrogen availability indices based on UV absorption of soil extracts (in Japanese with English summary). Jpn J Soil Sci Plant Nutr 59:493-495

Saito M, Ishii K (1987) Estimation of soil nitrogen mineralization in corn-grown fields based on mineralization parameters. Soil Sci Plant Nutr 33:555-566

Sakuta C, Oda A, Yamakawa S, et al (1998) Root-specific expression of genes for novel glycine-rich proteins cloned by use of an antiserum against xylem sap proteins of cucumber. Plant Cell Physiol 39:1330-1336

Satoh S, Iizuka C, Kikuchi A, et al (1992) Proteins and carbohydrates in xylem sap from squash root. Plant Cell Physiol 33:841-847

Smith EL, Stockel A (1954) Amino acid composition of crystalline carboxipeptidase. J Biol Chem 61:586-589

Stanford G, Ayres AS, Doi M (1965) Mineralizable soil nitrogen in relation to fertilizer needs of sugarcane in Hawaii. Soil Sci 99:132-137

Stevenson FJ (1958) Distribution of the forms of nitrogen in some soil profiles. Soil Sci Soc Am Proc 22:283-287

Stevenson FJ, Braids OC (1968) Variation in the relative distribution of amino sugar with depth in some soil profiles. Soil Sci Soc Am Proc 32:598-590

Urquiaga S, Cadisch G, Alves BJR, et al (1998) Influence of decomposition of roots of tropical forage species on the availability of soil nitrogen. Soil Biol Biochem 30:2099-2106

Wagner GH, Mutatkar VK (1968) Amino components of soil organic matter formed during huminification of ^{14}C glucose. Soil Sci Soc Am Proc 32:683-686

Wright DE (1962) Amino acid uptake by plant roots. Ar Biochem Biophys 97:174-180

Yamagata M, Ae N (1996) Nitrogen uptake response of crops to organic nitrogen. Soil Sci Plant Nutr 42:389-394

Yamagata M, Ae N (1997) Preferential use of organic nitrogen by upland rice compared to maize. In: Ando T, Fujita K, Mae T, et al (Eds) Plant nutrition for sustainable food production and environment. Kluwer Academic, Netherlands, pp 509-510

Yamagata M, Nakagawa K, Ae N (1997a) Difference of protease activity between crops in rhizosphere (in Japanese with English summary). Jpn Soil Sci Plant Nutr 68:295-300

Yamagata, M, Nakagawa K, Ae N (1997b) Differences among crops on nitrogen uptake from rice bran using the ^{15}N tracer technique (in Japanese with English summary). Jpn Soil Sci Plant Nutr 68:291-294

Yamamuro S (1986) Behavior of nitrogen in paddy soils. JARQ 20:100-107

Direct Uptake of Macro Organic Molecules

Naoko K. Nishizawa and Satoshi Mori

Summary. Agricultural intensification has had environmental consequences such as the leaching of nitrate and pesticides into freshwater and marine systems, and emission of environmentally hazardous nitrogen-containing gases to the atmosphere. The intensified application of chemical fertilizers is based on the assumption that plants do not absorb and use organic nitrogen directly. It has long been believed that plants cannot take up and use organic nitrogen as a nutrient. Therefore, it is assumed that the mineralization of organic nitrogen is a prerequisite for plant nitrogen acquisition. Here, we would like to show that plants can take up and use organic nitrogen directly, and that they preferentially absorb it for growth in some instances. This has major implications for sustainable agriculture in industrial and developing countries.

To examine the possibility that plants, especially crop plants, use organic nitrogen as a nutrient, we conducted a series of hydroponic experiments starting in 1974 (Mori et al. 1977). We water-cultured rice and barley plants with various nitrogen sources, including NH_4, NO_3, 22 amino acids and macro organic molecules.

Key words. Endocytosis, Heterophagy, Rice root, Protein uptake, Vauole formation, Lysosome, Vesicular transport

1. Uptake of Small Organic Molecules

Most of the amino acids tested had a positive effect on the growth of barley and rice. Of these, arginine (Arg) produced the highest grain yield in barley (Table 1). Glutamine and citrulline also had positive effects on growth. In contrast, methionine, isoleucine, and histidine had negative effects resulting in no grain yield. Therefore, barley and rice can grow using some amino acids as the sole source of nitrogen (Mori et al 1977). Moreover, barley plants pre-cultured with an amino acid preferentially absorbed the amino acid from a culture medium containing both organic and inorganic nitrogen (Mori and Nishizawa 1979). This result suggested that barley might absorb some amino acids from the rhizosphere where both organic and inorganic nitrogen exist. Using Arg as a nitrogen source in culture medium alleviated the reduced growth and grain yields caused by adverse climatic conditions such as low solar energy or low temperature (Mori et al. 1985). Unlike inorganic nitrogen, Arg consists of a carbon skeleton containing four nitrogen atoms in one molecule. Therefore, nitrogen and carbon are absorbed and supplied more efficiently with Arg than with either NH_4 or NO_3. The high effi-

Table 1. Effects of twenty-two amino acids as the sole source of nitrogen on the growth of barley

Nitrogen source	Grain yield (g)	1000-kernel weight (g)	Total weight (g)	No. of productive culms	Shoot height (cm)	Root length (cm)
Arginine	60.3	27.4	165	40.5	72	30.5
Glutamine	54.2	28.3	154	36	69.5	28.5
Citrulline	53.8	31.2	142	34	80.5	32.5
NO3	53.5	28.0	151	29.5	82	31.5
Asparagine	48.2	28.4	130	30.5	70	28.5
Ornithine	47.7	37.0	129	25.5	73.5	25
Alanine	42.2	32.0	108	22.5	67	20.5
Glycine	37.5	31.5	96	21	73	29
Lysine	33.6	33.5	89	23.5	68	24
Serine	33.5	38.6	81	15.5	70	21
Proline	29.8	27.1	82	18	73.5	25
Phenylalanine	26.2	32.9	74	19	73	20
Threonine	24.2	31.0	83	25.5	61.5	25
Cysteine	23.5	32.6	62	16.5	60.5	28
Valine	21.4	32.6	87	23.5	61	35
Glutamic acid	21.3	31.6	58	15	61	20
Tyrosine	21.2	28.6	53	13	58.5	18
Leucine	19.8	32.5	72	19.5	60.5	22
Aspartic acid	18.9	31.5	44	11	61.5	15
Tryptophan	14.7	30.0	53	16.5	72	10
NH_4	13.0	35.6	29	11	60	16
Histidine	0	0	74	—	47.5	31
Isoleucine	0	0	33	—	35.5	29
Methionine	0	0	—	—	—	—

Each value is a mean (*n*=2)

ciency of using amino acids as a nutrient source may work effectively, especially under adverse conditions.

Recently, the transport of amino acids and sucrose in plants has been clarified at the molecular level. Using the yeast functional complementation method, Frommer et al. (1993) first isolated an amino acid transporter, AAP1, from *Arabidopsis*. Yeast mutants lacking uptake systems for certain amino acids, which therefore cannot grow on media containing those amino acids as the sole nitrogen source, were chosen as the complementation system. The mutants were transformed with plant cDNA libraries in order to screen for transporters that were able to suppress the growth-deficient phenotype. Using various yeast amino acid transport mutants, many plant amino acid transporter genes (>13 genes from *Arabidopsis*) have now been identified (Fischer et al. 1995, 1998; Frommer et al. 1994b). We presume that among these plant amino acid transporters there must be amino acid transporters that participate in the direct uptake by roots of amino acid from the rhizosphere. In *Arabidopsis*, di-, tri-, and oligo-peptide transporter genes have also been isolated (Frommer et al. 1994a; Steiner et al.1994; Rentsch et al. 1995). A peptide transporter that was expressed during the early stage of germination was also isolated (West et al. 1998). Similarly, roots may use these peptide transporters to absorb small peptides in addition to amino acids. Sucrose transporter genes in plants have been isolated using the same complementation method (Riesmeier et al. 1992). Plant roots may also take up sucrose present in the rhizosphere through the sucrose transporter. Moreover, as described in the chapter by Mori S, the mugineic acid family phytosiderophores (MAs) secreted from iron-deficient graminaceous plants are incorporated in the root cells as an MAs-Fe complex. This uptake system is a mechanism by which roots can take up low molecular weight organic materials directly.

2. Uptake of Macro Organic Molecules

As mentioned above, we water-cultured rice and barley plants with macro organic molecules, such as hemoglobin, ovalbumin, peptone, and ribonucleic acid (RNA). As shown in Tables 2 and 3, differences in the nitrogen sources affected the growth of rice and barley plants. Both plants grew very well in hydroponic medium containing proteins, peptone, or RNA (Mori 1986) as the sole nitrogen source. We assumed that macro organic molecules must be incorporated into the cells by endocytosis and utilized as a nutrient. Therefore, we used electron microscopy to examine rice plants that were water-cultured with hemoglobin. As a result, we found that the exogenous macromolecule is incorporated and utilized through a mechanism of heterophagy in rice roots, which is induced by a supply of macromolecular nitrogen sources (Nishizawa and Mori 1977, 1978, 1980ab, 1982). The term heterophagy means that endocytosed extracellular molecules are degraded in the lytic compartment within the cells.

In addition to the transport mechanism utilizing the transporter or channels in the plasma membrane, there is a vesicular transport mechanism that brings extracellular substances into cells. Endocytosis, the uptake into cells by vesicular

Table 2. Yield component of rice plants water-cultured with proteins and inorganic nitrogen as the sole source of nitrogen (in 1975)

Nitrogen source	Seed weight (g)	Number of seeds	Weight of a head (g)	Grain number per head	Thousand grain wt. (g)	Number of ears
NH_4	64.0	3415	1.0	52	18.7	66
NO_3	63.8	3602	1.2	70	17.7	52
NH_4+Hem	76.0	4783	1.1	71	15.9	67
NO_3+Hem	94.7	5413	1.8	98	17.5	56
Hem	76.0	3741	1.7	82	20.3	46
Peptone	90.3	4796	1.4	73	18.8	66

Each value is a mean (n=2)
1/2000 a pot
Hem, hemoglobin.

Table 3. Yield of barley plants water-cultured with ovalbumin and inorganic nitrogen as the sole source of nitrogen (in 1974-1975)

Nitrogen source	Seed weight (g)	Root dry wt. (g)	Shoot dry wt. (g)
NH_4	3.57	0.39	3.99
NH_4+Ovalbumin	20.39	2.22	33.91
Ovalbumin	72.03	11.10	87.83
NO_3	31.03	2.47	40.94

Each value is a mean (n=2)
1/2000 a pot

transport, is well known in amoebae and macrophage cells, and this mechanism was once thought to be a special system in these particular cells. However, it has recently become clear that the vesicular transport mechanism is ubiquitous in animal cells. Material to be ingested is progressively enclosed by a small portion of the plasma membrane, which first invaginates and then pinches off to form an intracellular vesicle containing the ingested substance. Two main types of endocytosis are distinguished on the basis of the size of the endocytotic vesicles formed: pinocytosis, which involves the ingestion of fluid and solutes via small vesicles ($\leq$ 150 nm in diameter), and phagocytosis, which involves the ingestion of large particles such as microorganisms or cell debris, via large vesicles, generally > 250 nm in diameter. Endocytosis occurs both constitutively and as a response triggered by extracellular signals. There are two forms of endocytosis. One is fluid-phase endocytosis, the non-specific incorporation of substances dissolved in the extracellular medium, and the other is receptor-mediated endocytosis, the incorporation of substances after they bind to specific receptor sites on the plasma membrane. Receptor-mediated endocytosis is an efficient pathway for taking up specific macromolecules, and its detailed mechanism in animal cells is now being studied (Fig. 1). In addition, this mechanism also mediates the incorporation of foreign substances such as viruses and peptide toxins, and the uptake of important functional compounds such as low-density lipoprotein (LDL), macroglobulin, asialoglycoprotein (ASGP), epidermal growth factor (EGF), and transferrin (ferrous ion transporter). These macromolecules bind to complementary cell surface receptors, which are transmembrane proteins, and enter the cells as receptor-macromolecule complexes in the form of coated vesicles through invagination of the plasma membrane. Occasionally, the receptor-binding foreign substances are gathered in a coated pit and then incorporated into the cells in the same manner. The coated vesicles are small and coated with several peptides, principally clathrin. Although these phenomena have been studied mainly in animal cells, coated vesicles and coated pits are also observed in plant cells. Therefore, the uptake of extracellular substances by endocytosis has also attracted much attention in plant cells (Robinson and Depta 1988). However, the possibility that plant cells use this mechanism to take up organic molecules as nutrients directly has been completely neglected until now. In the rest of this chapter, this uptake mechanism for exogenous macromolecules in rice root cells cultured with proteins as the sole source of nitrogen will be described in detail (Nishizawa 1995).

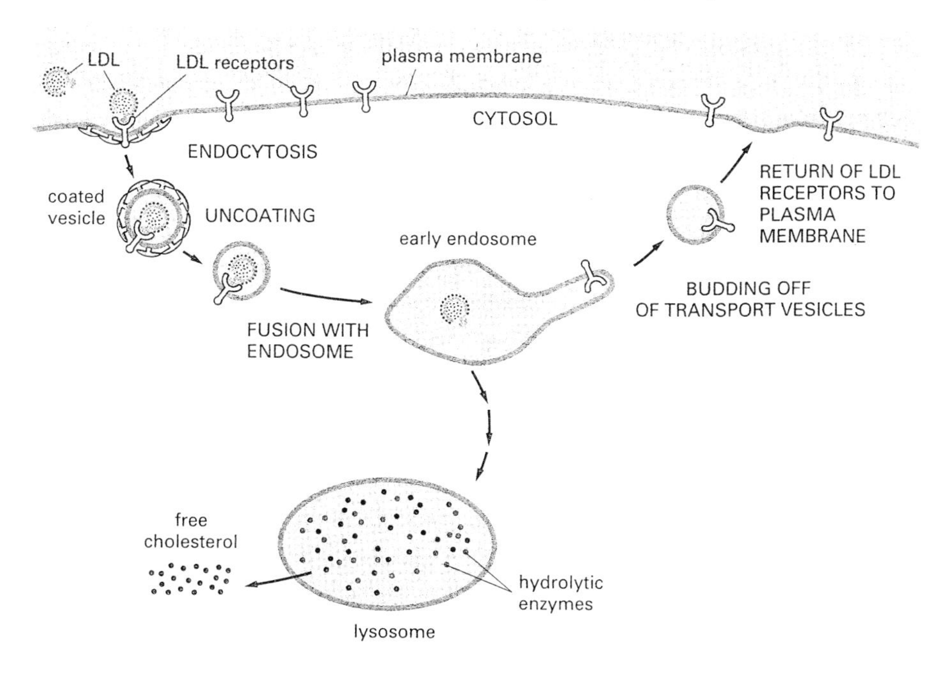

Fig. 1. Receptor-mediated endocytosis of low-density lipoproteins (*LDL*). LDL particles bound to LDL receptors in the coated pits in plasma membrane are rapidly internalized in the clathrin coated vesicles. After uncoating of clathrin, vesicles fuse to early endosome. The LDL dissociates from its receptors in the acidic environment of the endosome. After a number of steps, the LDL ends up in lysosomes, where it is degraded to release free cholesterol. In contrast, the LDL receptor proteins are returned to the plasma membrane via transport vesicles

2.1 Invagination of the Plasma Membrane (PL-Invagination)

Rice plants (*Oryzae sativa* L. cv. Honenwase) were water-cultured with hemoglobin, ovalbumin, bovine serum albumin (BSA), or peptone as the sole nitrogen source until the panicle formation stage. Although the culture conditions were not aseptic, the hydroponic medium was renewed every day to avoid degradation of the macromolecules. Fresh crown root tips were cut and prefixed in 4% paraformaldehyde, 5% glutaraldehyde, 0.1 M $CaCl_2$, and 0.1 M cacodylate buffer (pH 7.2) for 3 h on ice. After washing with the same buffer for 1.5 h, post-fixation was carried out in 2% osmium tetroxide and 0.1 M cacodylate buffer, for 1 h on ice. The samples were dehydrated serially in ethanol and propylene oxide, and embedded in Epon 812. Ultrathin sections were cut, stained with uranium and lead, and then observed with an electron microscope.

Many PL-invaginations were observed in the cortex cells of crown roots of these rice plants (Figs. 2, 3). The diameters of the PL-invaginations varied from 1 μm (Fig. 2) to over 12 μm. Occasionally they were large enough to fill the cytoplasm of one cell. In many cases, electron-dense fibrous materials or particles were at-

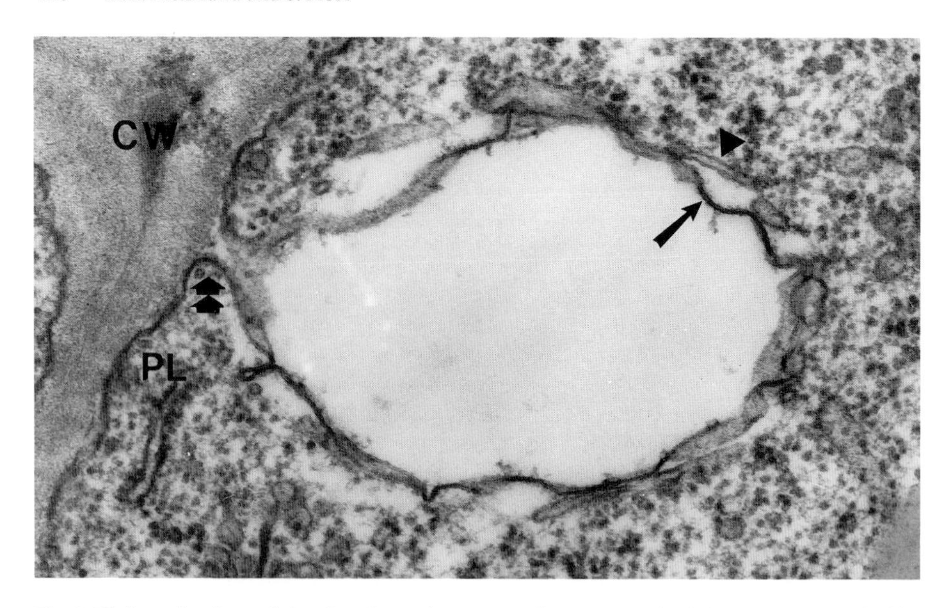

Fig. 2. PL-invagination originating from plasma membrane. A PL-invagination surrounded by tubular structures and vesicles. The limiting membrane of the invagination (*long arrow*) is thicker than that of the surrounding structures (*arrowhead*). Plasma membrane is invaginated into the cytoplasm (*double arrows*). *CW* indicates the cell wall. Transmission electron micrographs of cortex cells of rice roots grown in water-culture with hemoglobin as the sole nitrogen source. *PL*, plasmalemma

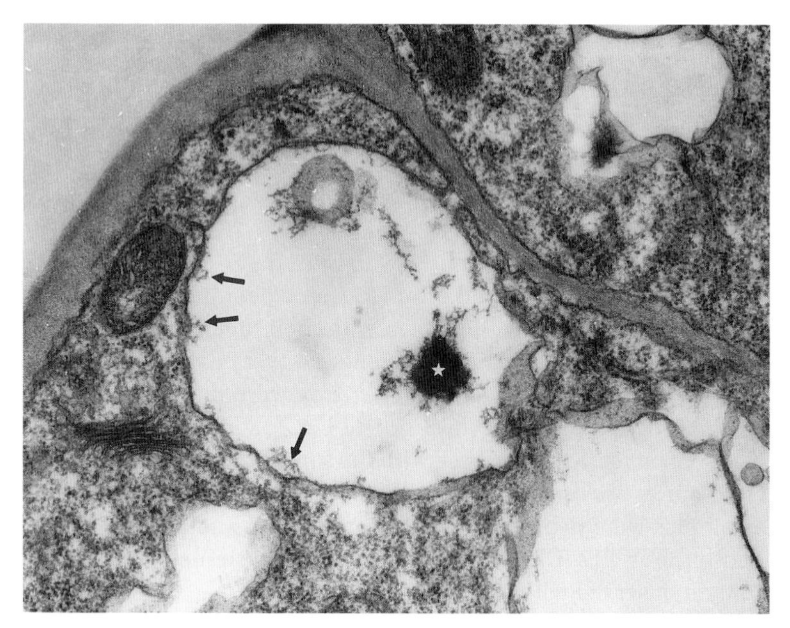

Fig. 3. PL-invagination in a cortex cell of a rice root, containing electron-dense deposits of fibrous materials (*white star*). Some fibrous materials are also attached to the surface of a limiting membrane (*arrows*). *Double arrows* show the site from where plasma membrane is invaginated into the cytoplasm

tached to the inner surface of the limiting membrane of the PL-invagination, corresponding to the extracellular surface of the plasma membrane. The electron-dense materials inside the PL-invagination were occasionally aggregated into a large mass (Fig. 3).

In less-differentiated cortex cells, where vacuoles did not occupy as large an area, vesicles or tubular structures frequently surrounded the PL-invaginations (Fig. 2). These surrounding structures included vesicles containing electron-dense materials, which looked like eyes. Therefore, they were named 'eye-like structures'. The limiting membranes of these structures were apparently thinner than the membranes of PL-invaginations or plasma membrane, and were similar to the endoplasmic reticulum (ER). The membranes in a cell can be classified into two types with respect to their thickness. Exoplasmic membranes, such as plasma membrane or Golgi vesicles, are thicker than endoplasmic membranes, such as

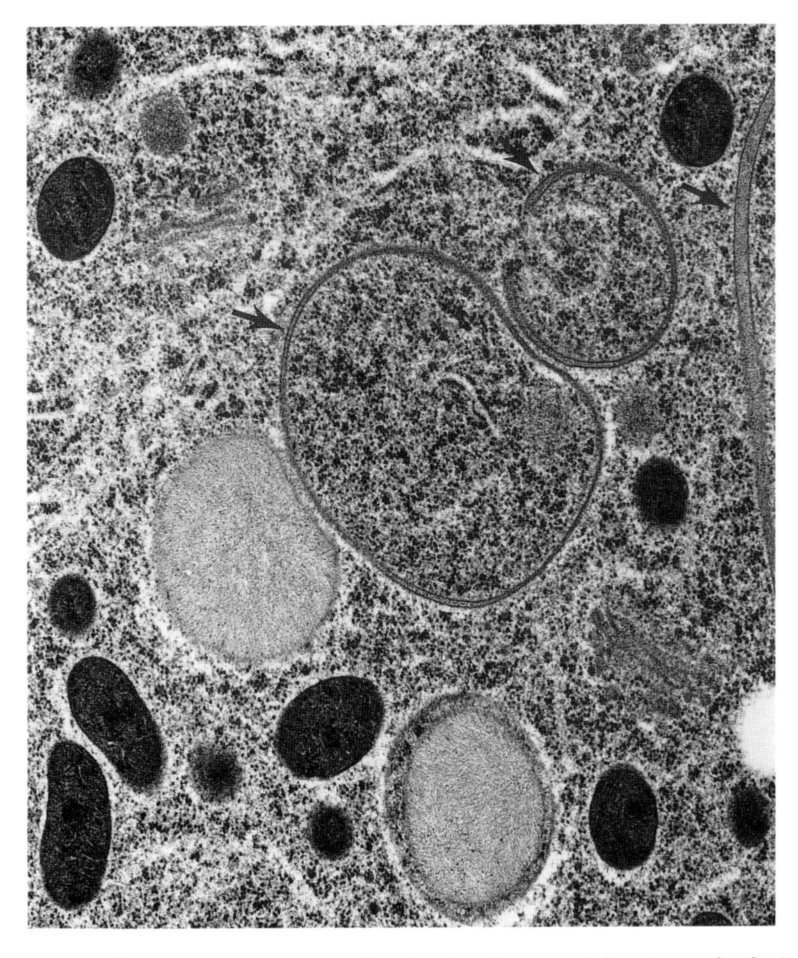

Fig. 4. A root cell of bean plants obtained by liquid helium rapid freezing and substitution fixation method. Lysosome-like structures (*arrows*) which are different from the smooth ER are surrounding a portion of cytoplasm

the nuclear envelope membrane or ER. The difference in membrane thickness observed here implies that the vesicles surrounding the PL-invaginations are not Golgi vesicles, but are possibly derived from the ER. Initially, it was thought that these tubular structures were the smooth ER (ER without ribosomes). However, results obtained using the liquid helium rapid freezing and substitution fixation method, which is a new method of fixing plant cells (Nishizawa and Mori 1989a), suggest that these structures are lysosome-like structures that are different from the smooth ER (Fig. 4).

After they are pinched off from the plasma membrane, PL-invaginations form invagination-derived vacuoles (endocytotic vacuoles, or endosomes) that are completely surrounded by ER-derived vesicles, including the 'eye-like structures' and lysosome-like structures (Fig 5). In rice cortex cells, a number of these endocytotic vacuoles surrounded by ER-derived vesicles was observed in addition to the PL-invaginations. The surrounding structures enveloping the endocytotic vacuoles joined together and enlarged, distorting the enclosed endocytotic vacuoles, which changed into filamentous structures in newly-formed vacuoles. In other words, new vacuoles derived from the surrounding structures formed after the disappearance of the endocytotic vacuoles. In the differentiating cortex cells, where

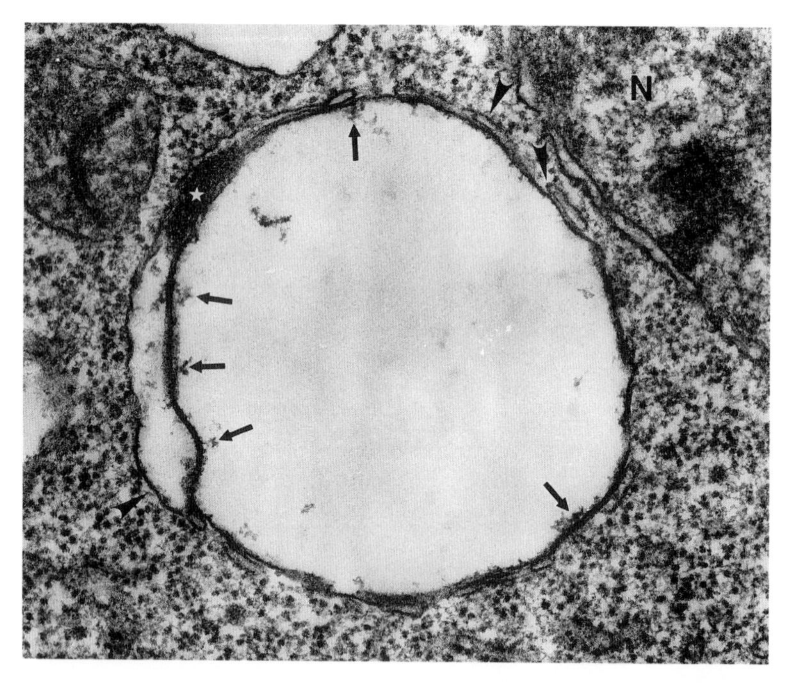

Fig. 5. An endocytotic vacuole derived from the PL-invagination. PL-invagination is pinched off from the plasma membrane into the cytoplasm and the endocytotic vacuole is formed. As in Fig. 3, electron-dense materials attached to the surface of the limiting membrane were observed (*arrows*). The endocytotic vacuole is surrounded with the ER (endoplasmic reticulum) -derived vesicles (*arrowheads*). One of the vesicles contains electron-dense deposits (*white star*) and is termed an "eye-like structure". *N* is the nucleus

the vacuoles fill the cytoplasm, the PL-invaginations were completely surrounded by pre-existing vacuoles. Pinched off endocytotic vacuoles were incorporated into the pre-existing vacuoles, and then distorted into filamentous residues.

2.2 PTA (Phosphotungstic Acid)-CrO_3 (Chromic Acid) Staining Studies of PL-Invagination

This interpretation of the fate of PL-invaginations was confirmed by PTA-CrO_3 staining, which is a specific staining technique for plasma membrane. The invaginations stained positively with this stain, proving that they were definitely derived from plasma membrane. Using a common staining method, the vacuoles in the cell were stained evenly. With this method, however, two types of vacuoles were distinguished in the cytoplasm: one was not stained (pre-existing vacuoles) and the other stained positively (endocytotic vacuoles) (Fig. 6). Moreover, the endocytotic vacuole itself was stained positively with this stain, but the limiting membrane of the vesicle and the tubular structures surrounding the stained endocytotic vacuole were not stained (Figs. 6, 7, 8, 9, 10). This proved that the limiting membrane of the surrounding structure is different from the plasma membrane or Golgi vesicles. This staining clearly showed that these surrounding structures fuse to each other, and then begin to inflate and concomitantly distort the invagination-derived vacuoles (endocytotic vacuoles) that they enclose. Similarly,

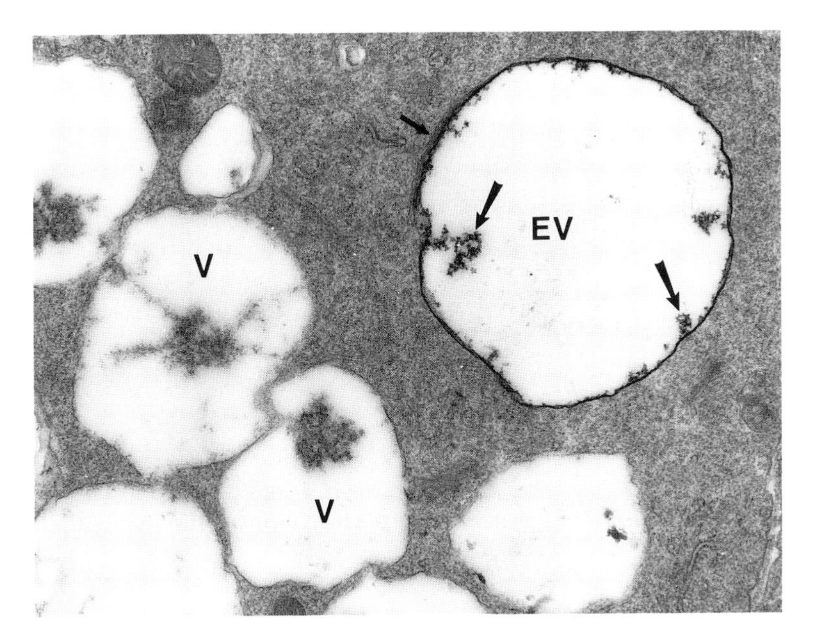

Fig. 6. An endocytotic vacuole (*EV*) stained by the PTA-CrO_3 staining method. Since the tonoplast of a pre-existing vacuole is not stained by this method, the endocytotic vacuole is distinguished from the pre-existing vacuole. The limiting membrane of the endocytotic vacuole is stained but that of the surrounding structures is not stained (*short arrow*). Particle materials attached to the limiting membrane were observed (*long arrows*). *V* is the vacuole

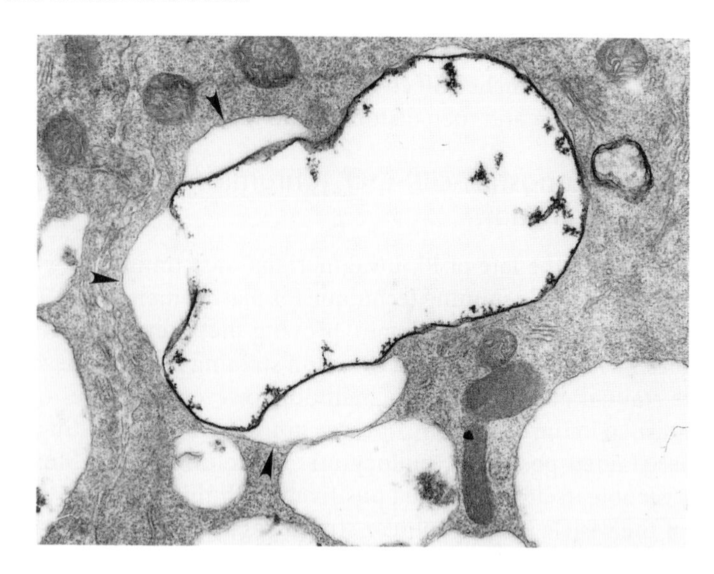

Fig. 7. An endocytotic vacuole stained by the PTA-CrO_3 staining method. The surrounding structures or ER-derived vesicles are beginning to inflate (*arrowheads*)

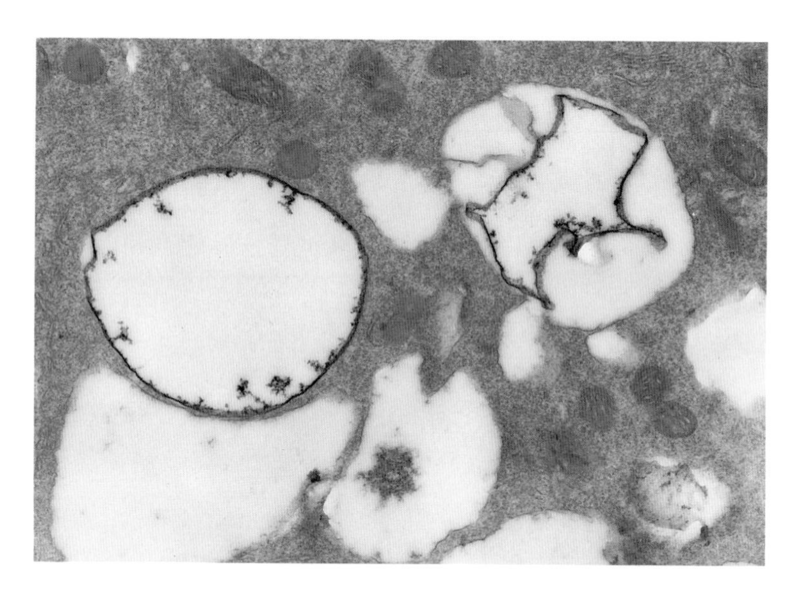

Fig. 8. An endocytotic vacuole stained by the PTA-CrO_3 staining method. The left vacuole is just pinched off from the plasma membrane and the right vacuole is distorted by the inflation of surrounding structures

the endocytotic vacuoles incorporated in the pre-existing vacuoles were seen to be distorted into filamentous structures in the vacuoles.

These results showed that the PL-invagination-derived endocytotic vacuoles are digested with lytic enzymes supplied by the surrounding vesicles or lysosome-like structures, and that new vacuoles formed with an endoplasmic-type limiting

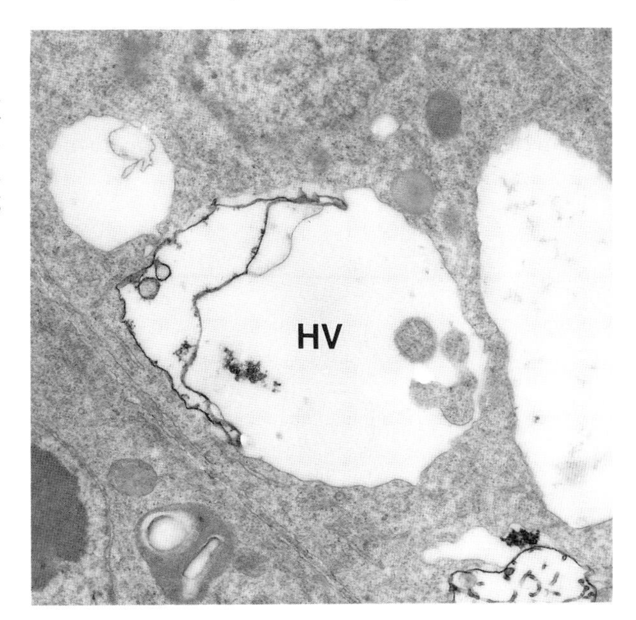

Fig. 9. Newly formed heterophagic vacuole (*HV*). The endocytotic vacuole is distorted and remains in the newly formed vacuole as a filamentous structure which is stained by the PTA-CrO_3 staining method

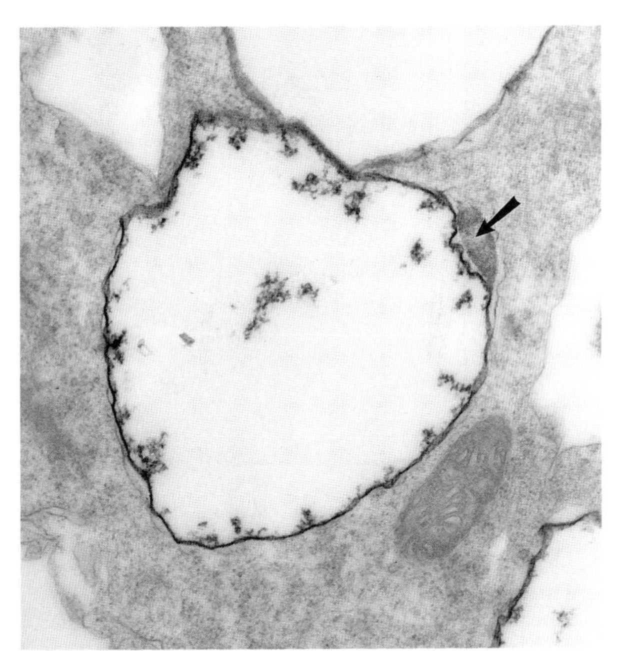

Fig. 10. The endocytotic vacuole stained by the PTA-CrO_3 staining method surrounded by the "eye-like structure" (*arrow*), which is not stained by this method

membrane were the same type of vacuole as the pre-existing vacuoles. When the endocytotic vacuoles were incorporated into the pre-existing vacuoles, the pre-existing vacuoles supplied lytic enzymes to digest the endocytotic vacuoles, and a new vacuole did not form.

As mentioned earlier, in animal cells it is well known that exogenous substances are incorporated into a cell by endocytosis and digested intracellularly through a lysosome system. This process is called heterophagy, in contrast to autophagy, which is a process involving digestion with endogenous substances, such as a portion of the cytoplasm or an organelle. The above-mentioned sequential process in rice root cells was actually thought to be the same process of heterophagy.

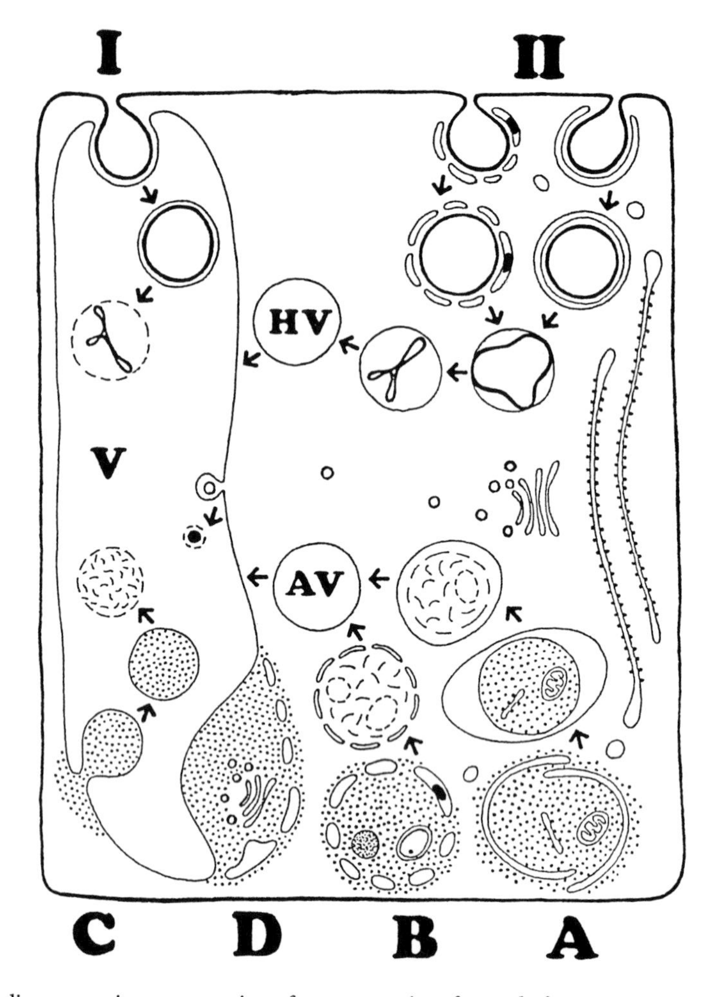

Fig. 11. A diagrammatic representation of two categories of vacuole formation and intracellular digestion in rice root cells. The process of heterophagy is classified into two types: *I* and *II*. In type *I* heterophagy, PL-invagination is pinched off from the plasma membrane, and then incorporated into the pre-existing vacuole. In type *II* heterophagy, PL-invagination is surrounded by ER or ER-derived vesicles, including eye-like structures. These surrounding structures are fused together, and then inflate to induce distortion and digestion of endocytotic vacuoles. Type II heterophagy results in the formation of a new vacuole, a heterophagic vacuole (*HV*). The process of autophagy is classified into four types: A, B, C and D. The ER cisterna in *type A* and the ER-derived vesicles in *type B* sequestrate a portion of the cytoplasm, and then digest the endogenous materials resulting in the formation of a new vacuole, an autophagic vacuole (*AV*). In *type C* autophagy, the tonoplast invaginates into the vacuole enclosing a portion of the cytoplasm, and then the vesicles containing cytoplasmic materials are pinched off into a vacuole. In *type D*, invagination of the tonoplast and ER-derived vesicles combine to sequestrate a portion of the cytoplasm. *Thicker lines* represent an exoplasmic type of membrane, and *thinner lines* represent an endoplasmic type

Therefore, two different sequences of heterophagy were identified, Type I and Type II. Type I heterophagy is the incorporation of endocytotic vacuoles into pre-existing vacuoles, where they are digested. In Type II heterophagy, endocytotic vacuoles are digested by the surrounding structures, and new vacuoles are formed with the endoplasmic type of limiting membrane. The new vacuoles formed by Type II heterophagy are called heterophagic vacuoles (Fig. 11).

2.3 Detection of AcPase (Acid Phosphatase) Activity

It was thought that lytic enzymes were released into the endocytotic vacuoles from pre-existing vacuoles in the case of Type I heterophagy, and from surrounding structures in the case of Type II heterophagy. Since hydrolytic enzymes are reported frequently in pre-existing vacuoles, it was readily believed that lytic enzymes in pre-existing vacuoles functioned in the case of Type I heterophagy. On the contrary, there was no knowledge of the existence of lytic enzymes in the surrounding structures, in the case of Type II heterophagy.

Therefore, in order to confirm the involvement of lytic enzymes in these heterophagic processes, the intracellular localization of AcPase, which is a lysosome marker enzyme, was investigated using a cytochemical detection method. To determine AcPase activity, samples were pre-fixed, washed, and incubated in medium consisting of 25 mg β-glycerophosphate sodium salt, 12 ml distilled water, 10 ml 0.5 M acetate buffer (pH 5.0), 3 ml 0.1% lead nitrate, and 1.25 g sucrose. After incubation, the samples were post-fixed in 1% osmium tetroxide.

AcPase activity was detected in the vesicles that surround the PL-invagination, including the 'eye-like structures' wrapping the pinched-off endocytotic vacuole (Fig. 12). The activity was especially vigorous in an electron-dense precipitation of the 'eye-like structures'. Thus, as expected, the surrounding vesicles, including the 'eye-like structures' were demonstrated to function as primary lysosomes carrying lytic enzymes. In addition, activity was also observed in the pre-existing vacuoles. Golgi vesicles showing AcPase activity were incorporated into the pre-existing vacuoles by invagination of the tonoplast.

The cytochemical detection of AcPase activity proved that lytic processes involving the lysosome system participated in both Type I and Type II heterophagy. Both types of heterophagy were observed in rice root cells cultured with hemoglobin as the sole nitrogen source, and both are quite different from the process of heterophagy in animal cells. In animal cells, the primary lysosomes fuse to the endocytotic vacuoles in the process of heterophagy. It is believed that the difference in the heterophagic processes of plant and animal cells stems from the fact that in plant cells the limiting membranes of the lysosome system are endoplasmic type membranes. Therefore, these two heterophagic processes observed in rice root cells are highly specific to plant cells.

2.4 Incorporation of Hemoglobin into Cells

It was demonstrated that the process of heterophagy occurred in the cortex cells of rice roots cultured with hemoglobin as the sole source of nitrogen. Therefore, in order to ascertain whether exogenously supplied hemoglobin was actually incorporated into cells by heterophagy, the uptake of ^{3}H-labeled hemoglobin was examined using electron microscope-autoradiography (Nishizawa and Mori 1980a, 1982).

Rice plants were water-cultured using hemoglobin as the sole source of nitrogen for 80 days after transplantation. ^{3}H-labeled hemoglobin was synthesized by the method of Bolton and Hunter, using N-succinimidyl (2, 3-^{3}H) propionate as the acylating agent. Fresh crown roots were dipped in a nutrient solution containing ^{3}H-labeled hemoglobin for four hours, then fixed for electron microscopic observations. Ultra-thin sections were cut, stained with uranium and lead, and then coated with carbon and Ilford L4 emulsion. Autoradiographs were exposed for eight to ten weeks in a refrigerator.

As previously mentioned, particles were frequently observed attached to the limiting membranes of the PL-invaginations and endocytotic vacuoles in the cortex cells of rice plants. These particles were thought to be hemoglobin molecules. Hemoglobin is a well-known protein that contains Fe. Therefore, it can be observed under an electron microscope. Similar particles attached to the limiting membrane were observed in the cortex cells of rice plants supplied with ^{3}H-labeled hemoglobin. The silver grains indicating ^{3}H radioactivity were coincident with these particles (Figs. 13, 14, 15), identifying the particles in the endocytotic vacuoles as hemoglobin. The structure of the hemoglobin in the endocytotic vacuoles suggested that hemoglobin was incorporated into the cells as a macromolecule and not as a smaller molecule after digestion in the rhizosphere.

Detailed studies of the distribution of incorporated hemoglobin in the endocytotic vacuoles indicated that hemoglobin was present inside the endocytotic vacuoles and also attached to the limiting membrane of the endocytotic vacuoles (Figs. 15, 16). This result suggests that hemoglobin is first attached or bound to the

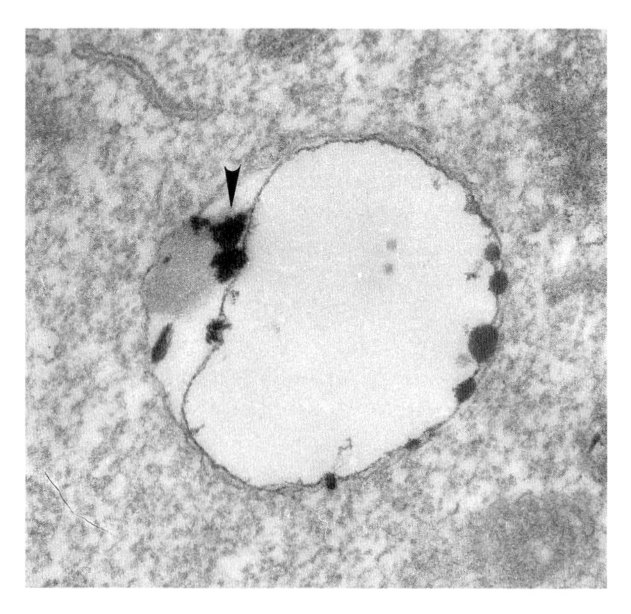

Fig. 12. Detection of AcPase (acid phosphatase) activity. The surrounding "eye-like structure" enveloping an endocytotic vacuole shows lead precipitate (*arrowhead*) which indicates AcPase activity. No staining

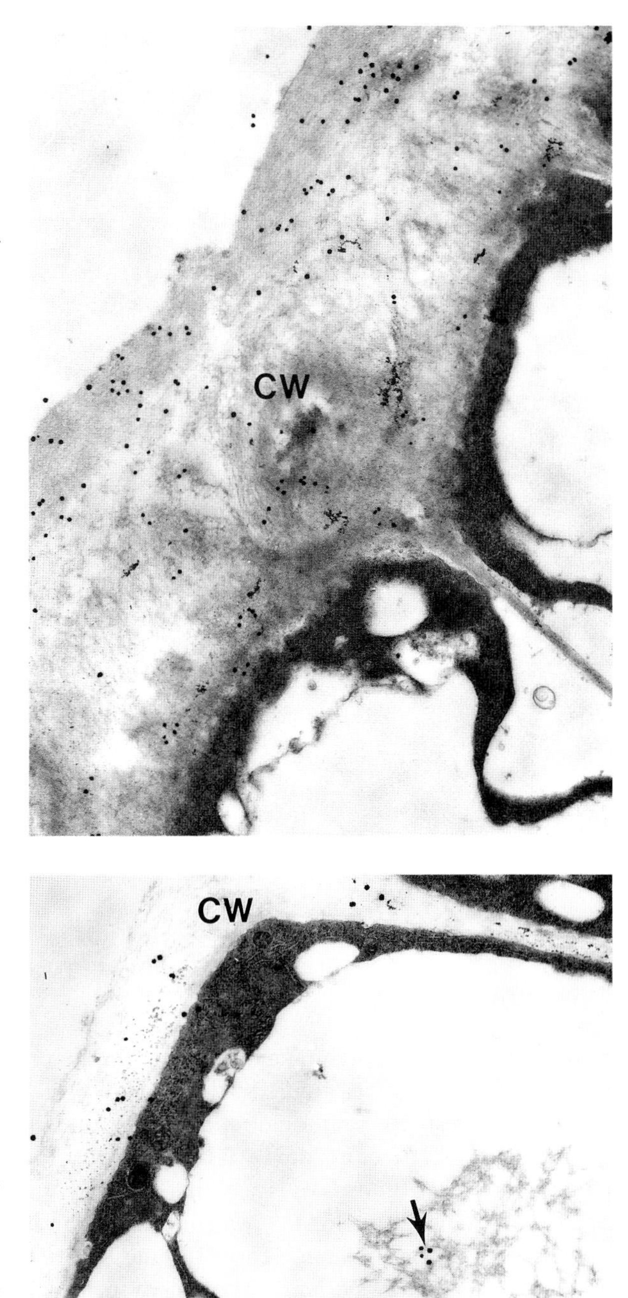

Fig. 13. An electron microscopic autoradiograph of rice roots fed with ^{3}H-labeled hemoglobin. Silver grains produced by the radioisotope (RI) are distributed inside the cell wall (*CW*). This result shows that ^{3}H-hemoglobin passes through the cell wall of epidermal cells

Fig. 14. An electron microscopic autoradiograph of rice roots fed with ^{3}H-labeled hemoglobin. Silver grains (*arrow*) were also observed in the precipitate in the vacuole

plasma membrane, and then the plasma membrane is invaginated into the cells. The incorporation of hemoglobin into rice root cells is a process similar to receptor-mediated endocytosis. In rice roots, some recognition mechanism is probably triggered when hemoglobin binds to the plasma membrane, before invagination of the plasma membrane.

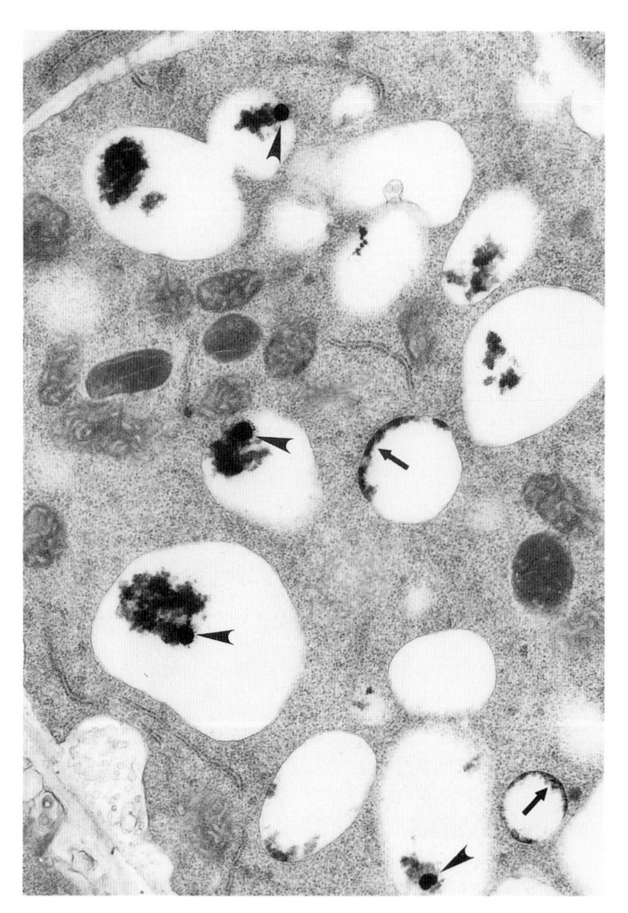

Fig. 15. An electron microscopic autoradiograph of cortex cells in rice roots fed with ^{3}H-labeled hemoglobin. Silver grains (*arrowheads*) are coincident with particle materials in the endocytotic vacuoles. The concentration of ^{3}H-labeled hemoglobin fed is higher than that of hemoglobin usually supplied as the sole nitrogen source in water-culture. Therefore, these particle materials are greater in amount than the particle materials observed in the endocytotic vacuoles indicated in Fig. 3. They are attached to the surface of the limiting membrane (*arrows*)

2.5 Detecting Incorporated Intracellular Proteins Using a Cytochemical Method

In order to confirm the uptake of hemoglobin macromolecules, the presence of hemoglobin in cells was studied using a cytochemical method. Non-enzymatic heme proteins such as hemoglobin or myoglobin also have peroxidatic activity, like peroxidase. Therefore, the DAB (diaminobenzidine) method should localize hemoglobin. The root tips of rice plants cultured with hemoglobin were cut and pre-fixed. After pre-fixation and washing, the samples were incubated for peroxidase activity in medium consisting of 10 mg DAB, 0.1 ml 3% hydrogen peroxide, and 10 ml 0.05 M AMP buffer. After incubation, the samples were post-fixed in 1% osmium tetroxide (Nishizawa and Mori 1980b). Since endogenous peroxidase activity is present in vacuoles and the cell wall, rice roots grown hydroponically with albumin were used as controls. The particles in the vacuoles reacted positively, proving the presence of hemoglobin in the vacuoles (Fig. 17). In addition, we examined the incorporation of ferritin into rice roots. Rice plants were water-cultured with albumin as the sole source of nitrogen, instead of hemoglobin. Then they were fed ferritin as an electron-dense marker protein. Incorporated ferritin

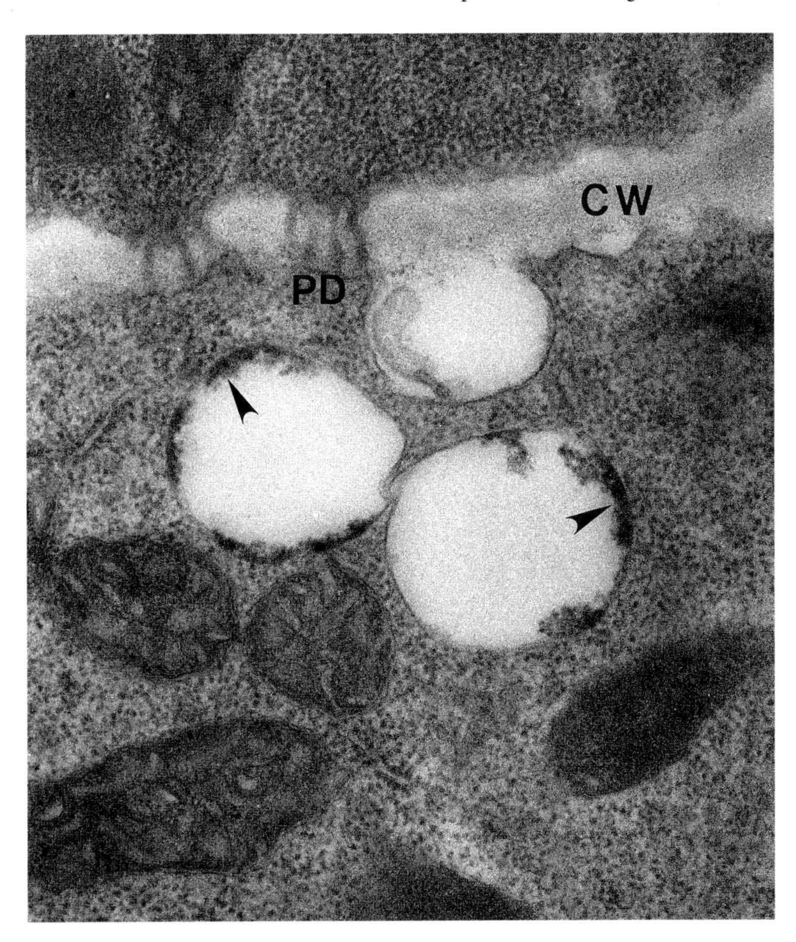

Fig. 16. Hemoglobin inside the endocytotic vacuoles. Endocytotic vacuoles near PL-invagination, containing hemoglobin (*arrowheads*) attached to their limiting membranes. *PD*, plasmodesmata; *CW*, cell wall

was observed in the vacuoles (Fig. 18). These cytochemical results confirmed that root cells take up hemoglobin and ferritin in the form of macromolecules.

In conclusion, rice roots incorporate, digest, and utilize exogenously supplied hemoglobin macromolecules as a nutrient source through heterophagy. This process is shown diagrammatically in Fig. 19.

2.6 Quantitative Analysis of Direct Protein Uptake in Rice Plants

To analyze the direct uptake of proteins by rice roots quantitatively, we examined the incorporation of ^{125}I- and ^{131}I-labeled bovine serum albumin (BSA) by rice roots (Mori 1979). ^{125}I- and ^{131}I-labeled BSA were prepared using Na^{125}I or Na^{131}I and BSA. Radio-labeled protein fractions were obtained after column chromatography using Sephadex G50. Rice plants cultured for one week after germina-

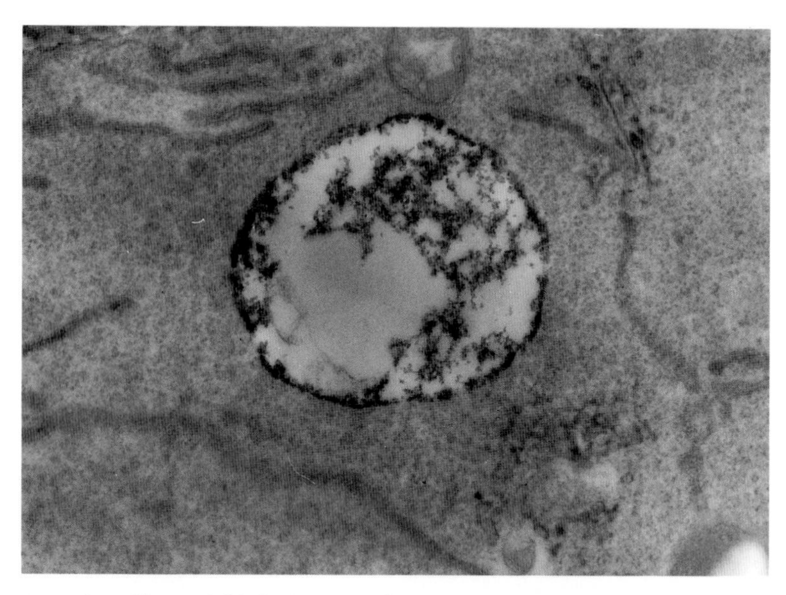

Fig. 17. Detection of hemoglobin by its peroxidase activity. Peroxidase activity in rice roots was detected by a cytochemical method using DAB. Particle substances in the vacuole as well as the Golgi body are positively stained. This implies that the particle substances in the vacuole are hemoglobin. The property of hemoproteins, such as hemoglobin, as characterized by its peroxidase activity is effectively used for detecting hemoglobin in the cell

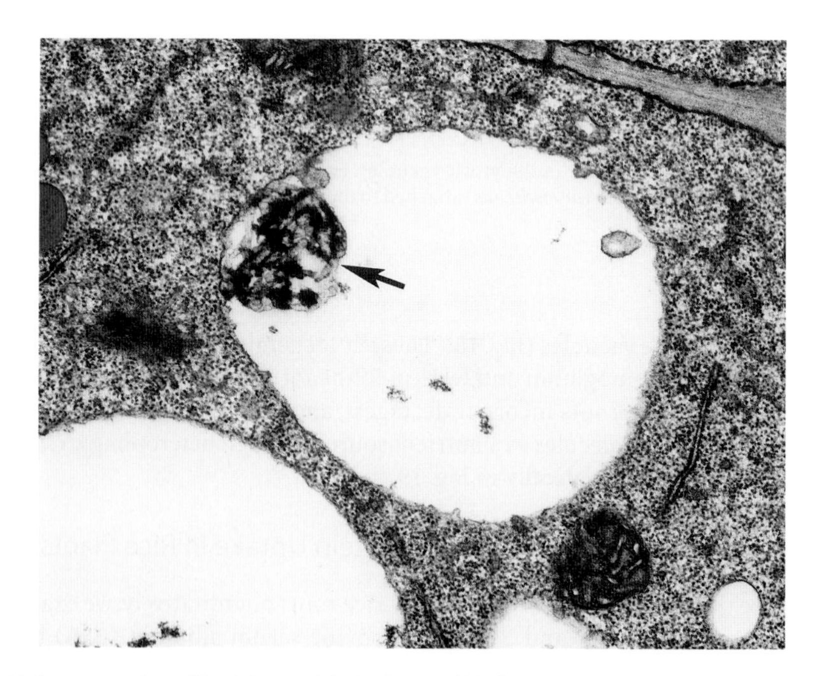

Fig. 18. Incorporation of ferritin. Ferritin is observed by the rice root water-cultured with albumin as the sole source of nitrogen. Ferritin particles (*arrow*) enveloped by a membrane were observed in the vacuoles in a cortex cell

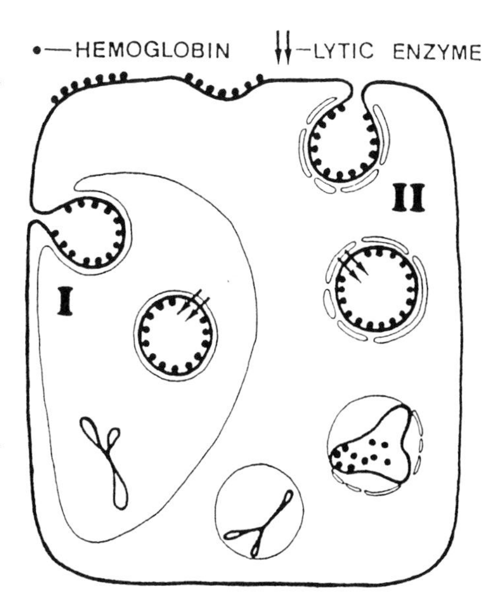

Fig. 19. A diagrammatic representation of the entry and digestion of exogenous hemoglobins in rice root cells through two types of heterophagy. The hemoglobin binding to a plasma membrane enters root cells through the invagination of the plasma membrane and subsequently forms an endocytotic vacuole, which is thereafter incorporated into a pre-existing vacuole. While being supplied with lytic enzymes, the hemoglobin is digested (Type I heterophagy). In the other case, the endocytotic vacuole containing hemoglobin is wrapped by ER-derived vesicles, including an "eye-like structure" or ER itself, which supply them with lytic enzymes (Type II heterophagy). After digestion, a new heterophagic vacuole is formed

tion were used in the following experiments. Twenty plants were treated with medium containing the radio-labeled BSA for one hour. After washing, the radioactivity of incorporated ^{125}I- or ^{131}I-labeled BSA was measured with a well-type scintillation counter. The measured radioactivity was assumed to be the total of the radio-labeled BSA incorporated into cells and adsorbed to the cell wall. As shown in Fig. 20, rice roots effectively absorbed ^{125}I-labeled BSA. In a one-hour treatment, only a small amount of radioactivity was transported to the shoots. Treatment with 2,4-dinitrophenol inhibited the absorption of ^{131}I-labeled BSA by rice roots (Fig. 21). Other metabolic inhibitors, such as KCN (10^{-4} M) and ouabain (10^{-4} M), also inhibited the absorption of ^{125}I-labeled BSA by rice plants. Therefore, the absorption of BSA by rice roots is energy-dependent. Moreover, as shown in Fig. 22, the addition of ATP to the medium enhanced the absorption of ^{125}I-labeled BSA remarkably. The enhancing effect of ATP on BSA absorption was accelerated with the addition of Mg. This effect was specific to ATP, since other nucleotides, including uridine triphosphate (UTP), cytidine triphosphate (CTP), guanosine triphosphate (GTP), and inositol triphosphate (ITP), lacked the enhancing effect (Fig. 23). The results of the experiments using radio-labeled BSA clearly show that rice plants can take up exogenously supplied protein.

2.7 Field Observations that Support the Direct Uptake of Organic Nitrogen

Recently, trees in boreal forest (Näsholm et al. 1998) and arctic tundra (Chapin III et al. 1993) were reported to use organic nitrogen for growth preferentially. Plant growth in the boreal forest and arctic tundra is generally nitrogen-limited, de-

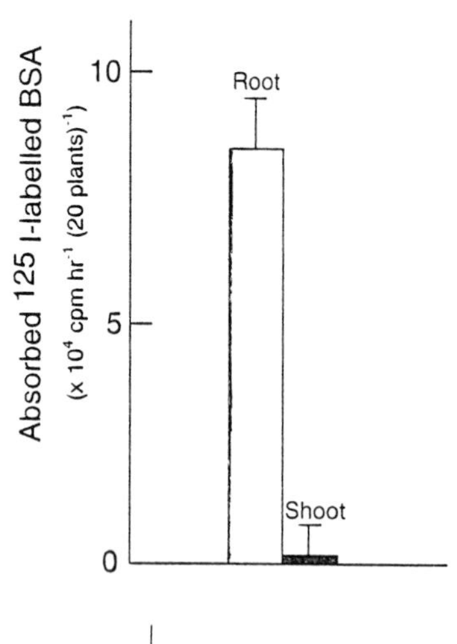

Fig. 20. Absorption of ^{125}I-labeled bovine serum albumin (BSA) by rice roots for one hour. Each value is a mean (n=3)

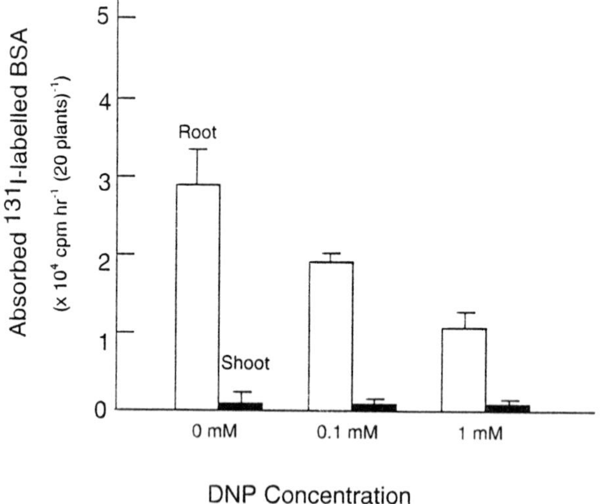

Fig. 21. Inhibitory effect of dinitrophenol (*DNP*) on the absorption of ^{131}I-labeled BSA by rice roots. Each value is a mean (n=3)

spite large pools of soil organic nitrogen. The presumed cause of this limitation is the slow mineralization of soil organic nitrogen because of low temperatures. In these areas, plants preferentially absorb and use amino acids such as glycine. The preferential absorption and use of amino acids by plants grown in northern ecosystems is consistent with our observation that arginine alleviates the growth reduction of rice caused by adverse conditions such as low temperature or low solar energy. Amino acids are a better nitrogen source than inorganic nitrogen in terms of saving energy. Therefore, plants growing in boreal areas use amino acids preferentially not only because of limited amounts of inorganic nitrogen, but also because they are a more advantageous nitrogen source. By absorbing and utilizing amino acids directly, these plants acquire an energetically favorable source of nitrogen and short-circuit the bottleneck in the arctic nitrogen cycle imposed by the temperature-limited mineralization process. The trees of the boreal forest are

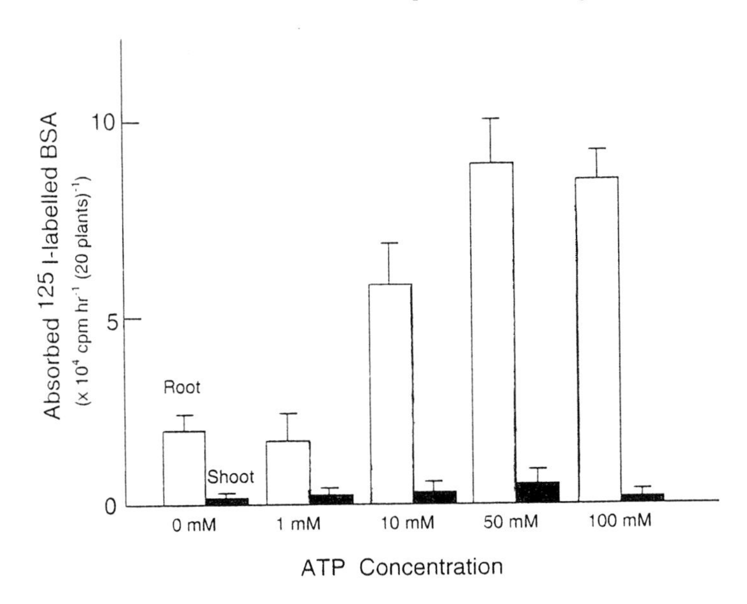

Fig. 22. Absorption of ^{125}I-labeled bovine serum albumin (*BSA*) by rice roots at various concentrations of ATP. Addition of ATP enhanced absorption of ^{125}I-labeled BSA by rice roots. Each value is a mean (n=3)

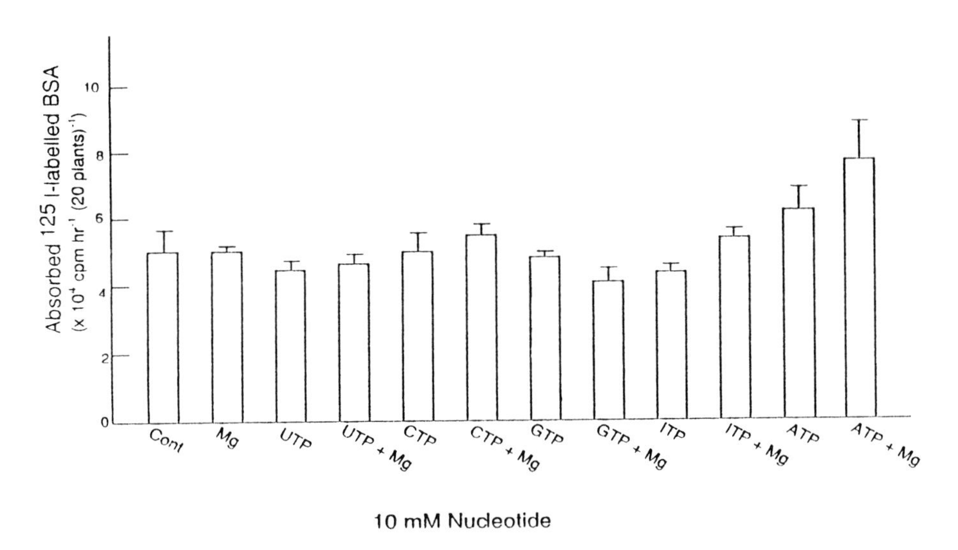

Fig. 23. Effects of various nucleotides (10 mM) and Mg ions on the absorption of ^{125}I-labeled BSA by rice roots. Each value is a mean (n=3)

reported to utilize protein as the sole nitrogen source after infection by ectomycorrhiza with proteolytic capability (Abuzinadah and Read 1986). In this case, trees may not absorb exogenous proteins directly. In field experiments, Yamagata and Ae (1996, 1997) showed the preferential use of organic nitrogen by upland rice, and suggested that upland rice took up protein directly from the soil where inorganic nitrogen coexisted, as described in the chapter by Yamagata M et al.

3. Future Aspects

In future, we have to clarify the molecular mechanism for the direct uptake of organic nitrogen by plant roots. Since many transporter genes have now been isolated from plants, the absorption of amino acids, small peptides, or sucrose by plant roots can be readily investigated at the molecular level. We have started to identify the amino acid transporters that function in the absorption of amino acids from the rhizosphere, in collaboration with a German group. However, nothing is known of the direct uptake of proteins by plant roots at the molecular level. In contrast, many proteins that participate in endocytosis are well-studied in animal cells. Dynamin is a 100 kDa GTPase that regulates late events in clathrin-coated vesicle formation. GTP hydrolysis is required to detach vesicles from the plasma membrane. Syntaxin is essential for the protrusion of the plasma membrane to start invagination. Plant counterparts homologous to animal dynamin genes have been isolated from soybean (Gu and Verma 1996) and *Arabidopsis* (Park et al. 1998). Genes homologous to syntaxin have been isolated from *Arabidopsis* (Lukowitz et al. 1996; Bassham et al. 1995). These homologous genes isolated in plants have been reported to function in the chloroplast development, or cell plate formation during cytokinesis. However, more genes homologous to them will be found in plants. Among these homologues, there may be proteins that function in the direct uptake of proteins from the rhizosphere through endocytosis. We are certain that in the near future we will be able to understand the molecular mechanism of protein uptake and use it to generate novel transgenic plants that absorb organic nitrogen primarily.

References

Abuzinadah RA, Read DJ (1986) The role of proteins in the nitrogen nutrition of ectomycorrhizal plants. III. Protein utilization by *Betula*, *Picea* and *Pinus* in mycorrhizal association with *Hebeloma crustuliniforme*. New Phytol 103:507-514

Bassham D, Gal S, da Silva Conceicao A, Raikhel N (1995) An *Arabidopsis* syntaxin homologue isolated by functional complementation of a yeast *pep*12 mutant. Proc Natl Acad Sci USA 92:7262-7266

Chapin FS, Moilanen L, Kielland K (1993) Preferential use of organic nitrogen for growth by a non-mycorrhizal arctic sedge. Nature 361:150-153

Fischer WN, Kwart M, Hummel S, Frommer WB (1995) Substrate specificity and expression profile of amino acid transporters (AAPs) in *Arabidopsis*. J Biol Chem 270:16315-16320

Fischer WN, André B, Rentsch D, Krolkiewicz S, Tegeder M, Breitkreuz K, Frommer WB (1998) Amino acid transport in plants. Trends Plant Sci 3:188-195

Frommer WB, Hummel S, Riesmeier JW (1993) Expression cloning in yeast of a cDNA encoding a broad specificity amino acid permease from *Arabidopsis thaliana*. Proc Natl Acad Sci USA 90:5944-5948

Frommer WB, Kwart M, Hirner B, Fischer WN, Hummel S, Ninnemann O (1994a) Transporters for nitrogenous compounds in plants. Plant Mol Biol 26:1651-1670

Frommer WB, Hummel S, Rentsch D (1994b) Cloning of an *Arabidopsis* histidine transporting protein related to nitrate and peptide transporters. FEBS Lett 347:185-189

Gu X, Verma DP (1996) Phragmoplastin, a dynamin-like protein associated with cell plate formation in plants. EMBO J 15:695-704

Lukowitz W, Mayer U, Jurgens G (1996) Cytokinesis in the *Arabidopsis* embryo involves the syntaxin-related KNOLLE gene product. Cell 84:61-71

Mori S (1979) Criticism to the mineral nutrition theory. Doctoral Thesis, University of Tokyo

Mori S (1986) Prominent positive effect of ribonucleic acid on the growth of barley. Jpn J Soil Sci Plant Nutr 57:171-178

Mori S, Nishizawa N (1979) Nitrogen absorption by plant root from the culture medium where organic and inorganic nitrogen coexist. II. Which nitrogen is preferentially absorbed among (U-^{14}C)gln, (2,3-^{3}H)Arg and $Na^{15}NO_3$? Soil Sci Plant Nutr 25:51-58

Mori S, Nishizawa N, Uchino H, Nishimura Y (1977) Utilization of organic nitrogen as the sole nitrogen source for barley. Proceedings of the international seminar on soil environment and fertility management in intensive agriculture, Tokyo, Japan, pp 612-617

Mori S, Uchino H, Sago F, Suzuki S, Nishikawa A (1985) Alleviation effect of arginine on artificially reduced grain yield of NH_4^+ or NO_3^--fed rice. Soil Sci Plant Nutr 31:55-67

Näsholm T, Ekblad A, Nordin A, Giesler R, Högberg M, Högberg P (1998) Boreal forest plants take up organic nitrogen. Nature 392:914-916

Nishizawa N (1995) Nutrient absorption of the rice plant. In: Matsuo T (Ed) Science of the rice plant. Volume 2: Physiology, pp 249-263

Nishizawa N, Mori S (1977) Invagination of the plasmalemma: its role in the absorption of macromolecules in rice roots. Plant Cell Physiol 18:767-782

Nishizawa N, Mori S (1978) Endocytosis (heterophagy) in plant cells: Involvement of ER and ER-derived vesicles. Plant Cell Physiol 19:717-730

Nishizawa N, Mori S (1980a) Electron microscope-autoradiographical evidence for the incorporation of exogenous protein in to rice root cells. Plant Cell Physiol. 21:493-496

Nishizawa N, Mori S (1980b) Vacuole formation as consequence of intracellular digestion: Acid phosphatase localization as associated with plasmalemma-invagination and vacuole formation. Soil Sci Plant Nutr 26:525-540

Nishizawa N, Mori S (1982) Utilization of exogenous hemoglobin by rice root cells through the mechanism of heterophagy. Proceedings of the 9th international plant nutrition colloquium, pp 431-436

Nishizawa N, Mori S (1989a) Ultrastructure of thylakoid membrane in tomato leaf chloroplast revealed by liquid helium rapid-freezing and substitution-fixation method. Plant Cell Physiol 30:1-7

Nishizawa N, Mori S (1989b) Microbodies containing crystalloid inclusions in rice root cells. Soil Sci Plant Nutr 35:485-490

Park JM, Cho JH, Kang SG, Jang HJ, Pih KT, X Piao HL, Cho MJ, Hwang I (1998) A dynamin-like protein in *Arabidopsis thaliana* is involved in biogenesis of thylakoid membranes. EMBO J 17:859-867

Rentsch D, Laloi M, Rouhara I, Schmelzer E, Delrot S, Frommer WB (1995) *NTR1* encodes a high affinity oligopeptide transporter in *Arabidopsis*. FEBS Lett 370:264-268

Riesmeier JW, Willmitzer L, Frommer WB (1992) Isolation and characterization of a sucrose carrier cDNA from spinach by functional expression in yeast. EMBO J 11:4705-4713

Robinson DG, Depta H (1988) Coated vesicles. Ann Rev Plant Mol Biol 39:53-99

Steiner HY, Song W, Zhang L, Naider F, Becker JM, Stacey G (1994) An *Arabidopsis* peptide transporter is a member of a new class of membrane transport proteins. Plant Cell 6:1289-1299

West CE, Waterworth WM, Stephens SM, Smith CP, Bray CM (1998) Cloning and functional characterisation of a peptide transporter expressed in the scutellum of barley grain during the early stages of germination. Plant J 15:221-229

Yamagata M, Ae N (1996) Nitrogen uptake response of crops to organic nitrogen. Soil Sci Plant Nutr 42:389-394

Yamagata M, Ae N (1997) Preferential use of organic nitrogen by upland rice compared to maize. Proc 13th Internatl Plant Nutr Colloq, Tokyo, Japan. Kluwer Academic Publishers, Dordrecht, The Netherlands, pp 509-510

Part VI.
Practical Implications

Implications of Soil-Acidity Tolerant Maize Cultivars to Increase Production in Developing Countries

Luis Narro, Shivaji Pandey, Carlos De León, Fredy Salazar, and Maria P. Arias

Summary. About 3950 million ha (30%) of the total ice-free land area in the world is under acid soils. Acid soils are characterized by low fertility caused by high levels of Al, Mn, and Fe, and deficiencies of P, Ca, Mg, K, S, and Zn. Of these, Al toxicity and P deficiency seem to be the most important causes for low maize yields (about 400 kg/ha for land race cultivars) in these soils. Maize is a staple for millions of people in developing countries where imports are growing up by 1.5 million tons (7%) per year. There are at least two alternatives to increasing maize production in acid soils. The first is to use amendments (lime, gypsum) to correct soil acidity. This is expensive and not available for small farmers. Another disadvantage is that only the upper 30 cm of soil is corrected making maize roots to be concentrated in that layer and not growing beyond. The second approach is to develop tolerant cultivars. This solution is relatively inexpensive, environmentally clean, permanent, and energy conserving. International Center for the Improvement of Wheat and Maize (CIMMYT), in collaboration with several National Agriculture Research NARS all over the world, has been developing soil-acidity tolerant maize cultivars to increase maize production. For this purpose, acid soil tolerant maize populations were formed and recurrent selection was used to improve these populations for grain yield under both acid and normal soils. Cultivars developed from these populations show a consistent increase in grain yield in both acid and non-acidic soils. Under acid soils, the average grain yield of genotypes used to form the base population in 1977, was below 0.4 t/ha. The average grain yield of acid soil tolerant open pollinated varieties (OPVs) developed in 1993 and evaluated across 13 acid soil environments was 3.2 t/ha, while that of non conventional hybrids developed in 1995 and evaluated across six acid soil environments was 3.84 t/ha. In acid soils, high parent heterosis of up to 42.5% has been observed in inter-variety crosses and up to 261% in single crosses. Also, although superiority of hybrids compared to OPVs has been reported, OPVs will continue to be more important than hybrids for the acid soils over the next years, due to the poor economic conditions of farmers cultivating them. From our agronomic research, response to P of Sikuani, an acid soil tolerant cultivar, is higher than that of Tuxpeño (a susceptible cultivar) both in acidic and non-acidic soils. Grain yield of maize was higher using amendments with Ca, Mg, and S than with Ca and Mg only, although there were no differences among methods of application of amendments. Studies of microelements in acid soils showed a highly significant response to Zn. Planting maize in association with pastures in savannas could be a profitable alternative for farmers allowing them to raise more animals/

ha with an associated gain of weight/animal. We are also looking for morphological and biochemical traits that may improve grain yield. Molecular and physiological studies are being continued to improve the efficiency of our breeding program. These activities are being developed in collaboration with prominent universities and research institutions worldwide.

Key words. Acid soils, Tropical maize, *Zea mays* L., Germplasm, Selection criteria, Genetic gain, Inheritance, Tolerant cultivars, Agronomy, Microelements, Amendments, Management, Phosphorus

1. Introduction

Historically, civilizations have tended to settle on high-base-status soils, but when populations increase they have migrated to other environments mainly with acid soils (Sanchez 1976). Acid soils cover about 30% of the total ice-free land area, equivalent to 3 950 million ha of the earth. About 16.7% of Africa, 6.1% of Australia and New Zealand, 9.9% of Europe, 26.4% of Asia, and 40.9% of America have acid soils (Von Uexkull and Mutert 1995). They cover a significant part of at least 48 developing countries located mainly in the tropical area, being more frequent in Ultisols and Oxisols in South America and in Oxisols in Africa. In Asia Gleisols, Podzoluvisols, and Ultizols are the most important acid soils.

Low fertility of acid soils is due to high levels of Al, Mn, and Fe and deficiencies of P, Ca, Mg, K, S, and Zn (Clark et al. 1988) although Al toxicity (Foy et al. 1978; Zeigler et al. 1995) and P deficiency seem to be the most important factors. Aluminum possibly affects between 40 and 70% of the world's arable land that is potentially usable for food and biomass production (Haug and Caldwell 1985). Al toxicity varies with its concentration, pH, temperature and concentration of Ca and Mg and interferes with cell division at the tips of the seminal and secondary roots. Al toxicity decreases the root length (Fig. 1), increases rigidity of the cell wall due to pectin formation and reduces DNA replication due to an increased rigidity of the DNA double helix. It also fixes P in forms non-available to the roots, decreases root respiration, and interferes with the enzymes controlling sugar phosphorylation and polysaccharide accumulation in the cell walls. It interferes with the assimilation, transport and utilization of essential elements including Ca, Mg, P, K, and Fe (Foy 1988). Usually pH per se does not have a direct effect on plant growth, although at pH lower than 4.2 the hydrogen ion concentration may hinder or even reverse cation uptake by plant roots (Jackson 1967).

Low fertility of acid soils is responsible for low maize yields, with an average of 400 kg/ha in conditions where local farmer technology is used in those soils. It is urgent to increase maize production in developing countries where maize is a staple for millions of people and maize imports grow by 1.5 million tons or 7% per year (CIMMYT 1992). To accomplish this, it is necessary to incorporate acid soils in intensive and profitable crop production.

For maize, there are at least two alternatives to increasing maize production in acid soils. First, to use amendments (lime, gypsum) to correct soil acidity and

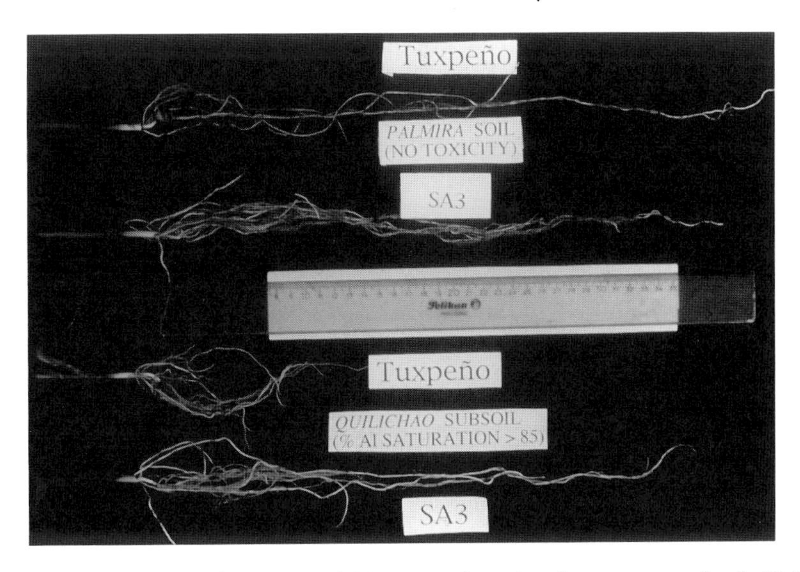

Fig. 1. Two maize cultivars (*Tuxpeño* and *SA-3* were planted under two types of soils (*Palmira* soil, a non-acidic soil, and *Quilichao* sub-soil, an acid soil). There was no difference in root length between Tuxpeño and SA-3 when grown in non-acidic soils; however, root length of Tuxpeño decreased dramatically when growing in acid soil. The effect of soil type was minimum for the root length of SA-3

second to develop tolerant cultivars. The first option is an expensive alternative not available for poor, small farmers. Also, liming subsoil deeper than 30 cm is difficult and consequently Al present below 30 cm in the subsoil does not allow roots to go deeper for nutrients and moisture. On the other hand, the use of amendments must be repeated every few years. Although both are complementary alternatives for better utilization of acid soils, the use of tolerant cultivars is a permanent solution to the problem and is relatively inexpensive for poor farmers. Additionally this is a clean and energy conserving solution (Pandey et al. 1994a).

CIMMYT, in collaboration with several NARS all over the world, has been developing soil-acidity tolerant maize cultivars to increase maize production for poor farmers in developing countries.

2. Development of Cultivars Tolerant to Acid Soils

2.1 Germplasm

The right choice of germplasm is critical in any maize breeding program for development of both hybrids and open pollinated varieties (OPVs).

CIMMYT started developing maize populations tolerant to acid soils in the 1970s when 192 cultivars from Bolivia, Colombia, Mexico, Peru, and Thailand were evaluated under acid soil conditions (pH = 4.8, 45% Al saturation, 6 mg kg^{-1} P) at Santander de Quilichao, Colombia. As a result, two full season populations (120 days to maturity) were developed: SA-3, a yellow population, and SA-8, formed by selecting the white kernels from SA-3. A complete description of the background of these populations is included in Granados et al. (1995). The focus of these populations was to serve as a source of open pollinated varieties.

Heterotic populations were developed in the 1980s in order to generate lines for hybrid development. Heterosis is known to increase yield in maize in the presence and absence of many biotic and abiotic stresses. Heterosis is exploited via hybrids where two or more genetically diverse genotypes are crossed. Genetic diversity is ensured by grouping genotypes into heterotic patterns. Heterosis is also a function of degree of dominance, which is more important in the performance of hybrids and less important in the performance of OPVs. Two heterotic groups have been formed, SA-4 (dent) and SA-5 (flint) constitute the yellow group, while SA-6 (dent) and SA-7 (flint) constitute the white group. The white and yellow color populations were developed to satisfy the food and feed needs in different parts of the world. Yellow maize is generally used as feed and as food in Asia and Latin America. White maize is used as food in Africa and is preferred as food in Latin America.

The method to develop each of these four populations was as follows:

- Extract S1 lines from different cultivars (mainly OPVs).
- Cross each yellow line with two heterotic lines (i.e. ICA-239 and ICA-240) as a tester. Heterotic testers for white lines were ICA-235 and ICA-236.
- Evaluate each of these top-crosses in replicated trials across locations, maintaining the corresponding grain color in each trial.
- Select the best lines for each tester to form the corresponding population.

SA-4 included 25 S1s showing high heterosis with line ICA-239 and low heterosis with line ICA-240. SA-5 included 31 S1 lines showing high heterosis with line ICA-240 and low heterosis with line ICA-239. SA-6 was formed with 16 S1 lines with high heterosis with line ICA-236 and low heterosis with line ICA-235. SA-7 was formed with 13 S1 lines showing high heterosis with line ICA-235 and low heterosis with line ICA-236. Pandey et al. (1995) present a detailed description of the formation of these populations.

Consequently, by the early 1990s, six acidity-tolerant maize populations were available. In order to release resources for other activities and to increase effectiveness in the hybrid program, it was decided to have only four populations (two yellow heterotic populations and two white heterotic populations). For this purpose, information of heterotic response between populations from diallel studies was used. Populations SA-3 and SA-5 are heterotic with SA-4, while population SA-7 and SA-8 are heterotic with SA-6 (Pandey et al. 1994b; Salazar et al. 1997). To merge SA-3 and SA-5 in only one population, S1s were obtained from each of these populations. Each S1 was pollinated with a bulk of pollen from population SA-4 and the top-crosses obtained were evaluated across locations in replicated trials. Based on the top-cross evaluation, selected lines were included in the new SA-3 population (yellow, flint). The corresponding heterotic population, SA-4 (yellow dent), was formed with the best S1 lines from the top-cross evaluation formed by the S1 lines from population SA-4 pollinated by SA-3. Similar criteria were used for the white populations. More information is presented by Narro et al.

(1997). As a consequence, four full-season heterotic populations are available in our program (Table 1).

In the recurrent selection scheme, new elite germplasm (lines and populations) from CIMMYT, NARS, private companies and native cultivars is being included. Lines are crossed to two heterotic testers (white or yellow) and after evaluation under acid and non-acidic soil conditions, selected ones are included in appropriate populations. For populations, S3 lines are included and crossed to heterotic testers as previously mentioned.

2.2 Improvement of Germplasm

After formation of base populations, they were improved following a recurrent selection scheme. Selection methods included half–sib, full sib, and S1 family selection or a combination of them. A half-sib family is obtained from seeds produced by the selected plant pollinated by a sample of pollen from the population (only the female parent is known). A full-sib family is produced by crossing a pair of selected plants from the population (both, the male and female parents are known). An S1 family is produced by selfing a selected plant. The genetic variance among progenies varies with the type of family in evaluation as follows: $1/4\,\sigma^2_A$ for half-sibs, $1/2\sigma^2_A + 1/4\,\sigma^2_D$ for full-sibs, and $\sigma^2_A + 1/4\,\sigma^2_D$ for S1, assuming no epistasis. These figures are important in predicting the expected genetic gain from selection.

Products of the first phase of our selection process were OPVs and synthetics. Inter-population methods are being used in the second phase of our selection program and the products are inbred lines, hybrids between lines from populations of different heterotic groups, and synthetics formed with high combining ability lines from each population.

2.3 Selection Criteria

Grain yield in acid and non-acidic soils is the main criterion in selecting the best progenies in our program. Indirect selection considering traits of easy evaluation (t_1) is used when the product of r_A (the genetic correlation between traits t_1 and t_2) and h_{t1} (h = the square root of heritability for t_1) is greater than h_{t2}, i.e., $(r_A)\,(h_{t1}) > h_{t2}$. Alternatively, indirect selection is more effective than direct selection for a trait when $[r_A(h^2 \text{ of alternate trait}/\ h^2 \text{ of trait of interest})^{1/2}]$ is greater than 1.0. In population SA-3, h^2 values for grain yield and ears per plant were 36.3 and 40.5%,

Table 1. Heterotic maize populations being improved for acid soil tolerance by CIMMYT in collaboration with NARS

Population	Grain color	Texture	Heterotic group
SA-3	Yellow	Flint	ETO
SA-4	Yellow	Dent	Tuxpeño
SA-6	White	Dent	Tuxpeño
SA-7	White	Flint	ETO

Table 2. Soil characteristics of locations in Colombia used for breeding and evaluation in the acid soil CIMMYT project

Location	pH	Al sat. (%)	P (mg kg^{-1})
S. de Quilichao	4.0	65	10
Villavicencio 1	4.6	50	10
Villavicencio 2	4.6	65	10
Carimagua	4.9	60	10
Matazul-1	4.4	70	10
Matazul-2	4.4	60	10
Palmira	7.1	0	60

respectively, and r_A between these traits was 0.84. Consequently, selection for number of ears per plant was only 88.7% [i.e. 0.84 $(40.5/36.3)^{1/2} = 88.7$] as effective at improving yield as direct selection for yield (Pandey et al. 1994a).

High variability, and consequently high experimental errors in field trials, are characteristics of acid soils, which reduce the precision of obtaining genetic differences among progenies in the selection process. Neither selection in nutrient solutions nor in greenhouse pots has been more efficient to improve grain yield for acid soils than direct selection using field plots considering grain yield as the selection criteria (Urrea-Gomez et al. 1996; Kasim et al. 1990; Magnavaca et al. 1987a).

Genetic and phenotypic correlations for grain yield were calculated to determine if selection in one location is more efficient than selection in another location. Results show that selection based on data from a given location is more efficient in improving grain yield at that location. Coefficients of genetic correlation between grain yield under acid and fertile soils were low but positive, which is an indication that selection for grain yield based on data from both types of soils is favorable for increasing grain yield in acid soils. Additionally, regeneration of progenies in fertile soil allows obtaining greater amounts of seed and, therefore, increases the number of trials for progeny evaluation (Pandey et al. 1994a). Therefore, seed increase and crosses are made on fertile soils at CIAT-Palmira, Colombia, while evaluation of progenies is done at CIAT-Palmira, three experimental stations with soil acidity in Colombia (Table 2), and other collaborative NARS in several countries.

2.4 Genetic Gain

Gain from selection (Gs) is product of heritability (h^2 = ratio of additive genetic variance and phenotypic variance) and selection differential (S = the difference between the mean of the selected portion of the population and the mean of the population). An increase in h^2 and/or in selection differential increases gain from selection. Accordingly:

$$Gs = h^2\, S,$$

being $h^2 = (\sigma^2_A/\sigma^2_{Ph})$, and $S = \bar{X}s - \bar{X}$.

If truncation selection is used, as usually occurs, S can be expressed in terms of phenotypic standard deviations, i.e., $k = S/\sigma_{Ph}$ and consequently $S = k\,\sigma_{Ph}$. We can rewrite the Gs equation as follows:

$$Gs = k\,\sigma^2_A/\sigma_{Ph}$$

It is important to consider in any recurrent selection scheme if selection includes only male or female gametes or both sexes and the time used to complete a cycle of selection; therefore, two additional terms (ci and Y, respectively) related to these concepts are added in the Gs formula. The new formula will be:

$$Gs = k\,ci\,\sigma^2_A/\sigma_{Ph}\,Y$$

Consequently, Gs increases selecting for both sexes, using a high selection intensity and few years per cycle, and allowing the estimation of large σ^2_A and low σ_{Ph}.

Several studies show that selection has been effective in increasing grain yield and other agronomic traits in maize for acid soils. Magnavaca et al. (1987a) reported gains after four cycles of full-sib selection in the field in the "Composto Amplo" population. Although grain yield was greater for cycle 4, root length under nutrient solution was greater for C0 than for C4. Lima et al. (1992) reported two cycles of divergent selection for Al tolerance in the IAC-Taiuba maize population based on radicle length. An average change of 5.2 cm (26.1%) was observed in radicle length after positive and negative selection. The effects of selection were also observed on yield (1.17 t/ha or 15.1%), plant height (0.10 m or 5.6%), and ear height (0.03 m or 3.6%). Granados et al. (1993) reported grain yield gains of 2 and 14% per cycle after 14 cycles of modified ear-to-row and two cycles of full-sib selection, respectively, in the population SA-3. Ceballos et al. (1995) reported yield improvements of 4.72% across five tropical maize populations evaluated at three locations. The advanced cycles of selection show better performance (grain yield and other agronomic traits) than initial cycles, both under acid and non-acidic soil conditions which suggests that high grain yield in acid soils does not mean low grain yield in fertile soils. Therefore, it is possible to simultaneously increase grain yield for both environments.

2.5 Inheritance Studies

Several researchers have reported on the inheritance of tolerance to acid soils in maize. Rhue et al. (1978), Stockmeyer and Everett (1979), Miranda et al. (1984), and Prioli (1987) concluded that one or few dominant genes control the tolerance to Al. Garcia et al. (1979) reported a single dominant gene with possible action of modifiers, as responsible for the Al tolerance in maize. On the other hand, Sawasaki and Furlani (1987), Magnavaca et al. (1987b), and Lima et al. (1992) have reported quantitative inheritance for yielding capacity of maize under acid soils. Studies in our program show the following results:

Duque-Vargas et al. (1994) studied the inheritance of tolerance to soil acidity in population SA-3 using North Carolina Design I. For this purpose, each male parent was mated to four females resulting in 256 full-sib progenies for evaluation (i.e. 64 males) grouped in 8 sets. Trials were planted in Colombia during 1990-1991 using two replications under one normal and three acidic soil environments. Across acid soils, additive genetic variance (σ^2_A) was similar to dominance variance (σ^2_D) for grain yield (0.15 vs. 0.12); σ^2_D was significantly higher than (σ^2_A) for days to silk (2.98 vs. 1.26), while the opposite was true for plant height (σ^2_A = 47.97, σ^2_D = 6.83), and ears per plant (σ^2_A = 0.01, σ^2_D = 0.00) (Table 3). Additive genetic variance is largely exploited in selection and is an important component in the performance of OPVs. Dominance genetic variance is largely exploited in hybridization and is an important component in the performance of hybrids. Magnitude of interaction of these components with environments is indicative of stability of performance of OPVs and hybrids. On the other hand, additive × environment interaction was an important component of genetic variance. Also, the magnitudes of additive and additive x environmental variances and of additive genetic correlations (r_A's) among the environments suggested that recurrent selection, based on multi-location testing, would be effective in improving grain yield under acid soils. Also, yield showed the highest positive correlation (r_A = 0.84**) with ears per plant under acid soils (Pandey et al. 1994a).

Borrero et al. (1995) reported the results of the North Carolina Design II study in the populations SA-4. One-hundred-sixty S1 lines were randomly assigned to 20 sets of eight lines. Within each set, four lines were arbitrarily designated as males and four as females. Each male was crossed with each female in the set, providing 16 full-sib families. The 20 sets, with 16 full-sib families each, were field planted during 1992 in Colombia and Brazil in two replications in one non-acidic and four acid soil environments. Across acid soils, σ^2_D was significantly greater than σ^2_A for yield (0.22 vs. 0.09) and days to silk (2.33 vs. 1.35) and lower for ear height (23.93 vs. 37.48), and ears per plant (4.86 vs. 13.68). Magnitudes of σ^2_A and σ^2_D and their interactions with environments indicated that reciprocal recurrent selection, using multilocation testing, would be effective for developing improved maize cultivars for acid soils (Table 3).

The differences between the genetic variances in SA-3 and SA-4 may be due to the different genetic make-up of the populations as well as to different matting designs and environments used in their estimation.

Pandey et al. (1994b) reported the results of a diallel study of six soil-acidity tolerant (CIMCALI 90: SA-3, SA-4, SA-5, SA-6, SA-7, and CMS-36) and two susceptible (Tuxpeño and Pool 26) parents evaluated in seven acid soil environments. The tolerant parents averaged higher yield than the susceptible parents (2.19 vs. 1.58 t/ha) (Table 4). Parents vs. crosses mean squares were highly significant indicating heterosis and non-additive gene effects for this trait. The crosses between tolerant parents averaged higher in yield (3.00 t/ha) than those between tolerant and susceptible parents (2.40 t/ha) and between susceptible parents (2.01 t/ha), indicating polygenic inheritance for this trait.

A diallel study, involving eight segregating populations and their 56 crosses (direct and reciprocal), was conducted in five acid soil environments to determine the relative importance of nuclear and cytoplasmic factors for yield, days to silk, ear height, ears per plant, and ear rot. Average and specific heterosis accounted for 65 and 31% of the total sum of squares for heterosis for yield. Population heterosis effects for yield were not significant suggesting that these effects would be of little value in selecting populations for developing superior hybrids. The absence of reciprocal differences for all traits indicated that tolerance to soil acidity was controlled by nuclear genes (Salazar et al. 1997).

Other diallel studies involved eight S6 lines with different reaction to acid soils (2 tolerant, 2 susceptible, 2 moderate tolerant, and 2 moderate susceptible) and their 28 crosses. Grain yield was the criterion used to classify reaction to acid soils of the lines and consequently there was a close association between the reaction to soil acidity and the general combining ability, i.e., the more tolerant lines had the highest combining ability. The high heterosis among these lines was the most important phenomenon in this study. Results of four acid soil environments showed a heterosis of 207% for line 6, averaged across the other seven lines, ranging from 166.7 with line 8 to 261% with line 5. Line 6 had the highest general

Table 3. Genetic parameters for grain yield, days to silk, plant height, and ears per plant in populations SA-3 and SA-4 evaluated across acid soil environments in Colombia

Parameters	Yield	Days to silk	Plant height	Ears per plant
Population SA-3				
σ^2_A	0.15 ± 0.08	1.26 ± 0.81	47.97 ± 17.94	0.010 ± 0.005
σ^2_D	0.12 ± 0.05	2.98 ± 0.65	6.83 ± 8.95	0.000 ± 0.002
$\sigma^2_A \times E$	0.37 ± 0.08	3.43 ± 0.71	66.60 ± 12.68	0.027 ± 0.005
$\sigma^2_D \times E$	-0.63 ± 0.08	-6.11 ± 0.74	-122.20 ± 11.81	-0.051 ± 0.004
h^2	0.36 ± 0.20	0.31 ± 0.20	0.52 ± 0.19	0.405 ± 0.198
Population SA-4				
σ^2_A	0.09 ± 0.03	1.35 ± 0.44	37.48 ± 17.56	13.68 ± 4.01
σ^2_D	0.22 ± 0.05	2.33 ± 0.58	23.93 ± 6.71	4.86 ± 5.30
$\sigma^2_A \times E$	0.13 ± 0.05	1.95 ± 0.75	17.63 ± 8.50	14.79 ± 8.55
$\sigma^2_D \times E$	-0.11 ± 0.08	-0.91 ± 1.06	-18.16 ± 12.48	-1.73 ± 13.20
h^2	0.39 ± 0.14	0.43 ± 0.14	0.66 ± 0.13	0.48 ± 0.14

σ^2_A, additive variance; σ^2_D, dominance variance; $\sigma^2_A \times E$, additive × environment interaction variance; $\sigma^2_D \times E$, dominance × environment interaction variance; and h^2, heritability

Table 4. Means of parents (*diagonal*), crosses (*above diagonal*) and high parent heterosis (*below diagonal*) for grain yield (t/ha) across seven acid soil environments

Pedigree	90SA-3	90SA-4	90SA-5	90SA-6	90SA-7	CMS-36	Tuxpeño	Pool 26
90SA-3	2.66	3.10	2.72	2.83	2.97	2.96	2.89	2.59
90SA-4	16.6*	2.33	2.65	2.81	2.67	3.02	2.49	2.38
90SA-5	2.4	13.4	2.07	2.49	2.35	2.74	2.29	2.31
90SA-6	6.4	20.3*	20.5*	2.02	2.40	2.90	2.13	2.26
90SA-7	11.9	14.3	13.8	18.8*	2.01	2.84	2.41	2.19
CMS-36	11.5	29.6*	32.7**	42.5**	39.4**	2.04	2.63	2.28
Tuxpeño	8.9	6.6	11.0	5.2	20.2	29.1	1.61	2.01
Pool 26	-2.4	2.1	11.7	11.8	9.1	11.9	24.7	1.55

* $P < 0.05$; ** $P < 0.01$.

combining ability and also the highest yield in combination with the other lines, ranging from 2.71 t/ha with line 8 to 3.67 t/ha with line 5 (Table 5). It is important to point out that line 6 is the only line from a different population (SA-5), the other lines being from population SA-4; consequently, difference in gene frequencies when combining with the other lines could be the reason for the observed heterosis.

2.6 Evaluation of Cultivars

Average yield of cultivars developed for acid soil conditions has increased dramatically since our project started. Although no yield data of cultivars used to form the base population in 1977 are available, Granados et al. (1993) reported that from 192 genotypes (about 8500 plants) initially screened, only 70 open pollinated ears were selected and 45 genotypes with good vegetative development, but poor ear development, were identified. This means that the average yield was too low, probably lower than 400 kg/ha, the average yield obtained with land race cultivars under acid soil conditions. Recurrent selection and simultaneous cultivar development was started after the base populations were developed. The first experimental cultivars were formed in 1988 recombining the 10 selected S1 progenies in the recurrent selection scheme. Cultivar formation and evaluation is a continuous process that allows furthering improvement of the germplasm each cycle. Different types of cultivars have been developed: OPVs, synthetics, non-conventional and conventional hybrids, inbred lines, etc. Cultivars generated from 1988 to 1991 were evaluated in four non-acidic and 20 acid soil environments. In non-acidic soil environments, there were no significant differences between our best experimental variety and the best check for the location where the trial was evaluated (6.64 and 6.73 t/ha, respectively). On the average of 20 acid soil environments, average yield of the best acid soil variety was 2.83, and 2.44 t/ha for the best check. In 15 acid soil environments, our best variety yielded more than the

Table 5. Means of inbred lines with different reaction to soil acidity (*diagonal*), crosses (*above diagonal*) and high parent heterosis (*below diagonal*) for grain yield (t/ha) across four acid soil environments

Line	1	2	3	4	5	6	7	8
1	1.19	2.44	2.46	3.31	1.22	3.53	1.06	2.42
2	106	0.72	1.20	0.98	2.65	3.29	2.62	1.45
3	107	67	0.52	0.78	2.72	3.37	2.82	1.66
4	179	36	49	0.52	2.65	2.85	2.60	1.08
5	3	180	188	180	0.95	3.67	1.04	2.37
6	198	222	231	180	261	1.02	3.40	2.71
7	-11	124	141	122	-11	191	1.17	2.52
8	104	102	221	106	150	167	115	0.40

Pedigree of lines: (1) SA-4-HC7-1-5-1-3-4-7BB (2) SA-4-HC7-1-4-1-2-1-7 BB (3) SA-4-HC7-1-4-1-2-5-2 BB (4) SA-4-HC7-1-4-1-2-10-5 BB (5) SA-4-HC7-1-5-1-3-7-5 BB (6) SA-5-HC1-3-9-2-1-3-4 BB (7) SA-4-HC7-1-5-1-3-7-3 BB (8) SA-4-HC7-1-4-1-1-1-1 BB

Table 6. Yield (t/ha) of experimental varieties evaluated in 13 acidic and two non-acidic environments during 1994-1995

ENTRY	Carimagua	Santander Quilichao 1	Santander Quilichao 2	Villavicencio 1	Villavicencio 2	Brazil	A. Mayo Peru	Guarico Venezuela	Thailand 1	Thailand 2	Ivory Coast	Bolivia 1	Bolivia 2	Palmira 1	Palmira 2	Mean acidic sites	Mean non-Acidic sites
Cimcali 93SA3	3.93	2.77	3.08	2.67	3.23	3.96	1.72	3.15	3.51	3.76	4.99	2.33	1.69	4.94	7.56	2.97	6.25
Cimcali 93SA4	3.99	1.82	2.42	2.52	2.74	4.85	1.54	3.34	4.89	4.79	5.93	2.96	1.74	5.53	7.71	3.21	6.62
Cimcali 93SA5	3.47	3.00	2.58	1.84	2.29	3.35	2.07	2.44	4.26	4.18	4.88	3.44	2.00	4.92	7.43	2.86	6.17
Cimcali 93SA6	4.15	2.36	2.48	1.92	2.21	4.76	1.98	1.54	4.64	4.71	6.20	3.15	2.03	6.92	8.05	3.11	7.48
Cimcali 93SA7	4.00	1.29	1.80	1.31	2.62	3.88	1.54	2.69	4.24	3.58	6.28	3.07	1.66	4.85	7.20	2.70	6.03
Cimcali 93SA8	3.42	2.66	2.42	2.32	2.51	3.08	1.35	2.63	3.61	4.10	5.51	2.96	1.70	4.66	7.99	2.82	6.32
Tuxpeño Seq.	3.53	1.92	0.94	2.32	1.64	2.92	0.62	2.17	3.30	3.75	4.91	2.35	1.47	3.92	6.43	2.31	5.17
Local Check 1	3.67	1.95	1.67	1.72	1.35	5.24	1.25	3.59	5.90	6.43	6.01	2.94	1.67	2.79	5.87	3.21	4.33
Local Check 2		0.65	1.78	0.94	1.70	5.57	1.51	3.05	5.48	5.46	5.93	2.07	2.04	4.59	5.59	2.95	5.09
Local Check 3	3.83	2.40	2.89	2.06	2.64	6.78	0.78	1.86	4.08	4.22	5.74	1.28	1.24	4.91	5.81	3.04	5.36
Mean Acidity tolerant EV's	3.83	2.32	2.46	2.09	2.60	3.98	1.7	2.63	4.19	4.19	5.63	2.98	1.80	5.30	7.66	2.95	6.48
Best tolerant EV/ Best check (%)	108	125	107	131	122	72	138	93	83	74	105	117	100	141	137		
Best tolerant EV/ Tuxpeño Seq. (%)	118	157	283	116	197	166	334	154	148	128	128	146	138	176	125		
CV(%)	21.0	32.0	22.6	17.5	17.3	21.3	23.8	17.3	12.8	19.2	11.4	27.9	22.7	13.3	10.9		
DMS (0.05)	1.15	0.99	0.72	0.50	0.58	1.38	0.53	0.66	0.81	1.26	0.93	1.08	0.57	0.93	1.10		

local check; for instance, in Vietnam yield of CIMCALI 91 SA-5 was 5.37 t/ha, while yield of the best local check was 4.62 t/ha (CIMMYT, 1995).

Cultivars generated in 1993 were evaluated during 1994 and 1995 in two non-acidic environments and 13 acidic environments using Tuxpeño, a susceptible cultivar, as the common check in all environments, in addition to local checks for each location (Table 6). In two trials at Palmira, Colombia, a non-acidic environment, the yield of the best acid soil tolerant cultivar (CIMCALI 93 SA-6 = 6.92 and 8.05 t/ha for each trial, respectively) was 76 and 25% higher than yield of Tuxpeño (3.92 and 6.43 t/ha). Grain yield of the best tolerant cultivar was higher than of the best check at nine acid soil environments (ranging from 7 to 38%). In three locations (Brazil, and two in Thailand) yield of the best checks were significantly higher than of the best acid soil experimental cultivar (Table 6). The main reason for these differences is that local checks were adapted hybrids with obvious higher yield potential than the acid soil tolerant OPVs evaluated in those trials. Based on this information, the first two new maize varieties tolerant to acid soils, Antasena

Table 7. Grain yield of open pollinated varieties (OPV), non-conventional hybrids (NCH), two common checks and two local checks, evaluated in six acid soil environments and three non-acid soil environments during 1996-97

Pedigree	Acid soil t/ha	Non-acid soil t/ha
OPV		
CIMCALI 95B SA-3	2.85	3.09
CIMCALI 95B SA-4	2.53	3.02
CIMCALI 95B SA-6	2.97	3.77
CIMCALI 95B SA-7	2.16	2.33
CIMCALI 95B DMR (10)	1.58	1.36
CIMCALI 95B DMR (20)	1.81	1.57
CIMCALI 95A SA-4	3.18	3.32
CIMCALI 95A SA-8	2.70	2.90
NCH		
SA-3 C6HC91-6 × CM93 SA-4	3.85	4.56
SA-5 C4HC3-10 × CM93 SA-4	3.37	4.42
SA-4 C6HC17-14 × CM93 SA-3	3.73	4.31
SA-4 C6HC17-6 × CM93 SA-3	3.74	4.30
SA-6 C4HC16-1 × CM93 SA-8	3.83	5.12
SA-6 C4HC24-9 × CM93 SA-8	3.65	4.89
SA-8 C3HC49-3 × CM93 SA-6	4.53	5.48
SA-8 C3HC114-14 × CM93 SA-6	4.01	4.55
Checks		
Sikuani	3.07	3.65
Tuxpeño	2.51	3.86
Local check 1	3.10	4.80
Local check 2	3.64	5.54

and ICA-Sikuani V 110 (Sikuani), were released in 1994 by the national programs of Indonesia and Colombia, respectively.

A trial including eight OPVs developed in 1995, eight new non-conventional hybrids (NCH = Line x variety), two common checks (Sikuani and TuxpeĖo), and two local checks was planted at six acid-soil environments and three non-acidic soil environments. Grain yield of NCH hybrids was greater than of the OPVs under acid and non-acidic soil environments (56 and 76%, respectively). In acid soil environments, there was no difference between the best OPV and Sikuani (3.18 and 3.07 t/ha, respectively); the best NCH yielded 4.53 t/ha or 48% more than Sikuani and 24 % more than the best local check. In non-acidic soil environments, grain yield of Sikuani, Tuxpeño, and the best OPV were similar (about 3.70 t/ha); the best NCH yielded 5.48 t/ha or 45% more than the best OPV. SA-8 C3HC49-3 x CIMCALI 93SA-6 was the best NCH in acid and non-acidic soils (Table 7). Based on the available information, we have concluded that heterosis can be exploited to increase maize yields in acidic soils.

3. Agronomic and Systems Research

In evaluations of tolerant OPVs under different levels of Ca and P, Sikuani, the tolerant OPV was compared with the susceptible cultivar, Tuxpeño. At 31% Al saturation and 0 kg/ha of P_2O_5, Sikuani yielded 3.98 t/ha compared to 3.57 t/ha of Tuxpeño (i.e., 11% more). At 31% Al saturation but 90 kg/ha of P_2O_5 Sikuani yielded 5.77 t/ha vs. 4.80 t/ha of Tuxpeño (i.e., 20% more). At 64% Al saturation and 0 kg/ha of P_2O_5, Sikuani yielded 3.05 t/ha and Tuxpeño yielded 1.70 t/ha (i.e., 79% more). At the same level of Al saturation and 90 kg/ha of P_2O_5, Sikuani yielded 4.19 t/ha vs. 3.20 t/ha of Tuxpeño (i.e., 31% more). These results show the superiority of Sikuani over a range of acidities and demonstrate its increasing superiority with

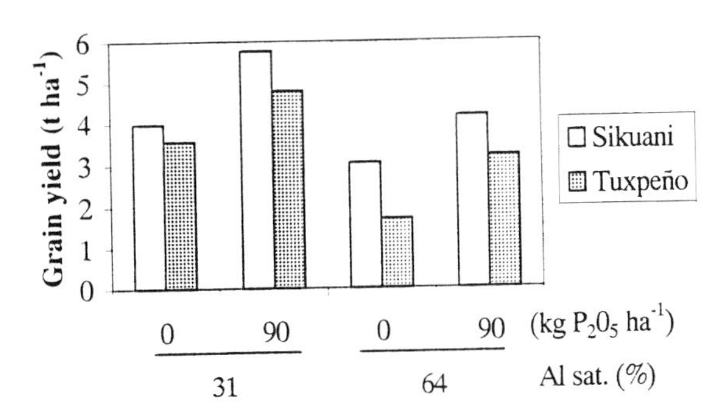

Fig. 2. Sikuani and Tuxpeño were evaluated under two levels of Al saturation (31% and 64%) and two levels of P_2O_5 (0 and 90 kg/ha). Grain yield of Sikuani was higher than that of Tuxpeño in all the treatments. On the average, grain yield increased 50% using 90 kg of P_2O_5 (3.07 vs. 4.49 t/ha for 0 and 90 P_2O_5, respectively) or applying lime to decrease Al saturation to 31% (3.04 vs. 4.53 t/ha for 64 and 31% Al saturation, respectively)

increasing levels of acidity stress (Fig. 2).

Comparisons of the effect of amendments, i.e., dolomitic lime (57% $CaCO_3$ + 35% Mg CO_3) and Sulcamag (25% CaO, 12% MgO, and 8% S) and four methods of application to reduce soil acidity were evaluated. The amendments were placed in hills, applied in bands, broadcasted, and banded + broadcasted. Maize grain yield with sulcamag and dolomitic lime was 3.5 and 1.6 t/ha, respectively, indicating that when sulcamag was used, grain yield was double that than when dolomitic lime was used. Sulfur particularly helps in sandy acidic soils. There were no differences among the methods and methods x amendments interactions were also non significant. Thus, broadcasting, which is the cheapest and easiest method was as good as any.

In another trial, an experimental hybrid and Sikuani were grown at 55% Al saturation. At 0 kg/ha of P_2O_5, the hybrid yielded 1.58 t/ha and Sikuani 1.47 t/ha (i.e. 7% more). At 80 kg/ha of P_2O_5, the hybrid yielded 4.60 t/ha and Sikuani 4.00 t/ha (i.e., 15% more). This shows superiority of the hybrid over the OPV and also the greater responsiveness of the hybrid over the OPV to higher fertility levels (Fig. 3).

A study of four micronutrients (B, Zn, Mn, and Cu) using the Sikuani variety was performed in the Colombian savanna (Matazul) during 1997. Constant to the trial were applications of dolomitic lime to correct Al saturation to 60%, application of 10 mg kg^{-1} of P, and fertilization with 100 and 60 kg/ha of N and P_2O_5, respectively.

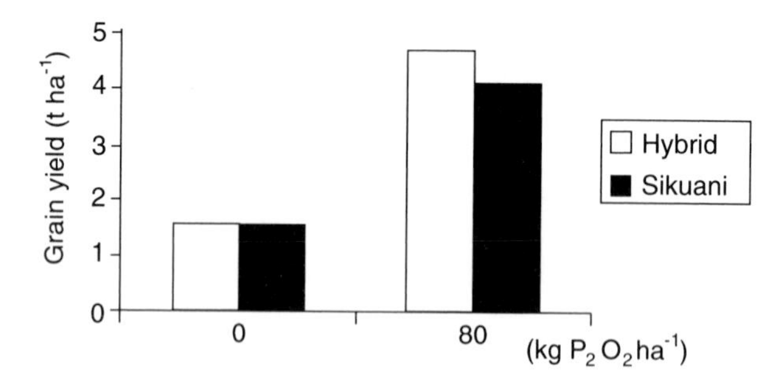

Fig. 3. An experimental hybrid and Sikuani were evaluated under two levels of P_2O_5 (0 and 80 kg/ha). On the average, grain yield was almost three fold when using P_2O_5 (1.52 vs. 4.30 t/ha for 0 and 80 kg/ha of P_2O_5, respectively)

Table 8. Levels of microelements evaluated on Sikuani cultivar in acid soils

	Microelement/levels (kg/ha)			
	B	Zn	Mn	Cu
Treatment	1.23	8.8	5.2	1.92
F	✔	✔	✔	✔
F-B	–	✔	✔	✔
F-Zn	✔	–	✔	✔
F-Mn	✔	✔	–	✔
F-Cu	✔	✔	✔	–
B	✔	–	–	–
Zn	–	✔	–	–
Mn	–	–	✔	–
Cu	–	–	–	✔

✔, microelement included in the treatment; –, microelement not included in the treatment

Doses for each micronutrient were 1.23, 8.80, 5.2, and 1.92 kg/ha of B, Zn, Mn, and Cu, respectively. The nine treatments of micronutrients are included in Table 8. A clear response to Zn application was observed; when the mixture excluded Zn (i.e. F-Zn) the grain yield was only 0.29 t/ha; in contrast, when only Zn was applied the grain yield increased to 1.70 t/ha. Grain yield was 2.36 t/ha when the four micronutrients were included (Fig. 4). Interactions between micro and/or macronutrients must be evaluated in the future to increase grain yields of maize. Certainly, enhancement of Zn-use efficiency will be a worthwhile goal for such soils.

Although low fertility is characteristic of acid soils, these vast areas have a large proportion of favorable topography for agriculture, adequate temperatures for plant growth throughout the year, sufficient moisture availability year-round in 70% of the region, and for 6-9 months in the remaining 30%. When the chemical constraints are eliminated by liming and using adequate amounts of fertilizers, the productivity of Oxisols and Ultisols are among the highest in the world (Sanchez and Salinas 1981). However, market conditions and transportation facilities are also limiting factors in increasing productivity of these soils and most of these areas are actually occupied by rain forest or native pastures. Much of the acid savannas are under degraded pastures that support one animal/10 ha/yr. With weight gains of about 300 g/day/ animal on such pastures, it gives a farmer 0.3 kg/ day $\times$ 365 days = 110 kg of meat or about 110 US dollars. Since this is the gain over 10 ha of pastures, the farmer really gains only 11 dollar/ha/yr. In comparison, improved grass pastures support up to two animals/ha, or 20 times more than the native pastures. Weight gain/animal is also higher at about 500 g/day. Thus, total weight gain/ha with improved pastures is 2 animals $\times$ 0.5 kg/day/animal $\times$ 365 days = 365 kg or 365 US dollars. The mixtures of grass and legume pastures also support two animals/ha but animal weight gains due to legumes are 10-15% higher, giving nearly 400 dollars of profit/ha/ year.

It is clear that planting improved pastures would help raise more animals/ha and increase farmer's income, but upgrading pastures is very expensive. One could plant maize with improved pastures. When ready, maize can be harvested and sold. The upgraded pasture would be ready for the animals. Would maize allow pastures to grow and pay for the upgrading costs of the pastures? To answer this question, maize was grown in monoculture, in association with grass pastures, and a mixture of grass and legume pastures. In every cropping system, maize was an economic crop for the acid savannas. In fact, the grass and legume mixtures,

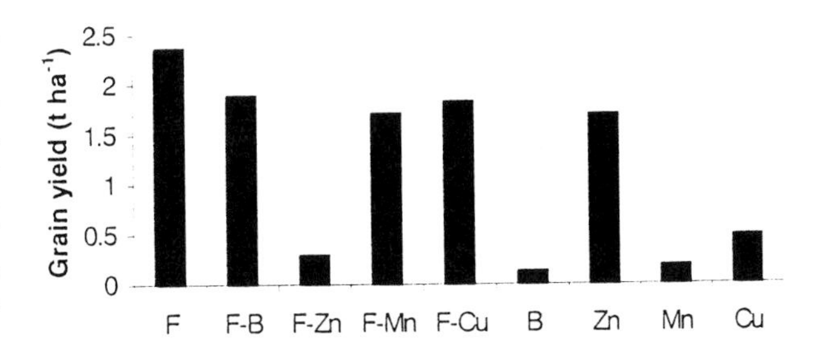

Fig. 4. Response of Sikuani to application of micronutrients (B, Zn, Mn, and Cu) was evaluated. A significant response to Zn was observed. Grain yield was 2.36 t/ha when the four microelements were included. Grain yield decreased to 0.29 t/ ha when Zn was not used. When only Zn was included, grain yield was 1.70 t/ha

when *Panicum* was used, supported up to 20 animals/ha the first year, increasing the first year profits by 10 times, finally the second year stabilizing to two animals/ha. A big advantage with maize is that, unlike rice or other crops, maize does not need to be transported for sale or processing. It can be fed to the animals and consumed right where it is produced. This is particularly useful in the acid savannas where there is limited infrastructure and roads are not available.

4. Future Studies

CIMMYT will continue to develop OPVs and hybrids to increase maize grain yields in both acid and non-acidic environments. Population improvement and pedigree selection will support the above activities. We will continue to look for molecular and physiological traits or markers that can enhance our ability and efficiency to develop superior cultivars. For such studies, we will seek collaboration from advanced research institutions. Since acid soils are low in fertility and fragile, work, in collaboration with agronomists and soil scientists, will continue to develop sustainable maize-based systems. Thus, future work will focus on efficiency and sustainability through development and deployment of acidity tolerant maize cultivars.

References

Borrero J, Pandey S, Ceballos H, Magnavaca R, Bahia Filho A (1995) Genetic variances for tolerance to soil acidity in a tropical maize population. Maydica 40: 283-288

Ceballos H, Pandey S, Knapp E, Duque J (1995) Progress from selection for tolerance to soil acidity in five tropical maize populations. In: Date R et al. (Ed) Plant soil interaction at low pH: Principles and management. Kluwer Academic, Dordrecht, The Netherlands, pp 419-124

CIMMYT (1992) 1991-92 CIMMYT World maize facts and trends: Maize research investment and impacts in developing countries. CIMMYT, Mexico, D.F

CIMMYT (1995) South American Maize Program. CIMMYT Annual Research Report. Palmira, Colombia

Clark R, Flores C, Gourley L (1988) Mineral element concentrations in acid soil tolerant and susceptible sorghum genotypes. Commun Soil Sci Plant Anal 19:1003-1017

Duque-Vargas J, Pandey S, Granados G, Ceballos H, Knapp E (1994) Inheritance of tolerance to soil acidity in tropical maize. Crop Sci 34:50-54

Foy C (1988) Plant adaptation to acid, aluminum-toxic soils. Commun Soil Sci Plant Anal 19:959-987

Foy C, Chancy R, While M (1978) The physiology of metal toxicity in plants. Ann Rev Plant Physiol 29:511-566

Garcia J, Silva W, Massel M (1979) An efficient method for screening maize inbreds for Aluminum tolerance. Maydica 23:75-82

Granados G, Pandey S, Ceballos H (1993) Response to selection for tolerance to acid soils in a tropical maize population. Crop Sci 33:936-940

Granados G, Pandey S, Ceballos H (1995) Registration of acid soil tolerant maize populations SA-3 and SA-8. Crop Sci 35:1236

Haug A, Caldwell C (1985) Aluminum toxicity in plants: The role of root plasma membrane and calmodulin. In: St. John J, Berlin E, Jackson P (Eds) Frontiers of membrane research in agriculture. Rowman and Allanheld, Totowa, pp 359-381

Jackson W (1967) Physiological effects of soil acidity. Agron Monogr 12: 3-124

Kasim F, Haag W, Wassom C (1990) Genotypic response of corn to aluminum stress. II Field performance of corn varieties in acid soils and its relationship with performance at seedling stage. Indonesian J Crop Sci 5:53-65

Lima M, Furlani P, Miranda Filho J (1992) Divergent selection for aluminum tolerance in a maize (*Zea mays* L.) population. Maydica 37:123-132

Magnavaca R, Gardner C, Clark R (1987a) Comparisons of maize populations for aluminum tolerance in nutrient solution. In: Gabelman H (Ed) Genetic aspects of plant mineral nutrition. Martinus Nijhoff, Dordrecht, The Netherlands, pp 189-199

Magnavaca R, Gardner C, Clark R (1987b) Inheritance of aluminum tolerance in maize. In: Gabelman H (Ed) Genetic aspects of plan mineral nutrition. Martinus Nijhoff, Dordrecht, The Netherlands, pp 201-212

Miranda L, Furlani P, Miranda L, Sawasaki E (1984) Genetics of environmental resistance and super genes: Latent aluminum tolerance. Maize Genetics Coop Newsl 58:46-48

Narro L, Pandey S, De Leon C, Perez J, Salazar F, Arias M (1997) Heterosis in acid-soil tolerant maize germplasm. In: CIMMYT 1997. Book of abstracts. The genetics and exploitation of heterosis in crops. An international symposium. Mexico D.F., Mexico, pp 290-293

Pandey S, Ceballos H, Granados G (1994a) Development of soil acidity tolerant cultivars for the tropics. Proc 15th World Congress Soil Sci, Acapulco, Mexico, pp579-592

Pandey S, Ceballos H, Magnavaca R, Bahia Filho A, Duque-Vargas J, Vinasco L (1994b) Genetics of tolerance to soil acidity in tropical maize. Crop Sci 34:1511-1514

Pandey S, Ceballos H, Granados, G (1995) Registration of four tropical maize populations with acid soil tolerance: SA-4, SA-5, SA6, and SA-7. Crop Sci 35:1230-1231

Prioli A (1987) Analise genetica da tolerancia a toxidez do aluminio em milho (*Zea mays* L.) Ph.D. thesis, Univ Campinas, Campinas, SP, Brazil. 182 pp

Rhue R, Gragan C, Stockmayer E, Everett H (1978) Genetic control of aluminum tolerance in corn. Crop Sci 18:1063-1067

Salazar F, Pandey S, Narro L, Perez J, Ceballos H, Parentoni S, Bahia Filho A (1997) Diallel analysis of acid-soil tolerant and intolerant tropical maize populations. Crop Sci 37:1457-1462

Sanchez P (1976) Properties and management of acid soils in the tropics. John Wiley and Sons, New York

Sanchez P, Salinas J (1981) Low-input technology for managing oxisols and ultisols in tropical America. Adv Agron 34:279-406

Sawasaki E, Furlani P (1987) Genetics of tolerance to aluminum in inbred lines of cateto maize (Genetica da tolerancia ao aluminio em linhagens de milho cateto). Bragantia 46:269-278

Stockmeyer E, Everett N (1979) Studies on the mechanisms of Al response in maize. Maize Genetics Coop Newsl 53: 47-48

Urrea-Gómez R, Ceballos H, Pandey S, Bahía Filho A, León L (1996) A greenhouse screening technique for acid soil tolerance in maize. Agron J 88:806-812

Von Uexkull H, Mutert E (1995) Global extent, development and economic impact of acid soils. Plant Soil 171:1-15

Zeigler R, Pandey S, Miles J, Gourley L, Sarkarung S (1995) Advances in the selection and breeding of acid-tolerant plants: Rice, maize, sorghum and tropical forages. In: Date R, Grundon N, Rayment G, Probert M (Eds) Plant-Soil interactions at low pH: Principles and management. Kluwer Academic. Dordrecht, The Netherlands, pp 391-406

A Marker-Based Approach to Improve Nutrient Acquisition in Rice

Matthias Wissuwa

Summary. Rice yields are frequently limited by soil deficiency of phosphorus (P). Increasing the P deficiency tolerance of rice cultivars may represent a more cost-effective solution to this problem than reliance on fertilizer application alone. Experimental evidence suggests that considerable genotypic variation for the ability to take up P from a low-P soil exists and that traditional varieties are likely to carry genes for high P uptake that have not been exploited in rice improvement. One reason this variation may have largely remained unnoticed is because traditional varieties, due to their low harvest index, may not show a yield advantage over modern varieties despite having superior P uptake from low-P soil. This chapter explores the possibility of transferring high P uptake ability from traditional rice cultivars to modern cultivars. The strategy followed here involves the identification of molecular markers linked to genes for P uptake. The advantage of such markers for selecting genotypes with superior P uptake, as well as in providing material for physiological studies, will be discussed.

A population of 98 backcross inbred lines derived from a cross of the modern *japonica* variety 'Nipponbare' with the traditional *indica* variety 'Kasalath' was grown on a highly P-deficient soil. Lines were genotyped at 245 RFLP marker loci and, by relating genotypic information to P uptake, four quantitative trait loci (QTL) for P uptake could be identified. One QTL linked to marker C 443 on chromosome 12 had a major effect, and lines carrying Kasalath alleles at this QTL had twice the P uptake compared to lines with Nipponbare alleles. Near isogenic lines (NILs) for QTL C443 (chromosome 12) and for QTL C498 (chromosome 6) were developed by three additional backcrosses to Nipponbare. These NILs genetically resembled Nipponbare to 90% but carried chromosomal segments derived from Kasalath at the respective QTLs. The effect of both QTLs could be confirmed using these NILs. Under P deficiency both NILs had more than twice the P uptake of Nipponbare, however, uptake was not different when genotypes were supplied with an optimum amount of P. Genes at both QTLs therefore appear to encode for mechanisms specific to P deficiency. Initial physiological studies imply that Kasalath alleles at C498 enhance root development under P deficiency.

These results show that genotypic variation for a quantitative trait such as P uptake can be attributed to several loci of known chromosomal location. Furthermore, it was demonstrated that it is possible to double the P uptake of a modern rice cultivar by transferring these loci from a donor of otherwise low agronomic value.

After fine mapping both QTLs, efforts will concentrate on cloning the underlying gene or genes. Once this has been accomplished, the transformation of modern cultivars promises to be far quicker than with traditional plant breeding methods.

Key words. Deficiency tolerance, Mapping, Near isogenic line, Phosphorus, Uptake, QTL, Rice, Traditional variety

1. Introduction

Rice yields have increased dramatically over the past three decades. This increase was mainly due to the extension and improvement of irrigation facilities, the development and widespread acceptance of modern high-yielding varieties (HYV), and the extended use of fertilizers (De Datta et al. 1990). For rainfed lowland and upland rice, however, yield improvements have not matched those of irrigated paddy rice. For example, in most Asian countries yields of lowland rainfed rice and upland rice are 60% and 30% of the level achieved under irrigation, respectively (IRRI, 1993).

The reason overall rice production could outpace population growth since the beginning of the green revolution has therefore mainly been due to productivity improvements for irrigated rice. In recent years, however, the growth in rice production seems to have slowed considerably (Swaminathan 1993; Hossain 1996). Two factors are responsible for this development. Paddy yields have apparently reached a plateau in many of the highly productive regions that had previously shown a steady increase in yields (Cassman and Pingali 1995). Farmland around urban centers, typically located in fertile river basins, has furthermore been diverted to urban and industrial uses at a growing rate (Swaminathan 1993; Brown 1995). To compensate for this trend towards stagnation of production of paddy rice and to keep up with the steadily increasing demand for rice that is driven by an ever increasing population, one of the main tasks in the future will be to find ways for improving yields under rainfed lowland and upland conditions–areas where yields have lagged far behind the genetic potential of modern varieties.

In order to devise a strategy to achieve this goal, factors that currently limit yields of rainfed lowland and upland rice need to be examined. Apart from an unreliable water supply, soil phosphorus (P) deficiency has been identified as a major factor limiting rice yields (Fageria 1980; Sanchez and Salinas 1981; Hedley et al. 1994). Application of phosphorus fertilizers can alleviate this problem and Fageria et al. (1982) showed that as little as 22 kg P ha^{-1} could dramatically increase rice yields on a Brazilian Latosol low in available P. However, the lack of locally available P sources and the high costs of importing and transporting P fertilizers frequently prevent resource-poor farmers from applying P to their deficient fields (Sanchez and Salinas 1981; Fageria et al. 1988).

In addition to low soil-P contents in absolute terms, P deficiency can arise from the high P fixing capacity of some soils (Sanchez and Salinas 1981; Dobermann et al. 1998). It has been estimated that in highly fixing soils, more than 90% of the added fertilizer-P is rapidly transformed to P forms such as aluminum and iron phosphates that are not readily available to plants. Soils of this kind are thus not P-deficient per se but contain a relatively high content of P in fixed form. Whereas these phosphate forms become partly plant-available under the anaerobic conditions prevailing with irrigation, they remain largely unavailable with aerobic soil conditions that predominate in upland rice and to a lesser extent in rainfed lowland rice soils. To overcome P deficiency on P-fixing upland soils, higher amounts of phosphorus need to be applied than on non-fixing soils to produce a significant effect on yield (Kirk et al. 1998). This low fertilizer efficiency further discourages the use of P fertilizers by resource-poor farmers.

Rather than relying on costly fertilizer inputs, the development of rice cultivars capable of producing acceptable yields on P-deficient soils may be an attractive and cost-effective alternative with positive effects on the long-term sustainability of agricultural systems. Cultivars with the ability to use fixed P sources would not only result in higher yields on P-deficient soils, but should also be more responsive to P fertilization (Kirk et al. 1998). A cultivar with an improved ability to access fixed P forms will be able to use a greater portion of the added but rapidly fixed fertilizer-P compared to unimproved cultivars. With higher fertilizer efficiency, the application of additional P may become more economical and thereby more attractive to farmers who generally have not been able to commit their limited resources to fertilizer inputs. In the long run that would be a necessary development because a sustainable system will require the replenishing of nutrients that have been removed with grains at harvest.

2. Breeding Rice with Improved Tolerance to P Deficiency

Tolerance to P deficiency can be defined in several ways. Yield produced by a cultivar under conditions of P deficiency is the most important and most commonly used measure of P deficiency tolerance. However, yield is only the end product of a series of physiological and developmental processes that are listed in schematic form in Fig. 1. In an environment where low soil-P availability is the major growth-limiting factor, P uptake is the first determinant of grain yield. Any cultivar able to take up more of the growth-limiting factor could therefore be classified as tolerant to P deficiency. Genotypic differences in P uptake can furthermore be attributed to two causal factors. Because of the low mobility of P in soil solution, P uptake will depend on root size and the amount of P taken up per unit root surface area or root efficiency (Itoh and Barber 1983; Jones et al. 1989; Otani and Ae 1996).

Yield will then depend on how efficiently a unit of the P taken up is used for grain formation. This "overall" P-use efficiency can subsequently be divided into the internal efficiency of P use for vegetative growth and for reproductive growth. The way total plant biomass is distributed between vegetative and reproductive growth, as expressed by the harvest index (HI), will have an additional major influence on yield and on the "overall" P-use efficiency.

Because of their effect on final grain yield, all factors listed in Fig. 1 can and have been used to define the P deficiency tolerance of cultivars (Fageria et al.

1988). The lack of a single well-defined definition for P deficiency tolerance may seem problematic; the use of several definitions can nevertheless be justified depending on their purpose. If the aim is to identify the best available cultivar under P-deficient conditions, grain yield will be the only meaningful definition for P deficiency tolerance. In developing new cultivars that exceed any presently available ones in yield potential on a low-P soil, however, yield alone would not provide all the desirable information to initiate a breeding cycle. It is unlikely that highest P uptake, most efficient P-use, and high HI can be found in a single cultivar. The classification of a wide variety of cultivars as P deficiency tolerant in terms of factors leading to grain yield on a low-P soil would allow choosing parents that compliment each other in components of P deficiency tolerance. Following a cross of such parents it should be possible to select lines in segregating generations that possess a more favorable combination of factors than any single cultivar. Consequently all components of P deficiency tolerance will be considered in this paper. Major emphasis will be placed on P uptake and P-use efficiency because these physiological properties represent specific adaptations to P deficiency whereas developmental factors such as HI are of more general importance and appear to be not greatly affected by P availability (Jones et al. 1989).

2.1 Looking Beyond Currently Used Modern High-Yielding Varieties

First reports of genotypic differences for P deficiency tolerance in rice appeared in the 1970's (IRRI 1971; Koyama et al. 1973). Efforts since have concentrated on screening available cultivars and advanced breeding lines for genotypes that show an above-average yield on P-deficient soil rather than on developing new cultivars that are specifically adapted to P deficiency (IRRI 1978; Fageria et al. 1988;

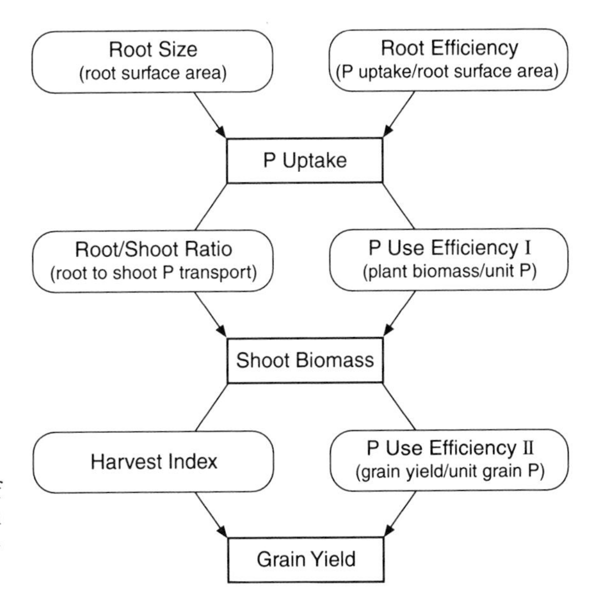

Fig. 1. Schematic presentation of factors that influence grain yield of rice under conditions of P deficiency

Hedley et al. 1994). Trials that compared modern high-yielding varieties (HYVs) to traditional varieties under moderate P deficiency generally showed that HYVs out yielded traditional varieties (Liu et al. 1987; Horst et al. 1996), but their superiority decreased with increasing P deficiency (Jones et al. 1989).

If traditional varieties do not show higher yields than HYVs over a wide range of P supply, do they justify any further attention? They may if one recalls that grain yield is only the end product of a series of processes (see Fig. 1). HYVs typically have a much higher HI compared to traditional varieties (Khush 1993; Hay 1995) and it is possible that the apparent P deficiency tolerance of HYVs was mainly caused by their high HI rather than by more specific adaptations to P deficiency. If yield is the only criterion used for evaluation, cultivars possessing efficient P uptake or P-use mechanisms may have been overlooked if they have an inherently low yield potential due to a low HI. Nonetheless, such cultivars could be used in a breeding scheme as donors of genes that would further improve the performance of modern HYVs. This approach is routinely used to incorporate resistance genes from otherwise unadapted germplasm such as traditional varieties and wild relatives (Khush 1993). Apparently such a strategy has not been pursued to improve P deficiency tolerance in rice despite some evidence from other crop species that such an approach may be worth serious consideration.

2.1.1 Case Studies in Wheat and Common Beans

Jones et al. (1989) compared modern semi-dwarf wheat cultivars to older cultivars under highly P-deficient and non-limiting P conditions. On average both groups did not differ significantly in grain yield but older cultivars had a 50% higher P uptake. In the low-P treatment, the cultivar 'Five' (released in 1840 and thereby the oldest of all cultivars tested), had more than twice the P uptake of the recently developed, highly responsive semi-dwarf type 'Sunstar' (Fig. 2a). Despite being the cultivar most tolerant to P deficiency in terms of the ability to extract P from a low-P soil, 'Five' was not able to transform high P uptake into high grain yield because of a very low HI. Consequently 'Sunstar' out-yielded 'Five' by 30% (Fig.2b,c).

Jones et al. reasoned that selection for high grain yield in modern cultivars automatically produced genotypes that are highly efficient in P use (Fig. 2d), a characteristic that is mainly due to their high HI and they saw little scope for further increases in P-use efficiency. Any further improvements in P deficiency tolerance may therefore have to rely on improving P uptake, and cultivars such as 'Five' may be valuable donors of genes associated with high P uptake.

Schettini et al. (1987) went one step further and crossed the modern bean cultivar 'Sanilac' to an exotic donor superior in P uptake. After two backcrosses to 'Sanilac' and several generations of self-pollination, they were able to produce a population of backcross inbred lines that resembled 'Sanilac' in appearance but showed considerable variation in P uptake. Sixteen percent of these lines had a significantly higher P uptake, being superior to 'Sanilac' by 50% (Fig. 3a). But whereas the donor parent was not able to transform high P uptake into seed yield, the best backcross inbred lines showed a yield increase of 30% over 'Sanilac' (Fig. 3b). Schettini et al. concluded that it was feasible to transfer a quantitative trait

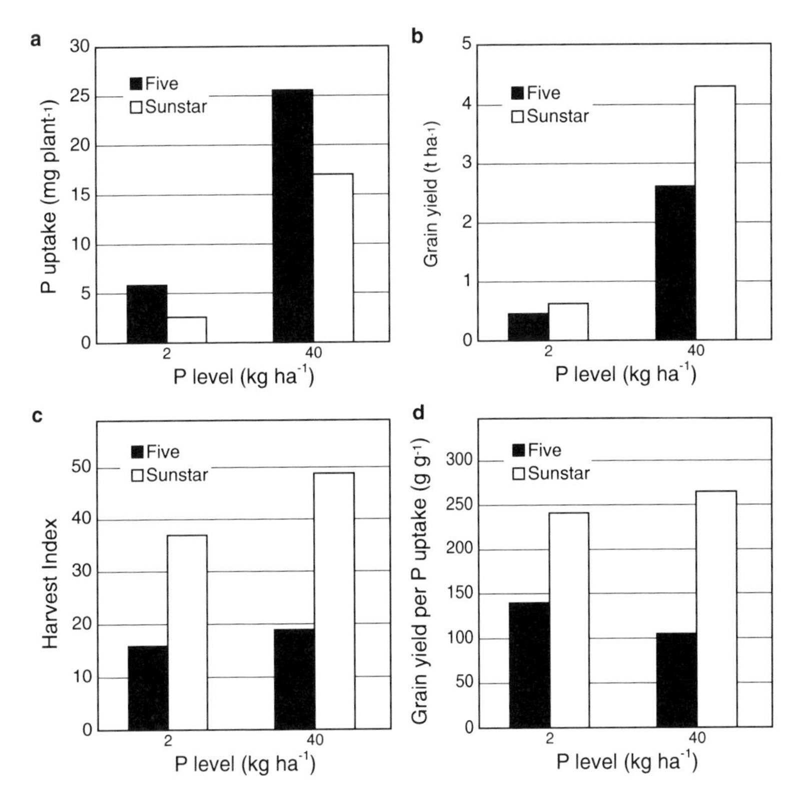

Fig. 2. Inability of the traditional wheat cultivar "Five" to transform high phosphorus (P) uptake (**a**) into grain yield (**b**) is due to a low harvest index (**c**), which leads to inefficient P use (**d**) (adapted from Jones et al. 1989)

such as high P uptake from exotic to modern cultivars, thereby increasing yield due to a combination of both high P uptake and efficient P use.

2.1.2 Variation for P Uptake from Low-P Soil in Rice

To follow a similar approach in rice potential donors of genes for high P uptake need to be identified. Genotypic variation for P uptake was assessed on more than 200 rice accessions of diverse origin and plant type that were obtained from the National Institute of Agro-Biological Resources, Tsukuba, Japan. Based on results of an initial screening experiment conducted in half-liter pots, 12 cultivars were selected for a field experiment at two P levels. Plots of the minus-P treatment had not received P fertilizers for 30 years whereas plus-P plots were supplied with 50 kg P ha^{-1} $year^{-1}$. Plants were harvested 120 days after sowing to determine the P content in aboveground plant tissue.

Among the two components of P deficiency tolerance assessed here, genotypic variation was much smaller for P use efficiency (CV = 25.6%) than for P uptake (CV = 83.1%). Similar results were reported by Fageria et al. (1988). The high variability for P uptake was reduced to a CV of 34.2% in the plus-P treatment.

Genotypic differences were highly significant in both treatments (Table 1) and the three outstanding genotypes were all traditional varieties. Interestingly their superiority increased in the minus-P treatment. Traditional varieties were thus more likely to be insensitive to P deficiency and their ability to maintain relatively high P uptake under P deficiency suggests that they possess some specific adaptation to limited P supply. These results show that considerable genotypic variation for P uptake exists among rice cultivars and that traditional rather than modern varieties may be carriers of potential donor genes for P uptake.

2.2 Developing Modern Cultivars with Higher Potential for P Uptake–the Task at Hand

The importance of improving P uptake and P-use efficiency for achieving superior rice yields under conditions of low P availability has been stressed repeat-

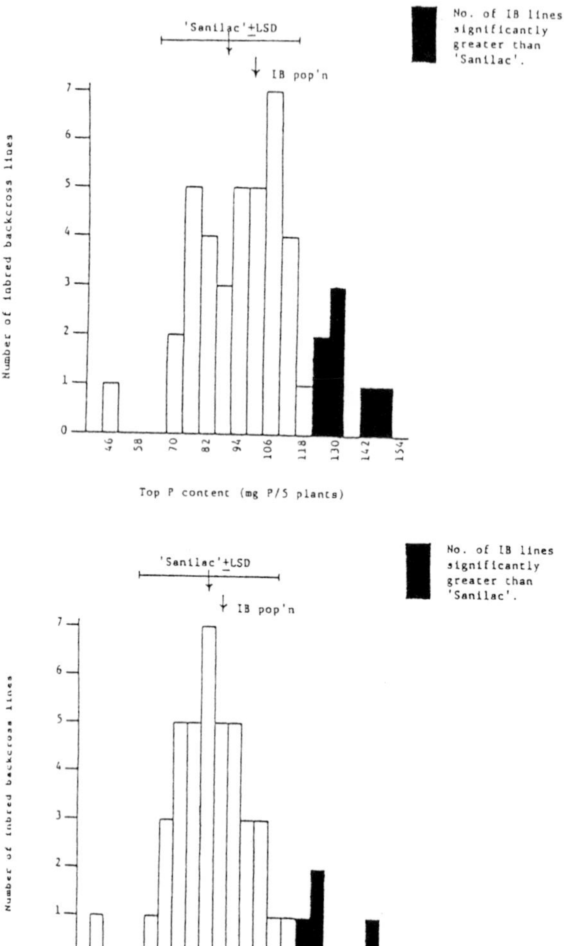

Fig. 3. Distribution of inbred backcross lines derived from a cross of Sanilac (recurrent parent) with PI 206002, a donor of genes for high P uptake, for top-P content and seed yield. Lines were grown in a field plot without added P (21 kg P ha-1, Bray-1) (Schettini et al. 1987). Reprinted with kind permission of Kluwer Academic Publishers

Table 1. P uptake from a P-deficient Andosol of 12 rice genotypes of diverse origin and plant type (Wissuwa unpublished data)

Genotype	Origin	Type of variety	P uptake[a] minus-P (mg P)	plus-P (mg P)	relative[b] (%)	P-use efficiency[c] (g dw mg P^{-1})
Morogsol	India	Traditional	13.5	36.0	37.6	1.28
Marss	Nepal	Traditional	11.3	28.9	39.1	1.35
Kasalath	India	Traditional	10.9	33.1	32.9	1.64
IAC 47	Brazil	Traditional	6.7	28.5	23.4	1.48
IR 36	IRRI	Modern	4.6	25.6	17.9	1.02
Progresso	Brazil	Modern	4.1	27.6	14.9	1.32
Basmoti	India	Traditional	3.9	27.9	14.1	1.52
Oryzica Sabana 6	Colombia	Modern	2.9	15.8	18.4	1.53
Yamadanishiki	Japan	Modern	2.8	19.6	14.4	1.86
Nipponbare	Japan	Modern	1.8	17.0	10.3	2.07
CNA 7013	Brazil	Traditional	0.5	17.6	2.6	2.07
CT 11891-2-2-7	CIAT	Modern	0.4	8.1	4.7	2.50
		HSD ($P \leq 0.05$)	2.9	13.7		0.81

P, phosphorus; IRRI, International Rice Research Institute; CIAT, Centro Internacional de Agricultura Tropical; HSD, Honestly significant difference

[a] Calculated by multiplying shoot dry weight with shoot-P concentrations of individual plants that were grown in the field in a randomized complete block design with 5 replications. The field was located in Tsukuba, Japan and had either not received P fertilizers for 30 years (minus-P) or was continuously supplied with 50 kg P ha^{-1}

[b] Shoot-P content of genotypes in the minus-P relative to the plus-P treatment

[c] Only for the minus-P treatment

edly. Studies cited above indicate the presence of considerable heritable genotypic variation for P uptake that has not yet been exploited by plant breeders. Improving the P uptake of modern cultivars therefore seems feasible. Chances of improving P-use efficiency, on the other hand, appear more modest because genotypic variation for P-use efficiency was found to be small compared to variation for P uptake (Table 1; Wade et al. 1998). In addition, limitations for further improvements of P-use efficiency do exist due to agronomic rather than physiological reasons. At harvest the largest plant-P pool is found in seeds (Dobermann et al. 1996). Improving P-use efficiency would therefore have to be mainly aimed at reducing grain-P concentrations. However, early crop establishment and ultimately grain yield of rice grown in low-P soil was found to be positively associated with high seed-P reserves (Hedley et al. 1994). To improve internal P-use efficiency by developing rice cultivars with a reduced seed-P content would therefore negatively affect subsequent field performance because seedlings would lack a sufficient amount of P reserves during early growth stages.

To increase the yield potential of rice cultivars on P-deficient soils, breeders may have to re-evaluate their current strategies and devote more resources to the evaluation of P uptake of a wide variety of genotypes including traditional varieties. In putting more emphasis on traditional varieties, the question is not if HYVs or traditional varieties are the preferable plant type in a stressful environment.

Rather than advocating the use of either plant type directly, the task at hand is to improve HYVs by using traditional varieties as donors of genes associated with superior P uptake. Such an approach may require the cooperation of breeders with plant physiologists and will take several generations of repeated backcrossing until results are finally seen in field trials. But by broadening the genetic base via the inclusion of genes for high P uptake that are currently not present in modern elite varieties it is likely that the extra effort will be justified.

3. Detection of Molecular Markers for P Uptake in Rice

The amount of time and resources required for transferring a trait from an exotic donor to modern genotypes depends on the genetic complexity of the trait and the ease of monitoring a successful transfer during backcrossing. The best-case scenario is a qualitative trait governed by a single gene that can be evaluated visibly and that is expressed regardless of environmental influences. Unfortunately, none of these conditions apply to P uptake. To identify backcross progeny carrying donor genes for high P uptake will require specific screening experiments on a P-deficient growth medium. In addition laborious laboratory analysis to determine tissue-P concentrations may be necessary unless a tight correlation between P uptake and another trait such as dry weight or grain yield can be established. P deficiency tolerance of rice and presumably P uptake are furthermore considered to be quantitative traits governed by several, mostly additive genes (Majumder et al. 1989; Chaubey et al. 1994). The difficulties of transferring quantitative traits that are hard to evaluate can be greatly reduced by identifying a molecular marker that is associated with the trait of interest and by selecting for this marker rather than for the trait that is difficult to evaluate. The availability of detailed molecular linkage maps in rice (Kurata et al. 1994) has allowed the dissection of quantitatively expressed traits into Mendelian factors referred to as quantitative trait loci (QTLs), each linked to molecular markers of known map position (Paterson et al. 1988; Yano and Sasaki 1997). The detection of putative QTLs is a crucial first step towards the identification of markers that are tightly linked to genes controlling traits like P uptake. Using such markers in a procedure called marker-assisted selection would allow breeders to monitor the progress in transferring genes for high P uptake from a donor without the need to conduct experiments under P deficiency during all stages of the backcrossing procedure. Eventually the detection of QTLs can also lead to the identification and cloning of genes controlling the trait of interest. Once this has been accomplished, genes for high P uptake could be directly transferred by molecular methods, replacing the time-consuming backcross procedure.

The objective of the study presented here was to take the first step towards this long-term goal by detecting and mapping QTLs for P uptake in a rice population derived from a wide cross of a modern *japonica* cultivar with an exotic *indica*-type traditional variety. Furthermore it will be demonstrated how results of the QTL analysis can be used for further studies in the field of plant breeding and plant physiology.

3.1 Materials and Methods

3.1.1 Plant Material

The parents used to form the rice population studied here differed greatly in P uptake from a deficient soil. Kasalath, a traditional *indica* variety of low agronomic value from Assam, India, had a six-fold higher P uptake compared to the Japanese variety Nipponbare of the *japonica* subspecies (Table 1). Kasalath was chosen as the exotic parent because other potential donors of high P uptake genes failed to produce viable offspring in a wide cross with *japonica* genotypes. After one backcross to Nipponbare to further improve the fertility of the *indica* x *japonica* hybrid, 98 backcross inbred lines (BILs) were developed by advancing BC_1F_1 lines for five generations by the single-seed descent method (Lin et al. 1998). Leaf material of the BC_1F_5 was used for DNA extraction and RFLP analysis, whereas phenotypic data were collected on BC_1F_6 lines.

3.1.2 Evaluation of Plant Performance Under Low-P Stress

Genotypic differences in P uptake are likely caused by differences in root size and by mechanisms increasing P uptake per unit root surface area (root efficiency). Screening experiments conducted in pots would more likely yield QTLs related to root efficiency because the limited volume of pots prevents the full realization of genotypic differences in root size (Otani and Ae 1996). Field screening experiments on the other hand are most likely to favor the detection of QTLs for root size due to the essentially unlimited soil volume available for root development. A compromise between pot and field was found by screening the rice population used here in a large fiberglass container of dimensions $11.60 \times 0.85 \times 0.22$ m (length $\times$ width $\times$ depth). The shallow depth of this container limits root growth but otherwise field conditions are simulated inasmuch as inter-plant competition for P by overlapping root zones is possible. The container was filled with topsoil (Humic haplic andosol) from the same field that was used to screen the rice genotypes described in Table 1. The concentration of plant-available P in the soil was very low at 1.0 mg P kg^{-1} (Truog-P) or 4.5 mg P kg^{-1} (BrayII-P) (Otani and Ae 1996). Hardly soluble Al and Fe-P were the predominant inorganic P forms and the more plant-available Ca-P was only detected in traces. Pre-germinated seeds were sown in a randomized complete block design with five replications (rows) that were spaced 15 cm apart and spacing within rows was 10 cm. Each row contained the 98 BILs plus both parents and an additional set of 12 cultivars used as standards. Plants were grown under upland conditions but drought stress was avoided by watering plants during dry spells in summer. Because lines of this population differed in maturity the experiment was harvested 125 days after planting when the majority of lines were at the early grain filling stage (before any line reached maturity).

3.1.3 RFLP Mapping and QTL Detection

A linkage map of 245 RFLP markers used for QTL detection was obtained from the Rice Genome Project, Japan. QTL analysis was performed by the composite interval mapping (CIM) method using the software package PLABQTL (Utz and Melchinger 1996). The significance threshold for QTL detection was set at a LOD score >2.80. Support intervals were determined using a LOD fall-off of 1.0. (For a more detailed description of the linkage map and the QTL detection method please refer to Lin et al. (1998) and Wissuwa et al. (1998), respectively.) In addition to P uptake as the trait of primary interest, a QTL analysis was conducted for shoot dry weight and P-use efficiency (the inverse of tissue-P concentration).

3.1.4 Development and Evaluation of Near Isogenic Lines

Near isogenic lines (NILs) were developed from selected BC_1F_2 lines of the QTL mapping population, which were then backcrossed to Nipponbare an additional three times. Genotypic monitoring and selection against Kasalath alleles was done using 118 RFLP markers. NILs were grown together with Nipponbare in 5 replications in 60-L buckets filled with the same P-deficient soil described above. After a 100-day growth period shoots were cut on July 30, 1998, oven-dried at 65°C for 5 days, and weighed. P analysis was done as described above. On the day of harvest roots of NILs were sampled by carefully separating and cleaning them under running water. Root length and surface area were determined on 1g of sample by using a digital scanner and the software package WinRhizo (Regent, Canada). In addition, the number of crown roots was determined.

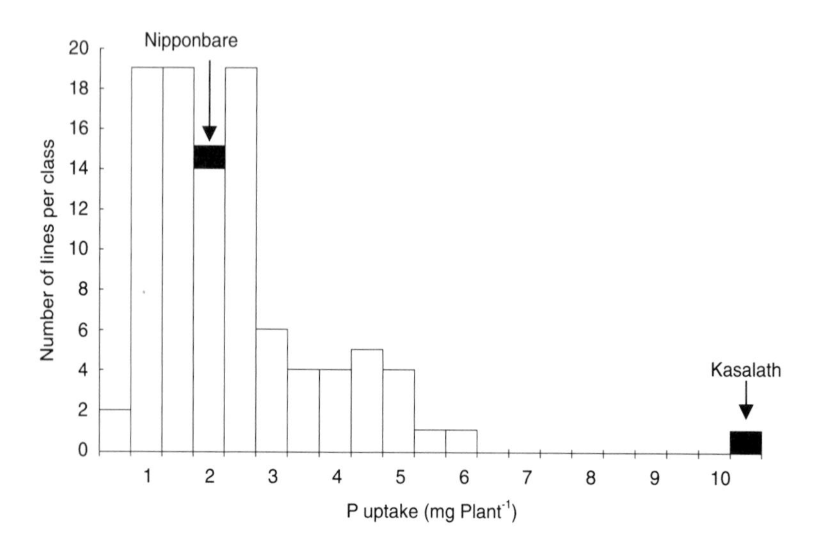

Fig. 4. Frequency distribution for phosphorus (P) uptake of 98 backcross inbred lines and their parents (Wissuwa et al. 1998)

3.2 Results

For the detection of QTLs, backcross inbred lines (BILs) were only grown in P-deficient soil without a control treatment that was fertilized with an adequate amount of P. That the low level of plant-available P was indeed the growth limiting factor on this soil had previously been shown (Table 1). P uptake was reduced by 81% and shoot dry weight by 58% relative to the plus-P control. That dry weight was less affected compared to P uptake can be attributed to increases in P-use efficiency (reduction of shoot-P concentrations).

P uptake of the 98 BILs ranged from 0.4 to 5.7 mg P plant^{-1} with a mean of 2.1 mg P plant^{-1} (Fig. 4). Nipponbare was slightly below average whereas Kasalath far exceeded any BIL with a P uptake of 10.1 mg P plant^{-1}. That none of the BILs was equal to or even exceeded Kasalath is not surprising considering that the F_1 of the original Nipponbare x Kasalath cross was once backcrossed to Nipponbare. BILs therefore genetically resembled Nipponbare, the parent with inferior P uptake, to 75% (on average).

Four putative QTLs for P uptake were detected on chromosomes 2, 6, 10, and 12 (Table 2) and except for the QTL on chromosome 10, the alleles increasing P uptake came from Kasalath. Together the four QTLs accounted for 54.5% of the variation for P uptake observed among the BILs. The QTL linked to marker C443 on chromosome 12 had a major effect. It accounted for half of the explained variation and lines carrying the Kasalath allele at this QTL had twice the P uptake compared to lines with the Nipponbare allele (Table 3). Three QTLs were detected for shoot dry weight and two of these were mapped to locations on chromosome 6 and 12 that were in agreement with QTLs for P uptake. Again the QTL linked to marker C443 had a major effect. QTLs for dry weight were not mapped to the minor P uptake QTLs on chromosomes 2 and 10, but peaks at those marker inter-

Table 2. Putative QTLs for P uptake, P use efficiency, and shoot dry weight under P deficiency

	Marker Interval[a]	Chromo some	Position[b]	LOD score	Variation[c] (%)	Positive allele[d]
P uptake	<u>G227</u> - C365	2	106	2.82	5.8	K
	<u>C498</u> - R1954	6	13	3.52	9.8	K
	<u>R1629</u> - R2447	10	32	4.70	7.7	N
	G2140 - <u>C443</u>	12	30	10.74	27.9	K
Shoot dry weight	C1488 - <u>C63</u>	3	86	3.08	6.4	N
	<u>C191</u> - C498	6	10	4.71	9.7	K
	G2140 - <u>C443</u>	12	30	10.50	26.5	K
P-use efficiency	<u>G227</u> - C365	2	106	5.22	9.8	N
	C946 - <u>R1854</u>	4	86	4.35	9.4	K
	G2140 - <u>C443</u>	12	28	6.57	19.1	N

QTL, quantitative trait loci; LOD, log 10 of the likelihood odds ratio
[a]Marker nearest to QTL is underlined
[b]Position on chromosome in cM
[c]Portion of phenotypic variation explained by QTL
[d]Allele increasing phenotype from Kasalath (K) or Nipponbare (N)

Table 3. Average P uptake of backcross inbred lines with Nipponbare or Kasalath alleles at putative QTLs

QTL	Chromosome	Nipponbare (mg)	Kasalath (mg)	Substitution effect[a] (%)
G 227	2	1.90	2.65	39.5
C 498	6	1.85	2.38	28.6
R 1629	10	2.18	1.66	-31.3
C 443	12	1.67	3.52	110.8

[a] Effect of substituting Nipponbare alleles with Kasalath alleles at respective QTL

vals were detected for dry weight as well, albeit with LOD scores below the significance threshold of 2.80. Similarly the QTL for dry weight on chromosome 3 coincided with a non-significant peak for P uptake. From the almost perfect overlap of QTLs for P uptake and dry weight it can be concluded that both traits were controlled by the same genes. Of the three putative QTLs discovered for P-use efficiency, the major QTL on chromosome 12 and the minor one on chromosome 2 coincided with QTLs for P uptake.However, whereas Kasalath alleles had increased P uptake, they reduced P-use efficiency.

3.3 Discussion

That two QTLs affected P uptake and P-use efficiency simultaneously but with opposite effects would be problematic in breeding if both traits were equally important for P deficiency tolerance. That was not the case in this study, however. Dry weight was tightly and positively correlated to P uptake (r = 0.96), whereas P-use efficiency was negatively correlated to both dry weight (r = -0.60) and P uptake (-0.72). If P-use efficiency is one of the traits leading to P deficiency tolerance, how could it be negatively correlated to dry weight, one of the tolerance indicators? Among the 98 BILs studied here, variability for P-use efficiency (CV of 17%) was much lower than for P uptake (CV of 64%). P uptake therefore was the more influential of the two parameters. While plants with superior dry weight could achieve this due to their high P uptake, inferior plants were in a state of severe P deprivation caused by insufficient P uptake. They produced their small amount of biomass at highly suboptimal tissue-P concentrations below 0.05% (Fig. 5). Concentrations so far below the deficiency threshold of 0.1% (IRRI, 1981) may represent the absolute minimum for survival. But because P-use efficiency is calculated as the inverse of tissue-P concentrations, plants in a state of P deprivation appear to be highly efficient in P use. Efficient use of P as a result of starvation should not be considered a mechanism of tolerance to P deficiency in a positive sense if it was caused by low tolerance to P deficiency in terms of insufficient P uptake. It can be expected that lines with higher P uptake and higher tissue-P concentrations (lower P-use efficiency) would have P concentrations comparable to such a survival level if they experienced an equivalent degree of P deprivation. The observed genotypic differences in P-use efficiency are therefore caused by the direct and opposite effect that genes for P uptake have on P-use efficiency and QTLs for P-use efficiency have to be considered artifacts of this analysis (Pseudo-QTLs).

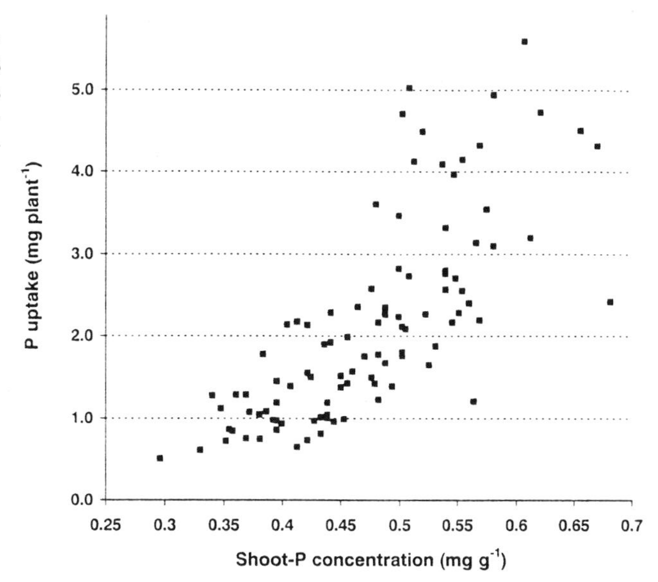

Fig. 5. Shoot-P concentration versus P uptake of 98 backcross inbred lines grown in highly P-deficient soil (Wissuwa unpublished data)

These results furthermore indicate that P-use efficiency should be used cautiously in screening experiments. Meaningful comparisons of genotypes will only be possible if P-use efficiency is independent of P uptake and if genotypic variation for P-use efficiency is equal to or greater than the variation for P uptake. These conditions will more likely be met at intermediate P deficiency levels where tissue-P concentrations of plants will vary around a mean closer to 0.1%, the level considered to be the threshold for P deficiency in rice (IRRI 1981). In this study severe P deprivation limited variation for P-use efficiency by pushing the tissue-P concentration of genotypes towards a physiological limit.

A different form of efficient use of P that was not assessed here because plants were harvested prior to maturity can be expressed in terms of grain yield produced per unit P taken up. As discussed earlier, this will depend largely on the harvest index (HI) and will be mostly under control of genes for plant architecture rather than of genes related to tolerance to P deficiency. Breeders have successfully increased the HI in rice during the past three decades and may continue to do so in the future (Khush 1993; Hay 1995). Independent of that development, the presence of QTLs for P uptake at putative chromosomal locations needs to be confirmed to achieve the goal of adding true P deficiency tolerance to HYVs.

4. Confirmation and First Characterization of Putative QTLs for P Uptake

Near isogenic lines (NIL) that would carry Nipponbare genes on all loci except for the chromosomal region containing a putative QTL are the type of plant material ideally suited to confirm that the QTL has the estimated effect. NILs can furthermore be used for physiological studies to investigate mechanisms responsible for superior P uptake associated with a QTL. Developing NILs, however, is a process that will take several years to accomplish if that process has to be started anew. Fortunately, the 98 lines used in the QTL study were already backcrossed to Nipponbare once before being genotyped at 245 RFLP marker loci. From this stock of lines several were chosen that together had chromosomal segments from Kasalath inserted at all loci. After three additional backcrosses to Nipponbare, a set of substitution lines was developed at the National Institute of Agro-Biological Resources (NIAR). These lines carried small chromosomal segments from Kasalath but otherwise were 85 to 97% identical to Nipponbare.

4.1 Near Isogenic Lines for QTL C443 and C498

Among these lines candidate NILs were found for the major QTL on chromosome 12 (C443) and for the minor one on chromosome 6 (C498). Even though lines were selected to contain only a small proportion of Kasalath alleles, three backcrosses were not sufficient to produce true NILs. For example, line NIL-C443-K carried undesired Kasalath segments on chromosomes 1, 8, and 10 in addition to the segment containing C443 on chromosome 12 (Fig. 6a). It can not be ruled out that these other segments contain genes with minor effects on P uptake that were not detected during the initial QTL analysis. It would thus be impossible to show with certainty that any superior P uptake by line NIL-C443-K was caused exclusively by the QTL on chromosome 12.

The set of substitution lines was therefore screened once more for lines that carried the same Kasalath inserts on chromosomes 1, 8, and 10 as NIL-C443-K but lacked the insert containing C443. One such line was found and designated NIL-C443-N (Fig. 6b). By using NIL-C443-N as a second negative check in addition to Nipponbare, three genotypes of high genetic similarity (Table 4) could be compared and since only NIL-C443-K carried Kasalath alleles at C443, the effect of that QTL could be estimated with high certainty. The same procedure was followed for choosing a set of NILs for the QTL C498 on chromosome six. NIL-C498-K carried small additional Kasalath segments on chromosomes 1, 6, and 9 but the effects of inserts on chromosomes 1 and 9 could be neutralized by including NIL-C498-N as well that contains both segments. The genetic similarity of this set of NILs was very high, approaching the status of true NILs (Table 4).

4.2 Evaluation of Near Isogenic Lines for P Uptake

To confirm the putative QTLs on chromosomes twelve and six, the sets of NILs were grown together with both parents in 60-liter buckets filled with the same P-deficient Andosol described earlier. A plus-P control received the equivalent of 60 kg P ha^{-1}. With adequate P the NILs and Nipponbare did not differ in total dry weight or P uptake (Fig. 7). On the P-deficient soil, however, differences between NILs with and without Kasalath alleles at QTLs were highly significant.

NIL-C498-K and NIL-C443-K had about twice the dry weight and P uptake of their isogenic lines with Nipponbare alleles at QTLs. Compared to Nipponbare, NILs with Kasalath alleles at QTLs even showed a four-fold increase for P uptake.

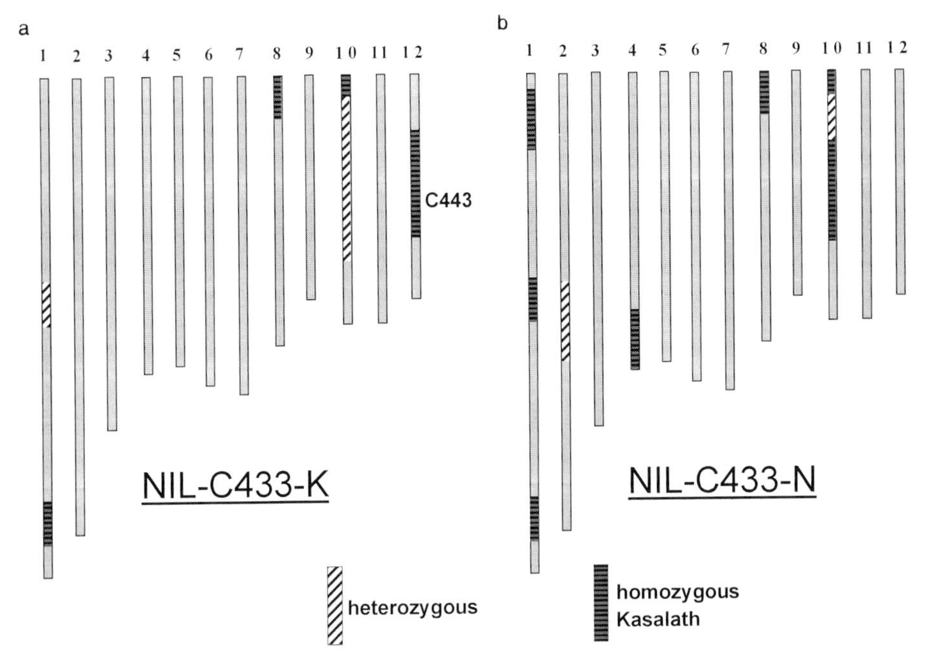

Fig. 6. Near isogenic lines (*NILs*) for a QTL increasing P uptake that is linked to marker C443 on chromosome 12 (From STAFF Institute and National Institute of Agro-Biological Resources, Tsukuba, Japan)

Table 4. Genetic similarity between Nipponbare and pairs of near isogenic lines for markers C498 and C443

Near isogenic line (NIL)	QTL	Genotype at QTL	Genetic Similarity (%)[a]	
			among pair	with Nipponbare
NIL-C498-K	C 498	Kasalath	96.2	93.7
NIL-C498-N	C 498	Nipponbare		96.9
NIL-C443-K	C 443	Kasalath	92.0	91.1
NIL-C443-N	C 443	Nipponbare		88.8

[a] Portion of the genome carrying identical alleles based on 118 maker loci

Differences between Nipponbare and control NILs with Nipponbare alleles at QTLs were not significant for P uptake, but for dry weight NIL-C443-N was superior to Nipponbare. Several important conclusions can be drawn from these results:

1. The presence of QTLs for P uptake near markers C498 and C443 could be confirmed. That NIL-C498-N and NIL-C443-N did not differ significantly from Nipponbare in P uptake further shows that effects of the additional Kasalath segments carried by NILs were small compared to the effect of the QTL in question.
2. The pair of NILs for C443 contained a higher portion of Kasalath alleles unrelated to known QTLs for P uptake compared to the pair of NILs for C498. That NIL-C443-N had a significantly higher dry weight than Nipponbare could indicate the presence of additional loci with minor effects on P deficiency tolerance.

3. Both QTLs did not increase P uptake in the plus-P treatment. Genes at the QTLs must therefore code for P uptake mechanisms that specifically increase P uptake from a low-P soil, thereby conveying tolerance to P deficiency.
4. The magnitude of the effect of the QTL linked to marker C443 on chromosome twelve was confirmed. As within the backcross inbred lines used for QTL detection, Kasalath alleles at C443 more than doubled P uptake.
5. The effect of C498 was greater than initially estimated. C498 had been estimated to increase P uptake by 29% in the population of BILs (Table 3) whereas the set of NILs for C498 differed by 111%. In a study of root characteristics in rice, Ray et al. (1996) identified a QTL for total root number on chromosome 6 that overlapped in position with C498. If both QTLs were identical the greater effect of C498 in the present study may have been the result of the large soil volume (60 l) available for root development that would allow NIL-C498-K to fully take advantage of its larger root system.
6. Genotypic differences in P uptake, a trait considered to be quantitatively inherited, could be attributed to several regions on the rice genome that can be successfully transferred from a donor parent to a modern semidwarf variety. Both NILs with Kasalath alleles at QTLs had a more than four-fold higher P uptake compared to Nipponbare, despite being identical to Nipponbare for more than 90% of all genes. Breeders could therefore use QTLs like the ones detected here to successfully increase P uptake in modern rice genotypes.

4.3 Effect of QTLs on Root Size and Root Efficiency

After having confirmed both QTLs, one of the questions that can be addressed is how the superior P uptake of lines carrying Kasalath alleles at these QTLs can be explained. This question is of more interest to a plant physiologist than to a plant breeder, yet breeders may benefit from a better understanding of the physiological aspects of P uptake. Based on such information rapid, gene-specific screening procedures could be developed that would simplify the screening and selection process during the development of varieties with improved tolerance to P deficiency. To establish a connection between a QTL and a physiological process would furthermore be of help in the process of isolating the gene or genes underlying a QTL.

P uptake is a process limited by the low diffusion rates of P in the soil solution. Uptake of such a nutrient is generally considered to be proportional to the root surface area in contact with the soil solution (Itoh and Barber 1983). More recently it has been postulated that some plant species possess additional mechanisms that increase the efficiency of P uptake by solubilizing soil phosphates that are otherwise not available (Ae et al. 1990; Kirk et al. 1998). To test if the superior P uptake of NILs with Kasalath alleles at QTLs can be attributed to the size of the root system or is caused by factors increasing root efficiency, roots of the NILs grown in 60-l buckets were sampled and analyzed after shoots had been harvested.

The comparison of Kasalath and Nipponbare indicates that the superior P uptake of Kasalath was due to both its larger root surface area and its higher root efficiency (Table 5). For both effects parental differences in root surface area were more pronounced, suggesting that differences in root size contributed a greater portion to the variation in P uptake than differences in root efficiency. Similar results were obtained for NILs. None of the NILs reached Kasalath in root size but both NILs with Kasalath alleles at QTLs had a significantly larger root surface area under P deficiency compared to Nipponbare (Table 5; Fig. 7c). With addition of P fertilizer, however, the root surface area of Nipponbare increased to a level not different from NILs. Kasalath was able to produce a root system more than twice the size of any NIL in the plus-P treatment which shows that root development of NILs was not limited by soil volume (60 L). Root size in the plus-P treatment can thus be considered to reflect the true genetic capacity of genotypes. If NILs and Nipponbare have an inherently similar potential for root development, the effect of genes at QTLs cannot be one of increasing root size per se. Rather than having genes encoding for a smaller root system, Nipponbare fails to realize its potential for root growth if grown in P-deficient soil.

The observation made in comparison to Nipponbare can be confirmed for the set of NILs for QTL C498. NIL-C498-K had a significantly greater root surface area than NIL-C498-N, both in absolute and relative terms (Table 5). That the QTL at marker C498 is identical to a QTL increasing root number (Ray et al. 1996) could not be confirmed here. Differences within pairs of NILs and in comparison to Nipponbare were not significant. The primary effect caused by Kasalath alleles at QTL C498 therefore lies in the capability of NIL-C498-K to expand the root system despite P deficiency. Differences in root size are consequently a secondary effect driven by genotypic differences in the ability to maintain root growth in low-P soil.

The effect of QTL C443 on chromosome twelve cannot be attributed to root size or root efficiency with certainty. NIL-C443-K had the highest relative root surface area and root efficiency among NILs and differed significantly from Nipponbare in both traits. But differences between NIL-C443-K and NIL-C443-N were not significant. It is therefore not entirely clear how much of the difference between Nipponbare and NIL-C443-K is due to QTL C443. Both NIL-C433-K and NIL-C443-N had significantly larger root systems than Nipponbare and may pos-

Table 5. P uptake and root parameters under P deficiency of NILs carrying Kasalath (C443-K/C498-K) or Nipponbare (C443-N/C498-N) alleles at QTLs

Genotype	P uptake[a] (mg P)	Root number[b]	Root surface area (cm^2)	Relative root surface area[c] (%)	Root efficiency ($\mu g\ P\ cm^{-2}$)
Kasalath	17.50 a	83.8 a	2329 a	40.1 ab	7.88 a
Nipponbare	1.09 c	54.2 a	366 d	28.8 b	2.98 c
NIL-C498-K	4.80 b	102.2 a	1106 b	59.0 a	4.36 bc
NIL-C498-N	2.27 c	77.9 a	577 cd	29.9 b	3.88 bc
NIL-C443-K	5.20 b	88.8 a	1004 bc	62.8 a	5.27 b
NIL-C443-N	2.67 c	76.8 a	737 bc	56.7 a	3.60 bc

[a] Values followed by different letters are significantly different at $P< 0.05$ (Tukey HSD)
[b] Root number was determined 1 cm from the crown
[c] Root surface area in the minus-P relative to the plus-P treatment

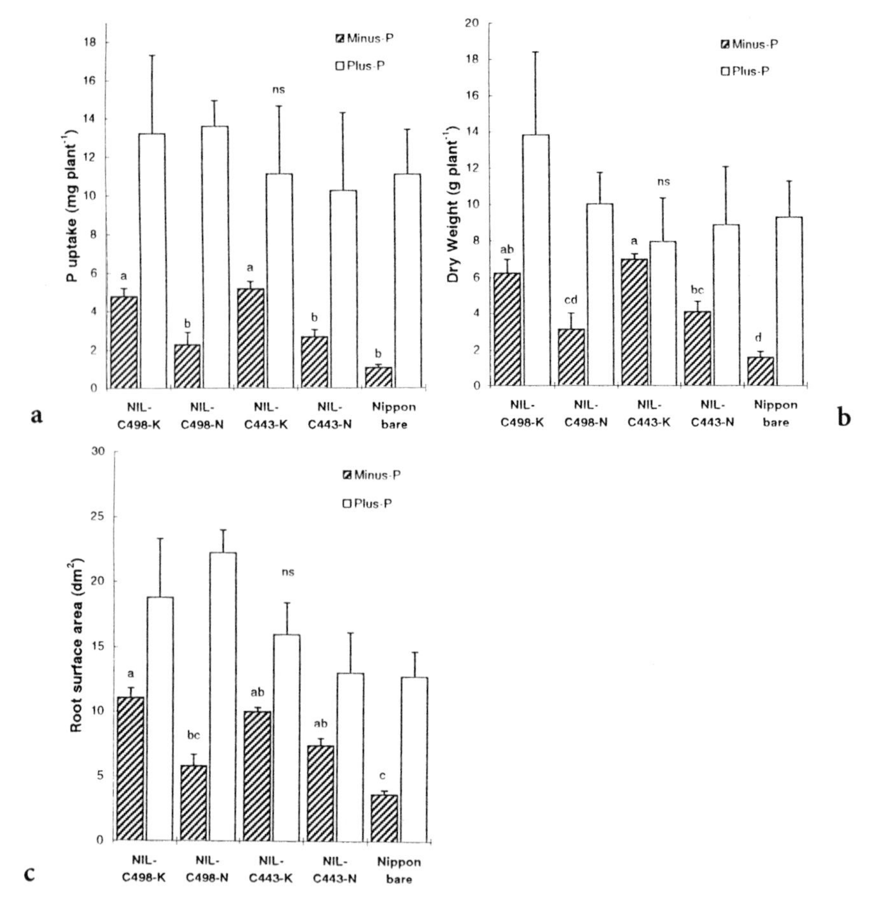

Fig. 7. The effect of Kasalath (-K) alleles at QTLs C443 and C498 on P uptake, total dry weight, and root surface area as illustrated by the performance of two sets of near isogenic lines (NILs). Letters above vertical bars represent significant genotypic differences (HSD ≤ 0.05) (Wissuwa unpublished data)

sess an additional QTL affecting the ability to maintain root growth in P-deficient soil that has not been detected in the QTL analysis. But this potential QTL appears to be not related to C443 and should be located on the additional Kasalath segments that the pair of NILs for C443 carry on chromosomes one, eight, or ten.

5. Conclusions

The starting point for the study presented here was the identification of rice genotypes that differed greatly in their ability to take up P from a P-deficient soil. The two genotypes subsequently used for further experiments varied for a great number of other traits in addition to having a six-fold difference in P uptake. The traditional *indica* variety Kasalath produced a large number of unproductive tillers and had small awned grains that shattered easily at maturity. Nipponbare on the other hand is a modern semi-dwarf type of the *japonica* subspecies. With that many differences between both genotypes it would amount to the proverbial search for the "needle in the haystack" to directly compare both genotypes with the aim of identifying genetic and physiological factors responsible for their contrasting P uptake. But this was made possible by the approach followed in this study. By using a segregating backcross population derived from a cross of both contrasting genotypes and by relating phenotypic variation observed in this population to genetic differences at molecular markers, several regions in the rice genome could be identified that accounted for a portion of the variation for P uptake. This approach allowed the separation of factors affecting P uptake from those responsible for variation in other traits. The presence of two QTLs for P uptake were then confirmed using sets of NILs differing for the QTL in question. Beyond confirming a QTL, the development of NILs is a crucial step for breeders and physiologists alike. This plant material allows the separate investigation of the effect that several chromosomal regions have on P uptake in an otherwise similar genetic background, a much more powerful tool compared to direct comparisons of different cultivars like Nipponbare and Kasalath.

5.1 Physiology of P Uptake

Kasalath alleles contained in NILs were found to increase root size under P deficiency, possibly by some mechanism that allowed carriers of these genes to maintain root growth despite a lack of readily plant-available P. Future studies should address the question of how NILs maintained root growth. Altering the root/shoot ratio due to a preferential supply of roots with assimilates and possibly P, a process that could be under hormonal control, is one potential mechanism. Alternatively roots of NILs may be less sensitive to sub-optimal soil-P concentrations or may require lower internal P concentrations for maintaining root growth.

Mechanisms mentioned above assume that P uptake is mainly driven by an expanding root system. In addition continued root growth may have been achieved by processes that would increase root efficiency. Roots that obtain P more efficiently would be able to expand further due to an increased P supply. Several crop species including rice are known to increase P availability by excretion of organic acids that solubilize P otherwise unavailable to plants (Ae et al. 1990; Kirk et al. 1998). Alternatively to actively solubilizing P in the rhizosphere, NILs may rely on associations with mycorrhiza for P uptake (Kirk et al. 1998). Genes at QTLs could potentially increase the rate of root colonization by these soil organisms. All the mechanisms discussed above are hypothetical explanations for the observed differences caused by QTLs for P uptake but the availability of NILs will make it possible to test these hypotheses in an effective way that has previously not been available for rice.

5.2 Breeding and Genetics

Even without any detailed knowledge of the physiology of P uptake, breeders can work on improving the P deficiency tolerance of rice cultivars. That the genes behind both QTLs confirmed so far can be transferred to a modern rice variety has been show successfully in this study. In both cases Kasalath genes increased P uptake four-fold compared to Nipponbare. A success of such magnitude is truly encouraging but before Kasalath can be widely recommended as a donor of genes for high P uptake, several questions should be addressed:

1. Part of the reason for the great effect that Kasalath alleles have in a Nipponbare background is the low P deficiency tolerance of Nipponbare. It remains to be seen how much of an effect QTLs C443 and C498 have in a different genetic background, specifically in cultivars that are currently used in areas characterized by soils of low P-availability.
2. All tests have up to now been conducted on a P-fixing volcanic ash soil rich in organic matter. To show that QTLs identified here are of universal significance, the superiority of NILs would have to be demonstrated on other soils such as the highly weathered mineral soils encountered throughout the tropics.
3. Currently the transfer of individual QTLs has been successfully accomplished but information about the effect a simultaneous transfer of both QTLs can achieve is not yet available. Crossing NIL-C443-K with NIL-C498-K should reveal if both QTLs act in a desirable additive manner. This would further enhance the prospect of producing cultivars with high tolerance to P deficiency.

Transferring genes in a conventional manner can be simplified by marker assisted selection. A prerequisite for the successful application of this method is the establishment of a very close association between a marker and the gene of interest. Due to the low number of backcross inbred lines (98 BILs) available in the mapping population used for QTL detection, QTLs identified here have not been mapped with a precision high enough to safely identify a marker in close proximity of the genes of interest. It will therefore be necessary to develop a population segregating for the chromosomal regions containing a QTL for the purpose of fine-mapping these QTLs. This is currently being done by backcrossing NIL-C443-K and NIL-C498-K to Nipponbare. In addition fine-mapping QTLs will resolve the question if the effect of a QTL was due to a single gene or multiple genes located in close proximity to each other. Furthermore, additional QTLs with effects on P uptake could be identified during this process. The set of NILs for C433 both had a significantly larger root system than Nipponbare. This could not be attributed to QTL C433 and may be due to genes located on the additional Kasalath segments that these NILs contain on chromosomes one, eight, or ten. Finally true near isogenic lines for QTLs C433 and C498 that do not carry undesirable Kasalath segments can be developed as a by-product of the fine-mapping procedure.

The final goal of the project described so far is the identification of genes causing differences in P uptake. Several important steps towards this goal have already been taken or will be accomplished in the near future. Once a gene is cloned it should theoretically be possible to transfer it with greater speed compared to traditional breeding methods. Furthermore more breeders might be willing to address the issue of P uptake if an easier method for gene transfer is available that does not involve a series of time consuming backcrosses.

Acknowledgments. The author is grateful to Dr. Masahiro Yano and his group at the National Institute of Agro-Biological Resources for generously providing most of the plant material used in this study.

References

Ae N, Arihara K, Okada K, Yoshihara T, Johansen TC (1990) Phosphorus uptake by pigeonpea and its role in cropping systems of the Indian subcontinent. Science. 248:477-480

Brown LR (1995) Who will feed China? A wake-up call for a small planet. The Worldwatch Environmental Series 163 pp

Cassman KG, Pingali PL (1995) Extrapolating trends from long-term experiments to farmers' fields: the case of irrigated rice systems in Asia. In: Barnett V, Payne R, Steiner R (Eds) Agricultural sustainability: economic, environmental, and statistical considerations. John Wiley & Sons, London, pp 63-84

Chaubey CN, Senadhira D, Gregorio GB (1994) Genetic analysis of tolerance for phosphorus deficiency in rice (*Oryza sativa* L.). Theor Appl Genet 89:313-317

De Datta SK, Biswas TK, Charoenchamratcheep C (1990) Phosphorus requirements and management for lowland rice. In: Phosphorus requirements for sustainable agriculture in Asia and Oceania. IRRI, Philippines, pp 307-323

Dobermann A, Cassman KG, Sta Cruz PC, Adviento MAA, Pampolino MF (1996) Fertilizer inputs, nutrient balance, and soil-nutrient supplying power in intensive, irrigated rice systems: II. Phosphorus. Nutrient Cycl Agroecosyst 46:111-125

Dobermann A, Cassman KG, Mamaril CP, Sheehy JE (1998) Management of phosphorus, potassium, and sulfur in intensive, irrigated lowland rice. Field Crop Res 56:113-138

Fageria NK (1980) Deficiencia hidrica em arroz de Cerrado e resposta ao fosforo. Pesq. Agropec Bras 15:259-265

Fageria NK, Barbosa Filho MP, Carvalho JRP (1982) Response of upland rice to phosphorus fertilization on an oxisol of central Brazil. Agron J 74:51-56

Fageria NK, Morais OP, Baligar VC, Wright RJ (1988) Response of rice cultivars to phosphorus supply on an oxisol. Fert Res 16:195-206

Hay RK (1995) Harvest index: a review of its use in plant breeding and crop physiology. Ann Appl Biol 126:197-216

Hedley MJ, Kirk GJD, Santos MB (1994) Phosphorus efficiency and the forms of soil phosphorus utilized by upland rice cultivars. Plant Soil 158:53-62

Horst WJ, Mohamed-Abdou, Wiesler F (1996) Differences between wheat cultivars in acquisition and utilization of phosphorus. Z. Pflanzenern Bodenk 159:155-161

Hossain M (1996) Recent developments in the Asian rice economy: challenges for rice research. In: Rice research in Asia: progress and priorities. CAB International, pp17-33

IRRI (1971) IRRI program report for 1970. International Rice Research Institute, Manila, Philippines

IRRI (1978) IRRI program report for 1977. International Rice Research Institute, Manila, Philippines.

IRRI (1981) Fundamentals of rice crop science. International Rice Research Institute, Manila, Philippines, p 186

IRRI (1993) IRRI Rice almanac. International Rice Research Institute, Manila, Philippines

Itoh S, Barber S A (1983) A numerical solution of whole plant nutrient uptake for soil-root systems with root hairs. Plant Soil 70:403-413

Jones GPD, Blair GJ, Jessop RS (1989) Phosphorus efficiency in wheat - a useful selection criterion? Field Crops Res 21:257-264

Khush GS (1993) Breeding rice for sustainable agricultural systems. In: International crop science, CSSA, Madison, WI, pp 189-199

Kirk GJD, George T, Courtois B, Senadhira D (1998) Opportunities to improve phosphorus efficiency and soil fertility in rainfed lowland and upland rice ecosystems. Field Crops Res 56: 73-92

Koyama T, Chammek C, Snitwonge P (1973) Varietal differences of Thai rice in the resistance to phosphorus deficiency. Trop Agr Res Cent Jpn Tech Bull 4:32 pp

Kurata N, Nagamura Y, Yamamoto K, Harushima Y, Sue N, Wu J, Antonio BA, Shomura A, Shimizu T, Lin S-Y, Inoue T, Fukuda A, Shimano T, Kuboki Y, Toyama T, Miyamoto Y, Krihara T, Hayasaka K, Miyao A, Monna L, Zhong HS, Tamura Y, Wang Z-X, Momma T, Umehara Y, Yano M, Sasaki T, Minobe Y (1994) A 300-kilobase interval genetic map of rice including 883 expressed sequences. Nature Genet 8:365-372

Lin SY, Sasaki T, Yano M (1998) Mapping quantitative trait loci controlling seed dormancy and heading date in rice, *Oryza sativa* L. using backcross inbred lines. Theor Appl Genet 96:997-1003

Liu HG, Liu ZX, Liu FX (1987) Screening of rice cultivars (lines) tolerant to low phosphorus and a study of their characteristics. Fujian Agric SciTechnol 6:9-11

Majumder ND, Borthakur DN, Rakshit SC (1989) Heterosis in rice under phosphorus stress. Indian J Genet 49:231-235

Otani T, Ae N (1996) Sensitivity of phosphorus uptake to changes in root length and soil volume. Agron J 88:371-375

Paterson AH, Lander ES, Hewitt JD, Peterson S, Lincoln SE, Tanksley SD (1988) Resolution of quantitative traits into Mendelian factors by using a complete linkage map of restriction fragment length polymorphisms. Nature 335:721-726

Ray JD, Yu L, McCouch SR, Champoux MC (1996) Mapping quantitative trait loci associated with root penetration ability in rice (*Oryza sativa* L.). Theor Appl Genet 92:627-636

Sanchez PA, Salinas JG (1981) Low-input technology for managing oxisols and ultisols in tropical America. Adv Agron 34:279-406

Schettini TM, Gabelman WH, Gerloff GC (1987) Incorporation of phosphorus efficiency from exotic germplasm into agriculturally adapted germplasm of common bean (*Phaseolus vulgaris* L.). Plant Soil 99:175-184

Swaminathan MS (1993) From nature to crop production. In: International crop science. CSSA, Madison, WI, pp 385-394

Utz HF, Melchinger AE (1996) PLABQTL: A program for composite interval mapping of QTL. J Quant Trait Loci http://probe.nalusda.gov:8000/otherdocs/jqtl

Wade LJ, George T, Ladha JK, Singh U, Bhuiyan SI, Pandey S (1998) Opportunities to manipulate nutrient-by-water interactions in rainfed lowland rice systems. Field Crop Res 56:93-112

Wissuwa M, Yano M, Ae N (1998) Mapping of QTLs for phosphorus-deficiency tolerance in rice (*Oryza sativa* L.). Theor Appl Genet 97:777-783

Yano M, Sasaki T (1997) Genetic and molecular dissection of quantitative traits in rice. Plant Mol Biol 35:145-153

Significance of Nutrient Uptake Mechanisms in Cropping Systems

Joji Arihara and Ancha Srinivasan

Summary. Given the widespread prevalence of nutrient stresses worldwide, a thorough understanding of acquisition, utilization and recycling of both organic and mineral forms of nutrients at the level of the cropping system is essential. This chapter outlines selected research advances in nutrient acquisition and management, and identifies a few priorities and strategies to increase the efficiency of nutrients mobilized from all sources. It is recognized that productivity levels of cropping systems cannot be increased and sustained if current practices, such as over or under application of nutrients and inefficient ways of utilizing crop residues and farm wastes, are continued. Although fertilization will continue to be important for alleviating nutrient limitations in many regions, research on biological and chemical processes aimed at optimizing nutrient cycling and recovery, minimizing indiscriminate application of fertilizers, and maximizing the efficiency of nutrient use is expected to provide innovative management alternatives. Further advances in soil fertility evaluation methods, nutrient acquisition models, and phytoremediation technologies will ultimately provide the most appropriate ways to manipulate nutrient availability and devise strategies for protecting and restoring environmental quality while sustaining farm yields and profitability. It is concluded that research initiatives for sustainable nutrient management must be conceived on the basis of a holistic understanding of location-specific interactions among production, environmental, biological components in cropping systems.

Key words. Cropping systems, Sustainability, Soil fertility evaluation, Nutrient uptake mechanism, Soil fauna, Indigenous cropping systems, Environmental quality, Organic matter, Fertility maintenance, Agronomic practices, Nutrient use efficiency

1. Introduction

Research on nutrient acquisition and management in cropping systems is entering a critical phase. While developed countries are mainly concerned about the adverse impacts of intensive cropping systems on environment (soil, water and air) and society (rural displacement, urban sprawl, etc.), developing countries are confronted with an ever-growing demand for intensifying agricultural production while sustaining their already fragile resource base (Sanchez et al. 1997; Smaling et al. 1997). Although the concept of sustainability originated mainly from pressures of environmentalists in the developed countries, concerns on the long-term viability of cropping systems have implications across all scales of farming and in all agroecological zones.

Nitrate contamination issues recently uncovered in north central Europe and the US resulted from what seemed to be optimal economic fertilization practices (Fedkiw 1991). Researchers estimate that 20% of the N that humans are putting into watersheds is consistently getting into rivers. In Japan, 670 000 t of N fertilizer is applied annually to agricultural fields, which is equal to the application of 130kg N ha^{-1} to each agricultural field. In addition to this, 700 000 t of N in animal manure is also applied, mainly to agricultural fields. This heavy N application is causing nitrate contamination of ground water and watersheds surrounding agroecosystems. In the future, 690 000 t of various forms of wastes from human activities is expected to be applied to agricultural fields. Technologies to efficiently utilize organic N for crop production have to be developed to reduce the application of chemical N fertilizers, and to reduce nitrate pollution. Improving organic N use efficiency of crops would be a key factor to achieve this goal.

In developing countries too, where the arable land per capita ratio is steadily decreasing, adoption of certain features of cropping systems from the Green Revolution era has systematically displaced local practices and knowledge which were inherently sustainable. For instance, the subsistence agriculture of the pre-chemical era efficiently sustained the N status of soils by maintaining a balance between N loss through grain harvest and N gain from biological N fixation. This was possible with less intensive cropping, adoption of rational crop rotations and intercropping systems, and use of legumes as green manure. The agriculture of the modern chemical era, however, concentrates on maximum output but overlooks input efficiency. Likewise, the traditional less intensive wet-dry rotation of rice culture was gradually replaced by intensive continuous wetland rice culture leading to a lower available soil N pool (Kundu and Ladha 1995). Depletion in organic carbon was also observed in field experiments conducted in Kenya in spite of fertilization at optimal rates (Smaling et al. 1997). The present challenge is to sustain soil fertility in cropping systems operating at high productivity levels.

Stresses (deficiencies and toxicities) due to availability, acquisition and utilization of nutrients are increasingly becoming widespread in many soils leading to low crop productivity. For example, yield potential of cropping systems on acid soils, which cover about 3 950 million hectares of the earth's surface, is restrained by deficiencies of P, Ca, Mg, and K, and toxicities of Al, Mn, and Fe (Salazar et al. 1997). Conventionally, fertilizers or soil amendments are used for curing such stresses. However, total dependence on fertilizers is neither economical nor pragmatic because of (a) the inability of many farmers to buy enough fertilizer, and (b) the capacity of many soils to fix applied nutrients into forms unavailable to plants (Sanchez and Uehara 1980). There is also increasing evidence to show that fertilization alone cannot sustain yields for long periods. For example, in continuous rice cropping with two to three crops grown annually, the use of fertilizer N increased with time but the yields often remained stagnant (Cassman and Pingali 1995). This can be due to the higher requirement of fertilizer to produce the same yield, implying a decline in yield response to nutrients, and/or overuse of fertilizer, but either situation is a reason for concern. As an alternative, tailoring plants to fit the soil through genetic improvement is considered more economical than changing the soil. It is believed that farmers would more easily adopt a genotype with useful traits than crop and soil management practices that are associated with extra costs. However, the magnitude of nutrient stresses is so severe and widespread that no single remedial measure can perhaps effectively solve this problem. For example, it is estimated that as much as 8-26 million hectares of maize, at least 60% of beans in Latin America, and approximately 44% of beans in Africa are grown on severely P-deficient soils (Yan et al. 1996). Likewise, it is estimated that at least 50% of the arable land used for crop production worldwide is low in availability of one or more of the essential micronutrients for current varieties (Ruel and Bouis 1997).

There are two approaches in research on nutrient acquisition and management in cropping systems. One approach is to analyze survival mechanisms of crops grown on nutrient-poor soils and results from such studies must be the foundation for basic process research. The other must emphasize basic research for a better understanding of the mechanisms of nutrient acquisition in cropping systems. The study methods for each approach are distinct and it is uncommon for researchers working in basic research to have the full competence needed for adaptive research, and vice versa. We believe, however, that to establish sustainable and productive cropping systems, two different but complementary strategies need to be conducted in a collaborative and highly focused manner.

2. Evaluation of Soil Fertility

Most soil fertility evaluation methods are based on the assumption that crop plants absorb nutrients already dissolved in soil solution. Crop activities to acquire nutrients from the soil are not usually taken into consideration for the evaluation. This is one of the most important reasons for the lack of explicit methods for evaluating soil fertility in cropping systems, which is perhaps a major factor that has led to over-fertilization in developed countries.

While soil inorganic nitrogen or NO_3^- or NH_4^+ can be measured rather easily, reliability of methods evaluating N mineralized from soil organic matter during the growing season is not satisfactory. It is necessary to improve our understanding of the mechanism of crop plants in utilizing organic forms of soil nutrients. Though some advances in this aspect of research are reported elsewhere in this book, further efforts need to intensively explore the mechanisms of crop N acquisition before establishing a reliable soil N test.

Methods for testing soil P fertility also need further investigation. For instance, use of such tests as Bray, Mehlich, or Olsen on soils different from those for which the tests were originally developed provides inaccurate available soil P estimates. Available P measured by Olsen's method, which uses calcium carbonate (pH 8.5) as an extractant, correlates well with P uptake. However, it gives a lower estimate of available phosphorus for alkaline soils (e.g., Vertisols) and a higher estimate for acidic to neutral soils (e.g., Alfisols; Ae et al. 1991).

The Truog method, which uses sulfuric acid as an extractant, underestimates the available P in acidic to neutral soils, especially when crops can utilize Fe-P or Al-P solubilized by organic chelating compounds exuded from their roots. In Andosols, for example, no linear relationship existed between soil available P measured with Truog method and P uptake by crops (Souma 1986). In recent years, the recommended rates of P fertilizer in cropping systems in some regions are much higher than the amount of P actually needed by crops, especially because of low crop response to applied P. Soil P content in some fields in Hokkaido, for example, was so high that eroded soil particles into rivers increased the P concentration of water to a critical level as to cause algal boom and hypoxia (Tachibana et al. 1992).

Sharpley et al. (1992) suggested that accounting for contributions of soil organic N and P to fertility was a major constraint for developing valid soil testing methods. This is especially relevant in the highly weathered soils of southern Africa and India, where mineralization of organic forms may be the main mechanism by which nutrients become available to crops. Methods involving the use of ion exchange resins and iron-oxide impregnated paper strips have been shown to provide more reliable estimates of plant available soil P (Sharpley et al. 1994) but the results need validation in a wider range of soils and cropping systems.

To establish cropping systems suitable for the soil conditions of different locations, a reliable soil test for standardization of soil nutrient availability is necessary. Further understanding of the dynamic mechanisms of crop plants in acquiring nutrients from the soil would be indispensable to improve soil tests for evaluating soil nutrient availability.

3. Research on Novel Mechanisms of Nutrient Uptake: Root Exudates, Cell Walls, Architecture

In India, chickpea is widely grown on alkaline soils such as Vertisols or Aridsols. Pigeonpea, on the other hand, is grown on acidic soils such as Alfisols or Oxisols. It has been recognized that chickpea and pigeonpea are crops less responsive to P application on those soils. A sizeable fraction of the inorganic P in Vertisols is associated with Ca, and this calcium-bound P (Ca-P) is considered to be a sparingly soluble compound, largely in the form of apatite, as the available P level of Vertisols as measured by Olsen's method is generally very low. Chickpea was found to solubilize otherwise insoluble Ca-P in alkaline soil by lowering the rhizosphere pH with the excretion of citric acid (Ae et al. 1991). Weak response of pigeonpea to P application in an Alfisol field suggested that pigeonpea was able to efficiently utilize iron-bound P (Fe-P). It was found that pigeonpea can acquire P from Alfisols by excretion of an organic acid, piscidic acid and its derivatives, which can specifically chelate Fe from Fe-P ligands (Ae et al. 1990). Pigeonpea later was confirmed to exude a significant amount of malonic and oxalic acids along with piscidic acid. Those acids were considered to release P from Fe-P and Al-P in soils of low P fertility (Otani and Ae 1996a). Roots of P-stressed white lupin are also known to excrete organic acids, which increases the availability of P. Johnson et al. (1996) reported that lateral root development was altered in P-stressed lupin. Clustered tertiary roots called proteoid roots formed over 60% of the root mass on P-stressed plants. The P-stressed plants exuded 25 fold more citrate and malate as compared with P-sufficient plants. Recently, Ae et al.(1996) showed that groundnut can solubilize Fe-P or Al-P with the chelating capability of the cell wall itself.

Researchers at the International Center for Tropical Agriculture (CIAT), Cali, Colombia, who observed promising genetic variation in P efficiency in bean germplasm, suspected that the variation might be due to a specific capacity of some of the genotypes to obtain P from recalcitrant organic matter. Such a specific adaptation was attributed to the release of phosphatases or other compounds capable of liberating P from organic complexes as reported in bean as well as in other plants (Helal 1990; Tarafdar and Jungk 1987). Intraspecific variation in excreted phosphatase has been correlated with depletion of organic P in the rhizosphere, and presumably, P uptake (Asmar et al. 1995). Since recalcitrant soil organic materials often undergo physicochemical interactions with soil minerals, it is imperative that organic matter must be freed from those minerals, perhaps through chelation. While phosphatase itself may have the ability to chelate ions, a combination of other natural chelating substances and phosphatases might release P much faster from recalcitrant soil organic matter. Higher P uptake of pigeonpea from an Andisol of high humic substances than other crops (Otani and Ae 1996a) seems to support this hypothesis. Studies in this direction have been limited so far.

Yan et al. (1996), however, reported that contrasting bean genotypes did not differ in their ability to mobilize P from organic, Al, or Fe sources. This suggests that beans do not have obvious soil specificity in response to low P availability in contrasting soils. However, large seeded Andean bean genotypes such as G19833 tended to utilize Ca-P better than other genotypes. The superior ability of some genotypes to utilize Ca-P thus opens up the possibility of developing bean cultivars that can be fertilized more efficiently with Ca-phosphate or rock phosphate. Lack of specific adaptation of bean to P sources can be interpreted as a positive result from the agronomic point of view, because this suggests that bean adaptation to P availability is stable across soil environments. This would make breeding for P efficiency relatively easier.

Nutrient-efficient genotypes may have better root growth and root architecture (Lynch and van Beem 1993; Lynch and Beebe 1995; Yan et al. 1995). Wissuwa et al. reported that an efficient rice genotype to take up P from low available P Andosol had a more vigorous root system than inefficient types in chapter 21. They also identified the position of genes concerned with efficiency of P uptake (Wissuwa et al. 1998). A vigorous, effective crop rooting system is essential for efficient nutrient acquisition, particularly for less mobile nutrients. The other general adaptive mechanisms could include better mycorrhizal symbiosis.

In highly P-depleted farms of Western Kenya, a combination of Minjingu rock phosphate at 250 kg ha^{-1} and 1.8 t ha^{-1} of *Tithonia diversifolia*, a common shrub planted in thousands of kilometers of farm boundaries, raised yields by 400%. *Tithonia* apparently helps solubilize P fixed by iron oxides in Oxisols. Both *Tithonia* and rock phosphate are indigenous nutrient sources (ICRAF 1997). Low-grade

Florida rock phosphate or BPL61 applied to 20-cm depth of Andosol was more effective for 3 years to increase dry matter production and P uptake of orchard grass and white clover than similarly applied high-grade Florida rock phosphate or BPL 72 and single super phosphate. The mechanism of efficient utilization of low-grade rock phosphate by those forage crops is not clear, although the involvement of arbuscular mycorrhizae (AM) for increased dry matter production and P uptake (Kondo et al. 1997) is suggested.

With the development of minimum or no tillage systems, an agro-pastoral system that rotates pasture and crop fields for a few years' interval is becoming a practical and promising method for productive agriculture, especially in Brazil and Colombia. In this system, the productivity of pasture is increased after field crops even though no fertilizer is applied to pasture. Pasture crops seem to have some mechanism to efficiently acquire less soluble P from the soil once pasture is established. For example, Kitagishi (1962) reported that orchard grass grown without P application on volcanic ash soil of very high P-fixing capability, higher than 2500 mg P kg^{-1} soil, did not show P-deficiency symptoms once the pasture was established. Ladino clover, however, decreased dry matter production without P application. Where rice plants were grown on an adjacent field to pasture they showed severe yield reduction under control conditions. In southern Thailand, tropical forage crops or stylosanthes, *Brachiaria humidicola*, and *B. ruziziensis* grown on a podsolic soil without P produced comparable dry matter to those grown with NPK (Hayashi 2000, personal communication). Efficient uptake of less soluble soil P by grass plants might be one of the important reasons for higher dry matter production of pasture in agro-pastoral systems.

Soybean seems to efficiently utilize soil N that other plants cannot take up easily. The better growth of maize following soybean than continuous maize has been attributed mainly to the residual effects of N fixed by soybean nodules. In an experiment conducted on an Andosol field, however, maize grown after non-nodulating soybean showed growth efficacy similar to that after nodulating soybean (Fig. 1).

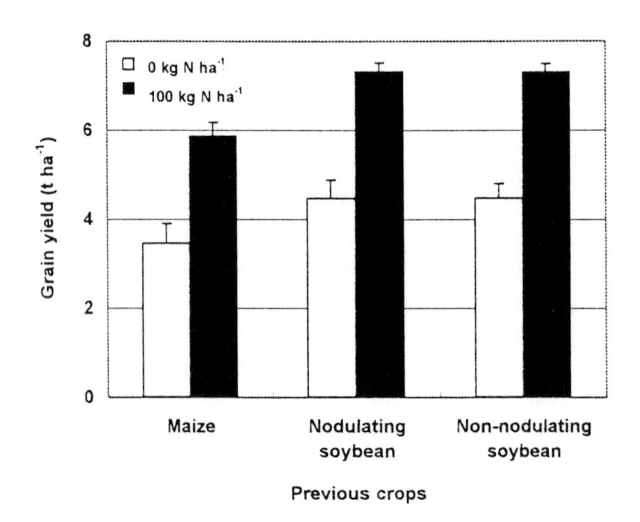

Fig. 1. Effect of cultivating maize and nodulating and non-nodulating soybeans on grain yield of succeeding maize with- (100 kg N ha^{-1}) and without-nitrogen fertilization. (Arihara unpublished data)

Based on a long-term crop rotation experiment, Vanotti and Bundy (1995) speculated that growth enhancement of maize after soybean was due to stimulation of soil N mineralizing microbes by soybean, which might gradually deplete readily available soil N. Also, alfalfa was reported to enhance soil N mineralization (Radke et al. 1988). It is suggested that some crops like flowering Chinese cabbage (*Brassica campestris* L spp. Chinensis (L)) and carrot absorbed organic N directly and/or solubilized insoluble forms of soil organic N (Matsumoto et al. 1999). Johnson and Damman (1996) reported that heath plants in N-deficient habitats such as peat bogs use amino acid N to compensate for deficiency of N available in inorganic forms. In bogs, the rate of soil N mineralization is among the lowest of any ecosystem, and the only source of N is precipitation. Organic N occurs as peat and in smaller quantities as amino acids. It is indicated that upland rice also can take up amino acid N (Yamagata and Ae 1996). Further studies are necessary to determine if crop plants also can take up novel forms of nutrients besides amino acids.

4. Role of Soil Fauna (Mycorrhizae, Bacteria, Nematodes, Endotrophs, etc.) in the Nutrient Balance of Cropping Systems

It is well established that soil fauna play a major role in increasing nutrient availability and uptake, especially in nutrient-poor soils. N-fixing systems, including free living symbiotic or associative organisms, contribute significant amounts of fixed N to cropping systems. Rhizobia-legume systems fix N at rates in the range of 50-300 kg N ha^{-1} $year^{-1}$. Cyanobacteria fix 15-25 kg N ha^{-1} $year^{-1}$ and azospirillum-grass associations 10-30 kg N ha^{-1} $year^{-1}$ (Ladha et al. 1992). The interaction between mineral fertilizers and N-fixing systems should be further studied with a view towards a better integration within plant nutrition systems.

It is also indicated that enhanced AM association of crops through cultivation of mycorrhizal crops in the previous season showed significant growth and yield promotion on soils of high P fixation capability (Thompson 1991; Arihara and Karasawa 2000). As AM inoculation is expensive and indigenous AM fungi usually dominate inoculated AM fungi, increasing indigenous AM fungi through proper cropping systems is a practical way to enhance growth and P uptake of mycorrhizal crops. Soil factors such as P status, soil type, and pH, and climatic variables such as precipitation and temperature determine the growth promoting effects of AM in a cropping system. AM fungal populations and colonization of roots by AM fungi, and their contribution to P uptake was higher under soils of lower P availability. However, the effect of preceding crops on the growth of following crops is different among soils even when P availability was low, thereby raising the possibility that differences in indigenous AM fungi in various soils mediate the effect of the preceding crops. High soil moisture also increases colonization of AM even in soils of low population of AM spores, which eliminates the effects of previous crops on AM colonization (Karasawa et al. 2000). Under water stress conditions, mycorrhizal wheat plants had greater acquisition of P and other nutrients compared to non-mycorrhizal plants (Clark and Zeto 1996b). Similarly, mycorrhizal corn could take up more Fe in alkaline soils than non-mycorrhizal

mycorrhizal corn could take up more Fe in alkaline soils than non-mycorrhizal corn (Clark and Zeto 1996 a,b). Further studies are, however, necessary to maximize the potential advantages from the mycorrhiza-crop symbiosis through a detailed understanding of mycorrhizal ecology in cropping systems. Effects of cropping systems on managing harmful soil microorganisms need to be examined now so as to realize the positive benefits from nutrient cycling.

5. Research on Nutrient Balance and Fertility in Indigenous Cropping Systems

The fertilizer-intensive cropping systems have evolved in only the last 30 to 40 years. To extend these into the next century, some modifications in nutrient management will be required. For this to happen, we must again look at nutrient equilibria in traditional cropping systems with particular attention to soil and crop health. For instance, studies in the semi-arid tropics of India revealed that addition of pigeonpea as a sole or an intercrop in a cropping system not only helps soil N fertility but also enables more P reserves to be available for subsequent crop growth (Ae et al. 1991; Arihara et. al. 1991a, b). Based on studies of a seven century old practice of Egyptian clover-rice rotation, which covers about 60%-70% of the entire rice acreage in Egypt, Yanni et al (1997) reported a unique natural endophytic association between *Rhizobium leguminosarum* bv. *trifoli* and rice roots. The N supplied by this rotation replaces 25%-33% of the recommended rate of fertilizer application to rice but such benefit cannot be explained solely by an increased availability of fixed N through mineralization of N-rich clover crop residues. Yanni et al. (1997) found that clover-nodulating rhizobia naturally invade rice roots and achieve an internal population density of approximately 1.1×10^6, endophytes per gram fresh weight of rice roots. They reported that inoculation with two endophytic strains (E11 and E12) of *R. leguminosarum* significantly

Table 1. Effect of N fertilization and inoculation with *Rhizobium leguminosarum* bv. Trifolii rice endophytes on production of Giza 175 rice under field conditions

Fertilization (kg N ha^{-1})	Grain yield (t ha^{-1})			Harvest index (%)			Fertilizer use efficiency		
	C	E11	E12	C	E11	E12	C	E11	E12
0	4.3	6.3	6.2	30.1	35.0	35.4	—	—	—
48	5.8	7.9	7.0	30.0	40.0	38.5	121.7	165.2	145.0
96	5.4	7.1	7.9	31.8	36.7	37.4	56.4	74.1	81.8
144	6.4	7.0	7.0	31.6	30.6	30.0	44.3	48.6	48.4
LSD (0.05)									
Fertilizer (N)			0.31			2.5			3.4
Inoculation (I)			0.32			1.9			4.4
N × I			0.65			3.9			8.8

From Yanni et al. 1997

increased grain yield, harvest index and fertilizer N use efficiency of field-grown Giza 175 hybrid rice (Table 1). Similar studies on natural associations of endophytic diazotrophs in rice roots under rice-sesbania rotation in the Philippines are in progress (Ladha et al. 1993).

6. Research on Nutrient Acquisition in Relation to Environmental Quality and Bioremediation of Contaminated Soils

It is widely recognized that high yields resulting from the use of high doses of fertilizers in modern cropping systems have been achieved at some cost to environmental quality. In developed countries, in some places, soil nitrate concentration becomes so high that nitrate leached from agricultural fields has increased the nitrate concentration of groundwater to more than 10ppm, which is critical for human health. It is especially important for areas where groundwater is used for drinking water. It is well documented that leaching of nitrate occurs during the period between fall to spring when the downward flow of water exceeds evapotranspiration. Cultivating crops during this season is effective in preventing leaching of nitrate from the soil profile. It is interesting to note that some crops belonging to the Brassicaceae family have been indicated to directly take up soil organic N, which would be beneficial for their growth under low temperature conditions where the mineralization rate of organic N is low (Matsumoto et al. 1999).

Soybean is usually cultivated in rotation with maize or other cereal crops in many countries because of its N fixing capability. However, soybean has been speculated to significantly decrease nitrate concentration in the soil profile compared to maize (Fig. 2). The difference in the amount of nitrate in soil profiles of 60cm between maize and soybean plots was estimated to be more than 100kg N ha^{-1} when heavy doses of N fertilizer were applied (Arihara, unpublished data). This surprisingly high figure suggests the possibility for soybean to be used as a cleaning crop in fields of high nitrate concentration, though the mechanisms involved in this phenomenon are still unclear. Dominique et al. (1997) recently reported that

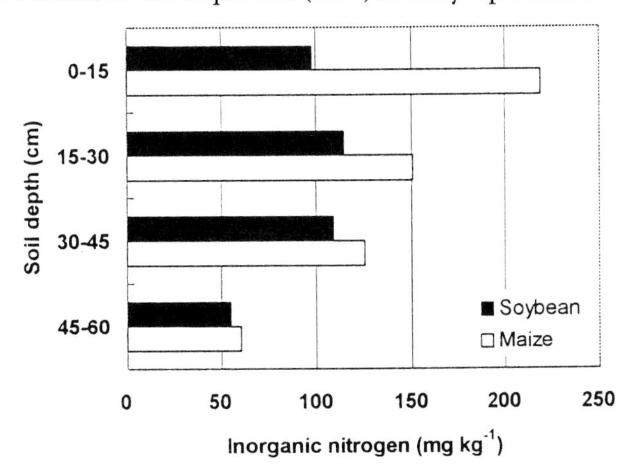

Fig. 2. Difference in soil inorganic nitrogen ($NO_3^{-1} + NH_4^{+1}$) concentration between soil profiles of maize and soybean plots on an Andosol field. (Arihara unpublished data)

plants colonized with AM fungi showed a greater ability to take up soil nitrate than those that are uncolonized. Based on this observation, they reported that induced colonization of crop plants with AM inoculum in the field could help alleviate nitrate contamination of groundwater.

Crops also vary in their ability to take up metals such as nickel from contaminated soils. Because of their rapid nickel transport rate, ryegrass and cabbage can accumulate high levels of nickel in the shoot as compared to maize and white clover which have low influx and transport rates (Chaney et al. 1995).

7. Crop Residue Management and its Relation to Fertility Maintenance

Maintaining indigenous soil fertility through judicious crop residue management is an urgent priority in tropical cropping systems. In highly productive fields, the soil nutrient supply matches the crop nutrient uptake pattern. The yield benefits from high native soil fertility are indeed difficult to replace with fertilizers. Cassman et al. (1995) reported that hybrid rice yields in China on a soil with low native N fertility, where 245 kg fertilizer N was applied, was 2. 2 t ha^{-1} less than that obtained on a soil with high native N fertility where only 54 kg fertilizer N ha^{-1} was applied. A study in Japan indicated that rice crops depend totally on the soil to meet N requirements from heading to maturity (Wada et al. 1986). Rice is preferably grown in flooded and puddle field. This practice has been thought ideal to conserve soil N. However, a decline in productivity in paddy rice caused by soil N reduction has been noticed in several places in Asia, especially where two to three crops of rice are grown annually.

Soil organic N is continually lost through plant removal, leaching, denitrification, and ammonia volatalization. Continuous rice cropping under wetland conditions, for instance, leads to excessively reduced soil N condition, if soil N was not replenished enough by biological N fixation. The decrease in soil N through continuous rice cropping, in turn, is thought to lower the available soil N pool in lowlands (Kundu and Ladha 1995). The decline in N supplying capacity of rice fields is attributed to degradation in the quality of soil organic matter under such a water regime (Cassman and Pengali 1995; Cassman et al. 1995). Therefore, shifting from continuous wetland rice cropping to judiciously managed multiple cropping systems with a wet-dry rotation which includes cultivation of leguminous crops may rectify such problems.

However, even a wet-dry rotation including legumes does not always conserve soil N. In Japan, more than 60% of soybean is grown on paddy fields in rotation with rice and wheat or other crops. Soybean yield has been slightly but steadily declining since the mid-1980s. As shown before, soybean exploits a heavy amount of soil N for the production of grain of which the N percentage is generally more than 8%. Sharp reduction of mineralized soil N caused by continuous cultivation of soybean is suspected to have occurred in many paddy fields with a wet-dry rotation. As grain legumes of good yield may exploit soil N more than rice, further amendments would be required to replenish soil N fertility.

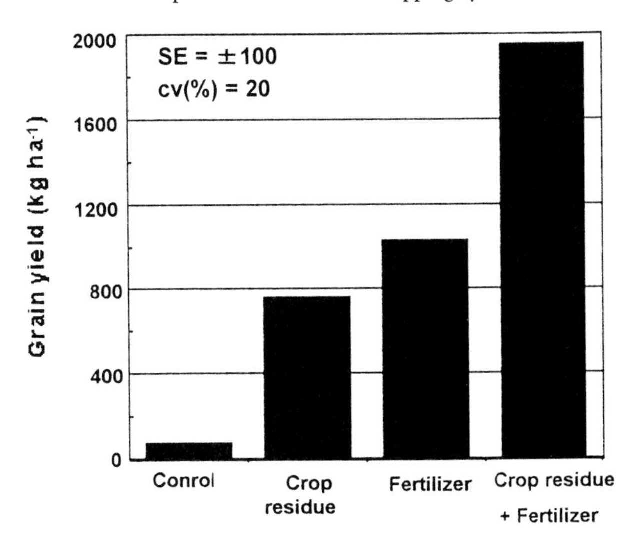

Fig. 3. Effect of fertilization and crop residue on pearl millet grain yield (Bationo et al. 1989)

Fertilizer-use efficiency is often low in tropical cropping systems due to the declining level of organic matter. Addition of small amounts of high quality organic matter (a narrow C/N ratio and a low proportion of lignin) to tropical soils can substantially improve fertilizer efficiency (Snapp 1995). Nearly four decades ago, two long-term cropping studies in Kenya and Nigeria indicated that organic plus inorganic inputs sustain fertility at a higher level than the expected additive effects of either input by itself (Dennison 1961). Nutrient effects alone do not explain the benefits derived from modest amounts of organic manure combined with inorganic fertilizers. High quality organic matter provides readily available N, energy (carbon), and nutrients to the soil ecosystem, and helps retain mineral nutrients (N, S, micronutrients) in the soil and makes them available to plants in small amounts over many years as it is mineralized. High-quality carbon and N provide substrate to support an active soil microbial community. Soil microbes are valuable not so much because they supply nutrients directly, but because they enhance the synchrony of plant nutrient demand with soil supply by reducing large pools of free nutrients (and consequent nutrient losses from leaching and denitrification). Thus microbes maintain a buffered, actively cycled nutrient supply (Snapp 1995). In addition, organic matter increases soil flora and fauna (associated with soil aggregation, improved infiltration of water and reduced soil erosion), complexes toxic Al and manganese (Mn) ions (leading to better rooting), increases the buffering capacity on low-activity clay soils, and increases water-holding capacity.

On-farm gains are usually lower because of inherently low soil fertility, water deficits, and management compromises. Drastic increase of pearl millet grain yield with incorporation of crop residue was reported from a field experiment conducted in Sadore, Niger (Fig. 3; Bationo et al. 1989). With application of crop residue alone grain yield increased from less than 0.2 t ha^{-1} to nearly 0.8 t ha^{-1}. Crop residue applied with fertilizer increased the grain yield up to 2 t ha^{-1} when it was about 1 t ha^{-1} with fertilizer only. Further research is necessary on how soil organic matter influences nutrient availability and uptake in various cropping systems.

8. Nutrient Acquisition and Management in Organic Cropping Systems

Recently, there has been a tremendous interest in organic agriculture, especially in developed countries. Soil fertility research in organic cropping systems is, however, still in its infancy and further studies are necessary to support this rapidly growing industry. For instance, research on organic onions at the University of California showed that organic farms have lower soil and plant nitrate levels even though their yields were comparable to those in conventional systems (Chaney 1996). Soil nitrate levels were similar at the beginning of the season (about 20 ppm), but by mid-June nitrate in the conventionally farmed soils increased to 76 ppm and remained significantly greater than in the organic system until the end of the season. Nitrate levels on the organic farms remained constant at about 15 ppm through the entire season. A similar pattern was found with soil ammonium. Nitrate-N in the onion roots and tops was also lower in the organic system than in the conventional system. The question is how organic onions obtained reasonably adequate supplies of N despite low nitrate levels in the soil and plants. The results suggest that nutrient acquisition and uptake patterns may be different in conventional and organic cropping systems. Although the pool of inorganic nutrients is small in such systems, the nutrient supply may be sufficient to support crop growth because of the continual re-supply of nutrients by microbially mediated turnover rates.

Moreover, if farmers rely on organic sources of N such as compost and cover crops for meeting the entire nutrient demands, it is important to determine how nutrient release from such materials and nutrient demands of crops can be synchronized. Synchronizing nutrient release with nutrient demand using organic residues is technically challenging. Laboratory incubation studies have shown that high-quality residues (e.g., narrow C/N ratio, low percent lignin) supplied in conjunction with low-quality residues (e.g., wide C/N ratio and/or high lignins) can provide a continuous nutrient supply, with N being released from high-quality residues first (Snapp et al. 1995). This synchrony may also be influenced by genotypic phenological traits and photothermal regimes (Ladha et al. 1993). Surface mulches in cold wet conditions, for example, increase soil-borne plant disease, promote weed infestation, impede tillage implements, and alter plant nutrient availability. Preliminary indications are that plants with high lignins, polyphenolics, and other anti-quality factors may not release nutrients for many growing seasons, and thus these residues may be undesirable to use as organic soil amendments (Myers et al. 1994). More research is necessary on how to increase plant access to nutrients in such situations. Humification parameters and electrophoretic procedures can be used to establish quality criteria of soil organic matter and to follow the stabilization and turnover rate of organic amendments.

Pallant et al. (1997) reported that corn root systems developed more fully under low-input/organic agriculture systems than in conventional systems. As a result, corn yields tended to be higher on average under organic systems, especially during years under drought conditions or under nutrient stress, because well-developed root systems are more efficient in extracting available soil water or nutrients. Moreover, the team found some of the differences between systems to be highly significant, e.g., root density (root length/ soil volume, cm cm^{-3}) under low-input/animal systems at mid-growth was 5.51 compared to 2.99 in conventionally fertilized corn/corn systems.

It must be borne in mind, however, that the increasing needs of agricultural production in developing countries could not be met neither by "low input" schemes or by "organic farming" only, because crop nutrient uptake tends to far exceed nutrient application through fertilizer.

9. Research on Agronomic Strategies to Improve Nutrient Use Efficiency in Cropping Systems

Because fertilization is not realistic in many regions for economic, logistic and other reasons, an integrated approach using germplasm efficient in nutrient acquisition and use seems appropriate. Adoption of genotypes with higher nutrient use efficiency (NUE) by farmers is considered easier as no additional costs are involved and no major changes in cropping systems are necessary. Also, nutrient efficient varieties may contribute to sustainability in many other ways (Graham and Welch 1996). They have a greater degree of disease resistance (thereby a reduced use of fungicides) due to enhanced membrane function and cell integrity, a greater ability to develop deep roots to penetrate subsoil in infertile soils, and a greater seedling vigor associated with higher seed yields.

NUE is defined in several ways such as efficiency of acquisition (plant nutrient content/available nutrient) or physiological efficiency with which a nutrient is used to produce biomass (plant biomass /plant nutrient content) or grain (grain yield/plant nutrient content). It is also defined as the amount of additional grain yield per unit of fertilizer applied. Efforts to improve it must be guided by a thorough understanding of the soil and plant processes that govern NUE. For example, an ideal and economical approach to improve NUE in acid soils might be the combination of liming practices for neutralizing soil acidity at the surface, coupled with selection for crops more tolerant to Al toxicity. In cropping systems where fertilizer use is already high, cost-effective technologies that improve NUE are necessary. In cropping systems with low fertilizer use, however, the most promising options to improve NUE are adding small amounts of high-quality organic matter and using crop varieties with a greater NUE.

Fertilizer use efficiency in many developing countries is generally low, seldom exceeding 25%-30%. Modifications in fertilization methods, use of biological resources (e.g., azolla and legumes as green manures), and genetic improvement can all help increase NUE. Although many fertilization technologies (effective formulations; split applications; specification of suitable timing, particle size, and placement methods; development of slow-release materials and nitrification inhibitors; combined use of organic and inorganic fertilizers, etc.) are long established, their acceptance and adoption by farmers is not always encouraging. Similarly, despite the knowledge that *Azolla* and *Sesbania* have high N supply potential to support higher yields, rice farmers are averse to using them extensively. Research must be initiated, therefore, to examine the causes and to explore the prospects for blending such technologies with farmer's practices.

The deficiency of secondary and micro-nutrients is another factor reducing N and P use efficiency and is becoming more widespread than before in cropping systems worldwide. Supply of secondary and micronutrients can improve the yield response to macronutrients considerably. In Malawi, for example, average maize yields was improved by 40% as compared with the standard N-P recommendation when the deficiencies of B, Zn, S, and K were satisfied (Kumwenda et al. 1996). Micronutrients can be included in common fertilizer blends, or applied directly as chelated powders and suspensions, but research on optimizing their supply from a cropping systems perspective has been limited. For example, the existence of a regulatory interplay between S and N acquisition has been known for long time but studies on integrating this knowledge in formulating S recommendations for various cropping systems are very few.

10. Conclusion

Sustainable crop production will obviously require enhanced flows of nutrients to crops, which in turn comprises higher amounts of nutrient reserves in soils and higher nutrient uptake and utilization by crops. Productivity of current cropping systems and protection of environmental quality cannot be sustained for long if practices such as excessive or under application of nutrients and inefficient utilization of crop residues and animal wastes are continued. To achieve this goal, it is necessary to quantify measurable sustainability indicators such as: levels of available nutrients, organic compounds, soil microflora and fauna, nutrients lost through runoff and leaching, and the rates of change in those variables as affected by specific nutrient management practices in cropping systems. It is not difficult to obtain those indicators with present technology. However, it is difficult to correctly evaluate the contributions of those indicators to the productivity of crops. Crop plants have recently been revealed to acquire plant nutrients with different mechanisms. Soils high in Fe-P may be suitable for pigeonpea with its special mechanism to utilize Fe-P. However, for crops without such a mechanism, like sorghum and maize, soils high in Fe-P are undesirable. Likewise, AM population density in the soil has significant meaning for mycorrhizal crops, but is meaningless for non-mycorrhizal crops. Such indicators and their responses are likely to vary with each cropping system due to a multitude of interactions among soils and crops in diverse climates. Design and development of comprehensive nutrient management plans for various cropping systems will also require not only knowledge on plant nutrition mechanisms, but also require an integrated knowledge to understand processes occurring in various cropping systems. Research initiatives for sustainable nutrient management at the level of the cropping system must be conceived, therefore, on the basis of a holistic understanding of interactions among production, environmental, and biological components.

References

Ae N, Arihara J, Okada K, Yoshihara T, Johansen C (1990) Phosphate uptake by pigeonpea and its role in cropping systems of the Indian subcontinent. Science 248:477-480

Ae N, Arihara J, Okada K (1991) Phosphorus response of chickpea and evaluation of phosphorus availability in Indian Alfisols and Vertisols. In: Johansen C et al. (Eds) Phosphorus nutrition of grain legumes in the semi-arid tropics. International Crops Research Institute for the Semi-Arid Tropics, Patancheru, India, pp 33-41

Ae N, Otani T, Makino T, Tazawa J (1996) Role of cell wall of groundnut roots in solubilizing sparingly soluble phosphorus in soil. Plant Soil 186:197-204

Arihara J, Ae N, Okada K (1991a) Improvement of soil productivity through legume-based cropping systems in Indian Alfisols and Vertisols under semi-arid environment. In: Tropical agriculture research series, Tropical Agriculture Research Center, Tsukuba, Japan, No 24, pp 157-173

Arihara J, Ae N, Okada K (1991b) Root development of pigeonpea and chickpea and its significance in different cropping systems. In: Johansen C et al. (Eds) Phosphorus nutrition of grain legumes in the semi-arid tropics. International Crops Research Institute for the Semi-Arid Tropics, Patancheru, India. pp 183-194

Arihara J, Karasawa T (2000) Effect of Previous Crops on Arbuscular Mycorrhizal Formation and Growth of Succeeding Maize. Soil Sci Plant Nutr 46: 43-51

Asmar F, Singh T, Gahoonia TS, Nielsen NE (1995) Barley genotypes differ in activity of soluble extracellular phosphatase and depletion of organic phosphorus in the rhizosphere soil. Plant Soil 172:117-122

Bationo A, Christianson CB, Mokwunye U (1989) Soil fertility management of the pearl millet-producing sandy soils of Sahelian west Africa: The Niger experience. In: Renard S et al. (Eds) Soil, crop, and water management in the Sudano-Sahelian zone. International Crops Research Institute for the Semi-Arid Tropics, Patancheru, India. pp159-168

Cassman KG, Pingali PL (1995) Extrapolating trends from long-term experiments to farmers fields: the case of irrigated rice systems in Asia. In: Barnett V (Ed) Agricultural sustainability in economic, environmental and statistical terms. Wiley, London pp 63-84

Cassman KG, De Datta SK, Olk DC, Alcantara JM, Samson MI, Descalsota JP, Dizon MA (1995) Yield decline and the nitrogen economy of long-term experiments on continuous irrigated rice systems in the tropics. In: Lal R et al. (Eds) Soil management: experimental basis for sustainability and environmental quality. Lewis/CRC Publishers, Boca Raton. pp 181-222

Chaney RL, Li Y, Bills DD (1995) Metal-scavenging plants to cleanse the soil. Agric Res, USDA-ARS Nov. pp 4-9

Chaney D (1996) Project update:Cover crops in central coast farming systems. University of California Sustainable Agriculture Research and Education Program (SAREP) Sustainable Agriculture Newsletter 8(4):24-25. at http://www.sarep.ucdavis.edu/NEWSLTR/v8n4/sa-3.htm

Clark RB, Zeto SK (1996a) Iron acquisition by mycorrhizal maize grown on alkaline soil. J Plant Nutr 19:247-264

Clark RB, Zeto SK (1996b) Mineral acquisition by mycorrhizal maize grown on acid and alkaline soil. Soil Biol Biochem 28:1495-1503

Dennison EB (1961) Value of farmyard manure in maintaining fertility in Northern Nigeria. Emp J Exp Agric 29:330-336

Dominique R, Pfeffer PE, Douds DD, Farrell HM, Shachar-Hill Y (1997) Effects of vesicular arbuscular mycorrhizal colonization on nitrogen metabolism in the host plant. USDA ARS Report No 83710

Fedkiw J (1991) Nitrate occurrence in U. S. waters: a reference summary of published sources from an agricultural perspective. USDA, Washington DC

Graham RD, Welch RM (1996) Breeding for staple food crops with high micronutrient density. Working papers on agricultural strategies for micronutrients 3. International Food Policy Research Institute, Washington DC

Helal HM (1990) Varietal differences in root phosphatase activity as related to the utilization of organic phosphates. Plant Soil 123:161-163

ICRAF (1997) Using the wild sunflower, *Tithonia*, in Kenya for soil fertility and crop yield improvement. International Center for Research in Agroforestry (ICRAF), Nairobi,12 pp

Johnson JF, Vance CP, Allan DL (1996) Phosphorus deficiency in *Lupinus albus*: Altered lateral root development and enhanced expression of phosphoenolpyruvate carboxylase. USDA ARS Report No 71471

Johnson LC, Damman A (1996) The role of amino acid uptake by mycorrhizal ericaceous shrubs in the nitrogen economy of nutrient-deficient peat bogs. NSF Ecosystems Prog Rep No 1

Karasawa T, Arihara J, Kasahara Y Effects of Previous Crops on Arbuscular Mycorrhizal Formation and Growth of Maize under Various Soil Moisture Conditions. Soil Sci Plant Nutr 46: 53-60

Kitagishi K (1962) Edaphological studies on intensive cropping of pasture crops on volcanic ash soils (in Japanese with English summary). Bull Tohoku Natl Agric Expt Stn 23:1-67

Kondo H, Shibuya T, Kimura T, Kojima M, Yamamoto K (1997) Effect of the application of low grade rock phosphate to subsoil on the growth and phosphorus uptake of pasture plants. Bull Natl Grassland Res Inst, Japan. 55:1-8

Kumwenda JDT, Waddington SR, Snapp SS, Jones RB, Blackie MJ (1996) Soil fertility management research for the maize cropping systems of smallholders in southern Africa: A review. CIMMYT, Mexico, DF, NRG Paper 96-102

Kundu DK, Ladha JK (1995) Efficient management of soil and biologically fixed nitrogen in intensively cultivated rice fields. Soil Biol Biochem 27:431-439

Ladha JK, George T, Bohlool BB (Eds) (1992) Biological nitrogen fixation for sustainable agriculture. Kluwer Academic, Boston

Ladha JK, Tirol-Padre A, Reddy K, Ventura W (1993) Prospects and problems of biological nitrogen fixation in rice production: A critical assessment. In: Palacios R et al. (Eds) New horizons in nitrogen fixation. Kluwer Academic, Dordrecht. pp 677-682

Lynch J, van Beem J (1993) Growth and architecture of seedling roots of common bean genotypes. Crop Sci 33:1253-1257

Lynch JP, Beebe SE (1995) Adaptation of beans (*Phaseolus vulgaris* L.) to low phosphorus availability. Hort Sci 30:1165-1171

Matsumoto S, Ae N, Yamagata M (1999) Nitrogen uptake response of vegetable crops to organic materials. Soil Sci Plant Nutr 45:269-278

Myers RJK, Palm CA, Cuevas E, Gunatilleke IUN, Brossard M (1994) The synchronization of nutrient mineralization and plant nutrient demand. In: Wooner PL et al. (Eds) The biological management of tropical soil fertility. John Wiley, Chichester, U.K. pp 81-116

Otani T, Ae N (1996a) Phosphorus (P) uptake mechanisms of crops grown in soils with low P status. I. Screening crops for efficient P uptake. Soil Sci Plant Nutr 42:155-163

Otani T, Ae N (1996b) Phosphorus (P) uptake mechanisms of crops grown in soils with low P status. II. Significance of organic acids in root exudates of pigeonpea. Soil Sci Plant Nutr 42:553-500

Pallant E, Lansky DL, Rio JE, Jacobs LD, Schuler GS, Whimpenny WG (1997) Growth of corn roots under low-input and conventional farming systems. Am J Alt Agr 12:173-177

Radke JK, Andrews RW, Janke RR, Peters SE (1988) Low-input cropping systems and efficiency of water and nitrogen use. In: Hargrove WC (Ed) Cropping strategies for efficient use of water and nitrogen. ASA, CSSA, SSSA, Madison, Wisconsin. Am Soc Agron Special Publ No 51, pp 193-218

Ruel MT, Bouis HE (1997) Plant breeding: A long-term strategy for the control of zinc deficiency in vulnerable populations. Int Food Policy Res Inst, Washington DC, FCND Discussion Paper No. 20

Salazar FS, Pandey S, Narro L, Perez JC, Ceballos H, Parentoni SN, Bahia Filho AFC (1997) Diallel analysis of acid-soil tolerant and intolerant tropical maize populations. Crop Sci 37:1457-1462

Sanchez PA, Uehara G (1980) Management considerations for acid soils with high phosphorus fixation capacity. In: Khasawheh FE et al. (Eds) The role of phosphorus in agriculture. ASA, CSSA, SSSA, Madison, Wisconsin, pp 471-514

Sanchez PA, Shepherd KD, Soule MJ, Place FM, Buresh RJ, Izac AMN (1997) Soil fertility replenishment in Africa: An investment in natural resource capital. In: Roland J et al. (Eds) Replenishing soil fertility in Africa. SSSA, ASA. Madison, Wisconsin. SSSA Special Publ No 51, pp 1-46

Sharpley AN, Meisinger JJ, Power JF, Suarez DL (1992) Root extraction of nutrients associated with long-term soil management. Adv Soil Sci 19:151-217

Sharpley AN, Sims JT, Pierzynsky GM (1994) Innovative soil phosphorus availability indices: assessing inorganic phosphorus. USDA ARS Rep No 50515

Smaling EMA, Nandwa SM, Jansse BH (1997) Soil fertility in Africa is at stake. In: Roland J et al. (Eds) Replenishing soil fertility in Africa. SSSA, ASA. Madison, Wisconsin. SSSA Special Publ No 51 pp 47-79

Snapp SS (1995) Improving fertilizer efficiency with small additions of high quality organic inputs. In: Waddington SR (Ed) Report on the first meeting of the network working group. Soil fertility research network for maize-based farming systems in selected countries of southern Africa. The Rockefeller Foundation. Southern Africa Agricultural Sciences Program and CIMMYT. Lilongwe, Malawi, and Harare, Zimbabwe. pp 60-65

Snapp SS, Phombeya HSK, Materechera S (1995) An agroforestry based test of new fractionation methods in soil organic matter analysis. Paper presented at the Crop Science Society of Zimbabwe 25th Anniversary Symposium, June, Harare, Zimbabwe

Souma S (1986) Management of soil and fertilizer application meeting needs of various vegetables and soil fertility of vegetable fields in Hokkaido. Rep Hokkaido Prefectural Agric Expt Stn 56:1-126

Tachibana H, Moriguchi A, Imaoka T, Yamada M (1992) Run-off of particulate phosphorus and its effect on algal growth. In: Hongliang L, Yutia Z, Haisheng L (Eds) Lake conservation and management. Proc 4th Int Conf Cons Managt Lakes, Hangzhou, China, pp 82-94

Tarafdar JC, Jungk A (1987) Phosphatase activity in the rhizosphere and its relation to the depletion of soil organic phosphorus. Biol Fertil Soils 3:199-204

Taylor GJ (1995) Overcoming barriers to understanding the cellular basis of aluminum resistance. In: Date RA et al (Eds) Plant-soil interaction at low pH: Principles and management. Kluwer Academic, Dordrecht, pp 255-269

Thompson JP (1991) Improving the mycorrhizal condition of the soil through cultural practices and effects on growth and phosphorus uptake by plants. In: Johansen C et al. (Eds) Phosphorus nutrition of grain legumes in the semi-arid tropics. ICRSAT, Patancheru, India, pp 117-137

Vanotti MB, Bundy LG (1995) Soybean effects on soil nitrogen availability in crop rotation. Agron. J 87:676-686

Wada G, Shoji S, Mae T (1986) Relationship between nitrogen absorption and growth and yield of rice plants. Jpn Agric Res Quart(JARQ) 20: 135-145

Wissuwa M, Yano M, Ae N (1998) Mapping of QTLs for phosphorus-deficiency tolerance in rice (Oryza sativa L.). Theor Appl Genet 97: 777-783

Yamagata M, Ae N (1996) Nitrogen uptake response of crops to organic nitrogen. Soil Sci Plant Nutr 42:389-394

Yan X, Lynch JP, Beebe SE (1995) Genetic variation for phosphorus efficiency of common bean in contrasting soil types: I. Vegetation response. Crop Sci 35:1086-1093

Yan X, Lynch JP, Beebe SE (1996) Utilization of phosphorus substrates by contrasting common bean genotypes. Crop Sci 36:936-941

Yanni YG, Rizk RY, Corich V, Squartini A, Ninke K, Philip-Hollingsworth S, Orgambide G, de Bruijn F, Stoltzfus J, Buckley D, Schmidt TM, Mateos PF, Ladha JK, Dazzo FB (1997) Natural endophytic association between *Rhizobium leguminosarum* bv. Trifoli and rice roots and assessment of its potential to promote rice growth. Plant Soil 194:99-114

Subject Index

a

b

d

e

h

i

j

k

l

m

n

q

r

Epilogue

Noriharu Ae, Joji Arihara, Kensuke Okada, and Ancha Srinivasan

Environmental pollution, characterized by contamination of groundwater and soils with nitrate-N and heavy metals respectively, is a major concern in developed countries. It is caused largely by indiscriminate use of chemical fertilizers, pesticides, herbicides, and manure from farm by-products and urban wastes. As most developed countries have surplus food production, consumers in these countries are now increasingly attracted to organically grown produce. Although organic farming is becoming popular, we do not yet know of any clear quantitative and qualitative differences between organic produce and food produced from conventional chemical fertilization.

At the same time, there are still chronic food shortages in many developing countries. With the world's population expected to rise by almost 90 million a year over the next decade, sustainable increases in food production will have to come through higher productivity of existing resources. It is important to note, however, that indigenous agricultural systems in developing countries are complex in many ways and have been sustained for a long time despite low input use. For instance, our studies on Indian Alfisols, in which the predominant form of phosphorus (P) is iron-bound (Fe-P), revealed that pigeonpea has a superior ability to take up such sparingly soluble P. Pigeonpea has long been cultivated as an intercrop with sorghum or pearl millet, neither of which has the ability to take up Fe-P. Therefore, introduction of pigeonpea into cropping systems on low-P Alfisols enhanced sustainability from the viewpoint of P utilization and food production in conditions of low fertilization or none at all. The results imply that Indian farmers had probably known the nutrient acquisition characteristics of various crop species in their traditional farming systems. However, following rapid developments in the fertilizer industry and associated economic changes, farmers ignored unique attributes of crop species and applied excess amounts of fertilizer and organic matter, thereby leading to environmental pollution.

What we must do now is to reveal the hidden knowledge in traditional practices and to probe various crop characteristics such as growth, nutrient uptake, and the relationship between root and soil microorganisms in order to identify and potentially manipulate physiological processes involved in nutrient acquisition and use. It is also important to integrate this knowledge in designing modern cropping systems. Authors contributing to the present volume did pioneering research in this complex field. For example, sedge in tundra—where soil pH and temperatures are low, soil organic matter and nitrogen mineralization are limited, and amino acids as N sources are rich—has been shown to respond better to amino acids than to inorganic nitrogen (Chapter 17). Likewise, Chingensai (a kind of Chinese cabbage) and carrot have been shown to respond better to soil organic nitrogen than to chemical fertilizer (Chapter 18).

Despite significant biotechnological advances in the recent past, especially in relation to genome analysis, we do not know enough about the biological nature and causes of genotype and environment interactions in a range of crops and pastures. Analysis of such interactions will help us to better understand the relationship between crop performance and environment, and to develop innovative and environment-friendly technologies for sustained improvement of crop productivity. To investigate new aspects of plant response to the environment, however, research in a particular field environment is most appropriate as plants respond in unique but diverse ways to similar nutrient environments under natural conditions. Unexpected responses in a field experiment also arouse interest in conducting further research and drawing new conclusions.

Postulating revolutionary and innovative concepts is often based on a commitment to questioning, contradicting, and evaluating existing knowledge. In general, most people are too conservative to come up with radical ideas from a known "textbook." We believe that the Plant Nutrient Acquisition Workshop provided a forum for participants to challenge existing concepts and exchange diverse views on mechanisms of plant nutrient acquisition. Workshop participants reviewed current research and described promising research approaches. They concluded that sustainable crop production will depend on explicit considerations of nutrient acquisition and that the concepts will need further review from time to time. We believe that this book, which is based on proceedings of the Workshop, provides the reader with a new outlook and balanced perspective on concepts of nutrient acquisition, although some of those are yet to be widely accepted by the scientific community. We believe that the results presented will provide researchers with a better understanding and appreciation of various concepts and will stimulate them to postulate and test new hypotheses in the near future.

We apologize for the delay in publishing this book due to circumstances beyond our control. We also take this opportunity to thank NIAES for supporting the Workshop and the publication of this volume.